HAESE & HARRIS PUBLICATIONS

Specialists in mathematics publishing

Mathematics

for the international student

Mathematical Studies SL

Mal Coad

Glen Whiffen

John Owen

Robert Haese

Sandra Haese

Mark Bruce

International Baccalaureate Diploma Programme

MATHEMATICS FOR THE INTERNATIONAL STUDENT
International Baccalaureate Mathematical Studies SL Course

Mal Coad	B.Ec., Dip.T.
Glen Whiffen	B.Sc., B.Ed.
John Owen	B.Sc., Dip.T.
Robert Haese	B.Sc.
Sandra Haese	B.Sc.
Mark Bruce	B.Ed.

Haese & Harris Publications
3 Frank Collopy Court, Adelaide Airport, SA 5950, AUSTRALIA
Telephone: +61 8 8355 9444, Fax: + 61 8 8355 9471
Email: info@haeseandharris.com.au
Web: www.haeseandharris.com.au

National Library of Australia Card Number & ISBN 1 876543 15 9

© Haese & Harris Publications 2004

Published by Raksar Nominees Pty Ltd
3 Frank Collopy Court, Adelaide Airport, SA 5950, AUSTRALIA

First Edition 2004

Cartoon artwork by John Martin. Artwork by Piotr Poturaj and David Purton
Cover design by Piotr Poturaj
Computer software by David Purton

Typeset in Australia by Susan Haese (Raksar Nominees). Typeset in Times Roman $10\frac{1}{2}/11\frac{1}{2}$

The textbook and its accompanying CD have been developed independently of the International Baccalaureate Organization (IBO). The textbook and CD are in no way connected with, or endorsed by, the IBO.

Acknowledgements: The publishers acknowledge the cooperation of Oxford University Press, Australia, for the reproduction of material originally published in textbooks produced in association with Haese & Harris Publications.

While every attempt has been made to trace and acknowledge copyright, the authors and publishers apologise for any accidental infringement where copyright has proved untraceable. They would be pleased to come to a suitable agreement with the rightful owner.

Disclaimer: All the internet addresses (URL's) given in this book were valid at the time of printing. While the authors and publisher regret any inconvenience that changes of address may cause readers, no responsibility for any such changes can be accepted by either the authors or the publisher.

FOREWORD

Mathematics for the International Student: Mathematical Studies SL has been written to embrace the syllabus for the new two-year Mathematical Studies SL Course, which is one of the courses of study in the International Baccalaureate Diploma Programme. It is not our intention to define the course. Teachers are encouraged to use other resources. We have developed the book independently of the International Baccalaureate Organization (IBO) in consultation with many experienced teachers of IB Mathematics. The text is not endorsed by the IBO.

This package is language rich and technology rich. The combination of textbook and interactive Student CD will foster the mathematical development of students in a stimulating way. Frequent use of the interactive features on the CD is certain to nurture a much deeper understanding and appreciation of mathematical concepts.

The book contains many problems from the basic to the advanced, to cater for a wide range of student abilities and interests. While some of the exercises are simply designed to build skills, every effort has been made to contextualise problems, so that students can see everyday uses and practical applications of the mathematics they are studying, and appreciate the universality of mathematics.

Emphasis is placed on the gradual development of concepts with appropriate worked examples, but we have also provided extension material for those who wish to go beyond the scope of the syllabus.

It is not our intention that each chapter be worked through in full. Time constraints will not allow for this. Teachers must select exercises carefully, according to the abilities and prior knowledge of their students, to make the most efficient use of time and give as thorough coverage of work as possible.

Investigations throughout the book will add to the discovery aspect of the course and enhance student understanding and learning.

Review sets appear at the end of each chapter and a suggested order for teaching the two-year course is given at the end of this Foreword.

The extensive use of graphics calculators and computer packages throughout the book enables students to realise the importance, application and appropriate use of technology. No single aspect of technology has been favoured. It is as important that students work with a pen and paper as it is that they use their calculator or graphics calculator, or use a spreadsheet or graphing package on computer.

The interactive features of the CD allow immediate access to our own specially designed geometry packages, graphing packages and more. Teachers are provided with a quick and easy way to demonstrate concepts, and students can discover for themselves and re-visit when necessary.

Instructions appropriate to each graphic calculator problem are on the CD and can be printed for students. These instructions are written for Texas Instruments and Casio calculators.

In this changing world of mathematics education, we believe that the contextual approach shown in this book, with the associated use of technology, will enhance the students' understanding, knowledge and appreciation of mathematics, and its universal application.

We welcome your feedback.

Email: info@haeseandharris.com.au
Web: www.haeseandharris.com.au

MC GAW JTO
RCH SHH MFB

Thank you

The authors and publishers would like to thank all those teachers who offered advice and encouragement. These teachers include: Marjut Mäenpää, Cameron Hall, Fran O'Connor, Glenn Smith, Anne Walker, Ian Hilditch, Phil Moore, Julie Wilson, Kestie Nelligan, Carolyn Farr.

TEACHING THE TWO-YEAR COURSE – A SUGGESTED ORDER

For the first year, it is suggested that students work progressively from Chapter 1 through to Chapter 11, although some teachers may prefer to include Chapter 18 'Two-variable statistics' in the first year as a basis for internal assessment. Other teachers may prefer to leave statistics until early in the second year and then have students work progressively from Chapter 12 through to Chapter 19.

However, it is acknowledged that there is no single best way for all teachers to work through the syllabus. Individual teachers have to consider particular needs of their students and other requirements and preferences that they may have.

We invite teachers to email their preferred order or suggestions to us. We can put these suggestions on our website to be shared with other teachers.

USING THE INTERACTIVE STUDENT CD

The CD is ideal for independent study. Frequent use will nurture a deeper understanding of Mathematics. Students can revisit concepts taught in class and undertake their own revision and practice. The CD also has the text of the book, allowing students to leave the textbook at school and keep the CD at home.

The icon denotes an active link on the CD. Simply 'click' the icon to access a range of interactive features:

CD LINK

- spreadsheets
- video clips
- graphing and geometry software
- graphics calculator instructions
- computer demonstrations and simulations
- background knowledge

Graphics calculators: Instructions for using graphics calculators are also given on the CD and can be printed. Instructions are given for Texas Instruments and Casio calculators. Click on the relevant icon (TI or C) to access printable instructions.

Examples in the textbook are not always given for both types of calculator. Where that occurs, click on the relevant icon to access the instructions for the other type of calculator.

NOTE ON ACCURACY

Students are reminded that in assessment tasks, including examination papers, unless otherwise stated in the question, all numerical answers must be given exactly or to three significant figures.

ERRATA

If you find an error in this book, please notify us by emailing errata@haeseandharris.com.au.

As a help to other teachers and students, we will include the correction on our website and correct the book at the first reprint opportunity.

TABLE OF CONTENTS

SYMBOLS AND NOTATION

N	the set of positive integers and zero, $\{0, 1, 2, 3, ...\}$
Z	the set of integers, $\{0, \pm1, \pm2, \pm3, ...\}$
Z^+	the set of positive integers, $\{1, 2, 3, ...\}$
Q	the set of rational numbers
Q^+	the set of positive rational numbers, $\{x \mid x \in Q, \quad x > 0\}$
R	the set of real numbers
R^+	the set of positive real numbers, $\{x \mid x \in R, \quad x > 0\}$
$\{x_1, x_2, ...\}$	the set with elements $x_1, x_2, ...$
$n(A)$	the number of elements in the finite set A
$\{x \mid ....$ or $\{x: ....$	the set of all x such that
$\in$	is an element of
$\notin$	is not an element of
$\varnothing$	the empty (null) set
U	the universal set
$\cup$	union
$\cap$	intersection
A'	the complement of the set A
$p \Rightarrow q$	implication: if p then q
$p \Leftarrow q$	implication: if q then p
$p \Leftrightarrow q$	equivalence: p is equivalent q
$\subset$	is a subset of
$p \wedge q$	conjunction: p and q
$p \vee q$	disjunction: p or q (or both)
$p \veebar q$	exclusive disjunction: p or q (**not** both)
$\neg p$	negation: not p
$a^{\frac{1}{n}}, \sqrt[n]{a}$	a to the power of $\frac{1}{n}$, nth root of a (if $a \geqslant 0$ then $\sqrt[n]{a} \geqslant 0$)
$a^{\frac{1}{2}}, \sqrt{a}$	a to the power $\frac{1}{2}$, square root of a (if $a \geqslant 0$ then $\sqrt{a} \geqslant 0$)
$\lvert x \rvert$	the modulus or absolute value of x, that is $\begin{cases} x \text{ for } x \geqslant 0 & x \in R \\ -x \text{ for } x < 0 & x \in R \end{cases}$
$\equiv$	identity or is equivalent to
$\approx$ or $\doteqdot$	is approximately equal to
$>$	is greater than
$\geq$ or $\geqslant$	is greater than or equal to
$<$	is less than
$\leq$ or $\leqslant$	is less than or equal to
$\ngtr$	is not greater than
$\nless$	is not less than
$[a, b]$	the closed interval $a \leqslant x \leqslant b$
$]a, b[$	the open interval $a < x < b$
u_n	the nth term of a sequence or series
d	the common difference of an arithmetic sequence
r	the common ratio of a geometric sequence
S_n	the sum of the first n terms of a sequence, $u_1 + u_2 + ... + u_n$
$\displaystyle\sum_{i=1}^{n} u_i$	$u_1 + u_2 + ... + u_n$
$f: \quad A \to B$	f is a function under which each element of set A has an image in set B
$f : x \mapsto y$	f is a function under which x is mapped to y
$f(x)$	the image of x under the function f
f^{-1}	the inverse function of the function f
$\displaystyle\lim_{x \to a} f(x)$	the limit of $f(x)$ as x tends to a
$\dfrac{dy}{dx}$	the derivative of y with respect to x
$f'(x)$	the derivative of $f(x)$ with respect to x
$\dfrac{d^2y}{dx^2}$	the second derivative of y with respect to x
$f''(x)$	the second derivative of $f(x)$ with respect to x
sin, cos, tan	the circular functions
A(x, y)	the point A in the plane with Cartesian coordinates x and y
[AB]	the line segment with end points A and B
AB	the length of [AB]
(AB)	the line containing points A and B
$\widehat{A}$	the angle at A
$\widehat{CAB}$ or $\angle CAB$	the angle between [CA] and [AB]
$\triangle ABC$	the triangle whose vertices are A, B and C
P(A)	probability of event A
P$'$(A)	probability of the event "not A"
P(A $\mid$ B)	probability of the event A given B
$x_1, x_2, ...$	observations of a variable
$f_1, f_2, ...$	frequencies with which the observations $x_1, x_2, x_3, ...$ occur
$\overline{x}$	sample mean
μ	population mean
s_n	standard deviation of the sample
σ	standard deviation of the population
r	Pearson's product-moment correlation coefficient
χ^2	chi-squared

USEFUL FORMULAE SUMMARY

STATISTICS

Mean

$$\overline{x} = \frac{\sum fx}{n} \quad \text{where} \quad n = \sum f$$

Standard deviation

$$s_n = \sqrt{\frac{\sum f(x - \overline{x})^2}{n}}, \quad \text{where} \quad n = \sum f$$

Pearson's product-moment correlation coefficient

$$r = \frac{s_{xy}}{s_x s_y}, \quad \text{where} \quad s_x = \sqrt{\frac{\sum(x - \overline{x})^2}{n}},$$

$$s_y = \sqrt{\frac{\sum(y - \overline{y})^2}{n}} \quad \text{and} \quad s_{xy} \text{ is the covariance.}$$

Equation of regression line for y on x

$$y - \overline{y} = \frac{s_{xy}}{s_x^2}(x - \overline{x})$$

The χ^2 test statistic

$$\chi_{calc}^2 = \sum \frac{(f_e - f_o)^2}{f_e}, \quad \text{where } f_o \text{ are the observed frequencies, } f_e \text{ are the expected frequencies.}$$

GEOMETRY

Equation of a straight line

$$y = mx + c; \quad ax + by + d = 0$$

Gradient formula

$$m = \frac{y_2 - y_1}{x_2 - x_1}$$

Equation of axis of symmetry

$$x = \frac{-b}{2a}$$

Distance between two points (x_1, y_1) and (x_2, y_2)

$$d = \sqrt{(x_1 - x_2)^2 + (y_1 - y_2)^2}$$

Coordinates of the midpoint of a line segment with endpoints (x_1, y_1) and (x_2, y_2)

$$\left(\frac{x_1 + x_2}{2}, \frac{y_1 + y_2}{2} \right)$$

ALGEBRA

Solution of a quadratic equation

$$ax^2 + bx + c = 0, \quad a \neq 0 \quad \Rightarrow \quad x = \frac{-b \pm \sqrt{b^2 - 4ac}}{2a}$$

TRIGONOMETRY

Sine rule

$$\frac{a}{\sin A} = \frac{b}{\sin B} = \frac{c}{\sin C}$$

Cosine rule

$$a^2 = b^2 + c^2 - 2bc \cos A; \quad \cos A = \frac{b^2 + c^2 - a^2}{2bc}$$

Area of a triangle

$$A = \tfrac{1}{2}ab \sin C, \quad \text{where } a \text{ and } b \text{ are adjacent sides, } C \text{ is the included angle}$$

PLANE AND SOLID FIGURES

Area of a parallelogram $A = (b \times h)$,   where b is the base, h is the height

Area of a triangle $A = \frac{1}{2}(b \times h)$,   where b is the base, h is the height

Area of a trapezium $A = \frac{1}{2}(a + b)h$,   where a and b are the parallel sides, h is the height

Area of a circle $A = \pi r^2$,   where r is the radius

Circumference of a circle $C = 2\pi r$,   where r is the radius

Volume of a pyramid $V = \frac{1}{3}(\text{area of base} \times \text{vertical height})$

Volume of a cuboid $V = l \times w \times h$,   where l is the length, w is the width, h is the height

Volume of a cylinder $V = \pi r^2 h$,   where r is the radius, h is the height

Area of the curved surface of a cylinder $A = 2\pi rh$,   where r is the radius, h is the height

Volume of a sphere $V = \frac{4}{3}\pi r^3$,   where r is the radius

Surface area of a sphere $A = 4\pi r^2$,   where r is the radius

Volume of a cone $V = \frac{1}{3}\pi r^2 h$,   where r is the radius, h is the height

Area of the curved surface of a cone πrl,   where r is the radius, l is the slant height

FINITE SEQUENCES

The nth term of an arithmetic sequence $u_n = u_1 + (n - 1)d$

The sum of n terms of an arithmetic sequence $S_n = \dfrac{n}{2}(2u_1 + (n - 1)d) = \dfrac{n}{2}(u_1 + u_n)$

The nth term of a geometric sequence $u_n = u_1 r^{n-1}$

The sum of n terms of a geometric sequence $S_n = \dfrac{u_1(r^n - 1)}{r - 1} = \dfrac{u_1(1 - r^n)}{1 - r}$, $r \neq 1$

PROBABILITY

Probability of an event A $P(A) = \dfrac{n(A)}{n(U)}$

Complementary events $P(A') = 1 - P(A)$

Combined events $P(A \cup B) = P(A) + P(B) - P(A \cap B)$

Mutually exclusive events $P(A \cup B) = P(A) + P(B)$

Independent events $P(A \cap B) = P(A)P(B)$

Conditional probability $P(A|B) = \dfrac{P(A \cap B)}{P(B)}$

FINANCIAL MATHEMATICS

Simple Interest

$$I = \frac{Crn}{100},$$

where C is the capital,
$r\%$ is the interest rate,
n is the number of time periods,
I is the interest

Compound Interest

$$I = C \times \left(1 + \tfrac{r}{100}\right)^n - C,$$

where C is the capital,
$r\%$ is the interest rate,
n is the number of time periods,
I is the interest

DIFFERENTIAL CALCULUS

Derivative of $f(x)$

$$y = f(x) \Rightarrow \frac{dy}{dx} = f'(x) = \lim_{h \to 0} \left(\frac{f(x+h) - f(x)}{h} \right)$$

Derivative of ax^n

$$f(x) = ax^n \Rightarrow f'(x) = nax^{n-1}$$

Derivative of a polynomial

$$f(x) = ax^n + bx^{n-1} + \dots$$
$$\Rightarrow f'(x) = nax^{n-1} + (n-1)bx^{n-2} + \dots$$

Chapter 1

Number sets and properties

Contents:

SOME SET LANGUAGE

The set of all digits we use to form numbers is   {0, 1, 2, 3, 4, 5, 6, 7, 8, 9}.

Notice that the ten digits have been enclosed within the brackets { and }.

Sets do not have to contain numbers. For example, the set of all vowels is   {a, e, i, o, u}.

> A **set** is a collection of numbers or objects.

We usually use capital letters to represent sets. Every item within the brackets is called **an element** (or **member**) of the set.

Notation:
- $\in$ reads *is an element of*.
- $\notin$ reads *is not an element of*.
- { } or $\varnothing$ is called *the empty set* and contains no elements.

So, for the set   A = {1, 2, 3, 4, 5, 6, 7}   we can say $4 \in A$, but $9 \notin A$.

SUBSETS

If P and Q are two sets then:

> $P \subset Q$ reads   'P is a **subset** of Q and means that every element in P is also an element of Q'.

For example   {2, 3, 5} $\subset$ {1, 2, 3, 4, 5, 6}   as every element in the first set is also in the second set.

UNION AND INTERSECTION

If P and Q are two sets then:

- $P \cup Q$ is the **union** of sets P and Q and is the set made up of all elements which are in P **or** Q.
- $P \cap Q$ is the **intersection** of sets P and Q and is the set made up of all elements which are in **both** P **and** Q.

For example, if P = {1, 3, 4} and Q = {2, 3, 5} then:

$$P \cup Q = \{1, 2, 3, 4, 5\}$$
$$\text{and} P \cap Q = \{3\}.$$

Every element in P or in Q must be in $P \cup Q$.

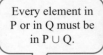

DEMO

Example 1

$M = \{2, 3, 5, 7, 8, 9\}$ and $N = \{3, 4, 6, 9, 10\}$
a   True or false?   **i** $4 \in M$ **ii** $6 \notin M$ **b**   List   **i** $M \cap N$ **ii** $M \cup N$
c   Is   **i** $M \subset N$ **ii** $\{9, 6, 3\} \subset N$?

---

a   **i** $4 \in M$ is false   **ii** $6 \notin M$ is true

b   **i** $M \cap N = \{3, 9\}$ ←   3 and 9 are in both sets

   ii $M \cup N = \{2, 3, 4, 5, 6, 7, 8, 9, 10\}$ ←   look at set M and add to it all ele-
                                                                    ments of N which are not in M.

c   **i**   No, as every element of M is not an element of N,   e.g., 2.

   ii   Yes, as 9, 6 and 3 are also in N.

EXERCISE 1A

1   Write using set language:
   a   5 is an element of set D   **b**   6 is not an element of set G
   c d is not an element of the set of all vowels
   d $\{2, 5\}$ is a subset of $\{1, 2, 3, 4, 5, 6\}$
   e $\{3, 8, 6\}$ is not a subset of $\{1, 2, 3, 4, 5, 6\}$

2   Find   **i** $A \cap B$ **ii** $A \cup B$ for:
   a $A = \{6, 7, 9, 11, 12\}$ and $B = \{5, 8, 10, 13, 9\}$
   b $A = \{1, 2, 3, 4\}$ and $B = \{5, 6, 7, 8\}$
   c $A = \{1, 3, 5, 7\}$ and $B = \{1, 2, 3, 4, 5, 6, 7, 8, 9\}$

3   If two sets have no common elements then their intersection is the empty set and we say
   that the sets are **disjoint**.
   a   Which of these sets are disjoint?
      i $A = \{3, 5, 7, 9\}$ and $B = \{2, 4, 6, 8\}$
      ii $P = \{3, 5, 6, 7, 8, 10\}$ and $Q = \{4, 9, 10\}$

   b   True or false?   If R and S are two sets and $R \cap S = \varnothing$ then R and S are disjoint.

B   NUMBER SETS

Having established some basic set notation, we can now define all of the components (subsets) of the set of all real numbers.

$N^* = \{1, 2, 3, 4, 5, 6, ...\}$ is the set of all **counting numbers**.

$N = \{0, 1, 2, 3, 4, 5, 6, ...\}$ is the set of all **natural numbers**.

$Z = \{0, \pm1, \pm2, \pm3, \pm4, ...\}$ is the set of all **integers**.

$Z^+ = \{1, 2, 3, 4, 5, 6, ...\}$ is the set of all **positive integers**.

$Z^- = \{-1, -2, -3, -4, -5, ...\}$ is the set of all **negative integers**.

$Q = \{\frac{p}{q}$ where p and q are integers and $q \neq 0\}$ is the set of all **rational numbers**.

$R = \{$real numbers$\}$ is the set of all real numbers,
        i.e., all numbers which can be placed on the number line.

Notice that: $Z = Z^- \cup \{0\} \cup Z^+$ which simply means that the set of all integers is
        made up of the set of all positive and negative integers and zero.

Example 2

Give an example of: **a**   why a positive integer is also a rational number
                      b   why -7 is a rational number.

a   Consider 5.   Now $5 = \frac{5}{1}$ (or $\frac{15}{3}$) $\therefore$ 5 is rational.

b $-7 = \frac{-7}{1}$ (or $\frac{-14}{2}$) $\therefore$ -7 is rational.

Example 3

Show that: **a**   0.45   **b**   0.135   are rational numbers.

a $0.45 = \frac{45}{100} = \frac{9}{20}$,   so 0.45 is rational.

b $0.135 = \frac{135}{1000} = \frac{27}{200}$,   so 0.135 is rational.

$\{0\}$ is the set containing 0 only. It is not the empty set.

Note: All terminating decimal numbers, such as those in **Example 3**, are rational numbers.

EXERCISE 1B

1 Show that: **a** 8   **b** -11 are rational numbers.

2 Why is $\frac{4}{0}$ not a rational number?

3 True or false?

    a $N^* \subset N$ **b** $N \subset Z$ **c** $N^* = Z^+$ **d** $Z^- \subset Z$

    e $Q \subset Z$ **f** $\{0\} \subset Z$ **g** $Z \subset Q$ **h** $Z^+ \cup Z^- = Z$

4 Show that the following are rational numbers:

    a   0.8   **b**   0.75   **c**   0.45   **d**   0.215   **e**   0.864

5 True or false?

    a $127 \in N$ **b** $\frac{138}{279} \in Q$ **c** $3\frac{1}{7} \notin Q$ **d** $-\frac{4}{11} \in Q$

EXTENSION

All **recurring decimal numbers** can be shown to be rational.

Example 4

Show that: **a** $0.7777777.....$ **b** $0.363636.....$ are rational.

a   Let $x = 0.7777777.....$
$\therefore \quad 10x = 7.7777777.....$
$\therefore \quad 10x = 7 + 0.777777.....$
$\therefore \quad 10x = 7 + x$
$\therefore \quad 9x = 7$
$\therefore \quad x = \frac{7}{9}$
So, $0.777777..... = \frac{7}{9}$

b   Let $x = 0.363636.....$
$\therefore \quad 100x = 36.363636.....$
$\therefore \quad 100x = 36 + x .....$
$\therefore \quad 99x = 36$
$\therefore \quad x = \frac{36}{99}$
$\therefore \quad x = \frac{4}{11}$
So, $0.363636..... = \frac{4}{11}$

---

6   Show that these are rational:

   a $0.444444.....$ **b** $0.212121......$ **c** $0.325325325.....$

IRRATIONAL NUMBERS

Either a real number is rational or irrational.

Irrational numbers cannot be written in the form $\dfrac{p}{q}$ where p and q are integers, $q \neq 0$.

Numbers such as $\sqrt{2}$, $\sqrt{3}$, $\sqrt{5}$, $\sqrt{7}$, $\sqrt[3]{2}$, $\sqrt[3]{11}$, etc., are all irrational.

This means that their decimal expansions do not terminate or recur.

Other irrationals include numbers like $\pi = 3.141593.....$ which is the ratio of a circle's circumference to its diameter, and exponential $e = 2.718281828235.....$ which is very important in SL and HL mathematics and beyond. The following shows the relationship between various number sets:

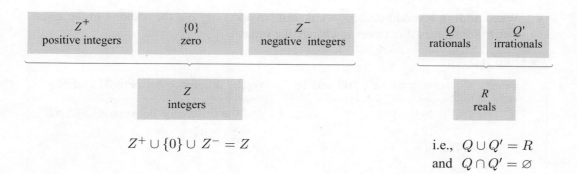

| Z^+ positive integers | {0} zero | Z^- negative integers | | Q rationals | Q' irrationals |

| Z integers | | R reals |

$Z^+ \cup \{0\} \cup Z^- = Z$

i.e., $Q \cup Q' = R$
and $Q \cap Q' = \varnothing$

C   WORDS USED IN MATHEMATICS

Many words used in mathematics have special meanings and we should not avoid using them. It is important to learn what each word (or phrase) means and to use it correctly. For example, when we write any number, we write some combination of the ten symbols: 1, 2, 3, 4, 5, 6, 7, 8, 9 and 0. These symbols are called **digits**.

There are four **basic operations** that are carried out with numbers:

Addition	$+$	to find the **sum**
Subtraction	$-$	to find the **difference**
Multiplication	$\times$	to find the **product**
Division	$\div$	to find the **quotient**

Here are some words which are frequently used with these operations:

- **terms**   numbers being added or subtracted
- **product**   the result of a multiplication
- **factors**   numbers which divide exactly into another number
- **quotient**   the result of a division
- **divisor**   the number by which we divide
- **dividend**   the number being divided

For example:

in $3 + 5$ 3 and 5 are terms
in $8 - 2 + 6$ 8, 2 and 6 are terms
in $2 \times 7 = 14$ 14 is the product of 2 and 7   in $6 \div 3 = 2$ $\begin{cases} \text{2 is the quotient} \\ \text{3 is the divisor} \\ \text{6 is the dividend} \end{cases}$
in $2 \times 3 = 6$ 2 and 3 are factors of 6

SUMS AND DIFFERENCES

- To find the **sum** of two or more numbers, we *add* them.
  For example, the sum of 3 and 16 is $3 + 16 = 19$.

- To find the **difference** between two numbers, we *subtract* the smaller from the larger.
  For example, the difference between 3 and 16 is $16 - 3 = 13$.

- When adding or subtracting **zero (0)**, the number remains unchanged.
  For example, $23 + 0 = 23$, $23 - 0 = 23$.

- When **adding** several numbers, we do not have to carry out the addition in the given order. Sometimes it is easier to change the order.

Example 5

Find: **a** the sum of 187, 369 and 13   **b** the difference between 37 and 82

a $187 + 369 + 13$	**b** the difference between 37 and 82
$= 187 + 13 + 369$	$= 82 - 37$
$= 200 + 369$	$= 45$
$= 569$	

EXERCISE 1C

1 Find:

    a the sum of 5, 7 and 8

    c the sum of the first 10 natural numbers

    b the difference between 19 and 56

    d by how much 639 exceeds 483

2 Solve the following problems:

    a What number must be increased by 374 to get 832?

    b What number must be decreased by 674 to get 3705?

3 Solve the following problems:

    a In a golf tournament Aaron won the first prize of \$163 700 and Sean came second with \$97 330. What was the difference between the two prizes? What would they each have won if they had tied?

    b My bank account balance was \$7667 and I withdrew amounts of \$1379, \$2608 and \$937. What is my bank balance now?

    c Agneta stands on some scales with a 15 kg dumbell in each hand. If the scales read 92 kg, what does she weigh?

PRODUCTS AND QUOTIENTS

- The word **product** is used to represent the result of a multiplication.
  For example, the product of 3 and 5 is $3 \times 5 = 15$.

- The word **quotient** is used to represent the result of a division.
  For example, the quotient of $15 \div 3$ is 5.

- Multiplying by **one (1)** does not change the value of a number.
  For example, $17 \times 1 = 17$, $1 \times 17 = 17$.

- Multiplying by **zero (0)** produces zero.
  For example, $17 \times 0 = 0$.

- Division by zero (0) is meaningless; we say it is **undefined**.
  For example, $0 \div 4 = 0$ but $4 \div 0$ is undefined.

- The order in which numbers are multiplied does **not** change the resultant number. For example, $3 \times 7 \times 2 = 2 \times 3 \times 7 = 42$.

4 Find:

    a the product of 17 and 32

    b the quotient of 437 and 19

    c the product of the first 5 natural numbers

5 Solve the following problems:

    a What must I multiply \$25 by to get \$1375?

    b What answer would I get if I start with 69 and add on 8, 31 times?

c I planted 400 rows of cabbages and each row contained 250 plants. How many cabbages were planted
altogether?

d Juan swims 4500 m in a training session. If the pool
is 50 m long, how many laps does he swim?

6 Solve the following problems:

a A contractor bought 34 loads of soil each weighing
12 tonnes at $13 per tonne. What was the total cost?

b All rooms of a motel cost $78 per day to rent. The
motel has 6 floors and 37 rooms per floor. What is
the total rental received per day if the motel is fully
occupied?

c How many 38-passenger buses are needed to transport 646 students to the athletics stadium?

D   EXPONENTIAL (INDEX) NOTATION

When numbers like 252 are written as a product of prime factors,
that is $252 = 2 \times 2 \times 3 \times 3 \times 7$, we notice that some of the factors are identical.

A convenient way to write a product of *identical factors* is to use **exponential notation**.

For example,   32 can be written as $2 \times 2 \times 2 \times 2 \times 2$ (five identical factors, each a 2).

We can write $2 \times 2 \times 2 \times 2 \times 2$ as 2^5.

The small 5 is called the **exponent** (or **index**) and 2 is called the **base**,
therefore 2^5 tells us that five 2's are multiplied together.

For example, consider 7^4 ⟵ exponent
⟵ base number

The symbol 7^4 tells us there are 4 factors of 7 multiplied together, or $7 \times 7 \times 7 \times 7$.

The following table shows how the exponent form relates to the factorised number.

Natural number	Factorised form	Exponent form	Spoken form
2	2	2^1	two
4	2×2	2^2	two squared
8	$2 \times 2 \times 2$	2^3	two cubed
16	$2 \times 2 \times 2 \times 2$	2^4	two to the fourth
32	$2 \times 2 \times 2 \times 2 \times 2$	2^5	two to the fifth

Example 6

Write in exponent form:
$2 \times 2 \times 2 \times 2 \times 3 \times 3 \times 3$

$2 \times 2 \times 2 \times 2 \times 3 \times 3 \times 3$
$= 2^4 \times 3^3$
{4 factors of 2, and 3 factors of 3}

EXERCISE 1D

1 Write each number in exponent form:

    a $2 \times 2 \times 3 \times 3 \times 3$ **b** $2 \times 5 \times 5$

    c $2 \times 3 \times 3 \times 3 \times 5$ **d** $5 \times 5 \times 7 \times 7$

    e $2 \times 2 \times 5 \times 5 \times 5 \times 7$ **f** $3 \times 3 \times 7 \times 7 \times 11 \times 11$

Example 7	
Write as a natural number: $2^3 \times 3^2 \times 5$	$2^3 \times 3^2 \times 5$ $= 2 \times 2 \times 2 \times 3 \times 3 \times 5$ $= 8 \times 9 \times 5$ $= 40 \times 9$ $= 360$

2 Convert each product into natural number form:

    a $2 \times 3 \times 5$ **b** $2^2 \times 5$ **c** $2^3 \times 7$

    d $2 \times 3^3 \times 5$ **e** $2^2 \times 3^2 \times 11$ **f** $2^3 \times 5^2 \times 11^2$

CALCULATOR USE

The **power key** of your calculator ( ) can be used to enter numbers into the calculator when they are in **index form**.

Example 8	
Convert $2^3 \times 3^4 \times 11^2$ into natural number form.	
Key in 2 3 $\boxed{\times}$ 3 $\boxed{\wedge}$ 4 $\boxed{\times}$ 11 $\boxed{\wedge}$ 2 **ENTER** *Answer*: 78 408	

3 Use your calculator to convert each product into natural number form:

    a $2^5 \times 3^7$ **b** $2^3 \times 3^4 \times 7^3$ **c** $2^3 \times 3^2 \times 11^4$

    d $2^5 \times 5^3 \times 7^2 \times 11$ **e** $3^4 \times 5^3 \times 13^2$ **f** $2^8 \times 5^2 \times 15^3$

E   FACTORS OF NATURAL NUMBERS

The **factors** of a natural number are the natural numbers which divide exactly into it.

For example, the factors of 10 are   1, 2, 5 and 10 since

$$10 \div 1 = 10$$
$$10 \div 2 = 5$$
$$10 \div 5 = 2$$
$$\text{and} \quad 10 \div 10 = 1.$$

As $63 \times 27 = 1701$,   then 63 and 27 are factors of 1701. There may be more.

A number may have many factors.

> When we write the number as a product of factors we say it is **factorised.**

10 may be factorised as a product of two factors in two ways: 1×10 or 2×5.

12 has factors   1, 2, 3, 4, 6, 12,   and can be factorised into two factors in three ways.

These are: 1×12, 2×6,   and 3×4.

EXERCISE 1E

1 **a** List all the factors of 9.   **b** List all the factors of 12.

c Copy and complete this equation: $12 = 2 \times$ ....

d Write another pair of factors which multiply to give 12.

2 List *all* the factors of each of the following numbers:

    a 10   **b** 18   **c** 30   **d** 35

    e 44   **f** 56   **g** 50   **h** 84

3 Complete the factorisations below:

    a $24 = 6 \times$ ....   **b** $25 = 5 \times$ ....   **c** $28 = 4 \times$ ....

    d $100 = 5 \times$ ....   **e** $88 = 11 \times$ ....   **f** $88 = 2 \times$ ....

    g $36 = 2 \times$ ....   **h** $36 = 3 \times$ ....   **i** $36 = 9 \times$ ....

4 Write the largest factor (not itself) of each of the following numbers:

    a 12   **b** 18   **c** 27   **d** 48

    e 44   **f** 75   **g** 90   **h** 39

EVEN AND ODD NUMBERS

> A natural number is **even** if it has at least one factor of 2.
>
> A natural number is **odd** if it does not have 2 as a factor.

5 **a** Beginning with 8, write three consecutive even numbers.

b Beginning with 17, write five consecutive odd numbers.

6 **a** Write two even numbers which are not consecutive and which add to 10.

b Write all the pairs of two non-consecutive odd numbers which add to 20.

7 Use the words "even" and "odd" to complete these sentences correctly:

    a The sum of two even numbers is always ......

    b The sum of two odd numbers is always ......

    c The sum of three even numbers is always ......

    d The sum of three odd numbers is always ......

    e The sum of an odd number and an even number is always ......

    f When an even number is subtracted from an odd number the result is ......

    g When an odd number is subtracted from an odd number the result is ......

    h The product of two odd numbers is always ....

    i The product of an even and an odd number is always ......

PRIMES AND COMPOSITES

Some numbers can be written as the product of *two factors* only.

For example,   the only two factors of 3 are 3 and 1, and for 11 are 11 and 1.

Numbers of this type are called **prime numbers**.

> A **prime** number is a natural number greater than 1 which has exactly two factors, 1 and itself.
>
> A **composite** number is a natural number greater than 1 which has more than two factors.

From the definition of prime and composite numbers we can see that *the number 1 is neither prime nor composite.*

6 is a **composite number** since it has 4 factors: 1, 6, 2, 3.

Notice that one pair of factors of 6, namely 2 and 3, are both prime numbers. In fact, *all composite numbers can be written as the product of prime number factors.*

The **Fundamental theorem of arithmetic** is:

Every composite number can be written as a product of prime factors in one and only one way (ignoring order).

So, although $252 = 2^2 \times 3^2 \times 7$ or $3^2 \times 7 \times 2^2$,   etc, the factors of 252 cannot involve different prime base numbers.

Note: If 1 was a prime number then there would not be one and only one factorisation for each composite number.

For example, $252 = 1^3 \times 2^2 \times 3^2 \times 7$ or $1^7 \times 2^2 \times 3^2 \times 7$ etc.

PRIME FACTORS

To find the prime factors of a composite number we systematically divide the number by the prime numbers which are its factors, starting with the smallest.

Example 9

Express 252 as the product of its prime factors.

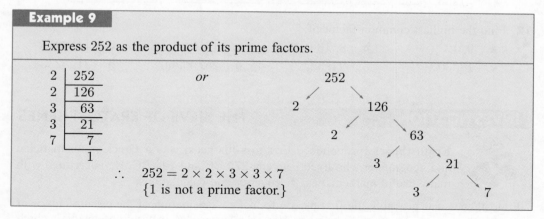

$$\begin{array}{c|c} 2 & 252 \\ \hline 2 & 126 \\ \hline 3 & 63 \\ \hline 3 & 21 \\ \hline 7 & 7 \\ \hline & 1 \end{array}$$

$\therefore$ $252 = 2 \times 2 \times 3 \times 3 \times 7$
{1 is not a prime factor.}

8   **a**   List all the prime numbers less than 30.

  b   Are there any prime numbers which are even?

9 Show that the following are composites by finding a factor other than 1 or itself:

   a 5485   **b** 8230   **c** 7882   **d** 999

10 Express each of the following numbers as a product of prime factors:

   a 24   **b** 28   **c** 63   **d** 72   **e** 136

11 Use your list of prime numbers to help you find the unknown number if:

   a I am the smallest one-digit odd prime.

   b Both of my digits are the same. The product of my digits is not a composite number.

   c I am the only odd two-digit composite number less than 20.

HIGHEST COMMON FACTOR

> A number which is a factor of two or more other numbers is called a **common factor** of these numbers.

For example, 7 is a common factor of 28 and 35.

We can use the method of finding prime factors to find the **highest common factor** (**HCF**) of two or more natural numbers.

Example 10

Find the highest common factor (HCF) of 18 and 24.

2	18
3	9
3	3
	1

2	24
2	12
2	6
3	3
	1

$18 = \mathbf{2} \times \mathbf{3} \times 3$
$24 = \mathbf{2} \times 2 \times 2 \times \mathbf{3}$

2×3 is common to both 18 and 24, and $2 \times 3 = 6$

$\therefore$ 6 is the highest common factor of 18 and 24.

---

12 Find the highest common factor of:

   a 9, 12   **b** 8, 16   **c** 18, 24   **d** 14, 42

   e 18, 30   **f** 24, 32   **g** 25, 50, 60   **h** 21, 42, 84

INVESTIGATION   THE SIEVE OF ERATOSTHENES

Eratosthenes (pronounced Erra-toss-tha-nees) was a Greek mathematician and geographer who lived between 275 BC and 194 BC. He is credited with many useful mathematical discoveries and calculations.

Eratosthenes was probably the first person to make a calculation of the circumference of the earth by using lengths of shadows. His calculation was in terms of 'stadia' which were the units of length in his day. When converted to metres, his calculation was found to be very close to modern day calculations.

Eratosthenes also found a method for 'sieving' out composite numbers from the set of naturals from 1 to 100 to leave only the primes.

The method was to

- cross out 1
- cross out all evens, except 2
- cross out all multiples of 3, except 3
- cross out all multiples of 5, except 5
- cross out all multiples of 7, except 7.

1	2	3	4	5	6	7	8	9	10
11	12	13	14	15	16	17	18	19	20
21	22	23	24	25	26	27	28	29	30
31	32	33	34	35	36	37	38	39	40
41	42	43	44	45	46	47	48	49	50
51	52	53	54	55	56	57	58	59	60
61	62	63	64	65	66	67	68	69	70
71	72	73	74	75	76	77	78	79	80
81	82	83	84	85	86	87	88	89	90
91	92	93	94	95	96	97	98	99	100

What to do:

1 Print the table above or write out as shown above, the naturals from 1 to 100 and use Eratosthenes' method to discover the primes between 1 and 100.

PRINTABLE
TABLE

2 Click on the icon to sieve for prime numbers by computer.

SIEVE OF
ERATOSTHENES

F   MULTIPLES OF NATURAL NUMBERS

The multiples of 10 are   10, 20, 30, 40, 50, ...

These are found by multiplying each of the natural numbers by 10,   i.e.,
$$1 \times 10 = 10$$
$$2 \times 10 = 20$$
$$3 \times 10 = 30$$
$$4 \times 10 = 40.$$

Likewise, the multiples of 15 are   15, 30, 45, 60, 75, ...

The number 30 is a multiple of both 10 and 15, so we say 30 is a **common multiple** of 10 and 15. Notice that 10 and 15 are both factors of 30.

In fact 30 has several factors including 2, 3, 5, 10 and 15.
30 is a *common multiple* of each of its factors.

Example 11

Find common multiples of 4 and 6 between 20 and 40.

Multiples of 4 are   4, 8, **12**, 16, 20, **24**, 28, 32, **36**, 40, ...
Multiples of 6 are   6, **12**, 18, **24**, 30, **36**, 42, ...
∴   the common multiples between 20 and 40 are   24 and 36.

EXERCISE 1F

1 List the first six multiples of:

    a 3   **b** 8   **c** 12

2 Find the:

    a fifth multiple of 6   **b** eighth multiple of 5

3 List the numbers from 1 to 30.

    a Put a circle around each multiple of 3.

    b Put a square around each multiple of 4.

    c List the common multiples of 3 and 4 which are less than 30.

4 Use this list of multiples of 15 to answer the following questions:

    15   30   45   60   75   90   105   120   135   150

    State the numbers which are common multiples of both

    a 15 and 10   **b** 15 and 9   **c** 20 and 30   **d** 4 and 30

LOWEST COMMON MULTIPLE

> The **lowest common multiple (LCM)** of two or more numbers is the smallest number which has *each* of these numbers as a *factor*.

Example 12

Find the lowest common multiple of 9 and 12.

Multiples of 9 are: 9, 18, 27, **36**, 45, 54, 63, **72**, 81, ...

Multiples of 12 are: 12, 24, **36**, 48, 60, **72**, 84, ...

∴   the common multiples are 36, 72, ... and 36 is the smallest of these

∴   the LCM is 36.

5 Find the lowest common multiples of the following sets:

    a 3, 6   **b** 4, 6   **c** 5, 8   **d** 12, 15

    e 6, 8   **f** 3, 8   **g** 3, 4, 9   **h** 5, 9, 12

6 Find the:

    a smallest multiple of 6 that is greater than 200

    b greatest multiple of 11 that is less than 500.

7 A piece of rope is either to be cut exactly into 12 metre lengths or exactly into 18 metre lengths. Find the shortest length of rope which can be used.

8 Three bells toll at intervals of 4, 6 and 9 seconds respectively. If they start to ring at the same instant, how long will it take before they will again ring together?

G | ORDER OF OPERATIONS

When two or more operations are carried out, different answers can result depending on the **order** in which the operations are performed.

For example: to find the value of $16 - 10 \div 2$, Sonia decided to subtract and then divide whereas Wei divided first and then subtracted.

Sonia's method: Subtract first, then divide.

$$16 - 10 \div 2$$
$$= 6 \div 2$$
$$= 3$$

Wei's method: Divide first then subtract.

$$16 - 10 \div 2$$
$$= 16 - 5$$
$$= 11$$

Which answer is correct, 3 or 11?

To avoid this problem, a set of rules, which states the **order of performing operations**, has been agreed upon by all mathematicians.

RULES FOR ORDER OF OPERATIONS

- Perform the operations within **B**rackets first, then
- calculate any part involving **E**xponents, then
- starting from the left, perform all **D**ivisions and **M**ultiplications as you come to them.
- Finally, work from the left, and perform all **A**dditions and **S**ubtractions.

The acronym **BEDMAS** may help you remember this order.

Note:
- If an expression contains more than one set of brackets, find the innermost brackets first.
- The division line of fractions behaves like a grouping symbol. This means that the numerator and denominator must be found before doing the division.

Using these rules, Wei's method is correct in the above example, and $16 - 10 \div 2 = 11$.

Example 13		
Evaluate: $35 - 10 \div 2 \times 5 + 3$	$35 - 10 \div 2 \times 5 + 3$	
	$= 35 - 5 \times 5 + 3$	{division first}
	$= 35 - 25 + 3$	{multiplication next}
	$= 10 + 3$	{subtraction next}
	$= 13$	{addition last}

EXERCISE 1G

1 Evaluate the following:

a $5 + 6 - 6$

b $7 + 8 \div 2$

c $8 \div 2 + 7$

d $9 \div 3 + 4$

e $100 + 6 - 7$

f $7 \times 9 \div 3$

g $30 \div 3 \div 5$

h $18 \div 3 + 11 \times 2$

i $6 + 3 \times 5 \times 2$

j $7 \times 4 - 3 \times 5$

k $8 + 6 \div 3 \times 4$

l $4 + 5 - 3 \times 2$

Example 14

Evaluate: $2 \times (3 \times 6 - 4) + 7$

$2 \times (3 \times 6 - 4) + 7$
$= 2 \times (18 - 4) + 7$ {inside brackets, multiply}
$= 2 \times 14 + 7$ {complete brackets}
$= 28 + 7$ {multiplication next}
$= 35$ {addition last}

If you do not follow the order rules, you are likely to get the wrong answer.

2 Evaluate the following: (Remember to complete the brackets first.)

a $(12 + 3) \times 2$

b $(17 - 8) \times 2$

c $(3 + 7) \div 10$

d $5 \times (7 + 3)$

e $36 - (8 - 6) \times 5$

f $18 \div 6 + 5 \times 3$

g $5 + 4 \times 7 + 27 \div 9$

h $(14 - 8) \div 2$

i $6 \times (7 - 2)$

j $17 - (5 + 3) \div 8$

k $(12 + 6) \div (8 - 5)$

l $5 \times (4 - 2) + 3$

m $36 - (12 - 4)$

n $52 - (10 + 2)$

o $25 - (10 - 3)$

Example 15

Simplify:

$5 + [13 - (8 \div 4)]$

$5 + [13 - (8 \div 4)]$
$= 5 + [13 - 2]$ {innermost brackets first}
$= 5 + 11$ {remaining brackets next}
$= 16$ {addition last}

3 Simplify:

a $[3 \times (4 + 2)] \times 5$

b $[(3 \times 4) - 5] \times 4$

c $[4 \times (16 - 1)] - 6$

d $[(3 + 4) \times 6] - 11$

e $5 + [6 + (7 \times 2)] \div 5$

f $4 \times [(4 \times 3) \div 2] \times 7$

Example 16

Simplify:

$\dfrac{16 - (4 - 2)}{14 \div (3 + 4)}$

$\dfrac{16 - (4 - 2)}{14 \div (3 + 4)}$

$= \dfrac{16 - 2}{14 \div 7}$ {brackets first}

$= \dfrac{14}{2}$ {evaluate numerator, denominator}

$= 7$ {do the division}

4 Simplify:

a $\dfrac{21}{16 - 9}$

b $\dfrac{18 \div 3}{14 - 11}$

c $\dfrac{(8 \times 7) - 5}{17}$

d $\dfrac{12 + 3 \times 4}{5 + 7}$

e $\dfrac{3 \times 7 - 5}{2}$

f $\dfrac{3 \times (7 - 5)}{2}$

5 Simplify:

    a 3×4^2

    b 2×3^3

    c $3^2 + 2^3$

    d $(5 - 2)^2 - 6$

    e $3 \times 4 + 5^2$

    f $4 \times 3^2 - (3 + 2)^2$

6 Simplify:

    a $3 \times -2 + 18$

    b $-3 \times -2 - 18$

    c $23 - 5 \times -3$

    d $\{3 - (-2 + 7)\} + 4$

    e $(18 \div 3) \times -2$

    f $2(7 - 13) - (6 - 12)$

    g $-6 \times (2 - 7)$

    h $-(14 - 8) \div -2$

    i $-18 - (8 - 15)$

    j $-52 \div (6 - 19)$

    k $\dfrac{38 - -4}{6 \times -7}$

    l $\dfrac{28 - (-3 \times 4)}{10 \times -2}$

USING A CALCULATOR

To calculate $15 \times 60 \div (8 + 7)$ press

$$15 \;\boxed{\times}\; 60 \;\boxed{\div}\; \boxed{(}\; 8 \;\boxed{+}\; 7 \;\boxed{)}\; \boxed{\textbf{ENTER}} \qquad \textit{Answer:}\quad 60$$

To calculate $\dfrac{27 + 13}{5 \times 4}$ first write as $\dfrac{(27 + 13)}{(5 \times 4)}$ then press

$$\boxed{(}\; 27 \;\boxed{+}\; 13 \;\boxed{)}\; \boxed{\div}\; \boxed{(}\; 5 \;\boxed{\times}\; 4 \;\boxed{)}\; \boxed{\textbf{ENTER}} \qquad \textit{Answer:}\quad 2$$

Notice that by pressing $27 \;\boxed{+}\; 13 \;\boxed{\div}\; 5 \;\boxed{\times}\; 4 \;\boxed{\textbf{ENTER}}$ we are finding the value of a

very different expression: $27 + \dfrac{13}{5} \times 4$ (*Answer:*   37.4)

7 Use your calculator to simplify:

    a $6 \times 8 - 18 \div (2 + 4)$

    b $10 \div 5 + 20 \div (4 + 1)$

    c $5 + (2 \times 10 - 5) - 6$

    d $18 - (15 \div 3 + 4) + 1$

    e $(2 \times 3 - 4) + (33 \div 11 + 5)$

    f $(18 \div 3 + 3) \div (4 \times 4 - 7)$

    g $(50 \div 5 + 6) - (8 \times 2 - 4)$

    h $(10 \times 3 - 20) + 3 \times (9 \div 3 + 2)$

    i $(7 - 3 \times 2) \div (8 \div 4 - 1)$

    j $(5 + 3) \times 2 + 10 \div (8 - 3)$

    k $\dfrac{27 - (18 \div 3) + 3}{3 \times 4}$

    l $\dfrac{620 - 224}{9 \times 4 \times 11}$

REVIEW SET 1A

1 $P = \{2, 3, 5, 7\}$ and $Q = \{5, 7, 9\}$

    a Find **i** $P \cap Q$ **ii** $P \cup Q$ **b** Are P and Q disjoint sets?   **c** Is $Q \subset P$?

2 Find $1834 - 712 + 78$.

3 By how much does 738 exceed 572?

4 How many times larger than 7×8 is 700×80?

5 Write down the 37th even number.

6 If 73 students have a total mass of 4161 kg, what is their average mass?

7 My bank account contains \$3621 and I make monthly withdrawals of \$78 for 12 months. What is my new bank balance?

8 List the first 6 powers of 2, starting with 2.

9 Write $2^2 \times 3 \times 5$ as a natural number.

10 Write 420 as the product of its prime factors in exponent form.

11 What is the largest prime number which will divide into 91?

12 List the factors of 162.

13 Find the largest number which divides exactly into both 63 and 84.

14 List the multiples of 6 which lie between 30 and 50.

15 Determine the LCM of 6 and 8.

16 Find the smallest multiple of 11 which is greater than 300.

17 Simplify $24 - 12 \div 2^2$

REVIEW SET 1B

1 a Show that 0.75 is a rational number. **b** Explain why $\sqrt{7}$ is an irrational number.

2 True or false: $Z \subset Q$?

3 a Find the sum of 8, 15 and 9.
  b What number must be increased by 293 to get 648?
  c Determine $5 \times 39 \times 20$. **d** Find the product of 14 and 38.
  e Find the quotient of 437 and 19.

4 Simplify the following:
  a $5 \times 6 - 4 \times 3$ **b** $3 + 18 \times 2^2$
  c $5 \times 8 - 18 \div (2 + 4)$ **d** $(5 + 4) \times 2 + 20 \div (8 - 3)$

5 a List all the factors of 84. **b** List the prime numbers between 20 and 30.
  c Express 124 as the product of prime factors in exponent form.
  d Find the highest common factor of 14 and 49.

6 a Write in exponent form: $3 \times 3 \times 5 \times 5 \times 5$
  b Convert into natural number form: $2^4 \times 3^2$

7 Find the lowest common multiple of 15 and 18.

8 What is the minimum number of sweets needed if they are to be shared exactly between either 5, 6 or 9 children?

9 Simplify:
  a {odd numbers} $\cup$ {even numbers}   **b** {odd numbers} $\cap$ {even numbers}

Chapter **2**

Measurement

TIME

In early civilisations, time was measured by regular changes in the sky. The recurring period of daylight and darkness came to be called a **day**. The Babylonians divided the day into hours, minutes and seconds.

By tracking the apparent movement of the stars at night, ancient astronomers also noticed a recurring pattern. The sun seemed to go full circle around the sky in 365 days (actually $365\frac{1}{4}$ days). This came to be known as a **year**.

The base unit of time in the International System of Units (SI) is the **second** (abbreviated **s**).

CONVERSION OF UNITS

Unlike most units within the SI system, conversions of time are **not** based on powers of 10. They are based on the external influence of the sun, moon and the earth's rotation.

1 minute = 60 seconds	1 hour = 60 minutes = 3600 seconds
1 day = 24 hours	1 week = 7 days
1 year = 12 months = $365\frac{1}{4}$ days	

For time **less than one second**, base 10 is used. We have introduced base 10 because it is easy to use. The use of such small units of time are becoming more common with the advances in computer technology.

For example, some of the units of time used in the computer world are:

$1 \text{ microsec} = \frac{1}{1\,000\,000} \text{ sec}$ and $1 \text{ nanosec} = \frac{1}{1\,000\,000\,000} \text{ sec}$.

For times **greater than one year**, base 10 is also used.

1 decade = 10 years
1 century = 100 years
1 millenium = 1000 years

Example 1

What is the time difference between 9.55 am and 1.25 pm?

9.55 am to 10.00 am = 5 min
10.00 am to 1.00 pm = 3 h
1.00 pm to 1.25 pm = 25 min
i.e., $\overline{\text{3 h 30 min}}$

EXERCISE 2A.1

1   Find the time difference between:
  a   3.20 am and 9.43 am
  b   11.17 am and 6.28 pm
  c   4.10 pm and 10.08 pm
  d   10.53 am and 3.29 pm

e   9.46 pm and 3.15 am   **f**   7.25 am and 9.45 am next day

2   Marni left home at 7.25 am and arrived at work at 9.03 am. How long did it take her to get to work?

Example 2

What is the time $3\frac{1}{2}$ hours after: **a**   10.40 am   **b**   before 1.15 pm?

a 10.40 am $+ 3\frac{1}{2}$ hours $= 10.40$ am $+ 3$ h $+ 30$ min $= 1.40$ pm $+ 30$ min $= 2.10$ pm	**b** 1.15 pm $- 3\frac{1}{2}$ hours $= 1.15$ pm $- 3$ h $- 30$ min $= 10.15$ am $- 30$ min $= 9.45$ am

---

3   Calculate the time:

  a   2 hours after 4.16 pm   **b**   4 hours before 10.15 am

  c $7\frac{1}{2}$ hours after 9.20 am   **d** $2\frac{1}{2}$ hours after 11.00 am

  e   5 hours after 10.15 pm   **f** $3\frac{1}{4}$ hours before 10.50 pm

4   Vinh caught a plane flight at 11.40 am. The flight was $5\frac{1}{2}$ hours. At what time did he arrive at his destination?

TIME ZONES

The Earth rotates from West to East about its axis. This rotation causes day and night on Earth.

As the sun rises on the east coast of the USA, the west coast is still in darkness. So it is earlier in the day in Los Angeles than in New York.

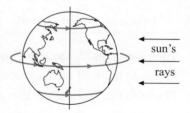
sun's
rays

As the sun rises in Melbourne in Australia, Wellington in New Zealand has already experienced about two hours of daylight. So it is later in the day in Wellington than in Melbourne.

TRUE LOCAL TIME

The earth rotates a full 360° every day.

That is 360° in   24 hours
or 15° in   1 hour $(360 \div 24 = 15)$
or 1° in   4 minutes. $(\frac{1}{15}$ of 60 min $= 4$ min$)$

Places on the same line of longitude share the same **true local time**. However, using true local time would cause problems. For example some states or counties stretch across almost 15° of longitude. This would mean time differences of up to one hour in the same state. The solution to these problems was to create **time zones**.

STANDARD TIME ZONES

The map following shows lines of longitude between the North and South Poles. These lines then follow the borders of countries, states or regions, or natural boundaries such as rivers and mountains.

The first line of longitude, 0°, passes through Greenwich near London. This first or **prime** line (meridian) is the starting point for 12 time zones west of Greenwich and 12 time zones east of Greenwich. Places which lie in the same time zone share the same **standard time**. Standard Time Zones are measured in 1 hour units.

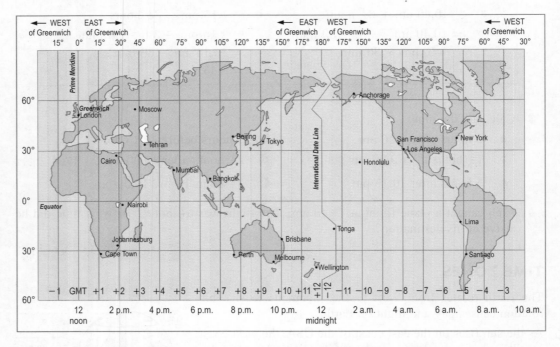

Time along the Prime Meridian is called **Greenwich Mean Time (GMT)**.

Places to the East of the Prime Meridian are **ahead** of GMT.

Places to the West of the Prime Meridian are **behind** GMT.

The numbers in the zones show how many hours have to be added or subtracted from Greenwich Mean Time to work out the **standard** time for that zone.

Example 3

If it is 12 noon in Greenwich, what is the standard time in:
a   Perth   **b**   San Francisco?

a   Perth is in a zone marked $+8$
∴   standard time in Perth is 8 hours ahead of GMT.
∴   standard time in Perth is 8 pm.

b   San Francisco is in a zone marked -8
∴   standard time in San Francisco is 8 hours behind GMT.
∴   standard time in San Francisco is 4 am.

Some cultures believe a new day begins at sunrise. Most countries use midnight as the beginning of a day.

EXERCISE 2A.2   Use the Standard Time Zone map to answer these questions.

1   If it is 12 noon in Greenwich, what is the standard time in:
   a   Moscow   **b**   Beijing   **c**   Tokyo   **d**   Santiago?

2   If it is 12 midnight Monday in Greenwich, what is the standard time in:
   a   Cairo   **b**   Mumbai   **c**   Tokyo   **d**   London?

3   If it is 11 pm on Tuesday in Greenwich, what is the standard time in:
   a   New York   **b**   San Francisco   **c**   Brisbane   **d**   Johannesburg?

4   If it is 3:15 am on Sunday in Greenwich, what is the standard time in:
   a   New Zealand   **b**   Mumbai   **c**   Santiago   **d**   Los Angeles?

5   If it is 8 pm standard time in Anchorage, what time is it in Lima?

6   Ravi lives in London. He wants to make a business call to Mumbai at 10 am Mumbai time. What (UK) time should he make the call?

7   Finland is 9 hours 'behind' New Zealand in time. James lives in Auckland, New Zealand and his girlfriend Johana lives in Helsinki, Finland.
   a   James wants to telephone Johana when she arrives home from school at 1700 hours Helsinki time. At what time in New Zealand should he make the call?
   b   Johana wants to send an email to James at the time he wakes up at 0830 hours in New Zealand. At what time in Helsinki should she send the email?

B   TEMPERATURE

The two most common scales used to measure temperature are degrees Celsius (SI unit) and degrees Fahrenheit (Imperial Unit).

The differences between these units are best illustrated with some common examples:

	Celsius (°C)	Fahrenheit (°F)
water freezes	0	32
water boils	100	212

Using this information we can construct a simple conversion graph (below) which allows us to change between these units.

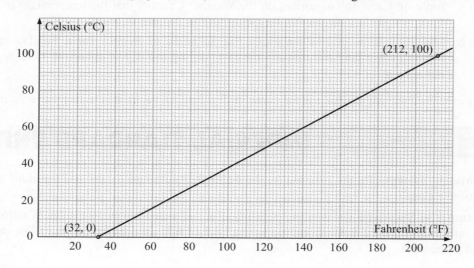

EXERCISE 2B

1   Use the temperature conversion graph to estimate:

   a   the temperature in $°C$ for: i $40°F$ ii $200°F$

   b   the temperature in $°F$ for: i $20°C$ ii $50°C$.

In order to calculate these more exactly, we can use the following formulae:

- To convert $°C$ to $°F$,   use $F = \frac{9}{5}C + 32$. F is in degrees Fahrenheit
- To convert $°F$ to $°C$,   use $C = \frac{5}{9}(F - 32)$. C is in degrees Celsius

The first formula can be rearranged to give the second.

Example 4

Convert: a $392°F$ to $°C$ b $30°C$ to $°F$

a	$C = \frac{5}{9}(F - 32)$	b	$F = \frac{9}{5}C + 32$
	$\therefore \quad C = \frac{5}{9}(392 - 32)$		$\therefore \quad F = \frac{9}{5} \times 30 + 32$
	$\therefore \quad C = 200$		$F = 86$
	i.e., $392°F \equiv 200°C$		i.e., $30°C \equiv 86°F$

2   Convert the following into $°C$:

   a $0°F$ b $100°F$ c $20°F$

3   Convert the following into $°F$:

   a $70°C$ b $-10°C$ c $400°C$

4   Check your results from question 1 using the appropriate formula.

5   Rearrange $F = \frac{9}{5}C + 32$ to show $C = \frac{5}{9}(F - 32)$.

6   a   Jonte, in Cape Town, notices on the Weather Channel that the temperature in New York is $-16°F$. Convert this temperature into degrees Celsius.

   b   Mary-Lou lives in Los Angeles. She is going to Sydney for a holiday and is told to expect temperatures around $25°C$ during the day. Convert $25°C$ to degrees Fahrenheit.

7   Find the temperature which is the same in degrees Celsius as it is in degrees Fahrenheit.

C | IMPERIAL STANDARD UNITS

The **Metric System** of units is based on the decimal system and is now used in most countries of the world.

The **Imperial Standard Units** were established in 1824 in England. This set of units was only one of many used throughout the world, and is still used by some countries; one of which is the USA.

The most commonly used units for:
- length are miles, yards, feet and inches
- weight are tons, pounds and ounces.

The basic units for **length** are:

	Imperial → Metric	Metric → Imperial
12 inches = 1 foot	1 inch ≐ 2.54 cm	1 cm ≐ 0.394 inches
3 feet = 1 yard	1 yard ≐ 91.44 cm	1 cm ≐ 0.0109 yards
1760 yards = 1 mile	1 mile ≐ 1.62 km	1 km ≐ 0.617 miles

The basic units for **weight** are:

	Imperial → Metric	Metric → Imperial
16 ounces = 1 pound	1 pound ≐ 0.454 kg	1 kg ≐ 2.20 pounds
2240 pounds = 1 ton	1 ton ≐ 0.982 tonnes	1 tonne ≐ 1.018 ton

UNIT CONVERSIONS

Example 5

Convert: **a** 4 yards to metres   **b** 2.5 tonnes to pounds

a   1 yard ≐ 91.44 cm
   ∴   4 yards ≐ 91.44 × 4 cm
              ≐ 365.76 cm
              ≐ 3.66 m

b   1 tonne ≐ 1.018 ton
   ∴   2.5 tonnes ≐ 2.5 × 1.018 ton
              ≐ 2.5 × 1.018 × 2240 pounds
              ≐ 5701 pounds

EXERCISE 2C

1   Use the table above to help perform these conversions:
   a 3.6 kg to pounds   **b** 5 miles to km   **c** 2.5 inches to cm
   d 1.5 metres to inches   **e** 8 ounces to kg   **f** 2 feet to cm

2   A hamburger contains a quarter of a pound of steak. Convert this mass to grams.

3   An American tourist in Germany noticed that his taxi was travelling at 180 km/h on the autobahn. Convert this speed to miles per hour.

4   The height of a horse is measured in 'hands'. This measurement has been used for centuries and is still used today in the horse racing and breeding industries. However, our hands vary in size so, 1 hand = 4 inches is used.

16 hands high

   a How high is the horse in inches?
   b How high is the horse to the nearest cm?

A horse which is 17 hands and 1 inch is said to be 17.1 hands high.

   c Explain why this 17.1 is not a decimal number.
   d How high in centimetres is a horse which is 15.3 hands high?

1 hand

Example 6

a   Ben is 6 feet 5 inches tall. Find his height in cm.

b   Stacey is 165 cm tall. Find her height in feet and inches.

a 1 foot = 12 inches	**b** 165 cm
$\therefore$ 6 feet 5 inches $= 6 \times 12 + 5$ inches	$\doteqdot 165 \times 0.394$ inches
$= 77$ inches	$\doteqdot 65$ inches
$\doteqdot 77 \times 2.54$ cm	$\doteqdot 5 \times 12 + 5$ inches
$\doteqdot 195.6$ cm	$\doteqdot 5$ feet 5 inches

---

5   **a**   Convert these heights to Metric units:

    i   5 feet 10 inches   **ii**   6 feet 1 inch

  b   Convert these heights to feet and inches:

    i   187 cm   **ii**   2 metres

6   Convert: **a**   6 kg to pounds and ounces   **b**   5 pounds 4 ounces to kilograms.

7   Brett works in Auckland making masts for yachts using a new light but very strong alloy. On January 7, 2002 he received a large order from the USA for masts. The USA company specified the masts required as "13 yards long and weighing 154 pounds".

  a   What do the masts weigh in kg?   **b**   How long must Brett make the masts?

8   Heidi has a business in Manheim which makes bits for electric drills. She has developed a technique for making drills which stay sharper longer than those produced by competitors. Her bits attract huge interest in the USA and one order she receives specifies drills to be $\frac{1}{8}$ inch, $\frac{3}{16}$ inch and $\frac{5}{32}$ inch diameters.

  a   What are these dimensions in millimetres to one decimal place?

  b   Which of the three imperial sizes is virtually the same as a metric bit given that the metric bits are made with diameters of 0.5 mm, 1 mm, 1.5 mm, 2 mm, ...

D   STANDARD FORM (SCIENTIFIC NOTATION)

Observe the pattern

$$10\,000 = 10^4 \quad \longleftarrow \text{ power is 4}$$
$$\div 10 \searrow \quad 1000 = 10^3 \quad \Big) -1$$
$$\div 10 \searrow \quad 100 = 10^2 \quad \Big) -1$$
$$\div 10 \searrow \quad 10 = 10^1 \quad \Big) -1$$
$$\div 10 \searrow \quad 1 = 10^0 \quad \Big) -1$$
$$\div 10 \searrow \quad \tfrac{1}{10} = 10^{-1} \quad \Big) -1$$
$$\div 10 \searrow \quad \tfrac{1}{100} = 10^{-2} \quad \Big) -1$$
$$\div 10 \searrow \quad \tfrac{1}{1000} = 10^{-3}, \quad ....\text{etc.}$$

As we divide by 10, the **exponent (power)** of 10 decreases by one.
If we continue this pattern, we get

We can use this pattern to simplify the writing of very large and very small numbers.

For example,

$$5\,000\,000$$
$$= 5 \times 1\,000\,000$$
$$= 5 \times 10^6$$

and

$$0.000\,003$$
$$= \frac{3}{1\,000\,000}$$
$$= \frac{3}{1} \times \frac{1}{1\,000\,000}$$
$$= 3 \times 10^{-6}$$

9.4488×10^{15}

This is a very large number!

5.304×10^{-11}

This is an extremely small number!

Many people in various fields of science work with very large numbers, or very small numbers, or both.

Scientific notation is used to write a number in the form $a \times 10^k$ that is, as a number between 1 and 10 multiplied by a power of 10, i.e., $1 \leqslant a < 10$ and k is an integer.

Example 7	
Write 9 448 800 000 000 000 in scientific notation.	$a = 9.4488$ To get from 9.4488 to 9 448 800 000 000 000 the decimal point is moved 15 places to the right. Hence, $k = 15$. $\therefore$ number required $= 9.4488 \times 10^{15}$

Example 8	
Write 0.000 000 000 000 053 04 in scientific notation.	$a = 5.304$ To get from 5.304 to 0.000 000 000 000 053 04 the decimal point is moved 14 places to the left. Hence $k = -14$. $\therefore$ number required $= 5.304 \times 10^{-14}$

Note: A number like 4.62 is already between 1 and 10, so it is written in standard form as 4.62×10^0 since $10^0 = 1$.

EXERCISE 2D

1   Which of the following numbers are *not* written in scientific notation?

    a 3.7×10^4 **b** 4.2×10^{-7} **c** 0.3×10^5 **d** 21×10^{11}

2 Copy and complete:

$$1000 = 10^3$$
$$100 = 10^2$$
$$10 =$$
$$1 =$$
$$0.1 =$$
$$0.01 = 10^{-2}$$
$$0.001 =$$
$$0.0001 =$$

3 Write as ordinary decimal numbers:

   a 8.2×10^4 **b** 3.6×10^1 **c** 8.7×10^0 **d** 4.9×10^2

   e 5.5×10^5 **f** 1.91×10^0 **g** 6.1×10^9 **h** 2.7×10^3

4 Write as ordinary decimal numbers:

   a 7.8×10^{-3} **b** 3.6×10^0 **c** 5.5×10^{-2} **d** 4.8×10^{-1}

   e 2.9×10^{-5} **f** 4.63×10^0 **g** 3.76×10^{-1} **h** 2.02×10^{-3}

5 Write as an ordinary decimal number:

   a The estimated population of the world in the year 2010 is 6.8×10^9 people.

   b 80 gsm photocopy paper has a thickness of 1.1×10^{-4} metres.

   c A small virus has greatest width of 9.8×10^{-4} mm.

   d The diameter of the Milky Way is 1.4×10^5 light years.

6 Express the following in scientific notation:

   a A dust particle is about 0.002 mm across.

   b The distance from the Earth to the Sun averages about 138 000 000 000 m.

   c The temperature of the sun is about 8.3 million degrees celsius.

   d Each minute, about 3 000 000 000 cells in the human body die.

LARGE AND SMALL MEASUREMENTS

Prefix	Symbol	means
giga	G	one billion (1 000 000 000) times the base unit
mega	M	one million (1 000 000) times the base unit
kilo	k	one thousand (1000) times the base unit
centi	c	one hundredth $\left(\frac{1}{100}\right)$ of the base unit
milli	m	one thousandth $\left(\frac{1}{1000}\right)$ of the base unit
micro	μ	one millionth $\left(\frac{1}{1\,000\,000}\right)$ of the base unit
nano	n	one billionth $\left(\frac{1}{1\,000\,000\,000}\right)$ of the base unit

7 A reservoir holds 2.78×10^4 megalitres of water. Write this quantity in:

   a litres   **b** gigalitres

8 Write one nanosecond in scientific notation.

E ROUNDING NUMBERS

Often we are not really interested in the exact value of a number. We just want a reasonable estimate of it. For example:

The number of people at a concert was 43 759. We are only interested in an approximate number, possibly to the nearest thousand, and $43\,759 \doteqdot 44\,000$.

Rules for rounding off are:

- If the digit after the one being rounded off is **less than 5** (i.e., 0, 1, 2, 3 or 4) we round **down.**
- If the digit after the one being rounded off is **5 or more** (i.e., 5, 6, 7, 8, 9) we round **up**.

Example 9

Round off the following to the nearest 10:

| a | 48 | b | 583 | c | 5705 |

DEMO

a	$48 \doteqdot 50$	{Round up, as 8 is greater than 5}
b	$583 \doteqdot 580$	{Round down, as 3 is less than 5}
c	$5705 \doteqdot 5710$	{Round up, halfway is always rounded up}

When a number is halfway between tens we always round up, so $255 \doteqdot 260$.

EXERCISE 2E.1

1 Round off to the nearest 10:

| a | 75 | b | 78 | c | 298 | d | 637 |
| e | 3994 | f | 1651 | g | 9797 | h | 1015 |

Example 10

Round off the following to the nearest 100: | a | 452 | b | 37 239 |

| a | $452 \doteqdot 500$ | {Round up for 5 or more} |
| b | $37\,239 \doteqdot 37\,200$ | {Round down, as 3 is less than 5} |

DEMO

2 Round off to the nearest 100:

| a | 78 | b | 468 | c | 923 | d | 954 |
| e | 5449 | f | 4765 | g | 13 066 | h | 43 951 |

3 Round off to the nearest 1000:

| a | 748 | b | 5500 | c | 9990 | d | 3743 |
| e | 65 438 | f | 123 456 | g | 434 576 | h | 570 846 |

4 Round off to the accuracy given:

  a the cost of an overseas holiday is $15 387 (to the nearest $1000)

  b the mass of a horse is 468 kg (to the nearest ten kg)

  c a weekly wage of $610 (to the nearest $100)

  d a distance of 5735 km (to the nearest 100 km)

  e the annual amount of water used in a household was 367 489 litres (to the nearest kilolitre)   {1 kL = 1000 L}

  f the population of a town is 46 495 (to nearest one thousand)

  g the population of a city is 997 952 (to nearest hundred thousand)

  h the box-office takings for a new movie were $6 543 722 (to nearest hundred thousand)

  i the area of a property is 32 457 hectares (to the nearest thousand)

  j the number of times the average heart will beat in one year is 35 765 280 times (to nearest million)

  k a year's loss by a large mining company was $1 322 469 175 (to nearest billion).

ROUNDING DECIMAL NUMBERS

If a traffic survey showed that 1852 cars carried 4376 people, it would not be sensible to give the average number per car as 2.362 850 972. An approximate answer of 2.4 is more appropriate.

There is clearly a need to shorten or **round off** some numbers which have more figures in them than are required.

We round off to a certain number of   • **decimal places**, or

  • **significant figures**.

Example 11

Round: **a** 3.27 to one decimal place   **b** 6.3829 to two decimal places.

---

  a 3.27 has 2 in the *first* decimal place
         and 7 in the *second* decimal place.

  Since 7 is in the second decimal place and is greater than 4, we increase the digit in the first decimal place by 1 and delete what follows.   So,   3.27 ≑ 3.3

  b 6.3829 has 8 in the *second* decimal place
         and 2 in the *third*

  Since 2 is less than 5, we retain the 8 and delete all digits after it.
  So,   6.3829 ≑ 6.38

EXERCISE 2E.2

1 Round the following to the number of decimal places stated in brackets.

    a 3.47   [1]   **b** 5.362   [2]   **c** 7.164   [1]

    d 15.234 [2]   **e** 9.0246 [3]   **f** 12.6234 [1]

    g 0.4372 [2]   **h** 9.276 43 [2]   **i** 0.0099 [2]

Example 12

Calculate, to 2 decimal places:

  a $(2.8 + 3.7)(0.82 - 0.57)$ **b** $18.6 - \dfrac{12.2 - 4.3}{5.2}$

---

a 

    Screen: 1.625   *Answer:* 1.63

b 18.6 ⊟ ⦅ 12.2 ⊟ 4.3 ⦆ ÷ 5.2 **ENTER**

    Screen: 17.080 769 23   *Answer:* 17.08

---

2 Find, giving your answers correct to 2 decimal places where necessary:

    a $(16.8 + 12.4) \times 17.1$ **b** $16.8 + 12.4 \times 17.1$ **c** $127 \div 9 - 5$

    d $127 \div (9 - 5)$ **e** $37.4 - 16.1 \div (4.2 - 2.7)$ **f** $\dfrac{16.84}{7.9 + 11.2}$

    g $\dfrac{27.4}{3.2} - \dfrac{18.6}{16.1}$ **h** $\dfrac{27.9 - 17.3}{8.6} + 4.7$ **i** $\dfrac{0.0768 + 7.1}{18.69 - 3.824}$

THE PROCEDURE FOR ROUNDING OFF TO SIGNIFICANT FIGURES

Rule: To round off to n significant figures, look at the $(n + 1)$th figure:

      • if it is 0, 1, 2, 3 or 4 do not change the nth figure,   DEMO

      • if it is 5, 6, 7, 8 or 9 increase the nth figure by 1,

    and delete all figures after the nth figure, replacing by 0's if necessary.

Converting to **standard form** provides us with a safe method of rounding off.

Example 13

Write **a** 278 463 correct to 3 significant figures

        b 0.007 6584 correct to 3 significant figures.

---

a $278\,463 = 2.784\,63 \times 10^5$

            $\doteqdot 2.78 \times 10^5$ {4th figure is 4, so 3rd stays as an 8}

            $\doteqdot 278\,000$

b $0.007\,6584 = 7.6584 \times 10^{-3}$

                  $\doteqdot 7.66 \times 10^{-3}$ {4th figure is 8 and so 3rd goes up by 1}

                  $\doteqdot 0.007\,66$

EXERCISE 2E.3

1   Write correct to 2 significant figures:

a	567	b	16 342	c	70.7	d	3.001	e	0.716
f	49.6	g	3.046	h	1760	i	0.040 9	j	45 600

2   Write correct to 3 significant figures:

a	43 620	b	10 076	c	$0.\overline{6}$	d	0.036 821	e	0.318 6
f	0.719 6	g	$0.6\overline{3}$	h	0.063 71	i	18.997	j	256 800

3   Write correct to 4 significant figures:

a	28.039 2	b	0.005 362	c	23 683.9	d	42 366 709
e	0.038 792	f	0.006 377 9	g	0.000 899 9	h	43.076 321

With practice converting into standard form becomes unnecessary.

Example 14

Round: a   5.371 to 2 significant figures   b   0.0086 to 1 significant figure

c   423 to 1 significant figure   d   4.053 to 3 significant figures

a $5.371 \doteqdot 5.4$ (2 s.f.)

This is the 2nd significant figure, so we look at the next figure which is 7.
The 7 tells us to round the 3 to a 4 and leave off the remaining digits.

b $0.0086 \doteqdot 0.009$ (1 s.f.)

These zeros at the front are place holders and so must stay. The first significant
figure is the 8. The next figure, 6 tells us to round the 8 to a 9 and leave off
the remaining digits.

c $423 \doteqdot 400$ (1 s.f.)

4 is the first significant figure so it has to be rounded. The second figure, 2,
tells us to keep the original 4 in the hundreds place. We convert the 23 into
00. These two zeros are place holders which are not significant but need to
be there as it would be silly to say $423 \doteqdot 4$.

d $4.053 \doteqdot 4.05$ (3 s.f.)

This 0 is significant as it lies between two non-zero digits.

---

4   How many significant figures are in:

a	62	b	620	c	0.062	d	0.602	e	0.060 02
f	137	g	0.137	h	13.7	i	1.3007	j	0.103 07

5   Round correct to the number of significant figures shown in brackets.

a	42.3	[2]	b	6.237	[3]	c	0.0462	[2]
d	0.2461	[2]	e	437	[2]	f	2064	[2]

g	31 009	[3]	**h**	10.27	[3]	**i**	0.999	[1]
j	0.999	[2]	**k**	264 003	[4]	**l**	0.037 642	[4]

6 The crowd at a rugby match was officially
26 247 people.

  a Round the crowd size to:
    i 1 significant figure
    ii 2 significant figures

  b Which of these figures would be used
    by the media to indicate crowd size?

USING SCIENTIFIC NOTATION ON A CALCULATOR

If you used your calculator in the previous exercise, you may have noticed that it gives answers in **scientific notation** when the answers are too large or too small to fit on the display.

If we find 47 000 $\boxed{\times}$ 500 000 we get $\boxed{2.35E\,10}$ displayed and this is actually 2.35×10^{10}.

If we find 0.000 036 $\boxed{\div}$ 5 000 000 we get $\boxed{7.2E\text{-}12}$ and this is actually 7.2×10^{-12}.

To set the calculator to use scientific notation, click on the appropriate icon.

EXERCISE 2E.4

1 Write these calculator display numbers in scientific notation:

  a $\boxed{4.5E\,07}$ **b** $\boxed{3.8E\text{-}04}$ **c** $\boxed{2.1E\,05}$ **d** $\boxed{4.0E\text{-}03}$

  e $\boxed{6.1E\,03}$ **f** $\boxed{1.6E\text{-}06}$ **g** $\boxed{3.9E\,04}$ **h** $\boxed{6.7E\text{-}02}$

2 Write the numbers displayed in question **1** as ordinary decimal numbers.

  For example, $\boxed{3.9E\,06}$ is 3.9×10^6

                         $= 3.900\,000$

                         $= 3\,900\,000$

3 Calculate the following giving answers in scientific notation correct to three significant figures (where necessary).

  a $0.002 \times 0.003 \div 15\,000$ **b** $70\,000 \times 500^3$ **c** $\sqrt{0.000\,078\,2}$

  d $1250^2 \times 650^3$ **e** $(0.000\,172)^4$ **f** $\dfrac{385 \times 1250}{0.000\,006\,7}$

4 A rocket travels at 2.8×10^4 kmph in space. Find how far it would travel in:

  a 5 hours   **b** a day   **c** a year

  (Give answers in standard form correct to 3 sig. figs. and assume that
  1 year $\doteqdot$ 365.25 days.)

5 Sheets of paper are 1.2×10^{-1} mm thick. Find how many sheets are needed to make a pile of paper:

    a   2 mm high   **b**   20 mm high   **c**   1 metre high

6 The average heart beats 4×10^3 times in one hour. Find how many beats there are in:

    a   5 hours   **b**   a week   **c**   a year

7 Evaluate the following, giving your answers in scientific form, correct to 3 sig. figs.

    a $234\,000\,000 \times 289\,800\,000$ **b** $998\,600\,000\,000 \times 122\,000\,000$

    c $0.4569 \times 0.000\,000\,399\,3$ **d** $0.000\,000\,988 \times 0.000\,000\,768$

    e $(394\,550\,000)^2$ **f** $(0.000\,000\,120\,34)^2$

8 **a** How many seconds are there in one minute?

    b How many minutes are there in an hour?

    c How many hours are there in a day?

    d If there are 365 days in a year, show that there are $31\,536\,000$ seconds in a year, expressing this value in scientific notation.

    e Light travels approximately 9.4488×10^{15} m in a year.

        How far does light travel in one second?

        Give your answer in scientific notation, correct to 3 significant figures.

Example 15

Use your calculator to simplify, correct to 3 significant figures:

    a $(3.2 \times 10^4) \times (8.6 \times 10^{-16})$ **b** $\dfrac{2.8 \times 10^5}{6.4 \times 10^{-8}}$

  a   Press: 3.2 $\boxed{\text{2nd}}$ $\boxed{\text{EE}}$ 4 $\boxed{\times}$ 8.6

                    $\boxed{\text{2nd}}$ $\boxed{\text{EE}}$ $\boxed{\text{(-)}}$ 16 $\boxed{\text{ENTER}}$

        Answer: 2.75×10^{-11}

```
3.2E4*8.6E-16
            2.75E-11
■
```

  b   Press: 2.8 $\boxed{\text{EXP}}$ 5 $\boxed{\div}$ 6.4 $\boxed{\text{EXP}}$ $\boxed{\text{(-)}}$ 8 $\boxed{\text{EXE}}$

        Answer: 4.38×10^{12}

```
2.8E5÷6.4E-8
            4.38E+12
```

EXERCISE 2E.5

1 Use your calculator to simplify, correct to 3 significant figures:

    a $(4.7 \times 10^5) \times (8.5 \times 10^7)$ **b** $(2.7 \times 10^{-3}) \times (9.6 \times 10^9)$

    c $(3.4 \times 10^7) \div (4.8 \times 10^{15})$ **d** $(7.3 \times 10^{-7}) \div (1.5 \times 10^4)$

    e $(2.83 \times 10^3)^2$ **f** $(5.96 \times 10^{-5})^2$

2 Use your calculator to answer the following:

There are 365.25 days in a year.

> **a** A rocket travels in space at 4×10^5 km/h. How far does it travel in:
>
> **i** 30 days   **ii** 20 years?
>
> **b** A satellite travels 5×10^3 km in 2×10^{-5} hours. Find its average speed in kilometres per hour.
>
> **c** A bullet travelling at an average speed of 2×10^3 km/h hits a target 500 m away. Find the time of flight of the bullet in seconds.
>
> **d** The planet Mars is 2.28×10^8 km from the sun whilst Mercury is 5.79×10^7 km from the sun. How many times further from the sun is Mars than Mercury?
>
> **e** Microbe C has mass 2.63×10^{-5} grams, whereas microbe D has mass 8×10^{-7} grams. Which microbe is heavier? How many times is it heavier than the other one?

F   RATES

A **rate** is an ordered comparison of quantities of different kinds.

Some examples of rates are:
- **rates of pay**   (dollars per hour)
- **petrol consumption**   (litres per 100 km)
- **annual rainfall**   (mm per year)
- **unit cost**   (dollars per kilogram)
- **population density**   (people per square km)

RESEARCH   RATE DATA

Obtain internet data on:
- the average annual rainfall rate of your city and the breakdown into average monthly rates. Compare these with rates obtained for other cities.
- comparative rates of petrol consumption of cars. Also obtain comparative rates between 4 cylinder and 6 cylinder cars.

SPEED

One of the most common rates we use is **speed**. Speed is actually a comparison between the distance travelled and the time taken. Because we are comparing quantities of different kinds (e.g., distance and time) the units cannot be omitted as they are with ratios.

Thus, speed is usually expressed in kilometres per hour (km/h).

For example,   a car travelling 400 km in 5 hours is moving at $\dfrac{400 \text{ km}}{5 \text{ h}} = 80$ km/h.

$$\text{Average speed} = \frac{\text{distance travelled}}{\text{time taken}} \qquad \text{Time taken} = \frac{\text{distance travelled}}{\text{average speed}}$$

$$\text{Distance travelled} = \text{average speed} \times \text{time taken}$$

Example 16

A car is travelling a distance of 325 km.
a   What is its average speed if the trip takes 4 h 17 min?
b   What is the time taken if the average speed was 93 km/h?

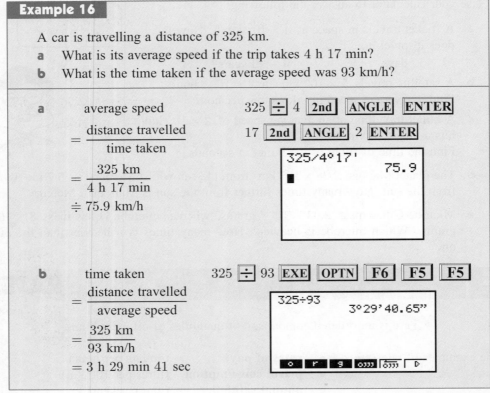

a   average speed

$$= \frac{\text{distance travelled}}{\text{time taken}}$$

$$= \frac{325 \text{ km}}{4 \text{ h } 17 \text{ min}}$$

$$\doteqdot 75.9 \text{ km/h}$$

325 ÷ 4 [2nd] [ANGLE] [ENTER]
17 [2nd] [ANGLE] 2 [ENTER]

```
325/4°17'
               75.9
■
```

b   time taken

$$= \frac{\text{distance travelled}}{\text{average speed}}$$

$$= \frac{325 \text{ km}}{93 \text{ km/h}}$$

$$= 3 \text{ h } 29 \text{ min } 41 \text{ sec}$$

325 ÷ 93 [EXE] [OPTN] [F6] [F5] [F5]

```
325÷93
          3°29'40.65"
```

EXERCISE 2F.1

1   Find the average speed of a car travelling:
    a   71.2 km in 51 minutes
    b   468 km in 5 hours 37 minutes.

2   An aeroplane flies a path from Adelaide to Melbourne which is 726 km and it takes 58 minutes. What is the average speed of the plane in km/h?

3   To convert km/h to m/s divide the km/h by 3.6 . Which is faster, 100 km/h or 30 m/s?

4   Find the distance travelled if you are travelling at an average speed of:
    a   95 km/h for 3 hours 23 minutes
    b   25.3 km/h for 1 hour 17.5 min

5   How long would it take to travel 42.3 km at an average walking speed of 5.7 km/h?

INVESTIGATION 1   STOPPING DISTANCES

A car does *not* stop the instant you decide to do so. Two factors control how far a car travels between you seeing a problem, and the car coming to a halt.

The first is called the **reaction time**, the time for the driver to react and hit the brake pedal, and the second is the **braking distance**, the distance the car travels when the brakes are being applied.

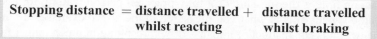

**Stopping distance = distance travelled + distance travelled
whilst reacting   whilst braking**

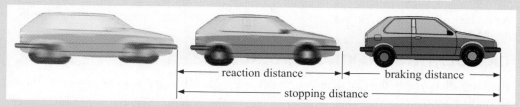

reaction distance ◄────► braking distance

◄──────────────── stopping distance ────────────────►

MEASURING REACTION TIME

Click on the reaction time demo *or*

REACTION
TIME DEMO

   use: •   Electronic timer (measuring to hundredths of
              a second) with remote switching capacity

           •   Pedal simulator

           •   Light to suit timer mechanism

The "driver" is to have his/her foot resting on the "accelerator"
pedal. When the light flashes the foot is moved rapidly to the
"brake" pedal to press it.

Do this test five times, and average the reaction times obtained.

CALCULATING REACTION DISTANCE

As you have a good estimate of your personal reaction time, you can now find how far
you have travelled in that time. To do this, multiply your reaction time by the speed in
metres per second.

> To convert **km/h** to **m/s** divide the km/h by **3.6**

For example, if you are travelling at 30 km/h and your reaction time is 0.5 seconds, then

      Step 1:   30 km/h = (30 ÷ 3.6) m/s = 8.333 m/s

      Step 2:   Distance travelled = 0.5 × 8.333 = 4.166 metres, *before* you hit
                  the brake pedal.

What to do:

1   Calculate the speed in m/sec for a speed of 60 km/h.

2   Using your own reaction time, repeat the calculations above to find your **reaction
distance** for 60 km/h.

3   Having calculated your reaction distance for 60 km/h in part **2**, now use your personal
reaction time from the test to calculate your reaction distances at each of the following
speeds:

    (Remember, this is the distance covered *before the car even begins to slow down!*)

Personal reaction distances

Speed in km/h	30	40	50	60	70	80	90	100	110	120
Speed in m/sec	8.33									
Reaction distance										

Simple reaction time is the measure of a fully prepared person waiting for a signal, and their reaction to the signal. This is what you measured in the above exercises. In the real world of driving, many things can distract or impair a driver, resulting in much longer reaction times.

As a class, discuss possible ways that your reaction time may be increased, and list these on the board.

CALCULATING BRAKING DISTANCES

Braking distance is the distance you travel *after* you have applied the brakes as the car slows to a stop.

On a good bitumen surface, with a car in perfect condition, it takes about 28 m to stop while travelling at 60 km/h. This gets longer on wet roads, or gravel surfaces.

We can calculate the effects of different road surfaces on braking distances using the following formula:

distance travelled (in m) = speed (in m/sec) × (2.08 + 0.96 × surface factor)

What to do:

Calculate the braking distances for each of the speed and surface conditions listed in the table alongside. Copy and complete the table which follows:

(You may like to set up a **spreadsheet** to do this.)

Surface factor	
dry asphalt	1.4
wet asphalt	1.7
gravel	2.1
hard snow	6.7
ice	14.4

Speed (km/h)	Speed (m/s)	Dry asphalt	Wet asphalt	Gravel	Snow	Ice
50						
60						
70						
⋮						
120						

CALCULATING STOPPING DISTANCE

Remember: **stopping distance = reaction distance + braking distance**

What to do:

Using your Personal reaction distances, find your total stopping distances for the conditions listed previously. Copy and complete a table like the one following or set up a **spreadsheet**:

Personal stopping distances

Condition	Personal reaction distance	Braking distance	Stopping distance
30 km/h, dry bitumen			
50 km/h, dry bitumen			
60 km/h, dry bitumen			
80 km/h, dry bitumen			
⋮			

OTHER RATES PROBLEMS

Example 17

Convert 35 apples bought for $9.45 to a rate of *cents per apple*.

35 apples bought for $9.45 is a rate of

$9.45 per 35 apples

$= 945$ cents / 35 apples

$= \dfrac{945 \text{ cents}}{35 \text{ apples}}$

$= 27$ cents per apple

EXERCISE 2F.2

1 Copy and complete:

   a If 24 kg of peas are sold for $66.24, they earn me $...... per kg.

   b My car uses 18 L of petrol every 261 km. The rate of petrol consumption is ...... km per litre.

   c 675 litres of water are pumped into a tank in 25 minutes. This is a rate of ...... L/minute.

   d Jasmin is paid $57 for 6 hours work. This is a rate of ...... /h.

   e A temperature rise of 14 degrees in $3\frac{1}{2}$ hours is a rate of ...... degrees per hour.

   f 38.5 kg of seed spread over 7 m^2 is a rate of ...... kg/m^2.

   g $173.47 for 1660 kwh of power is a rate of ...... cents/kilowatt hour.

   h Dominic types 220 words in 4 minutes. This rate is ...... words/minute.

2 A worker in a local factory earns $11.67 per hour.

   a How much does he earn in a 40 hour week?

   b If he receives the same weekly wage but works only a 35 hour week, what will be his new hourly rate?

Example 18

Suburb A covers 6.3 km^2 and has a population of 28 700 people while suburb B covers 3.9 km^2 and has a population of 16 100 people. Which suburb is more heavily populated?

Suburb A has $\dfrac{28\,700 \text{ people}}{6.3 \text{ km}^2} \doteqdot 4556$ people per km^2.

Suburb B has $\dfrac{16\,100 \text{ people}}{3.9 \text{ km}^2} \doteqdot 4128$ people per km^2.

∴   Suburb A is more heavily populated.

3 A farmer harvested 866 bags of wheat from a 83 hectare paddock (A) and 792 bags from a 68 hectare paddock (B). Which paddock yielded the better crop per hectare?

4 When a netball club decided the winner of the trophy for the highest number of goals

thrown per match, the two contenders were: Pat, who threw 446 goals in 18 matches, and Jo, who threw 394 goals in 15 matches. Who won the trophy?

5 A family uses 96 kilolitres of water in 90 days.

   a Find the rate in litres per day.

   b If the water board charges 65 cents per kilolitre, how much do they pay for the water used:

      i   over the 90 day period   **ii**   per day?

6 Phillipa types at a rate of 50 words per minute.

   a How long would it take her to type a 500 word essay at this rate?

   b How much longer would it take Kurt to type this essay if he types at 35 words per minute?

7 The cost for electricity is 13.49 cents/kwh for the first 300 kwh, then 10.25 cents/kwh for the remainder. How much does 2050 kwh of power cost?

8 The temperature at 2.30 pm was 11°C and fell steadily until it reached −2°C at 1.45 am. Find the

   a decrease in temperature

   b rate of decrease per hour (to 2 decimal places).

9 Convert: **a** $12 per hour into cents per minute

           b   240 000 litres per min into kL per second

           c   30 mL/second into litres per hour

           d $2.73 per gram to dollars per kilogram

           e   1 death every 10 min to deaths per year.

G MEASURING DEVICES AND THEIR ACCURACY

We measure quantities with various instruments and devices.

INVESTIGATION 2   MEASURING DEVICES

Examine a variety of measuring instruments at school and at home. Make a list of the names of these instruments, what they measure, what the units are and the degree of accuracy they can measure too.

For example

A ruler measures length. In the Metric System it measures in centimetres and millimetres and can measure to the nearest millimetre. Record your findings in a table.

Record your findings in a table.

READING MEASURING INSTRUMENTS

An electricity meter measures the amount of electricity used by a consumer in **kilowatt-hours**.

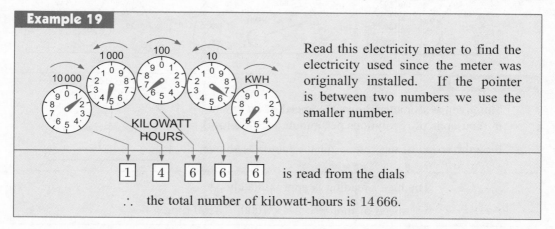

Example 19

Read this electricity meter to find the electricity used since the meter was originally installed. If the pointer is between two numbers we use the smaller number.

| 1 | 4 | 6 | 6 | 6 | is read from the dials

∴ the total number of kilowatt-hours is 14 666.

Note:
- Observe the reverse directions of alternating dials due to the cog driven system.
- Many new meters are now digital.

EXERCISE 2G

1 Read the following electricity meters:

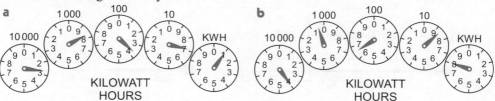

a

b

2 Find the readings on the following meter scales:

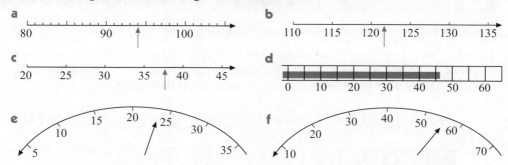

a

b

c

d

e

f

3 Find the fraction of fuel remaining in the following:

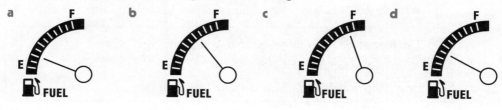

a   b   c   d

4   Find the speed from the following speedometers:

a
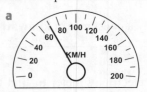

b
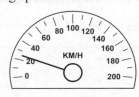

c
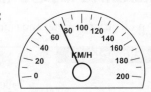

Example 20

The given tachometer shows the speed of an engine in thousands of revolutions per minute ($\times 1000$ rpm).

Find the speed of the engine for the motor as shown.

The meter reading is approximately $3\frac{1}{4}$

$\therefore$ speed of motor $= 3\frac{1}{4} \times 1000 = 3250$ rpm.

5   Find the speed of the following engines:

a

b
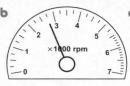

c

d

6   Read the following thermometers:

a
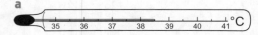

b
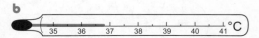

H   ACCURACY OF MEASUREMENTS

When we take measurements, we are usually reading some sort of scale.

The scale of a ruler may have millimetres (mm) marked on it, but when we measure the length of an object, it is likely to fall between two divisions. So, we **estimate** to the nearest mm say. The ruler is really only accurate to the nearest half a mm.

A measurement is only accurate to $\pm\frac{1}{2}$ of the smallest division on the scale.

Example 21

Rod used a tape measure with graduations in cm to measure his height. What is Rod's height range if he estimates his height to be 188 cm?

The tape measure is only accurate to $\pm\frac{1}{2}$ cm

$\therefore$ range of heights is $188 \pm \frac{1}{2}$ cm

i.e.,   Rod is between 187.5 cm and 188.5 cm tall.

EXERCISE 2H

1   State the accuracy of the following measuring devices:

    a   a ruler marked in mm   **b**   a set of scales marked in kg

    c   a tape measure marked in cm

    d   a measuring cup marked with 100 mL graduations.

2   Su-Lin used a tape measure marked with cm to measure the following lengths. State the range of lengths possible for each.

    a   155 cm   **b**   210 cm   **c**   189 cm

3   Stefan used a set of scales marked with kg to weigh some friends. State the range of weights possible for each person:

    a   Kylie 53 kg   **b**   Nev 95 kg   **c**   Wei 79 kg

Example 22

Find the range of values for a measurement of 18.7 cm.

18.7 cm is 187 mm, so the measuring device must be accurate to the nearest half mm.

$\therefore$ range of values is $187 \pm \frac{1}{2}$ mm

         i.e., $186\frac{1}{2}$ mm to $187\frac{1}{2}$ mm   i.e.,   18.65 cm to 18.75 cm.

---

4   Find the range of values for the following measurements:

    a   27 mm   **b**   38.3 cm   **c**   4.8 m

    d   1.5 kg   **e**   25 g   **f**   3.75 kg

5   Four students measured the width of their classroom using the same tape measure. The measurements were 6.1 m, 6.4 m, 6.0 m, 5.9 m.

    a   Which measurement is likely to be incorrect?

    b   What answer would you give for the width of the classroom?

    c   What graduations do you think were on the tape measure?

Example 23

A rectangular block of wood was measured at 78 cm by 24 cm. What are the boundary limits of its perimeter?

The length of 78 cm could be from $77\frac{1}{2}$ cm to $78\frac{1}{2}$ cm.

The width of 24 cm could be from $23\frac{1}{2}$ cm to $24\frac{1}{2}$ cm.

$\therefore$ perimeter's lower boundary is $2 \times 77\frac{1}{2} + 2 \times 23\frac{1}{2} = 202$ cm

$\therefore$ perimeter's upper boundary is $2 \times 78\frac{1}{2} + 2 \times 24\frac{1}{2} = 206$ cm

i.e.,   the perimeter is between 202 cm and 206 cm.   i.e., 204 ± 2 cm.

---

6   A rectangular bath mat was measured as 86 cm by 38 cm to the nearest centimetre. What are the boundary values of its perimeter?

7 A garden bed is measured to have dimensions 252 cm by 143 cm. Between what two values could the length of edging required to border the garden bed be?

8 A rectangle is measured to be 6 cm by 8 cm. What is:

    a the largest area it could have   **b** the smallest area it could have?

Example 24

A paver is measured as 18 cm × 10 cm. What are the boundary values of its actual area?

The length of 18 cm could be from $17\frac{1}{2}$ cm to $18\frac{1}{2}$ cm.

The width of 10 cm could be from $9\frac{1}{2}$ cm to $10\frac{1}{2}$ cm.

$\therefore$ lower boundary of the area is $17\frac{1}{2} \times 9\frac{1}{2} = 166.25$ cm^2

$\therefore$ upper boundary of the area is $18\frac{1}{2} \times 10\frac{1}{2} = 194.25$ cm^2

i.e., the area is between 166.25 cm^2 and 194.25 cm^2.

(This could also be represented as $\dfrac{166.25 + 194.25}{2} \pm 14$ cm^2

$\doteqdot 180 \pm 14$ cm^2)

---

9 Find the boundary values for the actual area of a glass window measured as 42 cm by 26 cm.

10 A triangle has its base measured as 9 cm and height as 8 cm. What are the boundary values for its actual area?

Example 25

A rectangular box container is measured as 10 cm by 6 cm by 12 cm. What are the boundary values of its actual volume?

The length measurement is between 9.5 cm and 10.5 cm. Likewise the width is between 5.5 cm and 6.5 cm and depth between 11.5 cm and 12.5 cm.

$\therefore$ lower boundary for the volume $= 9.5 \times 5.5 \times 11.5$
$$= 600.875 \text{ cm}^3$$

and the upper boundary for the volume $= 10.5 \times 6.5 \times 12.5$
$$= 853.125 \text{ cm}^3$$

$\therefore$ the actual volume is between 600.875 cm^3 and 853.125 cm^3.

---

11 Find the boundary values for the volume of a box measuring 4 cm by 8 cm by 6 cm.

12 Find the boundary values for the actual volume of a house brick measuring 21.3 cm by 9.8 cm by 7.3 cm.

13 A cylinder has radius 5 cm and height 15 cm. Find the boundary values for the cylinder's volume. $\{V = \pi r^2 h\}$

14 A cone has a measured radius of 8.4 cm and height 4.6 cm. Find the boundary values for its volume. $\{V = \frac{1}{3}\pi r^2 h\}$

❙ ERROR AND PERCENTAGE ERROR

APPROXIMATING AND ESTIMATING

An **approximation** is a value given to a number which is close to, but not equal to, its true value.

For example,   36.428 97   is approximately   36.4

An **estimation** of a quantity is an approximation which has been found by judgement instead of carrying out a more accurate determination.

For example, 38.7×5.1 is estimated to be $40 \times 5 = 200$ whereas its true value

is   197.37   and a good approximation using this true value is 197.

In order to make reasonable estimation we appeal to our previous experience.

Here are some things to help you visualise some of the units:

INVESTIGATION 3   **A GRAM IN THE HAND IS WORTH .....**

What to do:

1 Measure in mm the length and breadth of a sheet of 80 gsm A4 photocopying paper.

2 What is its area in m^2 and how many sheets make up 1 m^2?

3 80 gsm means 80 grams per square metre. What is the mass of one sheet of A4 paper?

4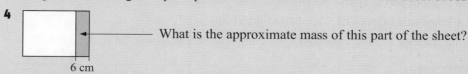
What is the approximate mass of this part of the sheet?

6 cm

5 Crumple the 6 cm strip in your hand and feel how heavy it is.

ERRORS IN ESTIMATING

On TV it was stated that the crowd size at Wimbledon was 27 000 whereas the actual number was 26 784.

The error in rounding $= 27\,000 - 26\,784 = 216$.

> **Error = estimated value − true value**

The **percentage error** is found by comparing the error with the actual value and expressing this as a percentage.

ABSOLUTE PERCENTAGE ERROR

$$\textbf{Absolute percentage error} = \frac{|\textbf{error}|}{\textbf{actual value}} \times \textbf{100\%}$$

Note: The vertical bars $|\ldots|$ means   *modulus of ....*

The **modulus** of a number is its absolute value, i.e., its size ignoring its sign.

For example, $|-2| = |2| = 2$.

For the crowd size at Wimbledon example,   absolute percentage error $= \dfrac{216}{26\,784} \times 100\%$

$\doteqdot 0.81\%$

Example 26

Suppose we are rounding a crowd size of $45\,471$ to $45\,000$.   Find:
a   the error   **b**   the absolute percentage error

a   Error
= estimated value $-$ true value
$= 45\,000 - 45\,471$
$= -471$

b   Absolute percentage error

$= \dfrac{|\text{error}|}{\text{actual value}} \times 100\%$

$= \dfrac{471}{45\,471} \times 100\%$

$\doteqdot 1.04\%$

EXERCISE 21

1 Find **i**   the error   **ii**   the absolute percentage error   in rounding:

a   the yearly profit of $\$1\,367\,540$ made by a company to $\$1.37$ million.

b   a population of $31\,467$ people to $31\,000$ people

c   a retail sales figure of $\$458\,110$ to $\$460\,000$

d   the number of new cars sold by a company in a year was 2811, to 3000.

Example 27

You estimate a fence's length to be 70 m and when you measure it, its true length is 78.3 m.   Find, correct to one decimal place:
a   the error   **b**   the absolute percentage error.

a $70 - 78.3 = -8.3$
$\therefore$ the error is -8.3 m

b   absolute percentage error

$= \dfrac{|\text{error}|}{\text{actual value}} \times 100\%$

$= \dfrac{8.3}{78.3} \times 100\%$

$\doteqdot 10.6\%$

2 Find the **i** error   **ii** absolute percentage error   if you estimate:

    a the mass of a brick to be 5 kg when its measured mass is 6.238 kg

    b the perimeter of a property to be 100 m when its measured length is 97.6 m

    c the capacity of a container to be 20 L when its measured capacity is 23.8 L

    d the time to solve a problem to be 50 hours when it actually took 72 hours.

3 Jon's lounge room is exactly a 10.3 m by 9.7 m rectangle.

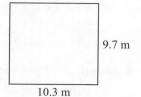

    a What is the floor area if he rounded each length to the nearest metre?

    b What is the actual area of the floor?

    c What is Jon's rounding error for the floor's area?

    d What absolute percentage error was made?

4 The cost of freight on a parcel is dependent on its volume. Justine lists the dimensions of a parcel as 24 cm by 15 cm by 9 cm on the consignment note. The actual dimensions are 23.9 cm × 14.8 cm × 9.2 cm.

    a Calculate the actual volume.

    b Calculate the volume given on the consignment note.

    c Find the rounding error in the calculation.

    d What absolute percentage error was made?

5 Luigi estimates that he can travel at an average of 70 km/h between his home and the nearest town, 87 km away. On one particular journey it took him 1 hour and 20 minutes.

    a Calculate his average speed for this journey.

    b Find the error in his estimate.

    c Find the absolute percentage error in his estimate.

ESTIMATION v MEASURING

Sometimes it is better to estimate rather than measure providing that our estimation is reasonably accurate. The factors which influence our decision to estimate or measure might be:

- the time taken to measure and calculate
- the difficulty of measuring
- the accuracy needed.

INVESTIGATION 4   ESTIMATING AND ACCURACY

The purpose of this investigation is to carry out estimates of angle size, length, mass, temperature and time and then to check your ability to estimate with actual measurement.

What to do:

1 **Angle size**   *Equipment needed:*   Protractor

    a Draw three angles, one acute, one reflex and one obtuse.

    b Estimate the size of each angle in degrees.

    c Cut out an angle of exactly $30°$ using a protractor and use this angle to help you re-estimate the three angle sizes.

d Measure each angle with a protractor.

e Copy and complete a table like the one given.

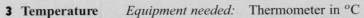

angle	estimate 1	estimate 2	measured size
i			
$\vdots$			

f Find the error in each case for both estimates and determine the absolute percentage errors.

g Did the 30° angle help to improve your estimate?

2 Length   *Equipment needed:*   Large tape measure, trundle wheel

a Estimate the length and width of *one* of the following:
   - your school tennis courts   (backstop to backstop)
   - your school gym   (wall to wall)
   - a rectangular school building   (wall to wall)

b Use paces to estimate the lengths more accurately.

c Check your estimates using a tape measure or trundle wheel.

d Find absolute percentage errors in each of your estimates.

3 Temperature   *Equipment needed:*   Thermometer in °C

a Make an estimate of the temperature of:

   i   air in the classroom   **ii**   air outside the classroom

   iii   water from the cold tap   **iv**   water from the hot tap (careful)

b Use an ordinary thermometer to measure the temperature of the air and water in **a**.

c Comment on the accuracy of your estimates.

4 Time   *Equipment needed:*   Stopwatch or watch with seconds

DEMO

a In pairs test each other at estimating a time interval of one minute.

b Can you find a way to improve your method of estimating?

REVIEW SET 2A

1 Find the time difference between 10.35 am and 4.52 pm.

2 What is the time:

   a $4\frac{1}{4}$ hours after 11.20 pm   **b** $3\frac{1}{2}$ hours before 1.20 pm

3 If it is 12 noon in Greenwich, what is the standard time in Tokyo if Tokyo is 9 hours ahead of GMT.

4 A racehorse weighs 540 kg. If 1 kg $\doteqdot$ 2.20 pounds, calculate its mass in pounds.

5 It is 35°C in a town in Mexico. Use the formula $F = \frac{9}{5}C + 32$ to calculate this temperature in degrees Fahrenheit.

6 Express the following quantities as an ordinary decimal number:

  a The sun has a diameter of 1.392×10^6 km.

  b A red blood cell is approximately 8.4×10^{-6} m   in diameter.

7 Use your calculator to answer the following. A rocket travels in space at 3.6×10^4 kmph.   Find how far it would travel in

  a a day   **b** a year   (assume 1 year = 365.25 days)

  Give your answer in standard form correct to 3 significant figures.

8 **a** Round 6.376 to: **i** 1 decimal place   **ii** 3 significant figures

  b Round 0.0473 to: **i** 2 decimal places   **ii** 2 significant figures

9 A cyclist is travelling a distance of 134 km.

  a What is her average speed, if the trip takes 5 hours and 18 minutes?

  b What is the time taken if her average speed is 24 km/h?

10 **a** How accurate is a tape measure marked in cm?

  b Find the range of values for a measurement of 36 cm.

  c If a square has sides measured to be 36 cm, between what two boundary values does its actual area lie?

11 Find the   **i** error   **ii** absolute percentage error if you:

  a estimate your credit card balance to be $2000 when it is $2590.

  b round 26.109 cm to 26 cm.

REVIEW SET 2B

1 Convert $2\frac{3}{4}$ hours to minutes.

2 Aniko left for work at 7.39 am and returned home at 6.43 pm. How long was she away from home?

3 Nairobi is 2 hours ahead of GMT and Honolulu is 10 hours behind GMT. Calculate the standard time in Nairobi when it is 2 pm in Honolulu.

4 Use the formula $C = \frac{5}{9}(F - 32)$ to convert $84°$F to degrees Celsius.

5 Bianca is 5 feet 11 inches tall. Calculate her height in cm, given 1 foot = 12 inches and 1 inch $\doteqdot$ 2.54 cm.

6 Write as ordinary decimal numbers:

  a 4.6×10^{11} **b** 1.9×10^0 **c** 3.2×10^{-3}

7 Write in scientific notation:

  a The diameter of the earth is approximately 12.76 million metres.

  b A bacterium has a diameter of 0.000 000 42 cm.

8 Use your calculator to answer the following: Sheets of paper are 3.2×10^{-4} m   in thickness. Find how many sheets are required to make a pile of paper 10 cm high.

9 **a** Round 59.397 to: **i** 1 decimal place   **ii** 4 significant figures.

  b Round 0.008 35 to: **i** 2 decimal places   **ii** 2 significant figures.

10 Paul works at the local supermarket and earns $13.75 per hour.

  a How much does he earn in a 37 hour week?

  b If he receives the same weekly wage but now works only a 35 hour week, what will be his new hourly rate?

11 The cost for electricity is 14.36 cents/kwh for the first 400 kwh, then 10.05 cents/kwh for the remainder. How much does 2125 kwh of power cost?

12 Jenny estimated the length of her front fence to be 32 m. When she measured it, its true length was 34.3 m. Find, correct to one decimal place:

  a   the error   **b**   the absolute percentage error.

13 A photograph was measured as 15 cm by 10 cm to the nearest cm. What are the boundary values of its perimeter?

Chapter 3

Sets and Venn diagrams

Contents:

OPENING PROBLEM

A city has three newspapers A, B and C.
Of the adult population, 1% read none of
these newspapers, 36% read A, 40% read
B, 52% read C, 8% read A and B, 11%
read B and C, 13% read A and C and 3% read all three
papers. What percentage of the adult population read:

a newspaper A only
b newspaper B or newspaper C
c newspaper A or B but not C?

The work we do in this chapter will enable us to organise the given figures and answer
questions like those in the Opening problem.

In **Chapter 1** we defined a **set** as a collection of numbers or objects and we used capital
letters to represent sets.

We defined a **subset** and used $\subset$ to represent *is a subset of*.

$\in$ reads *is an element of* or *is in* and $\notin$ reads *is not an element of* or *is not in*.

$\{\ \}$ or $\varnothing$ represents the empty set containing no members.

Examples: $3 \in \{1, 2, 3, 5, 9\}$ reads "3 is an element of the set 1, 2, 3, 5, 9".

$5 \notin \{1, 3, 6, 8\}$ reads "5 is not an element of the set 1, 3, 6, 8".

$\{1, 3, 5\} \subset \{1, 2, 3, 4, 5\}$ as every element of the first set is also in the
second set.

COUNTING IN A SET

The set $A = \{a, e, i, o, u\}$ has 5 elements, so we write $n(A) = 5$.

The set $B = \{2, \quad 4, \quad 6, \quad 8, \quad ...., \quad 98, \quad 100\}$ has 50 elements, so we write $n(A) = 50$.

1st 2nd 3rd 4th

> $n(A)$ is used to represent the number of elements in set A.

FINITE AND INFINITE SETS

> A set can have a finite number of elements and so is said to be a **finite set**.

The set of positive integers less than 10^9 is a finite set as 10^9, although very large, is
finite. It also has a largest member, 99 999 999.

> **Infinite sets** are sets which have infinitely many members.

For example, $Z = \{0, \pm 1, \pm 2, \pm 3, \pm 4, ...\}$ is an infinite set

$R = \{$real numbers$\}$ as an infinite set

$R = \{$all real numbers between 2.1 and 2.2$\}$ is also infinite.

We could write $n(Z) = \infty$.   Notice that $n(\varnothing) = 0$

A   SET BUILDER NOTATION

$A = \{x \mid x \in Z, \; -2 \leqslant x \leqslant 4\}$ reads   "the set of all x such that x is an integer between
-2 and 4, including -2 and 4."

such that
the set of all

$A = \{x: \; x \in Z, \; -2 \leqslant x \leqslant 4\}$ reads the same as above.

Example 1

Given $A = \{x \mid x \in Z, \; 3 < x \leqslant 10\}$ write down:
a   the meaning of the set builder notation.
b   List the elements of set A.   **c**   Find $n(A)$.

a   The set of all x such that x is an integer between 3 and 10, including 10.
b $A = \{4, 5, 6, 7, 8, 9, 10\}$ **c** $n(A) = 7$,   as there are 7 elements

---

EXERCISE 3A

1   Finite or infinite?

   a $\{x \mid x \in Z, \; -2 \leqslant x \leqslant 1\}$ **b** $\{x \mid x \in R, \; -2 \leqslant x \leqslant 1\}$

   c $\{x \mid x \in Z, \; x \geqslant 5\}$ **d** $\{x \mid x \in Q, \; 0 \leqslant x \leqslant 1\}$

2   For the following sets:

   i   Write down the meaning of the set builder notation.

   ii   If possible, list the elements of A.   **iii** Find $n(A)$.   **iv** Is A infinite?

   a $A = \{x \mid x \in Z, \; -1 \leqslant x < 7\}$ **b** $A = \{x \mid x \in N, \; -2 < x < 8\}$
   c $A = \{x \mid x \in R, \; 0 \leqslant x \leqslant 1\}$ **d** $A = \{x \mid x \in Q, \; 5 \leqslant x \leqslant 6\}$

3   Write in set builder notation:

   a   The set of all integers between -100 and 100.
   b   The set of all real numbers greater than 1000.
   c   The set of all rational numbers between 2 and 3, inclusive.

4   The subsets of $\{a\}$ are $\varnothing$ and $\{a\}$, i.e., two of them. The subsets of $\{a, b\}$ are $\varnothing$, $\{a\}$,
   $\{b\}$ and $\{a, b\}$, i.e., four of them.

   a   List the subsets of **i** $\{a, b, c\}$ **ii** $\{a, b, c, d\}$ and state the number of them
      in each case.

   b   Copy and complete: "If a set has n elements it has ...... subsets."

5   Is $A \subset B$ in these cases?

   a $A = \varnothing$ and $B = \{2, 5, 7, 9\}$ **b** $A = \{2, 5, 8, 9\}$ and $B = \{8, 9\}$
   c $A = \{x \mid x \in R, \; 2 \leqslant x \leqslant 3\}$ and $B = \{x \mid x \in R\}$
   d $A = \{x \mid x \in Q, \; 3 \leqslant x \leqslant 9\}$ and $B = \{x \mid x \in R, \; 0 \leqslant x \leqslant 10\}$
   e $A = \{x \mid x \in Z, \; -10 \leqslant x \leqslant 10\}$ and $B = \{z \mid z \in Z, \; 0 \leqslant z \leqslant 5\}$
   f $A = \{x \mid x \in Q, \; 0 \leqslant x \leqslant 1\}$ and $B = \{y \mid y \in Q, \; 0 < y \leqslant 2\}$

B   COMPLEMENTS OF SETS

UNIVERSAL SETS

If we are only interested in the natural numbers from 1 to 20 and we want to consider subsets of this set then $U = \{x \mid x \in N, \ 1 \leqslant x \leqslant 20\}$ is the universal set in this situation.

The symbol U is used to represent a universal set.

COMPLEMENTARY SETS

If the universal set is $U = \{1, 2, 3, 4, 5, 6, 7, 8\}$ and $A = \{1, 3, 5, 7, 8\}$ then the complement of A, denoted A' is $A' = \{2, 4, 6\}$.

The **complement** of A, denoted A' is the set of all elements of U which are **not in** A.

Three obvious relationships are observed connecting A and A'. These are:

- $A \cap A' = \varnothing$ as A' and A have no common members
- $A \cup A' = U$ as all elements of A and A' combined make up U
- $n(A') = n(U) - n(A)$ {see the example above where $n(A') = 3$ and $n(U) - n(A) = 8 - 5 = 3$}

Example 2

Find C' given that:
a $U = \{\text{all integers}\}$ and $C = \{\text{all even integers}\}$
b $C = \{x \mid x \in Z, \ x \geqslant 2\}$ and $U = Z$

a $C' = \{\text{all odd integers}\}$ b $C' = \{x \mid x \in Z, \ x \leqslant 1\}$

EXERCISE 3B

1   Find C', the complement of C given that:

a $U = \{\text{letters of the English alphabet}\}$ and $C = \{\text{vowels}\}$
b $U = \{\text{integers}\}$ and $C = \{\text{negative integers}\}$
c $U = Z$ and $C = \{x \mid x \in Z, \ x \leqslant -5\}$
d $U = Q$ and $C = \{x \mid x \in Q, \ x \leqslant 2 \ \text{or} \ x \geqslant 8\}$

Example 3

If $U = \{x \mid x \in Z, \ -5 \leqslant x \leqslant 5\}$, $A = \{x \mid x \in Z, \ 1 \leqslant x \leqslant 4\}$ and $B = \{x \mid x \in Z, \ -3 \leqslant x < 2\}$ list the elements of these sets:

a   A   b   B   c A' d B' e $A \cap B$ f $A \cup B$ g $A' \cap B$
h $A' \cup B'$

> **a** $A = \{1, 2, 3, 4\}$
>
> **b** $B = \{-3, -2, -1, 0, 1\}$
>
> **c** $A' = \{-5, -4, -3, -2, -1, 0, 5\}$
>
> **d** $B' = \{-5, -4, 2, 3, 4, 5\}$
>
> **e** $A \cap B = \{1\}$
>
> **f** $A \cup B = \{-3, -2, -1, 0, 1, 2, 3, 4\}$
>
> **g** $A' \cap B = \{-3, -2, -1, 0\}$
>
> **h** $A' \cup B' = \{-5, -4, -3, -2, -1, 0, 2, 3, 4, 5\}$

2 If $U = \{x \mid x \in Z, \ 0 \leqslant x \leqslant 8\}$, $A = \{x \mid x \in Z, \ 2 \leqslant x \leqslant 7\}$ and
   $B = \{x \mid x \in Z, \ 5 \leqslant x \leqslant 8\}$ list the elements of:

   a A   **b** A' **c** B   **d** B'

   e $A \cap B$ **f** $A \cup B$ **g** $A \cap B'$

3 If $n(U) = 15$, $n(P) = 6$ and $n(Q') = 4$ where P and Q are subsets of U, find:

   a $n(P')$ **b** $n(Q)$.

4 True or false?

   a If $n(U) = a$ and $n(A) = b$ then $n(A') = b - a$.

   b If Q is a subset of U then $Q' = \{x \mid x \notin Q, \ x \in U\}$.

5 If $U = \{x \mid 0 < x \leqslant 12, \ x \in Z\}$, $A = \{x \mid x \in Z, \ 2 \leqslant x \leqslant 7\}$,
   $B = \{x \mid 3 \leqslant x \leqslant 9, \ x \in Z\}$ and $C = \{x \mid x \in Z, \ 5 \leqslant x \leqslant 11\}$ list the
elements of:

   a B' **b** C' **c** A' **d** $A \cap B$

   e $(A \cap B)'$ **f** $A' \cap C$ **g** $B' \cup C$ **h** $(A \cup C) \cap B'$

Example 4

If $U = \{$positive integers$\}$, $P = \{$multiples of 4 less than 50$\}$
   $Q = \{$multiples of 6 less than 50$\}$:

   a list P and Q   **b** find $P \cap Q$ **c** find $P \cup Q$.

   d Verify that $n(P \cup Q) = n(P) + n(Q) - n(P \cap Q)$

   a $P = \{4, 8, 12, 16, 20, 24, 28, 32, 36, 40, 44, 48\}$
   $Q = \{6, 12, 18, 24, 30, 36, 42, 48\}$

   b $P \cap Q = \{12, 24, 36, 48\}$

   c $P \cup Q = \{4, 6, 8, 12, 16, 18, 20, 24, 28, 30, 32, 36, 40, 42, 44, 48\}$

   d $n(P \cup Q) = 16$ and $n(P) + n(Q) - n(P \cap Q) = 12 + 8 - 4 = 16$
   i.e., $n(P \cup Q) = n(P) + n(Q) - n(P \cap Q)$ is verified.

6 $U = Z^+$, $P = \{$prime numbers $< 25\}$ and
$Q = \{2, 4, 5, 11, 12, 15\}$

  a   List P.   **b**   Find $P \cap Q$.   **c**   Find $P \cup Q$.
  d   Verify that $n(P \cup Q) = n(P) + n(Q) - n(P \cap Q)$.

7 $U = Z^+$, $P = \{$multiples of 3 less than $30\}$ and
$Q = \{$multiples of 4 less than $30\}$

  a   List P.   **b**   Find $P \cap Q$.   **c**   Find $P \cup Q$.
  d   Verify that $n(P \cup Q) = n(P) + n(Q) - n(P \cap Q)$.

8 $U = Z^+$, $M = \{$multiples of 4 between 30 and $60\}$ and
$N = \{$multiples of 6 between 30 and $60\}$

  a   List M and N.   **b**   Find $M \cup N$.   **c**   Find $M \cap N$.
  d   Verify that $n(M \cup N) = n(M) + n(N) - n(M \cap N)$,

9 $U = Z$, $R = \{x \mid x \in Z, \ -2 \leqslant x \leqslant 4\}$ and
$S = \{x \mid x \in Z, \ 0 \leqslant x < 7\}$

  a   List R and S.   **b**   Find $R \cap S$.   **c**   Find $R \cup S$.
  d   Verify that $n(R \cup S) = n(R) + n(S) - n(R \cap S)$.

10 $U = Z$, $C = \{y \mid y \in Z, \ -4 \leqslant y \leqslant -1\}$ and
$D = \{y \mid y \in Z, \ -7 \leqslant y \leqslant 0\}$

  a   List C and D.   **b**   Find $C \cap D$.   **c**   Find $C \cup D$.
  d   Verify that $n(C \cup D) = n(C) + n(D) - n(C \cap D)$.

11 $U = Z^+$, $A = \{$multiples of 4 less than $40\}$,
$B = \{$multiples of 6 less than $40\}$ and
$C = \{$multiples of 12 less than $40\}$

  a   List the sets A, B and C.
  b   Find: **i** $A \cap B$ **ii** $B \cap C$ **iii** $A \cap C$ **iv** $A \cap B \cap C$
  c   Find: $A \cup B \cup C$
  d   Verify that $n(A \cup B \cup C) = n(A) + n(B) + n(C) - n(A \cap B) - n(B \cap C)$
$- n(A \cap C) + n(A \cap B \cap C)$

12 $U = Z^+$, $A = \{$multiples of 6 less than $31\}$,
$B = \{$factors of $30\}$ and
$C = \{$primes $< 30\}$

  a   List the sets A, B and C.
  b   Find: **i** $A \cap B$ **ii** $B \cap C$ **iii** $A \cap C$ **iv** $A \cap B \cap C$
  c   Find: $A \cup B \cup C$
  d   Verify that $n(A \cup B \cup C) = n(A) + n(B) + n(C) - n(A \cap B) - n(B \cap C)$
$- n(A \cap C) + n(A \cap B \cap C)$

C   VENN DIAGRAMS

Venn diagrams are diagrams used to represent sets of objects, numbers or things.

The universal set is usually represented by a rectangle whereas sets within it are usually represented by circles or ellipses.

For example: 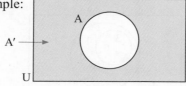 is a Venn diagram which shows set A within the universal set U.

A′, the complement of A is shaded.

In this case U could be $\{2, 3, 5, 7, 8\}$, $A = \{2, 7, 8\}$ and $A' = \{3, 5\}$.

These elements can be shown as:

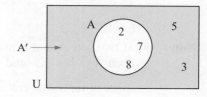

SUBSETS

As $B \subset A$ then B is contained within A as every element of B is also in A:

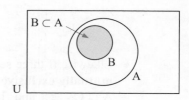

INTERSECTION

As $A \cap B$ consists of all elements common to both A and B, then $A \cap B$ is the shaded region:

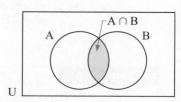

UNION

As $A \cup B$ consists of all elements in A or B (or both A and B) then $A \cup B$ is the shaded region:

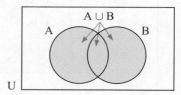

DISJOINT (OR MUTUALLY EXCLUSIVE) SETS

Disjoint sets do not have common elements.

For example, $A = \{2, 3, 8\}$ and $B = \{4, 5, 9\}$
i.e., $A \cap B = \varnothing$.

These are represented by non-overlapping circles.

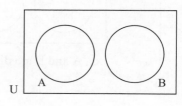

Example 5

Given that   U = {1, 2, 3, 4, 5, 6, 7, 8},   illustrate on a Venn diagram the sets:

a   A = {1, 3, 6, 8}   and   B = {2, 3, 4, 5, 8}

b   A = {1, 3, 6, 7, 8}   and   B = {3, 6, 8}

a   A ∩ B = {3, 8}

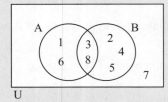

b   A ∩ B = {3, 6, 8},   B ⊂ A

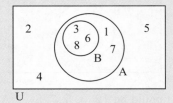

Note: • If two sets A and B are **disjoint** and **exhaustive** then A ∩ B = ∅ and U = A ∪ B.

We can represent this situation without using circles for the sets, as shown:

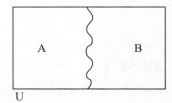

• Likewise if three sets A, B and C are **disjoint** and **mutually exclusive** then A ∩ B = ∅, B ∩ C = ∅, A ∩ C = ∅ and U = A ∪ B ∪ C.

We can represent this situation:

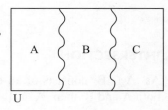

Example 6

Given U = {1, 2, 3, 4, 5, 6, 7, 8, 9}, illustrate on a Venn diagram the sets:

a   A = {2, 4, 8}   and   B = {1, 3, 5, 9}

b   A = {2, 4, 6, 8}   and   B = {1, 3, 5, 7, 9}

a   A ∪ B ≠ U

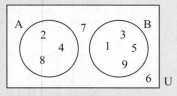

i.e., A and B are disjoint

b   A ∪ B = U

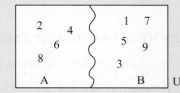

i.e., A and B are disjoint and mutually exclusive.

EXERCISE 3C

1 Represent sets A and B on a Venn diagram, given:

   a   U = {2, 3, 4, 5, 6, 7},   A = {2, 4, 6},   B = {5, 7}
   b   U = {2, 3, 4, 5, 6, 7},   A = {2, 4, 6},   B = {3, 5, 7}
   c   U = {1, 2, 3, 4, 5, 6, 7},   A = {2, 4, 5, 6}   B = {1, 4, 6, 7}
   d   U = {3, 4, 5, 7},   A = {3, 4, 5, 7}   B = {3, 5}

2 Given that   U = $\{x \mid x \in Z, \ 1 \leqslant x \leqslant 10\}$,
                    A = {odd numbers < 10},   B = {primes < 10}:

   a   List sets A and B.   **b**   Find A ∩ B and A ∪ B.
   c   Represent the sets A and B on a Venn diagram.

3 Given that   U = {even numbers between 0 and 30},
                    A = {multiples of 4 less than 30}, B = {multiples of 6 less than 30}:

   a   List sets A and B.   **b**   Find A ∩ B and A ∪ B.
   c   Represent the sets A and B on a Venn diagram.

4 Given that   U = $\{x \mid x \in Z^{+}, \ x \leqslant 30\}$,
                    R = {primes less than 30}, S = {composites less than 30}:

   a   List sets R and S.   **b**   Find R ∩ S and R ∪ S.
   c   Represent the sets R and S on a Venn diagram.

5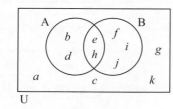

List the letters in set:

   a A   **b** B   **c** A′   **d** B′
   e A ∩ B   **f** A ∪ B   **g** (A ∪ B)′
   h A′ ∪ B′

6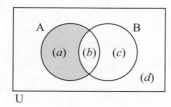

(a) means that there are a elements in the shaded region.

Notice that $n(A) = a + b$.

Find:

   a $n(B)$ **b** $n(A')$ **c** $n(A \cap B)$
   d $n(A \cup B)$ **e** $n((A \cap B)')$ **f** $n((A \cup B)')$

7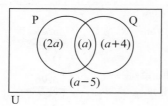

The Venn diagram shows us that $n(P \cap Q) = a$ and $n(P) = 3a$.

   a   Find:
      i $n(Q)$ **ii** $n(P \cup Q)$
      iii $n(Q')$ **iv** $n(U)$
   b   Find a if:
      i $n(U) = 29$ **ii** $n(U) = 31$ Comment!

8

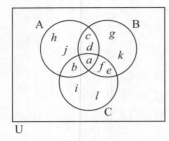

U

This Venn diagram consists of three overlapping circles A, B and C.

a List the letters in set:

    i A   **ii** B   **iii** C

    iv A ∩ B   **v** A ∪ B   **vi** B ∩ C

    vii A ∩ B ∩ C   **viii** A ∪ B ∪ C

b Find:

    i $n(A \cup B \cup C)$

    ii $n(A) + n(B) + n(C) - n(A \cap B) - n(A \cap C) - n(B \cap C) + n(A \cap B \cap C)$

D   VENN DIAGRAM REGIONS

We use shading to show various sets being considered. For example, for two intersecting sets:

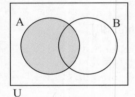

U

A is shaded

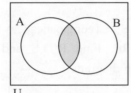

U

A ∩ B is shaded

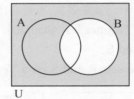

U

B′ is shaded

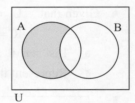

U

A ∩ B′ is shaded

Example 7

On separate Venn diagrams shade these regions for two overlapping sets A and B:

a A ∪ B   **b** A′ ∩ B   **c** (A ∩ B)′

a

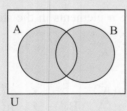

U

(in A, B or both)

b

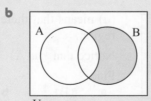

U

(outside A intersected with B)

c

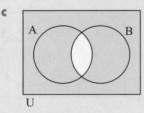

U

(outside A ∩ B)

Click on the icon to **find the shaded region** representing various subsets for two intersecting sets.

DEMO

Click on the icon to **practice shading** regions representing various subsets. If you are correct you will be informed of this.

DEMO

This demo has two parts:
- two intersecting sets
- three intersecting sets.

EXERCISE 3D

1 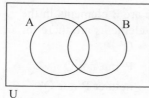 Draw separate Venn diagrams with shading of:

    a $A \cap B$ **b** $A \cap B'$ **c** $A' \cup B$

    d $A \cup B'$ **e** $(A \cap B)'$ **f** $(A \cup B)'$

PRINTABLE
VENN DIAGRAMS
(OVERLAPPING)

2 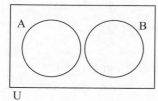 A and B are two disjoint sets. Shade on separate Venn diagrams:

    a A **b** B **c** A'

    d B' **e** $A \cap B$ **f** $A \cup B$

    g $A' \cap B$ **h** $A \cup B'$ **i** $(A \cap B)'$

PRINTABLE
VENN DIAGRAMS
(DISJOINT)

3 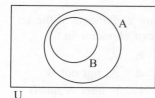 $B \subset A$ in the given Venn diagram. Shade on separate Venn diagrams:

    a A **b** B **c** A'

    d B' **e** $A \cap B$ **f** $A \cup B$

    g $A' \cap B$ **h** $A \cup B'$ **i** $(A \cap B)'$

PRINTABLE
VENN DIAGRAMS
(SUBSET)

4 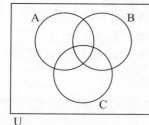 This Venn diagram consists of three overlapping circles (sets). On separate Venn diagrams shade:

    a A **b** B' **c** $B \cap C$

    d $A \cup B$ **e** $A \cap B \cap C$ **f** $A \cup B \cup C$

    g $(A \cap B \cap C)'$ **h** $(A \cup B) \cap C$ **i** $(B \cap C) \cup A$

PRINTABLE
VENN DIAGRAMS
(3 SETS)

E   NUMBERS IN REGIONS

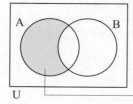 There are four regions on a Venn diagram which contains two overlapping sets A and B.

 is 'in A, but not B'

 is 'in B but not A'

 is 'in both A and B'

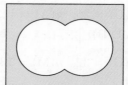 is 'neither in A nor in B'

Example 8

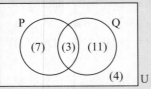

If (3) means that there are 3 elements in the set P ∩ Q, how many elements are there in:

a P   b Q′
c P ∪ Q   d P, but not Q
e Q, but not P   f neither P nor Q?

a $n(P) = 7 + 3 = 10$

b $n(Q') = 7 + 4 = 11$

c $n(P \cup Q) = 7 + 3 + 11$
   $\qquad\qquad\quad = 21$

d $n(P, \text{ but not } Q) = 7$

e $n(Q, \text{ but not } P) = 11$

f $n(\text{neither P nor Q}) = 4$

EXERCISE 3E

1

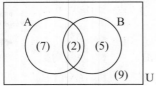

If (2) means that there are 2 elements in the set A ∩ B, give the number of elements in:

a B   b A′
c A ∪ B   d A, but not B
e B, but not A   f neither A nor B

2

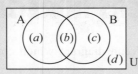

Give the number of elements in:

a X′   b X ∩ Y
c X ∪ Y   d X, but not Y
e Y, but not X   f neither X nor Y

Example 9

Given $n(U) = 30$, $n(A) = 14$, $n(B) = 17$ and $n(A \cap B) = 6$,
find: a $n(A \cup B)$ b $n(A, \text{ but not } B)$.

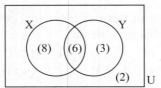

We see that $b = 6$ {as $n(A \cap B) = 6$}
$\qquad\qquad\quad a + b = 14$ {as $n(A) = 14$}
$\qquad\qquad\quad b + c = 17$ {as $n(B) = 17$}
$\qquad a + b + c + d = 30$ {as $n(U) = 30$}

Solving these, $b = 6$ ∴ $a = 8$, $c = 11$, $d = 5$

a $n(A \cup B) = a + b + c$
   $\qquad\qquad\quad = 25$

b $n(A, \text{ but not } B) = a$
   $\qquad\qquad\qquad\qquad = 8$

3 Given $n(U) = 26$, $n(A) = 11$, $n(B) = 12$ and $n(A \cap B) = 8$, find:

a $n(A \cup B)$ b $n(B, \text{ but not } A)$

4 Given $n(U) = 32$, $n(M) = 13$, $n(M \cap N) = 5$, $n(M \cup N) = 26$, find:

a $n(N)$ b $n((M \cup N)')$

5 Given $n(U) = 50$, $n(S) = 30$, $n(R) = 25$ and $n(R \cup S) = 48$, find:

    a $n(R \cap S)$ **b** $n(S, \text{ but not } R)$

Note:

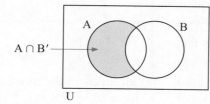

$$n(A \cap B') = n(A) - n(A \cap B)$$
$$n(A' \cap B) = n(B) - n(A \cap B)$$

are useful identities which could be used in questions like **3**, **4** and **5** above.

Example 10

A squash club has 27 members. 19 have black hair, 14 have brown eyes and 11 have both black hair and brown eyes.

  a Place this information on a Venn diagram.

  b Hence find the number of members with:

    i black hair or brown eyes   **ii** black hair, but not brown eyes.

---

  a Let Bl represent the black hair set and Br represent the brown eyes set.

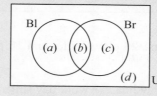

$$a + b + c + d = 27$$
$$a + b = 19$$
$$b + c = 14$$
$$b = 11$$
$$\therefore \quad a = 8, \quad c = 3, \quad d = 6$$

  b i.e.,

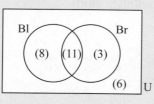

    i $n(\text{Bl} \cup \text{Br}) = 8 + 11 + 3$
                    $= 22$

    ii $n(\text{Bl, but not Br}) = 8$

---

6 Pele has 14 cavies as pets. Five have long hair and 8 are brown. Two are both brown and have long hair.

  a Place this information on a Venn diagram.

  b Hence find the number of cavies that:

    i are short haired   **ii** have short hair and are brown

    iii have short hair and are not brown.

7 During a 2 week period, Murielle took her umbrella with her on 8 days, it rained on 9 days, and Murielle took her umbrella on five of the days when it rained.

  a Display this information on a Venn diagram.

  b Hence find the number of days that:

    i Murielle did not take her umbrella and it rained

    ii Murielle did not take her umbrella and it did not rain.

Example 11

A platform diving squad of 25 has 18 members who dive from 10 m and 17 who dive from 4 m. How many dive from both platforms?

Let   T represent those who dive from 10 m and
      F represent those who dive from 4 m.

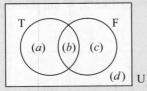

$d = 0$ {as all divers in the squad dive
                from 10 m or 4 m}

$a + b = 18$
$b + c = 17$ $\therefore$ $c = 7,\ a = 8,\ b = 10$
$a + b + c = 25$

i.e.,

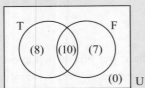

$n(\text{both T and F})$
$= n(\text{T} \cap \text{F})$
$= 10$

---

8   A badminton club has 31 playing members. 28 play singles and 16 play doubles. How many play both singles and doubles?

9   In a factory, 56 people work on the assembly line. 47 work day shifts and 29 work night shifts. How many work both day shifts and night shifts?

In an earlier exercise we verified that:

$$n(A \cup B) = n(A) + n(B) - n(A \cap B)$$

10

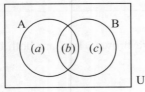

Use the figure given to show that:
$$n(A) + n(B) - n(A \cap B) = n(A \cup B).$$

Example 12   (The Opening Problem)

A city has three newspapers A, B and C. Of the adult population, 1% read none of these newspapers, 36% read A, 40% read B, 52% read C, 8% read A and B, 11% read B and C, 13% read A and C and 3% read all three papers. What percentage of the adult population read:

a   newspaper A only   **b**   newspaper B or newspaper C

c   newspaper A or B but not C.

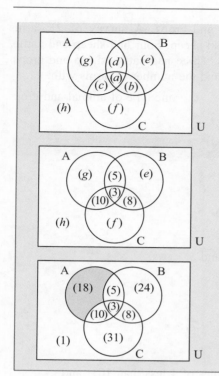

From the given information, working in %'s

$$a = 3$$
$$a + d = 8 \quad \{8\% \text{ read A and B}\}$$
$$a + b = 11 \quad \{11\% \text{ read B and C}\}$$
$$a + c = 13 \quad \{13\% \text{ read A and C}\}$$
$$\therefore \quad d = 5, \quad b = 8 \quad \text{and} \quad c = 10$$

$$g + 3 + 5 + 10 = 36 \quad \therefore \quad g = 18$$
$$e + 3 + 5 + 8 = 40 \quad \therefore \quad e = 24$$
$$f + 3 + 8 + 10 = 52 \quad \therefore \quad f = 31$$
$$\text{and} \quad h = 1$$

a P(reads A only) = 18%   {shaded}

b $n(\text{B or C}) = n(\text{B} \cup \text{C})$
$$= 100\% - 1\% - 18\%$$
$$= 81\%$$

c $n(\text{A or B, but not C}) = 18\% + 5\% + 24\%$
$$= 47\%$$

11 In a year group of 63 students, 22 study Biology, 26 study Chemistry and 25 study Physics. 18 study both Physics and Chemistry, four study both Biology and Chemistry and three study both Physics and Biology. One studies all three subjects. How many students study:

  a Biology only
  b Physics or Chemistry
  c none of Biology, Physics or Chemistry
  d Physics but not Chemistry?

12 36 students participated in the mid-year adventure trip. 19 students went paragliding, 21 went abseiling and 16 went white water rafting. 7 did abseiling and rafting, 8 did paragliding and rafting and 11 did paragliding and abseiling. 5 students did all three activities. Find the number of students who:

  a went paragliding or abseiling
  b only went white water rafting
  c did not participate in any of the activities mentioned
  d did exactly two of the activities mentioned.

13 There were 32 students available for the woodwind section of the school orchestra. 11 students could play the flute, 15 could play the clarinet and 12 could play the saxophone. Two could play the flute and the saxophone, two could play the flute and the clarinet and 6 could play the clarinet and the saxophone. One student could play all three instruments. Find the number of students who could play:

  a woodwind instruments other than the flute, clarinet and saxophone
  b only the saxophone
  c the saxophone and clarinet but not the flute
  d only one of the clarinet, saxophone or flute.

14 In a particular region, most farms had livestock and crops. A survey of 21 farms showed that 15 grew crops, 9 had cattle and 11 had sheep. Four had sheep and cattle, 7 had cattle and crops and 8 had sheep and crops. Three had cattle, sheep and crops. Two specialised in goats and did not grow crops. Find the number of farms with:

    a only crops   **b** only animals   **c** one type of animal and crops.

REVIEW SET 3A

1 If $P = \{x \mid x \in Z, \ 2 \leqslant x < 9\}$:

    a list the elements of P   **b** find $n(P)$

2   **a** Write in set builder notation:

        i the set of all integers greater than 3

        ii the set of rational numbers between -5 and 5

    b **i** Is the set in **a i** finite or infinite?

        ii Is the set in **a ii** finite or infinite?

3 If $P = \{x \mid x \in R, \ 0 \leqslant x \leqslant 10\}$ and $Q = \{2, 5, 7\}$:

    a Is $Q \subset P$?   **b** Is $\{\} \subset Q$?

4 If $U = \{x : x \in Z, \ 0 \leqslant x \leqslant 15\}$, $A = \{$multiples of 4 less than 16$\}$ and $B = \{$prime numbers less than 15$\}$, list the elements of:

    a A   **b** B   **c** B' **d** $A \cup B$ **e** $A \cap B'$

5

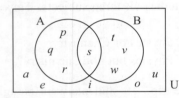

    a List the letters in:

        i A   **ii** B   **iii** $A \cap B$

        iv $(A \cup B)'$ **v** $A \cup B'$ **vi** $A' \cap B$

    b Find: **i** $n(A \cup B)$ **ii** $n(A \cap B')$

6 Of the 13 people in an elevator, eight were company executives, 10 of the people wore suits, and 7 of the people wearing suits were company executives.

    a Display this information on a Venn diagram.

    b Hence find the number of people who:

        i were not company executives and wore suits

        ii were not company executives and did not wear suits

7 All members of a class of 27 students participated in at least one event on sports day. 20 students participated in individual events and 12 students took part in team events. How many students took part in both individual and team events?

8 In a class of 24 students, 13 study economics, 13 study history and 12 study information technology (IT). Five study both economics and IT, six study both history and IT and six study both economics and history. Two study all three subjects. Find how many students study:

    a economics or history   **b** economics and history but not IT

    c exactly one of economics, history or IT.

REVIEW SET 3B

1 For the set $A = \{x \in Z^-, \ -4 < x < 0\}$:
   a write down the meaning of the set builder notation
   b list the elements of A.
   c Is A finite or infinite?

2 If $S = \{x \mid x \in Q, \ 0 \leqslant x \leqslant 5\}$ and $T = \{x \mid x \in Z, \ 0 \leqslant x < 6\}$, is $T \subset S$?

3 If $U = \{\text{positive integers}\}$, $P = \{\text{multiples of 2 less than 20}\}$ and $Q = \{\text{multiples of 4 less than 20}\}$:
   a list P and Q
   b find $P \cap Q$
   c find $P \cup Q$
   d Is $Q \subset P$?
   e Verify that $n(P \cup Q) = n(P) + n(Q) - n(P \cap Q)$

4 **a** Represent sets A and B on a Venn diagram given $U = \{x \mid x \in Z, \ -10 < x < 0\}$, $A = \{x \mid x \in Z, \ -5 < x < 0\}$, $B = \{-2, -4, -6, -8\}$.
   b If $U = \{x \mid x \in Z^+, \ x \leqslant 20\}$,
   $R = \{\text{even numbers less than 20}\}$, $S = \{\text{square numbers less than 20}\}$:
   i list sets R and S
   ii find $n(S)$
   iii list S′
   iv Is $S \subset R$?

5 Given $n(U) = 32$, $n(A) = 17$, $n(B) = 16$ and $n(A \cup B) = 25$, find:
   a $n(A \cap B)$
   b $n(A, \text{but not } B)$

6 The local library has 84 members. 52 members are female, 18 members are more than 50 years old, and 13 of the female members are more than 50 years old.
   a Display this information on a Venn diagram.
   b Hence find the number of members who are:
   i male and 50 or less years old
   ii 50 or less years old.

7 At an adventure camp, all children participated in at least one of swimming and kayaking. Of the 23 children who went to camp, 20 went swimming and 14 went kayaking. How many went swimming and kayaking?

8 The local Arts and Drama Society has 58 members. During the past year 23 members attended plays, 29 attended concerts and 25 attended films. 5 members attended plays, concerts and films. 14 members attended plays and concerts, and 8 members attended plays and films. 14 members only attended concerts. Find the number of members who:
   a only attended films
   b did not attend any plays, concerts or films
   c attended only one type of entertainment.

Chapter 4

The rule of Pythagoras

Contents:

Right angles (90° angles) are used in the construction of buildings and in the division of areas of land into rectangular regions.

The ancient **Egyptians** used a rope with 12 equally spaced knots to form a triangle with sides in the ratio 3 : 4 : 5. This triangle had a right angle between the sides of length 3 and 4 units.

This triangle is, in fact, the simplest right angled triangle with sides of integer length.

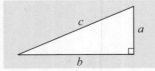

take hold of knots at arrows

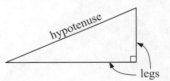

make rope taut

corner

line of one side of building

Pythagoras' Rule connects the lengths of the sides of any right angled triangle. It can be used to find unknown distances and has practical uses in many situations.

Square root numbers like $\sqrt{7}$ are called **surds** and are frequently seen when using Pythagoras' Rule.

A | THE RULE OF PYTHAGORAS (REVIEW)

A **right angled triangle** is a triangle which has a right angle as one of its angles.

The side **opposite** the **right angle** is called the **hypotenuse** and is the **longest** side of the triangle.

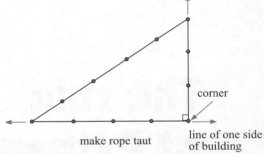

hypotenuse

legs

The other two sides are called the **legs** of the triangle.

Around 500 BC, the Greek mathematician **Pythagoras** formulated a rule which connects the lengths of the sides of all right angled triangles. It is thought that he discovered the rule while studying tessellations of tiles on bathroom floors. Such patterns, like the one illustrated, were common on the walls and floors of bathrooms in ancient **Greece**.

THE RULE OF PYTHAGORAS

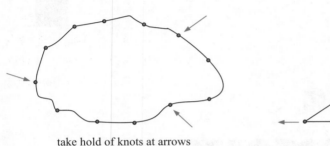

In a right angled triangle, with hypotenuse c and legs a and b,

$$c^2 = a^2 + b^2.$$

DEMO

Reminder: The **hypotenuse** is always the *longest side* and is *opposite the right angle.*

This theorem, known to the ancient Greeks, is particularly valuable in that:

- if we know the lengths of any two sides of a right angled triangle then we can calculate the length of the third side
- if we know the lengths of the three sides then we can determine whether or not the triangle is right angled.

The second statement here relies on the **converse of Pythagoras' Rule**, which is:

> If a triangle has sides of length a, b and c units and $a^2 + b^2 = c^2$ say, then the triangle is right angled and its hypotenuse is c units long.

Example 1

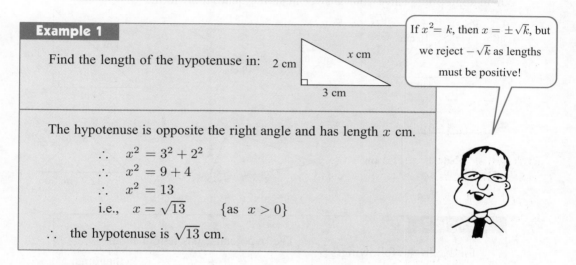

Find the length of the hypotenuse in:

2 cm x cm   3 cm

If $x^2 = k$, then $x = \pm\sqrt{k}$, but we reject $-\sqrt{k}$ as lengths must be positive!

The hypotenuse is opposite the right angle and has length x cm.

$$\therefore \quad x^2 = 3^2 + 2^2$$
$$\therefore \quad x^2 = 9 + 4$$
$$\therefore \quad x^2 = 13$$
$$\text{i.e.,} \quad x = \sqrt{13} \qquad \{\text{as } x > 0\}$$

$\therefore$ the hypotenuse is $\sqrt{13}$ cm.

EXERCISE 4A

1 Find the length of the hypotenuse in the following triangles, leaving your answer in surd (square root) form if appropriate:

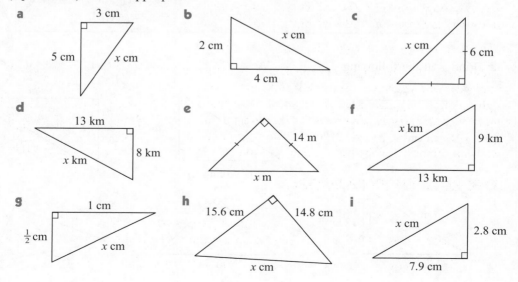

a 3 cm, 5 cm, x cm

b 2 cm, x cm, 4 cm

c x cm, 6 cm

d 13 km, 8 km, x km

e 14 m, x m

f x km, 9 km, 13 km

g 1 cm, $\frac{1}{2}$ cm, x cm

h 15.6 cm, 14.8 cm, x cm

i x cm, 2.8 cm, 7.9 cm

Example 2

Find the length of the
third side of:

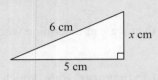

The hypotenuse has length 6 cm.

$\therefore \quad x^2 + 5^2 = 6^2 \quad$ {Pythagoras}

$\therefore \quad x^2 + 25 = 36$

$\therefore \quad x^2 = 11$

$\therefore \quad x = \sqrt{11} \quad$ {as $x > 0$}

$\therefore$ third side is $\sqrt{11}$ cm long.

---

2 Find the length of the third side of the following right angled triangles.
Where appropriate leave your answer in surd (square root) form.

a

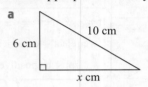

b

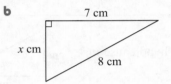

c

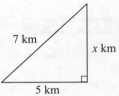

d

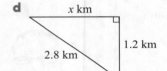

e

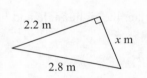

f

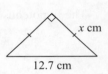

Example 3

Find x in the following:

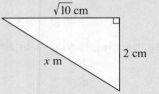

The hypotenuse has length x cm.

$\therefore \quad x^2 = 2^2 + (\sqrt{10})^2 \quad$ {Pythagoras}

$\therefore \quad x^2 = 4 + 10$

$\therefore \quad x^2 = 14$

$\therefore \quad x = \pm\sqrt{14}$

But $x > 0$, $\therefore$ $x = \sqrt{14}$.

Remember that
$(\sqrt{x})^2 = x$.

---

3 Find x in the following:

a

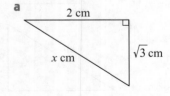

b

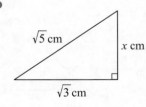

c

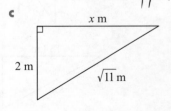

d

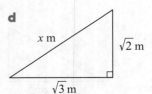

e

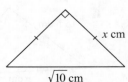

f

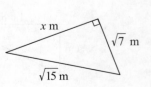

Example 4

Solve for x:

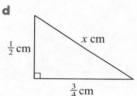

$$x^2 + \left(\tfrac{1}{2}\right)^2 = 1^2 \quad \{\text{Pythagoras}\}$$

$$\therefore \quad x^2 + \tfrac{1}{4} = 1$$

$$\therefore \quad x^2 = \tfrac{3}{4}$$

$$\therefore \quad x = \pm\sqrt{\tfrac{3}{4}}$$

$$\therefore \quad x = \sqrt{\tfrac{3}{4}} \quad \{\text{as } x > 0\}$$

4 Solve for x:

a

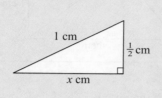

b

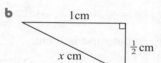

c

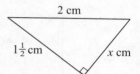

d

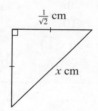

e

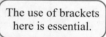

f

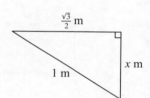

Example 5

Find the value of x:

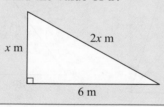

$$(2x)^2 = x^2 + 6^2 \quad \{\text{Pythagoras}\}$$

$$\therefore \quad 4x^2 = x^2 + 36$$

$$\therefore \quad 3x^2 = 36$$

$$\therefore \quad x^2 = 12$$

$$\therefore \quad x = \pm\sqrt{12}$$

But $x > 0$, $\therefore \quad x = \sqrt{12}$.

The use of brackets here is essential.

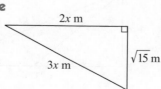

5 Find the value of x:

a

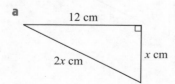

b

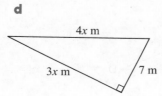

c

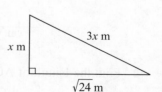

d

e

f

Example 6

Find the value of any unknowns:

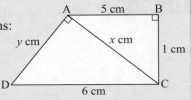

x is the hypotenuse in triangle ABC.

$$\therefore \quad x^2 = 5^2 + 1^2 \qquad \text{\{Pythagoras\}}$$
$$\therefore \quad x^2 = 26$$
$$\therefore \quad x = \pm\sqrt{26}$$
$$\therefore \quad x = \sqrt{26} \qquad \{x > 0\}$$

In $\triangle$ACD, the hypotenuse is 6 cm.

$$\therefore \quad y^2 + (\sqrt{26})^2 = 6^2 \qquad \text{\{Pythagoras\}}$$
$$\therefore \quad y^2 + 26 = 36$$
$$\therefore \quad y^2 = 10$$
$$\therefore \quad y = \pm\sqrt{10}$$
$$\therefore \quad y = \sqrt{10} \qquad \{y > 0\}$$

6 Find the value of any unknown:

a

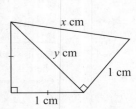

b

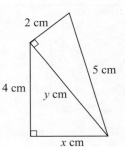

c

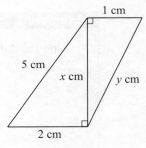

7 Find x:

a

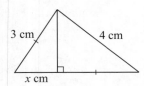

b

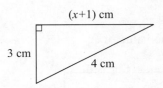

8 Find the length of AC in:

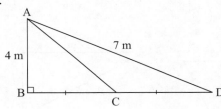

9 Find the distance AB, in the following figures.
  (**Hint:** It is necessary to draw an additional line or two on the figure in each case.)

a
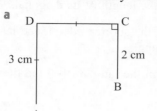

b
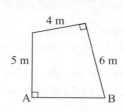

c
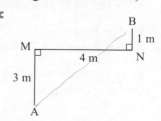

B PYTHAGORAS AND GEOMETRICAL FIGURES

SPECIAL GEOMETRICAL FIGURES

All of these figures contain right angled triangles where Pythagoras' Rule applies:

rectangle

In a **rectangle**, a right angle exists between adjacent sides.

Construct a **diagonal** to form a right angled triangle.

square

rhombus

In a **square** and a **rhombus**, the *diagonals bisect each other at right angles*.

In an **isosceles triangle** and an **equilateral triangle**, the *altitude bisects the base at right angles*.

isosceles triangle

equilateral triangle

Things to remember

- Draw a neat, clear diagram of the situation.
- Mark on known lengths and right angles.
- Use a symbol, such as x, to represent the unknown length.
- Write down Pythagoras' Rule for the given information.
- Solve the equation.
- Write your answer in sentence form (where necessary).

Example 7

The longer side of a rectangle is 12 cm and its diagonal is 13 cm. Find:
a the length of the shorter side   **b** the area of the rectangle.

a

12 cm

x cm

13 cm

Let the shorter side be x cm.

$$\therefore \quad x^2 + 12^2 = 13^2$$
$$\therefore \quad x^2 + 144 = 169$$
$$\therefore \quad x^2 = 25$$
$$\therefore \quad x = \pm 5$$

But $x > 0$, $\therefore$ $x = 5$ i.e., the shorter side is 5 cm.

b Area $= l \times w$
$= 12 \times 5$
$= 60$ cm^2

EXERCISE 4B

1 A rectangle has sides of length 7 cm and 3 cm. Find the length of its diagonals.

2 The longer side of a rectangle is double the length of the shorter side. If the length of the diagonal is 10 cm, find the dimensions of the rectangle.

3 A rectangle with diagonals of length 30 cm has sides in the ratio 3 : 1. Find the
   a perimeter   **b** area of the rectangle.

Example 8

A rhombus has diagonals of length 6 cm and 8 cm.
Find the length of its sides.

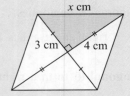

The diagonals of a rhombus *bisect at right angles.*
Let a side be x cm.

$\therefore \quad x^2 = 3^2 + 4^2 \qquad$ {Pythagoras}

$\therefore \quad x^2 = 9 + 16$

$\therefore \quad x^2 = 25$

$\therefore \quad x = \pm\sqrt{25}$

$\therefore \quad x = 5 \qquad \qquad \{x > 0\}$

i.e.,   the sides are 5 cm in length.

---

4 A rhombus has sides of length 7 cm. One of its diagonals is 10 cm long. Find the length of the other diagonal.

5 A square has diagonals of length 8 cm. Find the length of its sides.

6 A rhombus has diagonals of length 4 cm and 6 cm. Find its perimeter.

Example 9

An isosceles triangular garden bed has equal sides 12 m long and the third side is 10 m.
Find:

   a the altitude of the triangle

   b the area of the triangle.

a

The altitude bisects the base at right angles. So,

$a^2 + 5^2 = 12^2$

$\therefore \quad a^2 = 12^2 - 5^2$

$\therefore \quad a = \sqrt{(12^2 - 5^2)}$

$\therefore \quad a \doteqdot 10.91$

i.e.,   the altitude is 10.9 m   (3 s.f.)

b   The area $= \frac{1}{2}$ base $\times$ height

$= \frac{1}{2} \times 10 \times 10.91$

$\doteqdot 54.6$ m^2 (3 s.f.)

Note: Always sketch a diagram and mark on it all given lengths, etc.

7   An equilateral triangle has sides of length 12 cm.

   a   Find the length of one of its altitudes.

   b   Find the area of the triangle.

8   An isosceles triangle has equal sides of length 8 cm and a base of length 6 cm.

   a   Find the altitude of the triangle.

   b   Find the area of the triangle.

9   A new reserve is an equilateral triangle with sides 200 m. It is to be surfaced with instant turf.

   a   Find the length of an altitude of the triangle (to 2 d.p.).

   b   Find the area of turf needed to grass it.

   c   If the turf costs $4.25 per m^2, fully laid, find the total cost of grassing the reserve.

C | THE CONVERSE OF PYTHAGORAS' RULE

In building, rectangular walls and floors are essential.

Here is a typical timber framed wall (sometimes called *the inner 'skin'*):

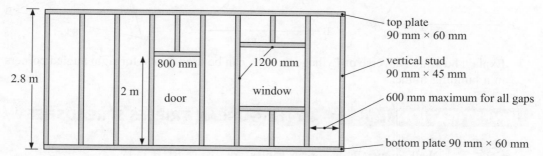

How do the workers make sure that they have right angled corners?

The **converse** of **Pythagoras' Rule** gives us a simple test as to whether (or not) a triangle is right angled when we are given the lengths of its three sides.

THE CONVERSE OF PYTHAGORAS' RULE

> If a triangle has sides of length a, b and c units and $a^2 + b^2 = c^2$, then the triangle is right angled.

GEOMETRY PACKAGE

Example 10

Is the triangle with sides 6 cm, 8 cm and 5 cm right angled?

The two shorter sides have lengths 5 cm and 6 cm, and

$$5^2 + 6^2$$
$$= 25 + 36$$
$$= 61$$

But $8^2 = 64$

$\therefore$ $5^2 + 6^2 \neq 8^2$ and hence the triangle is not right angled.

EXERCISE 4C

1 The following figures are not drawn accurately. Which of the triangles are right angled?

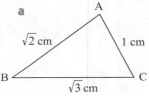

a   6 cm   5 cm   4 cm

b   6 cm   8 cm   10 cm

c   9 cm   3 cm   8 cm

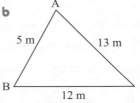

d   2 cm $\sqrt{7}$ cm $\sqrt{12}$ cm

e $\sqrt{5}$ m $\sqrt{11}$ m   4 m

f   22.4 cm   16.8 cm   28.0 cm

2 If any of the following triangles (not drawn accurately) is right angled, find the right angle:

a   A $\sqrt{2}$ cm   1 cm   B $\sqrt{3}$ cm   C

b   A   5 m   13 m   B   12 m   C

c   C   5 km $\sqrt{39}$ km   A   8 km   B

3 Explain how the converse of Pythagoras' rule can be used to test for right angled corners in a building frame.

INVESTIGATION 1   PYTHAGOREAN TRIPLES SPREADSHEET

Well known Pythagorean triples are $\{3, 4, 5\}$, $\{5, 12, 13\}$, $\{7, 24, 25\}$ and $\{8, 15, 17\}$. Formulae can be used to generate Pythagorean triples.

SPREADSHEET

An example is $2n + 1$, $2n^2 + 2n$, $2n^2 + 2n + 1$ where n is a positive integer.

A **spreadsheet** will quickly generate sets of Pythagorean triples using such formulae.

What to do:

1 Open a new spreadsheet and enter the following:

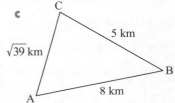

	A	B	C	D
1	n	a	b	c
2	1	=2*A2+1	=2*A2^2+2*A2	=C2+1
3	=A2+1			
4	↓	↓	↓	↓
5	↓		fill down	

2 Highlight the formulae in B2, C2 and D2 and **fill down** to Row 3. You should now have generated two sets of triples:

3 Highlight the formulae in Row 3 and **fill down** to Row 11 to generate 10 sets of triples.

4 Check that each set of numbers is actually a triple by adding two more columns to your spreadsheet.

	A	B	C	D
1	n	a	b	c
2	1	3	4	5
3	2	5	12	13
4				
5				

In E1 enter the heading 'a^2+b^2' and in F1 enter the heading 'c^2'.
In E2 enter the formula $= B2^2+C2^2$ and in F2 enter the formula $= D2^2$.

	A	B	C	D	E	F
1	n	a	b	c	$a\wedge2+b\wedge2$	$c\wedge2$
2	1	3	4	5	=B2^2+C2^2	=D2^2
3	2	5	12	13		
4	3	7	24	25		
5						fill down

5 Highlight the formulae in E2 and F2 and **fill down** to Row 11. Is each set of numbers a Pythagorean triple? [**Hint:** Does $a^2 + b^2 = c^2$?]

6 Your task is to **prove** that the formulae $\{2n+1, \ 2n^2+2n, \ 2n^2+2n+1\}$ will produce sets of Pythagorean triples for positive integer values of n.

We let $a = 2n+1$, $b = 2n^2+2n$ and $c = 2n^2+2n+1$.

Simplify $c^2-b^2 = (2n^2+2n+1)^2 - (2n^2+2n)^2$ using *the difference of two squares* factorisation, and hence show that it equals $(2n+1)^2 = a^2$.

D   PROBLEM SOLVING

Right angled triangles occur frequently in **problem solving** and often the presence of right angled triangles indicates that **Pythagoras' Rule** is likely to be used.

EXERCISE 4D

1 Find, correct to 3 significant figures, the value of x in:

a

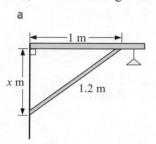

b

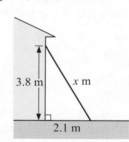

c

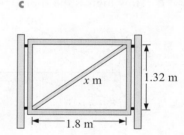

Example 11

A rectangular gate is 3 m wide and has a 3.5 m diagonal. How high is the gate?

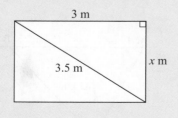

Let x m be the height of the gate.

Now $(3.5)^2 = x^2 + 3^2$ {Pythagoras}

$\therefore \quad 12.25 = x^2 + 9$

$\therefore \quad 3.25 = x^2$

$\therefore \quad x = \sqrt{3.25}$ {as $x > 0$}

$\therefore \quad x \doteqdot 1.803$

Thus the gate is approximately 1.80 m high.

2 The size of a television screen is the length across its diagonal. If a television screen is 55 cm long and 40 cm high, what size is it?

3

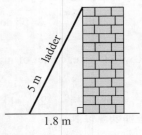

A 5 m long ladder leans against a brick wall. The feet of the ladder are 1.8 m from the base of the wall. How far up the wall does the ladder reach?

4 Julia finds that it is quicker to go to school across the diagonal of a rectangular paddock, than walking around its sides. The diagonal is 85 metres long. One side of the paddock is 42 m.

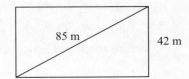

  a Find the length of the other side of the paddock.

  b By how much is it shorter to walk across the diagonal?

5 Find the length of the truss AB for the following roof structures:

  a
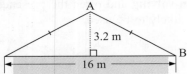

  b
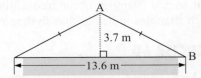

6 How high is the roof above the walls in the following roof structures?

  a
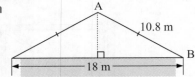

  b
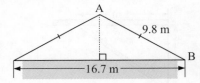

7

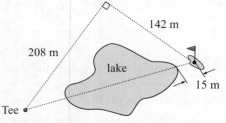

Bob is about to tee off on the sixth, a par 4 at the Royal Golf Club. If he chooses to hit over the lake, directly at the flag, how far must he hit the ball to clear the lake, given that the pin is 15 m from the water's edge?

8 A power station P, supplies two towns A and B with power. New underground power lines to each town are required. The towns are connected by a straight highway through A and B and the power station is 42 km from this highway.

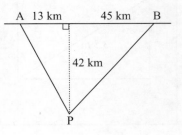

  a Find the length of power line required from P to each town.

b Find the total cost of the new power line given that each kilometre will cost $2350.

9 A surveyor makes the following measurements of a field. Calculate the perimeter of the field to the nearest metre.

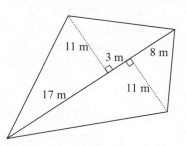

10

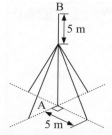

A pole AB is 16 m tall above the ground. At a point 5 m below B, four wires are connected from the pole to the ground.

Each wire is pegged to the ground 5 m from the base of the pole. What is the total length of the wire needed given that a total of 2 m extra is needed for tying?

Example 12

Find the total length of metal to make 15 000 right angled brackets like the one shown.

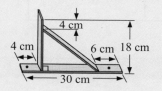

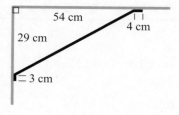

Sketch of a bracket
(measurements in cm)

By Pythagoras
$$x^2 = 14^2 + 20^2$$
$$\therefore \quad x = \sqrt{(14^2 + 20^2)}$$
$$\therefore \quad x \doteqdot 24.4 \text{ cm}$$

Length needed for one bracket
$$= 30 + 18 + 24.4 \text{ cm}$$
$$= 72.4 \text{ cm}$$

$\therefore$ total length needed $= 15\,000 \times 0.724$ m
$$\doteqdot 10\,900 \text{ m} (3 \text{ s.f.})$$

---

11 8 steel brackets (as shown) are used to support the roof of a shelter shed. The steel weighs 2.4 kg per metre. Find:

a the length of one bracket

b the total length of the brackets used

c the total weight of the brackets used.

12 A lawn with dimensions as shown is bordered with timber. Find:

a the length of AB

b the total length of timber required to make the border.

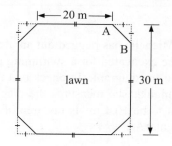

13 5 square-section metal trusses are used to make the roof of a garage. Find:

   a the length of PQ

   b the total length of metal to make one truss

   c the total length of metal to make all five trusses.

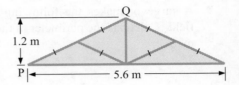

14 Find the total length of metal required to make these right angle brackets.

Dimensions		Number Required
a	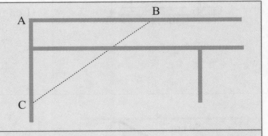	2500
b		8450

Example 13

Bjorn suspects that the corner A of a tennis court is not a right angle. With a measuring tape he finds that AB = 3.72 m, BC = 4.56 m and AC = 2.64 m. Is Bjorn's suspicion well founded?

$$BC^2 = 4.56^2 \doteq 20.8$$

$$\text{and} \quad AB^2 + AC^2 = 3.72^2 + 2.64^2 \doteq 20.8 \quad (3 \text{ s.f.})$$

So, it appears that the angle at A is a right angle (within the limitations of accuracy of the measurements).

---

15 Michelle has pegged out an area of ground to be excavated for a swimming pool, as shown in the diagram. To check that the area is rectangular she measures the diagonal and finds it to be 10.4 m. Is the area of ground rectangular?

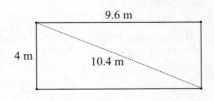

16 Ivar believes he has cut out a perfect rectangular canvas covering which has adjacent sides 8.6 m and 5.4 m. The opposite sides are equal. He measures a diagonal to be 10.155 m. Was Ivar's rectangle right angled?

17 Satomi says she has just cut out a triangular sail for her boat. The lengths of the sides are 6.23 m, 3.87 m and 4.88 m. The sail is supposed to be right angled. Is it?

Due to rounding $a^2 + b^2$ and c^2 may not be exactly equal for large numbers. If they are close we say that the triangle is right angled or approximately so.

E TRUE BEARINGS AND NAVIGATION

True bearings measure the direction of travel by comparing it with the **true north direction**. Measurements are always taken in the **clockwise** direction.

Imagine you are standing at point A, facing north. You turn **clockwise** through an angle until you face B.

The **bearing of B from A** is the angle through which you have turned. So, the bearing of B from A is the measure of the angle between AB and the 'north' line through A.

The bearing of B from A is $72°$ from true north. We write this as $72°T$ or $072°$.

To find the true **bearing of A from B**, we place ourselves at point B and face north.

We measure the clockwise angle through which we have to turn so that we face A. The true bearing of A from B is $252°$.

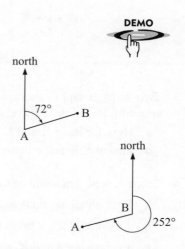

Example 14

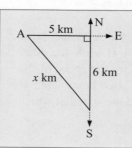

Kimi leaves point A and heads east for 5 km. He then heads south for 6 km. How far is he now from point A?

By Pythagoras,
$$x^2 = 5^2 + 6^2$$
$$\therefore \quad x = \sqrt{(5^2 + 6^2)}$$
$$\therefore \quad x \doteqdot 7.81$$

So, John is 7.81 km from A.

EXERCISE 4E

1   A sailing ship sails 46 km north then 74 km east.

    a   Draw a fully labelled diagram of the ship's course.

    b   How far is the ship from its starting point?

2   A runner is 22 km east and 15 km south of her starting point.

    a   How far is she from her starting point?

    b   How long would it take her to return to her starting point in a direct line if she can run at 10 km/h?

Example 15

A man travels due east by bicycle at 16 km/h. His son travels due south on his bicycle at 20 km/h. How far apart are they after 4 hours, if they both leave point A at the same time?

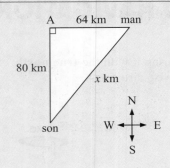

After 4 hours the man has travelled $4 \times 16 = 64$ km and his son has travelled $4 \times 20 = 80$ km.

Thus $x^2 = 64^2 + 80^2$ {Pythagoras}

i.e., $x^2 = 4096 + 6400$

$\therefore \quad x^2 = 10\,496$

$\therefore \quad x = \sqrt{10\,496}$ {as $x > 0$}

$\therefore \quad x \doteqdot 102.4$

$\therefore$ they are 102 km apart after 4 hours.

---

3   Two ships B and C leave a port A at the same time. B travels due west at a constant speed of 16 km/h. C travels due south at a constant speed of 18 km/h.

    a   How far have B and C each travelled in two hours?

    b   Find the distance between them after 2 hours.

4   Town A is 80 km south of town B. Town C is 150 km east of town B.

    a   Find how long it takes to travel directly from A to C by car at 90 km/h.

    b   Find how long it takes to travel from A to C via B in a train travelling at 130 km/h.

    c   Comparing **a** and **b**, which is faster?

5   Two runners set off from town A at the same time. One ran due east to town B while the other ran due south to town C at twice the speed of the first. They arrived at B and C two hours later. If B and C are 60 km apart, find the speed at which each runner travelled.

Example 16

A helicopter travels from base station S on a true bearing of $074°$ for 112 km to outpost A. It then travels 134 km on a true bearing of $164°$ to outpost B. How far is outpost B from base station S?

From the diagram alongside, in triangle SAB
$\angle SAB = 90°$.   Let   SB $= x$ km.

By Pythagoras, $x^2 = 112^2 + 134^2$

$$\therefore \quad x = \sqrt{(112^2 + 134^2)}$$

$$\therefore \quad x \doteqdot 174.6$$

i.e.,   outpost B is 175 km from base station S.

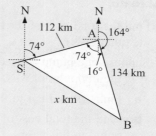

6 Jan walks 450 m from home on a true bearing of $127°$. She then walks 710 m on a true bearing of $217°$.

    a Draw a fully labelled sketch of Jan's path as in **Example 16**.

    b Find how far Jan is from her starting point.

7 Two rally car drivers set off from town C at the same time. A travels in a direction $063°$ at 120 km/h and B travels in a direction $333°$ at 135 km/h. How far apart are they after one hour?

8 Two ships B and C leave a port A at the same time. B travels in a direction $67°$T at a constant speed of 36 km/h. C travels in a direction $157°$ at a constant speed of 28 km/h. Two hours go by.

    a How far does B travel in the two hours?

    b How far does C travel in the two hours?

    c Draw a fully labelled sketch of the situation.

    d Find the distance between them after two hours.

INVESTIGATION 2   SHORTEST DISTANCE

A and B are two homesteads which are 4 km and 3 km away from a water pipeline. M and N are the nearest points (on the pipeline) to A and B respectively, and MN $= 6$ km.

The cost of running a spur pipeline across country from the main pipe line is $3000 per km and the cost of a pump is $8000. Your task is to determine the most economic way of pumping the water from the pipeline to A and B. Should you use two pumps (located at M and N) or use one pump located somewhere between M and N knowing that one pump would be satisfactory to pump sufficient water to meet the needs of both homesteads?

What to do:

1 Find the total cost of the pumps and pipelines if two pumps are used (one at M and the other at N).

2 Suppose one pump is used and it is located at P, the midpoint of MN.

    a Find AP and PB to the nearest metre.

    b Find the total cost of the pipeline and pump in this case.

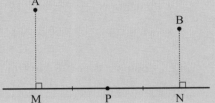

F   CIRCLE PROBLEMS

There are certain properties of circles where right angles occur and so Pythagoras' Rule can be used.

A CHORD OF A CIRCLE

> The line drawn from the centre of a circle at right angles to a chord bisects the chord.

Notice the use of the **isosceles triangle theorem**. The construction of radii from the centre of the circle to the end-points of the chord produces two right angled triangles.

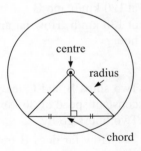

Example 17

A circle has a chord of length 10 cm, and the radius of the circle is 8 cm.
Find the shortest distance from the centre of the circle to the chord.

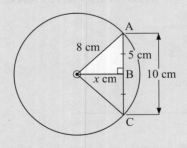

The shortest distance is the 'perpendicular distance'. Since the line drawn from the centre of a circle, perpendicular to a chord, bisects the chord, then
$$AB = BC = 5 \text{ cm}.$$

In $\triangle AOB$, $\quad 5^2 + x^2 = 8^2 \quad$ {Pythagoras}

$$\therefore \quad x^2 = 64 - 25 = 39$$

$$\therefore \quad x = \sqrt{39} \qquad \{\text{as } x > 0\}$$

$$\therefore \quad x \doteqdot 6.24 \qquad \{\text{to 3 sf.}\}$$

Thus the shortest distance is 6.24 cm.

EXERCISE 4F.1

1 A chord of a circle has length 2 cm and the circle has radius 3 cm. Find the shortest distance from the centre of the circle to the chord.

2 A chord of length 8 cm is 4 cm from the centre of a circle. Find the length of the circle's radius.

3 A chord is 6 cm from the centre of a circle of radius 10 cm. Find the length of the chord.

TANGENT-RADIUS PROPERTY

A tangent to a circle and a radius at the point of contact meet at right angles.

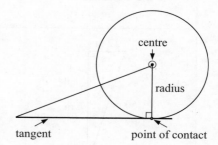

The line from the centre of the circle to the point of contact of the tangent completes a right angled triangle.

Example 18

A tangent of length 10 cm is drawn to a circle with radius 7 cm. How far is the centre of the circle from the end point of the tangent?

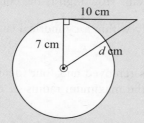

Let the distance be d cm.

$$\therefore \quad d^2 = 7^2 + 10^2 \qquad \{\text{Pythagoras}\}$$
$$\therefore \quad d^2 = 49 + 100$$
$$\therefore \quad d^2 = 149$$
$$\therefore \quad d = \sqrt{149} \qquad \{\text{as } d > 0\}$$
$$\therefore \quad d \doteqdot 12.2 \qquad \{\text{to 3 s.f.}\}$$

i.e., the centre is 12.2 cm from the endpoint of the tangent.

EXERCISE 4F.2

1 A circle has radius 2 cm. A tangent is drawn to the circle from point P which is 9 cm from O, the circle's centre. How long is the tangent?

2 Find the radius of a circle for which a tangent of length 10 cm has its end point 15 cm from the circle's centre.

3 If the earth has a radius of 6400 km and you are in a rocket 50 km directly above the earth's surface, determine the distance to the horizon.

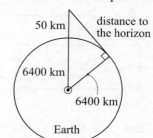

G   THREE-DIMENSIONAL PROBLEMS

Pythagoras' Rule is often used when finding lengths in **three-dimensional solids**.

Example 19

A 50 m rope is attached inside an empty cylindrical wheat silo of diameter 12 m as shown. How high is the wheat silo?

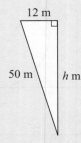

Let the height be h m.

$$\therefore \quad h^2 + 12^2 = 50^2 \qquad \{\text{Pythagoras}\}$$
$$\therefore \quad h^2 + 144 = 2500$$
$$\therefore \quad h^2 = 2356$$
$$\therefore \quad h = \sqrt{2356} \qquad \{\text{as } h > 0\}$$
$$\therefore \quad h \doteqdot 48.5 \qquad \{\text{to 3 s.f.}\}$$

i.e.,   the wheat silo is 48.5 m high.

EXERCISE 4G

1   A cone has a slant height of 13 cm and a base radius of 5 cm. How high is the cone?

2   Find the length of the longest nail that could be put entirely within a cylindrical can of radius 3 cm and height 10 cm.

3   A 68 cm nail just fits inside a cylindrical can. The nail is removed and four identical spherical balls are fitted exactly within the can. What is the maximum radius of each ball?

The **rule of Pythagoras** is often used *twice* in three-dimensional problem solving questions. We look for right angled triangles which contain two sides of given length.

Example 20

A room is 6 m by 4 m at floor level and the floor to ceiling height is 3 m. Find the distance from a floor corner point to the opposite corner point on the ceiling.

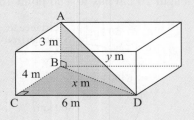

The required distance is AD. We join BD.

In $\triangle$BCD, $x^2 = 4^2 + 6^2$ $\{\text{Pythagoras}\}$

In $\triangle$ABD, $y^2 = x^2 + 3^2$

$$\therefore \quad y^2 = 4^2 + 6^2 + 3^2$$
$$\therefore \quad y^2 = 61$$
$$\therefore \quad y = \pm\sqrt{61}$$

But $y > 0$ $\therefore$ the required distance is $\sqrt{61} \doteqdot 7.81$ m. $\{\text{to 3 s.f.}\}$

4 A cube has sides of length 2 cm. Find the length of a diagonal of the cube.

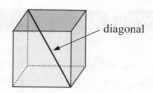

diagonal

5 A room is 6 m by 5 m and has a height of 3 m. Find the distance from a corner point on the floor to the opposite corner of the ceiling.

6 A rectangular box is 2 cm by 3 cm by 4 cm (internally). Find the length of the longest toothpick that can be placed within the box.

7 Determine the length of the longest piece of timber which could be stored in a rectangular shed 8 m by 5 m by 3 m high.

Example 21

A pyramid of height 40 m has a square base with edges 50 m.
Determine the length of the slant edges.

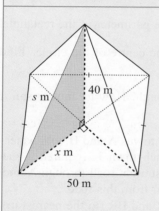

Let a slant edge have length s m.

Let half a diagonal have length x m.

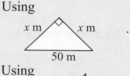

Using

$$x^2 + x^2 = 50^2 \qquad \text{\{Pythagoras\}}$$
$$\therefore \quad 2x^2 = 2500$$
$$\therefore \quad x^2 = 1250$$

Using

$$s^2 = x^2 + 40^2 \qquad \text{\{Pythagoras\}}$$
$$\therefore \quad s^2 = 1250 + 1600$$
$$\therefore \quad s^2 = 2850$$
$$\therefore \quad s = \sqrt{2850} \qquad \text{\{as } s > 0\text{\}}$$
$$\therefore \quad s \doteq 53.4 \qquad \text{\{to 1 dec. place\}}$$

i.e.,   each slant edge is 53.4 m long.

---

8 ABCD is a square-based pyramid. E, the apex of the pyramid is vertically above M, the point of intersection of AC and BD.

If an Egyptian Pharoah wished to build a square-based pyramid with all edges 200 m, how high (to the nearest metre) would the pyramid reach above the desert sands?

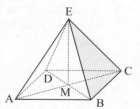

9 A symmetrical square-based pyramid has height 10 cm and slant edges of 12 cm. Find the dimensions of its square base.

10 A cube has sides of length 1 m. B is at the centre of one face, and A is an opposite vertex. Find the direct distance from A to B.

REVIEW SET 4A

1 A young tree has a 2 m support rope tied to it 1.6 m above ground level. The other end is tied to a peg in the ground. How far is the peg from the base of the tree?

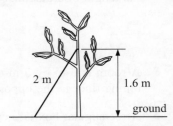

2 A city park is 320 metres square. The council decides to build a new walkway from one corner to the opposite corner. How long will the walkway be?

3

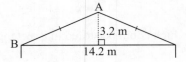

Find the length of the truss AB for the roof structure shown.

4 Is the following triangle right angled? Give evidence.

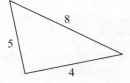

5 A rectangle has diagonal 15 cm and one side 9 cm. Find the perimeter of the rectangle.

6 A chord of a circle is 14 cm in length and is 9 cm from the centre of the circle. Find the radius of the circle.

7 A fishing boat leaves port and sails 12 km east then 5 km south. How far must it sail if it returns to port by the shortest distance?

8

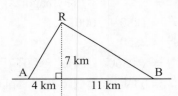

A reservoir R supplies two towns A and B with water. New piping from R to each town is required. The towns are connected by a straight road through A and B and the reservoir is 7 km from this road.

  a Find the distances AR and BR (to the nearest m).

  b Find the total cost of the new pipelines given that each 100 m will cost $2550.

9 A bracket for hanging flower baskets is shown in the diagram. Find the length of steel needed to make this bracket.

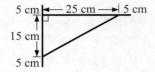

10 Kate and Ric leave home at the same time. Kate walks north at 5 km/h and Ric walks west at 6.5 km/h.

  a How far do they each walk in 4 hours?

  b How far apart are they at that time?

11 Can a 10 m long piece of timber be placed in a rectangular shed of dimensions 8 m by 5 m by 3 m? Give evidence.

REVIEW SET 4B

1 A television screen size is advertised as 34 cm. If the screen is 20 cm high, how wide must it be?

2 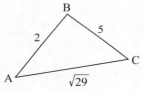 How high is the roof above the walls in the roof structure shown?

3 A ladder is 2 m long. It leans against a wall so that it is 90 cm from the base of the wall.

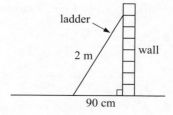

   a Write 90 cm as metres.

   b How far up the wall does the ladder reach?

4

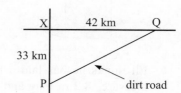

Show that the following triangle is right angled and state which vertex is the right angle:

5 If the diameter of a circle is 24 cm, find the shortest distance from a chord of length 18 cm to the centre of the circle.

6 If a softball diamond has sides of length 30 m, determine the distance the catcher must throw the ball from the home base to reach second base.

7 Two country roads meet at right angles at X. A motorist can travel from P to X to Q on a sealed road, travelling at 100 km/h.

   a **i** How far does he travel from P to X to Q?

   ii How long does it take him (in minutes)?

   b He could also travel from P to Q along a dirt road.

   i Find the distance from P to Q along the dirt road.

   ii If he can travel at 60 km/h on the dirt road, how long does it take him?

   c Which is the quicker route?

8 An aeroplane departs A and flies on a course of 143°T for 368 km. It then changes direction to a 233°T course and flies a further 472 km to town C.

   a Draw a fully labelled sketch of the flight.

   b Find the distance of C from A.

9 A pyramid of height 40 m has a square base with edges 50 m. Determine the length of the slant edges.

REVIEW SET 4C

1   A farmer has a paddock which is 350 m long and 280 m wide. She walks along two sides checking the electric fence then returns by walking diagonally across the paddock. How far has she walked?

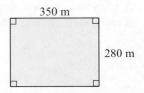

2   A circle has a chord of length 10 cm. The shortest distance from the circle's centre to the chord is 4 cm. Find the radius of the circle.

3   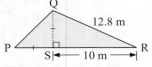   For the roof structure given:

    a   Find the height QS of the roof above the walls.

    b   Find the length of the roof truss PQ.

4   A pole XY is 18 metres tall. Four wires from the top of the pole X connect it to the ground.

Each wire is pegged 6 metres from the base of the pole. Find the total length of wire needed if a total of 2 m extra is needed for tying.

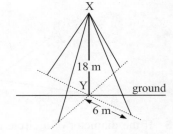

5   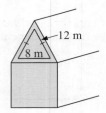   A church roof is shaped as shown. There is a triangular window at the front.

    a   Find the height of the window.

    b   Find the area of glass for the window.

6   Eli thinks he has laid a rectangular slab of concrete for the floor of his toolshed. It measures 3.2 m long and 2.1 m wide. The diagonal measures 3.83 m. Check that Eli's concrete is rectangular.

7   A boat leaves port and sails 14.8 km in a direction 125°T. It then turns and sails 5.3 km in a direction 215°T. It then returns to port.

    a   Draw a fully labelled sketch of the boat trip.

    b   How far did the boat sail on this trip?

8   A cube has sides of length 10 cm. Find the length of a diagonal of the cube.

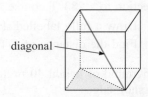

9   A room is 7 m by 4 m and has a height of 3 m. Find the distance from a corner point on the floor to the opposite corner of the ceiling.

Chapter 5

Descriptive statistics

Contents:

STATISTICS

Statistics is the art of solving problems and answering questions by collecting and analysing data.

Statistics are used by governments, businesses and sports organisations so that they can make informed decisions when they are providing services such as in health, transport and commerce or developing new tactics. They are also interested in using statistics as a means of analysing the effects of certain changes that may have been made, or in predicting what may happen in the future.

 # DESCRIBING DATA

TYPES OF DATA

Data are individual observations of a **variable**. A variable is a quantity that can have a value recorded for it or to which we can assign an attribute or quality.

There are two types of variable that we commonly deal with:

CATEGORICAL VARIABLES

A **categorical variable** is one which describes a particular quality or characteristic. It can be divided into **categories**. The information collected is called **categorical data**.

Examples of categorical variables are:
- *Getting to school:*   the categories could be train, bus, car and walking.
- *Colour of eyes:*   the categories could be blue, brown, hazel, green, grey.
- *Gender:*   male and female.

QUANTITATIVE (NUMERICAL) VARIABLES

A **quantitative (numerical) variable** is one which has a numerical value and is often called a numerical variable. The information collected is called **numerical data**.

Quantitative variables can be either discrete or continuous.

A **quantitative discrete variable** takes exact number values and is often a result of **counting.**

Examples of discrete quantitative variables are:
- *The number of people in a household:*   the variable could take the values 1, 2, 3, ...
- *The score out of* 30 *on a test:*   the variable could take the values 0, 1, 2, 3, ... 30.

A **quantitative continuous variable** takes numerical values within a certain continuous range. It is usually a result of **measuring**.

Examples of quantitative continuous variables are:
- *The weight of newborn babies:*   the variable could take any value on the number line but is likely to be in the range 0.5 kg to 8 kg.

- *The heights of Year 8 students:* the variable would be measured in centimetres. A student whose height is recorded as 145 cm could have exact height between 144.5 cm and 145.5 cm.

Example 1

Classify these variables as categorical, quantitative discrete or quantitative continuous:
  a   the number of heads when 3 coins are tossed
  b   the brand of toothpaste used by the students in a class
  c   the heights of a group of 15 year old children.

  a   The values of the variables are obtained by counting the number of heads. The result can only be one of the values 0, 1, 2 or 3. It is quantitative discrete data.

  b   The variable describes the brands of toothpaste. It is categorical data.

  c   This is numerical data obtained by measuring. The results can take any value between certain limits determined by the degree of accuracy of the measuring device. It is quantitative continuous data.

EXERCISE 5A

1   For each of the following possible investigations, classify the variable as categorical, quantitative discrete or quantitative continuous:

  a   the number of goals scored each week by a basketball ball team
  b   the heights of the members of a football team
  c   the most popular radio station
  d   the number of children in a Japanese family
  e   the number of loaves of bread bought each week by a family
  f   the pets owned by students in a year 8 class

  g   the number of leaves on the stems of plants
  h   the amount of sunshine in a day
  i   the number of people who die from cancer each year in the USA
  j   the amount of rainfall in each month of the year
  k   the countries of origin of immigrants
  l   the most popular colours of cars
  m   the gender of school principals
  n   the time spent doing homework
  o   the marks scored in a class test
  p   the items sold at the school canteen
  q   the number of matches in a box
  r   the reasons people use taxis
  s   the sports played by students in high schools
  t   the stopping distances of cars doing 60 km/h
  u   the pulse rates of a group of athletes at rest.

2 **a** For the categorical variables in question **1**, write down two or three possible categories. (In all cases but one, there will be more than three categories possible.) Discuss your answers.

  b For each of the quantitative variables (discrete and continuous) identified in question **1**, discuss as a class the range of possible values you would expect.

INVESTIGATION 1   STATISTICS FROM THE INTERNET

In this investigation you will be exploring the web sites of a number of organisations to find out the topics and the types of data that they collect and analyse.

Note that the web addresses given here were operative at the time of writing but there is a chance that they will have changed in the meantime. If the address does not work, try using a search engine to find the site of the organisation.

What to do:

Visit the site of a world organisation such as the United Nations (www.un.org) or the World Health Organisation (www.who.int) and see the available types of data and statistics.

B   PRESENTING AND INTERPRETING DATA

ORGANISING CATEGORICAL DATA

A **tally and frequency table** can be used to organise categorical data.

For example, a survey was conducted on the type of fuel used by 50 randomly selected vehicles.

The variable 'type of fuel' is a categorical variable because the information collected for each vehicle can only be one of the four categories: Unleaded, Lead Replacement, LPG or Diesel. The data has been tallied and organised in the given frequency table:

Fuel type	Tally	Freq.
Unleaded	ǁꞀꞀ ǁꞀꞀ ǁꞀꞀ ǁꞀꞀ ǁꞀꞀ ǁǁǁ	28
Lead Rep	ǁꞀꞀ ǁꞀꞀ ǁǁ	12
LPG	ǁꞀꞀ ǁǁǁ	8
Diesel	ǁǁ	2
	Total	50

DISPLAYING CATEGORICAL DATA

Acceptable graphs to display the 'type of fuel' categorical data are:

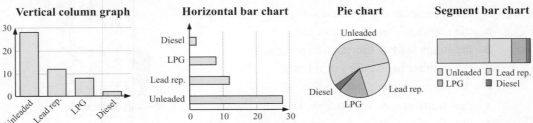

For categorical data, the **mode** is the category which occurs most frequently.

ORGANISING DISCRETE NUMERICAL DATA

Discrete numerical data can be organised:

- in a **tally and frequency table**
- using a **dot plot**
- using a **stem-and-leaf plot** (also called a **stemplot**).

Stemplots are used when there are many possible data values. The stemplot is a form of grouping of the data which displays frequencies but retains the actual data values.

Examples:

- **frequency table**

Number	Tally	Freq.
3	||	2
4	﹊﹊ ||||	9
5	﹊﹊ ﹊﹊ |||	13
6	﹊﹊	5
7	|	1

- **dot plot**

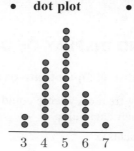

- **stemplot**

  Example:

Stem	Leaf
0	9
1	7 1
2	8 3 6 7 6 4
3	9 3 5 5 6 8 2 1
4	7 9 3 4 2
5	1

As data is collected it can be entered directly into a carefully set up tally table, dot plot or stemplot blank sheet.

THE PEA PROBLEM

A farmer wishes to investigate the effect of a new organic fertiliser on his crops of peas. He is hoping to improve the crop yield by using the fertiliser. He set up a small garden which was subdivided into two equal plots and planted many peas. Both plots were treated the same except for the use of the fertiliser on one, but not the other. All other factors such as watering were as normal.

A random sample of 150 pods was harvested from each plot at the same time and the number of peas in each pod counted. The results were:

Without fertiliser

4 6 5 6 5 6 4 6 4 9 5 3 6 8 5 4 6 8 6 5 6 7 4 6 5 2 8 6 5 6 5 5 5 4 4 4 6 7 5 6
7 5 5 6 4 8 5 3 7 5 3 6 4 7 5 6 5 7 5 7 6 7 5 4 7 5 5 5 6 6 5 6 7 5 8 6 8 6 7 6
6 3 7 6 8 3 3 4 4 7 6 5 6 4 5 7 3 7 7 6 7 7 4 6 6 5 6 7 6 3 4 6 6 3 7 6 7 6 8 6
6 6 6 4 7 6 6 5 3 8 6 7 6 8 6 7 6 6 6 8 4 4 8 6 6 2 6 5 7 3

With fertiliser

6 7 7 4 9 5 5 5 8 9 8 9 7 7 5 8 7 6 6 7 9 7 7 7 8 9 3 7 4 8 5 10 8 6 7 6 7 5 6 8
7 9 4 4 9 6 8 5 8 7 7 4 7 8 10 6 10 7 7 7 9 7 7 8 6 8 6 8 7 4 8 6 8 7 3 8 7 6 9 7
6 9 7 6 8 3 9 5 7 6 8 7 9 7 8 4 8 7 7 7 6 6 8 6 3 8 5 8 7 6 7 4 9 6 6 6 8 4 7 8
9 7 7 4 7 5 7 4 7 6 4 6 7 7 6 7 8 7 6 6 7 8 6 7 10 5 13 4 7 7

For you to consider:

- Can you state clearly the problem that the farmer wants to solve?
- How has the farmer tried to make a fair comparison?
- How could the farmer make sure that his selection is at random?
- What is the best way of organising this data?
- What are suitable methods of display?
- Are there any abnormally high or low results and how should they be treated?
- How can we best indicate the most typical pod size?
- How can we best indicate the spread of possible pod sizes?
- What is the best way to show 'typical pod size' and the spread?
- Can a satisfactory conclusion be made?

ORGANISATION AND DISPLAY OF DISCRETE DATA

In **The Pea Problem**, the **discrete quantitative variable is**: *The number of peas in a pod.*

To organise the data a tally/frequency table could be used. We count the data systematically and use a '|' to indicate each data value.

Remember that 𝍸 represents 5.

Below is the table for *Without fertiliser*:

Number of peas/pod	*Tally*	*Frequency*
1		0
2	||	2
3	𝍸 𝍸 |	11
4	𝍸 𝍸 𝍸 ||||	19
5	𝍸 𝍸 𝍸 𝍸 𝍸 ||||	29
6	𝍸 𝍸 𝍸 𝍸 𝍸 𝍸 𝍸 𝍸 𝍸 𝍸 |	51
7	𝍸 𝍸 𝍸 𝍸 𝍸	25
8	𝍸 𝍸 ||	12
9	|	1

A **dot plot** could be used to organise and display the results, or a **column graph** could be used to display the results.

Column graph of *Without fertiliser*

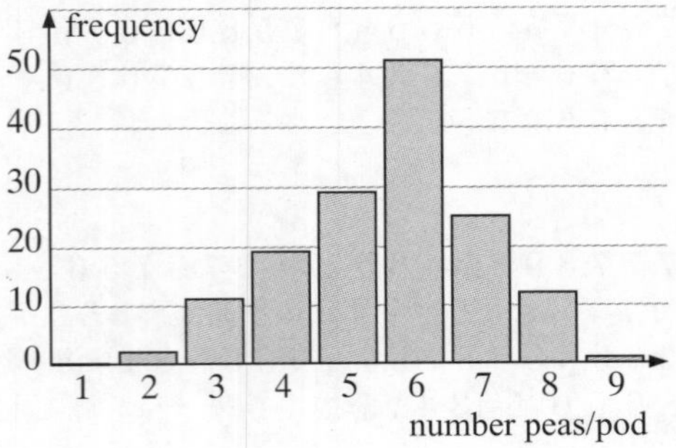

Dot plot of *Without fertiliser*

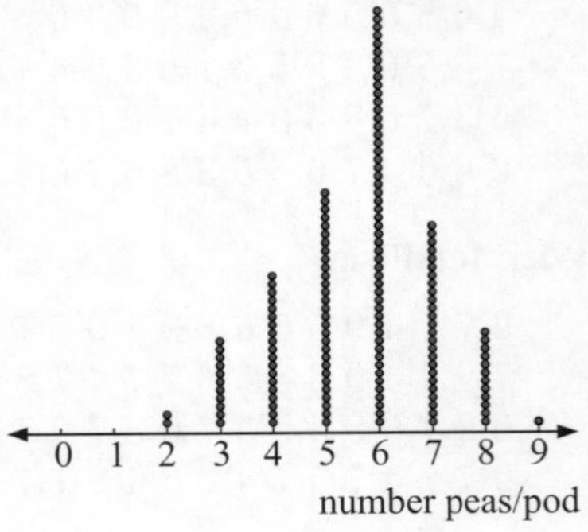

DESCRIPTIVE STATISTICS (Chapter 5)   115

Are there any advantages/disadvantages in using a dot plot rather than a column graph?

From both graphs we can make observations and calculations such as:

- 6 peas per pod is the mode of the *Without fertiliser* data.
- 8.7% of the pods had fewer than 4 peas in them, etc.

DESCRIBING THE DISTRIBUTION OF THE DATA SET

Many data sets show **symmetry** or **partial symmetry** about the mode.

If we place a curve over the column graph (or dot plot) we see that this curve shows symmetry and we say that we have a **symmetrical distribution** of the data.

For the *Without fertiliser* data we have: This distribution is said to be **negatively skewed** as if we compare it with the symmetrical distribution it has been 'stretched' on the left (or negative) side of the mode.

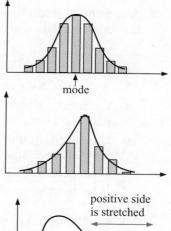

So we have:

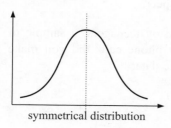

symmetrical distribution

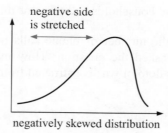

negative side is stretched

negatively skewed distribution

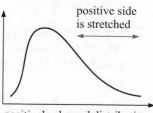

positive side is stretched

positively skewed distribution

OUTLIERS

Outliers are data values that are either much larger or much smaller than the general body of data. Outliers appear separated from the body of data on a frequency graph.

For example, if the farmer in **The Pea Problem** (page **113**) found one pod in the *Without fertiliser* sample contained 13 peas, then the data value 13 would be considered an outlier. It is much larger than the other data in the sample. On the column graph it would appear separated.

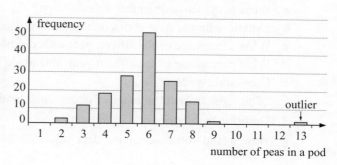

EXERCISE 5B

1 State whether the following quantitative (or numerical) variables are discrete or continuous:

    a the time taken to run 100 metres

    b the maximum temperature reached on a January day

    c the number of matches in a box

    d the weight of luggage taken on an aircraft

    e the time taken for a battery to run down

    f the number of bricks needed to build a house

    g the number of passengers on a bus

    h the number of minutes spent on the internet per day

2 A class of 20 students was asked "How many pets do you have in your household?" and the following data was collected:

    0 1 2 2 1 3 4 3 1 2 0 0 1 0 2 1 0 1 0 1

    a What is the variable in this investigation?

    b Is the data discrete or continuous? Why?

    c Construct a dotplot to display the data. Use a heading for the graph, and scale and label the axes.

    d How would you describe the distribution of the data? (Is it symmetrical, positively skewed or negatively skewed? Are there any outliers?)

    e What percentage of the households had no pets?

    f What percentage of the households had three or more pets?

3 For an investigation into the number of phone calls made by teenagers, a sample of 50 fifteen-year-olds were asked the question "How many phone calls did you make yesterday?" The following dotplot was constructed from the data:

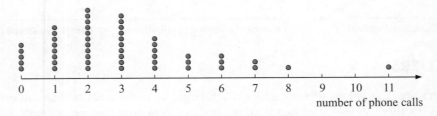

    a What is the variable in this investigation?

    b Explain why the data is discrete numerical data.

    c What percentage of the fifteen-year-olds did not make any phone calls?

    d What percentage of the fifteen-year-olds made 5 or more phone calls?

    e Copy and complete:

       "The most frequent number of phone calls made was ...."

    f Describe the distribution of the data.

    g How would you describe the data value '11'?

4   A randomly selected sample of households has been asked, 'How many people live in
    your household?' A column graph has been constructed for the results.

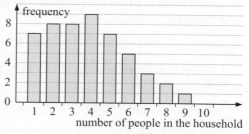

a   How many households gave data in the
    survey?

b   How many of the households had only
    one or two occupants?

c   What percentage of the households had
    five or more occupants?

d   Describe the distribution of the data.

5   The number of matches in a box is stated as 50 but the actual number of matches has
    been found to vary. To investigate this, the number of matches in a box has been counted
    for a sample of 60 boxes:

    51 50 50 51 52 49 50 48 51 50 47 50 52 48 50 49 51 50 50 52
    52 51 50 50 52 50 53 48 50 51 50 50 49 48 51 49 52 50 49 50
    50 52 50 51 49 52 52 50 49 50 49 51 50 50 51 50 53 48 49 49

a   What is the variable in this investigation?

b   Is the data continuous or discrete numerical data?

c   Construct a frequency table for this data.

d   Display the data using a bar chart.

e   What percentage of the boxes contained exactly 50 matches?

6   Revisiting **The Pea Problem**.   For the *With fertiliser* data:
    a   Organise the data in a tally-frequency table.
    b   Draw a column graph of the data.
    c   Are there any outliers?
    d   What evidence is there that the fertiliser 'increases the number of peas in a pod'?
    e   Can it be said that the fertiliser will increase the farmer's pea crop and his profits?

C   GROUPED DISCRETE DATA

A local kindergarten is concerned about the number of vehicles passing by between 8.45 am
and 9.00 am. Over 30 consecutive week days they recorded data.

The results were: 27, 30, 17, 13, 46, 23, 40, 28, 38, 24, 23, 22, 18, 29, 16,
                      35, 24, 18, 24, 44, 32, 52, 31, 39, 32, 9, 41, 38, 24, 32

In situations like this we group the data into
class intervals.

It seems sensible to use class intervals of
length 10 in this case.

The tally/frequency table is:

Number of cars	Tally	Frequency
0 to 9			1
10 to 19	卌	5
20 to 29	卌 卌	10
30 to 39	卌 ||||	9
40 to 49	||||	4
50 to 59			1
	Total	30

STEM-AND-LEAF PLOTS

A **stem-and-leaf plot** (often called a **stemplot**) is a way of writing down the data in groups. It is used for small data sets.

A stemplot shows actual data values. It also shows a comparison of frequencies.

For numbers with two digits, the first digit forms part of the **stem** and the second digit forms a **leaf**.

For example, for the data value 18,   1 is recorded on the stem, 8 is a leaf value.

For the data value 123, 12 is recorded on the stem and 3 is the leaf.

The **stem-and-leaf plot** is:

Stem	Leaf
0	9
1	73868
2	7384329444
3	085219282
4	6041
5	2

The **ordered stem-and-leaf plot** is:

Stem	Leaf
0	9
1	36788
2	2334444789
3	012225889
4	0146
5	2

Note: 1 | 7   means 17

The ordered stemplot arranges all data from smallest to largest.

Notice that:

- all the actual data is shown
- the minimum (smallest) data value is 9
- the maximum (largest) data value is 52
- the 'twenties' interval (20 to 29) occurred most often.

BACK TO BACK STEM-AND-LEAF PLOTS

This back to back stemplot represents the times for the 100 metre freestyle recorded by members of a swimming squad.

Note: The fastest time for a girl is 33.4 seconds and the slowest time for a girl is 41.1 seconds.

Girls	Stem	Boys
	32	1
4	33	0227
763	34	13448
87430	35	024799
8833	36	788
7666	37	0
6	38	
0	39	
	40	
1	41	

leaf unit: 0.1 seconds.

COLUMN GRAPHS

A vertical column graph can be used to display grouped discrete data.

For example, consider the local kindergarten data of page **117**.

The **frequency table** is:

Number of cars	Frequency
0 to 9	1
10 to 19	5
20 to 29	10
30 to 39	9
40 to 49	4
50 to 59	1

The **column graph** for this data is:

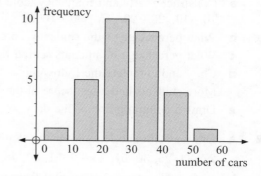

INVESTIGATION 2   TAXI SIR?

Two taxi drivers, Peter and Ivan, are friendly rivals. Each claims that he is the more successful driver. They agree to randomly select 25 days on which they work and record the daily fare totals. The data collected to the nearest dollar was:

Peter   194   99   188   208   95   168   205   196   233
116   132   153   205   191   182   118   140   270
183   155   93   154   190   223   147

Ivan   260   152   127   163   180   161   110   153   139
139   142   161   97   116   129   215   241   160
110   159   147   174   162   158   223

What to do:

1   Produce back-to-back stem plots of this data.

2   Explain why "the amount of money collected per hour" would have been a better variable to use rather than "the daily fare totals".

3   Produce back-to-back stem plots for "the amount of money collected per hour" given below.

Peter ($ per hour)   17.27   11.31   15.72   18.92   9.55   12.98   19.12   16.69   11.68
15.84   12.81   24.03   15.03   12.95   20.09   18.64   18.94   13.92
11.69   15.52   15.21   18.26   12.25   18.59   22.79

Ivan ($ per hour)   23.70   13.30   12.18   14.20   15.74   14.01   10.05   10.05   12.20
13.50   18.64   13.29   12.65   13.54   8.83   11.09   12.29   18.94
20.08   13.84   14.57   13.34   13.44   13.63   14.18

4   Present a brief written report (of about 250 words) to summarise your findings and present your conclusion.

EXERCISE 5C

1 The data set below is the test scores (out of 100) for a Maths test for 50 students.

56   29   78   67   68   69   80   89   92   58   66   56   88   81   70   73   63   74
67   64   62   55   56   75   90   92   47   59   64   89   62   51   87   89   76   59
72   80   95   68   80   64   53   43   61   71   38   44   88   39

    a Construct a tally and frequency table for this data using class intervals 0 - 9,
       10 - 19, 20 - 29, ..., 90 - 100.

    b What percentage of the students scored 80 or more for the test?

    c What percentage of students scored less than 50 for the test?

    d Copy and complete the following:
       More students had a test score in the interval .... than in any other interval.

    e Draw a column graph of the data.

2   **a** Draw a stem-and-leaf plot using stems 2, 3, 4, and 5 for the following data:
       27, 34, 25, 36, 57, 34, 42, 51, 50, 48, 29, 27, 33, 30, 46, 40, 35, 24, 21, 58

    b Redraw the stem-and-leaf plot from **a** so that it is ordered.

3 For the ordered stem-and-leaf plot given find:

Stem	Leaf
0	2 3 7
1	0 4 4 7 8 9 9
2	0 0 1 1 2 2 3 5 5 6 8 8
3	0 1 2 4 4 5 8 9
4	0 3 7
5	5
6	2

    a the minimum value

    b the maximum value

    c the number of data with a value greater than 25

    d the number of data with a value of at least 40

    e the percentage of the data with a value less than 15.

    f How would you describe the distribution of the data? **Hint**: Turn your stemplot on its side.

4 The test score, out of 50 marks, is recorded for a group of 45 students.

25   28   35   42   44   28   24   49   29   33   33   34   38   28   26
32   34   39   41   46   35   35   43   45   50   30   22   20   35   48
36   25   20   18   9   40   32   33   28   33   34   34   36   25   42

    a Construct a stem-and-leaf plot for this data using 0, 1, 2, 3, 4, and 5 as the stems.

    b Redraw the stem-and-leaf plot so that it is ordered.

    c What advantage does a stem-and-leaf plot have over a frequency table?

    d What is the **i** highest **ii** lowest mark scored for the test?

    e If an 'A' was awarded to students who scored 40 or more for the test, what percentage of students scored an 'A'?

    f What percentage of students scored less than half marks for the test?

5 The students of class 11C and class 11P were asked to record the number of hours they spent watching television during a one week period. The results were as follows:

Class 11C							*Class 11P*					
21	10	18	27	32	13	5	22	23	29	25	18	30
12	14	26	21	20	37	14	15	14	19	29	7	16
9	15	11	27	17	30	12	24	11	12	14	15	6

Draw a back to back stemplot to represent this data.

CONTINUOUS DATA

Continuous data is numerical data which has values within a continuous range.

For example, if we consider the weights of students in a rugby training squad we might find that all weights lie between 50 kg and 100 kg.

Suppose   2 students lie in the 50 kg up to but not including 60 kg,
           7 students lie in the 60 kg up to but not including 70 kg,
           9 students lie in the 70 kg up to but not including 80 kg,
           5 students lie in the 80 kg up to but not including 90 kg,
           3 students lie in the 90 kg up to but not including 100 kg.

The frequency table would be: We could use a histogram to represent the data graphically.

Weight interval	Frequency
50 -	2
60 -	7
70 -	9
80 -	5
90 -	3

Weights of the students in the rugby squad

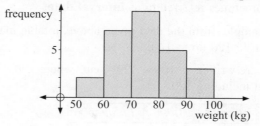

HISTOGRAMS

A **histogram** is a vertical column graph used to represent continuous grouped data.

There are no gaps between the columns in a histogram as the data is continuous.

The bar widths must be equal and each bar height must reflect the frequency.

Example 2

The time, in minutes (ignoring any seconds) for students to get to school on a given day is as follows:

   5   12   32   26   41   37   22   27   41   52   49   38   41   62   69
  37   21   4   7   12   32   36   39   14   24   27   29   22   21   25

a   Organise this data on a frequency table. Use time intervals of 0 -, 10 -, 20 -, etc.
b   Draw a histogram to represent the data.

a

Time int.	Tally	Freq.
0 -	|||	3
10 -	|||	3
20 -	++++ ++++	10
30 -	++++ ||	7
40 -	||||	4
50 -			1
60 -	||	2

b

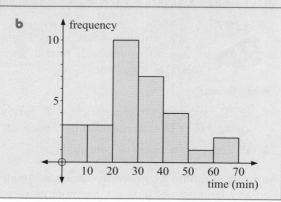

COLLATING CONTINUOUS DATA

The following data are results (in grams) of weighing 50
laboratory rats:

261	133	173	295	265	142	140	271	185
251	166	100	292	107	201	234	239	159
153	263	195	151	156	117	144	189	234
171	233	182	165	122	281	149	152	289
168	260	256	156	239	203	101	268	241
217	254	240	214	221				

Continuous data can also be grouped.

Each group is called a **class**.

The size of the grouping is called the **class interval**. Because continuous data can be grouped
it is sometimes referred to as **interval data**.

For example, from the data given above, a table may be designed so that the classes are 100
up to 120, 120 up to 140, ...

These are written as 100 -, 120 -, etc. where 100 - means all of the numbers from 100 up to
but not including 120.

Class interval	Tally	Frequency
100 -	||||	4
120 -	||	2
140 -	|||| ||||	10
160 -	||||	5
180 -	||||	4
200 -	||||	4
220 -	|||| |	6
240 -	||||	5
260 -	|||| |	6
280 -	||||	4
Total		50

INVESTIGATION 3   CHOOSING CLASS INTERVALS

When dividing data values into intervals, the choice of
how many intervals to use, and hence the width of each
class, is important.

DEMO

What to do:

1 Click on the icon to experiment with various data sets. You can change the number of
classes. How does the number of classes alter the way in which we can read the data?

2 Write a brief account of your findings.

As a rule of thumb we generally use approximately $\sqrt{n}$ classes for a data set of n individuals.
For very large sets of data we use more classes rather than less.

EXERCISE 5D

1 The initial weights of students (in kg) in a fitness class were:

69   90   61   58   94   68   77   80   64   87   59   81   73   56   72   66
62   75   78   75   87   73   81   70   73   58   85   73   76   92   82   86
82   71   79   80   85   70   60   75

a Using classes 50 - , 60 - , 70 - , 80 - , 90 - , tabulate the data using columns of score, tally, frequency.

b How many students are in the 70 -   class?

c How many students weighed less than 80 kg?

d Find the percentage of students who weighed 70 kg or more.

2 A group of young athletes were invited to participate in a javelin throwing competition. The following results were obtained:

Distance (metres)	0 -	10 -	20 -	30 -	40 -
No. of athletes	3	26	41	14	6

a How many athletes threw less than 20 metres?

b What percentage of the athletes were able to throw at least 30 metres?

3

Height (mm)	Frequency
300 -	12
325 -	18
350 -	42
375 -	28
400 -	14
425 -	6

A plant inspector takes a random sample of two week old seedlings from a nursery and measures their height to the nearest mm.

The results are shown in the table alongside.

a How many of the seedlings are 400 mm or more?

b What percentage of the seedlings are between 349 and 400 mm?

c The total number of seedlings in the nursery is 1462. Estimate the number of seedlings which measure:

   i less than 400 mm   **ii** between 374 and 425 mm.

E   FREQUENCY DISTRIBUTION TABLES

Frequency tables may include other columns which show **relative frequency**, **cumulative frequency**, and **cumulative relative frequencies**.

The **relative frequency** of an event is the frequency of that event expressed as a fraction (or decimal equivalent) of the total frequency.

The **cumulative frequency** of an event is the accumulation (sum) of the frequencies up to and including that event.

The **cumulative relative frequency** of an event is the accumulation (sum) of the relative frequencies up to and including that event.

The data listed below gives the lengths of steel rods in metres as produced by a machine:

1.95 3.5   2.5   2.8   1.25 4.75 5.5   5.5   5.75 5.75 4.5   2.25 4.75 0.65   3.5
2.8   4.5   4.75 5.5   5.5   4.75 4.5   4.75 1.25 2.25 4.5   4.75 5.5   5.5   5.75
5.5   5.5   4.75 2.8   2.5   3.5   1.4   4.75 5.5   5.75 5.75 5.75 5.5   5.5   4.75
4.5   2.25 1.4   5.75 5.75

More sense can be made of this data when it is organised according to size and frequencies and presented in a tabular format.

The various rod sizes can be listed in the table from lowest to highest and a tally column can be used to record the number of times each value (length) occurs.

See **Example 3** below.

CLASS INTERVAL

Another way that this data could be organised is by grouping the rod lengths together into various ranges of lengths called class intervals.

These class intervals are usually of **equal length** and must be constructed so that each value is assigned to one class interval only.

The size of the class intervals devised will depend on the data being organised, but it is generally thought that about 6 or 12 class intervals is the right number for a frequency table.

See **Example 4** on the next page.

Example 3

Prepare a frequency distribution table for the steel rod data. Include columns for length, tally, frequency, cumulative frequency, relative frequency and cumulative relative frequency.

Frequency distribution table for the Steel Rod production

Rod Length (m)	Tally	Frequency	Cumulative Frequency	Relative Frequency	Cumu. Rel. Frequency
0.65	|	1	1	0.02	0.02
1.25	||	2	3	0.04	0.06
1.40	||	2	5	0.04	0.10
1.95	|	1	6	0.02	0.12
2.25	|||	3	9	0.06	0.18
2.50	||	2	11	0.04	0.22
2.80	|||	3	14	0.06	0.28
3.50	|||	3	17	0.06	0.34
4.50	卌	5	22	0.10	0.44
4.75	卌 ||||	9	31	0.18	0.62
5.50	卌 卌 |	11	42	0.22	0.84
5.75	卌 |||	8	50	0.16	1.00
Total		50			

Example 4

Rework the frequency distribution table using the class intervals:
0-0.99,   1.00-1.99,   2.00-2.99,   etc.

Rod Length (metres)	Tally	Frequency	Cumulative Frequency	Relative Frequency	Cumulative Relative Frequency
0 - 0.99	|	1	1	0.02	0.02
1.00 - 1.99	⫴⫴	5	6	0.10	0.12
2.00 - 2.99	⫴⫴ ⫴⫴	8	14	0.16	0.28
3.00 - 3.99	⫴⫴	3	17	0.06	0.34
4.00 - 4.99	⫴⫴ ⫴⫴ ⫴⫴	14	31	0.28	0.62
5.00 - 5.99	⫴⫴ ⫴⫴ ⫴⫴ ⫴⫴	19	50	0.38	1.00
Total		50			

HISTOGRAMS

Once data has been summarised in a frequency table, it can be represented diagrammatically by a **histogram**.

Example 5

Draw a frequency histogram of the steel rod data from **Example 4**.

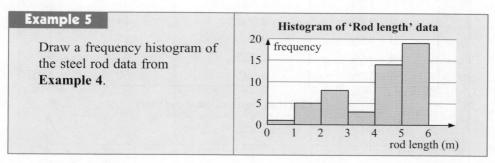

FREQUENCY POLYGON

A **frequency polygon** is a line graph which, like the histogram, gives a good visual appreciation of the shape of the frequency distribution.

Instead of drawing the bars, the **midpoint** of each bar is found and is used to represent the whole interval. These points are then joined by straight lines.

Example 6

Draw a frequency polygon of the steel rod data from **Example 4**.

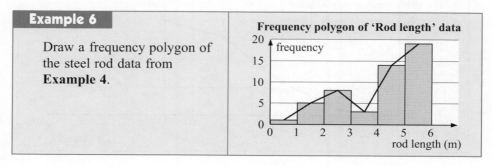

LESS THAN OGIVE (CUMULATIVE FREQUENCY DIAGRAM)

An **ogive** or **cumulative frequency diagram** is used to represent **cumulative frequencies**.

Since cumulative frequency is the sum of the frequencies up to and including the value being examined, we plot the cumulative frequencies against the **upper end points** of each class interval. These points are then joined by straight lines to produce the ogive.

The **less than ogive** is a polygonal graph which is based on **increasing accumulations** and one is illustrated in the example which follows. The graph begins at the lowest boundary.

In this case the curve slopes upwards as we look from left to right on the diagram. It is called a **less than ogive** as it gives all points less than any particular value. For instance, on the ogive which follows, if you wanted to know how many rods were less than, say, 4 metres, you could easily find the answer by reading off the graph. There are 17 such rods less than 4 metres.

Example 7

Draw a less than ogive for the steel rod data from **Example 4**.

Rod length (metres)	Upper End Point	Cumulative Frequency
0 - 0.99	0.995	1
1 - 1.99	1.995	6
2 - 2.99	2.995	14
3 - 3.99	3.995	17
4 - 4.99	4.995	31
5 - 5.99	5.995	50

Less than Ogive of 'Steel rod' data

EXERCISE 5E

1 Copy and complete the table alongside including columns of cumulative frequency, relative frequency and cumulative relative frequency.

Value	Frequency
32	3
35	7
36	8
39	11
41	15
44	12
45	8
46	7
49	5
50	4
total	80

2 A bottle manufacturer records the following number of rejected bottles each day over a 30 day period:

112 127 92 147 134 131 104 99 116 122
125 118 109 96 142 127 106 100 132 138
113 123 118 131 115 94 144 124 117 124

a Organise the data into six class intervals starting with 91 - 100 and ending with 141 - 150. Hence, complete a frequency table including columns of cumulative frequency, relative frequency and cumulative relative frequency.

b From your table find:
  i how many times there were less than 120 rejections
  ii what percentage of time there were between 91 and 100 (inclusive) rejections
  iii what percentage of time there were less than 130 rejections.

c Construct a histogram and ogive from your table.

3 A country motel has room for 80 people.
The manager keeps records of the number of guests staying at the motel over the summer period of 90 days. The results are shown alongside:

No. of guests	Freq. (days)
1 - 10	8
11 - 20	11
21 - 30	14
31 - 40	17
41 - 50	20
51 - 60	9
61 - 70	7
71 - 80	4

a Construct a frequency table for this data and show cumulative frequencies for each class interval.

b Construct a histogram and frequency polygon of the data.

c Prepare a 'less than' ogive of the data and hence find:
  i the number of nights when the motel had 30 or less guests
  ii the number of nights when the motel had less than 45 guests.

d The motel will 'break even' if it has less than 30 guests for no more than 25% of the time. Did it break even?

4   A market research company conducts a survey on the number of times people eat out at restaurants each year. The results from the 100 people surveyed are as follows:

No. of times	Freq.
1 - 20	15
21 - 40	21
41 - 60	24
61 - 80	18
81 - 100	12
101 - 120	6
121 - 140	4

   a   Construct a frequency table which also shows cumulative frequencies.

   b   Construct a frequency polygon.

   c   Construct a 'less than' ogive and use it to answer the following:

      i   What percentage of people ate at restaurants less than 81 times a year?

      ii   If a person was in the lower 25% of the distribution, what is the maximum number of times they would have eaten at restaurants during the year?

5   The manager of a fast food restaurant is concerned that customers are waiting too long for their food. She decides to gather some statistics on customer waiting times and the following times (in minutes) are recorded:

1.25   2.5   8.5   1.8   4.6   10.5   3.4   7.0   6.25   4.10   5.15   5.95   7.35   5.8   2.9
0.9   3.4   6.0   8.8   2.7   10.2   4.5   5.2   4.1   2.5   7.7   3.8   2.1   5.5   6.25
4.3   1.8   8.0   9.9   3.7   4.4   6.2   3.3   7.2   8.6   3.45   6.55   2.85   9.4   4.25
5.6   11.9 6.4   4.8   5.8

   a   Group this data into classes of 0 - 1.99, 2 - 3.99, etc. and construct a frequency table which also shows cumulative frequencies.

   b   Construct a histogram of the data.

   c   Construct a 'less than' ogive and answer the following questions:

      i   How many customers have to wait less than 4 minutes for their food?

      ii   What percentage of customers have to wait more than 5 minutes for their food?

      iii   If the restaurant's goal is for 90% of the customers to be given their food within 8 minutes, are they achieving this goal?

F   SUMMARISING THE DATA

MEASURES OF THE MIDDLE OF A DISTRIBUTION

After collecting and presenting statistical data, you can now attempt to interpret the data. One way of doing this is to find the value of the **centre or middle** of the distribution.

There are three commonly used measures for the middle of a distribution; the **mean**, the **mode** and the **median**.

However, before proceeding further we will define some of the terms that will be used from now on:

- **Ungrouped data** comprises of single values which have not been put into groups or classes.

   For example, the heights of five children are 1.23 m, 1.56 m, 1.34 m, 1.09 m, 1.71 m.

- **Grouped data** has been grouped together according to the number of times each value occurs (i.e., the frequency). There are two types of grouped data:
  - ▸ **Grouped discrete data** which can be precisely determined and has been grouped together according to the number of times each value occurs.
  - ▸ **Grouped continuous data** which cannot be precisely determined and has been grouped together into classes.

MEAN, MODE AND MEDIAN

The **mean** of a set of scores is their arithmetic average obtained by adding all the scores and dividing by the total number of scores.

The **mode(s)** of a set of scores is the score(s) which occurs most frequently.

The **median** of a set of scores is the middle score after they have been placed in order of size from smallest to largest.

A set of scores is **bimodal** if it has two modes. If it has more than two modes we do not use them as a measure of the centre.

MEDIAN FINDER

UNGROUPED DATA

Example 8

Find the mean, mode and median of the following distributions:
  a   3, 6, 5, 6, 4, 5, 5, 6, 7   **b**   13, 12, 15, 13, 18, 14, 16, 15, 15, 17.

a   mean $= \dfrac{3+6+5+6+4+5+5+6+7}{9}$

$= \dfrac{47}{9}$

$\doteqdot 5.2$

modes are 5 and 6 i.e., is bimodal {both 5 and 6 occur with highest frequency }

median $= 5$ {In order of size: 3, 4, 5, 5, 5, 6, 6, 6, 7 }

↑

middle score

b   mean $= \dfrac{13+12+15+13+18+14+16+15+15+17}{10}$

$= \dfrac{148}{10}$

$\doteqdot 14.8$

mode $= 15$ {occurs most frequently}

median $= 15$ {In order of size: 12, 13, 13, 14, 15, 15, 15, 16, 17, 18}

↑

middle scores

∴   take average, which is 15.

Note: For a sample containing:

- an **odd number** of scores, n say, the **median** score is the $\left(\frac{n+1}{2}\right)$th score

- an **even number** of scores, n say, the **median** score is the average of the $\left(\frac{n}{2}\right)$th and $\left(\frac{n}{2}+1\right)$th scores.

Be sure that you distinguish between the position of a score and its value.

The table below shows the rank score from smallest to largest and the position of each score, for data in **Example 8**, part **b**.

position	1	2	3	4	5	6	7	8	9	10
score	12	13	13	14	15	15	15	16	17	18

└ i.e., the 4th score is 14

EXERCISE 5F.1

1   Below are the points scored by two basketball teams over a 12 match series:

Team A: 91, 76, 104, 88, 73, 55, 121, 98, 102, 91, 114, 82
Team B: 87, 104, 112, 82, 64, 48, 99, 119, 112, 77, 89, 108

Which team had the higher mean score?

2   Select the mode(s) for the following sets of numbers:

a   44, 42, 42, 49, 47, 44, 48, 47, 49, 41, 45, 40, 49

b   148, 144, 147, 147, 149, 148, 146, 144, 145, 143, 142, 144, 147

c   25, 21, 20, 24, 28, 27, 25, 29, 26, 28, 22, 25

3   Calculate the median value for the following data:

a   21, 23, 24, 25, 29, 31, 34, 37, 41

b   105, 106, 107, 107, 107, 107, 109, 120, 124, 132

c   173, 146, 128, 132, 116, 129, 141, 163, 187, 153, 162, 184

4   A survey of 50 students revealed the following number of siblings per student:

1, 1, 3, 2, 2, 2, 0, 0, 3, 2, 0, 0, 1, 3, 3, 4, 0, 0, 5, 3, 3, 0, 1, 4, 5,
1, 3, 2, 2, 0, 0, 1, 1, 5, 1, 0, 0, 1, 2, 2, 1, 3, 2, 1, 4, 2, 0, 0, 1, 2

a   What is the modal number of siblings per student?

b   What is the mean number of siblings per student?

c   What is the median number of siblings per student?

5   The following table shows the average monthly rainfall for a city.

Month	J	F	M	A	M	J	J	A	S	O	N	D
Av. rainfall (mm)	16	34	38	41	98	172	166	159	106	71	52	21

Calculate the mean average monthly rainfall for this city.

6 The selling prices of the last 10 houses sold in a certain district were as follows:

$146 400, $127 600, $211 000, $192 500,
$256 400, $132 400, $148 000, $129 500,
$131 400, $162 500

a Calculate the mean and median selling prices of these houses and comment on the results.

b Which measure would you use if you were:
  i a vendor wanting to sell your house
  ii looking to buy a house in the district?

7 Find x if 5, 9, 11, 12, 13, 14, 17 and x have a mean of 12.

8 Towards the end of season, a basketballer had played 14 matches and had an average of 16.5 goals per game. In the final two matches of the season the basketballer threw 21 goals and 24 goals. Find the basketballer's new average.

9 A sample of 12 measurements has a mean of 16.5 and a sample of 15 measurements has a mean of 18.6. Find the mean of all 27 measurements.

10 15 of 31 measurements are below 10 cm and 12 measurements are above 11 cm. Find the median if the other 4 measurements are 10.1 cm, 10.4 cm, 10.7 cm and 10.9 cm.

11 The mean and median of a set of 9 measurements are both 12. If 7 of the measurements are 7, 9, 11, 13, 14, 17 and 19, find the other two measurements.

12 Seven sample values are: 2, 7, 3, 8, 4, a and b where $a < b$. These have a mean of 6 and a median of 5. Find **a** a and b **b** the mode.

MEASURES OF THE CENTRE FROM OTHER SOURCES

When the same data appear several times we often summarise the data in table form. Consider the data of the **given table**:

We can find the measures of the centre directly from the table.

The mode

The mode is 7. There are 15 occurances of this data value which is more than any other data value.

Data value	Frequency	$f \times x$
3	1	$3 \times 1 = 3$
4	1	$4 \times 1 = 4$
5	3	$5 \times 3 = 15$
6	7	$6 \times 7 = 42$
7	15	$7 \times 15 = 105$
8	8	$8 \times 8 = 64$
9	5	$9 \times 5 = 45$
Total	40	278

The mean

Adding an $f \times x$ **column** to the table helps to add all scores. For example, there are 15 occurances of the data value 7, these add to $15 \times 7 = 105$.

So,   mean $= \dfrac{278}{40} = 6.95$ i.e.,   mean $= \dfrac{\sum f \times x}{\sum f}$.

$\sum$ is the Greek letter sigma, which we use to represent '*the sum of*'.

The median

There are 40 data values, an even number, so there are *two middle* data values. What are they? How do we find them from the table?

As the sample size $n = 40$, $\dfrac{n+1}{2} = \dfrac{41}{2} = 20.5$

$\therefore$ the median is the average of the 20th and 21st data values.

In the table, the blue numbers show us accumulated values.

Data Value	Frequency	Cumulative frequency	
3	1	1	← one number is 3
4	1	2	← two numbers are 4 or less
5	3	5	← five numbers are 5 or less
6	7	12	← 12 numbers are 6 or less
7	15	27	← 27 numbers are 7 or less
8	8	35	← 35 numbers are 8 or less
9	5	40	← 40 numbers are 9 or less
Total	40		

We can see that the 20th and 21st data values (in order) are both 7's,

$\therefore$ median $= \dfrac{7+7}{2} = 7$

Notice that in this example the distribution is clearly skewed even though the mean, median and mode are nearly equal. So, we must be careful in saying that equal values of these measures of the middle enable us to say with certainty that the distribution is symmetric.

GROUPED DISCRETE DATA

The mean, $\overline{x}$, is calculated by:
- multiplying each score (x) by its frequency (f)
- finding the sum of all the values of $f \times x$,
- using the formula, $\overline{x} = \dfrac{\text{total of all } f \times x}{\text{total frequency}} = \dfrac{\sum f \times x}{\sum f}$

Example 9

For the following distribution find:

a the mean

b mode

c median

Score	Frequency
1	6
2	9
3	4
4	7

Score (x)	Frequency (f)	$f \times x$	Cumu. freq.
1	6	6	6
2	9	18	15
3	4	12	19
4	7	28	26
	26	64	

Note:

6 scores of 1.

15 scores of 1 or 2

$\therefore$ 7th, 8th, ..., 15th are all 2's.

a mean $= \dfrac{\sum f \times x}{\sum f}$

$= \dfrac{64}{26}$

$\doteqdot 2.46$

b mode $= 2$ {occurs 9 times}

c Since there are 26 scores, there are two 'middle' scores

$\therefore$ median $= \dfrac{\text{13th score } + \text{ 14th score}}{2}$

$= \dfrac{2 + 2}{2}$

$= 2$

Example 10

The distribution obtained by counting the contents of 25 match boxes is shown:

Find the:

   a mean

   b mode

   c median number of matches per box.

Number of matches	Frequency
47	2
48	4
49	7
50	8
51	3
53	1

Number of matches (x)	Frequency (f)	$f \times x$	Cumulative frequency
47	2	94	2
48	4	192	6
49	7	343	13
50	8	400	21
51	3	153	24
53	1	53	25
Total	25	1235	-

Note:

6 scores are 47 or 48.

13 scores are 47, 48 or 49.

$\therefore$ 7th, 8th, ..., 13th are all 49s.

a mean $= \dfrac{\sum f \times x}{\sum f}$

$= \dfrac{1235}{25}$

$= 49.4$

b mode $= 50$

c median is the 13th score

$= 49$

$\{\frac{25+1}{2} = 13,$ i.e., 13th$\}$

EXERCISE 5F.2

1 A hardware store maintains that packets contain 60 nails. To test this, a quality control inspector tested 100 packets and found the following distribution:

Number of nails	Frequency
56	8
57	11
58	14
59	18
60	21
61	8
62	12
63	8
Total	100

    a Find the mean, mode and median number of nails per packet.

    b Comment on these results in relation to the store's claim.

    c Which of these three measures is most reliable? Comment on your answer.

2 51 packets of chocolate almonds were opened and their contents counted. The following table gives the distribution of the number of chocolates per packet sampled.

Find the mean, mode and median of the distribution.

Number in packet	Frequency
32	6
33	8
34	9
35	13
36	10
37	3
38	2

3 The table alongside compares the mass at birth of some guinea pigs with their mass when they were two weeks old.

    a What was the mean birth mass?

    b What was the mean mass after two weeks?

    c What was the mean increase over the two weeks?

Guinea Pig	Mass (g) at birth	Mass (g) at 2 weeks
A	75	210
B	70	200
C	80	200
D	70	220
E	74	215
F	60	200
G	55	206
H	83	230

GROUPED CONTINUOUS DATA

When information has been gathered in classes we use the **midpoint** of the class to represent all scores within that interval.

> The **midpoint** of a class interval is the average of its endpoints.

For example,   the midpoint of $0 - < 50$ is $\dfrac{0 + 50}{2} = 25$.

We are assuming that the scores within each class are evenly distributed throughout that interval. The mean calculated will therefore be an **approximation** to the true value.

The mode is simply the **modal class** (the class which occurs most frequently).

Example 11

For the following distribution find:

a mean

b mode

Class Interval	Freq.
0 - 49.99	12
50 - 99.99	20
100 - 149.99	24
150 - 199.99	23
200 - 249.99	17
250 - 299.99	6

Class interval	Freq. (f)	Midpt. (x)	$f \times x$
0 - < 50	12	25	300
50 - < 100	20	75	1500
100 - < 150	24	125	3000
150 - < 200	23	175	4025
200 - < 250	17	225	3825
250 - < 300	6	275	1650
Total	102		14 300

Note: midpoint of 150 - < 200 is $\dfrac{150 + 200}{2}$

i.e., 175

a mean $= \dfrac{\sum f \times x}{\sum f}$

$ \doteqdot \dfrac{14300}{102}$

$ \doteqdot 140$

b modal class is 100 - < 150

Note: In **Example 11**, the **true class boundary** of, say, the class interval 50-99.99 is 49.995 to 99.995 and the class interval length is 50. The lower limit of this class is therefore 49.995, not 50.

EXERCISE 5F.3

1 Find the approximate mean for each of the following distributions:

a

Score (x)	Frequency (f)
1-5	7
6-10	12
11-15	15
16-20	10
21-25	11

b

Score (x)	Frequency (f)
40-42	2
43-45	1
46-48	5
49-51	6
52-54	12
55-57	3

2 50 students sit a mathematics test and the results are as follows:

Score	0-9	10-19	20-29	30-39	40-49
Frequency	2	5	7	27	9

Find an estimate of the mean score.

3 Following is a record of the number of goals Chloë has scored in her basketball matches.

15 8 6 10 0 9 2 16 11 23 14 13 17 16 20 12 13
12 10 3 13 5 18 14 19 4 15 15 19 19 14 6 11 29
8 9 3 20 9 25 7 15 19 21 23 12 17 22 14 26

a Find the mean number of goals per match.

b Estimate the mean by grouping the data into:
  i intervals 0-4, 5-9, 10-14, etc.   **ii** intervals 0-8, 9-16, 17-24, 25-30.
c Comment on your answers from **a** and **b**.

4 The table shows the length of newborn babies at a hospital over a one week period.
Find the approximate mean length of the newborn babies.

Length (mm)	frequency
400 to 424	2
425 to 449	7
450 to 474	15
475 to 499	31
500 to 524	27
525 to 549	12
550 to 574	4
575 to 599	1

5 The table shows the petrol sales in one day by a number of city service stations.
  a How many service stations were involved in the survey?
  b Estimate the total amount of petrol sold for the day by the service stations.
  c Find the approximate mean sales of petrol for the day.

Thousands of litres (l)	frequency
2000 to 2999	4
3000 to 3999	4
4000 to 4999	9
5000 to 5999	14
6000 to 6999	23
7000 to 7999	16

6 This histogram illustrates the results of an aptitude test given to a group of people seeking positions in a company.
  a How many people sat for the test?
  b Find an estimate of the mean score for the test.

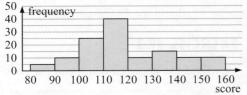

  c What fraction of the people scored less than 100 for the test?
  d If the top 20% of the people are offered positions in the company, estimate the minimum mark required.

7

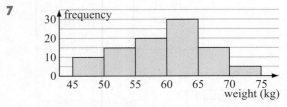

The histogram shows the weights (in kg) of a group of year 10 students at a country high school.

  a How many students were involved in the survey?
  b Calculate the mean weight of the students.
  c How many students weigh less than 56 kg?
  d What percentage of students weigh between 50 and 60 kg?
  e If a student was selected at random, what would be the chance that the student weighed less than 60 kg?

For grouped continuous data, the **median** can be determined by either of two methods:

- by drawing an ogive and finding the 50th percentile, or
- by examining the cumulative frequency table, then
  - ▶ finding the interval in which the median lies, (this interval is called the median class)
  - ▶ using the formula,

$$\textbf{median} = \boldsymbol{L} + \left(\frac{\boldsymbol{i}}{\boldsymbol{f}} \times \boldsymbol{C}\right),$$

where L is the **lower limit of the median class**

i is the **number of scores in the median class needed to arrive at the middle score**

f is the **number of scores in the median class**

C is the **length of the class interval**.

Example 12

Use the formula to find the median of the mathematics test marks:

Test Mark	Number of students
30 - 39	6
40 - 49	20
50 - 59	64
60 - 69	87
70 - 79	51
80 - 89	19
90 - 99	10

Test mark	Lower boundary	Number of students	Cumulative frequency	
30 - < 39	29.5	6	6	
40 - < 49	39.5	20	26	
50 - < 59	49.5	64	90	── 90 scores $\leqslant$ 59
60 - < 69	59.5	87	177	── 177 scores $\leqslant$ 69
70 - < 79	69.5	51	228	
80 - < 89	79.5	19	247	
90 - < 99	89.5	10	257	
	Total	257		

Since there are 257 scores, the median is the 129th score $\left\{\frac{257+1}{2} = 129\right\}$

∴ median class is 60 - 69.

Now median $= L + \dfrac{i}{f} \times C$

$\qquad\quad = 59.5 + \frac{39}{87} \times 10$

$\qquad\quad \doteqdot 64.0$

$L = 59.5$ halfway between 59 and 60
$i = 129 - 90 = 39$
$f = 87$
$C = 10$

PERCENTILES

A 'less than ogive' can be used to find the percentage of the data **below** a certain value.

A 'more than ogive' can be likewise used to find the percentage of the data **above** a particular value. These values are known as **percentiles**.

Percentiles are those values of the data below/above which certain percentages of the frequencies lie.

For instance, the 25th percentile is the value that has 25% of the total frequencies below it.

The 50th percentile is the middle value or **median value** of the data.

The 25th percentile is referred to as the **lower quartile** of the data.

The 75th percentile is referred to as the **upper quartile** of the data.

Example 13

From the less than ogive in **Example 7**, find:

a the 75th percentile   **b** the 30th percentile

c the median value   **d** the 25th percentile.

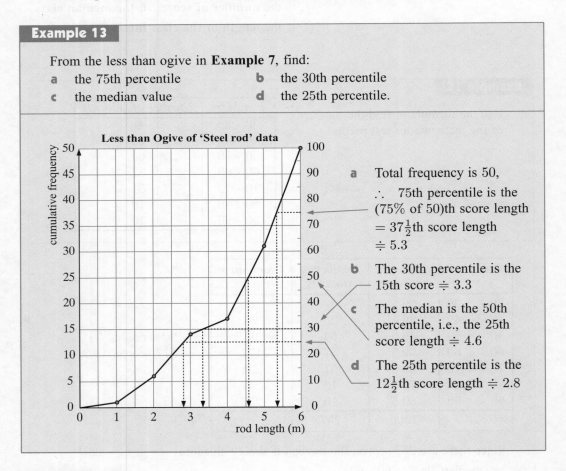

a Total frequency is 50,

∴ 75th percentile is the (75% of 50)th score length

$= 37\frac{1}{2}$th score length

$\doteqdot 5.3$

b The 30th percentile is the 15th score $\doteqdot 3.3$

c The median is the 50th percentile, i.e., the 25th score length $\doteqdot 4.6$

d The 25th percentile is the $12\frac{1}{2}$th score length $\doteqdot 2.8$

EXERCISE 5F.4

1 Calculate the median of the following distributions:

a

score	1	2	3	4	5	6
frequency	25	11	8	5	4	1

b

score	5	6	7	8	9	10
frequency	1	3	11	12	8	2

2 This table indicates the number of errors in randomly chosen pages of a telephone directory:

number of errors	0	1	2	3	4	5	6
frequency	67	35	17	8	11	2	1

Find the median number of errors.

3 The following data shows the lengths of 30 trout caught in a lake during a fishing competition. Measurements are to the nearest centimetre.

31   38   34   40   24   33   30   36   38   32   35   32   36   27   35
40   34   37   44   38   36   34   33   31   38   35   36   33   33   28

   a Construct a cumulative frequency table for trout lengths, x cm, using the following intervals $24 \leqslant x < 27$, $27 \leqslant x < 30$, ... etc.

   b Draw a cumulative frequency graph.

   c Use **b** to find the median length.

   d Use the original data to find its median and compare your answer with **c**. Comment!

4 In an examination the following scores were achieved by a group of students:

Draw a less than ogive of the data and use it to find:

   a the median examination mark

   b how many students scored less than 65 marks

   c how many students scored between 50 and 70 marks

   d how many students failed, given that the pass mark was 45

   e the credit mark, given that the top 16% of students were awarded credits.

Score	frequency
$10 \leqslant x < 20$	2
$20 \leqslant x < 30$	5
$30 \leqslant x < 40$	7
$40 \leqslant x < 50$	21
$50 \leqslant x < 60$	36
$60 \leqslant x < 70$	40
$70 \leqslant x < 80$	27
$80 \leqslant x < 90$	9
$90 \leqslant x < 100$	3

5 In a cross-country race, the times (in minutes) of 80 competitors were recorded as follows:

Draw a less than ogive of the data and use it to find:

   a the median time

   b the 75th percentile

   c the 30th percentile

Score	frequency
$20 \leqslant t < 25$	15
$25 \leqslant t < 30$	33
$30 \leqslant t < 35$	21
$35 \leqslant t < 40$	10
$40 \leqslant t < 45$	1

6 The following table gives the age groups of car drivers involved in an accident in a city for a given year.

Draw a cumulative frequency graph of the data and use it to find:

   a the median age of the drivers involved in the accidents

   b the percentage of drivers, with ages of 23 or less, involved in accidents.

   c Estimate the probability that a driver involved in an accident is:

     i aged less than or equal to 27 years

     ii aged 27 years.

Age (in years)	No. of accidents
$16 \leqslant x < 20$	59
$20 \leqslant x < 25$	82
$25 \leqslant x < 30$	43
$30 \leqslant x < 35$	21
$35 \leqslant x < 40$	19
$40 \leqslant x < 50$	11
$50 \leqslant x < 60$	24
$60 \leqslant x < 80$	41

INVESTIGATION 4   EFFECTS OF OUTLIERS

 In a set of data an **outlier**, or **extreme value**, is a value which is much greater than, or much less than, the other values.

Your task: Examine the effect of an outlier on the three measures of central tendency.

What to do:

1 Consider the following set of data: 4, 5, 6, 6, 6, 7, 7, 8, 9, 10. Calculate:

     a the mean   **b** the mode   **c** the median.

2 Now introduce an extreme value, say 100, to the data. The data set is now 4, 5, 6, 6, 6, 7, 7, 8, 9, 10, 100. Calculate:

     a the mean   **b** the mode   **c** the median.

3 Comment on the effect that this extreme value has on:

     a the mean   **b** the mode   **c** the median.

4 Which of the three measures of central tendency is most affected by the inclusion of an outlier? Discuss your findings with your class.

CHOOSING THE APPROPRIATE MEASURE

The mean, mode and median can be used to indicate the centre of a set of numbers. Which of these values is the most appropriate measure to use, will depend upon the type of data under consideration.

For example, when reporting on shoe size stocked by a shoe store, the average or mean size would be a useless measure of the stock. In this case, the mode would be the most useful measure. In real estate values the median is used.

When selecting which of the three measures of central tendency to use as a representative figure for a set of data, you should keep the following **advantages and disadvantages** of each measure in mind.

▶ **Mean**
  - The mean's main advantage is that it is commonly used, easy to understand and easy to calculate.
  - Its main disadvantage is that it is affected by extreme values within a set of data and so may give a distorted impression of the data.
    For example, consider the following data: 4, 6, 7, 8, 19, 111. The total of these 6 numbers is 155, and so the mean is approximately 25.8. Is 25.8 a representative figure for the data? The **extreme value (or outlier)** of 111 has distorted the mean in this case.

▶ **Median**
  - The median's main advantage is that it is easily calculated and is the middle value of the data.
  - Unlike the mean, it is not affected by extreme values.
  - The main disadvantage is that it ignores all values outside the middle range and so its representativeness is questionable.

► **Mode**

- The mode's main advantage as a representative figure is that it is the most usual value within a set of data.

- The mode has an advantage over the mean in that it is not affected by extreme values contained in the data.

- The main disadvantage is that it does not take into account all values within the data and this makes its representativeness questionable.

G   MEASURING THE SPREAD OF DATA

If, in addition to having measures of the middle of a data set, we also have an indication of the **spread** of the data, then a more accurate picture of the data set is possible.

For example,   1, 4, 5, 5, 6, 7, 8, 9, 9   has a mean value of 6 and so does

4, 4, 5, 6, 6, 7, 7, 7, 8.   However, the first data set is more widely spread than the second one.

Two commonly used statistics that indicate the spread of a set of data are:

- the range   • the interquartile range.

THE RANGE

The **range** is the difference between the **maximum** (largest) data value and the **minimum** (smallest) data value.

range = maximum data value − minimum data value

Example 14

Find the range of the data set: 4, 7, 5, 3, 4, 3, 6, 5, 7, 5, 3, 8, 9, 3, 6, 5, 6

Searching through the data we find: minimum value = 3   maximum value = 9

∴   range = 9 − 3 = 6

THE UPPER AND LOWER QUARTILES AND THE INTERQUARTILE RANGE

The median divides the ordered data set into two halves and these halves are divided in half again by the **quartiles**.

The middle value of the lower half is called the **lower quartile** (Q_1). One-quarter, or 25%, of the data have a value less than or equal to the lower quartile. 75% of the data have values greater than or equal to the lower quartile.

The middle value of the upper half is called the **upper quartile** (Q_3). One-quarter, or 25%, of the data have a value greater than or equal to the upper quartile. 75% of the data have values less than or equal to the upper quartile.

interquartile range = upper quartile − lower quartile

The interquartile range is the range of the middle half (50%) of the data.

The data set has been divided into quarters by the lower quartile (Q_1), the median (Q_2) and the upper quartile (Q_3).

So,   the **interquartile range**, is $\boxed{IQR = Q_3 - Q_1}$.

Example 15

For the data set   6, 7, 3, 7, 9, 8, 5, 5, 4, 6, 6, 8, 7, 6, 6, 5, 4, 5, 6   find the:

  a   median   **b**   lower quartile

  c   upper quartile   **d**   interquartile range

The ordered data set is:

  3 4 4 5 5 5 5 6 6 6 6 6 6 7 7 7 8 8 9   (19 of them)

  a   The median $= \left(\frac{19+1}{2}\right)$th score $=$ 10th score $= 6$

  b/c   As the median is a data value we ignore it and split the remaining data into two groups.

         3 4 4 5 5 5 5 6 6   6 6 6 7 7 7 8 8 9

         $Q_1 =$ median of lower half $Q_3 =$ median of upper half

              $= 5$ $= 7$

  d $IQR = Q_3 - Q_1 = 2$

Example 16

For the data set   9, 8, 2, 3, 7, 6, 5, 4, 5, 4, 6, 8, 9, 5, 5, 5, 4, 6, 6, 8   find the:

  a   median   **b**   lower quartile

  c   upper quartile   **d**   interquartile range

The ordered data set is:

  2 3 4 4 4 5 5 5 5 5 6 6 6 6 7 8 8 8 9 9   (20 of them)

  a   As $n = 20$, $\frac{n+1}{2} = \frac{21}{2} = 10.5$

       $\therefore$ median $= \dfrac{\text{10th value } + \text{ 11th value}}{2} = \dfrac{5+6}{2} = 5.5$

  b/c   As the median is not a data value we split the original data into two equal groups of 10.

         2 3 4 4 4 5 5 5 5 5   6 6 6 6 7 8 8 8 9 9

         $\therefore$ $Q_1 = 4.5$ $\therefore$ $Q_3 = 7.5$

  d $IQR = Q_3 - Q_1 = 3$

EXERCISE 5G

1 For each of the following sets of data find:

    i the upper quartile   **ii** the lower quartile
    iii the interquartile range   **iv** the range

  a 2, 3, 4, 7, 8, 10, 11, 13, 14, 15, 15

  b 35, 41, 43, 48, 48, 49, 50, 51, 52, 52, 52, 56

  c

Stem	Leaf
1	3 5 7 7 9
2	0 1 3 4 6 7 8 9
3	0 1 2 7
4	2 6
5	1 *Scale:* 4 | 2 means 42

  d

Score	0	1	2	3	4	5
Frequency	1	4	7	3	3	1

2 The time spent (in minutes) by 24 people in a queue at a bank, waiting to be attended by a teller, has been recorded as follows:

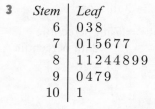

0	3.2	0	2.4	3.2	0	1.3	0
1.6	2.8	1.4	2.9	0	3.2	4.8	1.7
3.0	0.9	3.7	5.6	1.4	2.6	3.1	1.6

  a Find the median waiting time and the upper and lower quartiles.

  b Find the range and interquartile range of the waiting time.

  c Copy and complete the following statements:

    i "50% of the waiting times were greater than .... minutes."

    ii "75% of the waiting times were less than ...... minutes."

    iii "The minimum waiting time was .... minutes and the maximum waiting time was ..... minutes. The waiting times were spread over ...... minutes."

3

Stem	Leaf
6	0 3 8
7	0 1 5 6 7 7
8	1 1 2 4 4 8 9 9
9	0 4 7 9
10	1

For the data set given, find:

  a the minimum value   **b** the maximum value
  c the median   **d** the lower quartile
  e the upper quartile   **f** the range
  g the interquartile range

Scale: 7 \| 5 means 7.5

INTERQUARTILE RANGE FROM OGIVES

The IQR of a distribution of grouped scores can also be obtained using an ogive.

Remember that: $\text{IQR} = Q_3 - Q_1$

                         = 75th percentile − 25th percentile

Example 17

Draw an ogive for the following distribution and hence determine the interquartile range.

Scores	Frequencies
1 - 9.99	3
10 - 19.99	11
20 - 29.99	25
30 - 39.99	36
40 - 49.99	31
50 - 59.99	14

Scores	Upper End point	Frequency	Cumulative Frequency
0 - 9.99	9.995	3	3
10 - 19.99	19.995	11	14
20 - 29.99	29.995	25	39
30 - 39.99	39.995	36	75
40 - 49.99	49.995	31	106
50 - 59.99	59.995	14	120
Total		120	

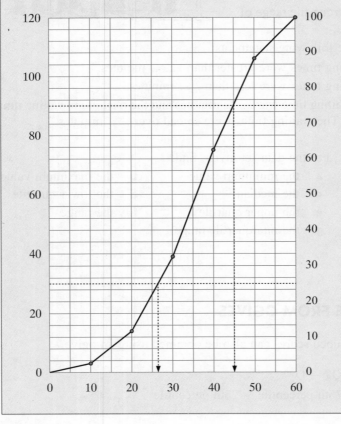

75% of 120 = 90

∴   75th percentile is
≑ 90th score.

25% of 120 = 30

∴   25th percentile is
≑ 30th score.

IQR = 75th percentile
        − 25th percentile
     ≑ 45 − 26
     ≑ 19

4   A botanist has measured the heights of 60 seedlings and has presented her findings on the ogive below.

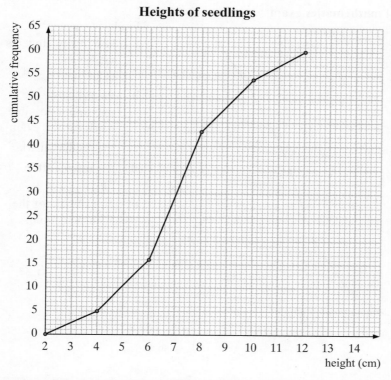

Heights of seedlings

a   How many seedlings have heights of 5 cm or less?

b   What percentage of seedlings are taller than 8 cm?

c   What is the median height?

d   What is the inter-quartile range for the heights?

e   Find the 90th percentile for the data and explain what your answer means.

5   The following ogive displays the performance of 80 competitors in a cross-country race.

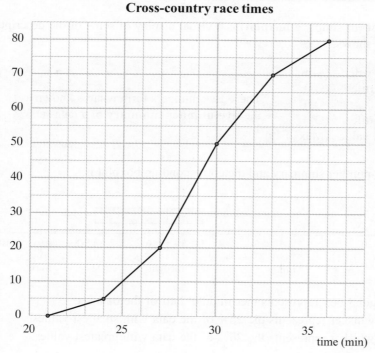

Cross-country race times

Find:

a   the lower quartile time

b   the median

c   the upper quartile

d   the interquartile range

e   an estimate of the 40th percentile.

6 The ogive below displays the marks scored by year 12 students from a cluster of schools in a common trial mathematics exam.

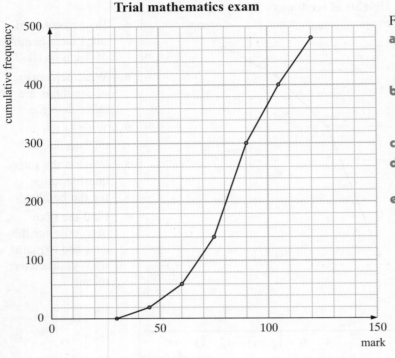

Trial mathematics exam

Find:

a how many students sat for the examination

b the probable maximum possible mark for the exam

c the median mark

d the interquartile range

e an estimate of the 85th percentile.

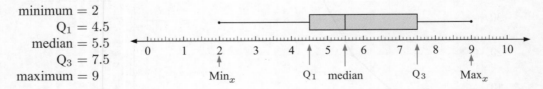

BOX-AND-WHISKER PLOTS

A **box-and-whisker plot** (or simply a **boxplot**) is a visual display of some of the descriptive statistics of a data set. It shows:

- the minimum value (Min_x)
- the lower quartile (Q_1)
- the median (Q_2) These five numbers form what is known as
- the upper quartile (Q_3) the **five-number summary** of a data set.
- the maximum value (Max_x)

In **Example 16** the five-number summary and the corresponding boxplot is:

minimum $= 2$
$Q_1 = 4.5$
median $= 5.5$
$Q_3 = 7.5$
maximum $= 9$

Note:
- The rectangular box represents the 'middle' half of the data set.
- The lower whisker represents the 25% of the data with smallest values.
- The upper whisker represents the 25% of the data with greatest values.

Example 18

For the data set: 5 6 7 6 2 8 9 8 4 6 7 4 5 4 3 6 6

a   construct the five-number summary   **b**   draw a boxplot
c   find the   **i** range   **ii** interquartile range
d   find the percentage of data values below 7

a   The ordered data set is:

2   3   4   4   4   5   5   6   6   6   6   6   7   7   8   8   9   (17 of them)

$Q_1 = 4$ median $= 6$ $Q_3 = 7$

So the **5-number summary** is:
$$\begin{cases} \text{Min}_x = 2 \\ Q_1 = 4 \\ \text{median} = 6 \\ Q_3 = 7 \\ \text{Max}_x = 9 \end{cases}$$

b

c   **i**   range $= \text{Max}_x - \text{Min}_x$
                   $= 9 - 2$
                   $= 7$

ii   IQR $= Q_3 - Q_1$
                   $= 7 - 4$
                   $= 3$

d   75% of the data values are less than or equal to 7.

BOXPLOTS AND OUTLIERS

Outliers are extraordinary data that are usually separated from the main body of the data. Outliers are either much larger or much smaller than most of the data.

There are several 'tests' that identify data that are outliers.

A commonly used test involves the calculation of 'boundaries':

- **The upper boundary =**   **upper quartile + 1.5 × IQR**.
  Any data larger than the upper boundary is an outlier.

- **The lower boundary =**   **lower quartile − 1.5 × IQR**.
  Any data smaller than the lower boundary is an outlier.

Outliers are marked with an asterisk on a boxplot and it is possible to have more than one outlier at either end.

The whiskers extend to the last value that is not an outlier.

Example 19

Draw a boxplot for the following data, testing for outliers and marking them, if they exist, with an asterisk on the boxplot:

3, 7, 8, 8, 5, 9, 10, 12, 14, 7, 1, 3, 8, 16, 8, 6, 9, 10, 13, 7

The ordered data set is:

1 3 3 5 6 7 7 7 8 8 8 8 8 9 9 10 10 12 13 14 16 $(n = 20)$

Min_x	Q_1	median	Q_3	Max_x
$= 1$	$= 6.5$	$= 8$	$= 10$	$= 16$

$IQR = Q_3 - Q_1 = 3.5$

Test for outliers:

upper boundary   and   lower boundary

= upper quartile + 1.5 × IQR   = lower quartile − 1.5 × IQR

= 10 + 1.5 × 3.5   = 6.5 − 1.5 × 3.5

= 15.25   = 1.25

As 16 is above the upper boundary it is an outlier.
As 1 is below the lower boundary it is an outlier.

> Notice that the whisker is drawn to the last value that is not an outlier.

So, the boxplot is:

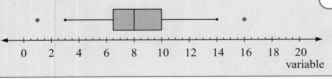

EXERCISE 5H

1   A boxplot has been drawn to show the distribution of marks (out of 100) in a test for a particular class.

   a   What was the   i   highest mark   ii   lowest mark   scored?

   b   What was the median test score for this class?

   c   What was the range of marks scored for this test?

   d   What percentage of students scored 60 or more for the test?

   e   What was the interquartile range for this test?

   f   The top 25% of students scored a mark between ..... and .....

   g   If you scored 70 for this test, would you be in the top 50% of students in this class?

   h   Comment on the symmetry of the distribution of marks.

2   A set of data has a lower quartile of 31.5, median of 37 and an upper quartile of 43.5.

   a   Calculate the interquartile range for this data set.

   b   Calculate the boundaries that identify outliers.

   c   Which of the data 22, 13.2, 60, 65 would be outliers?

3 Julie examines a new variety of bean and does a count on the number of beans in 33 pods. Her results were:

5, 8, 10, 4, 2, 12, 6, 5, 7, 7, 5, 5, 5, 13, 9, 3, 4, 4, 7, 8, 9, 5, 5, 4, 3, 6, 6, 6, 6, 9, 8, 7, 6

   a Find the median, lower quartile and upper quartile of the data set.

   b Find the interquartile range of the data set.

   c What are the lower and upper boundaries for outliers?

   d Are there any outliers according to **c**?   **e** Draw a boxplot of the data set.

4 Andrew counts the number of bolts in several boxes and tabulates the data as shown below:

Number of bolts	33	34	35	36	37	38	39	40
Frequency	1	5	7	13	12	8	0	1

   a Find the five-number summary for this data set.

   b Find the **i** range **ii** IQR for the data set.

   c Are there any outliers? Test for them.   **d** Construct a boxplot for the data set.

I   THE STANDARD DEVIATION

The standard deviation is the most widely used measure of the spread of a sample.

The **standard deviation** measures the **deviation between scores and the mean**, i.e., is a measure of the **dispersal** of the data.

The differences between the scores and the mean are squared, and the average of these squares is then found. The standard deviation is the square root of this average.

The standard deviation gives insight into how the data is **dispersed**.

The larger the standard deviation, the more widely spread the data would be and vice versa. The standard deviation provides a better measure of the spread than either the range or the interquartile range because it considers all scores in the data.

UNGROUPED DATA

The standard deviation, s, can be determined using the formula: $s = \sqrt{\dfrac{\sum(x - \overline{x})^2}{n}}$

    where x is any **score**

            $\overline{x}$ is the **mean** of the distribution

            n is the **total number of scores**.

Example 20

Calculate the standard deviation for the sample: 2, 5, 4, 6, 7, 5, 6.

$$\overline{x} = \frac{2+5+4+6+7+5+6}{7}$$

$$= \frac{35}{7}$$

$$= 5$$

$$s = \sqrt{\frac{\sum(x-\overline{x})^2}{n}}$$

$$= \sqrt{\frac{16}{7}}$$

$$\doteqdot 1.512$$

Score (x)	$x - \overline{x}$	$(x - \overline{x})^2$
2	-3	9
4	-1	1
5	0	0
5	0	0
6	1	1
6	1	1
7	2	4
35		16

EXERCISE 51.1

1   A company recorded the following weekly petrol usage (in litres) by its salespersons:

    62, 40, 52, 48, 64, 55, 44, 75, 40, 68, 60, 42, 70, 49, 56

Find the mean and standard deviation of the petrol used.

2   The weights of a sample of cooking chickens to the nearest kilogram are:

    1.5, 1.8, 1.7, 1.4, 1.7, 1.8, 2.0, 1.5, 1.6, 1.6, 1.9, 1.7, 1.4, 1.7, 1.8, 2.0

Find the mean and standard deviation of weights.

GROUPED DISCRETE DATA

The standard deviation for grouped data is: $s = \sqrt{\dfrac{\sum f \times (x - \overline{x})^2}{n}}$

where x is any **score**,

   $\overline{x}$ is the **mean**,

   f is the **frequency**

   n is the **number of scores**.

Note: A large population of many thousands or even millions has a mean which we denote as μ, a Greek letter called *mu*, and a standard deviation denoted σ, called *sigma*.

Often μ and σ are unknown and are too difficult and/or too expensive to find.

We therefore take a sample of sufficient size and find its mean $\overline{x}$ and standard deviation s.

If we believe that the sample is a true reflection of the whole population then we use $\overline{x}$ as an estimate of μ and s as an estimate of σ.

Example 21

Find the standard deviation of the following distribution of scores:

Score	Frequency
11	3
12	4
13	5
14	2
15	8
16	9
17	5
18	8
19	6
Total	50

Score (x)	Frequency (f)	$f \times x$	$x - \overline{x}$	$(x - \overline{x})^2$	$f \times (x - \overline{x})^2$
11	3	33	-4.62	21.3444	64.0332
12	4	48	-3.62	13.1044	52.4176
13	5	65	-2.62	6.8644	34.3220
14	2	28	-1.62	2.6244	5.2488
15	8	120	-0.62	0.3844	3.0752
16	9	144	0.38	0.1444	1.2996
17	5	85	1.38	1.9044	9.522
18	8	144	2.38	5.6644	45.3152
19	6	114	3.38	11.4244	68.5464
	50	781			283.7800

The mean, $\overline{x} = \dfrac{\sum f \times x}{n}$

$$= \frac{781}{50}$$

$$= 15.62$$

The standard deviation is $s = \sqrt{\dfrac{\sum f \times (x - \overline{x})^2}{n}}$

$$= \sqrt{\frac{283.7800}{50}}$$

$$\doteqdot 2.38$$

3 The local Health and Fitness Centre recorded the following number of clients per week during the last year:

Calculate the average number of clients per week during the year and the standard deviation from this number.

Number of clients	Frequency
36	2
39	5
44	9
45	11
46	15
48	5
50	4
52	1
Total	52

GROUPED CONTINUOUS DATA

For the grouped continuous data we use exactly the same method as we did for grouped discrete data with the exception that **class midpoint** is used to represent all scores in a particular class.

Example 22

Find the standard deviation for the following distribution of examination scores:

Mark	Frequency	Mark	Frequency
0 - 9	1	50 - 59	16
10 - 19	1	60 - 69	24
20 - 29	2	70 - 79	13
30 - 39	4	80 - 89	6
40 - 49	11	90 - 99	2

Class int.	Midpt. (x)	Freq. (f)	$f \times x$	$x - \overline{x}$	$(x - \overline{x})^2$	$f \times (x - \overline{x})^2$
0 - 9	4.5	1	4.5	-55.25	3052.56	3052.56
10 - 19	14.5	1	14.5	-45.25	2047.56	2047.56
20 - 29	24.5	2	49.0	-35.25	1242.56	2485.12
30 - 39	34.5	4	138.0	-25.25	637.56	2550.25
40 - 49	44.5	11	489.5	-15.25	232.56	2558.19
50 - 59	54.5	16	872.0	-5.25	27.56	441.00
60 - 69	64.5	24	1548.0	4.75	22.56	541.50
70 - 79	74.5	13	968.5	14.75	217.56	2828.31
80 - 89	84.5	6	507.0	24.75	612.56	3675.38
90 - 99	94.5	2	189.0	34.75	1207.56	2415.12
		80	4780			22594.99

$$\overline{x} = \frac{\sum f \times x}{n} \qquad \text{and} \qquad s = \sqrt{\frac{\sum f \times (x - \overline{x})^2}{n}}$$

$$= \frac{4780}{80} \qquad\qquad\qquad\qquad = \sqrt{\frac{22\,594.99}{80}}$$

$$= 59.75 \qquad\qquad\qquad\qquad\quad \doteqdot 16.8$$

VARIANCE

Another measure of the spread of a set of data is the variance. The **variance** of a set of scores is simply the square of the standard deviation, s^2 (or σ^2 for a population).

4 The hours worked last week by 40 employees of a local clothing factory were as follows:

38 40 46 32 41 39 44 38 40 42 38 40 43 41 47 36 38 39 34 40
48 30 49 40 40 43 45 36 35 39 42 44 48 36 38 42 46 38 39 40

a Without using class intervals, group this data and hence calculate the mean and standard deviation correct to 2 decimal places.

b Group the data into classes 30 - 33, 34 - 37, etc., and recalculate the mean and standard deviation correct to 2 decimal places. Examine any differences in the 2 sets of answers.

c Draw an ogive of the data and determine the interquartile range.

d Represent this data on a box and whisker plot.

5 A traffic survey by the highways department revealed the following number of vehicles at a suburban intersection at various 15 minute intervals during the day.

Calculate the mean, the variance and the standard deviation correct to 2 decimal places. Draw an ogive of the data and determine the interquartile range.

Number of vehicles	Frequency
1 - 5	4
6 - 10	16
11 - 15	22
16 - 20	28
21 - 25	14
26 - 30	9
31 - 35	5
36 - 40	2

6 A coffee shop owner wanted to know how long customers spent in the shop and so conducted a survey of 50 customers and recorded these results (in whole minutes):

13 38 24 32 42 15   8   12 19 36 25 22 18 14 26 29 33   6 11 24
 9 20 26 40 52 38 31 22 16   9 50 37 28 19   7   14 26   4 32 44
40 40 51 59 18 24 21 34 13 18

Group this data into class intervals 1 - 10,   11 - 20, ...,   51 - 60.   Hence, calculate the mean, variance and standard deviation of the data. Draw an ogive of the data and determine the interquartile range.

7 The following exam results were recorded by two classes of students studying Spanish:

Class A: 64 69 74 67 78 88 76 90 89 84 83 87 78 80 95 75 55 78 81
Class B: 94 90 88 81 86 96 92 93 88 72 94 61 87 90 97 95 77 77 82 90

a Determine the median and upper and lower quartiles for each class scores.

b Construct a box and whisker plot for the scores of each class and ensure that any extreme scores (outliers) are correctly shown.

c Which class records the better results? Why?

COMPARING THE SPREAD

To compare the spread of two sets of data of different sizes or units, we can calculate the **coefficients of variation** for each of them.

The **coefficient of variation** is the ratio of the standard deviation to the mean of a set of data, expressed as a percentage.

In symbols, $v = \dfrac{s}{\overline{x}} \times 100\%$ where v is the **coefficient of variation**

s is the **standard deviation** of the data

$\overline{x}$ is the **mean** of the data.

Example 23

Corner store A has average daily sales of \$500 and a standard deviation of \$50. Store B, on the opposite corner, has average daily sales of \$800 and a standard deviation of \$68. Which store has the greater dispersion in daily sales?

For store A, coefficient of variation $= \frac{50}{500} \times 100\% = 10\%$

For store B, coefficient of variation $= \frac{68}{800} \times 100\% = 8.5\%$

Since store A has the greater coefficient of variation, it has a greater relative dispersion, even though it has a smaller absolute dispersion.

EXERCISE 51.2

1 A factory is producing and filling 1 kg packets of soap powders using two different machines. Samples are taken from the production of the two machines and the following results are recorded: machine A - average weight 0.98 kg with a standard deviation of 0.07;   machine B - average weight 1.04 kg with a standard deviation of 0.08. Which machine produces greater variability in sizes?

2 The quality control department of a towel manufacturer has found that, over time, the average number of rejected towels per day is 88 with a standard deviation of 8:5. However, over the past six months, the average number of rejections has fallen to 80 with a standard deviation of 8:0 towels. What does this change signify to the management of the company?

3 The average wage of males working in the factory is \$495 per week with a standard deviation of \$40. The average wage of females in the same factory is \$450 with a standard deviation of \$35. Whose wage is more widely dispersed, the males or the females?

INVESTIGATION 5   HEART STOPPER

A new drug that is claimed to lower the cholesterol level in humans has been developed.

A heart specialist was interested to know if the claims made by the company selling the drug were accurate. He enlisted the help of 50 of his patients.

They agreed to take part in an experiment in which 25 of them would be randomly allocated to take the new drug and the other 25 would take an identical looking pill that was actually a placebo (a sugar pill that would have no effect at all).

All participants had their cholesterol level measured before starting the course of pills and then at the end of two months of taking the drug, they had their cholesterol level measured again.

The data collected by the doctor is given below.

cholesterol levels of **all** participants **before** the experiment	7.1 8.2 8.4 6.5 6.5 7.1 7.2 7.1 6.1 6.0 8.5 5.0 6.3 6.7 7.3 8.9 6.2 6.3 7.1 8.4 7.4 7.6 7.5 6.6 8.1 6.2 6.2 7.0 8.1 8.4 6.4 7.6 8.6 7.5 7.9 6.2 6.8 7.5 6.0 5.0 8.3 7.9 6.7 7.3 6.0 7.4 7.4 8.6 6.5 7.6
cholesterol levels of the 25 participants who took the drug	4.8 5.6 4.7 4.2 4.8 4.6 4.8 5.2 4.8 5.0 4.7 5.1 4.4 4.7 4.9 6.2 4.7 4.7 4.4 5.6 3.2 4.4 4.6 5.2 4.7
cholesterol levels of the 25 participants who took the placebo	7.0 8.4 8.8 6.1 6.6 7.6 6.5 7.9 6.2 6.8 7.5 6.0 8.2 5.7 8.3 7.9 6.7 7.3 6.1 7.4 8.4 6.6 6.5 7.6 6.1

What to do:

1 Produce a single stem plot for the cholesterol levels of all participants before and after the experiment. Present the stem plot so that this data can be simply compared to all the measurements before the experiment began.

2 Use technology to calculate the relevant statistical data.

3 Use the data to complete the table:

Cholesterol Level	Before Experiment	25 participants taking the drug	25 participants taking the placebo
4.0 -			
4.5 -			
5.0 -			
5.5 -			
6.0 -			
6.5 -			
7.0 -			
7.5 -			
8.0 -			
8.5 -			

4 Calculate the mean and standard deviation for each group in the table using the formula for continuous data in classes.

5 Write a report presenting your data and findings based on that data.

J   STATISTICS USING TECHNOLOGY

GRAPHICS CALCULATOR

A **graphics calculator** can be used to find descriptive statistics and to draw some types of graphs.

(You will need to change the **viewing window** as appropriate.)

Consider the data set: 5 2 3 3 6 4 5 3 7 5 7 1 8 9 5

No matter what brand of calculator you use you should be able to:

- Enter the data as a **list**.
- Enter the **statistics calculation** part of the menu and obtain the descriptive statistics like these shown.

$\overline{x}$ is the mean

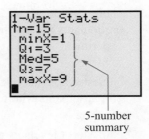

5-number summary

- Obtain a box-and-whisker plot such as:

(These screen dumps are from a TI-83.)

- Obtain a vertical barchart if required.

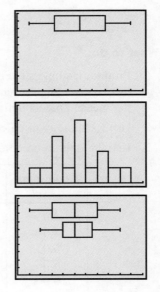

- Enter a second data set into another list and obtain a side-by-side boxplot for comparison with the first one.

  Use: 9 6 2 3 5 5 7 5 6 7 6 3 4 4 5 8 4

Now you should be able to create these by yourself.

EXERCISE 5J.1

1 **a** Using your calculator enter the data set: 5 2 3 3 6 4 5 3 7 5 7 1 8 9 5   and obtain the mean and the 5-number summary. This is the first graphics calculator example shown above and you should check your results from it.

   b Obtain the boxplot for question **a**.

   c Obtain the vertical bar chart for question **a**.

   d Enter this data set: 9 6 2 3 5 5 7 5 6 7 6 3 4 4 5 8 4   into a second list. Find the mean and 5-number summary. Now create a side-by-side boxplot for both sets of data.

STATISTICS FROM A COMPUTER PACKAGE

Click on the icon to enter the **statistics package** on the CD.

Enter data set 1:   5 2 3 3 6 4 5 3 7 5 7 1 8 9 5

Enter data set 2:   9 6 2 3 5 5 7 5 6 7 6 3 4 4 5 8 4

STATISTICS
PACKAGE

Examine the side-by-side column graphs.

Click on the Box-and-Whisker spot to view the side-by-side boxplots.

Click on the Statistics spot to obtain the descriptive statistics.

Click on Print to obtain a print-out of all of these on one sheet of paper.

EXERCISE 5J.2 (Computer package and/or graphics calculator)

1 Shane and Brett play in the same cricket team and are fierce but friendly rivals when it comes to bowling. During a season the number of wickets per innings taken by each bowler was recorded as:

Shane: 1 6 2 0 3 4 1 4 2 3 0 3 2 4 3 4 3 3
        3 4 2 4 3 2 3 3 0 5 3 5 3 2 4 3 4 3

Brett: 7 2 4 8 1 3 4 2 3 0 5 3 5 2 3 1 2 0
        4 3 4 0 3 3 0 2 5 1 1 2 2 5 1 4 0 1

   a Is the data discrete or continuous?

   b Enter the data into a graphics calculator or statistics package.

   c Produce a vertical column graph for each data set.

   d Are there any outliers? Should they be deleted before we start to analyse the data?

   e Describe the shape of each distribution.

   f Compare the measures of the centre of each distribution.

   g Compare the spreads of each distribution.

   h Obtain side-by-side boxplots.

   i If using the statistics package, print out the graphs, boxplots and relevant statistics.

   j What conclusions, if any, can be drawn from the data?

2 A manufacturer of light globes claims that the newly invented type has a life 20% longer than the current globe type. Forty of each globe type are randomly selected and tested. Here are the results to the nearest hour.

Old type:	103	96	113	111	126	100	122	110	84	117
	111	87	90	121	99	114	105	121	93	109
	87	127	117	131	115	116	82	130	113	95
	103	113	104	104	87	118	75	111	108	112

New type:	146	131	132	160	128	119	133	117	139	123
	191	117	132	107	141	136	146	142	123	144
	133	124	153	129	118	130	134	151	145	131
	109	129	109	131	145	125	164	125	133	135

   a Is the data discrete or continuous?

   b Enter the data into a graphics calculator or statistics package.

   c Are there any outliers? Should they be deleted before we start to analyse the data?

   d Compare the measures of centre and spread.

   e Obtain side-by-side boxplots.

   f Use **e** to describe the shape of each distribution.

   g What conclusions, if any, can be drawn from the data?

PARALLEL BOXPLOTS

Parallel boxplots enable us to make a *visual comparison* of the distribution of the data and the descriptive statistics (median, range and interquartile range). Parallel boxplots could be horizontal or vertical. For example:

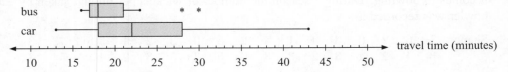

It is suggested that you use appropriate technology in this section.

Example 24

An office worker has the choice of travelling to work by car or bus and has collected data giving the travel times from recent journeys using both of these types of transport. He is interested to know which type of transport is the quickest to get him to work and which is the most reliable.

Car travel times (m): 21, 25, 18, 13, 33, 27, 28, 14, 18, 43, 19, 22, 30, 22, 24
Bus travel times (m): 23, 18, 16, 16, 30, 20, 21, 18, 18, 17, 20, 21, 28, 17, 16

Prepare parallel boxplots for the datasets and use them to compare the two methods of transport for quickness and reliability.

The 5-number summaries are:

For car travel:		For bus travel:	
	min. $= 13$		min. $= 16$
	$Q_1 = 18$		$Q_1 = 17$
	median $= 22$		median $= 18$
	$Q_3 = 28$		$Q_3 = 21$
	max. $= 43$		max. $= 30$

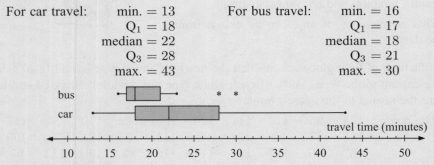

Using the median, 50% of the time, bus travel takes 18 minutes or less, compared with car travel at 22 minutes or less.
i.e., bus travel is generally *quicker*.

Comparing spread: range for car $= 43 - 13$ range for bus $= 30 - 16$
$$= 30 \qquad\qquad = 14$$

$$\text{IQR} = Q_3 - Q_1 \qquad\qquad \text{IQR} = Q_3 - Q_1$$
$$= 28 - 18 \qquad\qquad = 21 - 17$$
$$= 10 \qquad\qquad = 4$$

When comparing these spread measures, they indicate that the bus travel times are less 'spread out' than the car travel times. They are *more predictable or reliable.*

EXERCISE 5K

1 The following boxplots compare the time students in years 10 and 12 spend on homework over a one week period.

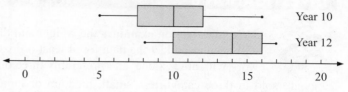

a Find the 5-number summaries for both the year 10 and year 12 students.

b Determine the i range ii interquartile range for each group.

2 Two classes have completed the same test. Boxplots have been drawn to summarise and display the results. They have been drawn on the same set of axes so that the results can be compared.

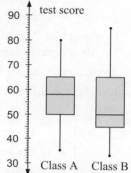

a In which class was:

   i the highest mark   ii the lowest mark

   iii there a larger spread of marks?

b Find:

   i the range of marks in class B

   ii the interquartile range for class A.

c If the pass mark was 50 for the test what percentage of students passed the test in:

   i class A   ii class B?

d Describe the distribution of marks in: i class A   ii class B.

e Copy and complete:

   The students in class .... generally scored higher marks.

   The marks in class ...... were more varied.

3 The batting averages for the Australian and Indian teams for the 2001 test series in India were as follows:

Australia 109.8, 48.6, 47.0, 33.2, 32.2, 29.8, 24.8, 20.0, 10.8, 10.0, 6.0, 3.4, 1.0

India   83.83, 56.33, 50.67, 28.83, 27.00, 26.00, 21.00, 20.00, 17.67, 11.33, 10.00, 6.00, 4.00, 4.00, 1.00, 0.00

a Record the 5-number summary for each country and test for outliers.

b Construct parallel boxplots for the data.

c Compare and comment on the centres and spread of the data sets.

d Should any outliers be discarded and the data be reanalysed?

4 The heights (to the nearest centimetre) of boys and girls in a school year are as follows:

Boys 164 168 175 169 172 171 171 180 168 168 166 168 170 165 171 173 187 179 181 175 174 165 167 163 160 169 167 172 174 177 188 177 185 167 160

Girls 165 170 158 166 168 163 170 171 177 169 168 165 156 159 165 164 154 170 171 172 166 152 169 170 163 162 165 163 168 155 175 176 170 166

a Find the five-number summary for each of the data sets. Test for outliers and construct parallel boxplots.

b Compare and comment on the distribution of the data.

INVESTIGATION 6   HOW DO YOU LIKE YOUR EGGS?

This investigation examines the weight and dimensions of eggs. Because you will need to collect the data for at least 5 dozen eggs, it is suggested that you work with at least one, and preferably three other people.

Eggs are sold in three categories: small, medium or large. Decide which category your group will use.

Using a set of electronic scales, measure the weight of at least five dozen eggs in the category of your choice.

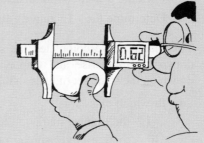

Use a set of electronic calipers to measure the length and maximum diameter of each egg.

Record your results in a spreadsheet.

Use the spreadsheet to organise the data in three separate ways.

By weight	By length	By width
• in classes of 0.1 g	• in classes of 0.1 mm	• in classes of 0.1 mm
• in classes of 0.2 g	• in classes of 0.2 mm	• in classes of 0.2 mm
• without classes	• without classes	• without classes

Present your results in a suitable table.

Draw side by side box and whisker plots for each set of data.

In a brief report, comment on your results. Discuss the characteristics of a typical egg in the category you have studied.

By chance, you hear a suggestion that hens of different breeds produce eggs of different sizes. Discuss how you would set about examining that conjecture.

On the basis of your work in this investigation, discuss the characteristics you would expect to find for a different category of eggs. How many eggs do you think it would take to test your conjecture?

If someone brought an egg to you and asked if it was of the category you had measured, how would you check?

A POSSIBLE PROJECT

The work you have covered in this chapter so far should give you sufficient knowledge to carry out your own statistical investigation.

Begin by choosing a problem or topic that interests you. Outline your view of the problem question and the data you need to answer it. Discuss your problem and proposed analysis with your teacher. If you need, refine both the problem and your proposed method of solving it.

Collect, in sufficient quantity, the data needed. Aim to ensure your data is randomly selected. Use the software available to produce any graphs and statistical calculations required.

Click on this icon to obtain additional material for projects (samples and surveys).

Prepare a report of your work. You may choose how you present your work. You may present it as:

- a newspaper or magazine article
- a video
- a powerpoint slide presentation
- a wordprocessed document.

In your report include:

- A description of the problem or issue that you are investigating.
- A simple account of the method you have employed to carry out the investigation.
- The analysis you carried out. This includes a copy of your data, any graphics and summary statistics you produced and the argument that you wrote to support your conclusion.
- Your conclusion.
- A discussion of any weaknesses in your method that may cause your conclusion to be suspect.

REVIEW SET 5A

1 Classify the following data as categorical, discrete numerical or continuous numerical:

a the number of pages in a daily newspaper

b the maximum daily temperature in the city

c the manufacturer of a car

d the preferred football code

e the position taken by a player on a soccer field

f the time it takes 15-year-olds to run one kilometre

g the length of feet

h the number of goals shot by a soccer player

i the amount spent weekly, by an individual, at the supermarket.

2 The following marks were scored for a test where the maximum score was 50:

47 32 32 29 36 39 40 46 43 39 44 18 38 45 35 46 7 44 27 48

a Construct an ordered stemplot for the data.

b What percentage of the students scored 40 or more marks?

c What percentage of the students scored less than 30 marks?

d If a score of 25 or more is a pass, what percentage of the students passed?

e Describe the distribution of the data.

3 A frequency table for the heights of a basketball squad is given below.

Height (cm)	Frequency
170 -	1
175 -	8
180 -	9
185 -	11
190 -	9
195 -	3
200 - < 205	3

a Explain why 'height' is a continuous variable.

b Construct a histogram for the data. The axes should be carefully marked and labelled and include a heading for the graph.

c What is the modal class? Explain what this means.

d Describe the distribution of the data.

4 Six scores have mean 8. What must the seventh score be to increase the mean by 1?

5 The data below shows the distance, in metres, Kapil was able to throw a cricket ball.

71.2	65.1	68.0	71.1	74.6	68.8	83.2	85.0	74.5	87.4
84.3	77.0	82.8	84.4	80.6	75.9	89.7	83.2	97.5	82.9
90.5	85.5	90.7	92.9	95.6	85.5	64.6	73.9	80.0	86.5

  a Determine the highest and lowest value for the data set.

  b Produce between 6 and 12 groups in which to place all the data values.

  c Prepare a frequency distribution table.

  d For this data, draw:

    i a frequency histogram   **ii** a relative frequency histogram

    iii a less than ogive.

  e Determine: **i** the mean   **ii** the median.

6 The given parallel boxplots represent the 100-metre sprint times
for the members of two athletics
squads.

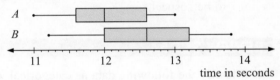

  a Determine the 5-number summaries for both A and B.

  b Determine the **i** range **ii** interquartile range   for each group.

  c Copy and complete:

    i The members of squad ...... generally ran faster times.

    ii The times in squad ...... were more varied.

7 Find, using your calculator, the mean and standard deviation of these sets of data:

  a 117, 129, 105, 124, 123, 128, 131, 124, 123, 125, 108

  b 6.1, 5.6, 7.2, 8.3, 6.6, 8.4, 7.7, 6.2

8 Find a, given that 3, 0, a, a, 4, a, 6, a and 3 have a mean of 4.

9 The back to back stemplot alongside
represents the times for the 100 metre freestyle recorded by members of a
swimming squad.

Girls		Boys
	32	1
4	33	0 2 2 7
7 6 3	34	1 3 4 4 8
8 7 4 3 0	35	0 2 4 7 9 9
8 8 3 3	36	7 8 8
7 6 6 6	37	0
6	38	
0	39	
	40	
1	41	leaf unit: 0.1 sec

  a Copy and complete the following table:

	Girls	Boys
outliers		
shape		
centre (median)		
spread (range)		

  b Write an argument that supports the conclusion you have drawn about the girls' and
boys' swimming times.

REVIEW SET 5B

1 A sample of lamp-posts were surveyed for the following data. Classify the data as categorical, discrete numerical or continuous numerical:

a the diameter of the lamp-post (in centimetres) measured 1 metre from its base

b the material from which the lamp-post is made

c the location of the lamp-post (inner, outer, North, South, East or West)

d the height of the lamp-post, in metres

e the time (in months) since the last inspection

f the number of inspections since installation

g the condition of the lamp-post (very good, good, fair, unsatisfactory).

2 The data supplied below is the diameter (in cm) of a number of bacteria colonies as measured by a microbiologist 12 hours after seeding.

```
0.4   2.1   3.4   3.9   4.7   3.7   0.8   3.6   4.1   4.9
2.5   3.1   1.5   2.6   1.3   3.5   0.9   1.5   4.2   3.5
2.1   3.0   1.7   3.6   2.8   3.7   2.8   3.2   4.0   3.3
```

a Produce a stemplot for this data.

b Find the: **i** median   **ii** range   of the data.

c Comment on the skewness of the data.

3 Consider the graph alongside.

a Name the type of graph.

b Is the data represented discrete or continuous?

c Find the modal class.

d Construct a table for the data showing class intervals, frequencies, midpoints and products.

e Find the approximate mean of the data.

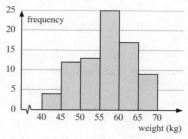

4 The ogive shows the cumulative frequency of the weights of a herd of 12-month old female alpacas.

a How many alpacas were the the herd?

b What percentage of alpacas have weights under 52 kg?

c What percentage of alpacas have weights over 55 kg?

d Read the cumulative frequencies from the graph and set up a cumulative frequency table.

e Hence set up a frequency table for classes 48-, 50-, 52-, 54-

f Draw the histogram.   **g** Find the modal class.

h Is the histogram symmetrical about the modal class?

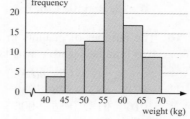

5 12 of 29 measurements are below 20 cm and 13 measurements are above 21 cm. Find the median if the other 4 measurements are 20.1 cm, 20.4 cm, 20.7 cm and 20.9 cm.

6 Eight sample values are: 6, a, 7, a, 4, b, 6 and 8 where a and b are single digit numbers and the mean is 7.

    a Show that a and b have two possible solutions.

    b If there is a single mode, what is the median?

7 Determine the five number summary and the interquartile range for each of the following data sets that have already been placed in rank order. Then draw a boxplot for each data set:

    a 4.0, 10.1, 13.4, 14.2, 15.0, 16.5, 22.2, 22.4, 23.1, 30.0

    b 11, 15, 17, 21, 23, 25, 25, 25, 27, 47, 49, 49

8 Anneka's golf scores for her last 20 rounds were:

        90, 106, 84, 103, 112, 100, 105, 81, 104, 98,
        107, 95, 104, 108, 99, 101, 106, 102, 98, 101

    a Find the: **i** median **ii** lower quartile **iii** upper quartile

    b Find the interquartile range of the data set.

    c Find the mean and standard deviation of her scores.

9 Find, using technology, the **i** mean **ii** standard deviation **iii** variance of the following data sets:

    a 3.4, 3.5, 3.6, 3.7, 3.8, 3.9, 4.0, 4.1

    b 12, 9, 3, 0, 1, 3, 6

10 The number of peanuts in a jar varies slightly from jar to jar. A sample of 30 jars for two brands X and Y was taken and the number of peanuts in each jar was recorded.

Brand X						Brand Y					
871	885	878	882	889	885	909	906	913	891	898	901
916	913	886	905	907	898	894	894	928	893	924	892
874	904	901	894	897	899	927	907	901	900	907	913
908	901	898	894	895	895	921	904	903	896	901	895
910	904	896	893	903	888	917	903	910	903	909	904

    a Produce a back-to-back stemplot for the data for each brand.

    b Complete the table below:

	Brand X	Brand Y
outliers		
shape		
centre (median)		
spread (range)		

    c Use the above information to compare Brand X and Brand Y.

Chapter 6

Linear and exponential algebra

Contents:

Algebra is a very powerful tool which is used to make problem solving easier. Algebra involves using **pronumerals** (letters) to represent **unknown** values or values which can **vary** depending on the situation.

Many worded problems can be converted to algebraic symbols to make algebraic **equations**. We learn how to **solve** equations in order to find the **solutions** to problems.

Algebra can also be used to construct **formulae**, which are equations that connect two or more **variables**. Many people use formulae as part of their jobs, so an understanding of how to **substitute** into and **rearrange** formulae is essential. Builders, nurses, pharmacists, engineers, financial planners and computer programmers all use formulae which rely on algebra.

OPENING PROBLEM

Holly is offered two different mobile phone plans. Plan A costs $25 for the monthly access fee and calls are billed at 17 cents per minute. Plan B costs only $10 for the monthly access fee but calls are billed at 23 cents per minute.

Consider the following:

* Find *formulae* that connect the total cost (C) with the number of minutes (m) used per month for *each* plan.

* Use the formulae to find the *cost* for each plan if Holly uses her phone for
  i 150 minutes per month   **ii** 300 minutes per month.

* For how many minutes of use per month would the cost of each plan work out to be the *same*?

* What advice would you give to Holly regarding her choice of plans?

ALGEBRAIC NOTATION

The ability to convert worded sentences and problems into algebraic symbols and to understand algebraic notation is essential in the problem solving process.

Notice that:
* $x^2 + 3x$ is an algebraic **expression**, whereas
* $x^2 + 3x = 8$ is an **equation**, and
* $x^2 + 3x > 28$ is an **inequality** (sometimes called an **inequation**).

Recall that when we simplify **repeated sums**, we use **product** notation:

i.e., $x + x$ and $x + x + x$
$= 2$ 'lots' of x $= 3$ 'lots' of x
$= 2 \times x$ $= 3 \times x$
$= 2x$ $= 3x$

Also, when we simplify **repeated products**, we use **index** notation:

i.e., $x \times x = x^2$ and $x \times x \times x = x^3$

A ALGEBRAIC SUBSTITUTION

Consider the number crunching machine alongside:

If we place any number x, into the machine, it calculates $5x - 7$, i.e., x is multiplied by 5 and then 7 is subtracted.

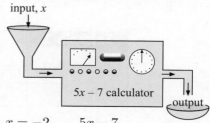

input, x

$5x - 7$ calculator

output

For example: if $x = 2$,

$$5x - 7$$
$$= 5 \times 2 - 7$$
$$= 10 - 7$$
$$= 3$$

and if $x = -2$,

$$5x - 7$$
$$= 5 \times -2 - 7$$
$$= -10 - 7$$
$$= -17$$

To **evaluate** an *algebraic expression*, we find its **value** for particular numerical *substitutions* of the *unknowns*.

Example 1

If $p = 4$, $q = -2$ and $r = 3$, find the value of:

a $3q - 2r$ **b** $2pq - r$ **c** $\dfrac{p - 2q + 2r}{p + r}$

a $3q - 2r$
$$= 3 \times -2 - 2 \times 3$$
$$= -12$$

b $2pq - r$
$$= 2 \times 4 \times -2 - 3$$
$$= -19$$

c $\dfrac{p - 2q + 2r}{p + r}$
$$= \frac{4 - 2 \times -2 + 2 \times 3}{(4 + 3)}$$
$$= 2$$

EXERCISE 6A

1 If $p = 4$, $q = -2$ and $r = 3$ find the value of:

a $5p$ **b** $4q$ **c** $3pq$ **d** pqr
e $3p - 2q$ **f** $5r - 4q$ **g** $4q - 2r$ **h** $2pr + 5q$

2 If $w = 2$, $x = -1$ and $y = 3$, evaluate:

a $\dfrac{y}{w}$ **b** $\dfrac{y + w}{x}$ **c** $\dfrac{3x - y}{w}$ **d** $\dfrac{5w - 2x}{y - x}$

Example 2

If $a = 3$, $b = -2$ and $c = -1$, evaluate:
a b^2 **b** $ab - c^3$

a b^2
$$= (-2)^2$$
$$= 4$$

b $ab - c^3$
$$= 3 \times -2 - (-1)^3$$
$$= -5$$

Notice the use of brackets!

3 If $a = 3$, $b = -2$ and $c = -1$, evaluate:

    a c^2 **b** b^3 **c** $a^2 + b^2$ **d** $(a+b)^2$

    e $b^3 + c^3$ **f** $(b+c)^3$ **g** $(2a)^2$ **h** $2a^2$

Example 3

If $p = 4$, $q = -3$ and $r = 2$, evaluate:

    a $\sqrt{p - q + r}$ **b** $\sqrt{p + q^2}$

a $\sqrt{p - q + r}$	**b** $\sqrt{p + q^2}$
$= \sqrt{4 - -3 + 2}$	$= \sqrt{4 + (-3)^2}$
$= 3$	$\doteqdot 3.61$ (3 s.f.)

---

4 If $p = 4$, $q = -3$ and $r = 2$, evaluate:

    a $\sqrt{p} + q$ **b** $\sqrt{p + q}$ **c** $\sqrt{r - q}$ **d** $\sqrt{p - pq}$

    e $\sqrt{pr - q}$ **f** $\sqrt{p^2 + q^2}$ **g** $\sqrt{p + r + 2q}$ **h** $\sqrt{2q - 5r}$

INVESTIGATION 1   SOLVING EQUATIONS

 Linear equations like $5x - 3 = 12$ can be solved using a table of values on a **graphics calculator**.

We try to find the value of x which makes the expression $5x - 3$ equal to 12 upon *substitution*. This is the **solution** to the equation.

What to do:

1 Enter the **function** $Y_1 = 5X - 3$ into your calculator.

2 Set up a **table** that calculates the value of $y = 5x - 3$ for x values from -5 to 5.

3 View the table and scroll down until you find the value of x that makes Y_1 equal to 12.

    As we can see, the solution is $x = 3$.

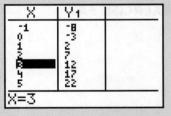

4 Use your calculator and the method given above to solve the following equations:

    a $7x + 1 = -20$ **b** $8 - 3x = -4$

    c $\dfrac{x}{4} + 2 = 1$ **d** $\frac{1}{3}(2x - 1) = 3$

5 The solutions to the following equations are *not integers*, so change your table to investigate x values from -5 to 5 in intervals of 0.5:

    a $2x - 3 = -6$ **b** $6 - 4x = 8$ **c** $x - 5 = -3.5$

6 Use a calculator to solve the following equations:

    a $3x + 2 = 41$ **b** $5 - 4x = 70$ **c** $\dfrac{2x}{3} + 5 = 2\frac{2}{3}$

B   LINEAR EQUATIONS

Many problems can be written as **equations** using algebraic notation. So, it is essential we are able to **solve** equations.

> **Linear equations** are equations which are in the form (or can be converted to the form) $ax + b = 0$, where x is the **unknown (variable)** and a, b are **constants**.

SOLVING EQUATIONS

The following steps should be followed when solving simple equations:

Step 1:   Decide how the expression containing the unknown has been '**built up**'.

Step 2:   **Isolate** the unknown by performing **inverse** operations on **both sides** of the equation to 'undo' the 'build up' in **reverse** order.

Step 3:   **Check** your solution by **substitution**.

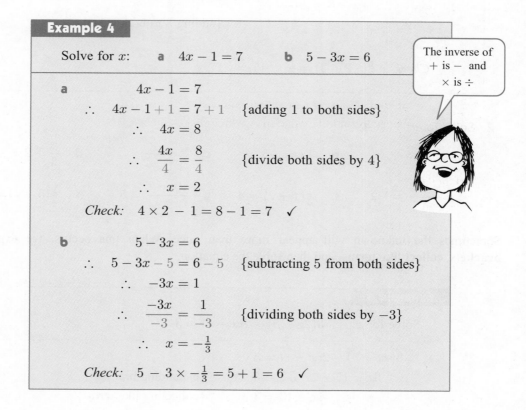

Example 4

Solve for x: **a** $4x - 1 = 7$ **b** $5 - 3x = 6$

The inverse of + is − and × is ÷

a
$$4x - 1 = 7$$
$$\therefore \quad 4x - 1 + 1 = 7 + 1 \quad \text{\{adding 1 to both sides\}}$$
$$\therefore \quad 4x = 8$$
$$\therefore \quad \frac{4x}{4} = \frac{8}{4} \quad \text{\{divide both sides by 4\}}$$
$$\therefore \quad x = 2$$

Check: $4 \times 2 - 1 = 8 - 1 = 7$ ✓

b
$$5 - 3x = 6$$
$$\therefore \quad 5 - 3x - 5 = 6 - 5 \quad \text{\{subtracting 5 from both sides\}}$$
$$\therefore \quad -3x = 1$$
$$\therefore \quad \frac{-3x}{-3} = \frac{1}{-3} \quad \text{\{dividing both sides by } -3\text{\}}$$
$$\therefore \quad x = -\tfrac{1}{3}$$

Check: $5 - 3 \times -\tfrac{1}{3} = 5 + 1 = 6$ ✓

EXERCISE 6B

1 Solve for x:

   a $x + 8 = 4$ **b** $5x = 35$ **c** $-22 = -2x$ **d** $5 - x = 11$

   e $2x + 3 = 7$ **f** $3x - 2 = -8$ **g** $3 - 4x = -11$ **h** $6 = 11 - 2x$

Example 5

Solve for x: **a** $\dfrac{x}{5} - 3 = -1$ **b** $\dfrac{1}{5}(x - 3) = -1$

a
$$\dfrac{x}{5} - 3 = -1$$

$$\therefore \quad \dfrac{x}{5} - 3 + 3 = -1 + 3 \qquad \{\text{adding 3 to both sides}\}$$

$$\therefore \quad \dfrac{x}{5} = 2$$

$$\therefore \quad \dfrac{x}{5} \times 5 = 2 \times 5 \qquad \{\text{multiplying both sides by 5}\}$$

$$\therefore \quad x = 10 \qquad Check: \ \tfrac{10}{5} - 3 = 2 - 3 = -1 \ \checkmark$$

b
$$\tfrac{1}{5}(x - 3) = -1$$

$$\therefore \quad \tfrac{1}{5}(x - 3) \times 5 = -1 \times 5 \qquad \{\text{multiplying both sides by 5}\}$$

$$\therefore \quad x - 3 = -5$$

$$\therefore \quad x - 3 + 3 = -5 + 3 \qquad \{\text{adding 3 to both sides}\}$$

$$\therefore \quad x = -2$$

$$Check: \ \tfrac{1}{5}(-2 - 3) = \tfrac{1}{5} \times -5 = -1 \ \checkmark$$

2 Solve for x:

a $\dfrac{x}{4} = 8$ **b** $\tfrac{1}{2}x = -5$ **c** $3 = \dfrac{x}{-2}$ **d** $\dfrac{x}{3} + 2 = -1$

e $\dfrac{x + 3}{4} = -2$ **f** $\tfrac{1}{3}(x + 2) = 4$ **g** $\dfrac{2x - 1}{3} = 5$ **h** $\tfrac{1}{2}(5 - x) = -3$

Sometimes the unknown will appear more than once. When this occurs, we **expand brackets**, **collect like terms**, and then **solve** the equation.

Example 6

Solve for x: $3(2x - 5) - 2(x - 1) = 3$

$$3(2x - 5) - 2(x - 1) = 3$$

$$\therefore \quad 6x - 15 - 2x + 2 = 3 \qquad \{\text{expanding brackets}\}$$

$$\therefore \quad 4x - 13 = 3 \qquad \{\text{collecting like terms}\}$$

$$\therefore \quad 4x - 13 + 13 = 3 + 13 \qquad \{\text{adding 13 to both sides}\}$$

$$\therefore \quad 4x = 16$$

$$\therefore \quad x = 4 \qquad \{\text{dividing both sides by 4}\}$$

$$Check: \ 3(2 \times 4 - 5) - 2(4 - 1) = 3 \times 3 - 2 \times 3 = 3 \ \checkmark$$

3   Solve for x:

    a $2(x + 8) + 5(x - 1) = 6$ **b** $2(x - 3) + 3(x + 2) = 0$

    c $3(x + 3) - 2(x + 1) = -2$ **d** $4(2x - 3) + 2(x + 2) = 2$

    e $3(4x + 1) - 2(3x - 4) = 1$ **f** $5(x + 2) - 2(3 - 2x) = 3$

If the unknown appears on **both sides** of the equation,

we   •   **expand** any brackets and **collect like terms**

      •   move the **unknown to one side** of the equation and the remaining terms to the other side

      •   **simplify** and **solve** the equation.

Example 7

Solve for x:

  a $3x - 4 = 2x + 6$ **b** $4 - 3(2 + x) = x$

---

a

$$3x - 4 = 2x + 6$$
$$\therefore \quad 3x - 4 - 2x = 2x + 6 - 2x \quad \text{\{subtracting } 2x \text{ from both sides\}}$$
$$\therefore \quad x - 4 = 6$$
$$\therefore \quad x - 4 + 4 = 6 + 4 \quad \text{\{adding 4 to both sides\}}$$
$$\therefore \quad x = 10$$

Check: LHS $= 3 \times 10 - 4 = 26$, RHS $= 2 \times 10 + 6 = 26$. ✓

b

$$4 - 3(2 + x) = x$$
$$\therefore \quad 4 - 6 - 3x = x \quad \text{\{expanding\}}$$
$$\therefore \quad -2 - 3x = x$$
$$\therefore \quad -2 - 3x + 3x = x + 3x \quad \text{\{adding } 3x \text{ to both sides\}}$$
$$\therefore \quad -2 = 4x$$
$$\therefore \quad \frac{-2}{4} = \frac{4x}{4} \quad \text{\{dividing both sides by 4\}}$$
$$-\tfrac{1}{2} = x$$
$$\text{i.e., } \quad x = -\tfrac{1}{2}$$

Check: LHS $= 4 - 3(2 + -\tfrac{1}{2}) = 4 - 3 \times \tfrac{3}{2} = 4 - 4\tfrac{1}{2} = -\tfrac{1}{2} =$ RHS ✓

---

4   Solve for x:

    a $2x - 3 = x + 6$ **b** $4x - 2 = 5 - x$

    c $4 - 5x = 3x - 7$ **d** $-x = x + 4$

    e $12 - 7x = 3x + 8$ **f** $5x - 9 = 1 + 6x$

    g $4 - x - 3(2 - x) = 6 + x$ **h** $5 - 3(1 - x) = 2 - x$

    i $3 - 2x - (2x + 1) = -1$ **j** $3(4x + 2) - 2x = -7 + x$

Sometimes when more complicated equations are expanded a linear equation results.

Example 8

Solve for x: $(x-3)^2 = (4+x)(2+x)$

$$(x-3)^2 = (4+x)(2+x)$$

$\therefore \quad x^2 - 6x + 9 = 8 + 4x + 2x + x^2$ {expanding each side}

$\therefore \quad x^2 - 6x + 9 - x^2 = 8 + 4x + 2x + x^2 - x^2$ {subtracting x^2 from both sides}

$\therefore \quad -6x + 9 = 8 + 6x$

$\therefore \quad -6x + 9 + 6x = 8 + 6x + 6x$ {adding $6x$ to both sides}

$\therefore \quad 9 = 12x + 8$

$\therefore \quad 9 - 8 = 12x + 8 - 8$ {subtracting 8 from both sides}

$\therefore \quad 1 = 12x$

$\therefore \quad \dfrac{1}{12} = \dfrac{12x}{12}$ {dividing both sides by 12}

$\therefore \quad x = \tfrac{1}{12}$

---

5 Solve for x:

  a $\quad x(x+5) = (x-4)(x-3)$
  b $\quad x(2x+1) - 2(x+1) = 2x(x-1)$

  c $\quad (x+3)(x-2) = (4-x)^2$
  d $\quad x^2 - 3 = (2+x)(3+x)$

  e $\quad (x+2)(2x-1) = 2x(x+3)$
  f $\quad (x+4)^2 = (x-1)(x+3)$

6 Solve for x:

  a $\quad 3(2x-1) + 2 = 6x - 1$
  b $\quad 4(3x+2) = 6(2x+1)$

  c Comment on your solutions to **a** and **b**.

C   FRACTIONAL EQUATIONS

Fractional equations can be simplified by finding the **least common denominator (LCD)** of the fractions. Each term is then multiplied by the fraction which makes the denominators the same (LCD) and then the numerators are equated.

Consider the following fractional equations:

$$\frac{x}{2} = \frac{x}{3}$$ LCD is 2×3 i.e.,   6

$$\frac{5}{2x} = \frac{3x}{5}$$ LCD is $2x \times 5$ i.e., $10x$

$$\frac{x-7}{3} = \frac{x}{2x-1}$$ LCD is $3 \times (2x-1)$ i.e., $3(2x-1)$

Example 9

Solve for x: $\dfrac{x}{2} = \dfrac{3+x}{5}$

$\dfrac{x}{2} = \dfrac{3+x}{5}$ has LCD $= 10$

$\therefore \quad \dfrac{x}{2} \times \dfrac{5}{5} = \dfrac{2}{2} \times \left(\dfrac{3+x}{5} \right)$ {to create a common denominator}

Notice the insertion of brackets here.

$\therefore \quad 5x = 2(3+x)$ {equating numerators}
$\therefore \quad 5x = 6 + 2x$ {expanding brackets}
$\therefore \quad 5x - 2x = 6 + 2x - 2x$ {taking $2x$ from both sides}
$\therefore \quad 3x = 6$
$\therefore \quad x = 2$ {dividing both sides by 3}

EXERCISE 6C.1

1 Solve for x:

a $\dfrac{x}{2} = \dfrac{3}{7}$

b $\dfrac{3}{5} = \dfrac{x}{6}$

c $\dfrac{x}{5} = \dfrac{x-2}{3}$

d $\dfrac{x+1}{3} = \dfrac{2x-1}{8}$

e $\dfrac{2x}{3} = \dfrac{5-x}{4}$

f $\dfrac{3x+2}{3} = \dfrac{2x-5}{2}$

g $\dfrac{2x-1}{3} = \dfrac{4-x}{12}$

h $\dfrac{4x+7}{11} = \dfrac{5-x}{2}$

i $\dfrac{3x+7}{6} = \dfrac{4x-1}{-2}$

Example 10

Solve for x: $\dfrac{4}{x} = \dfrac{3}{4}$

$\dfrac{4}{x} = \dfrac{3}{4}$ has LCD $= 4x$

$\therefore \quad \dfrac{4}{x} \times \dfrac{4}{4} = \dfrac{3}{4} \times \dfrac{x}{x}$ {to create a common denominator}

$\therefore \quad 16 = 3x$ {equating numerators}

$\therefore \quad x = \dfrac{16}{3}$ {dividing both sides by 3}

2 Solve for x:

a $\dfrac{5}{x} = \dfrac{3}{5}$

b $\dfrac{6}{x} = \dfrac{2}{7}$

c $\dfrac{4}{3} = \dfrac{7}{x}$

d $\dfrac{3}{2x} = \dfrac{5}{4}$

e $\dfrac{3}{2x} = \dfrac{5}{3}$

f $\dfrac{2}{3x} = -\dfrac{1}{6}$

g $\dfrac{5}{4x} = -\dfrac{11}{12}$

h $\dfrac{4}{7x} = \dfrac{3}{8x}$

Example 11

Solve for x: $\dfrac{2x+1}{3-x} = \dfrac{3}{4}$.

$$\dfrac{2x+1}{3-x} = \dfrac{3}{4} \qquad \text{has LCD } 4(3-x)$$

$\therefore \quad \dfrac{4}{4} \times \left(\dfrac{2x+1}{3-x}\right) = \dfrac{3}{4} \times \left(\dfrac{3-x}{3-x}\right)$ {to create a common denominator}

$\therefore \quad 4(2x+1) = 3(3-x)$ {equating numerators}

$\therefore \quad 8x+4 = 9-3x$ {expanding the brackets}

$\therefore \quad 8x+4+3x = 9-3x+3x$ {adding $3x$ to both sides}

$\therefore \quad 11x+4 = 9$

$\therefore \quad 11x+4-4 = 9-4$ {subtracting 4 from both sides}

$\therefore \quad 11x = 5$

$\therefore \quad x = \frac{5}{11}$ {dividing both sides by 11}

Make sure you use brackets.

3 Solve for x:

a $\dfrac{2x+3}{x+1} = \dfrac{4}{3}$

b $\dfrac{x+1}{1-2x} = \dfrac{3}{5}$

c $\dfrac{2x-1}{4-3x} = -\dfrac{2}{3}$

d $\dfrac{x+3}{2x-1} = \dfrac{1}{2}$

e $\dfrac{4x+3}{2x-1} = 6$

f $\dfrac{3x-2}{x+4} = -5$

g $\dfrac{6x-1}{3-2x} = 10$

h $\dfrac{5x-1}{x+4} = 5$

i $\dfrac{2x+5}{x-1} = -4$

USING TECHNOLOGY

GRAPHING PACKAGE

You could use a **graphing package** (click on the icon) or a **graphics calculator** to solve linear equations.

For example: to solve $3.2x - 4.2 = 7.4$ we:

- Enter the **functions** $Y_1 = 3.2x - 4.2$ and $Y_2 = 7.4$ into the calculator.
- Adjust the viewing **window** settings for x values between -10 and 10 and y values between -10 and 10.
- **Graph** the two straight lines.
- Find the point of **intersection** of the two lines using the built in function.

  So, $x = 3.625$.

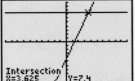

EXERCISE 6C.2

1 Use technology to solve the following:

a $5.4x + 7.2 = 15.6$

b $0.05x - 9.6 = 3.5$

c $23.24 - 13.08x = 8.94$

d $1234.32 + 37.56x = 259.04$

2   The scale reading, R (mm) when weights w (grams) are placed on a spring balance is given by the formula $R = 0.4w + 5$.   Find w when: **a** $R = 27$ **b** $R = 42$

3   In the United States of America temperature is measured in degrees fahrenheit ($^\circ$F) rather than in degrees Celsius ($^\circ$C). The rule showing the relationship between these two temperature scales is given by $F = 1.8C + 32$.   What temperature in $^\circ$C corresponds to a temperature of: **a** 40°F   **b** 0°F   **c** 200°F?

4   The total cost $\$C$, to sink a bore is given by the rule $C = 15d + 350$ where d is the depth in metres. How deep a bore can a farmer obtain for a cost of:
   a   \$2000   **b**   \$3200?

5   Use technology, as in question **1** to solve:
   a $3x + 2 = 5x - 17$ **b** $3.6x - 1.8 = 2.7x + 4.1$
   c $3.56x + 13.67 = 1.05x + 39.97$ **d** $21.67 + 3.67x = 5.83x - 58.88$

6   Find out how to use the **equation solver** function on your **graphics calculator**. Use this method to solve question **1** of this exercise.

D   PROBLEM SOLVING

Many problems can be translated into **algebraic equations**. When problems are solved using algebra, we follow these steps:

Step 1:   Decide the unknown quantity and allocate a pronumeral.
Step 2:   Decide which operations are involved.
Step 3:   Translate the problem into a linear equation and check your translation is correct.
Step 4:   Solve the linear equation by isolating the pronumeral.
Step 5:   Check that your solution does satisfy the original problem.
Step 6:   Write your answer in sentence form. Remember, there is usually no pronumeral in the original problem.

Example 12

When a number is trebled and subtracted from 7 the result is -11. Find the number.

Let x be the number.
$\therefore$ $3x$ is the number trebled.
$\therefore$ $7 - 3x$ is this number subtracted from 7.

So, $7 - 3x = -11$
$\therefore$ $7 - 3x - 7 = -11 - 7$ {subtracting 7 from both sides}
$\therefore$ $-3x = -18$
$\therefore$ $x = 6$ {dividing both sides by -3}

So, the number is 6.
Check: $7 - 3 \times 6 = 7 - 18 = -11$ ✓

EXERCISE 6D

1 When seven times a certain number is subtracted from 15, the result is -4. Find the number.

2 Four times a certain number, minus 5, is equal to 7 more than twice the number. What is the number?

3 Three times the result of subtracting a certain number from 5 gives the same answer as dividing the number by 3. Find the number.

Example 13

What number must be added to both the numerator and the denominator of the fraction $\frac{1}{3}$ to get a fraction equalling $\frac{7}{8}$?

Let x be the number.

$$\therefore \quad \frac{1+x}{3+x} = \frac{7}{8} \qquad \text{where the LCD is } 8(3+x)$$

$$\therefore \quad \frac{8}{8} \times \left(\frac{1+x}{3+x}\right) = \frac{7}{8} \times \left(\frac{3+x}{3+x}\right) \qquad \text{\{to get a common denominator\}}$$

$$\therefore \quad 8(1+x) = 7(3+x) \qquad \text{\{equating numerators\}}$$

$$\therefore \quad 8 + 8x = 21 + 7x \qquad \text{\{expanding brackets\}}$$

$$\therefore \quad 8 + 8x - 7x = 21 + 7x - 7x \qquad \text{\{subtracting } 7x \text{ from both sides\}}$$

$$\therefore \quad 8 + x = 21$$

$$\therefore \quad x = 13 \qquad \text{So, 13 is added to both.}$$

---

4 What number must be added to both the numerator and the denominator of the fraction $\frac{1}{5}$ to get a fraction equalling $\frac{3}{4}$?

5 What number must be subtracted from both the numerator and the denominator of the fraction $\frac{2}{3}$ to get a fraction equalling $\frac{1}{4}$?

Example 14

Sarah's age is one third her father's age and in 13 years time her age will be a half of her father's age. How old is Sarah now?

Let Sarah's present age be x years.

$\therefore$ father's present age is $3x$ years.

Table of ages:

	Now	13 years time
Sarah	x	$x + 13$
Father	$3x$	$3x + 13$

So, $3x + 13 = 2(x + 13)$

$\therefore \quad 3x + 13 = 2x + 26$

$\therefore \quad 3x - 2x = 26 - 13$

$\therefore \quad x = 13$

$\therefore$ Sarah's present age is 13.

6 Jake is now one-quarter of his father's age and in 7 years time his age will be one-third the age of his father. How old is Jake now?

7 When Heidi was born her mother was 21 years old. At present Heidi's age is 30% of her mother's age. How old is Heidi now?

Example 15

Brittney has only 2-cent and 5-cent stamps with a total value of $1.78 and there are two more 5-cent stamps than there are 2-cent stamps. How many 2-cent stamps are there?

If there are x 2-cent stamps then there are $(x + 2)$ 5-cent stamps

Type	Number	Value
2-cent	x	$2x$ cents
5-cent	$x + 2$	$5(x + 2)$ cents

$$\therefore \quad 2x + 5(x + 2) = 178 \quad \text{\{equating values in cents\}}$$
$$\therefore \quad 2x + 5x + 10 = 178$$
$$\therefore \quad 7x + 10 = 178$$
$$\therefore \quad 7x = 168$$
$$\therefore \quad x = 24$$

There are 24 2-cent stamps.

---

8 Yong has 50 stamps which are either 2-cent or 5-cent values. If the total value of the stamps is $2.02, how many 2-cent stamps does he have?

9 Jasmin has 50 old coins in a jar and they are either 1-cent, 2-cent or 5-cent pieces and their value is $1.20. If she has twice as many 2-cent coins as 1-cent coins, how many of each type does she have?

10 Carlo has 35 stamps. Some are valued at 2-cents, others at 5-cents and the remainder at 10-cents. He has three times as many 5-cent stamps as he has 2-cent stamps and the total value of all stamps is $1.89. How many of each stamp does he have?

11 Tickets at a football match cost $6, $10 or $15 each. The number of $10 tickets sold was 3 times the number of $6 tickets sold and 500 less than the number of $15 tickets sold. If the total gate receipts were $84 450, how many of each type of ticket was sold?

E FORMULA SUBSTITUTION

In previous questions we used the **formula** $\quad s = \dfrac{d}{t} \quad$ in problems concerned with speed (s), distance travelled (d) and time taken (t).

A **formula** is an equation which connects two or more variables. (The plural of formula is **formulae** or **formulas**.)

In a formula it is common for one of the variables to be on one side of the equation and the other variable(s) and constants to be on the other side.

The variable on its own is called the **subject** of the formula.

If a formula contains two or more variables and we know the value of all but one of them, we can use the formula to find the value of the unknown variable.

The Method:	*Step 1:*	Write down the formula.
	Step 2:	State the values of the known variables.
	Step 3:	Substitute into the formula to form a one variable equation.
	Step 4:	Solve the equation for the unknown variable.

Example 16

When a stone is dropped down a well the total distance fallen, D metres, is given by the formula $D = \frac{1}{2}gt^2$ where t is the time of fall (in seconds) and g is the gravitational constant of 9.8 . Find:

a   the distance fallen after 5 seconds

b   the time (to the nearest $\frac{1}{100}$th second) taken for the stone to fall 100 metres.

a $D = \frac{1}{2}gt^2$ where $g = 9.8$ and $t = 5$

∴ $D = \frac{1}{2} \times 9.8 \times 5^2$ *Calculator:*

∴ $D = 122.5$ 0.5 ✕ 9.8 ✕ 5 x^2 ENTER

∴   the stone has fallen 122.5 m.

b $D = \frac{1}{2}gt^2$ where $D = 100$, $g = 9.8$

> Remember that if $x^2 = k$, then $\sqrt{x} = \pm k$!

∴ $\frac{1}{2} \times 9.8 \times t^2 = 100$

∴ $4.9t^2 = 100$

∴ $t^2 = \dfrac{100}{4.9}$

∴ $t = \pm\sqrt{\dfrac{100}{4.9}}$ *Calculator:*

∴ $t = 4.517$ $\{t > 0\}$ 2nd $\sqrt{\ }$ 100 ÷ 4.9) ENTER

∴   time taken is approx. 4.52 seconds.

EXERCISE 6E

1   The formula for finding the circumference, C, of a circle of diameter d, is $C = \pi d$ where π is the constant with value approximately 3.14159. Find:

a   the circumference of a circle of diameter 11.4 cm

b   the diameter of a circle with circumference 250 cm

c   the radius of a circle of circumference 100 metres.

2

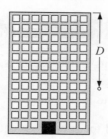

When a cricket ball is dropped from the top of a building the total distance fallen is given by the formula $D = \frac{1}{2}gt^2$ where D is the distance in metres and t is the time taken in seconds. Given that $g = 9.8$, find:

a the total distance fallen in the first 3 seconds of fall

b the height of the building, to the nearest metre, when the time of fall to hit the ground is 5.13 seconds.

3 When a car travels a distance d kilometres in time t hours, the average speed s km/h for the journey is given by the formula $s = \dfrac{d}{t}$. Find:

a the average speed of a car which travels 200 km in $2\frac{1}{2}$ hours

b the distance travelled by a car in $3\frac{1}{4}$ hours if its average speed is 80 km/h

c the time taken, to the nearest minute, for a car to travel 865 km at an average speed of 110 km/h.

4 A circle's area A, is given by $A = \pi r^2$ where r is its radius length. Find:

a the area of a circle of radius 5.6 cm

b the radius of a circular swimming pool which must have an area of 200 m².

5 A cylinder of radius r, and height h, has volume given by $V = \pi r^2 h$. Find:

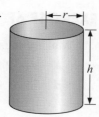

a the volume of a cylindrical tin can of radius 12 cm and height 17.5 cm

b the height of a cylinder of radius 4 cm given that its volume is 80 cm³

c the radius, in mm, of copper wire of volume 100 cm³ and length 0.2 km.

6

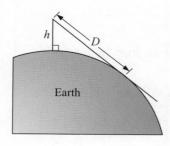

The formula $D = 3.56\sqrt{h}$ gives the approximate distance (D km) to the horizon which can be seen by a person with eye level h metres above the level of the sea. Find:

a the distance to the horizon when a person's eye level is 10 m above sea level

b how far above sea level a person's eye must be if the person wishes to see the horizon at a distance of 30 km.

7 The formula for calculating the total surface area of a sphere of radius r is given by $A = 4\pi r^2$. Find:

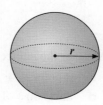

a the total surface area of a sphere of radius 6.9 cm

b the radius (in cm) of a spherical balloon which is to have a surface area of 1 m².

F   FORMULA REARRANGEMENT

For the formula $D = xt + p$ we say that D is the **subject**. This is because D is expressed in terms of the other variables, x, t and p.

In formula rearrangement we require one of the other variables to be the subject.

> To **rearrange** a formula we use the same processes as used for solving an equation for the variable we wish to be the subject.

Example 17

Make y the subject of $2x + 3y = 8$.

$$
\begin{aligned}
\text{If} \quad & 2x + 3y = 8 \\
\text{then} \quad & 2x + 3y - 2x = 8 - 2x \quad \{\text{subtract } 2x \text{ from both sides}\} \\
\therefore \quad & 3y = 8 - 2x \\
\therefore \quad & \frac{3y}{3} = \frac{8 - 2x}{3} \quad \{\text{divide both sides by 3}\} \\
\therefore \quad & y = \frac{8 - 2x}{3}
\end{aligned}
$$

EXERCISE 6F

1   Make y the subject of:

   a $x + 2y = 4$ **b** $2x + 6y = 7$ **c** $3x + 4y = 11$
   d $5x + 4y = 8$ **e** $7x + 2y = 20$ **f** $11x + 15y = 38$

Example 18

Make y the subject of $3x - 7y = 22$.

$$
\begin{aligned}
\text{If} \quad & 3x - 7y = 22 \\
\text{then} \quad & 3x - 7y - 3x = 22 - 3x \quad \{\text{subtract } 3x \text{ from both sides}\} \\
\therefore \quad & -7y = 22 - 3x \\
\therefore \quad & 7y = 3x - 22 \quad \{\times \text{ both sides by } -1\} \\
\therefore \quad & \frac{7y}{7} = \frac{3x - 22}{7} \quad \{\text{divide both sides by 7}\} \\
\therefore \quad & y = \frac{3x - 22}{7}
\end{aligned}
$$

2   Make y the subject of:

   a $x - 2y = 4$ **b** $2x - 6y = 7$ **c** $3x - 4y = -12$
   d $4x - 5y = 18$ **e** $7x - 6y = 42$ **f** $12x - 13y = -44$

3 Make x the subject of:

a $a + x = b$	**b** $ax = b$	**c** $2x + a = d$
d $c + x = t$	**e** $7x + 3y = d$	**f** $ax + by = c$
g $mx - y = c$	**h** $c - 2x = p$	**i** $a - 3x = t$
j $n - kx = 5$	**k** $a - bx = n$	**l** $p = a - nx$

Example 19

Make x the subject of $c = \dfrac{m}{x}$.

$$c = \frac{m}{x}$$

$$c \times x = \frac{m}{x} \times x \qquad \{\text{multiply both sides by } x\}$$

$$\therefore \quad cx = m$$

$$\therefore \quad \frac{cx}{c} = \frac{m}{c} \qquad \{\text{divide both sides by } c\}$$

$$\therefore \quad x = \frac{m}{c}$$

---

4 Make x the subject of:

a $a = \dfrac{x}{b}$	**b** $\dfrac{a}{x} = d$	**c** $p = \dfrac{2}{x}$
d $\dfrac{x}{2} = n$	**e** $\dfrac{b}{x} = s$	**f** $\dfrac{m}{x} = \dfrac{x}{n}$

INVESTIGATION 2   THE CYCLES PROBLEM

In the display window of the local cycle store there are bicycles and tricycles. Altogether there are 9 cycles and 21 wheels which can be seen. How many bicycles and tricycles are in the display?

What to do:

1 Copy and complete the following table:

Number of bicycles	0	1	2	3	4	5	6	7	8	9
Number of wheels										
Number of tricycles	9	8								
Number of wheels										
Total wheels										

2 Use the table to find the solution to the problem.

3 Suppose there were x bicycles and y tricycles in the window.

   a By considering the number of cycles seen, explain why $x + y = 9$.

b By considering the total number of wheels seen, explain why $2x + 3y = 21$.

4 You should have found that there were six bicycles and three tricycles in the window.

   a Substitute $x = 6$ and $y = 3$ into $x + y = 9$.   What do you notice?

   b Substitute $x = 6$ and $y = 3$ into $2x + 3y = 21$.   What do you notice?

We say that $x = 6$, $y = 3$ is the only solution to the **simultaneous equations**
We found the solution to the problem by trial and error.
However, the solution can also be found by algebraic means.

$$\begin{cases} x + y = 9 \\ 2x + 3y = 21. \end{cases}$$

G | LINEAR SIMULTANEOUS EQUATIONS

Simultaneous equations are two equations containing two unknowns, e.g., $\begin{cases} x + y = 9 \\ 2x + 3y = 21. \end{cases}$

When we solve these problems we are trying to find the solution which is common to both equations.

Notice that if $x = 6$ and $y = 3$ then:

- $x + y = 6 + 3 = 9$ ✓   i.e.,   the equation is satisfied
- $2x + 3y = 2 \times 6 + 3 \times 3 = 12 + 9 = 21$ ✓   i.e.,   the equation is satisfied.

So, $x = 6$ and $y = 3$ is the **solution** to the simultaneous equations $\begin{cases} x + y = 9 \\ 2x + 3y = 21. \end{cases}$

The solutions to **linear simultaneous equations** can be found by **trial and error** (a little tedious) or **graphically** (which can be inaccurate if solutions are not integers).

However, because of the limitations of these methods, other methods are used.

ACTIVITY

The *first* candle is 20 cm long when new and burns at a rate of 2.5 mm per hour and the *second* is 24.5 cm long when new and burns at a rate of 4 mm per hour.

What to do:

1 Show that the first candle's height after t hours is given by $h_1 = 200 - 2.5t$ millimetres and for the second candle $h_2 = 245 - 4t$ mm   where t is the time in hours.

2 Use the equations in **1** to determine how long each candle will last.

3 On the same set of axes graph each equation.

4 At what time do the candles have the same height?

5 If you want the candles to 'go out' together which candle would you light first and how long after this would you light the second one?

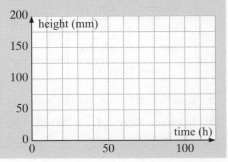

SOLUTION BY SUBSTITUTION

The method of **solution by substitution** is used when at least one equation is given with
either x or y as the **subject** of the formula.

Example 20

Solve simultaneously, by substitution: $y = 9 - x$
$$2x + 3y = 21$$

> Notice that
> $9 - x$ is substituted for y
> in the other equation.

$$y = 9 - x \quad ......(1)$$
$$2x + 3y = 21 \quad ......(2)$$

Since $y = 9 - x$,   then $2x + 3(9 - x) = 21$
$$\therefore \quad 2x + 27 - 3x = 21$$
$$\therefore \quad 27 - x = 21$$
$$\therefore \quad -x = 21 - 27$$
$$\therefore \quad -x = -6$$
$$\therefore \quad x = 6$$

and so, when $x = 6$, $y = 9 - 6$ {substituting $x = 6$ into (1)}
$$\therefore \quad y = 3.$$

Solution is: $x = 6$, $y = 3$.

Check: (1) $3 = 9 - 6$ ✓   (2) $2(6) + 3(3) = 12 + 9 = 21$ ✓

Example 21

Solve simultaneously, by substitution: $2y - x = 2$
$$x = 1 + 8y$$

$$2y - x = 2 \quad ...... (1)$$
$$x = 1 + 8y \quad ...... (2)$$

Substituting (2) into (1) gives
$$\therefore \quad 2y - (1 + 8y) = 2$$
$$\therefore \quad 2y - 1 - 8y = 2$$
$$\therefore \quad -6y - 1 = 2$$
$$\therefore \quad -6y = 3$$
$$\therefore \quad y = -\tfrac{1}{2}.$$

Substituting $y = -\tfrac{1}{2}$ into (2) gives
$$x = 1 + 8 \times -\tfrac{1}{2} = -3.$$

The solution is $x = -3$, $y = -\tfrac{1}{2}$.

Check: (1) $2(-\tfrac{1}{2}) - (-3) = -1 + 3 = 2$ ✓
(2) $1 + 8(-\tfrac{1}{2}) = 1 - 4 = -3$ ✓

EXERCISE 6G.1

1 Solve simultaneously, using substitution:

    a $x = 8 - 2y$
         $2x + 3y = 13$

    b $y = 4 + x$
         $5x - 3y = 0$

    c $x = -10 - 2y$
         $3y - 2x = -22$

    d $x = -1 + 2y$
         $x = 9 - 2y$

    e $3x - 2y = 8$
         $x = 3y + 12$

    f $x + 2y = 8$
         $y = 7 - 2x$

2 Use the substitution method to solve simultaneously:

    a $x = -1 - 2y$
         $2x - 3y = 12$

    b $y = 3 - 2x$
         $y = 3x + 1$

    c $x = 3y - 9$
         $5x + 2y = 23$

    d $y = 5x$
         $7x - 2y = 3$

    e $x = -2 - 3y$
         $3x - 2y = -17$

    f $3x - 5y = 26$
         $y = 4x - 12$

3 **a** Use the method of substitution to try to solve the equations $y = 3x + 1$ and $y = 3x + 4$.

    b What is the simultaneous solution for the equations in **a**?

4 **a** Use the method of substitution to try to solve the equations $y = 3x + 1$ and $2y = 6x + 2$.

    b How many simultaneous solutions do the equations in **a** have?

SOLUTION BY ELIMINATION

In many problems which require the simultaneous solution of linear equations, each equation will be of the form $ax + by = c$. Solution by substitution is often tedious in such situations and the method of elimination of one of the variables is preferred.

In the method of **elimination**, we eliminate (remove) one of the variables by making the coefficients of x (or y) the **same size** but **opposite in sign** and then **adding** the equations. This has the effect of **eliminating** one of the variables.

Example 22

Solve simultaneously, by elimination: $4x + 3y = 2$ .... (1)
                                    $x - 3y = 8$ .... (2)

We **sum** the LHS's and the RHS's to get an equation which contains x only.

$$\begin{array}{r} 4x + 3y = 2 \\ +\quad x - 3y = 8 \\ \hline 5x\qquad = 10 \end{array}$$ {on adding the equations}

$$\therefore\quad x = 2 \quad \text{\{dividing both sides by 5\}}$$

Let $x = 2$ in (1) $\therefore \quad 4 \times 2 + 3y = 2$

$$\therefore \quad 8 + 3y = 2$$

$$\therefore \quad 3y = 2 - 8 \quad \text{\{subtracting 8 from both sides\}}$$

$$\therefore \quad 3y = -6$$

$$\therefore \quad y = -2 \quad \text{\{dividing by 3 on both sides\}}$$

i.e., $x = 2$ and $y = -2$

Check: in (2): $(2) - 3(-2) = 2 + 6 = 8$ ✓

The method of elimination uses the fact that: If $a = b$ and $c = d$ then $a + c = b + d$.

EXERCISE 6G.2

1 What equation results when the following are added vertically?

a $5x + 3y = 12$
$x - 3y = -6$

b $2x + 5y = -4$
$-2x - 6y = 12$

c $4x - 6y = 9$
$x + 6y = -2$

d $12x + 15y = 33$
$-18x - 15y = -63$

e $5x + 6y = 12$
$-5x + 2y = -8$

f $-7x + y = -5$
$7x - 3y = -11$

2 Solve the following using the method of elimination:

a $2x + y = 3$
$3x - y = 7$

b $4x + 3y = 7$
$6x - 3y = -27$

c $2x + 5y = 16$
$-2x - 7y = -20$

d $3x + 5y = -11$
$-3x - 2y = 8$

e $4x - 7y = 41$
$3x + 7y = -6$

f $-4x + 3y = -25$
$4x - 5y = 31$

In problems where the coefficients of x (or y) are **not** the **same size** or **opposite in sign**, we may have to **multiply** each equation by a number to enable us to **eliminate** one variable.

Example 23	Solve simultaneously, by elimination: $3x + 2y = 7$

$$3x + 2y = 7 \quad ......(1)$$
$$2x - 5y = 11 \quad ......(2)$$

We can eliminate y by multiplying (1) by 5 and (2) by 2.

$$\therefore \quad 15x + 10y = 35$$
$$+ \quad 4x - 10y = 22$$

$$\therefore \quad 19x \qquad = 57 \qquad \text{\{on adding the equations\}}$$
$$\therefore \quad x = 3 \qquad \text{\{dividing both sides by 19\}}$$

Substituting $x = 3$ into equation (1) gives

$$3(3) + 2y = 7$$
$$\therefore \quad 9 + 2y = 7$$
$$\therefore \quad 2y = -2$$
$$\therefore \quad y = -1$$

So, the solution is: $x = 3$, $y = -1$.   *Check:* $3(3) + 2(-1) = 9 - 2 = 7$ ✓
$2(3) - 5(-1) = 6 + 5 = 11$ ✓

EXERCISE 6G.3

1 Give the equation that results when both sides of the equation:

a $3x + 4y = 2$ is multiplied by 3

b $x - 4y = 7$ is multiplied by -2

c $5x - y = -3$ is multiplied by 5

d $7x + 3y = -4$ is multiplied by -3

e $-2x - 5y = 1$ is multiplied by -4

f $3x - y = -1$ is multiplied by -1

Example 24

Solve by elimination: $3x + 4y = 14$
$4x + 5y = 17$

$$3x + 4y = 14 \quad ....(1)$$
$$4x + 5y = 17 \quad ....(2)$$

To eliminate x, multiply both sides of

(1) by 4: $12x + 16y = 56 \quad ....(3)$
(2) by -3: $\underline{-12x - 15y = -51 \quad ......(4)}$

$\qquad\qquad\qquad y = 5 \quad$ {on adding (3) and (4)}

and substituting $y = 5$ into (2) gives

$4x + 5(5) = 17$
$\therefore \quad 4x + 25 = 17$
$\therefore \quad 4x = -8$
$\therefore \quad x = -2$

Check:
(1) $3(-2) + 4(5) = (-6) + 20 = 14 \quad \checkmark$
(2) $4(-2) + 5(5) = (-8) + 25 = 17 \quad \checkmark$

Thus $x = -2$ and $y = 5$.

WHAT TO ELIMINATE

There is always a choice whether to eliminate x or y, so our choice depends on which variable is easier to eliminate.

Solve the problem in **Example 24** by multiplying (1) by 5 and (2) by -4. This eliminates y rather than x. The final solutions should be the same.

2 Solve the following using the method of elimination:

a $4x - 3y = 6$
$-2x + 5y = 4$

b $2x - y = 9$
$x + 4y = 36$

c $3x + 4y = 6$
$x - 3y = -11$

d $2x + 3y = 7$
$3x - 2y = 4$

e $4x - 3y = 6$
$6x + 7y = 32$

f $7x - 3y = 29$
$3x + 4y + 14 = 0$

g $2x + 5y = 20$
$3x + 2y = 19$

h $3x - 2y = 10$
$4x + 3y = 19$

i $3x + 4y + 11 = 0$
$5x + 6y + 7 = 0$

USING TECHNOLOGY

Graphics calculators can be used in a variety of ways to solve a system of linear equations. While we consider a graphical method in the Coordinate Geometry chapter (see the **Investigation**: "Finding where lines meet using technology" on page **228**), a non-graphical method also exists on most graphics calculators.

Click on the icon for specific instructions.

Investigate how your calculator can be used to solve the following system:

$$4x + 3y = 10 \qquad \text{(Answer:} \quad x = 1, \ y = 2)$$
$$x - 2y = -3$$

For casio users, the simultaneous equation solver for two unknowns requires the coefficients of each equation to be entered as shown.

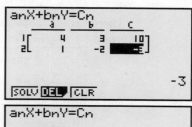

For TI-83 users, a 2×3 matrix is reduced to simplest form and the solutions derived from it.

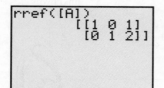

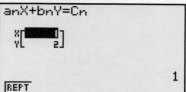

Note:
The top row says
$1x + 0y = 1$
i.e., $x = 1$

EXERCISE 6G.4

1 Use technology to find the point of intersection of:

a $y = x + 4$
$5x - 3y = 0$

b $x + 2y = 8$
$y = 7 - 2x$

c $x - y = 5$
$2x + 3y = 4$

d $2x + y = 7$
$3x - 2y = 1$

e $y = 3x - 1$
$x - y = 6$

f $y = -\dfrac{2x}{3} + 2$
$2x + y = 6$

H   PROBLEM SOLVING

Many problems can be described mathematically by a **pair of linear equations**, i.e., two equations of the form $ax + by = c$, where x and y are the two variables (unknowns).

We have already seen an example of this in **Investigation 2** on **page 181**.

Once the equations are formed, they can then be solved simultaneously and the original problem can be solved. The following method is recommended:

Step 1:   Decide on the two unknowns; call them x and y, say. Do not forget the units.

Step 2:   Write down **two** equations connecting x and y.

Step 3:   Solve the equations simultaneously (either algebraically or with technology).

Step 4:   Check your solutions with the original data given.

Step 5:   Give your answer in sentence form.

(**Note:**   The form of the original equations will help you decide whether to use the substitution method, or the elimination method. Of course you could use a calculator.)

Example 25

Two numbers have a sum of 45 and a difference of 13. Find the numbers.

Let x and y be the unknown numbers, where $x > y$.

Then	$x + y = 45$ ......(1)	{'sum' means add}
and	$x - y = 13$ ......(2)	{'difference' means subtract}
	$2x \quad = 58$	{adding (1) and (2)}
	$\therefore \quad x = 29$	{dividing both sides by 2}

and substituting into (1),   *Check:*

$$29 + y = 45$$ (1) $29 + 16 = 45$ ✓

$$\therefore \quad y = 16$$ (2) $29 - 16 = 13$ ✓

The numbers are   29   and   16.

> When solving problems with simultaneous equations we must find two equations containing two unknowns.

EXERCISE 6H

1   The sum of two numbers is 47 and their difference is 14. Find the numbers.

2   Find two numbers with sum 28 and half their difference 2.

3   The larger of two numbers is four times the smaller and their sum is 85. Find the two numbers.

Example 26

5 oranges and 14 bananas cost me \$1.30, and 8 oranges and 9 bananas cost \$1.41. Find the cost of each orange and each banana.

Let each orange cost x cents and $\therefore \quad 5x + 14y = 130$ .... (1)
each banana cost y cents. $8x + 9y = 141$ .... (2)

{**Note:**   Units must be the same on both sides of each equation   i.e., cents}

From technology, $x = 12$, $y = 5$.

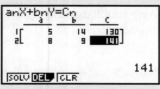

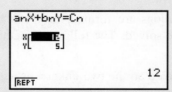

Thus oranges cost 12 cents, bananas cost 5 cents each.

4   Five pencils and 6 biros cost a total of \$4.64, whereas 7 pencils and 3 biros cost a total of \$3.58. Find the cost of each item.

5   Seven toffees and three chocolates cost a total of \$1.68, whereas four toffees and five chocolates cost a total of \$1.65. Find the cost of each of the sweets.

Example 27

In my pocket I have only 5-cent and 10-cent coins. How many of each type of coin do I have if I have 24 coins altogether and their total value is $1.55?

Let x be the number of 5-cent coins and y be the number of 10-cent coins.

$\therefore$ $x + y = 24$ ......(1)   {the total number of coins}

and $5x + 10y = 155$ ......(2)   {the total value of coins}

From technology, $x = 17$ and $y = 7$

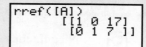

Check: $17 + 7 = 24$ ✓

$5 \times 17 + 10 \times 7 = 85 + 70 = 155.$ ✓

Thus   I have 17 five cent coins and 7 ten cent coins.

6 I collect only 50-cent and $1 coins. My collection consists of 43 coins and their total value is $35. How many of each coin type do I have?

7 Amy and Michelle have $29.40 between them and Amy's money is three quarters of Michelle's. How much money does each have?

8 Margarine is sold in either 250 g or 400 g packs. A supermarket manager ordered 19.6 kg of margarine and received 58 packs. How many of each type did the manager receive?

9 Given that the triangle alongside is equilateral, find a and b.

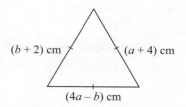

$(b+2)$ cm $(a+4)$ cm

$(4a - b)$ cm

10 A rectangle has perimeter 32 cm. If 3 cm is taken from the length and added to the width, the rectangle becomes a square. Find the dimensions of the original rectangle.

I   INDEX NOTATION (REVIEW)

Rather than write $2 \times 2 \times 2 \times 2 \times 2$,   we write such a product as 2^5.

2^5 reads "two to the power of five" or "two with index five".

Thus $5^3 = 5 \times 5 \times 5$ and $3^6 = 3 \times 3 \times 3 \times 3 \times 3 \times 3$.

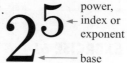

power, index or exponent

base

If n is a positive integer, then a^n is the product of n factors of a

i.e., $a^n = \underbrace{a \times a \times a \times a \times .... \times a}_{n \text{ factors}}$

EXERCISE 6I

1 Copy and complete the values of these powers. Try to become familiar with them.

a $2^1 = ...,$ $2^2 = ...,$ $2^3 = ...,$ $2^4 = ...,$ $2^5 = ...,$ $2^6 = ...$

b $3^1 = ...,$ $3^2 = ...,$ $3^3 = ...,$ $3^4 = ...$

c $5^1 = ...,$ $5^2 = ...,$ $5^3 = ...,$ $5^4 = ...$

d $7^1 = ...,$ $7^2 = ...,$ $7^3 = ...$

HISTORICAL NOTE

Nicomachus who lived around 100 AD discovered an interesting number pattern involving cubes and sums of odd numbers.

$$1 = 1^3$$
$$3 + 5 = 8 = 2^3$$
$$7 + 9 + 11 = 27 = 3^3 \quad \text{etc.}$$

J   NEGATIVE BASES

So far we have only considered **positive** bases raised to a power.

We will now briefly look at **negative** bases. Consider the statements below:

$(-1)^1 = -1$ $(-2)^1 = -2$

$(-1)^2 = -1 \times -1 = 1$ $(-2)^2 = -2 \times -2 = 4$

$(-1)^3 = -1 \times -1 \times -1 = -1$ $(-2)^3 = -2 \times -2 \times -2 = -8$

$(-1)^4 = -1 \times -1 \times -1 \times -1 = 1$ $(-2)^4 = -2 \times -2 \times -2 \times -2 = 16$

Note: A **negative** base raised to an **odd** power is **negative**; whereas a **negative** base raised to an **even** power is **positive**.

> Notice the effect of the brackets in these examples.

Example 28

Evaluate: **a** $(-2)^4$ **b** -2^4 **c** $(-2)^5$ **d** $-(-2)^5$

a $(-2)^4$	**b** -2^4	**c** $(-2)^5$	**d** $-(-2)^5$
$= 16$	$= -1 \times 2^4$	$= -32$	$= -1 \times (-2)^5$
	$= -16$		$= -1 \times -32$
			$= 32$

EXERCISE 6J

1 Simplify:

a $(-1)^3$ **b** $(-1)^4$ **c** $(-1)^{12}$ **d** $(-1)^{17}$

e $(-1)^6$ **f** -1^6 **g** $-(-1)^6$ **h** $(-2)^3$

i -2^3 **j** $-(-2)^3$ **k** $-(-5)^2$ **l** $-(-5)^3$

CALCULATOR USE

Although different calculators vary in the appearance of keys, they
all perform operations of raising to powers in a similar manner.

Power keys x^2 squares the number in the display.

$\boxed{\wedge}$ 3   raises the number in the display to the power 3.

$\boxed{\wedge}$ 5   raises the number in the display to the power 5.

$\boxed{\wedge}$ $\boxed{(-)}$ 4   raises the number in the display to the power -4.

Example 29

Find, using your calculator:

a 6^5 **b** $(-5)^4$ **c** -7^4

		Answer
a	Press: 6 $\boxed{\wedge}$ 5 $\boxed{\textbf{ENTER}}$	7776
b	Press: $\boxed{(}$ $\boxed{(-)}$ 5 $\boxed{)}$ $\boxed{\wedge}$ 4 $\boxed{\textbf{ENTER}}$	625
c	Press: $\boxed{(-)}$ 7 $\boxed{\wedge}$ 4 $\boxed{\textbf{ENTER}}$	-2401

Note: You will need to check if your calculator uses the same key sequence as in the
examples. If not, work out the sequence which gives you the correct answers.

2 Use your calculator to find the value of the following, recording the entire display:

a 2^9 **b** $(-5)^5$ **c** -3^5 **d** 7^5 **e** 8^3
f $(-9)^4$ **g** -9^4 **h** 1.16^{11} **i** -0.981^{14} **j** $(-1.14)^{23}$

Example 30

Find using your
calculator, and
comment on:

a 5^{-2} **b** $\dfrac{1}{5^2}$

		Answer
a	Press: 5 $\boxed{\wedge}$ $\boxed{(-)}$ 2 $\boxed{\textbf{ENTER}}$	0.04
b	Press: 1 $\boxed{\div}$ 5 $\boxed{\wedge}$ 2 $\boxed{\textbf{ENTER}}$	0.04

The answers indicate that $5^{-2} = \dfrac{1}{5^2}$.

---

3 Use your calculator to find the values of the following:

a 7^{-1} **b** $\dfrac{1}{7^1}$ **c** 3^{-2} **d** $\dfrac{1}{3^2}$

e 4^{-3} **f** $\dfrac{1}{4^3}$ **g** 13^0 **h** 172^0

What do you notice?

4 By considering 3^1, 3^2, 3^3, 3^4, 3^5 .... and looking for a pattern, find the last digit of 3^{33}.

5 What is the last digit of 7^{77}?

K   INDEX LAWS

Recall the following **index laws** where the bases a and b are both positive and the indices m and n are integers.

$a^m \times a^n = a^{m+n}$	To **multiply** numbers with the **same base**, keep the base and **add** the indices.
$\dfrac{a^m}{a^n} = a^{m-n}$	To **divide** numbers with the same base, keep the base and **subtract** the indices.
$(a^m)^n = a^{m \times n}$	When **raising** a **power** to a **power**, keep the base and **multiply** the indices.
$(ab)^n = a^n b^n$	The power of a product is the product of the powers.
$\left(\dfrac{a}{b}\right)^n = \dfrac{a^n}{b^n}$	The power of a quotient is the quotient of the powers.
$a^o = 1, \quad a \neq 0$	Any non-zero number raised to the power of zero is **1**.
$a^{-n} = \dfrac{1}{a^n}$ and	$\dfrac{1}{a^{-n}} = a^n$ and in particular $a^{-1} = \dfrac{1}{a}$.

Example 31

Simplify using $a^m \times a^n = a^{m+n}$:

a $11^5 \times 11^3$ **b** $a^4 \times a^5$ **c** $x^4 \times x^a$

a $\quad 11^5 \times 11^3$	**b** $\quad a^4 \times a^5$	**c** $\quad x^4 \times x^a$
$= 11^{5+3}$	$= a^{4+5}$	$= x^{4+a}$
$= 11^8$	$= a^9$	$(= x^{a+4})$

EXERCISE 6K

1 Simplify using $a^m \times a^n = a^{m+n}$:

   a $\quad 7^3 \times 7^2$ **b** $\quad 5^4 \times 5^3$ **c** $\quad a^7 \times a^2$ **d** $\quad a^4 \times a$

   e $\quad b^8 \times b^5$ **f** $\quad a^3 \times a^n$ **g** $\quad b^7 \times b^m$ **h** $\quad m^4 \times m^2 \times m^3$

Example 32

Simplify using $\dfrac{a^m}{a^n} = a^{m-n}$:   **a** $\dfrac{7^8}{7^5}$ **b** $\dfrac{b^6}{b^m}$

a $\dfrac{7^8}{7^5}$

$= 7^{8-5}$

$= 7^3$

b $= \dfrac{b^6}{b^m}$

$= b^{6-m}$

2 Simplify using $\dfrac{a^m}{a^n} = a^{m-n}$:

a $\dfrac{5^9}{5^2}$ **b** $\dfrac{11^{13}}{11^9}$ **c** $7^7 \div 7^4$ **d** $\dfrac{a^6}{a^2}$

e $\dfrac{b^{10}}{b^7}$ **f** $\dfrac{p^5}{p^m}$ **g** $\dfrac{y^a}{y^5}$ **h** $b^{2x} \div b$

Example 33

Simplify using $(a^m)^n = a^{m \times n}$:
a $(2^4)^3$ **b** $(x^3)^5$ **c** $(b^7)^m$

a $(2^4)^3$

$= 2^{4 \times 3}$

$= 2^{12}$

b $(x^3)^5$

$= x^{3 \times 5}$

$= x^{15}$

c $(b^7)^m$

$= b^{7 \times m}$

$= b^{7m}$

3 Simplify using $(a^m)^n = a^{m \times n}$:

a $(3^2)^4$ **b** $(5^3)^5$ **c** $(2^4)^7$ **d** $(a^5)^2$
e $(p^4)^5$ **f** $(b^5)^n$ **g** $(x^y)^3$ **h** $(a^{2x})^5$

Example 34

Express in simplest form with a prime number base:

a 9^4 **b** 4×2^p **c** $\dfrac{3^x}{9^y}$ **d** 25^{x-1}

a 9^4

$= (3^2)^4$

$= 3^{2 \times 4}$

$= 3^8$

b 4×2^p

$= 2^2 \times 2^p$

$= 2^{2+p}$

c $\dfrac{3^x}{9^y}$

$= \dfrac{3^x}{(3^2)^y}$

$= \dfrac{3^x}{3^{2y}}$

$= 3^{x-2y}$

d 25^{x-1}

$= (5^2)^{x-1}$

$= 5^{2(x-1)}$

$= 5^{2x-2}$

4 Express in simplest form with a prime number base:

a 8	**b** 25	**c** 27	**d** 4^3
e 9^2	**f** $3^a \times 9$	**g** $5^t \div 5$	**h** $3^n \times 9^n$
i $\dfrac{16}{2^x}$	**j** $\dfrac{3^{x+1}}{3^{x-1}}$	**k** $(5^4)^{x-1}$	**l** $2^x \times 2^{2-x}$
m $\dfrac{2^y}{4^x}$	**n** $\dfrac{4^y}{8^x}$	**o** $\dfrac{3^{x+1}}{3^{1-x}}$	**p** $\dfrac{2^t \times 4^t}{8^{t-1}}$

Example 35

Remove the brackets of: **a** $(2x)^3$ **b** $\left(\dfrac{3c}{b}\right)^4$

a $(2x)^3$
$= 2^3 \times x^3$
$= 8x^3$

b $\left(\dfrac{3c}{b}\right)^4$
$= \dfrac{3^4 \times c^4}{b^4}$
$= \dfrac{81c^4}{b^4}$

Remember that each factor within the brackets has to be raised to the power outside them.

---

5 Remove the brackets of:

a $(ab)^3$	**b** $(ac)^4$	**c** $(bc)^5$	**d** $(abc)^3$
e $(2a)^4$	**f** $(5b)^2$	**g** $(3n)^4$	**h** $(2bc)^3$
i $(4ab)^3$	**j** $\left(\dfrac{a}{b}\right)^3$	**k** $\left(\dfrac{m}{n}\right)^4$	**l** $\left(\dfrac{2c}{d}\right)^5$

Example 36

Express the following in simplest form, without brackets:

a $(3a^3b)^4$ **b** $\left(\dfrac{x^2}{2y}\right)^3$

a $(3a^3b)^4$
$= 3^4 \times (a^3)^4 \times b^4$
$= 81 \times a^{3\times4} \times b^4$
$= 81a^{12}b^4$

b $\left(\dfrac{x^2}{2y}\right)^3$
$= \dfrac{(x^2)^3}{2^3 \times y^3}$
$= \dfrac{x^6}{8y^3}$

6 Express the following in simplest form, without brackets:

a $(2b^4)^3$

b $\left(\dfrac{3}{x^2 y}\right)^2$

c $(5a^4 b)^2$

d $\left(\dfrac{m^3}{2n^2}\right)^4$

e $\left(\dfrac{3a^3}{b^5}\right)^3$

f $(2m^3 n^2)^5$

g $\left(\dfrac{4a^4}{b^2}\right)^2$

h $(5x^2 y^3)^3$

Example 37

Simplify using the index laws:

a $3x^2 \times 5x^5$

b $\dfrac{20a^9}{4a^6}$

c $\dfrac{b^3 \times b^7}{(b^2)^4}$

a $3x^2 \times 5x^5$
    $= 3 \times 5 \times x^2 \times x^5$
    $= 15 \times x^{2+5}$
    $= 15x^7$

b $\dfrac{20a^9}{4a^6}$
    $= \frac{20}{4} \times a^{9-6}$
    $= 5a^3$

c $\dfrac{b^3 \times b^7}{(b^2)^4}$
    $= \dfrac{b^{10}}{b^8}$
    $= b^{10-8}$
    $= b^2$

---

7 Simplify the following expressions using one or more of the index laws:

a $\dfrac{a^3}{a}$

b $4b^2 \times 2b^3$

c $\dfrac{m^5 n^4}{m^2 n^3}$

d $\dfrac{14a^7}{2a^2}$

e $\dfrac{12a^2 b^3}{3ab}$

f $\dfrac{18m^7 a^3}{4m^4 a^3}$

g $10hk^3 \times 4h^4$

h $\dfrac{m^{11}}{(m^2)^8}$

i $\dfrac{p^2 \times p^7}{(p^3)^2}$

> Notice that
> $\left(\dfrac{a}{b}\right)^{-2} = \left(\dfrac{b}{a}\right)^2$

Example 38

Simplify, giving answers in simplest rational form:

a 7^0 **b** 3^{-2} **c** $3^0 - 3^{-1}$ **d** $\left(\frac{5}{3}\right)^{-2}$

a 7^0
    $= 1$

b 3^{-2}
    $= \dfrac{1}{3^2}$
    $= \frac{1}{9}$

c $3^0 - 3^{-1}$
    $= 1 - \frac{1}{3}$
    $= \frac{2}{3}$

d $\left(\frac{5}{3}\right)^{-2}$
    $= \left(\frac{3}{5}\right)^2$
    $= \frac{9}{25}$

---

8 Simplify, giving answers in simplest rational form:

a 5^0

b 3^{-1}

c 6^{-1}

d 8^0

e 2^2

f 2^{-2}

g 2^3

h 2^{-3}

i 5^2

j 5^{-2}

k 10^2

l 10^{-2}

9   Simplify, giving answers in simplest rational form:

a $\left(\frac{2}{3}\right)^0$ **b** $\dfrac{4^3}{4^3}$ **c** $3y^0$ **d** $(3y)^0$

e 2×3^0 **f** 6^0 **g** $\dfrac{5^2}{5^4}$ **h** $\dfrac{2^{10}}{2^{15}}$

i $\left(\frac{1}{3}\right)^{-1}$ **j** $\left(\frac{2}{5}\right)^{-1}$ **k** $\left(\frac{4}{3}\right)^{-1}$ **l** $\left(\frac{1}{12}\right)^{-1}$

m $\left(\frac{2}{3}\right)^{-2}$ **n** $5^0 - 5^{-1}$ **o** $7^{-1} + 7^0$ **p** $2^0 + 2^1 + 2^{-1}$

Example 39

Write the following without brackets or negative indices:

a $(5x)^{-1}$ **b** $5x^{-1}$ **c** $(3b^2)^{-2}$

In $5x^{-1}$ the index -1 refers to the x only.

a $(5x)^{-1}$ **b** $5x^{-1}$ **c** $(3b^2)^{-2}$

$= \dfrac{1}{5x}$ $= \dfrac{5}{x}$ $= \dfrac{1}{(3b^2)^2}$

$= \dfrac{1}{3^2 b^4}$

$= \dfrac{1}{9b^4}$

---

10   Write the following without brackets or negative indices:

a $(2a)^{-1}$ **b** $2a^{-1}$ **c** $3b^{-1}$ **d** $(3b)^{-1}$

e $\left(\frac{2}{b}\right)^{-2}$ **f** $(2b)^{-2}$ **g** $(3n)^{-2}$ **h** $(3n^{-2})^{-1}$

i ab^{-1} **j** $(ab)^{-1}$ **k** ab^{-2} **l** $(ab)^{-2}$

Example 40

Write the following as powers of 2, 3 and/or 5:

a $\dfrac{1}{8}$ **b** $\dfrac{1}{9^n}$ **c** $\dfrac{25}{5^4}$

a $\dfrac{1}{8}$ **b** $\dfrac{1}{9^n}$ **c** $\dfrac{25}{5^4}$

$= \dfrac{1}{2^3}$ $= \dfrac{1}{(3^2)^n}$ $= \dfrac{5^2}{5^4}$

$= 2^{-3}$ $= \dfrac{1}{3^{2n}}$ $= 5^{2-4}$

$= 3^{-2n}$ $= 5^{-2}$

11 Write the following as powers of 2, 3 and/or 5:

a $\dfrac{1}{3}$ b $\dfrac{1}{2}$ c $\dfrac{1}{5}$ d $\dfrac{1}{4}$

e $\dfrac{1}{27}$ f $\dfrac{1}{25}$ g $\dfrac{1}{8^x}$ h $\dfrac{1}{16^y}$

i $\dfrac{1}{81^a}$ j $\dfrac{9}{3^4}$ k 25×5^{-4} l $\dfrac{5^{-1}}{5^2}$

12 The water lily *Growerosa Veryfasterosa* doubles its size every day. From the time it was planted until it completely covered the pond took 26 days.

How many days did it take to cover half the pond?

13 Suppose you have the following six coins in your pocket: 5 cents, 10 cents, 20 cents, 50 cents, \$1, \$2.   How many different sums of money can you make?

(**Hint:**   Simplify the problem to a smaller number of coins and look for a pattern.)

14 Read about Nicomachus' pattern on page **190** and find the sequence of odd numbers for:

a 5^3 b 7^3 c 12^3

Example 41

Write in non-fractional form: a $\dfrac{x^2 + 3x + 2}{x}$ b $\dfrac{x^3 + 5x - 3}{x^2}$

a $\dfrac{x^2 + 3x + 2}{x}$

$= \dfrac{x^2}{x} + \dfrac{3x}{x} + \dfrac{2}{x}$

$= x + 3 + 2x^{-1}$

b $\dfrac{x^3 + 5x - 3}{x^2}$

$= \dfrac{x^3}{x^2} + \dfrac{5x}{x^2} - \dfrac{3}{x^2}$

$= x - 5x^{-1} - 3x^{-2}$

---

15 Write in non-fractional form:

a $\dfrac{x + 3}{x}$ b $\dfrac{5 - x}{x^2}$ c $\dfrac{x + 2}{x^3}$

d $\dfrac{x^2 + 5}{x}$ e $\dfrac{x^2 + x - 2}{x}$ f $\dfrac{x^3 - 3x + 5}{x^2}$

g $\dfrac{5 - x - x^2}{x}$ h $\dfrac{16 - 3x + x^3}{x^2}$ i $\dfrac{5x^4 - 3x^2 + x + 6}{x^2}$

L   EXPONENTIAL EQUATIONS

An **exponential equation** is an equation in which the unknown occurs as part of the index or exponent.

For example: $2^x = 8$ and $30 \times 3^x = 7$ are both exponential equations.

If $2^x = 8$, then $2^x = 2^3$.   Thus $x = 3$, and this is the only solution.

Hence:

> If $a^x = a^k$, then $x = k$,
>
> i.e., if the base numbers are the same, we can **equate indices**.

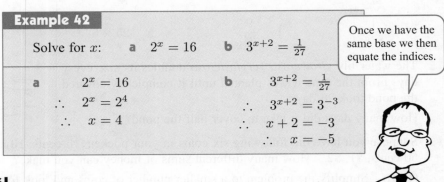

Example 42

Solve for x: **a** $2^x = 16$ **b** $3^{x+2} = \frac{1}{27}$

a $\qquad 2^x = 16$

$\therefore \quad 2^x = 2^4$

$\therefore \quad x = 4$

b $\qquad 3^{x+2} = \frac{1}{27}$

$\therefore \quad 3^{x+2} = 3^{-3}$

$\therefore \quad x + 2 = -3$

$\therefore \quad x = -5$

> Once we have the same base we then equate the indices.

EXERCISE 6L

1 Solve for x:

  a $2^x = 2$ **b** $2^x = 4$ **c** $3^x = 27$ **d** $2^x = 1$

  e $2^x = \frac{1}{2}$ **f** $3^x = \frac{1}{3}$ **g** $2^x = \frac{1}{8}$ **h** $2^{x+1} = 8$

  i $2^{x-2} = \frac{1}{4}$ **j** $3^{x+1} = \frac{1}{27}$ **k** $2^{x+1} = 64$ **l** $2^{1-2x} = \frac{1}{2}$

> Remember to use the index laws correctly!

Example 43

Solve for x: **a** $4^x = 8$ **b** $9^{x-2} = \frac{1}{3}$

a $\qquad 4^x = 8$

$\therefore \quad (2^2)^x = 2^3$

$\therefore \quad 2^{2x} = 2^3$

$\therefore \quad 2x = 3$

$\therefore \quad x = \frac{3}{2}$

b $\qquad 9^{x-2} = \frac{1}{3}$

$\therefore \quad (3^2)^{x-2} = 3^{-1}$

$\therefore \quad 3^{2(x-2)} = 3^{-1}$

$\therefore \quad 2x - 4 = -1$

$\therefore \quad 2x = 3$

$\therefore \quad x = \frac{3}{2}$

2 Solve for x:

  a $4^x = 32$ **b** $8^x = \frac{1}{4}$ **c** $9^x = \frac{1}{3}$ **d** $49^x = \frac{1}{7}$

  e $4^x = \frac{1}{8}$ **f** $25^x = \frac{1}{5}$ **g** $8^{x+2} = 32$ **h** $8^{1-x} = \frac{1}{4}$

  i $4^{2x-1} = \frac{1}{2}$ **j** $9^{x-3} = 3$ **k** $(\frac{1}{2})^{x+1} = 2$ **l** $(\frac{1}{3})^{x+2} = 9$

  m $4^x = 8^{-x}$ **n** $(\frac{1}{4})^{1-x} = 8$ **o** $(\frac{1}{7})^x = 49$ **p** $(\frac{1}{2})^{x+1} = 32$

SOLVING EXPONENTIAL EQUATIONS WITH TECHNOLOGY

Often we cannot write both sides of an exponential equation with the same base. In these cases a **graphics calculator solver** solution can be used.

Limitations of the Solver function

In order to use the solver mode effectively on a calculator, an approximate value of the solution must be known. If the quess is too far from the actual solution, be prepared to retry your guess.

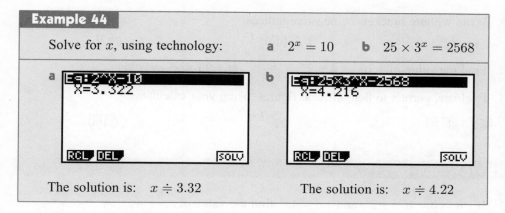

Example 44

Solve for x, using technology: **a** $2^x = 10$ **b** $25 \times 3^x = 2568$

a

Eq:2^X=10
X=3.322

RCL DEL   SOLV

The solution is: $x \doteqdot 3.32$

b

Eq:25×3^X=2568
X=4.216

RCL DEL   SOLV

The solution is: $x \doteqdot 4.22$

3 Solve the following exponential equations using technology:

   a $2^x = 20$ **b** $2^x = 100$ **c** $2^x = 5000$

   d $2^x = 80\,000$ **e** $3^x = 30$ **f** $5^x = 0.567$

   g $12^x = 23\,000$ **h** $(1.04)^x = 4.238$ **i** $3 \times 2^x = 93$

   j $5 \times 2^x = 420$ **k** $8 \times 3^x = 120$ **l** $21 \times (1.05)^x = 34$

REVIEW SET 6A

1 If $p = 5$, $q = -2$ and $r = -3$,   find the value of $\dfrac{p - 2q}{r}$.

2 Solve for x: **a** $5 - 2x = 3x + 4$ **b** $\dfrac{x - 1}{2} = \dfrac{2 - 3x}{7}$

3 Make x the subject of the formula $3x - 7y = 25$.

4 Solve the following simultaneous equations: **a** $y = 2x - 5$ **b** $3x + 5y = 1$
                                                    $3x - 2y = 11$ $4x - 3y = 11$

5 The period of a pendulum (the time for one complete swing) is approximately given by
$T = \frac{1}{5}\sqrt{l}$ seconds where l cm is the length of the pendulum.
Find the period if the pendulum has length 74 cm.

6 Solve the following problems:

   a When a number is decreased by 7 and this result is divided by the sum of the number and 11, the answer is $-\frac{3}{4}$. Find the number.

b Flour is sold in 5 kg and 2 kg packets. The 5 kg packets cost $2.75 and the 2 kg packets cost $1.25 each. If I bought 67 kg of flour and the total cost was $38.50, how many of each kind of packet did I buy?

7 Simplify:

    a $-(-1)^{10}$ **b** $-(-3)^3$ **c** $3^0 - 3^{-1}$

8 Simplify using the index laws:

    a $a^4b^5 \times a^2b^2$ **b** $6xy^5 \div 9x^2y^5$ **c** $\dfrac{5(x^2y)^2}{(5x^2)^2}$

9 Write without brackets or negative indices:

    a b^{-3} **b** $(ab)^{-1}$ **c** ab^{-1}

10 Find the value of x if: **a** $2^x = \frac{1}{32}$ **b** $25 \times 2^x = 7.234$

11 Evaluate, correct to 3 significant figures, using your calculator:

    a $3^{\frac{3}{4}}$ **b** $27^{-\frac{1}{5}}$ **c** $\sqrt[4]{100}$

REVIEW SET 6B

1 If $a = 3$, $b = -2$ and $c = 4$, find the value of $\dfrac{b-c}{2a}$.

2 Solve for x: **a** $2(x-3) + 4 = 5$ **b** $\dfrac{2x+3}{3} = 2$

3 Make y the subject of the formula: **a** $3x - 8y = 5$ **b** $a = \dfrac{4}{y}$

4 Solve the following simultaneous equations: **a** $3x - 2y = 16$ **b** $3x - 5y = 11$
                                                  $y = 2x - 10$ $4x + 3y = 5$

5 The volume of a cylinder is given by the formula $V = \pi r^2 h$, where r is its radius and h is its height. Find the volume, in cm^3, of a cylinder of height 23 cm and base radius 10 cm.

6 If a number is increased by 5 and then trebled, the result is six more than two thirds of the number. Find the number.

7 A bus company uses two different sized buses. If the company uses 7 small buses and 5 large buses to transport 331 people, but needs 4 small buses and 9 large buses to carry 398 people, determine the number of people each bus can carry.

8 Simplify: **a** $-(-2)^3$ **b** $5^{-1} - 5^0$

9 Simplify using the index laws:

    a $(a^7)^3$ **b** $pq^2 \times p^3q^4$ **c** $\dfrac{8ab^5}{2a^4b^4}$

10 Write without brackets or negative indices:

    a $x^{-2} \times x^{-3}$ **b** $2(ab)^{-2}$ **c** $2ab^{-2}$

11 Solve for x if:

    a $2^{x+1} = 32$ **b** $18 \times 4^x = 0.0317$

12 Use your calculator to evaluate, correct to 3 significant figures:

    a $4^{\frac{2}{3}}$ **b** $20^{-\frac{1}{2}}$ **c** $\sqrt[3]{30}$

REVIEW SET 6C

1 If $x = -3$, $y = 2$ and $z = -4$,   find the value of $5y - xz$.

2 Solve for x: **a** $4.03x + 3.1 = 11.54$ **b** $\dfrac{2 - 3x}{5} = -1$

3 Make y the subject of the formula: **a** $4x - 3y = 10$ **b** $m = \dfrac{c}{y}$.

4 Solve using the method of 'substitution': $y = 11 - 3x$
                                                           $4x + 3y = -7$

5 What number must be added to the numerator and denominator of $\frac{3}{4}$ in order to finish with a fraction equal to $\frac{1}{3}$?

6 Orange juice can be purchased in 2 L cartons or in 600 mL bottles. The 2 L cartons cost \$1.50 each and the 600 mL bottles cost \$0.60 each. A consumer purchases 73 L of orange juice and his total cost was \$57. How many of each container did the consumer buy?

7   **a** Write 4×2^n as a power of 2.

    b Evaluate $7^{-1} - 7^0$.

    c Write $(\frac{2}{3})^{-3}$ in simplest fractional form.

    d Simplify $\left(\dfrac{2a^{-1}}{b^2}\right)^2$.   Do not have negative indices or brackets in your answer.

    e Simplify $\dfrac{2^{x+1}}{2^{1-x}}$.

8 Write as powers of 5 in simplest form:

    a 1 **b** $5\sqrt{5}$ **c** $\dfrac{1}{\sqrt[4]{5}}$ **d** 25^{a+3}

9 Simplify:

    a $-(-2)^2$ **b** $(-\frac{1}{2}a^{-3})^2$ **c** $(-3b^{-1})^{-3}$

10 Evaluate, correct to 3 significant figures, using your calculator:

    a $10^{\frac{3}{4}}$ **b** $24^{-\frac{1}{3}}$ **c** $\sqrt[5]{40}$

11 Find x if:

    a $3^x = \frac{1}{27}$ **b** $(1.038)^x = 2.068$

Chapter 7

Coordinate geometry

Contents:

THE NUMBER PLANE

The position of any point in the **number plane** can be specified in terms of an **ordered pair** of numbers (x, y), where:

x is the **horizontal step** from a fixed point O, and y is the **vertical step** from O.

Once an **origin** O, has been given, two perpendicular axes are drawn. The x-**axis** is horizontal and the y-**axis** is vertical.

The **number plane** is also known as either:

- the **2-dimensional plane**, or
- the **Cartesian plane**, (named after **René Descartes**).

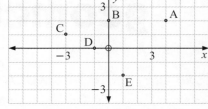

Note: (a, b) is called an **ordered pair**, where a and b are often referred to as **the coordinates** of the point.

   a is called the x-**coordinate** and

   b is called the y-**coordinate**.

Examples: The coordinates of the given points are

   A(4, 2)

   B(0, 2)

   C(−3, 1)

   D(−1, 0)

   E(1, −2).

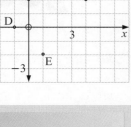

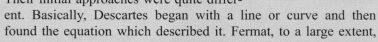

HISTORICAL NOTE

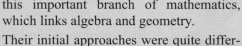

History now shows that the two Frenchmen René Descartes and Pierre de Fermat seem to have arrived at the idea of **analytical geometry** at about the same time. Descartes' work "*La Geometrie*", however, was published first (in 1637) and Fermat's "*Introduction to Loci*" was not published until after his death. Today, they are considered the co-founders of this important branch of mathematics, which links algebra and geometry.

Pierre de Fermat

René Descartes

Their initial approaches were quite different. Basically, Descartes began with a line or curve and then found the equation which described it. Fermat, to a large extent, started with an equation and investigated the shape of the curve it described. This interaction between algebra and geometry shows the power of **analytical geometry** as a branch of mathematics.

Analytical geometry and its use of coordinates, provided the mathematical tools which enabled Isaac Newton to later develop another important branch of mathematics called calculus.

Newton humbly stated: "*If I have seen further than Descartes, it is because I have stood on the shoulders of giants.*"

A DISTANCE BETWEEN TWO POINTS

Consider the points A(1, 3) and B(4, 1). The distance from A to B can be found by joining A to B using a straight line then drawing a right angled triangle with hypotenuse AB and with sides parallel to the axes.

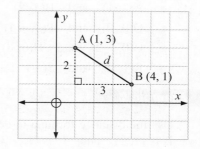

It is clear that $d^2 = 3^2 + 2^2$ {Pythagoras' Rule}

$$\therefore \quad d^2 = 13$$

$$\therefore \quad d = \sqrt{13} \qquad \{\text{as } d > 0\}$$

$\therefore$ the distance from A to B is $\sqrt{13}$ units.

EXERCISE 7A.1

1   If necessary, use the rule of Pythagoras to find the distance between:

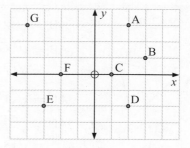

    a   A and B   **b**   A and D

    c   C and A   **d**   F and C

    e   G and F   **f**   C and G

    g   E and C   **h**   E and D

THE DISTANCE FORMULA

To avoid drawing a diagram each time we wish to find a distance, a **distance formula** can be developed.

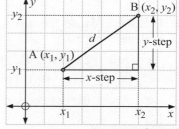

In going from A to B, x-step $= x_2 - x_1$,   and

$$y\text{-step} = y_2 - y_1.$$

Now, using Pythagoras' Rule,

$$(AB)^2 = (x\text{-step})^2 + (y\text{-step})^2$$

$$AB = \sqrt{(x\text{-step})^2 + (y\text{-step})^2}$$

$$\therefore \quad d = \sqrt{(x_2 - x_1)^2 + (y_2 - y_1)^2}.$$

If A(x_1, y_1) and B(x_2, y_2) are two points in a plane, then the **distance** between these points is given by $\mathbf{AB = \sqrt{(x_2 - x_1)^2 + (y_2 - y_1)^2}}$

This distance formula saves us having to graph the points each time we want to find a distance.

Example 1

Find the distance between A$(-2, 1)$ and B$(3, 4)$.

$$\begin{array}{ll}
\text{A}(-2, 1) \quad \text{B}(3, 4) & AB = \sqrt{(3 - -2)^2 + (4 - 1)^2} \\
\quad\uparrow\;\;\uparrow \qquad\;\; \uparrow\;\uparrow & \quad\;\; = \sqrt{5^2 + 3^2} \\
\quad x_1\, y_1 \qquad\; x_2\, y_2 & \quad\;\; = \sqrt{25 + 9} \\
 & \quad\;\; = \sqrt{34} \text{ units}
\end{array}$$

EXERCISE 7A.2

1   Find the distance between the following pairs of points:

    a   A(3, 1)   and   B(4, 2)

    b   C(-1, 2)   and   D(5, 2)

    c   O(0, 0)   and   P(3, -4)

    d   E(3, 0)   and   F(7, 0)

    e   G(0, -2)   and   H(0, 3)

    f   I(2, 0)   and   J(0, -4)

    g   R(1, 2)   and   S(-1, 1)

    h   W(3, -2)   and   Z(-1, -4)

Example 2

Use the distance formula to determine if the triangle ABC, where A is $(-2, 0)$, B is $(2, 1)$ and   C is $(1, -3)$,   is equilateral, isosceles or scalene.

A(-2, 0)
B(2, 1)
C(1, -3)

$$AB = \sqrt{(2--2)^2 + (1-0)^2}$$
$$\therefore \ AB = \sqrt{4^2 + 1^2}$$
$$= \sqrt{17} \text{ units}$$

$$BC = \sqrt{(1-2)^2 + (-3-1)^2}$$
$$\therefore \ BC = \sqrt{(-1)^2 + (-4)^2}$$
$$= \sqrt{17} \text{ units}$$

$$AC = \sqrt{(1--2)^2 + (-3-0)^2}$$
$$\therefore \ AC = \sqrt{3^2 + (-3)^2}$$
$$= \sqrt{18} \text{ units}$$

As   AB = BC,   triangle ABC is isosceles.

---

2   Use the distance formula to classify triangle ABC, as either equilateral, isosceles or scalene:

    a   A(5, -3),   B(1, 8),   C(-6, 1)

    b   A(1, 0),   B(3, 1),   C(7, 3)

    c   A(-2, -1),   B(0, 3),   C(4, 1)

    d   A($\sqrt{2}$, 0),   B($-\sqrt{2}$, 0),   C(0, $\sqrt{6}$)

    e   A(-2, 0), B(1, $\sqrt{3}$), C(1, $-\sqrt{3}$)

    f   A(a, b),   B(a, $-b$),   C(1, 0)

Example 3

Use the distance formula to show that triangle ABC is right angled if A is (1, 2),   B is (2, 5)   and   C is (4, 1).

$$AB = \sqrt{(2-1)^2 + (5-2)^2}$$
$$\therefore \ AB = \sqrt{1^2 + 3^2}$$
$$= \sqrt{10} \text{ units}$$

$$BC = \sqrt{(4-2)^2 + (1-5)^2}$$
$$\therefore \ BC = \sqrt{2^2 + (-4)^2}$$
$$= \sqrt{20} \text{ units}$$

$$AC = \sqrt{(4-1)^2 + (1-2)^2}$$
$$\therefore \ AC = \sqrt{3^2 + (-1)^2}$$
$$= \sqrt{10} \text{ units}$$

Now $AB^2 + AC^2 = 10 + 10 = 20$

and $BC^2 = 20$

$\therefore$ triangle ABC is right angled at A.

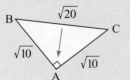

The right angle is opposite the longest side.

3 Use the distance formula to show that the following triangles are right angled and in each case state the right angle:

    a   A$(-2, -1)$, B$(-1, -1)$, C$(-2, 3)$ **b**   A$(-2, 2)$, B$(4, 2)$, C$(4, -6)$

    c   A$(1, -2)$, B$(3, 0)$, C$(-2, 1)$ **d**   A$(3, -4)$, B$(-2, -5)$, C$(1, 6)$

Example 4

Find b given that A$(3, -2)$ and B$(b, 1)$ are $\sqrt{13}$ units apart.

From A to B x-step $= b - 3$

                    y-step $= 1 - -2 = 3$

$\therefore\quad \sqrt{(b-3)^2 + 3^2} = \sqrt{13}$

$\therefore\quad (b-3)^2 + 9 = 13$ {squaring both sides}

$\therefore\quad (b-3)^2 = 4$ {subtracting 9}

$\therefore\quad b - 3 = \pm 2$ {if $X^2 = k$ then $X = \pm\sqrt{k}$}

$\therefore\quad b = 3 \pm 2$

i.e., $b = 5$ or 1.

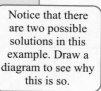

> Notice that there are two possible solutions in this example. Draw a diagram to see why this is so.

4 Find b given that:

    a A$(4, 10)$ and B$(b, 7)$ are 5 units apart

    b A$(3, b)$ and B$(-1, 2)$ are $\sqrt{20}$ units apart

    c A(b, b) is $\sqrt{18}$ units from the origin

    d P$(b, 2)$ is equidistant from A$(5, 4)$ and B$(2, 5)$

B   GRADIENT

When looking at line segments drawn on a set of axes, it is clear that different line segments are inclined to the horizontal at different angles, i.e., some appear to be steeper than others.

> The **gradient** of a line is a measure of its steepness.

If we choose any two distinct (different) points on the line, a **horizontal step** and a **vertical step** may be determined. We can always make the horizontal step in the positive direction.

Case 1:   *Case 2:*

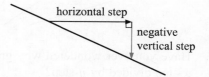

The **gradient** of a line may be determined by the fraction $\dfrac{\textbf{vertical step}}{\textbf{horizontal step}}$, i.e., $\dfrac{y\text{-step}}{x\text{-step}}$.

Example 5

Find the gradient of each line segment:

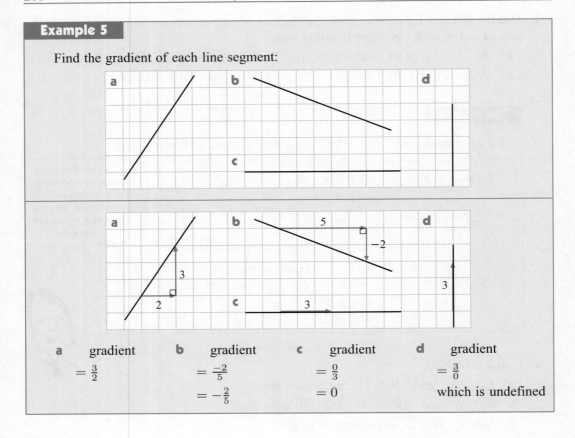

a gradient	**b** gradient	**c** gradient	**d** gradient
$= \frac{3}{2}$	$= \frac{-2}{5}$	$= \frac{0}{3}$	$= \frac{3}{0}$
	$= -\frac{2}{5}$	$= 0$	which is undefined

Note:

- The gradient of a **horizontal** line is **0**, since the vertical step (i.e., the numerator) is 0.
- The gradient of a **vertical** line is **undefined**, since the horizontal step (i.e., the denominator) is 0.

Lines like

are forward sloping and have **positive gradients**,

whereas lines like

are backwards sloping and have **negative gradients**.

Have you ever wondered why gradient is measured by y-step divided by x-step rather than x-step divided by y-step?

Horizontal lines have no gradient and zero (0) should represent this. As the gradient of a line increases we would want its numerical representation to increase.

EXERCISE 7B.1

1 Find the gradient
of each line segment:

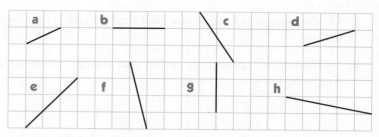

2 On grid paper draw a line segment with gradient:

    a $\frac{2}{5}$ **b** $-\frac{1}{2}$ **c** 3   **d** 0   **e** -4 **f** $-\frac{2}{7}$

THE GRADIENT FORMULA

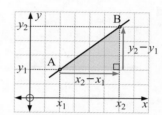

> If A is (x_1, y_1) and B is (x_2, y_2) then
>
> the **gradient** of AB is $\dfrac{y_2 - y_1}{x_2 - x_1}$

Example 6

Find the gradient of the line through $(3, -2)$ and $(6, 4)$.	$(3, -2) \quad (6, 4)$ $\uparrow \quad \uparrow \quad\; \uparrow \;\; \uparrow$ $x_1 \;\; y_1 \quad x_2 \; y_2$	gradient $= \dfrac{y_2 - y_1}{x_2 - x_1}$ $= \dfrac{4 - -2}{6 - 3}$ $= \dfrac{6}{3}$ $= 2$

EXERCISE 7B.2

1 Find the gradient of the line segment joining the following pairs of points:

    a $(2, 3)$ and $(5, 6)$ **b** $(3, 5)$ and $(1, 6)$ **c** $(1, -2)$ and $(3, 4)$

    d $(2, 5)$ and $(-1, 5)$ **e** $(3, -1)$ and $(3, 7)$ **f** $(3, -1)$ and $(-2, -3)$

    g $(-3, 1)$ and $(2, 0)$ **h** $(0, -1)$ and $(-3, -2)$

Example 7

Through $(2, 4)$ draw a line with gradient $-\frac{2}{3}$.

> It is a good idea to use a positive x-step here.

Plot the point $(2, 4)$

gradient $= \dfrac{y\text{-step}}{x\text{-step}} = \dfrac{-2}{3}$

$\therefore$ let y-step $= -2$, x-step $= 3$.

Use these steps to find another point
and draw the line through these points.

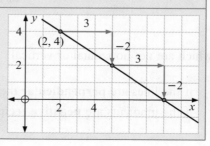

2 On the same set of axes draw lines through (2, 3) with gradients of $\frac{2}{5}$, $\frac{1}{2}$, 1, 2 and 4.

3 On the same set of axes draw lines through (−3, 1) with gradients of 0, $−\frac{1}{3}$, −1 and −3.

4 By finding an appropriate y-step and x-step, determine the gradient of each of the following lines:

a

b
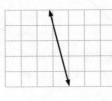

c
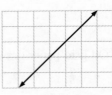

d
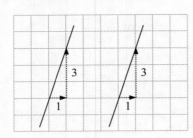

e
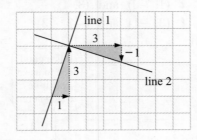

f
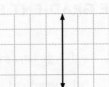

PARALLEL LINES

Notice that the given lines are parallel and both of them have a gradient of 3.

In fact:

- if two lines are **parallel**, then they have **equal gradient**,
- if two lines have **equal gradient**, then they are **parallel**.

PERPENDICULAR LINES

Notice that *line 1* and *line 2* are perpendicular.

Line 1 has gradient $\frac{3}{1} = 3$

Line 2 has gradient $\frac{-1}{3} = -\frac{1}{3}$

We see that the gradients are *negative reciprocals* of each other and their product is $3 \times -\frac{1}{3} = -1$.

For lines which are not horizontal or vertical:

- if the lines are **perpendicular** their gradients are **negative reciprocals**
- if the gradients are **negative reciprocals** the lines are **perpendicular**.

Proof:

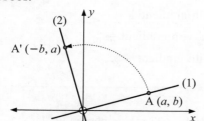

Suppose the two perpendicular lines are translated so that they intersect at the origin O. If $A(a, b)$ lies on one line, under an anti-clockwise rotation about O of $90°$ it finishes on the other line and its coordinates are $A'(-b, a)$.

The gradient of line (1) is $\dfrac{b - 0}{a - 0} = \dfrac{b}{a}$.

The gradient of line (2) is $\dfrac{a - 0}{-b - 0} = -\dfrac{a}{b}$.

The negative reciprocal of $\dfrac{b}{a}$ is $-\dfrac{a}{b}$.

Example 8

If a line has gradient $\frac{2}{3}$, find the gradient of all lines :

a parallel to the given line

b perpendicular to the given line.

a Since the original line has gradient $\frac{2}{3}$, the gradient of all parallel lines is also $\frac{2}{3}$.

b The gradient of all perpendicular lines is $-\frac{3}{2}$.   {the negative reciprocal}

EXERCISE 7B.3

1 Find the gradient of all lines perpendicular to a line with a gradient of:

  a $\frac{1}{3}$ **b** $\frac{3}{5}$ **c** 2 **d** 5 **e** $-\frac{3}{4}$ **f** $-2\frac{1}{2}$ **g** -4 **h** -1

2 The gradients of several pairs of lines are listed below.
Which of the line pairs are perpendicular?

  a $\frac{1}{4}, 4$ **b** $3, -3$ **c** $\frac{2}{7}, -3\frac{1}{2}$ **d** $5, -\frac{1}{5}$ **e** $7, -\frac{2}{7}$ **f** $\frac{3}{4}, -\frac{4}{3}$

Example 9

Find a given that the line joining $A(a, -1)$ to $B(2, 3)$ has gradient -2.

gradient of AB $= -2$ {parallel lines have equal gradient}

$\therefore \quad \dfrac{3 - -1}{2 - a} = \dfrac{-2}{1}$ {gradient formula}

$\therefore \quad \dfrac{4}{2 - a} = \dfrac{-2}{1}$

$\therefore \quad 4 \times 1 = -2(2 - a)$ {achieving a common denominator}

$\therefore \quad 4 = -4 + 2a$ {equating numerators}

$\therefore \quad 8 = 2a$

$\therefore \quad a = 4$

3 Find a given that the line joining:

   a A(1, 3)   to   B(3, a)   is parallel to a line with gradient 4

   b P(a, −3)   to   Q(4, −1)   is parallel to a line with gradient $\frac{1}{2}$

   c M(3, a)   to   N(a, 5)   is parallel to a line with gradient $-\frac{2}{3}$.

Example 10

Find t given that the line joining D(−1, −3) to C(1, t) is perpendicular to a line with gradient 2.

gradient of CD $= -\frac{1}{2}$ {perpendicular to line of gradient 2}

$\therefore \quad \dfrac{t - -3}{1 - -1} = -\frac{1}{2}$ {gradient formula}

$\therefore \quad \dfrac{t + 3}{2} = \dfrac{-1}{2}$ {simplifying}

$\therefore \quad t + 3 = -1$ {equating numerators}

$\therefore \quad\quad t = -4$

---

4 Find t given that the line joining:

   a A(1, −3)   to   B(−2, t)   is perpendicular to a line with gradient $1\frac{1}{2}$

   b P(t, −2)   to   Q(5, t)   is perpendicular to a line with gradient $-\frac{1}{3}$.

5 Given the points   A(1, 3),   B(−1, 0),   C(6, 4)   and   D(t, −1),   find t if:

   a AB is parallel to CD   **b** AC is parallel to DB

   c AB is perpendicular to CD   **d** AD is perpendicular to BC

6 P(3, a) lies on a semi-circle as shown.

   a Find a.

   b Using this value of a, find the gradient
        of: **i** AP   **ii** BP.

   c Use **b** to show that angle APB is a
        right angle.

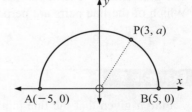

COLLINEAR POINTS

Three or more points are **collinear** if they lie on the same straight line.

i.e.,   if three points A, B and C are collinear,
the gradient of AB is equal to the gradient of BC.
(and the gradient of AC)

Example 11

Show that the following points are collinear: A(1, −1),   B(6, 9),   C(3, 3)

gradient of AB $= \dfrac{9 - -1}{6 - 1}$ gradient of BC $= \dfrac{3 - 9}{3 - 6}$

$\qquad\qquad\quad = \dfrac{10}{5}$ $\qquad\qquad\qquad = \dfrac{-6}{-3}$

$\qquad\qquad\quad = 2$ $\qquad\qquad\qquad = 2$

∴   AB is parallel to BC and as point B is common to both line segments, then A, B and C are *collinear*.

7 Determine whether or not the following sets of three points are collinear:

   a   A(1, 2), B(4, 6) and C(−5, −6)   **b**   P(−6, −6), Q(−1, 0) and R(9, 12)

   c   R(3, 4), S(−6, 5) and T(0, −4)   **d**   A(0, −2), B(−1, −5) and C(4, 10)

8 Find c given that:

   a   A(−5, −3), B(−1, 1) and C(c, 5) are collinear

   b   P(3, −2), Q(4, c) and R(0, 7) are collinear.

C   APPLICATIONS OF GRADIENT

In the previous exercise we considered the gradients of straight lines or gradients between points. In real life gradients occur in many situations and can be interpreted in a variety of ways.

For example, the sign alongside would indicate to motor vehicle drivers that there is an uphill climb ahead.

There are however, many less obvious situations where the gradient of a line can be interpreted in a way that we understand.

Consider the situation in the graph alongside where a motor vehicle travels at a constant speed for a distance of 600 km in 8 hours.

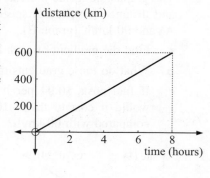

Clearly, the gradient of the line $= \dfrac{\text{vertical step}}{\text{horizontal step}}$

$\qquad\qquad = \dfrac{600}{8}$

$\qquad\qquad = 75$

However,   speed $= \dfrac{\text{distance}}{\text{time}} = \dfrac{600 \text{ km}}{8 \text{ hours}} = 75$ km/h.

So, we would expect that in a graph of distance against time the *gradient* can be interpreted as the *speed*.

EXERCISE 7C

1

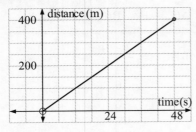

The graph alongside indicates the distance run by a sprinter in a number of seconds.

  a Find the gradient of the line.

  b Interpret the gradient found in **a**.

  c Is the speed of the runner constant or variable? What evidence do you have for your answer?

2 The graph alongside indicates the distances travelled by a lorry driver. Determine:

  a the average speed for the whole trip

  b the average speed from

    i O to A   **ii** B to C

  c the time interval over which the average speed is greatest.

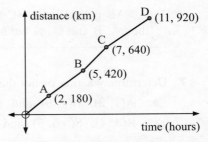

3

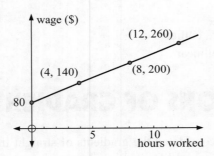

The graph alongside indicates the wages paid to sales assistants.

  a What does the intercept on the vertical axis mean?

  b Find the gradient of the line. What does this gradient mean?

  c Determine the wages for working:

    i 6 hours   **ii** 18 hours.

  d If no payment was made for not working but the same payment was made for 8 hours work, what would be the new rate of pay?

4 The graphs alongside indicate the fuel consumption and distance travelled at speeds of 60 km/h (graph A) and 90 km/h (graph B).

  a Find the gradient of each line.

  b What do these gradients mean?

  c If fuel costs \$0.94 per litre, how much more would it cost to travel 1000 km at 90 km/h compared with 60 km/h?

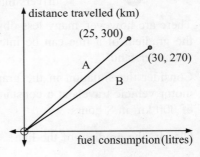

5

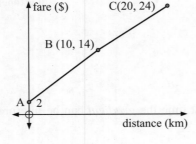

The graph alongside indicates the taxi-fare charged for distance travelled.

  a What does the value at A indicate?

  b Find the gradients of the line segments AB and BC. What do these gradients indicate?

  c If a straight line segment was drawn from A to C, find its gradient. What would this gradient mean?

D MIDPOINTS

If point M is halfway between points A and B then M is the **midpoint** of AB.

Consider the points A(1, 2) and B(5, 4).

It is clear from the diagram at right that the midpoint M, of AB is (3, 3).

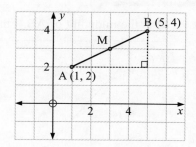

We notice that: $\dfrac{1+5}{2} = 3$ and $\dfrac{2+4}{2} = 3$.

So, the x-coordinate of M is the *average* of the x-coordinates of A and B,

and the y-coordinate of M is the *average* of the y-coordinates of A and B.

THE MIDPOINT FORMULA

In general,

if A(x_1, y_1) and B(x_2, y_2) are two points then the **midpoint** M of AB has coordinates

$$\left(\frac{x_1 + x_2}{2}, \frac{y_1 + y_2}{2} \right)$$

Example 12

Find the coordinates of the midpoint of AB for A(−1, 3) and B(4, 7).

x-coordinate of midpoint	y-coordinate of midpoint
$= \dfrac{-1+4}{2}$	$= \dfrac{3+7}{2}$
$= \dfrac{3}{2}$	$= 5$
$= 1\frac{1}{2}$	

$\therefore$ the midpoint of AB is $(1\frac{1}{2}, 5)$

EXERCISE 7D

1

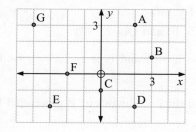

Use this diagram only to find the coordinates of the midpoint of the line segment:

a GA   b ED

c AC   d AD

e CD   f GF

g EG   h GD

2 Find the coordinates of the midpoint of the line segment joining the pairs of points:

a (6, 3) and (2, 5)   **b** (4, −1) and (0, 1)

c (3, 0) and (0, 5)   **d** (−1, 6) and (1, 6)

e (3, −1) and (−1, 0)   **f** (−1, 3) and (3, −1)

g (6, 8) and (−3, −4)   **h** (−2, 3) and (−5, 1)

Example 13

M is the midpoint of AB. Find the coordinates of B if A is (1, 3) and M is (4, −2).

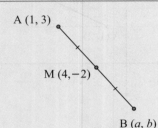

Let B be (a, b)

$\therefore \quad \dfrac{a+1}{2} = 4$ and $\dfrac{b+3}{2} = -2$

$\therefore \quad a + 1 = 8$ and $b + 3 = -4$

$\therefore \quad a = 7$ and $b = -7$

i.e., B is (7, −7)

---

3 M is the midpoint of AB. Find the coordinates of B if:

a A(4, 2) and M(3, −1)   **b** A(−3, 0) and M(0, −1)

c A(5, −1) and M($1\frac{1}{2}$, 2)   **d** A(−2, −1) and M($-\frac{1}{2}$, $2\frac{1}{2}$)

e A(3, −1) and M(0, 0)   **f** A(5, −2) and M(0, $-\frac{1}{2}$)

Example 14

Find the coordinates of B using *equal steps* if M is the midpoint of AB, A is (1, 3) and M is (4, −2).

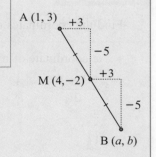

x-step: $1 \xrightarrow{+3} 4 \xrightarrow{+3} 7$

y-step: $3 \xrightarrow{-5} -2 \xrightarrow{-5} -7$

$\therefore$ B is (7, −7)

---

4 Check your answers to questions **3a** and **3b** using the equal steps method given in **Example 14**.

5 If T is the midpoint of PQ, find the coordinates of P for:

a T(−3, 4) and Q(5, −1)   **b** T(5, 0) and Q(−2, −3)

6 AB is the diameter of a circle, centre C. If A is (5, −6) and B is (−1, −4), find the coordinates of C.

7 PQ is a diameter of a circle, centre (5, $-\frac{1}{2}$). Find the coordinates of P given that Q is (−1, 2).

8 Line segments AB and CD bisect each other at T.
Find the coordinates of C.

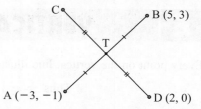

9 Triangle ABC has   A(−1, 4),   B(2, −1),   and   C(5, 2)   as vertices. Find the length
of the line segment from A to the midpoint of BC.

Example 15

Use midpoints to find the fourth
vertex of the given parallelogram:

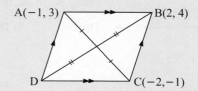

Since ABCD is a parallelogram, the diagonals bisect each other.

∴ the midpoint of DB is the same as the midpoint of AC, and if D is (a, b),

$$\frac{a+2}{2} = \frac{-1+-2}{2} \quad \text{and} \quad \frac{b+4}{2} = \frac{3+-1}{2}$$

$$\therefore \quad a+2 = -3 \quad \text{and} \quad b+4 = 2$$

$$\therefore \quad a = -5 \quad \text{and} \quad b = -2$$

$$\therefore \quad \text{D is } (-5, -2)$$

---

10 Use midpoints to find the fourth vertex of the given parallelograms:

a

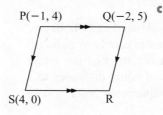

b

c

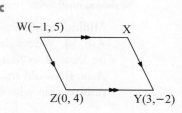

11 An inaccurate sketch of quadrilateral ABCD is
given. P, Q, R and S are the midpoints of AB,
BC, CD and DA respectively.

  a Find the coordinates of:
      i P **ii** Q **iii** R **iv** S

  b Find the gradient of:
      i PQ   **ii** QR
      iii RS   **iv** SP

  c What can be deduced about quadrilateral
     PQRS from **b**?

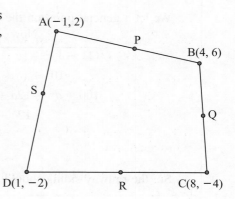

E   VERTICAL AND HORIZONTAL LINES

Every point on the vertical line illustrated has an x-coordinate of 3.

Thus $x = 3$ is the equation of this line.

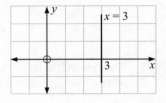

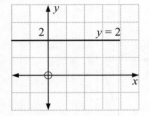

Every point on the horizontal line illustrated has a y-coordinate of 2.

Thus $y = 2$ is the equation of this line.

In general:

> All **vertical lines** have equations of the form $x = a$, (a is a constant).
> All **horizontal lines** have equations of the form $y = c$, (c is a constant).

EXERCISE 7E

1   Find the equation of the:

   a   horizontal line through $(3, -4)$
   b   vertical line with x-intercept 5
   c   vertical line through $(-1, -3)$
   d   horizontal line with y-intercept 2
   e x-axis
   f y-axis

Example 16

Mining towns are situated at B(1, 6) and A(5, 2). Where should a railway siding S be located so that ore trucks from either A or B would travel equal distances to a railway line with equation $x = 11$?

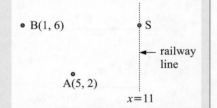

We let a general point on the line $x = 11$ have coordinates $(11, a)$ say.

Now $BS = AS$

$\therefore \quad \sqrt{(11-1)^2 + (a-6)^2} = \sqrt{(11-5)^2 + (a-2)^2}$

$\therefore \quad 10^2 + (a-6)^2 = 6^2 + (a-2)^2$ {squaring both sides}

$\therefore \quad 100 + a^2 - 12a + 36 = 36 + a^2 - 4a + 4$

$\therefore \quad -12a + 4a = 4 - 100$

$\therefore \quad -8a = -96$

$\therefore \quad a = 12$

So, the railway siding should be located at (11, 12).

2 A(5, 5) and B(7, 10) are houses and $y = 8$ is a gas pipeline. Where should the one outlet from the pipeline be placed so that it is the same distance from both houses so they pay equal service costs?

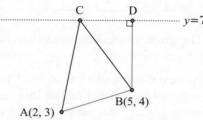

3 CD is a water pipeline. A and B are two towns. A pumping station is to be located on the pipeline to pump water to A and B. Each town is to pay for their own service pipes and insist on equality of costs.

   a Where should C be located for equality of costs to occur?

   b What is the total length of service pipe required?

   c If the towns agree to pay equal amounts, would it be cheaper to install the service pipeline from D to B to A?

4 Jason's girlfriend lives in a house on Clifton Highway which has equation $y = 8$. The distance 'as the crow flies' from Jason's house to his girlfriend's house is 11.73 km. If Jason lives at (4, 1), what are the coordinates of his girlfriend's house?

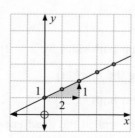

F   EQUATIONS OF LINES

The **equation of a line** is an equation which connects the x and y values for every point on the line.

Recall that:

   $y = mx + c$ is the equation of a line with gradient m and y-intercept c.

For example:

The illustrated line has

$$\text{gradient} = \frac{y\text{-step}}{x\text{-step}} = \tfrac{1}{2}$$

and the y-intercept is 1

$\therefore$ its equation is $y = \tfrac{1}{2}x + 1$.

Consider the equation $y = 2x - 3$.

We could set up a table of values which satisfy this equation:

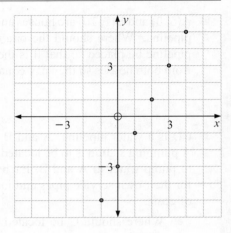

x	-1	0	1	2	3	4
y	-5	-3	-1	1	3	5

The graph alongside shows only discrete points for integer values of x.

However, x can take any value. Click on the demonstration to see the complete line being generated by using $\frac{1}{2}$, $\frac{1}{4}$, $\frac{1}{8}$, etc. values between the integer values of x.

GRAPHING PACKAGE

Experiment with other linear equations of your choosing.

Every point on the line has coordinates which satisfy the equation.

For example, if $x = -0.316$ in $y = 2x - 3$ there is a corresponding value of y which could be found by substitution.

FINDING THE EQUATION OF A LINE

Consider the illustrated line which has gradient $\frac{1}{2}$ and passes through the point $(2, 3)$

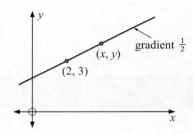

Suppose (x, y) is any point on the line.

The gradient between $(2, 3)$ and (x, y) is $\dfrac{y - 3}{x - 2}$.

Equating gradients gives us $\dfrac{y - 3}{x - 2} = \dfrac{1}{2}$ {gradient formula}

Consider two rearrangements of this equation of the line.

$$\frac{y - 3}{x - 2} = \frac{1}{2}$$

$\therefore \quad y - 3 = \frac{1}{2}(x - 2) \quad$ {multiplying both sides by $x - 2$}

$\therefore \quad y - 3 = \frac{1}{2}x - 1 \quad$ {expanding the bracket}

$\therefore \quad y = \frac{1}{2}x + 2 \quad$ {adding 3 to both sides}

and this is in the form $y = mx + c$ called the **gradient-intercept** form.

$$or \qquad \frac{y-3}{x-2} = \frac{1}{2}$$

$$\therefore \quad \frac{2}{2}\left(\frac{y-3}{x-2}\right) = \frac{1}{2}\left(\frac{x-2}{x-2}\right) \qquad \{\text{as LCD is } \ 2(x-2)\}$$

$$\therefore \quad 2(y-3) = 1(x-2) \qquad \{\text{equating numerators}\}$$

$$\therefore \quad 2y - 6 = x - 2 \qquad \{\text{expanding brackets}\}$$

$$\therefore \quad x - 2y = -4 \qquad \{\text{rearranging}\}$$

and this is in the form $\ Ax + By = C\ $ called the **general** form.

$$(A = 1, \quad B = -2, \quad C = -4)$$

So, to find the equation of a line we need to:

- know (or be able to find) the **gradient** and
- the coordinates of any **point** on the line.

Summary:

If a straight line has gradient m and passes through $(a,\,b)$ then it has equation $\dfrac{y-b}{x-a} = m$

which can be rearranged into $\qquad \boldsymbol{y = mx + c} \qquad \{\textbf{gradient-intercept} \text{ form}\}$

$$or \quad \boldsymbol{Ax + By = C} \quad \{\textbf{general} \text{ form}\}$$

Example 17

Find, in *gradient-intercept form*, the equation of the line through $(-1,\,3)$ with a gradient of 5.

The equation of the line is $\qquad \dfrac{y-3}{x--1} = 5$

> To find the equation of a line we need to know its gradient and a point on it.

i.e., $\qquad \dfrac{y-3}{x+1} = 5$

$$\therefore \quad y - 3 = 5(x+1)$$

$$\therefore \quad y - 3 = 5x + 5$$

$$\therefore \quad y = 5x + 8$$

EXERCISE 7F

1 Find, in *gradient-intercept form*, the equation of the line through:

a $(2,\,-5)$ having a gradient of 4

b $(-1,\,-2)$ having a gradient of -3

c $(7,\,-3)$ having a gradient of -5

d $(1,\,4)$ having a gradient of $\frac{1}{2}$

e $(-1,\,3)$ having a gradient of $-\frac{1}{3}$

f $(2,\,6)$ having a gradient of 0

Example 18

Find, in *general form*, the equation of the line with gradient $\frac{3}{4}$ and passing through $(5, -2)$.

The equation of the line is $\quad \dfrac{y - -2}{x - 5} = \dfrac{3}{4}$

$$\text{i.e.,} \quad \dfrac{y + 2}{x - 5} = \dfrac{3}{4}$$

$$\therefore \quad 4(y + 2) = 3(x - 5)$$

$$\therefore \quad 4y + 8 = 3x - 15$$

$$\therefore \quad 3x - 4y = 23$$

2 Find, in *general form*, the equation of the line through:

  a $(2, 5)$ having gradient $\frac{2}{3}$

  b $(-1, 4)$ having gradient $\frac{3}{5}$

  c $(5, 0)$ having gradient $-\frac{1}{3}$

  d $(6, -2)$ having gradient $-\frac{2}{7}$

  e $(-3, -1)$ having gradient 4

  f $(5, -3)$ having gradient -2

Example 19

Find the equation of the line which passes through the points $A(-1, 5)$ and $B(2, 3)$.

The gradient of the line is $\quad \dfrac{3 - 5}{2 - -1} = \dfrac{-2}{3}$

$\therefore$ using point A the equation is $\quad \dfrac{y - 5}{x - -1} = \dfrac{-2}{3} \qquad \{\text{or} \quad \dfrac{y - 3}{x - 2} = \dfrac{-2}{3}$

$$\therefore \quad \dfrac{y - 5}{x + 1} = \dfrac{-2}{3} \qquad \text{using point B}\}$$

$$\therefore \quad 3(y - 5) = -2(x + 1)$$

$$\therefore \quad 3y - 15 = -2x - 2$$

$$\therefore \quad 2x + 3y = 13$$

> Check that you get the same final answer using point B instead of A.

3 Find the equation of the line (in general form) which passes through the points:

  a $A(2, 3)$ and $B(4, 8)$

  b $A(0, 3)$ and $B(-1, 5)$

  c $A(-1, -2)$ and $B(4, -2)$

  d $C(-3, 1)$ and $D(2, 0)$

  e $P(5, -1)$ and $Q(-1, -2)$

  f $R(-1, -3)$ and $S(-4, -1)$

4 Find the equations of the lines (in general form) through:

  a $(0, 1)$ and $(3, 2)$

  b $(1, 4)$ and $(0, -1)$

  c $(2, -1)$ and $(-1, -4)$

  d $(0, -2)$ and $(5, 2)$

  e $(3, 2)$ and $(-1, 0)$

  f $(-1, -1)$ and $(2, -3)$

GRADIENT FROM THE EQUATION OF THE LINE

From equations of lines such as $y = 2x - 3,$ $y = \frac{1}{3}x + \frac{2}{3},$ and $y = 5 - 2x,$ we can easily find the gradient by looking at the coefficient of x. But, how do we find the gradients of equations of lines in the general form?   One method is to rearrange them.

Example 20	
Find the gradient of the line $2x + 5y = 17$.	$2x + 5y = 17$ $\therefore \quad 5y = 17 - 2x$ $\qquad$ {subtracting $2x$, both sides} $\therefore \quad y = \dfrac{17}{5} - \dfrac{2x}{5}$ $\qquad$ {dividing both sides by 5} $\therefore \quad y = -\frac{2}{5}x + \frac{17}{5}$ $\quad$ and so, the gradient is $-\frac{2}{5}$.

5 Find the gradient of the line with equation:

     a $\quad y = 3x + 2$
     b $\quad y = 3 - 2x$
     c $\quad y = 0$

     d $\quad x = 5$
     e $\quad y = \dfrac{2x + 1}{3}$
     f $\quad 3x + y = 7$

     g $\quad 2x - 7y = 8$
     h $\quad 2x + 7y = 8$
     i $\quad 3x - 4y = 11$

DOES A POINT LIE ON A LINE?

A point lies on a line if its coordinates satisfy the equation of the line.

Example 21
Does $(3, -2)$ lie on the line with equation $5x - 2y = 20$?
Substituting $(3, -2)$ into $5x - 2y = 20$ gives $\qquad\qquad\qquad 5(3) - 2(-2) = 20$ $\qquad\qquad$ i.e., $19 = 20$ which is false $\therefore \quad (3, -2)$ does not lie on the line.

6   **a**   Does $(3, 4)$ lie on the line with equation $3x - 2y = 1$?

     b   Does $(-2, 5)$ lie on the line with equation $5x + 3y = -5$?

     c   Does $(6, -\frac{1}{2})$ lie on the line $3x - 8y = 22$?

7 Find k if:

     a $(3, 4)$ lies on the line with equation $3x - 2y = k$

     b $(-1, 3)$ lies on the line with equation $5x - 2y = k.$

8 Find a given that:

     a $(a, 3)$ lies on the line with equation $y = 2x - 1$

     b $(-2, a)$ lies on the line with equation $y = 1 - 3x.$

9   A straight road is to pass through points A(5, 3) and B(1, 8).

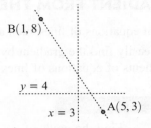

  a   Find where this road meets the road given by:
   i $x = 3$ **ii** $y = 4$

  b   If we wish to refer to points on road AB, but between A and B, how can we indicate this?

  c   Does C(23, −20) lie on the road?

EQUATIONS FROM GRAPHS

Provided that a graph contains sufficient information we can determine its equation.

Remember that we must have at least one point and we must be able to determine its gradient.

Example 22

Find the equation of the line with graph:

a

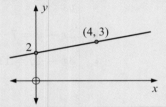

b

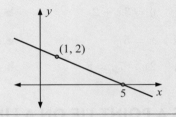

---

a   Two points on the line are (0, 2) and (4, 3)

  $\therefore$ gradient $m = \dfrac{3 - 2}{4 - 0} = \frac{1}{4}$

  and the y-intercept, $c = 2$

  $\therefore$ the equation is

   $y = \frac{1}{4}x + 2$.

  (gradient-intercept form)

b   Two points on the line are (1, 2) and (5, 0)

  $\therefore$ gradient $m = \dfrac{0 - 2}{5 - 1} = \dfrac{-2}{4} = \dfrac{-1}{2}$

  As we do not know the y-intercept we use

  equation is $\dfrac{y - 2}{x - 1} = \dfrac{-1}{2}$

   $\therefore$ $2(y - 2) = -1(x - 1)$

   $\therefore$ $2y - 4 = -x + 1$

   $\therefore$ $x + 2y = 5$ (general form)

---

10   Find the equation (in gradient-intercept form) of the line with:

  a   gradient 2 and y-intercept 7
  b   gradient 4 and y-intercept -6

  c   gradient -3 and y-intercept -1
  d   gradient $-\frac{1}{2}$ and y-intercept 2

  e   gradient 0 and y-intercept 8
  f   undefined gradient, through (2, 5)

11 Find the equations (in gradient-intercept form) of the illustrated lines:

a

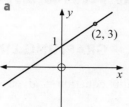

b

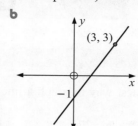

c

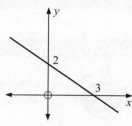

d

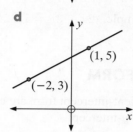

e

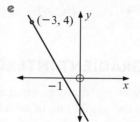

f

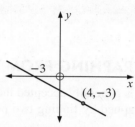

Example 23

Find the equation connecting
the variables in:

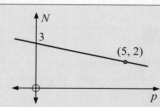

(0, 3) and (5, 2) lie on the straight line

$\therefore$ gradient $m = \dfrac{2-3}{5-0} = -\frac{1}{5}$ and as the y-intercept is $c = 3$,

the equation is of the form $Y = mX + c$ where $Y = N$, $X = p$

$\therefore$ is $N = -\frac{1}{5}p + 3$.

---

12 Find the equation (in gradient-intercept form) connecting the variables given:

a

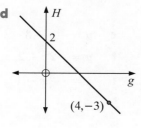

b

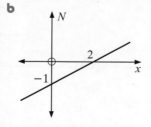

c

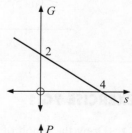

d

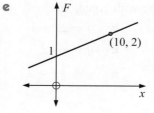

e

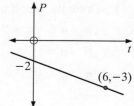

f

GRAPHING LINES

DISCUSSION   GRAPHING LINES

   Discuss the easiest way to graph a line when its equation is given in the form:

- $y = mx + c$ for example, $y = 2x + 3$
- $Ax + By = C$ for example, $2x + 3y = 12$.

GRAPHING FROM THE GRADIENT-INTERCEPT FORM

It is generally accepted that lines with equations given in the gradient-intercept form are easily graphed by finding two points on the graph, one of which is the y-intercept.

The other could be found by substitution or using the gradient.

Example 24

Graph the line with equation $y = \frac{1}{3}x + 2$.

Method 1:

The y-intercept is 2

when $x = 3$, $y = 1 + 2 = 3$

$\therefore$ $(0, 2)$ and $(3, 3)$ lie on the line.

Method 2:

The y-intercept is 2

and the gradient $= \frac{1}{3}$ ← y-step ← x-step

So we start at $(0, 2)$ and move to another point by moving across 3, then up 1.

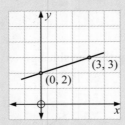

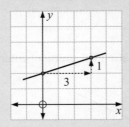

EXERCISE 7G.1

1 Draw the graph of the line with equation:

  a $y = \frac{1}{2}x + 2$ **b** $y = 2x + 1$ **c** $y = -x + 3$

  d $y = -3x + 2$ **e** $y = -\frac{1}{2}x$ **f** $y = -2x - 2$

  g $y = \frac{3}{2}x$ **h** $y = \frac{2}{3}x + 2$ **i** $y = -\frac{3}{4}x - 1$

GRAPHING FROM THE GENERAL FORM

The easiest method used to graph lines given in the
general form $Ax + By = C$ is to use axis intercepts.

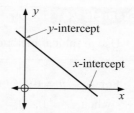

> The x-intercept is found by letting $y = 0$.
> The y-intercept is found by letting $x = 0$.

Example 25

Graph the line with equation $2x - 3y = 12$ using axis intercepts.	For $2x - 3y = 12$, when $x = 0$, $\quad -3y = 12$ $\qquad \therefore \quad y = -4$ when $y = 0$, $\quad 2x = 12$ $\qquad \therefore \quad x = 6$	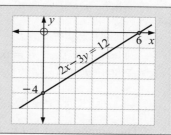

2 Use axis intercepts to draw sketch graphs of:

a	$x + 2y = 8$	**b**	$3x - y = 6$	**c**	$2x - 3y = 6$
d	$4x + 3y = 12$	**e**	$x + y = 5$	**f**	$x - y = -5$
g	$2x - y = -4$	**h**	$9x - 2y = 9$	**i**	$3x + 4y = -15$

3 Copy and complete:

	Equation of line	Gradient	x-intercept	y-intercept
a	$2x - 3y = 6$			
b	$4x + 5y = 20$			
c	$y = -2x + 5$			
d	$x = 8$			
e	$y = 5$			
f	$x + y = 11$			

> If a line has equation $y = mx + c$ then the gradient of the line is 'm'.

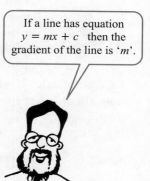

WHERE GRAPHS MEET

There are three possible situations which may occur. These are:

Case 1:

The lines meet in a
single **point of
intersection.**

Case 2:

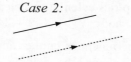

The lines are **parallel**
and **never meet**.
So, there is no point
of intersection.

Case 3:

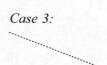

The lines are **coincident**
(the same line) and so
there are infinitely many
points of intersection.

Example 26

Use graphical methods to find where the lines $x + y = 6$ and $2x - y = 6$ meet.

For $x + y = 6$

  when $x = 0$, $y = 6$

  when $y = 0$, $x = 6$

x	0	6
y	6	0

For $2x - y = 6$

  when $x = 0$, $-y = 6$, $\therefore$ $y = -6$

  when $y = 0$, $2x = 6$, $\therefore$ $x = 3$

x	0	3
y	-6	0

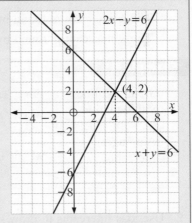

The graphs meet at $(4, 2)$.

Check: $4 + 2 = 6$ ✓ and $2 \times 4 - 2 = 6$ ✓

INVESTIGATION   FINDING WHERE LINES MEET USING TECHNOLOGY

Graphing packages and **graphics calculators** can be used to plot straight line graphs and hence find the point of intersection of the straight lines. This can be useful if the solutions are not integer values, although an algebraic method can also be used. However, most graphing packages and graphics calculators require the equation to be entered in the form $y = mx + c$.

Consequently, if an equation is given in **general form**, it must be rearranged into **gradient-intercept form**.

For example, if we wish to use technology to find the point of intersection of $4x + 3y = 10$ and $x - 2y = -3$:

Step 1: We **rearrange** each equation into the form $y = mx + c$, i.e.,

$$4x + 3y = 10 \qquad\qquad \text{and} \quad x - 2y = -3$$
$$\therefore \quad 3y = -4x + 10 \qquad\qquad \therefore \quad -2y = -x - 3$$
$$\therefore \quad y = -\tfrac{4}{3}x + \tfrac{10}{3} \qquad\qquad \therefore \quad y = \tfrac{x}{2} + \tfrac{3}{2}$$

GRAPHING PACKAGE

Step 2: If you are using the **graphing package**, click on the icon to open the package and enter the two equations.

If you are using a **graphics calculator**, enter the functions $Y_1 = -4X/3 + 10/3$ and $Y_2 = X/2 + 3/2$.

Step 3: Draw the **graphs** of the functions on the same set of axes. (You may have to change the viewing **window** if using a graphics calculator.)

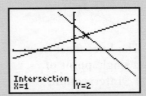

Step 4: Use the built in functions to calculate the point of **intersection**.

Thus, the point of intersection is $(1, 2)$.

What to do:

1   Use technology to find the point of intersection of:

   a $y = x + 4$
   $5x - 3y = 0$

   b $x + 2y = 8$
   $y = 7 - 2x$

   c $x - y = 5$
   $2x + 3y = 4$

   d $2x + y = 7$
   $3x - 2y = 1$

   e $y = 3x - 1$
   $3x - y = 6$

   f $y = -\dfrac{2x}{3} + 2$
   $2x + 3y = 6$

2   Comment on the use of technology to find the point(s) of intersection in **1 e** and **1 f**.

EXERCISE 7G.2

1   Use graphical methods to find the point of intersection of:

   a $y = x - 3$
   $y = 1 - x$

   b $x - y = 1$
   $y = 2x$

   c $4x + 3y = 12$
   $x - 2y = 3$

   d $3x + y = -3$
   $2x - 3y = -24$

   e $3x + y = 9$
   $3x - 2y = -12$

   f $x - 3y = -9$
   $2x - 3y = -12$

   g $2x - y = 6$
   $x + 2y = 8$

   h $y = 2x - 4$
   $2x - y = 2$

   i $y = -x - 5$
   $2x + 2y = -10$

2   How many points of intersection do the following pairs of lines have?
   Explain, but **do not** graph them.

   a $3x + y = 5$
   $3x + y = 8$

   b $3x + y = 5$
   $6x + 2y = 10$

   c $3x - y = 5$
   $3x - y = k$ (k takes all values)

3   Use technology to find where the following lines meet:

   a $x + 2y = 8$
   $y = 2x - 6$

   b $y = -3x - 3$
   $3x - 2y = -12$

   c $3x + y = -3$
   $2x - 3y = -24$

   d $2x - 3y = 8$
   $3x + 2y = 12$

   e $x + 3y = 10$
   $2x + 6y = 11$

   f $5x + 3y = 10$
   $10x + 6y = 20$

H   MIDPOINTS AND PERPENDICULAR BISECTORS

Recall that the **midpoint** of a line segment connecting (x_1, y_1) to (x_2, y_2) is

average of
x-coordinates
$$\left(\underbrace{\frac{x_1 + x_2}{2}}, \underbrace{\frac{y_1 + y_2}{2}} \right)$$
average of
y-coordinates

PERPENDICULAR BISECTORS

The perpendicular bisector of two points A and B divides the plane into two regions. On one side of the line points are closer to B than to A, and vice versa on the other side.

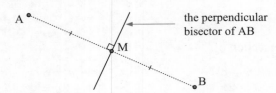

the perpendicular
bisector of AB

We observe that the midpoint of line segment AB must lie on the perpendicular bisector of AB.

Points on the perpendicular bisector of AB are **equidistant** to A and B.

Example 27

Find the equation of the perpendicular bisector of AB for A$(-1, 2)$ and B$(3, 4)$.

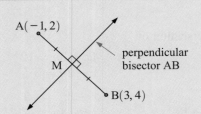

M is $\left(\dfrac{-1+3}{2}, \dfrac{2+4}{2} \right)$ i.e., M$(1, 3)$

gradient of AB is $\dfrac{4-2}{3--1} = \dfrac{2}{4} = \dfrac{1}{2}$

$\therefore$ gradient of perpendicular is $-\dfrac{2}{1}$

$\therefore$ equation of perpendicular bisector is $\dfrac{y-3}{x-1} = -2$ {using M$(1, 3)$}

$\therefore$ $y - 3 = -2(x - 1)$

$\therefore$ $y - 3 = -2x + 2$

$\therefore$ $y = -2x + 5$

Note: We could have let P on the perpendicular bisector be (x, y) and as AP $=$ PB then

$$\sqrt{(x--1)^2 + (y-2)^2} = \sqrt{(x-3)^2 + (y-4)^2}$$
$$\therefore \quad (x+1)^2 + (y-2)^2 = (x-3)^2 + (y-4)^2$$
$$\therefore \quad x^2 + 2x + 1 + y^2 - 4y + 4 = x^2 - 6x + 9 + y^2 - 8y + 16$$
$$\therefore \quad 4y = -8x + 20$$
$$\text{or} \quad y = -2x + 5$$

EXERCISE 7H

1 Find the equation of the perpendicular bisector of AB for:

   a   A$(3, -3)$ and B$(1, -1)$ **b**   A$(1, 3)$ and B$(-3, 5)$

   c   A$(3, 1)$ and B$(-3, 6)$ **d**   A$(4, -2)$ and B$(4, 4)$

2 Two Post Offices are located at P$(3, 8)$ and Q$(7, 2)$ on a Council map. What is the equation of the line which should form the boundary between the two regions being serviced by the Post Offices?

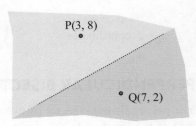

3 The perpendicular bisector of a chord of a circle passes through its centre.

Find the centre of a circle passing through points P(5, 7), Q(7, 1) and R(−1, 5) by finding the perpendicular bisectors of PQ and QR and solving them simultaneously.

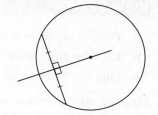

4 Triangle ABC has vertices as shown. Find:

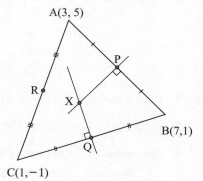

a the coordinates of P, Q and R, the midpoints of AB, BC and AC respectively

b the equation of the perpendicular bisector of

    i AB   **ii** BC   **iii** AC

c the coordinates of X, the point of intersection of the perpendicular bisector of AB and the perpendicular bisector of BC.

d Does the point X lie on the perpendicular bisector of AC?

e What does your result from **d** suggest about the perpendicular bisectors of the sides of a triangle?

f What is special about the point X in relation to the vertices of the triangle ABC?

3-DIMENSIONAL COORDINATE GEOMETRY (EXTENSION)

Click on the icon to obtain a printable introduction to 3-dimensional coordinate geometry.

In this section you should discover:

- 3-dimensional coordinate representation
- how to find distances in space
- how to find midpoints of intervals in space.

3D GEOMETRY INTRODUCTION

REVIEW SET 7A

1 **a** Find the distance between the points A(-3, 2) and B(1, 5).

   b Find the gradient of the line perpendicular to a line with gradient $\frac{3}{4}$.

   c Find the midpoint of the line segment joining C(-3, 1) to D(5, 7).

2 Find the axis intercepts and gradient of the line with equation $5x - 2y = 10$.

3 Determine the equation of the illustrated line:

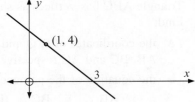

4 Find a given that   P(-3, 4), Q(2, 6) and R(5, a) are collinear.

5 Find c if $(-1, c)$ lies on the line with equation $3x - 2y = 7$.

6 Determine the equations of the following lines:

   a   gradient -3, y-intercept 4   **b**   through the points $(-3, 4)$ and $(3, 1)$.

7 Use graphical methods to find the point of intersection of
   $y = 2x - 9$ and $x + 4y = 36$.

8 Find the distance between   P(-4, 7)   and   Q(-1, 3).

9 Find the equation of the line:

   a   through P(1, -5) which is parallel to the line with equation $4x - 3y = 6$

   b   through Q(-2, 1) which is perpendicular to the line with equation $y = -4x + 7$.

10 The Circular Gardens are bounded by East Avenue and Diagonal Road. Diagonal Road intersects North Street at C and East Avenue at D. Diagonal Rd is tangential to the Circular Gardens at B.

   a Find the equation of:
      i East Avenue
      ii North Street
      iii Diagonal Road.

   b Where does Diagonal Road intersect
      i East Avenue
      ii North Street?

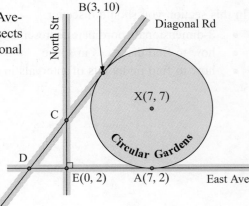

11 Use midpoints to find the fourth vertex of the given parallelogram:

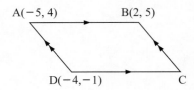

12 Find k given that $(-3, k)$ is 7 units away from $(2, 4)$.

REVIEW SET 7B

1 Determine the midpoint of the line segment joining K(3, 5) to L(7, −2).

2 Find in gradient-intercept form, the equation of the line through:

    a (2, −1) with gradient −3   **b** (3, −2) and (−1, 4)

3 Find where the following lines cut the axes:

    a $y = -\frac{3}{2}x + 7$ **b** $5x - 3y = 12$

4 Does (2, −5) lie on the line with equation $3x + 4y = -14$?

5 State the equation of the line:

    a parallel to $3x + 5y = 7$ through (−1, 3)

    b perpendicular to $2x - 7y = 5$ through (4, 2).

6 If $3x + ky = 7$ and $y = 3 - 4x$ are the equations of two lines, find k if:

    a the lines are parallel   **b** the lines are perpendicular.

7 Find the coordinates of the point where the line through A(−3, 2) and B(1, 7) meets the line with equation $3x + 2y = 6$.

8 A(3, 2) and B(5, 7) are beach shacks and $y = 5$ is a power line. Where should the one power outlet be placed so that it is an equal distance to both shacks and they pay equal service costs?

        • B(5, 7)

        $y = 5$

        • A(3, 2)

9 Given P(1, 5), Q(5, 7), R(3, 1):

    a Show that triangle PQR is isosceles.

    b Find the midpoint M of QR.

    c Use the gradient to verify that PM is perpendicular to QR

    d Draw a sketch to illustrate what you have found in **a**, **b** and **c**.

10 Given A(−1, 1), B(1, 5) and C(5, 1), where M is the midpoint of AB and N is the midpoint of BC:

    a show that MN is parallel to AC, using gradients

    b show that MN has half the length of AC.

REVIEW SET 7C

1 Find the equation of the line in general form, through:

a $(1, -5)$ with gradient $\frac{2}{3}$ **b** $(2, -3)$ and $(-4, -5)$

2 If $5x - 7y = 8$ and $3x + ky = -11$ are the equations of two lines, find the value of k for which:

  a   the lines are parallel   **b**   the lines are perpendicular.

3 A point T on the y-axis, is 3 units from the point A$(-1, 2)$. Find:

  a   the coordinates of T (there are two points T_1, T_2 say)

  b   the equation of the line AT_1, given that T_1 is above T_2.

4 If P(x, y) is equidistant from A$(-1, 4)$ and B$(3, -2)$:

  a   draw a sketch of the possible positions of P

  b   find the equation connecting x and y.

5 Determine the nature of the triangle KLM for K$(-5, -2)$, L$(0, 1)$ and M$(3, -4)$.

6 Use midpoints to find the fourth vertex, K, of parallelogram HIJK for H$(3, 4)$, I$(-3, -1)$, and J$(4, 10)$.

7 Find the equation of the perpendicular bisector of the line segment joining P$(7, -1)$ to Q$(-3, 5)$.

8 Two primary schools are located at P$(5, 12)$ and Q$(9, 4)$ on a council map. If the Local Education Authority wishes to zone the region so that children must attend that school which is closer to their place of residence, what is the equation of the line which should form this boundary?

9 Given A$(6, 8)$, B$(14, 6)$, C$(-1, -3)$ and D$(-9, -1)$:

  a   Use the gradient to show that:

     i   AB is parallel to DC   **ii**   BC is parallel to AD.

  b   What kind of figure is ABCD?

  c   Check that AB = DC and BC = AD using the distance formula.

  d   Find the midpoints of diagonals: **i** AC **ii** BD.

  e   What property of parallelograms has been checked in **d**?

For figures named ABCD, etc. the labelling is in cyclic order.

or

10 Given A$(1, 3)$, B$(6, 3)$, C$(3, -1)$ and D$(-2, -1)$:

  a   show that ABCD is a rhombus, using the distance formula

  b   find the midpoints of AC and BD

  c   show that AC and BD are perpendicular, using gradients.

Chapter 8

Quadratic algebra

Contents:

PRODUCTS AND EXPANSIONS

A **product** of two (or more) *factors* is the result obtained when *multiplying* them together.

Consider the *factors* $-3x$ and $2x^2$. Their product $-3x \times 2x^2$ can be simplified by following the steps below:

Step 1: Find the product of the **signs**.

Step 2: Find the product of the **numerals** (numbers).

Step 3: Find the product of the **pronumerals** (letters).

> In $-3x$,
> the sign is $-$,
> the numeral is 3,
> and the pronumeral is x.

So, $$-3x \times 2x^2 = -6x^3$$

$$- \times + = - \qquad \qquad x \times x^2 = x^3$$
$$2 \times 3 = 6$$

Example 1

Simplify the following products:

a $-3 \times 4x$	**b** $2x \times -x^2$	**c** $-4x \times -2x^2$

a $-3 \times 4x$	**b** $2x \times -x^2$	**c** $-4x \times -2x^2$
$= -12x$	$= -2x^3$	$= 8x^3$

EXERCISE 8A.1

1 Simplify the following:

a $2 \times 3x$	**b** $4x \times 5$	**c** $-2 \times 7x$	**d** $3 \times -2x$
e $2x \times x$	**f** $3x \times 2x$	**g** $-2x \times x$	**h** $-3x \times 4$
i $-2x \times -x$	**j** $-3x \times x^2$	**k** $-x^2 \times -2x$	**l** $3d \times -2d$
m $(-a)^2$	**n** $(-2a)^2$	**o** $2a^2 \times a^2$	**p** $a^2 \times -3a$

Other simplifications of two factors are possible using the **expansion** (or **distributive**) rules:

$$a(b+c) = ab + ac \qquad \text{and} \qquad a(b-c) = ab - ac$$

Geometric Demonstration:

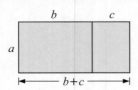

The overall area is $a(b+c)$,

and could also be found by adding the areas of the two small rectangles, i.e., $ab + ac$.

Hence, $a(b+c) = ab + ac$. {equating areas}

Example 2

Expand the following:

a $3(4x + 1)$ **b** $2x(5 - 2x)$ **c** $-2x(x - 3)$

a $3(4x + 1)$
$= 3 \times 4x + 3 \times 1$
$= 12x + 3$

b $2x(5 - 2x)$
$= 2x \times 5 - 2x \times 2x$
$= 10x - 4x^2$

c $-2x(x - 3)$
$= -2x \times x - -2x \times 3$
$= -2x^2 + 6x$

2   Expand and simplify:

a $3(x + 2)$ **b** $2(5 - x)$ **c** $-(x + 2)$ **d** $-(3 - x)$
e $-2(x + 4)$ **f** $-3(2x - 1)$ **g** $x(x + 3)$ **h** $2x(x - 5)$
i $a(a + b)$ **j** $-a(a - b)$ **k** $x(2x - 1)$ **l** $2x(x^2 - x - 2)$

> Notice that the minus sign in front of $2x$ in **b** affects both terms inside the following bracket.

Example 3

Expand and simplify:

a $2(3x - 1) + 3(5 - x)$ **b** $x(2x - 1) - 2x(5 - x)$

a $2(3x - 1) + 3(5 - x)$
$= 6x - 2 + 15 - 3x$
$= 3x + 13$

b $x(2x - 1) - 2x(5 - x)$
$= 2x^2 - x - 10x + 2x^2$
$= 4x^2 - 11x$

3   Expand and simplify:

a $3(x - 4) + 2(5 + x)$ **b** $2a + (a - 2b)$ **c** $2a - (a - 2b)$
d $3(y + 1) + 6(2 - y)$ **e** $2(y - 3) - 4(2y + 1)$ **f** $3x - 4(2 - 3x)$
g $2(b - a) + 3(a + b)$ **h** $x(x + 4) + 2(x - 3)$ **i** $x(x + 4) - 2(x - 3)$
j $x^2 + x(x - 1)$ **k** $-x^2 - x(x - 2)$ **l** $x(x + y) - y(x + y)$
m $-4(x - 2) - (3 - x)$ **n** $5(2x - 1) - (2x + 3)$ **o** $4x(x - 3) - 2x(5 - x)$

THE PRODUCT $(a + b)(c + d)$

Consider the **factors** $(a + b)$ and $(c + d)$.

The product $(a + b)(c + d)$ can be found using the distributive law several times.

$$(a + b)(c + d) = a(c + d) + b(c + d)$$
$$= ac + ad + bc + bd$$

i.e., $(a + b)(c + d) = ac + ad + bc + bd$

Notice that the final result contains four terms:

ac is the product of the **F**irst terms of each bracket.
ad is the product of the **O**uter terms of each bracket.
bc is the product of the **I**nner terms of each bracket.
bd is the product of the **L**ast terms of each bracket.

Remember, this is sometimes called the FOIL rule.

EXERCISE 8A.2

1 Consider the figure alongside:
Give an expression for the area of:

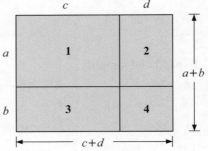

   a rectangle 1   **b** rectangle 2
   c rectangle 3   **d** rectangle 4
   e the overall rectangle.

What can you conclude?

Example 4

Expand and simplify:
$(x+3)(x+2)$

$(x+3)(x+2)$
$= x \times x + x \times 2 + 3 \times x + 3 \times 2$
$= x^2 + 2x + 3x + 6$
$= x^2 + 5x + 6$

In practice we do not include the second line of these worked examples.

Example 5

Expand and simplify:
$(2x+1)(3x-2)$

$(2x+1)(3x-2)$
$= 2x \times 3x - 2x \times 2 + 1 \times 3x - 1 \times 2$
$= 6x^2 - 4x + 3x - 2$
$= 6x^2 - x - 2$

---

2 Use the rule $(a+b)(c+d) = ac + ad + bc + bd$ to expand and simplify:

  a $(x+3)(x+7)$ **b** $(x+5)(x-4)$ **c** $(x-3)(x+6)$
  d $(x+2)(x-2)$ **e** $(x-8)(x+3)$ **f** $(2x+1)(3x+4)$
  g $(1-2x)(4x+1)$ **h** $(4-x)(2x+3)$ **i** $(3x-2)(1+2x)$
  j $(5-3x)(5+x)$ **k** $(7-x)(4x+1)$ **l** $(5x+2)(5x+2)$

Example 6

What do you notice about the two middle terms?

Expand and simplify: **a** $(x+3)(x-3)$ **b** $(3x-5)(3x+5)$

  a $(x+3)(x-3)$
     $= x^2 - 3x + 3x - 9$
     $= x^2 - 9$

  b $(3x-5)(3x+5)$
     $= 9x^2 + 15x - 15x - 25$
     $= 9x^2 - 25$

3 Expand and simplify:

a $(x+2)(x-2)$ b $(a-5)(a+5)$ c $(4+x)(4-x)$

d $(2x+1)(2x-1)$ e $(5a+3)(5a-3)$ f $(4+3a)(4-3a)$

Example 7

Expand and simplify:

a $(3x+1)^2$ b $(2x-3)^2$

> What do you notice about the two middle terms?

a $(3x+1)^2$
$= (3x+1)(3x+1)$
$= 9x^2+3x+3x+1$
$= 9x^2+6x+1$

b $(2x-3)^2$
$= (2x-3)(2x-3)$
$= 4x^2-6x-6x+9$
$= 4x^2-12x+9$

---

4 Expand and simplify:

a $(x+3)^2$ b $(x-2)^2$ c $(3x-2)^2$

d $(1-3x)^2$ e $(3-4x)^2$ f $(5x-y)^2$

DIFFERENCE OF TWO SQUARES

a^2 and b^2 are perfect squares and so a^2-b^2 is called the **difference of two squares**.

Notice that $(a+b)(a-b) = a^2 \underbrace{-ab+ab} -b^2 = a^2-b^2$

the middle
two terms
add to zero

Geometric Demonstration:

Consider the figure drawn alongside:

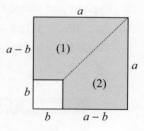

The shaded area
= area of large square − area of small square
$= a^2 - b^2$

Cutting along the dotted line, and flipping (2) over, we form a rectangle.

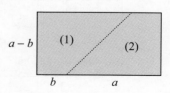

Notice that the rectangle's area is $(a+b)(a-b)$.

Thus, $(a+b)(a-b) = a^2 - b^2$

DEMO

Example 8

Expand and simplify:

a $(x+5)(x-5)$ b $(3-y)(3+y)$

a $(x+5)(x-5)$ b $(3-y)(3+y)$

$= x^2 - 5^2$ $= 3^2 - y^2$

$= x^2 - 25$ $= 9 - y^2$

EXERCISE 8A.3

1 Expand and simplify using the rule $(a+b)(a-b) = a^2 - b^2$:

a $(x+2)(x-2)$ b $(x-2)(x+2)$ c $(2+x)(2-x)$

d $(2-x)(2+x)$ e $(x+1)(x-1)$ f $(1-x)(1+x)$

g $(x+7)(x-7)$ h $(c+8)(c-8)$ i $(d-5)(d+5)$

j $(x+y)(x-y)$ k $(4+d)(4-d)$ l $(5+e)(5-e)$

Example 9

Expand and simplify:

a $(2x-3)(2x+3)$ b $(5-3y)(5+3y)$

a $(2x-3)(2x+3)$ b $(5-3y)(5+3y)$

$= (2x)^2 - 3^2$ $= 5^2 - (3y)^2$

$= 4x^2 - 9$ $= 25 - 9y^2$

2 Expand and simplify using the rule $(a+b)(a-b) = a^2 - b^2$:

a $(2x-1)(2x+1)$ b $(3x+2)(3x-2)$ c $(4y-5)(4y+5)$

d $(2y+5)(2y-5)$ e $(3x+1)(3x-1)$ f $(1-3x)(1+3x)$

g $(2-5y)(2+5y)$ h $(3+4a)(3-4a)$ i $(4+3a)(4-3a)$

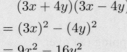

Example 10

Expand and simplify: $(3x+4y)(3x-4y)$

$(3x+4y)(3x-4y)$

$= (3x)^2 - (4y)^2$

$= 9x^2 - 16y^2$

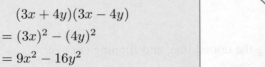

Your answers should
be a **difference** of
two squares!

3 Expand and simplify using the rule $(a+b)(a-b) = a^2 - b^2$:

a $(2a+b)(2a-b)$ b $(a-2b)(a+2b)$ c $(4x+y)(4x-y)$

d $(4x+5y)(4x-5y)$ e $(2x+3y)(2x-3y)$ f $(7x-2y)(7x+2y)$

INVESTIGATION 1 THE PRODUCT OF THREE CONSECUTIVE INTEGERS

Con was trying to multiply $19 \times 20 \times 21$ without a calculator. Aimee told him to 'cube the middle integer and then subtract the middle integer' to get the answer.

What to do:

1 Find $19 \times 20 \times 21$ using a calculator.

2 Find $20^3 - 20$ using a calculator. Does Aimee's rule seem to work?

3 Check that Aimee's rule works for the following products:

    a $4 \times 5 \times 6$ **b** $9 \times 10 \times 11$ **c** $49 \times 50 \times 51$

4 Try to prove Aimee's rule using algebra. By letting the middle integer be x, the other integers must be $(x - 1)$ and $(x + 1)$.

Find the product $(x - 1) \times x \times (x + 1)$ by expanding and simplifying. Have you proved Aimee's rule?

(**Hint**: Rearrange as $x \times (x - 1) \times (x + 1)$ and use the difference between two squares expansion.)

PERFECT SQUARES EXPANSION

$(a + b)^2$ and $(a - b)^2$ are called **perfect squares**.

Notice that:
$$(a + b)^2$$
$$= (a + b)(a + b)$$
$$= a^2 \underbrace{+\, ab + ab}_{\substack{\text{the middle} \\ \text{two terms} \\ \text{are identical}}} + b^2 \quad \{\text{using 'FOIL'}\}$$
$$= a^2 + 2ab + b^2$$

Also,
$$(a - b)^2$$
$$= (a - b)(a - b)$$
$$= a^2 \underbrace{-\, ab - ab}_{\substack{\text{the middle} \\ \text{two terms} \\ \text{are identical}}} + b^2$$
$$= a^2 - 2ab + b^2$$

Thus, we can state the perfect square expansion rules:

$$(a + b)^2 = a^2 + 2ab + b^2$$
$$(a - b)^2 = a^2 - 2ab + b^2$$

The following is a useful way of remembering the perfect square expansion rules:

    Step 1:   Square the *first term*.

    Step 2:   Add or subtract twice the product of the *first* and *last terms* depending on the sign between the terms.

    Step 3:   Add on the square of the *last term*.

EXERCISE 8A.4

1 Consider the figure alongside:

Give an expression for the area of:

    a square 1   **b** rectangle 2   **c** rectangle 3

    d square 4   **e** the overall square.

What can you conclude?

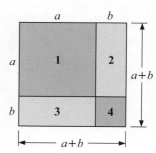

Example 11

Expand and simplify:

a $(x+3)^2$

b $(x-5)^2$

a $(x+3)^2$
$= x^2 + 2 \times x \times 3 + 3^2$
$= x^2 + 6x + 9$

b $(x-5)^2$
$= x^2 - 2 \times x \times 5 + 5^2$
$= x^2 - 10x + 25$

2 Use the rule $(a+b)^2 = a^2 + 2ab + b^2$ to expand and simplify:

a $(x+5)^2$ b $(x+4)^2$ c $(x+7)^2$

d $(a+2)^2$ e $(3+c)^2$ f $(5+x)^2$

3 Use the rule $(a-b)^2 = a^2 - 2ab + b^2$ to expand and simplify:

a $(x-3)^2$ b $(x-2)^2$ c $(y-8)^2$

d $(a-7)^2$ e $(5-x)^2$ f $(4-y)^2$

Example 12

Expand and simplify using perfect square expansion rules:

a $(5x+1)^2$

b $(4-3x)^2$

a $(5x+1)^2$
$= (5x)^2 + 2 \times 5x \times 1 + 1^2$
$= 25x^2 + 10x + 1$

b $(4-3x)^2$
$= 4^2 - 2 \times 4 \times 3x + (3x)^2$
$= 16 - 24x + 9x^2$

4 Expand and simplify using perfect square expansion rules:

a $(3x+4)^2$ b $(2a-3)^2$ c $(3y+1)^2$

d $(2x-5)^2$ e $(3y-5)^2$ f $(7+2a)^2$

g $(1+5x)^2$ h $(7-3y)^2$ i $(3+4a)^2$

Example 13

Expand and simplify: a $(2x^2+3)^2$ b $5-(x+2)^2$

a $(2x^2+3)^2$
$= (2x^2)^2 + 2 \times 2x^2 \times 3 + 3^2$
$= 4x^4 + 12x^2 + 9$

b $5-(x+2)^2$
$= 5 - [x^2 + 4x + 4]$
$= 5 - x^2 - 4x - 4$
$= 1 - x^2 - 4x$

Notice the use of square brackets in the second line. These remind us to change the signs inside them when they are removed.

5 Expand and simplify:

    a $(x^2 + 2)^2$ **b** $(y^2 - 3)^2$ **c** $(3a^2 + 4)^2$

    d $(1 - 2x^2)^2$ **e** $(x^2 + y^2)^2$ **f** $(x^2 - a^2)^2$

6 Expand and simplify:

    a $3x + 1 - (x + 3)^2$ **b** $5x - 2 + (x - 2)^2$

    c $(x + 2)(x - 2) + (x + 3)^2$ **d** $(x + 2)(x - 2) - (x + 3)^2$

    e $(3 - 2x)^2 - (x - 1)(x + 2)$ **f** $(1 - 3x)^2 + (x + 2)(x - 3)$

    g $(2x + 3)(2x - 3) - (x + 1)^2$ **h** $(4x + 3)(x - 2) - (2 - x)^2$

    i $(1 - x)^2 + (x + 2)^2$ **j** $(1 - x)^2 - (x + 2)^2$

B   FURTHER EXPANSION

Consider the expansion of $(a + b)(c + d + e)$.

    Now $(a + b)(c + d + e)$ Compare: $p(c + d + e)$

         $= (a + b)c + (a + b)d + (a + b)e$ $= pc + pd + pe$

         $= ac + bc + ad + bd + ae + be$

Notice that there are 6 terms in this expansion and that each term within the first bracket is multiplied by each term in the second,

  i.e.,   2 terms in first bracket $\times$ 3 terms in second bracket $\longrightarrow$ 6 terms in expansion.

Example 14

Expand and simplify: $(2x + 3)(x^2 + 4x + 5)$

     $(2x + 3)(x^2 + 4x + 5)$

     $= 2x^3 + 8x^2 + 10x$ {all terms of 2nd bracket $\times$ $2x$}

        $+ 3x^2 + 12x + 15$ {all terms of 2nd bracket $\times$ 3}

     $= 2x^3 + 11x^2 + 22x + 15$ {collecting like terms}

EXERCISE 8B

1 Expand and simplify:

    a $(x + 3)(x^2 + x + 2)$ **b** $(x + 4)(x^2 + x - 2)$

    c $(x + 2)(x^2 + x + 1)$ **d** $(x + 5)(x^2 - x - 1)$

    e $(2x + 1)(x^2 + x + 4)$ **f** $(3x - 2)(x^2 - x - 3)$

    g $(x + 2)(2x^2 - x + 2)$ **h** $(2x - 1)(3x^2 - x + 2)$

When expanding, each term of the first bracket is multiplied by each term of the second bracket.

Example 15

Expand and simplify: $(x+2)^3$

$$(x+2)^3 = (x+2)^2 \times (x+2)$$

$$= (x^2 + 4x + 4)(x+2)$$

$$= x^3 + 4x^2 + 4x \qquad \{\text{all terms in 1st bracket} \times x\}$$
$$\quad + 2x^2 + 8x + 8 \qquad \{\text{all terms in 1st bracket} \times 2\}$$
$$= x^3 + 6x^2 + 12x + 8 \qquad \{\text{collecting like terms}\}$$

2 Expand and simplify:

  a $(x+1)^3$ **b** $(x+3)^3$ **c** $(x-1)^3$

  d $(x-3)^3$ **e** $(2x+1)^3$ **f** $(3x-2)^3$

Example 16

Expand and simplify:

  a $x(x+1)(x+2)$ **b** $(x+1)(x-2)(x+2)$

  a $\quad x(x+1)(x+2)$

$$= (x^2 + x)(x+2) \qquad \{\text{all terms in first bracket} \times x\}$$
$$= x^3 + 2x^2 + x^2 + 2x \qquad \{\text{expand remaining factors}\}$$
$$= x^3 + 3x^2 + 2x \qquad \{\text{collect like terms}\}$$

  b $\quad (x+1)(x-2)(x+2)$

$$= (x^2 - 2x + x - 2)(x+2) \qquad \{\text{expand first two factors}\}$$
$$= (x^2 - x - 2)(x+2) \qquad \{\text{collect like terms}\}$$
$$= x^3 + 2x^2 - x^2 - 2x - 2x - 4 \qquad \{\text{expand remaining factors}\}$$
$$= x^3 + x^2 - 4x - 4 \qquad \{\text{collect like terms}\}$$

3 Expand and simplify:

  a $x(x+2)(x+3)$ **b** $x(x-4)(x+1)$ **c** $x(x-3)(x-2)$

  d $2x(x+3)(x+1)$ **e** $2x(x-4)(1-x)$ **f** $-x(3+x)(2-x)$

  g $-3x(2x-1)(x+2)$ **h** $x(1-3x)(2x+1)$

4 Expand and simplify:

  a $(x+3)(x+2)(x+1)$ **b** $(x-2)(x-1)(x+4)$

  c $(x-4)(x-1)(x-3)$ **d** $(2x-1)(x+2)(x-1)$

  e $(3x+2)(x+1)(x+3)$ **f** $(2x+1)(2x-1)(x+4)$

  g $(1-x)(3x+2)(x-2)$ **h** $(x-3)(1-x)(3x+2)$

5 State how many terms would be in the expansion of the following:

    a $(a+b)(c+d)$ **b** $(a+b+c)(d+e)$

    c $(a+b)(c+d+e)$ **d** $(a+b+c)(d+e+f)$

    e $(a+b+c+d)(e+f)$ **f** $(a+b+c+d)(e+f+g)$

    g $(a+b)(c+d)(e+f)$ **h** $(a+b+c)(d+e)(f+g)$

INVESTIGATION 2   THE EXPANSION OF $(a+b)^3$

The purpose of this investigation is to discover the binomial expansion for $(a+b)^3$.

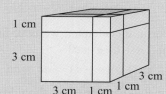

What to do:

1 Find a large potato and cut it to obtain a 4 cm by 4 cm by 4 cm cube.

2 Now, by making 3 cuts parallel to the cube's surfaces, divide the cube as shown in the given figure into 8 rectangular prisms.

3 Find how many prisms are:

    a   3 by 3 by 3   **b**   3 by 3 by 1

    c   3 by 1 by 1   **d**   1 by 1 by 1

4 Now instead of 3 cm and 1 cm dimensions, suppose the original cube has dimensions a cm and b cm. How many prisms are

    a a by a by a **b** a by a by b **c** a by b by b **d** b by b by b?

5 Why is the volume of the cube in **2** in terms of a and b given by $(a+b)^3$?

By adding the volumes of the 8 rectangular prisms, find an expression for the total volume, and hence write down the expansion formula for $(a+b)^3$.

C   FACTORISATION OF QUADRATIC EXPRESSIONS

Many of the previous **expansions** have resulted in expressions of the form $ax^2 + bx + c$.

For example: $x^2 + 5x + 6$, $4x^2 - 9$, $9x^2 + 6x + 1$.

These are known as **quadratic expressions**.

> A **quadratic expression** in x is an expression of the form $ax^2 + bx + c$ where x is the variable and a, b and c represent constants with $a \neq 0$.
>
> $$ax^2 \quad + \quad bx \quad + \quad c$$
> $$\uparrow \qquad\qquad \uparrow \qquad\qquad \uparrow$$
> the x^2 term   the x term   the constant term

[**Note:**   If $a = 0$,   then the expression becomes $bx + c$ which is **linear** in x.]

In the following sections, you will learn several techniques for **factorising** quadratic expressions and **solving** quadratic equations. **Factorisation** is critical in the solution of problems that convert to quadratic equations.

> **Factorisation** is the process of writing an expression as a *product* of *factors*.

For example: given the product of two factors such as $(x + 2)(x + 3)$ we can expand to get $(x + 2)(x + 3) = x^2 + 3x + 2x + 6 = x^2 + 5x + 6$.

Notice that: **Factorisation** is the reverse process of expansion.

For example:

$$\overbrace{(x + 2)\,(x + 3)}^{\text{expansion}} = \underbrace{x^2 + 5x + 6}_{\text{factorisation}}$$

As $x^2 + 5x + 6 = (x + 2)(x + 3)$,   we say that

$(x + 2)$ and $(x + 3)$ are *factors* of $x^2 + 5x + 6$.

So far in this chapter, we have used the **distributive law** to **expand** binomial products of the form $(a + b)(c + d)$

i.e., $(a + b)(c + d) = ac + ad + bc + bd$,   using the FOIL method.

Special expansions you should be familiar with include:

$(x + p)(x + q) = x^2 + (p + q)x + pq$ **sum and product** expansion

$\left. \begin{array}{l} (x + a)^2 = x^2 + 2ax + a^2 \\ (x - a)^2 = x^2 - 2ax + a^2 \end{array} \right\}$ **perfect square** expansion

$(x + a)(x - a) = x^2 - a^2$ **difference of two squares** expansion.

These statements are called **identities** because they are true for all values of the variable x.

Notice that the RHS of each identity is a quadratic expression which has been formed by **expanding** the LHS.

The LHS of the identities above can be obtained by **factorising** the RHS.

REMOVAL OF COMMON FACTORS

Many quadratic expressions are easily factorised by removing the Highest Common Factor (HCF). In fact, we should always look to remove the HCF before proceeding with any other factorisation.

Example 17

Factorise by removing a common factor:

a $2x^2 + 3x$ b $2x^2 - 6x$

---

a $2x^2 + 3x$ has HCF x b $2x^2 - 6x$ has HCF $2x$

$= x(2x + 3)$ $= 2x(x - 3)$

EXERCISE 8C.1

1 Fully factorise:

    a $3x^2 + 5x$ **b** $2x^2 - 7x$ **c** $3x^2 + 6x$

    d $4x^2 - 8x$ **e** $2x^2 - 9x$ **f** $3x^2 + 15x$

    g $4x + 8x^2$ **h** $5x - 10x^2$ **i** $12x - 4x^2$

Example 18

Fully factorise by removing a common factor:

    a $(x-5)^2 - 2(x-5)$ **b** $(x+2)^2 + 2x + 4$

    a $(x-5)^2 - 2(x-5)$

        $= (x-5)(x-5) - 2(x-5)$ has HCF $(x-5)$

        $= (x-5)[(x-5) - 2]$

        $= (x-5)(x-7)$ {simplifying}

    b $(x+2)^2 + 2x + 4$

        $= (x+2)(x+2) + 2(x+2)$ has HCF $(x+2)$

        $= (x+2)[(x+2) + 2]$

        $= (x+2)(x+4)$

Check your factorisations by expansion! Notice the use of the square brackets.

---

2 Fully factorise by removing a common factor:

    a $(x+2)^2 - 5(x+2)$ **b** $(x-1)^2 - 3(x-1)$ **c** $(x+1)^2 + 2(x+1)$

    d $(x-2)^2 + 3x - 6$ **e** $x+3+(x+3)^2$ **f** $(x+4)^2 + 8 + 2x$

    g $(x-3)^2 - x + 3$ **h** $(x+4)^2 - 2x - 8$ **i** $(x-4)^2 - 5x + 20$

THE DIFFERENCE OF TWO SQUARES FACTORISATION

We know the **expansion** of $(a+b)(a-b)$ is $a^2 - b^2$

Thus, the **factorisation** of $a^2 - b^2$ is $(a+b)(a-b)$

i.e., $\quad a^2 - b^2 = (a+b)(a-b)$

Example 19

Use the rule $\quad a^2 - b^2 = (a+b)(a-b)\quad$ to factorise fully:

    a $9 - x^2$ **b** $4x^2 - 25$

    a $9 - x^2$ **b** $4x^2 - 25$

        $= 3^2 - x^2$ $= (2x)^2 - 5^2$

        $= (3+x)(3-x)$ $= (2x+5)(2x-5)$

The difference between a^2 and b^2 is $a^2 - b^2$ which is the difference of two squares.

EXERCISE 8C.2

1 Use the rule $a^2 - b^2 = (a + b)(a - b)$ to fully factorise:

    a $x^2 - 4$ **b** $4 - x^2$ **c** $x^2 - 81$ **d** $25 - x^2$

    e $4x^2 - 1$ **f** $9x^2 - 16$ **g** $4x^2 - 9$ **h** $36 - 49x^2$

Example 20

Fully factorise: **a** $2x^2 - 8$ **b** $3x^2 - 48$

a $2x^2 - 8$

     $= 2(x^2 - 4)$ {HCF is 2}

     $= 2(x^2 - 2^2)$ {difference of squares}

     $= 2(x + 2)(x - 2)$

b $3x^2 - 48$

     $= 3(x^2 - 16)$ {HCF is 3}

     $= 3(x^2 - 4^2)$ {difference of squares}

     $= 3(x + 4)(x - 4)$

---

2 Fully factorise:

    a $3x^2 - 27$ **b** $-2x^2 + 8$ **c** $3x^2 - 75$

    d $-5x^2 + 5$ **e** $8x^2 - 18$ **f** $-27x^2 + 75$

We notice that $x^2 - 9$ is the difference between two squares and therefore we can factorise it using $a^2 - b^2 = (a + b)(a - b)$.

Even though 7 is not a perfect square we can still factorise $x^2 - 7$ if we use $7 = (\sqrt{7})^2$.

$$\text{So,} \quad x^2 - 7 = x^2 - (\sqrt{7})^2$$
$$= (x + \sqrt{7})(x - \sqrt{7}).$$

We say that $x + \sqrt{7}$ and $x - \sqrt{7}$ are the linear factors of $x^2 - 7$.

Example 21

Factorise into linear factors:

a $x^2 - 11$ **b** $(x + 3)^2 - 5$

a $x^2 - 11$ **b** $(x + 3)^2 - 5$

     $= x^2 - (\sqrt{11})^2$ $= (x + 3)^2 - (\sqrt{5})^2$

     $= (x + \sqrt{11})(x - \sqrt{11})$ $= [(x + 3) + \sqrt{5}][(x + 3) - \sqrt{5}]$

                                         $= [x + 3 + \sqrt{5}][x + 3 - \sqrt{5}]$

Note: The **sum of two squares** does not factorise into two real linear factors.

3 If possible, factorise into linear factors:

a $x^2 - 3$	**b** $x^2 + 4$	**c** $x^2 - 15$
d $3x^2 - 15$	**e** $(x + 1)^2 - 6$	**f** $(x + 2)^2 + 6$
g $(x - 2)^2 - 7$	**h** $(x + 3)^2 - 17$	**i** $(x - 4)^2 + 9$

Example 22

Factorise using the difference between two squares:

a $(3x + 2)^2 - 9$ **b** $(x + 2)^2 - (x - 1)^2$

a $\quad (3x + 2)^2 - 9$

$= (3x + 2)^2 - 3^2$

$= [(3x + 2) + 3][(3x + 2) - 3]$

$= [3x + 5][3x - 1]$

b $\quad (x + 2)^2 - (x - 1)^2$

$= [(x + 2) + (x - 1)][(x + 2) - (x - 1)]$

$= [x + 2 + x - 1][x + 2 - x + 1]$

$= [2x + 1][3]$

$= 3(2x + 1)$

---

4 Factorise using the difference between two squares:

a $(x + 1)^2 - 4$	**b** $(2x + 1)^2 - 9$	**c** $(1 - x)^2 - 16$
d $(x + 3)^2 - 4x^2$	**e** $4x^2 - (x + 2)^2$	**f** $9x^2 - (3 - x)^2$
g $(2x + 1)^2 - (x - 2)^2$	**h** $(3x - 1)^2 - (x + 1)^2$	**i** $4x^2 - (2x + 3)^2$

PERFECT SQUARE FACTORISATION

We know the **expansion** of $(x + a)^2$ is $x^2 + 2ax + a^2$ and $(x - a)^2$ is $x^2 - 2ax + a^2$.

Thus the **factorisation** of $x^2 + 2ax + a^2$ is $(x + a)^2$ and $x^2 - 2ax + a^2$ is $(x - a)^2$,

i.e., $\quad \mathbf{x^2 + 2ax + a^2 = (x + a)^2}$
$\quad\quad \mathbf{x^2 - 2ax + a^2 = (x - a)^2}$

Example 23

Use perfect square rules to fully factorise:

a $x^2 + 10x + 25$ **b** $x^2 - 14x + 49$

> $(x + a)^2$ and $(x - a)^2$ are **perfect squares!**

a $\quad x^2 + 10x + 25$

$= x^2 + 2 \times x \times 5 + 5^2$

$= (x + 5)^2$

b $\quad x^2 - 14x + 49$

$= x^2 - 2 \times x \times 7 + 7^2$

$= (x - 7)^2$

EXERCISE 8C.3

1 Use perfect square rules to fully factorise:

a $x^2 + 6x + 9$	**b** $x^2 + 8x + 16$	**c** $x^2 - 6x + 9$
d $x^2 - 8x + 16$	**e** $x^2 + 2x + 1$	**f** $x^2 - 10x + 25$
g $y^2 + 18y + 81$	**h** $m^2 - 20m + 100$	**i** $t^2 + 12t + 36$

Example 24

Fully factorise:

a $9x^2 - 6x + 1$

b $8x^2 + 24x + 18$

a $9x^2 - 6x + 1$
$= (3x)^2 - 2 \times 3x \times 1 + 1^2$
$= (3x - 1)^2$

b $8x^2 + 24x + 18$
$= 2(4x^2 + 12x + 9)$ {as HCF is 2}
$= 2([2x]^2 + 2 \times 2x \times 3 + 3^2)$
$= 2(2x + 3)^2$

2 Fully factorise:

a $9x^2 + 6x + 1$

b $4x^2 - 4x + 1$

c $9x^2 + 12x + 4$

d $25x^2 - 10x + 1$

e $16x^2 + 24x + 9$

f $25x^2 - 20x + 4$

g $x^2 - 2x + 1$

h $2x^2 + 8x + 8$

i $3x^2 + 30x + 75$

QUADRATIC TRINOMIAL FACTORISATION

A **quadratic trinomial** expression has an x^2 term and three terms overall.

For example: $x^2 + 7x + 6$ and $3x^2 - 13x - 10$.

Notice that $(x + p)(x + q) = x^2 + qx + px + pq = x^2 + (p + q)x + pq$

and so $x^2 + (p + q)x + pq = (x + p)(x + q)$

So, in order to factorise $x^2 + 7x + 6$, we look for two numbers with a product of 6 and a sum of 7. These numbers are 1 and 6, and so $x^2 + 7x + 6 = (x + 1)(x + 6)$.

Example 25

Use the sum and product method of factorisation to fully factorise:

a $x^2 + 5x + 4$

b $x^2 - x - 12$

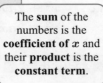

The **sum** of the numbers is the **coefficient of x** and their **product** is the **constant term**.

a $x^2 + 5x + 4$ has $p + q = 5$ and $pq = 4$.
So p and q are 1 and 4.
$\therefore \quad x^2 + 5x + 4 = (x + 1)(x + 4)$

b $x^2 - x - 12$ has $p + q = -1$ and $pq = -12$.
So p and q are -4 and 3.
$\therefore \quad x^2 - x - 12 = (x - 4)(x + 3)$

EXERCISE 8C.4

1 Use the $x^2 + (p + q)x + pq = (x + p)(x + q)$ factorisation to fully factorise:

a $x^2 + 3x + 2$

b $x^2 + 5x + 6$

c $x^2 - x - 6$

d $x^2 + 3x - 10$

e $x^2 + 4x - 21$

f $x^2 + 8x + 16$

g $x^2 - 14x + 49$

h $x^2 + 3x - 28$

i $x^2 + 7x + 10$

j	$x^2 - 11x + 24$	**k**	$x^2 + 15x + 44$	**l**	$x^2 + x - 42$
m	$x^2 - x - 56$	**n**	$x^2 - 18x + 81$	**o**	$x^2 - 4x - 32$

Example 26

Fully factorise by first removing a common factor:

a $3x^2 - 9x + 6$ 　　　　　　 **b** $2x^2 - 2x - 12$

a   　$3x^2 - 9x + 6$
　　$= 3(x^2 - 3x + 2)$　　{removing 3 as a common factor}
　　$= 3(x - 2)(x - 1)$　　{as $p + q = -3$ and $pq = 2$ gives $p = -2$, $q = -1$}

b   　$2x^2 - 2x - 12$
　　$= 2(x^2 - x - 6)$　　{removing 2 as a common factor}
　　$= 2(x - 3)(x + 2)$　　{as $p + q = -1$ and $pq = -6$ gives $p = -3$, $q = 2$}

2　Fully factorise by first removing a common factor:

a	$2x^2 - 6x - 8$	**b**	$3x^2 + 9x - 12$	**c**	$5x^2 + 10x - 15$
d	$4x^2 + 4x - 80$	**e**	$2x^2 - 4x - 30$	**f**	$3x^2 + 12x - 63$
g	$2x^2 - 2x - 40$	**h**	$3x^2 - 12x + 12$	**i**	$7x^2 + 21x - 28$
j	$5x^2 + 15x - 50$	**k**	$2x^2 + 8x - 42$	**l**	$-4x + x^2 - 32$

MISCELLANEOUS FACTORISATION

Use the following steps in order to **factorise quadratic expressions**:

Step 1:　Look at the quadratic expression to be factorised.

Step 2:　If there is a **common factor**, take it out.

Step 3:　Look for **perfect square** factorisation:　$x^2 + 2ax + a^2 = (x + a)^2$
　　　　　　　　　　　　　　　　　　　　　　$x^2 - 2ax + a^2 = (x - a)^2$

　　　　or　look for the **difference of two squares**:　$x^2 - a^2 = (x + a)(x - a)$

　　　　or　look for the **sum and product** type:　$x^2 + (p + q)x + pq$
　　　　　　　　　　　　　　　　　　　　　　$= (x + p)(x + q)$

EXERCISE 8C.5

1　Where possible, fully factorise the following miscellaneous expressions:

a	$3x^2 + 9x$	**b**	$4x^2 - 1$	**c**	$5x^2 - 15$
d	$3x - 5x^2$	**e**	$x^2 + 3x - 40$	**f**	$2x^2 - 32$
g	$x^2 + 9$	**h**	$x^2 + 10x + 25$	**i**	$x^2 - x - 6$
j	$x^2 - 16x + 39$	**k**	$x^2 - 7x - 60$	**l**	$x^2 - 2x - 8$
m	$x^2 + 11x + 30$	**n**	$x^2 + 6x - 16$	**o**	$x^2 - 5x - 24$
p	$3x^2 + 6x - 72$	**q**	$4x^2 - 8x - 60$	**r**	$3x^2 - 42x + 99$

D | FACTORISATION OF ax^2+bx+c $(a \neq 1)$

In the previous section we revised techniques for factorising quadratic expressions in the form $ax^2 + bx + c$ where: • $a = 1$ • a was a common factor

For example:

$$x^2 + 5x + 6$$
$$= (x + 3)(x + 2)$$

For example:

$$2x^2 + 10x + 12$$
$$= 2(x^2 + 5x + 6)$$
$$= 2(x + 3)(x + 2)$$

Factorising a quadratic expression such as $3x^2 + 11x + 6$ appears to be more complicated because the coefficient of x^2 is not 'one' and is not a common factor.

We need to develop a method for factorising this type of quadratic expression.

Two methods for factorising $ax^2 + bx + c$ where $a \neq 1$ are commonly used.

These are   • trial and error   • by 'splitting the x-term'

TRIAL AND ERROR

For example: consider the quadratic $3x^2 + 13x + 4$.

Since 3 is a prime number, $3x^2 + 13x + 4 = (3x\quad)(x\quad)$

To fill the gaps we are seeking two numbers with a product of 4 and a net result of inners and outers being $13x$.

As the product is 4 we will try   2 and 2,   4 and 1,   and   1 and 4.

$$(3x + 2)(x + 2) = 3x^2 + 6x + 2x + 4 \quad \text{fails}$$
$$(3x + 4)(x + 1) = 3x^2 + 3x + 4x + 4 \quad \text{fails}$$
$$(3x + 1)(x + 4) = 3x^2 + 12x + x + 4 \quad \text{is successful}$$

So, $3x^2 + 13x + 4 = (3x + 1)(x + 4)$

Note: We could set these trials out in table form:

$3x$	2	4	1
x	2	1	4
	$8x$	$7x$	$13x$

This entry is $3x \times 2 + x \times 2$ (as shown).

Now, if a and c are not prime in $ax^2 + bx + c$ there can be many possibilities.

For example,   consider $8x^2 + 22x + 15$.

By simply using trial and error the possible factorisations are:

$(8x + 5)(x + 3)$	$\times$	$(4x + 5)(2x + 3)$	$\checkmark$ this is correct
$(8x + 3)(x + 5)$	$\times$	$(4x + 3)(2x + 5)$	$\times$
$(8x + 1)(x + 15)$	$\times$	$(4x + 15)(2x + 1)$	$\times$
$(8x + 15)(x + 1)$	$\times$	$(4x + 1)(2x + 15)$	$\times$

We could set these trials out in table form:

$8x$	5	3	1	15
x	3	5	15	1

$$29x \qquad 43x \qquad 121x \qquad 23x$$

or

$4x$	5	3	1	15
$2x$	3	5	15	1

$$22x \qquad 26x \qquad 62x \qquad 34x$$

As you can see, this process can be very tedious and time consuming.

FACTORISATION BY 'SPLITTING' THE x-TERM

Using the distributive law to expand we see that

$$(2x + 3)(4x + 5)$$
$$= 8x^2 + 10x + 12x + 15$$
$$= 8x^2 + 22x + 15$$

We will now *reverse* the process to factorise the quadratic expression $8x^2 + 22x + 15$

Notice that:

$$8x^2 + 22x + 15$$
$$= 8x^2 + 10x + 12x + 15 \qquad \text{\{splitting the middle term\}}$$
$$= (8x^2 + 10x) + (12x + 15) \qquad \text{\{grouping in pairs\}}$$
$$= 2x(4x + 5) + 3(4x + 5) \qquad \text{\{factorising each pair separately\}}$$
$$= (4x + 5)(2x + 3) \qquad \text{\{completing the factorisation\}}$$

But how do we correctly 'split' the middle term? That is, how do we determine that $22x$ must be written as $+ 10x + 12x$?

When looking at $8x^2 + 10x + 12x + 15$ we notice that $8 \times 15 = 120$ and $10 \times 12 = 120$
and also $10 + 12 = 22$.

So, in $8x^2 + 22x + 15$,   we are looking for two numbers such that their *sum* is 22 and their *product* is $8 \times 15 = 120$ and these numbers are 10 and 12.

Likewise in $6x^2 + 19x + 15$ we seek two numbers of sum 19 and product $6 \times 15 = 90$.

These numbers are 10 and 9.   So $6x^2 + 19x + 15$
$$= 6x^2 + 10x + 9x + 15$$
$$= (6x^2 + 10x) + (9x + 15)$$
$$= 2x(3x + 5) + 3(3x + 5)$$
$$= (3x + 5)(2x + 3)$$

Rules for **splitting the x-term**:

> The following procedure is recommended for factorising $ax^2 + bx + c$:
>
> *Step 1:*   Find ac and then the factors of ac which add to b.
>
> *Step 2:*   If these factors are p and q replace bx by $px + qx$.
>
> *Step 3:*   Complete the factorisation.

Example 27

Show how to split the middle term of the following so that factorisation can occur:

a $3x^2 + 7x + 2$ **b** $10x^2 - 5x + 12$

a   In $3x^2 + 7x + 2$, $ac = 6$ and $b = 7$,   so we are looking for
two numbers with a product of 6 and a sum of 7. These are 1 and 6.
So, the split is $7x = x + 6x$.

> **b** In $10x^2 - 23x - 5$, $ac = -50$ and $b = -23$, so we are looking for
>   two numbers with a product of -50 and a sum of -23. These are -25 and 2.
>   So, the split is $-23x = -25x + 2x$.

Example 28

Factorise, using the 'splitting method':

a $6x^2 + 19x + 10$ **b** $3x^2 - x - 10$

---

a $6x^2 + 19x + 10$ has $ac = 60$ and $b = 19$. So we are looking for
    two numbers with product 60 and sum 19.

Searching amongst the factors of 60, only 4 and 15 have a sum of 19.

$\therefore \quad 6x^2 + 19x + 10$

$= 6x^2 + 4x + 15x + 10$ {splitting the x-term}

$= 2x(3x + 2) \; + \; 5(3x + 2)$ {factorising in pairs}

$= (3x + 2)(2x + 5)$ {taking out the common factor}

b $3x^2 - x - 10$ has $ac = -30$ and $b = -1$. So we are looking for
    two numbers with product -30 and sum -1.

Searching amongst the factors of -30, only 5 and -6 have a sum of -1.

$\therefore \quad 3x^2 - x - 10$

$= 3x^2 + 5x - 6x - 10$ {splitting the x-term}

$= x(3x + 5) \; - \; 2(3x + 5)$ {factorising in pairs}

$= (3x + 5)(x - 2)$ {common factor}

> Remember to check
> your factorisations
> by expansion!

EXERCISE 8D

1 Fully factorise:

a $2x^2 + 5x + 3$ **b** $2x^2 + 7x + 5$ **c** $7x^2 + 9x + 2$

d $3x^2 + 7x + 4$ **e** $3x^2 + 13x + 4$ **f** $3x^2 + 8x + 4$

g $8x^2 + 14x + 3$ **h** $21x^2 + 17x + 2$ **i** $6x^2 + 5x + 1$

j $6x^2 + 19x + 3$ **k** $10x^2 + 17x + 3$ **l** $14x^2 + 37x + 5$

2 Fully factorise:

a $2x^2 - 9x - 5$ **b** $3x^2 + 5x - 2$ **c** $3x^2 - 5x - 2$

d $2x^2 + 3x - 2$ **e** $2x^2 + 3x - 5$ **f** $5x^2 - 14x - 3$

g $5x^2 - 8x + 3$ **h** $11x^2 - 9x - 2$ **i** $3x^2 - 7x - 6$

j $2x^2 - 3x - 9$ **k** $3x^2 - 17x + 10$ **l** $5x^2 - 13x - 6$

m $3x^2 + 10x - 8$ **n** $2x^2 + 17x - 9$ **o** $2x^2 + 9x - 18$

p $2x^2 + 11x - 21$ **q** $15x^2 + x - 2$ **r** $21x^2 - 62x - 3$

s $9x^2 - 12x + 4$ **t** $12x^2 + 17x - 40$ **u** $16x^2 + 34x - 15$

E   QUADRATIC EQUATIONS

A **quadratic equation** in x is an equation of the form $ax^2 + bx + c = 0$ where x is the unknown and a, b and c are constants with $a \neq 0$.

Quadratic equations are examples of second degree equations, since the *highest power* of x in the equation is two.

QUADRATIC EQUATIONS OF THE FORM $x^2 = k$

Consider the equation $x^2 - 7 = 0$

i.e., $x^2 = 7$ (adding 7 to both sides)

Notice that $\sqrt{7} \times \sqrt{7} = 7$, so $x = \sqrt{7}$ is one solution

and $(-\sqrt{7}) \times (-\sqrt{7}) = 7$, so $x = -\sqrt{7}$ is also a solution.

Thus, if $x^2 = 7$, then $x = \pm\sqrt{7}$.

> $\pm\sqrt{7}$ is read as 'plus or minus the square root of 7'

SOLUTION OF $x^2 = k$

$$\text{If } x^2 = k \text{ then } \begin{cases} x = \pm\sqrt{k} & \text{if } k > 0 \\ x = 0 & \text{if } k = 0 \\ \text{there are no real solutions} & \text{if } k < 0 \end{cases}$$

Example 29

Solve for x: **a** $2x^2 + 1 = 15$ **b** $2 - 3x^2 = 8$

a $\quad 2x^2 + 1 = 15$

$\therefore \quad 2x^2 = 14$ {take 1 from both sides}

$\therefore \quad x^2 = 7$ {divide both sides by 2}

$\therefore \quad x = \pm\sqrt{7}$

b $\quad 2 - 3x^2 = 8$

$\therefore \quad -3x^2 = 6$ {take 2 from both sides}

$\therefore \quad x^2 = -2$ {dividing both sides by -3}

which has no solutions as x^2 cannot be < 0.

EXERCISE 8E.1

1 Solve for x:

   a $x^2 = 16$ **b** $2x^2 = 18$ **c** $3x^2 = 27$

   d $12x^2 = 72$ **e** $3x^2 = -12$ **f** $4x^2 = 0$

   g $2x^2 + 1 = 19$ **h** $1 - 3x^2 = 10$ **i** $2x^2 + 7 = 13$

Example 30

Solve for x: **a** $(x-3)^2 = 16$ **b** $(x+2)^2 = 11$

Did you notice that for equations of the form $(x \pm a)^2 = k$ we did not expand the LHS?

a $(x-3)^2 = 16$

$\therefore \quad x-3 = \pm\sqrt{16}$

$\therefore \quad x-3 = \pm 4$

$\therefore \quad x = 3 \pm 4$

$\therefore \quad x = 7 \text{ or } -1$

b $(x+2)^2 = 11$

$\therefore \quad x+2 = \pm\sqrt{11}$

$\therefore \quad x = -2 \pm \sqrt{11}$

2 Solve for x:

   a $(x-2)^2 = 9$
   b $(x+4)^2 = 25$
   c $(x+3)^2 = -1$

   d $(x-4)^2 = 2$
   e $(x+3)^2 = -7$
   f $(x+2)^2 = 0$

   g $(2x+5)^2 = 0$
   h $(3x-2)^2 = 4$
   i $\frac{1}{3}(2x-1)^2 = 8$

OTHER QUADRATIC EQUATIONS

Linear equations like $2x + 3 = 11$ have **only one solution**, $x = 4$ in this case.

However, as we have seen in the questions above, quadratic equations may have:

- **no real** solutions
- **one** solution   (two equal solutions)

or
- **two** solutions.

For example, consider $x^2 + 10x + 21 = 0$.

    If $x = -3,$ $x^2 + 10x + 21$ If $x = -7,$ $x^2 + 10x + 21$

                 $= 9 - 30 + 21$ $= 49 - 70 + 21$

                 $= 0$ $= 0$

    So, $x^2 + 10x + 21 = 0$ has solutions $x = -3 \text{ or } -7$.

How do we find these solutions without using 'trial and error'?

One method is to use **factorisation** and the **Null Factor law**.

THE NULL FACTOR LAW

When the product of two (or more) numbers is zero, then at least one of them must be zero,

i.e.,   if $ab = 0$ then $a = 0$ or $b = 0$.

Example 31

Solve for x using the Null Factor law:

a $3x(x-5) = 0$
      b $(x-4)(3x+7) = 0$

a $3x(x-5) = 0$ **b** $(x-4)(3x+7) = 0$

$\therefore \quad 3x = 0 \text{ or } x-5 = 0$ $\therefore \quad x-4 = 0 \text{ or } 3x+7 = 0$

$\therefore \quad x = 0 \text{ or } 5$ $\therefore \quad x = 4 \text{ or } 3x = -7$

                                         $\therefore \quad x = 4 \text{ or } -\frac{7}{3}$

EXERCISE 8E.2

1 Solve for x using the Null Factor law:

a $\quad x(x + 3) = 0$

b $\quad 2x(x - 5) = 0$

c $\quad (x - 1)(x - 3) = 0$

d $\quad 4x(2 - x) = 0$

e $\quad 3x(2x + 1) = 0$

f $\quad 5(x + 2)(2x - 1) = 0$

g $\quad (2x + 1)(2x - 1) = 0$

h $\quad 11(x + 2)(x - 7) = 0$

i $\quad 6(x - 5)(3x + 2) = 0$

j $\quad x^2 = 0$

k $\quad 2(x - 3)^2 = 0$

l $\quad 4(2x - 1)^2 = 0$

Example 32

Solve for x: $\quad x^2 = 3x$

$$x^2 = 3x$$
$$\therefore \quad x^2 - 3x = 0 \qquad \{\text{'equating to zero', i.e., 0 on RHS by itself}\}$$
$$\therefore \quad x(x - 3) = 0 \qquad \{\text{factorising the LHS}\}$$
$$\therefore \quad x = 0 \ \text{ or } \ x - 3 = 0 \qquad \{\text{Null Factor law}\}$$
$$\therefore \quad x = 0 \ \text{ or } \ 3$$

ILLEGAL CANCELLING

Let us reconsider the equation $x^2 = 3x$.

You may be tempted to divide both sides by x in order to solve this equation.

If we then try to simplify both sides by cancelling then $\dfrac{x^2}{x} = \dfrac{3x}{x}$ would finish as $x = 3$.

But, from **Example 32** we see that there are two solutions to this equation, $x = 0$ or 3. Consequently, we have 'lost' the solution $x = 0$.

From this example we conclude that:

> We must never cancel a variable common factor from both sides of an equation unless we know that the factor cannot be zero.

STEPS FOR SOLVING QUADRATIC EQUATIONS

Since cancellation by a variable is illegal we adopt the following method:

Step 1: If necessary rearrange the equation with one side being **zero**.

Step 2: Fully **factorise** the other side (usually the LHS).

Step 3: Use the **Null Factor law.**

Step 4: **Solve** the resulting simple equations.

Step 5: **Check** at least one of your solutions.

EXERCISE 8E.3

1 Factorise and hence solve for x:

a $\quad 3x^2 + 6x = 0$

b $\quad 2x^2 + 5x = 0$

c $\quad 4x^2 - 3x = 0$

d $\quad 4x^2 = 5x$

e $\quad 3x^2 = 9x$

f $\quad 4x = 8x^2$

Example 33

Solve for x: $x^2 + 3x = 28$

$$x^2 + 3x = 28$$
$$\therefore \quad x^2 + 3x - 28 = 0 \qquad \text{\{one side must be 0\}}$$
$$\therefore \quad (x+7)(x-4) = 0 \qquad \text{\{fully factorise the LHS\}}$$
$$\therefore \quad x + 7 = 0 \ \text{ or } \ x - 4 = 0 \qquad \text{\{Null Factor law\}}$$
$$\therefore \quad x = -7 \ \text{ or } \ 4 \qquad \text{\{solving linear equations\}}$$

2 Solve for x:

a $x^2 + 9x + 14 = 0$	**b** $x^2 + 11x + 30 = 0$	**c** $x^2 + 2x = 15$
d $x^2 + x = 12$	**e** $x^2 + 6 = 5x$	**f** $x^2 + 4 = 4x$
g $x^2 = x + 6$	**h** $x^2 = 7x + 60$	**i** $x^2 = 3x + 70$
j $10 - 3x = x^2$	**k** $x^2 + 12 = 7x$	**l** $9x + 36 = x^2$

Example 34

Solve for x: $5x^2 = 3x + 2$

$$5x^2 = 3x + 2$$
$$\therefore \quad 5x^2 - 3x - 2 = 0 \qquad \text{\{making the RHS} = 0\}$$
$$\therefore \quad 5x^2 - 5x + 2x - 2 = 0 \qquad \{ac = -10, \quad b = -3$$
$$\therefore \quad 5x(x-1) + 2(x-1) = 0 \qquad \therefore \ \text{ numbers are } -5 \text{ and } +2\}$$
$$\therefore \quad (x-1)(5x+2) = 0 \qquad \text{\{factorising\}}$$
$$\therefore \quad x - 1 = 0 \ \text{ or } \ 5x + 2 = 0 \qquad \text{\{Null Factor law\}}$$
$$\therefore \quad x = 1 \ \text{ or } \ -\tfrac{2}{5} \qquad \text{\{solving the linear equations\}}$$

3 Solve for x:

a $2x^2 + 2 = 5x$	**b** $3x^2 + 8x = 3$	**c** $3x^2 + 17x + 20 = 0$
d $2x^2 + 5x = 3$	**e** $2x^2 + 5 = 11x$	**f** $2x^2 + 7x + 5 = 0$
g $3x^2 + 13x + 4 = 0$	**h** $5x^2 = 13x + 6$	**i** $2x^2 + 17x = 9$
j $2x^2 + 3x = 5$	**k** $3x^2 + 2x = 8$	**l** $2x^2 + 9x = 18$

Example 35

Solve for x: $10x^2 - 13x - 3 = 0$

$$10x^2 - 13x - 3 = 0 \qquad \{ac = -30, \quad b = -13$$
$$\therefore \quad 10x^2 - 15x + 2x - 3 = 0 \qquad \therefore \ \text{ numbers are } -15 \text{ and } 2\}$$
$$\therefore \quad 5x(2x-3) + 1(2x-3) = 0 \qquad \text{\{factorising in pairs\}}$$
$$\therefore \quad (2x-3)(5x+1) = 0 \qquad \text{\{common factor factorisation\}}$$
$$\therefore \quad 2x - 3 = 0 \ \text{ or } \ 5x + 1 = 0 \qquad \text{\{using the Null Factor law\}}$$
$$\therefore \quad x = \tfrac{3}{2} \ \text{ or } \ -\tfrac{1}{5} \qquad \text{\{solving the linear equations\}}$$

4 Solve for x:

    a $6x^2 + 13x = 5$ **b** $6x^2 = x + 2$ **c** $6x^2 + 5x + 1 = 0$

    d $21x^2 = 62x + 3$ **e** $10x^2 + x = 2$ **f** $10x^2 = 7x + 3$

5 Solve for x by first expanding brackets and then equating to zero:

    a $x(x + 5) + 2(x + 6) = 0$ **b** $x(1 + x) + x = 3$

    c $(x - 1)(x + 9) = 8x$ **d** $3x(x + 2) - 5(x - 3) = 17$

    e $4x(x + 1) = -1$ **f** $2x(x - 6) = x - 20$

6 Solve for x by first eliminating the algebraic fractions:

    a $\dfrac{x}{3} = \dfrac{2}{x}$ **b** $\dfrac{4}{x} = \dfrac{x}{2}$ **c** $\dfrac{x}{5} = \dfrac{2}{x}$

    d $\dfrac{x - 1}{4} = \dfrac{3}{x}$ **e** $\dfrac{x - 1}{x} = \dfrac{x + 11}{5}$ **f** $\dfrac{x}{x + 2} = \dfrac{1}{x}$

    g $\dfrac{2x}{3x + 1} = \dfrac{1}{x + 2}$ **h** $\dfrac{2x + 1}{x} = 3x$ **i** $\dfrac{x + 2}{x - 1} = \dfrac{x}{2}$

F   COMPLETING THE SQUARE

Try as much as we like, but we will not be able to solve quadratic equations such as $x^2 + 4x - 7 = 0$ by using the factorisation methods already practised. To solve this equation we need a different technique.

Consider the solution to the equation

$$(x + 2)^2 = 11$$
$$\therefore \quad x + 2 = \pm\sqrt{11}$$
$$\therefore \quad x = -2 \pm \sqrt{11}$$

However,   if $(x + 2)^2 = 11$

           then $x^2 + 4x + 4 = 11$

           and $x^2 + 4x - 7 = 0$

and so,   the solutions to $x^2 + 4x - 7 = 0$ are the solutions to $(x + 2)^2 = 11$.

Consequently, an approach for solving such equations could be to reverse the above argument.

$$\text{Consider} \quad x^2 + 4x - 7 = 0$$
$$\therefore \quad x^2 + 4x = 7$$
$$\therefore \quad x^2 + 4x + 4 = 7 + 4$$
$$\therefore \quad (x + 2)^2 = 11$$
$$\therefore \quad x + 2 = \pm\sqrt{11}$$
$$\therefore \quad x = -2 \pm \sqrt{11}$$

Hence the solutions to $x^2 + 4x - 7 = 0$ are $x = -2 \pm \sqrt{11}$.

From the above example it can be seen that a **perfect square** needs to be created on the left hand side.

The process used is called **completing the square**.

From our previous study of perfect squares we observe that:

$$(x + 3)^2 = x^2 + 2 \times 3 \times x + 3^2 \quad \longleftarrow \quad \text{notice that} \quad 3^2 = \left(\frac{2 \times 3}{2}\right)^2$$

$$(x - 5)^2 = x^2 - 2 \times 5 \times x + 5^2 \quad \longleftarrow \quad \text{notice that} \quad 5^2 = \left(\frac{2 \times 5}{2}\right)^2$$

$$(x + p)^2 = x^2 + 2 \times p \times x + p^2 \quad \longleftarrow \quad \text{notice that} \quad p^2 = \left(\frac{2 \times p}{2}\right)^2$$

i.e.,   the constant term is **"the square of half the coefficient of x"**.

For
- $x^2 + 4x - 7$,   half the coefficient of x is 2,
  so the constant term is $2^2 = 4$

- $x^2 - 4x - 7$,   half the coefficient of x is -2,
  so the constant term is $(-2)^2 = 4$

Example 36

To create a perfect square on the LHS, what must be added to both sides
of the equation   **a** $x^2 + 8x = -5$ **b** $x^2 - 6x = 13$?
What does the equation become in each case?

a   In $x^2 + 8x = -5$, half the coefficient of x is $\dfrac{8}{2} = 4$

so, we add 4^2 to both sides

and the equation becomes
$$x^2 + 8x + 4^2 = -5 + 4^2$$
$$(x + 4)^2 = -5 + 16$$
$$(x + 4)^2 = 11$$

b   In $x^2 - 6x = 13$, half the coefficient of x is $\dfrac{-6}{2} = -3$

so, we add $(-3)^2 = 3^2$ to both sides

and the equation becomes
$$x^2 - 6x + 3^2 = 13 + 3^2$$
$$(x - 3)^2 = 13 + 9$$
$$(x - 3)^2 = 22$$

Notice that we keep the
equation balanced by
adding the same to both
sides of the equation.

EXERCISE 8F

1   For each of the following equations:

    i   find what must be added to both sides of the equation to create a perfect square
on the LHS

    ii   write each equation in the form $(x + p)^2 = k$

a $x^2 + 2x = 5$	**b** $x^2 - 2x = -7$	**c** $x^2 + 6x = 2$
d $x^2 - 6x = -3$	**e** $x^2 + 10x = 1$	**f** $x^2 - 8x = 5$
g $x^2 + 12x = 13$	**h** $x^2 + 5x = -2$	**i** $x^2 - 7x = 4$

Example 37

Solve for x by completing the square, leaving answers in surd form:

a $x^2 + 2x - 2 = 0$ **b** $x^2 - 4x + 6 = 0$

a $x^2 + 2x - 2 = 0$

$\therefore \quad x^2 + 2x \quad\quad = 2$ {remove the constant term to the RHS}

$\therefore \quad x^2 + 2x + 1^2 = 2 + 1^2$ {add $(\frac{2}{2})^2 = 1^2$ to both sides}

$\therefore \quad (x + 1)^2 = 3$ {factorise the LHS, simplify the RHS}

$\therefore \quad x + 1 = \pm\sqrt{3}$

$\therefore \quad x = -1 \pm \sqrt{3}$

So, solutions are $x = -1 + \sqrt{3}$ or $-1 - \sqrt{3}$.

> Remember that if $x^2 = k$, where $k > 0$ then $x = \pm\sqrt{k}$.

b $x^2 - 4x + 6 = 0$

$\therefore \quad x^2 - 4x \quad\quad = -6$ {remove the constant term to the RHS}

$\therefore \quad x^2 - 4x + 2^2 = -6 + 2^2$ {add $(-\frac{4}{2})^2 = 2^2$ to both sides}

$\therefore \quad (x - 2)^2 = -2$ {factorise the LHS, simplify the RHS}

which is impossible as no perfect square can be negative

$\therefore$ no real solutions exist.

2 If possible, solve for x using 'completing the square', leaving answers in surd form:

a $x^2 - 4x + 1 = 0$ **b** $x^2 - 2x - 2 = 0$ **c** $x^2 - 4x - 3 = 0$

d $x^2 + 2x - 1 = 0$ **e** $x^2 + 2x + 4 = 0$ **f** $x^2 + 4x + 1 = 0$

g $x^2 + 6x + 3 = 0$ **h** $x^2 - 6x + 11 = 0$ **i** $x^2 + 8x + 14 = 0$

3 Using the method of 'completing the square' solve for x, leaving answers in surd form.

a $x^2 + 3x + 2 = 0$ **b** $x^2 = 4x + 8$ **c** $x^2 - 5x + 6 = 0$

d $x^2 + x - 1 = 0$ **e** $x^2 + 3x - 1 = 0$ **f** $x^2 + 5x - 2 = 0$

GRAPHICS CALCULATOR SOLUTION OF QUADRATIC EQUATIONS

To solve $2x^2 + 3x - 7 = 0$,   click on the icon for instructions on how to solve quadratic equations for your particular calculator.

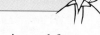

Example 38

Use technolngy to solve: $3x^2 + 2x - 11 = 0$.

Using a CASIO calculator, select the equation menu option and then choose to solve a polynomial of degree 2. Enter the quadratic coefficents and press   **F1**   to solve.

 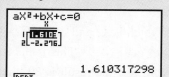

So the solutions are:
$x \doteqdot -2.28$ and 1.61
(Correct to 3 s.f.)

4 Use technology to solve these quadratic equations (to 3 s.f.):

a $x^2 + 6x + 3 = 0$ **b** $x^2 - 6x + 11 = 0$ **c** $x^2 + 8x + 14 = 0$

d $2x^2 + 4x - 1 = 0$ **e** $3x^2 - 12x + 7 = 0$ **f** $5x^2 - 10x + 3 = 0$

INVESTIGATION 3   THE QUADRATIC FORMULA (Extension)

What to do:

1 Use factorisation techniques and sometimes 'completing the square' where necessary to solve:

a $x^2 + 3x = 10$ **b** $2x^2 = 7x + 4$ **c** $x^2 = 4x + 3$

d $4x^2 - 12x + 9 = 0$ **e** $x^2 + 6x + 11 = 0$ **f** $4x^2 + 1 = 8x$

2 A formula for solving $ax^2 + bx + c = 0$ is $x = \dfrac{-b \pm \sqrt{b^2 - 4ac}}{2a}$.

Check that this formula gives the correct answer for each question in **1**.

3 What is the significance of $b^2 - 4ac$ in determining the solutions of the quadratic equation $ax^2 + bx + c = 0$? You should be able to answer this question from observations in **1** and **2**.

4 Establish the quadratic formula by 'completing the square' on $ax^2 + bx + c = 0$.

[**Hint:** Do not forget to divide each term by a to start with.]

HISTORICAL NOTE   BABYLONIAN ALGEBRA

The mathematics used by the **Babylonians** was recorded on clay tablets in cuneiform. One such tablet which has been preserved is called *Plimpton 322* (around 1600 BC).

The Ancient Babylonians were able to solve difficult linear equations using the rules we use today, such as transposing terms and multiplying both sides by like quantities to remove fractions. They were familiar with factorisation.

They could, for example, add $4xy$ to $(x - y)^2$ to obtain $(x + y)^2$. This was all achieved without the use of letters for unknown quantities. However, they often used words for the unknown.

Consider the following example from about 4000 years ago.

Problem: *"I have subtracted the side of my square from the area and the result is 870. What is the side of the square?"*

Solution: Take half of 1, which is $\frac{1}{2}$, and multiply $\frac{1}{2}$ by $\frac{1}{2}$ which is $\frac{1}{4}$;

add this to 870 to get $870\frac{1}{4}$. This is the square of $29\frac{1}{2}$.

Now add $\frac{1}{2}$ to $29\frac{1}{2}$ and the result is 30, the side of the square.

Using our modern symbols: the equation is $x^2 - x = 870$ and the solution is

$x = \sqrt{(\frac{1}{2})^2 + 870} + \frac{1}{2} = 30.$

Note this is the positive root of the equation using the quadratic formula.

G | PROBLEM SOLVING WITH QUADRATICS

Contained in this section are problems which when converted to algebraic form result in a **quadratic equation**.

It is essential that you are successful at solving these equations using **factorisation** (and sometimes 'completing the square'). You may need to revise this process.

Below is a method recommended for solving these problems:

Problem solving method

Step 1:	Carefully read the question until you understand the problem. A rough sketch may be useful.
Step 2:	Decide on the unknown quantity, calling it x, say.
Step 3:	Find an equation which connects x and the information you are given.
Step 4:	Solve the equation using one of the methods you have learnt.
Step 5:	Check that any solutions satisfy the original problem.
Step 6:	Write your answer to the question in sentence form.

Example 39

The sum of a number and its square is 42. Find the number.

Let the number be x.   Therefore its square is x^2.

$$x + x^2 = 42$$
$$\therefore \quad x^2 + x - 42 = 0 \quad \text{\{rearranging\}}$$
$$\therefore \quad (x + 7)(x - 6) = 0 \quad \text{\{factorising\}}$$
$$\therefore \quad x = -7 \quad \text{or} \quad x = 6$$

Check: If $x = -7$, $-7 + (-7)^2 = -7 + 49 = 42$ ✓

If $x = 6$, $6 + 6^2 = 6 + 36 = 42$ ✓

So, the number is -7 or 6.

EXERCISE 8G

1   The sum of a number and its square is 72. Find the number.

2   The product of a number and the number increased by 4 is 96. Find the two possible answers for the number.

3   When 24 is subtracted from the square of a number the result is five times the original number. Find the number.

4 The sum of two numbers is 9 and the sum of their squares is 261. Find the numbers.

5 Two numbers differ by 5 and the sum of their squares is 13. Find the numbers.

Example 40

A rectangle has length 5 cm greater than its width. If it has an area of 84 cm^2, find the dimensions of the rectangle.

If x cm is the width, then $(x+5)$ cm is the length.

$$\text{Now}\quad \text{area} = 84 \text{ cm}^2$$
$$\therefore\quad x(x+5) = 84$$
$$\therefore\quad x^2 + 5x = 84$$
$$\therefore\quad x^2 + 5x - 84 = 0$$
$$\therefore\quad (x+12)(x-7) = 0 \qquad \{\text{on factorisation}\}$$
$$\therefore\quad x = -12 \text{ or } 7.$$

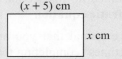

But $x > 0$ as lengths are positive quantities, $\therefore$ $x = 7$
$\therefore$ the rectangle is 7 cm by 12 cm.

---

6 A rectangle has length 6 cm greater than its width. Find its width given that its area is 91 cm^2.

7 A triangle has base 2 cm more than its altitude. If its area is 49.5 cm^2, find its altitude.

8 A rectangular enclosure is made from 40 m of fencing. The area enclosed is 96 m^2. Find the dimensions of the enclosure.

9 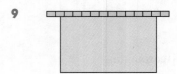 A rectangular pig pen is built against an existing brick fence. 24 m of fencing was used to enclose 70 m^2.

Find the dimensions of the pen.

10 Use the theorem of Pythagoras to find x given:

a 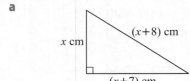   **b**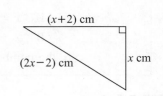

11 A right angled triangle has sides 2 cm and 9 cm respectively less than its hypotenuse. Find the length of each side of the triangle.

12 A gardener plants 600 cabbages in rows. If the number of cabbages in each row is 10 more than twice the number of rows, how many rows did the gardener plant?

13 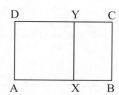 ABCD is a rectangle in which $AB = 21$ cm.

The square AXYD is removed and the remaining rectangle has area 80 cm^2.

Find the length of BC.

14 A rectangular sheet of tin plate is 20 cm by 16 cm and is to be made into an open box with a base having an area of 140 cm^2, by cutting out equal squares from the four corners and then bending the edges upwards.

Find the size of the squares cut out.

15 A rectangular swimming pool is 12 m long by 6 m wide. It is surrounded by a pavement of uniform width, the area of the pavement being $\frac{7}{8}$ of the area of the pool.

   a If the pavement is x m wide, show that the area of the pavement is $4x^2 + 36x$ m^2.

   b Hence, show that $4x^2 + 36x - 63 = 0$.

   c How wide is the pavement?

REVIEW SET 8A

1 Expand and simplify:

   a $2(x - 5)$ **b** $3(1 - 2x) - (x - 4)$

2 Expand and simplify:

   a $(3x - 2)(x + 2)$ **b** $(2x + 1)^2$ **c** $(3x + 1)(3x - 1)$

3 Fully factorise:

   a $x^2 - 4x - 21$ **b** $4x^2 - 25$ **c** $3x^2 - 6x - 72$ **d** $6x^2 + x - 2$

4 Solve for x:

   a $x^2 + 24 = 11x$ **b** $10x^2 - 11x - 6 = 0$

5 Solve by 'completing the square': $x^2 + 6x + 11 = 0$

6 The width of a rectangle is 7 cm less than its length and its area is 260 cm^2. Find its dimensions.

7 Expand and simplify:

   a $5 - 2x - (x + 3)^2$ **b** $(3x - 2)(x^2 + 2x + 7)$

REVIEW SET 8B

1 Expand and simplify:

   a $2(x - 3) + 3(2 - x)$ **b** $-x(3 - 4x) - 2x(x + 1)$

2 Expand and simplify:

   a $(2x - 5)(x + 3)$ **b** $(3x - 4)^2$ **c** $(3x + 2)(3x - 2)$

3 Fully factorise:

    a $x^2 - 8x - 33$ **b** $x^2 + 4x - 32$ **c** $x^2 - 10x + 25$

4 Solve for x:

    a $x^2 + 5x = 24$ **b** $2x^2 - 18 = 0$ **c** $8x^2 + 2x - 3 = 0$

5 Solve by 'completing the square': $x^2 - 2x = 100$

6 A rectangle has its length 3 cm greater than its width. If it has an area of 108 cm², find the dimensions of the rectangle.

7 When the square of a number is increased by 10, the result is seven times the original number. Find the number.

8 Expand and simplify:

    a $(2x - 1)^2 - (x + 2)(3 - x)$ **b** $(x^2 - 2x + 3)(2x + 1)$

REVIEW SET 8C

1 Expand and simplify:

    a $-7(2x - 5)$ **b** $3(x - 2) - 4(3 - x)$

2 Expand and simplify:

    a $(5x - 1)(x + 3)$ **b** $(3x - 1)^2$ **c** $(1 + 5x)(1 - 5x)$

3 Factorise:

    a $x^2 - 7x - 18$ **b** $4x^2 - 49$ **c** $2x^2 - 14x - 60$ **d** $5x^2 + 12x + 4$

4 Solve for x:

    a $x^2 + 6 = 5x$ **b** $x^2 + 16 = 8x$ **c** $3x^2 = 2x + 21$

5 Solve by 'completing the square': $x^2 - 14x + 7 = 0$

6 When the square of a number is increased by one the result is four times the original number. Find the number.

7 A right angled triangle has its hypotenuse one centimetre more than twice the length of the shortest side, while the other side is 7 cm longer than the shortest side. Find the length of each of the sides of the triangle.

8 Expand and simplify:

    a $3(x - 2) - (x + 3)^2$ **b** $(x^2 - x + 3)(2x + 5)$

Chapter **9**

Function notation and quadratic functions

RELATIONS AND FUNCTIONS

The charges for parking a car in a short-term car park at an Airport are given in the table shown alongside.

Car park charges	
Period (h)	Charge
0 - 1 hours	$5.00
1 - 2 hours	$9.00
2 - 3 hours	$11.00
3 - 6 hours	$13.00
6 - 9 hours	$18.00
9 - 12 hours	$22.00
12 - 24 hours	$28.00

There is an obvious relationship between time spent and the cost. The cost is dependent on the length of time the car is parked.

Looking at this table we might ask: How much would be charged for exactly one hour? Would it be $5 or $9?

To make the situation clear, and to avoid confusion, we could adjust the table and draw a graph. We need to indicate that 2-3 hours really means for time over 2 hours up to and including 3 hours   i.e., $2 < t \leqslant 3$.

So, we
now have

Car park charges	
Period	Charge
$0 < t \leqslant 1$ hours	$5.00
$1 < t \leqslant 2$ hours	$9.00
$2 < t \leqslant 3$ hours	$11.00
$3 < t \leqslant 6$ hours	$13.00
$6 < t \leqslant 9$ hours	$18.00
$9 < t \leqslant 12$ hours	$22.00
$12 < t \leqslant 24$ hours	$28.00

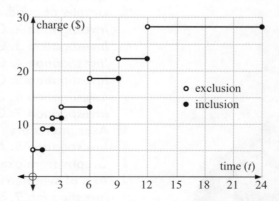

In mathematical terms, because we have a relationship between two variables, time and cost, the schedule of charges is an example of a **relation**.

A relation may consist of a finite number of ordered pairs, such as $\{(1, 5), (-2, 3), (4, 3), (1, 6)\}$ or an infinite number of ordered pairs.

The parking charges example is clearly the latter as any real value of time (t hours) in the interval $0 < t \leqslant 24$ is represented.

The set of possible values of the variable on the horizontal axis is called the **domain** of the relation.

For example: • $\{t: \ 0 < t \leqslant 24\}$ is the domain for the car park relation

• $\{-2, 1, 4\}$ is the domain of $\{(1, 5), (-2, 3), (4, 3), (1, 6)\}$.

The set which describes the possible y-values is called the **range** of the relation.

For example:
- the range of the car park relation is $\{5, 9, 11, 13, 18, 22, 28\}$
- the range of $\{(1, 5), (-2, 3), (4, 3), (1, 6)\}$ is $\{3, 5, 6\}$.

We will now look at relations and functions more formally.

RELATIONS

> A **relation** is any set of points on the Cartesian plane.

A relation is often expressed in the form of an **equation** connecting the **variables** x and y.

For example $y = x + 3$ and $x = y^2$ are the equations of two relations.

These equations generate sets of ordered pairs.

Their graphs are:

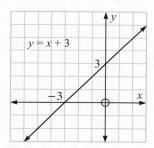
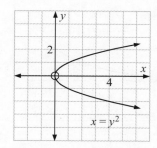

However, a relation may not be able to be defined by an equation. Below are two examples which show this:

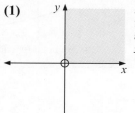

(1) All points in the first quadrant are a relation. $x > 0, y > 0$

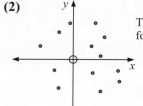

(2) These 13 points form a relation.

FUNCTIONS

> A **function** is a relation in which no two different ordered pairs have the same x-coordinate (first member).

We can see from the above definition that a function is a special type of relation.

TESTING FOR FUNCTIONS

Algebraic Test:

> If a relation is given as an equation, and the substitution of any value for x results in one and only one value of y, we have a function.

For example: • $y = 3x - 1$ is a function, as for any value of x there is only one value of y

• $x = y^2$ is not a function since if $x = 4$, say, then $y = \pm 2$.

Geometric Test   ("Vertical Line Test"):

> If we draw all possible vertical lines on the graph of a relation, the relation:
> • is a function if each line cuts the graph no more than once
> • is not a function if one line cuts the graph more than once.

DEMO

Example 1

Which of the following relations are functions?

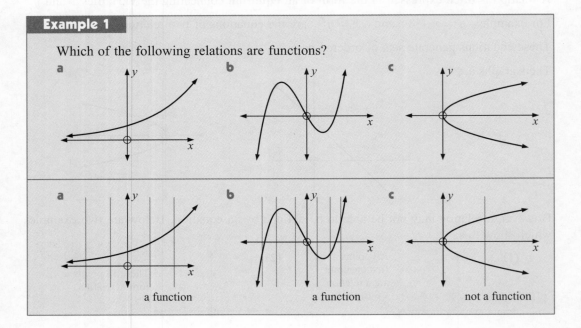

GRAPHICAL NOTE

• If a graph contains a small **open circle** end point such as ——o , the end point is **not included**.

• If a graph contains a small **filled-in circle** end point such as ——• , the end point **is included**.

• If a graph contains an **arrow head** at an end such as ——→ then the graph continues indefinitely in that general direction, or the shape may repeat as it has done previously.

EXERCISE 9A

1 Which of the following sets of ordered pairs are functions? Give reasons.

    a (1, 3), (2, 4), (3, 5), (4, 6)   **b** (1, 3), (3, 2), (1, 7), (−1, 4)

    c (2, −1), (2, 0), (2, 3), (2, 11)   **d** (7, 6), (5, 6), (3, 6), (−4, 6)

    e (0, 0), (1, 0), (3, 0), (5, 0)   **f** (0, 0), (0, −2), (0, 2), (0, 4)

2   Use the vertical line test to determine which of the following relations are functions:

a

b

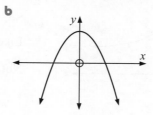

c

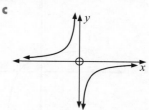

d

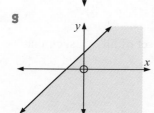

e

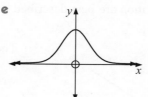

f

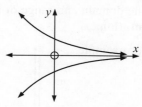

g

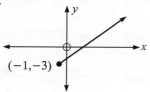

h

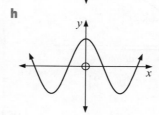

i

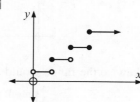

3   Will the graph of a straight line always be a function? Give evidence.

4   Give algebraic evidence to show that the relation $x^2 + y^2 = 9$ is not a function.

B | INTERVAL NOTATION, DOMAIN AND RANGE

DOMAIN AND RANGE

> The **domain** of a relation is the set of permissible values that x may have.
> The **range** of a relation is the set of permissible values that y may have.

For example:

(1)

All values of $x \geqslant -1$ are permissible.

So, the domain is $\{x: \ x \geqslant -1\}$.

All values of $y \geqslant -3$ are permissible.

So, the range is $\{y: \ y \geqslant -3\}$.

(2)

x can take any value.

So, the domain is $\{x: \ x \text{ is in } R\}$.

y cannot be > 1

$\therefore$ range is $\{y: \ y \leqslant 1\}$.

(3)

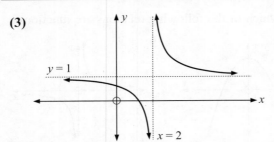

x can take all values except $x = 2$.

So, the domain is $\{x : x \neq 2\}$.

Likewise, the range is $\{y : y \neq 1\}$.

Note: R represents the set of all real values, i.e., all numbers on the number line.

The domain and range of a relation are best described where appropriate using **interval notation**.

For example:

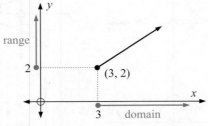

The domain consists of all real x such that $x \geqslant 3$ and we write this as

$$\{x : \ x \geqslant 3\}.$$

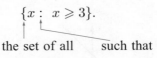

the set of all   such that

Likewise the range would be $\{y : \ y \geqslant 2\}$.

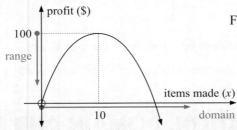

For this profit function:

- the domain is $\{x : \ x \geqslant 0\}$
- the range is $\{y : \ y \leqslant 100\}$.

Intervals have corresponding graphs.

For example:

$\{x : \ x \geqslant 3\}$ is read "the set of all x such that x is greater than or equal to 3" and has number line graph

$\{x : \ x < 2\}$ has number line graph

$\{x : \ -2 < x \leqslant 1\}$ has number line graph

$\{x : \ x \leqslant 0 \ \text{ or } \ x > 4\}$ has number line graph

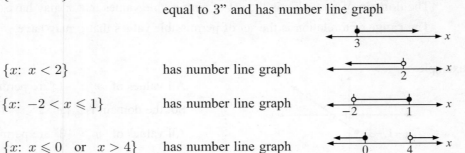

Note: for numbers *between* a and b we write $a < x < b$

for numbers '*outside*' a and b we write $x < a$ or $x > b$

Example 2

For each of the following graphs state the domain and range:

a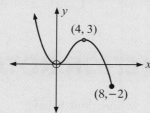

b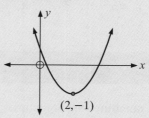

a	Domain is $\{x: x \leqslant 8\}$.	b	Domain is $\{x: x \text{ is in } R\}$.
	Range is $\{y: y \geqslant -2\}$.		Range is $\{y: y \geqslant -1\}$.

EXERCISE 9B

1 For each of the following graphs find the domain and range:

a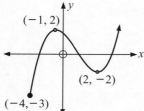

b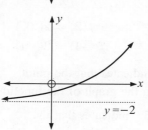

c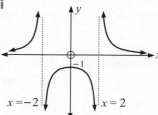

d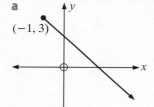

e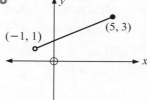

f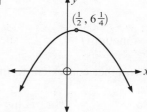

g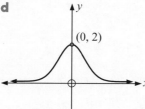

h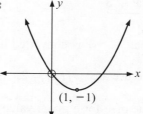

i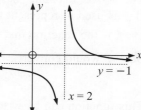

2 Use a graphics calculator to help sketch carefully the graphs of the following functions and find the domain and range of each:

a $y = \sqrt{x}$

b $y = \dfrac{1}{x^2}$

c $y = \sqrt{4 - x}$

d $y = x^2 - 7x + 10$

e $y = 5x - 3x^2$

f $y = x + \dfrac{1}{x}$

g $y = \dfrac{x+4}{x-2}$ **h** $y = x^3 - 3x^2 - 9x + 10$ **i** $y = \dfrac{3x-9}{x^2-x-2}$

j $y = x^2 + x^{-2}$ **k** $y = x^3 + \dfrac{1}{x^3}$ **l** $y = x^4 + 4x^3 - 16x + 3$

C   FUNCTION NOTATION

Function machines are sometimes used to illustrate how functions behave.

For example:

So, if 4 is fed into the machine, $2(4) + 3 = 11$ comes out.

The above 'machine' has been programmed to perform a particular function.
If f is used to represent that particular function we can write:

f is the function that will convert x into $2x + 3$.

So, f would convert 2 into $2(2) + 3 = 7$ and
                        -4 into $2(-4) + 3 = -5$.

This function can be written as:

$$f : \; x \longmapsto 2x + 3$$

function f such that x is converted into $2x + 3$

Two other equivalent forms we use are: $f(x) = 2x + 3$ or $y = 2x + 3$

So, $f(x)$ is the value of y for a given value of x,   i.e., $y = f(x)$.

Notice that for $f(x) = 2x + 3$, $f(2) = 2(2) + 3 = 7$ and
                                $f(-4) = 2(-4) + 3 = -5$.

Consequently, $f(2) = 7$ indicates that the point
                 (2, 7) lies on the graph of the function.

Likewise $f(-4) = -5$ indicates that the
                point $(-4, -5)$ also lies on the graph.

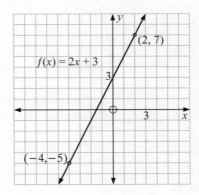

Note:
- $f(x)$ is read as "f of x" and is the value of the function at any value of x.
- If (x, y) is any point on the graph then $y = f(x)$.
- f is the function which converts x into $f(x)$, i.e., $f : x \longmapsto f(x)$.
- $f(x)$ is sometimes called the **image** of x.

Example 3

If $f : x \longmapsto 2x^2 - 3x$, find the value of: **a** $f(5)$ **b** $f(-4)$

$f(x) = 2x^2 - 3x$

a $\quad f(5) = 2(5)^2 - 3(5)$ {replacing x by (5)}
$\qquad\qquad = 2 \times 25 - 15$
$\qquad\qquad = 35$

b $\quad f(-4) = 2(-4)^2 - 3(-4)$ {replacing x by (-4)}
$\qquad\qquad\ = 2(16) + 12$
$\qquad\qquad\ = 44$

The **table mode** on your calculator can be used to find $f(x)$ given x. Click on the icon for instructions.

EXERCISE 9C

1 If $f : x \longmapsto 3x + 2$, find the value of:

a $f(0)$ **b** $f(2)$ **c** $f(-1)$ **d** $f(-5)$ **e** $f(-\frac{1}{3})$

2 If $g : x \longmapsto x - \dfrac{4}{x}$, find the value of:

a $g(1)$ **b** $g(4)$ **c** $g(-1)$ **d** $g(-4)$ **e** $g(-\frac{1}{2})$

3 If $f : x \longmapsto 3x - x^2 + 2$, use the **table mode** of your calculator to find:

a $f(0)$ **b** $f(3)$ **c** $f(-3)$ **d** $f(-7)$ **e** $f(\frac{3}{2})$

Example 4

If $f(x) = 5 - x - x^2$, find in simplest form: **a** $f(-x)$ **b** $f(x + 2)$

a $\quad f(-x) = 5 - (-x) - (-x)^2$ {replacing x by $(-x)$}
$\qquad\qquad\ = 5 + x - x^2$

b $\quad f(x + 2) = 5 - (x + 2) - (x + 2)^2$ {replacing x by $(x + 2)$}
$\qquad\qquad\quad\ = 5 - x - 2 - [x^2 + 4x + 4]$
$\qquad\qquad\quad\ = 3 - x - x^2 - 4x - 4$
$\qquad\qquad\quad\ = -x^2 - 5x - 1$

---

4 If $f(x) = 7 - 3x$, find in simplest form:

a $f(a)$ **b** $f(-a)$ **c** $f(a + 3)$ **d** $f(b - 1)$ **e** $f(x + 2)$

5 If $F(x) = 2x^2 + 3x - 1$,   find in simplest form:

     a $F(x + 4)$ **b** $F(2 - x)$ **c** $F(-x)$ **d** $F(x^2)$ **e** $F(x^2 - 1)$

6 If $G(x) = \dfrac{2x + 3}{x - 4}$:   **a**   evaluate   **i** $G(2)$ **ii** $G(0)$ **iii** $G(-\frac{1}{2})$

                         b   find a value of x where $G(x)$ does not exist

                         c   find $G(x + 2)$ in simplest form

                         d   find x if $G(x) = -3$.

7 f represents a function. What is the difference in meaning between f and $f(x)$?

8 If $f(x) = 2^x$,   show that $f(a)f(b) = f(a + b)$.

9 Given $f(x) = x^2$ find in simplest form: **a** $\dfrac{f(x) - f(3)}{x - 3}$ **b** $\dfrac{f(2 + h) - f(2)}{h}$

10 If the value of a photocopier t years after purchase is given by $V(t) = 9650 - 860t$ dollars:

     a   find $V(4)$ and state what $V(4)$ means

     b   find t when $V(t) = 5780$ and explain what this represents

     c   find the original purchase price of the photocopier.

11 On the same set of axes draw the graphs of three different functions $f(x)$ such that $f(2) = 1$ and $f(5) = 3$.

12 Find $f(x) = ax + b$,   a linear function, in which $f(2) = 1$ and $f(-3) = 11$.

13 Find constants a and b where $f(x) = ax + \dfrac{b}{x}$ and $f(1) = 1$, $f(2) = 5$.

14 Given $T(x) = ax^2 + bx + c$,   find a, b and c if $T(0) = -4$, $T(1) = -2$ and $T(2) = 6$.

D   FUNCTIONS AS MAPPINGS

In the previous section, we introduced functions as 'machines' which convert x values into $f(x)$ values.

Hence, $f: x \longmapsto f(x)$ can be thought of as a number crunching machine which **maps** elements in the domain (input) to elements in the range (output).

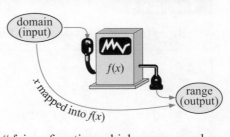

For example, if $f: x \longmapsto 5x + 2$,   we say this as "f is a function which maps x values into two more than five lots of the x values".

If we specify a domain (i.e., we limit the input possibilities) as $\{x: x \in Z, \ 1 \leqslant x \leqslant 4\}$, then the permissible values of $f(x)$ are restricted to: $f(1) = 5(1) + 2 = 7$

                                                           $f(2) = 5(2) + 2 = 12$

                                                             $f(3) = 5(3) + 2 = 17$

                                                             $f(4) = 5(4) + 2 = 22$

Hence, we can represent $f : x \longmapsto 5x + 2$
on the domain specified using a mapping diagram:

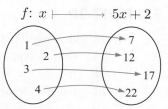

Notice that each element of the domain $\{1, 2, 3, 4\}$ corresponds to a single element within the range $\{7, 12, 17, 22\}$.

For function mappings (as opposed to relation mappings):

- It is possible that multiple elements in the domain correspond to the **same** element within the range.

- It is not possible for a single element in the domain to have more than one corresponding element within the range.

Example 5

Consider $f : x \longmapsto x^2$ for the domain $\{-2, -1, 0, 1, 2\}$

x	-2	-1	0	1	2
$f(x)$	4	1	0	1	4

$f(-2) = f(2) = 4$
$f(-1) = f(1) = 1$

Hence, we have a resultant mapping diagram showing some elements within the domain mapping to the same element within the range.

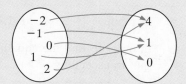

We can also represent this on a set of axes.

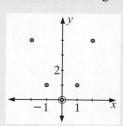

Note: We do not join the elements using a straight line.

Example 6

Consider: $p : x \longmapsto \pm\sqrt{x}$ for the domain $\{0, 1, 4, 9\}$

$f(0) = 0$
$f(1) = \pm 1$
$f(4) = \pm 2$
$f(9) = \pm 3$

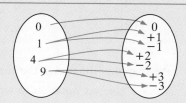

Hence out mapping diagram shows single elements of the domain mapping to **multiple** elements within the range. Hence, this is a relation mapping.

If we represent this function on a set of axes it is clear p is not a function.

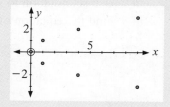

EXERCISE 9D

1 For the following functions of $f : x \longmapsto f(x)$ on $-2 \leqslant x \leqslant 2$ where $x \in Z$:

  i   draw a mapping diagram to represent $f(x)$

  ii list the elements of the domain of $f(x)$ using set notation

  iii list the elements of the range of $f(x)$ using set notation.

  a $f(x) = 3x - 1$ **b** $f(x) = x^2 + 1$ **c** $f(x) = 3 - 4x$

  d $f(x) = 2x^2 - x + 1$ **e** $f(x) = x^3$ **f** $f(x) = 3^x$

  g $f(x) = \dfrac{1}{x + 3}$ **h** $f(x) = \dfrac{x + 3}{x}; \; x \neq 0$

Example 7

The diagram (right) shows a function f mapping members of a set X to members of set Y.

a Using set notation, write down the members of the domain and range.

b Find the equation of the function f.

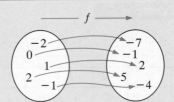

a Domain $= \{-2, -1, 0, 1, 2\}$ Range $= \{-7, -4, -1, 2, 5\}$

b In order to find the equation $f(x)$, it is very useful to use a table of values (put in ascending order),

i.e.,

x	-2	-1	0	1	2
$f(2)$	-7	-45	-1	2	5

A scatterplot also helps to determine the type of function (i.e., linear, quadratic, etc.). It is obvious from this scatterplot that $f(x)$ is linear.

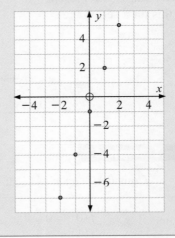

$$\therefore \quad f(x) = mx + c$$

where $m = \dfrac{5 - 2}{2 - 1}$ {using (2, 5) and (1, 2)}

$$\therefore \quad m = 3$$

$$\therefore \quad f(x) = 3x - 1 \quad \text{or} \quad f : x \longmapsto 3x - 1$$

---

2 For these mapping diagrams, the function f maps elements of X to elements of Y.

  i   Use set notation to write down the domain of f.

  ii Use set notation to write down the range of f.

  iii Find the equation of the function f.

a

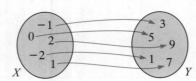

b

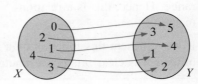

c 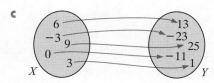   **d**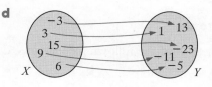

3 For the following mapping diagrams, where f maps X to Y:

    i sketch the domain (x) and range (y) on a set of axes

    ii find the equation of the function f.

a 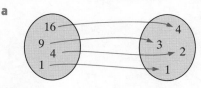   **b**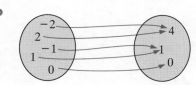

c 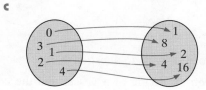   **d**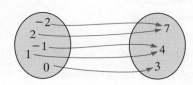

INVESTIGATION 1   FLUID FILLING FUNCTIONS

When water is added at a **constant rate** to a cylindrical container the depth of water in the container is a function of time.

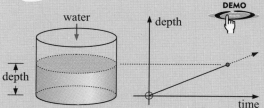

This is because the volume of water added is directly proportional to the time taken to add it. If water was not added at a constant rate the direct proportionality would not exist.

The depth-time graph for the case of a cylinder would be as shown alongside.

The question arises: 'What changes in appearance of the graph occur for different shaped containers?' Consider a vase of conical shape.

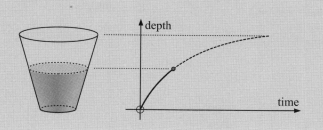

What to do:

1 For each of the following containers, draw a 'depth v time' graph as water is added:

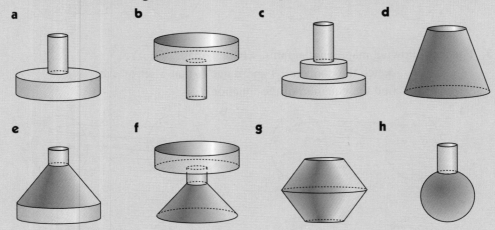

2 Use the water filling demonstration to check your answers to question **1**.

3 Write a brief report on the connection between the shape of a vessel and the corresponding shape of its depth-time graph. You may wish to discuss this in parts. For example, first examine cylindrical containers, then conical, then other shapes. Slopes of curves must be included in your report.

4 Draw possible containers as in question **1** which have the following 'depth v time' graphs:

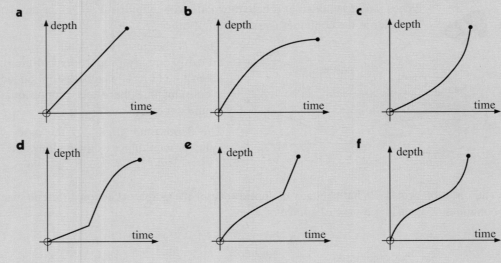

E   LINEAR FUNCTIONS

> A **linear function** is a function of the form $f(x) = ax + b$ or $f : x \longmapsto ax + b$
> where $a \neq 0$.

The *variables* are x and y whilst a and b are *constants*.

If we graph y against x we obtain a **straight line**, with *gradient* a and *y-intercept* b.

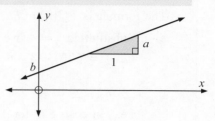

A GEOMETRICAL MODEL

Consider all rectangles with length twice their width. This information does not tell us how long or how wide the rectangle actually is.

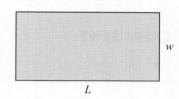

However, it does give us a connection between these two variables. This is $L = 2w$ or using functional notation $L(w) = 2w$.

The graph is:

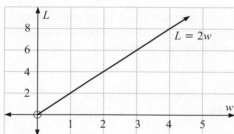

Notice in this case that the *domain* is $\{w : \; w > 0\}$ as sides of rectangles have positive lengths.

Clearly w and L are continuous variables and so the **full line** (not separate discrete points) is used to illustrate the relationship.

If a rectangle has width 5.23 m, then its length is
$$L(5.23)$$
$$= 2 \times 5.23$$
$$= 10.46 \text{ m}$$

KEY FEATURES OF LINEAR FUNCTION GRAPHS (REVIEW)

- The x-**intercept:**   The x-value where the graph cuts the x-axis.

  Let $y = 0$ or $f(x) = 0$ and solve for x.

- The y-**intercept:**   The y-value where the graph cuts the y-axis.

  Let $x = 0$,   solve $f(0) = b$.

  b is the y-intercept.

- The **gradient**:   As $f(1) = a + b$ then
  $(1, \, a + b)$ lies on the line.

  So,   gradient $= \dfrac{y\text{-step}}{x\text{-step}} = a$.

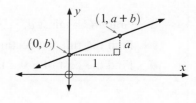

Example 8

The cost of hiring a tennis court is given by the formula $C(h) = 5h + 8$ where C is the cost in $ and h is the number of hours the court is hired for. Find the cost of hiring the tennis court for: **a**   4 hours.   **b**   10 hours.

The formula is $C(h) = 5h + 8$

a   Substituting $h = 4$ we get

$$C(4) = C(5) + 3$$
$$= 20 + 8$$
$$= 28$$

i.e., it costs $28 for 4 hours.

b   Substituting $h = 10$ we get

$$C(10) = 5(10) + 8$$
$$= 50 + 8$$
$$= 58$$

i.e., it costs $58 for 10 hours.

EXERCISE 9E

1   The cost of staying at a hotel is given by the formula, $C(d) = 50d + 20$ where C is the cost in $ and d is the number of days a person stays. Find the cost of staying for:

   a   3 days   **b**   6 days   **c**   2 weeks

2   The thermometer in a kitchen oven was designed using the Celsius (T_c) scale. However many recipe books give the required temperatures on the Fahrenheit (F) scale. The formula which links the two temperature scales is:

$$T_c(F) = \tfrac{5}{9}(F - 32)$$

Convert the following Fahrenheit temperatures into Celsius:

   a   212°F   **b**   32°F   **c**   104°F   **d**   374°F

3   If the value of a car t years after purchase is given by $V(t) = 25\,000 - 3000t$ dollars:

   a   find $V(0)$ and state the meaning of $V(0)$

   b   find $V(3)$ and state the meaning of $V(3)$

   c   find t when $V(t) = 10\,000$ and explain what this represents.

4   Find $f(x) = ax + b$, a linear function, in which $f(2) = 7$ and $f(-1) = -5$.

Example 9

Ace taxi services charge $3.30 for stopping to pick up a passenger and then $1.75 per km travelled thereafter.

   a   Copy and complete:

Distance (km)	0	2	4	6	8	10
Cost ($C)						

   b   Graph C against d.

   c   Find the rule connecting the variables.

   d   Find the cost of a 9.4 km trip.

a

Distance (d km)	0	2	4	6	8	10
Cost ($\$C$)	3.30	6.80	10.30	13.80	17.30	20.80

adding $2 \times \$1.75$ each time

b

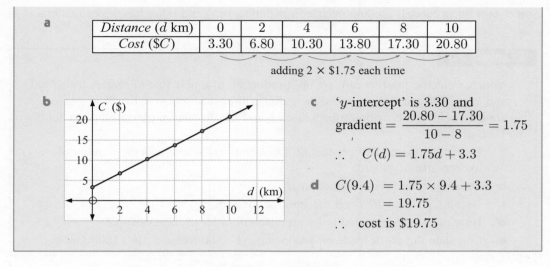

c 'y-intercept' is 3.30 and

$$\text{gradient} = \frac{20.80 - 17.30}{10 - 8} = 1.75$$

$$\therefore \quad C(d) = 1.75d + 3.3$$

d $C(9.4) = 1.75 \times 9.4 + 3.3$
$$= 19.75$$

$\therefore$ cost is $\$19.75$

---

5 An electrician charges $60 for calling and $45 per hour thereafter.

 a From a table of values, plot the charge ($\$C$) against the hours worked (t hours) for $t = 0, 1, 2, 3, 4$ and 5.

 b Use your graph to determine the cost function C, in terms of t.

 c Use the cost function to determine the electrician's total cost for a job lasting $6\frac{1}{2}$ hours. Use your graph to check your answer.

6 A rainwater tank contains 265 litres. The tap is left on and 11 litres escape per minute.

 a Construct a table of values of volume (V litres) left in the tank after time t (minutes) for $t = 0, 1, 2, 3, 4$ and 5.

 b Use your table to graph V against t.

 c Use your graph to determine the rule connecting V and t.

 d Use your rule to determine:

  i how much water is left in the tank after 15 minutes

  ii the time taken for the tank to empty.

 e Use your graph to check your answers to **d i** and **d ii**.

7 The cost of running a truck is $158 plus $365 per one thousand kilometres thereafter.

 a Without graphing, determine the cost ($\$C$) in terms of the number of thousands of kilometres (n).

 b Find the cost of running the truck a distance of 3750 km.

 c How far could the truck travel if $5000 was available?

8 A salesperson's wage is determined from the graph alongside.

 a Determine the weekly wage ($\$W$) in terms of the sales ($\$s$ thousand dollars).

 b Find the weekly wage when sales were $33\,500.

 c Determine the sales necessary for a weekly wage of $830.

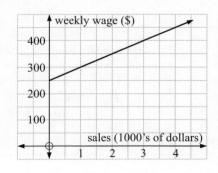

Sometimes linear models are appropriate in manufacturing and other business situations.

Example 10

Nonchar electric toasters can 'set up' production of a new line of toaster for $3000 and every 100 toasters produced thereafter will cost $1000 to make. The toasters are then sold to a distributor for $25 each and the distributor places an order for 1500 of them.

a Determine the cost of production function $C(n)$ where n is the number of toasters manufactured.

b Determine the revenue function $R(n)$.

c Graph $C(n)$ and $R(n)$ on the same set of axes.

d How many toasters need to be produced in order to 'break even'?

e Calculate the profit made on producing: **i** 800 toasters **ii** 1500 toasters.

a Each toaster costs $10 to produce and the fixed cost is $3000.
∴ $C(n) = 10n + 3000$ dollars   {gradient = cost/item = $10}

b $R(n) = n \times \$25$ i.e., $R(n) = 25n$ dollars

c
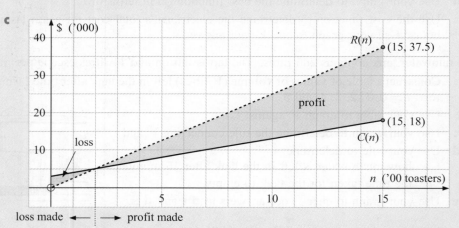

d We see that to the left of the point of intersection $C(n) > R(n)$ so a loss is made.

$$\text{So 'break even' is where}\quad C(n) = R(n)$$
$$\text{i.e.,}\quad 10n + 3000 = 25n$$
$$\therefore\quad 3000 = 15n$$
$$\therefore\quad n = 200$$

200 toasters must be produced to 'break even'.

e   Profit $= R(n) - C(n)$
∴ $P(n) = 25n - (10n + 3000)$
$= 15n - 3000$

i $P(800) = 15 \times 800 - 3000$
$= \$9000$

ii $P(1500) = 15 \times 1500 - 3000$
$= \$19\,500$

9 Self adhesive label packs are produced with a cost function, $C(n) = 3n + 20$ dollars and a revenue function, $R(n) = 5n + 10$ dollars and n is the number of packs produced.

  a Graph each function on the same set of axes, clearly labelling each graph.

  b Determine the 'break even' production number. Check your answer algebraically.

  c For what values of n is a profit made?

  d How many self adhesive label packs should be produced to make a profit of $100?

10 Two way adaptors sell for $7 each. The adaptors cost $2.50 each to make with fixed costs of $300 per day regardless of the number made.

  a Find revenue and cost functions in terms of the number manufactured, n.

  b On the same set of axes graph the revenue and cost functions.

  c Determine the 'break even' level of production and check your answer algebraically.

  d What level of production will guarantee a profit of $1000 per day?

11 A new novel is being printed. The production costs are $6000 (fixed) and $3250 per thousand books thereafter. The books are to sell at $9.50 each with an unlimited market.

  a Determine cost and revenue functions for the production of the novel.

  b On the same set of axes graph the cost and revenue functions.

  c How many books must be sold in order to 'break even'? Check your answer algebraically.

  d What level of production will produce a $10 000 profit?

12 Waverley Manufacturing produces carburettors for motor vehicles. At the beginning of each week it costs $2100 to start up the factory. Each carburettor costs $13.20 in materials and $14.80 in labour to produce. Waverley is able to sell the carburettors to the motor vehicle manufacturers at $70 each.

  a Determine Waverley's cost and revenue functions in terms of the number (n) manufactured per week.

  b Draw graphs of both of the functions in **a** on the same axes and use your graph to find the "break even" point.

  c Find an expression for the profit function and check your answer for the "break even" value of n.

  d Use your profit function to find:

   i the weekly profit for producing 125 carburettors

   ii the number of carburettors required to make a profit of at least $1300.

USING TECHNOLOGY

Use a **graphics calculator** or **graphing package** to check some answers to the previous exercises.

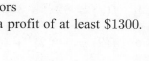

F   QUADRATIC FUNCTIONS

A **quadratic function** is a relationship between two variables which can be written in the form $y = ax^2 + bx + c$ where x and y are the variables and a, b, and c represent constants with $a \neq 0$.

Using function notation, $f(x) = ax^2 + bx + c$ can be written as $f : x \longmapsto ax^2 + bx + c$.

As with linear functions, for any value of x a corresponding value of y can be found by substituting into the function equation.

For example,   if $y = 2x^2 - 3x + 5$,   and $x = 3$,   then $y = 2 \times 3^2 - 3 \times 3 + 5$
$$= 14$$

Hence, the ordered pair (3, 14) satisfies the function $y = 2x^2 - 3x + 5$.

Similarly, using function notation we could write,

if $f(x) = 2x^2 - 3x + 5$ and $x = 3$ then $f(3) = 2 \times 3^2 - 3 \times 3 + 5$
$$= 14$$

Example 11

If $y = -2x^2 + 3x - 4$ find the value of y when: **a** $x = 0$ **b** $x = 3$

a   When $x = 0$
$$y = -2(0)^2 + 3(0) - 4$$
$$= 0 + 0 - 4$$
$$= -4$$

b   When $x = 3$
$$y = -2(3)^2 + 3(3) - 4$$
$$= -2(9) + 9 - 4$$
$$= -18 + 9 - 4$$
$$= -13$$

EXERCISE 9F

1   Which of the following are quadratic functions?

   a $y = 3x^2 - 4x + 1$ **b** $y = 5x - 7$ **c** $y = -x^2$

   d $y = \frac{2}{3}x^2 + 4$ **e** $2y + 3x^2 - 5 = 0$ **f** $y = 5x^3 + x - 6$

2   For each of the following functions, find the value of y for the given value of x:

   a $y = x^2 + 5x - 4$ $\{x = 3\}$ **b** $y = 2x^2 + 9$ $\{x = -3\}$

   c $f : x \longmapsto -2x^2 + 3x - 5$ $\{x = 1\}$ **d** $f : x \longmapsto 4x^2 - 7x + 1$ $\{x = 4\}$

Example 12

If $f(x) = -2x^2 + 3x - 4$ find: **a** $f(2)$ **b** $f(-4)$

a $f(2) = -2(2)^2 + 3(2) - 4$
$$= -2(4) + 6 - 4$$
$$= -8 + 6 - 4$$
$$= -6$$

b $f(-4) = -2(-4)^2 + 3(-4) - 4$
$$= -2(16) - 12 - 4$$
$$= -32 - 12 - 4$$
$$= -48$$

3 For each of the following functions find the value of $f(x)$ given in brackets:

Remember that $f(2) = -6$ means that the point with coordinates $(2, -6)$ lies on the graph of $y = f(x)$.

    a $f(x) = x^2 - 2x + 3$ $\{f(2)\}$

    b $f(x) = 4 - x^2$ $\{f(-3)\}$

    c $f(x) = \frac{1}{2}x^2 + 3x$ $\{f(2)\}$

Example 13

State whether the following quadratic functions are satisfied by the given ordered pairs:

  a $y = 3x^2 + 2x$ $(2, 16)$ **b** $f(x) = -x^2 - 2x + 1$ $(-3, 1)$

a $y = 3(2)^2 + 2(2)$ **b** $f(-3) = -(-3)^2 - 2(-3) + 1$

       $= 12 + 4$ $= -9 + 6 + 1$

       $= 16$ $= -2$

   i.e.,   when $x = 2$, $y = 16$ i.e., $f(-3) \neq 1$

   $\therefore$ $(2, 16)$ does satisfy $\therefore$ $(-3, 1)$ does not satisfy

       $y = 3x^2 + 2x$ $f(x) = -x^2 - 2x + 1$

---

4 State whether the following quadratic functions are satisfied by the given ordered pairs:

    a $f(x) = 5x^2 - 10$ $(0, 5)$ **b** $y = 2x^2 + 5x - 3$ $(4, 9)$

    c $y = -2x^2 + 3x$ $(-\frac{1}{2}, 1)$ **d** $f : x \longmapsto -7x^2 + 8x + 15$ $(-1, 16)$

FINDING x GIVEN y

It is also possible to substitute a value for y to find a corresponding value for x. However, unlike linear functions, with quadratic functions there may be 0, 1 or 2 possible values for x for any one value of y.

Example 14

If $y = x^2 - 6x + 8$ find the value(s) of x when: **a** $y = 15$ **b** $y = -1$

a   If $y = 15$, $x^2 - 6x + 8 = 15$

                   $\therefore$ $x^2 - 6x - 7 = 0$

             $\therefore$ $(x + 1)(x - 7) = 0$ {factorising}

           $\therefore$ $x = -1$ or $x = 7$ i.e.,   2 solutions.

b   If $y = -1$, $x^2 - 6x + 8 = -1$

            $\therefore$ $x^2 - 6x + 9 = 0$

          $\therefore$ $(x - 3)^2 = 0$ {factorising}

              $\therefore$ $x = 3$ i.e.,   only one solution

5 Find the value(s) of x for the given value of y for each of the following quadratic functions:

    a $y = x^2 + 6x + 10$ $\{y = 1\}$ **b** $y = x^2 + 5x + 8$ $\{y = 2\}$

    c $y = x^2 - 5x + 1$ $\{y = -3\}$ **d** $y = 3x^2$ $\{y = -3\}$

Example 15

If $f(x) = x^2 + 4x + 11$ find x when **a** $f(x) = 23$ **b** $f(x) = 7$

a   If $f(x) = 23$ **b**   If $f(x) = 7$

$\therefore \quad x^2 + 4x + 11 = 23$ $\therefore \quad x^2 + 4x + 11 = 7$

$\therefore \quad x^2 + 4x - 12 = 0$ $\therefore \quad x^2 + 4x + 4 = 0$

$\therefore \quad (x + 6)(x - 2) = 0$ {factorising} $\therefore \quad (x + 2)^2 = 0$ {factorising}

$\therefore \quad x = -6 \text{ or } 2$ $\therefore \quad x = -2$

i.e., 2 solutions.   i.e., one solution only.

---

6 Find the value(s) of x given that:

    a $f(x) = 3x^2 - 2x + 5$ and $f(x) = 5$

    b $f(x) = x^2 - x - 5$ and $f(x) = 1$

    c $f(x) = -2x^2 - 13x + 3$ and $f(x) = -4$

    d $f(x) = 2x^2 - 12x + 1$ and $f(x) = -17$

Example 16

A stone is thrown into the air and its height in metres above the ground is given by the function $h(t) = -5t^2 + 30t + 2$ where t is the time (in seconds) from when the stone is thrown.

    a How high above the ground is the stone at time $t = 3$ seconds?

    b How high above the ground was the stone released?

    c At what time was the stone's height above the ground 27 m?

a $h(3) = -5(3)^2 + 30(3) + 2$ **b**   The stone is released when

             $= -45 + 90 + 2$ $t = 0$ sec

             $= 47$ $\therefore \quad h(0) = -5(0)^2 + 30(0) + 2 = 2$

      i.e., 47 m above ground. $\therefore$ released 2 m above ground level.

c When $h(t) = 27$

          $-5t^2 + 30t + 2 = 27$

$\therefore \quad -5t^2 + 30t - 25 = 0$

$\therefore \quad t^2 - 6t + 5 = 0$ {dividing each term by -5}

$\therefore \quad (t - 1)(t - 5) = 0$ {factorising}

$\therefore \quad t = 1 \text{ or } 5$

i.e., after 1 sec and after 5 sec.   Can you explain the two answers?

7   An object is projected into the air with a velocity of 30 m/s. Its height in metres, after t seconds is given by the function $h(t) = 30t - 5t^2$.

   a   Calculate the height after: **i**   1 second   **ii**   5 seconds   **iii**   3 seconds.

   b   Calculate the time(s) at which the height is: **i**   40 m   **ii**   0 m.

   c   Explain your answers in part **b**.

8   A cake manufacturer finds that the profit in dollars, of making x cakes per day is given by the function $P : x \longmapsto -\frac{1}{4}x^2 + 16x - 30$.

   a   Calculate the profit if:

      i   0 cakes   **ii**   10 cakes   are made per day.

   b   How many cakes per day are made if the profit is \$57?

G GRAPHS OF QUADRATIC FUNCTIONS

The graphs of all quadratic functions are **parabolas**. The parabola is one of the conic sections.

Conic sections are curves which can be obtained by cutting a cone with a plane. The Ancient Greek mathematicians were fascinated by conic sections.

You may like to find the conic sections for yourself by cutting an icecream cone.

Cutting parallel to the side produces a parabola,   i.e.,

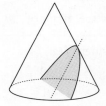

There are many examples of parabolas in every day life. The name parabola comes from the Greek word for **thrown** because when an object is thrown its path makes a parabolic shape.

Parabolic mirrors are used in car headlights, heaters, radar discs and radio telescopes because of their special geometric properties.

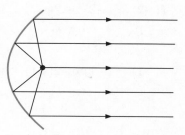

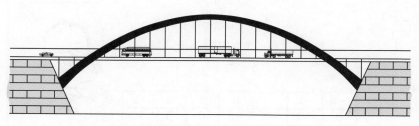

Alongside is a single span parabolic bridge.

Other suspension bridges, such as the Golden Gate bridge in San Francisco, also form parabolic curves.

THE SIMPLEST QUADRATIC FUNCTION

The simplest quadratic function is $y = x^2$ and its graph can be drawn from a table of values.

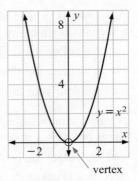

x	-3	-2	-1	0	1	2	3
y	9	4	1	0	1	4	9

Note:

* The curve is a **parabola** and it opens upwards.
* There are no negative y values,   i.e., the curve does not go below the x-axis.
* The curve is **symmetrical** about the y-axis because, for example, when $x = -3$, $y = (-3)^2$ and when $x = 3$, $y = 3^2$ have the same value.
* The curve has a **turning point** or **vertex** at (0, 0).

The **vertex** is the point where the graph is at its maximum or minimum.

Special note:

It is essential that you can draw the graph of $y = x^2$ without having to refer to a table of values.

Example 17

Draw the graph of $y = x^2 + 2x - 3$ from a table of values.

Consider $f(x) = x^2 + 2x - 3$

$f(-3) = (-3)^2 + 2(-3) - 3$
$\quad = 9 - 6 - 3$
$\quad = 0$

$f(-2) = (-2)^2 + 2(-2) - 3$
$\quad = 4 - 4 - 3$
$\quad = -3$

$f(-1) = (-1)^2 + 2(-1) - 3$
$\quad = 1 - 2 - 3$
$\quad = -4$

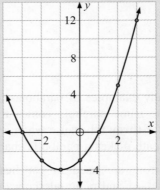

$f(0) = 0^2 + 2(0) - 3$
$\quad = 0 + 0 - 3$
$\quad = -3$

$f(1) = 1^2 + 2(1) - 3$
$\quad = 1 + 2 - 3$
$\quad = 0$

$f(2) = 2^2 + 2(2) - 3$
$\quad = 4 + 4 - 3$
$\quad = 5$

$f(3) = 3^2 + 2(3) - 3$
$\quad = 9 + 6 - 3$
$\quad = 12$

The easiest way to do these calculations is in **table mode**.

Tabled values:

x	-3	-2	-1	0	1	2	3
y	0	-3	-4	-3	0	5	12

EXERCISE 9G.1

1 From a table of values for $x = -3, -2, -1, 0, 1, 2, 3$ draw the graph of:

a $y = x^2 - 2x + 8$ **b** $y = -x^2 + 2x + 1$ **c** $y = 2x^2 + 3x$

d $y = -2x^2 + 4$ **e** $y = x^2 + x + 4$ **f** $y = -x^2 + 4x - 9$

2 Use your **graphics calculator** or **graphing package** to check your graphs in question **1**.

INVESTIGATION 2   **GRAPHS OF THE FORM** $y = x^2 + k$

What to do:

1 Using your **graphing package** or **graphics calculator**:

     i graph the two functions on the same set of axes

     ii state the coordinates of the vertex of each function.

GRAPHING PACKAGE

a $y = x^2$ and $y = x^2 + 2$ **b** $y = x^2$ and $y = x^2 - 2$

c $y = x^2$ and $y = x^2 + 4$ **d** $y = x^2$ and $y = x^2 - 4$

2 These functions are all members of the family $y = x^2 + k$ where k is a constant term. What effect does the value of k have on:

     a the position of the graph   **b** the shape of the graph?

3 To graph $y = x^2 + k$ from $y = x^2$, what transformation is used?

Graphs of the form $y = x^2 + k$ have exactly the same shape as the graph of $y = x^2$. In fact k is the **vertical translation** factor. Every point on the graph of $y = x^2$ is translated k units vertically to give the graph of $y = x^2 + k$.

Example 18

Sketch $y = x^2$ on a set of axes and hence sketch $y = x^2 + 3$. Mark the vertex of $y = x^2 + 3$.

Draw $y = x^2$ and translate it 3 units upwards.

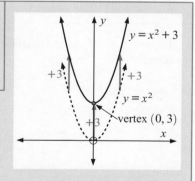

EXERCISE 9G.2

1 Without using your graphics calculator or graphing package, sketch the graph of $y = x^2$ on a set of axes and hence sketch each of the following functions, stating the coordinates of the vertex: (Use separate sets of axes for each part.)

a $y = x^2 - 3$ **b** $y = x^2 - 1$ **c** $y = x^2 + 1$

d $y = x^2 - 5$ **e** $y = x^2 + 5$ **f** $y = x^2 - \frac{1}{2}$

2   Use your **graphics calculator** or **graphing package** to check your graphs in question **1**.

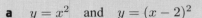

INVESTIGATION 3   **GRAPHS OF THE FORM** $y = (x - h)^2$

What to do: GRAPHING PACKAGE

1 Using your **graphing package** or **graphics calculator**:

    i   graph the two functions on the same set of axes

    ii state the coordinates of the vertex of each function.

  a $y = x^2$ and $y = (x - 2)^2$ **b** $y = x^2$ and $y = (x + 2)^2$

  c $y = x^2$ and $y = (x - 4)^2$ **d** $y = x^2$ and $y = (x + 4)^2$

2   These functions are all members of the family of $y = (x - h)^2$ where h is constant. Discuss the effect the value of h has on:

  a the position of the graph   **b**   the shape of the graph?

3   What transformation of $y = x^2$ is used to graph $y = (x - h)^2$?

Graphs of the form $y = (x - h)^2$ have exactly the same shape as the graph of $y = x^2$.

In fact, h is the **horizontal translation** factor. Every point on the graph of $y = x^2$ is translated h units horizontally to give the graph of $y = (x - h)^2$.

Example 19

Sketch $y = x^2$ on a set of axes and hence sketch $y = (x + 3)^2$. Mark the vertex of $y = (x + 3)^2$.

Draw $y = x^2$ and translate it 3 units left.

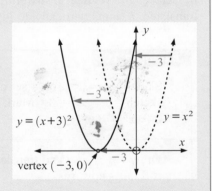

EXERCISE 9G.3

1   Without using your graphics calculator or graphing package, sketch the graph of $y = x^2$ on a set of axes and hence sketch each of the following functions, stating the coordinates of the vertex: (Use a separate set of axes for each part.)

  a $y = (x - 3)^2$ **b** $y = (x + 1)^2$ **c** $y = (x - 1)^2$

  d $y = (x - 5)^2$ **e** $y = (x + 5)^2$ **f** $y = (x - \frac{3}{2})^2$

2   Use your **graphics calculator** or **graphing package** to check your graphs in question **1**.

INVESTIGATION 4   GRAPHS OF THE FORM $y = (x - h)^2 + k$

What to do:

1 *Without using the assistance of technology* sketch the graph of $y = (x - 2)^2 + 3$, stating the coordinates of the vertex and commenting on the shape of the graph. (You should take into consideration what you have learnt from the two previous investigations.)

2 Use your **graphing package** or **graphics calculator** to draw, on the same set of axes, the graphs of $y = x^2$ and $y = (x - 2)^2 + 3$.

GRAPHING
PACKAGE

3 Compare the two graphs you have drawn commenting on their shape and position.

4 Repeat steps **1**, **2** and **3** for $y = (x + 4)^2 - 1$.

5 Copy and complete:

- The graph of $y = (x - h)^2 + k$ is the same shape as the graph of ......
- The graph of $y = (x - h)^2 + k$ is a .... of the graph of $y = x^2$ through ....

Graphs of the form $y = (x - h)^2 + k$ have the same shape as the graph of $y = x^2$ and can be obtained from $y = x^2$ by using a **horizontal shift** of h units and a **vertical shift** of k units.

The **vertex** is at (h, k).

Example 20

Sketch $y = x^2$ on a set of axes, then sketch the following, stating the coordinates of the vertex:

  a $y = (x - 2)^2 + 3$ **b** $y = (x + 2)^2 - 5$

a Draw $y = x^2$ and translate it 2 units right 3 units up.	**b** Draw $y = x^2$ and translate it 2 units left 5 units down

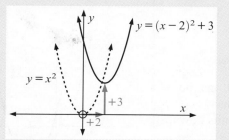

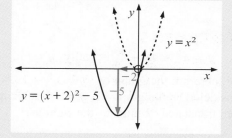

      The vertex is at $(2, 3)$.   The vertex is at $(-2, -5)$.

EXERCISE 9G.4

1 On separate sets of axes sketch $y = x^2$ and each of the following, stating the coordinates of the vertex:

   a $y = (x-1)^2 + 3$ **b** $y = (x-2)^2 - 1$ **c** $y = (x+1)^2 + 4$

   d $y = (x+2)^2 - 3$ **e** $y = (x+3)^2 - 2$ **f** $y = (x-3)^2 + 3$

2 Use your **graphics calculator** or **graphing package** to check your graphs in question 1.

INVESTIGATION 5   **GRAPHS OF THE FORM** $y = x^2 + bx + c$

What to do:

1 Using your **graphing package** or **graphics calculator**:

   a graph $y = x^2$ and $y = x^2 - 4x + 1$ on the same set of axes

   b state the coordinates of the vertex of each function

   c comment on the shape of each function.

GRAPHING PACKAGE

2 Write down the equation $y = x^2 - 4x + 1$ in the form $y = (x-h)^2 + k$, using your results from the previous exercise.

3 Expand and simplify $y = (x-2)^2 - 3$. What do you notice?

From the above investigation it appears that if we could write quadratics of the form $y = x^2 + bx + c$ in the form $y = (x-h)^2 + k$ we could easily sketch such graphs by hand *without needing* a table of values.

COMPLETING THE SQUARE

Recall that in **Chapter 8** we used the process of completing the square to assist us in solving quadratic equations which did not factorise.

This same process can be used here to convert quadratics into the form $y = (x-h)^2 + k$.

Consider $y = x^2 - 4x + 1$

$\therefore \quad y = x^2 - 4x + 2^2 + 1 - 2^2$ {keeping the equation balanced}

$\therefore \quad y = x^2 - 4x + 2^2 - 3$

$\therefore \quad y = (x-2)^2 - 3$

So, $y = x^2 - 4x + 1$ is really $y = (x-2)^2 - 3$

and therefore the graph of $y = x^2 - 4x + 1$ can be considered as the graph of $y = x^2$ after it has been translated 2 units to the right and 3 units down.

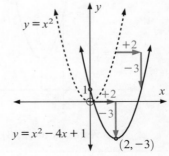

Example 21

Write $y = x^2 + 4x + 3$ in the form $y = (x - h)^2 + k$ using completing the square and hence sketch $y = x^2 + 4x + 3$, stating the coordinates of the vertex.

$y = x^2 + 4x + 3$

$\therefore \quad y = x^2 + 4x + 2^2 + 3 - 2^2$

$\therefore \quad y = (x + 2)^2 - 1$

shift 2	shift 1
units left	unit down

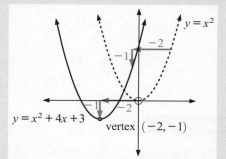

EXERCISE 9G.5

1 Write the following quadratics in the form $y = (x - h)^2 + k$ using 'completing the square' and hence sketch each function, stating the vertex:

   a $y = x^2 - 2x + 3$ **b** $y = x^2 - 6x + 4$ **c** $y = x^2 + 4x - 2$

   d $y = x^2 - 2x + 5$ **e** $y = x^2 - 4x$ **f** $y = x^2 + 3x$

   g $y = x^2 + 5x - 2$ **h** $y = x^2 - 3x + 2$ **i** $y = x^2 - 5x + 1$

2 Use your **graphing package** or **graphics calculator** to check your graphs in question **1**.

3 By using your **graphing package** or **graphics calculator**, graph each of the following functions, and hence write each function in the form $y = (x - h)^2 + k$:

   a $y = x^2 - 4x + 7$ **b** $y = x^2 + 6x + 3$ **c** $y = x^2 + 4x + 5$

   d $y = x^2 + 2x - 4$ **e** $y = x^2 - 3x + 1$ **f** $y = x^2 - 9x - 5$

INVESTIGATION 6   **GRAPHS OF THE FORM** $y = ax^2$

What to do:

1 Using your **graphing package** or **graphics calculator**:

   i graph the two functions on the same set of axes

   ii state the coordinates of the vertex of each function.

   a $y = x^2$ and $y = 2x^2$ **b** $y = x^2$ and $y = 4x^2$

   c $y = x^2$ and $y = \frac{1}{2}x^2$ **d** $y = x^2$ and $y = -x^2$

   e $y = x^2$ and $y = -2x^2$ **f** $y = x^2$ and $y = -\frac{1}{2}x^2$

GRAPHING PACKAGE

2 These functions are all members of the family $y = ax^2$ where a is the coefficient of the x^2 term. What effect does a have on:

   a the position of the graph   **b** the shape of the graph

   c the direction in which the graph opens?

If $a > 0$, $y = ax^2$ opens upwards   i.e.,

If $a < 0$, $y = ax^2$ opens downwards   i.e.,

Graphs of the form $y = ax^2$ are 'wider' or 'thinner' than $y = x^2$.

If $a < -1$ or $a > 1$, $y = ax^2$ is 'thinner' than $y = x^2$.
If $-1 < a < 1$ $(a \neq 0)$, $y = ax^2$ is 'wider' than $y = x^2$.

Example 22

Sketch $y = x^2$ on a set of axes and hence sketch:
a $y = 3x^2$ **b** $y = -3x^2$

---

a $y = 3x^2$ is 'thinner' than $y = x^2$.

b $y = -3x^2$ is the same shape as
$y = 3x^2$ but opens downwards.

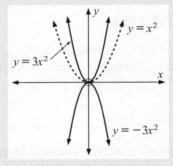

EXERCISE 9G.6

1 On separate sets of axes sketch $y = x^2$ and then sketch the following, commenting
on: **i**   the shape of the graph   **ii**   the direction in which the graph opens.

a $y = 5x^2$ **b** $y = -5x^2$ **c** $y = \frac{1}{3}x^2$

d $y = -\frac{1}{3}x^2$ **e** $y = -4x^2$ **f** $y = \frac{1}{4}x^2$

2 Use your **graphics calculator** or **graphing package** to check your graphs in question **1**.

INVESTIGATION 7   GRAPHS OF THE FORM $y = a(x - h)^2 + k$

What to do:

1 *Without the assistance of technology*, on the same set of axes, sketch the
graph of $y = 2x^2$ and hence the graph of $y = 2(x - 1)^2 + 3$ stating
the coordinates of the vertex and commenting on the shape of the two
graphs. (You should take into consideration what you have learnt from
previous investigations.)

2 Use your **graphing package** or **graphics calculator** to check your graphs in step **1**.

3 Compare the two graphs you have drawn commenting on their shape and position.

4 Repeat steps **1**, **2** and **3** for

    a $y = -x^2$ and $y = -(x+2)^2 + 3$ **b** $y = \frac{1}{2}x^2$ and $y = \frac{1}{2}(x-2)^2 - 4$

5 Copy and complete:

- The graph of $y = a(x-h)^2 + k$ is the same shape and opens in the same direction as the graph of ......
- The graph of $y = a(x-h)^2 + k$ is a .... of the graph of $y = ax^2$ through ....

Summary:

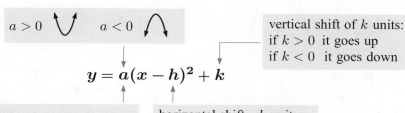

$a > 0$ $a < 0$

vertical shift of k units:
if $k > 0$ it goes up
if $k < 0$ it goes down

$$y = a(x - h)^2 + k$$

affects the width
$a < -1$ or $a > 1$, thinner than $y = x^2$
$a \neq 0$, $-1 < a < 1$, wider than $y = x^2$

horizontal shift h units:
$h > 0$ right, $h < 0$ left

Example 23

Sketch the graph of $y = -(x-2)^2 - 3$ from the graph of $y = x^2$ and hence state the coordinates of the vertex.

$y = -(x-2)^2 - 3$

reflect in x-axis

vertical shift 3 units down

horizontal shift $+2$ units

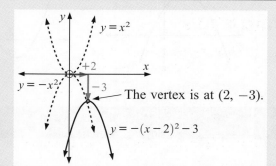

The vertex is at $(2, -3)$.

EXERCISE 9G.7

1 Sketch the graphs of the following functions without using tables of values and state the coordinates of the vertex:

    a $y = -(x-1)^2 + 3$ **b** $y = 2x^2 + 4$ **c** $y = -(x-2)^2 + 4$

    d $y = 3(x+1)^2 - 4$ **e** $y = \frac{1}{2}(x+3)^2$ **f** $y = -\frac{1}{2}(x+3)^2 + 1$

    g $y = -2(x+4)^2 + 3$ **h** $y = 2(x-3)^2 + 5$ **i** $y = \frac{1}{2}(x-2)^2 + 1$

    j $y = -\frac{1}{2}(x+1)^2 - 4$ **k** $y = 3(x+2)^2$ **l** $y = -\frac{1}{3}(x+3)^2 + 1\frac{2}{3}$

2 Use your **graphics calculator** or **graphing package** to check your graphs in question **1**.

3 Match each quadratic function with its corresponding graph:

- **a** $y = -1(x+1)^2 + 3$
- **b** $y = -2(x-3)^2 + 2$
- **c** $y = x^2 + 2$
- **d** $y = -1(x-1)^2 + 1$
- **e** $y = (x-2)^2 - 2$
- **f** $y = \frac{1}{3}(x+3)^2 - 3$
- **g** $y = -x^2$
- **h** $y = -\frac{1}{2}(x-1)^2 + 1$
- **i** $y = 2(x+2)^2 - 1$

A
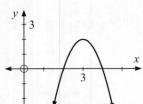

B
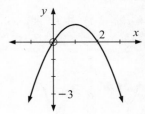

C
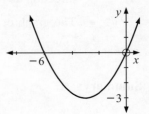

D
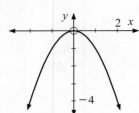

E
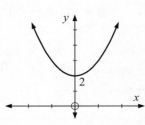

F
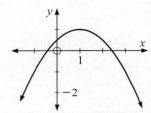

G
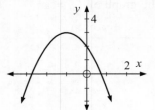

H
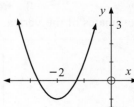

I
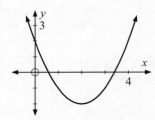

H | AXES INTERCEPTS

Given the equation of any curve:

An x-**intercept** is a value of x where the graph meets the x-axis,

A y-**intercept** is a value of y where the graph meets the y-axis.

x-intercepts are found by letting y be 0 in the equation of the curve.

y-intercepts are found by letting x be 0 in the equation of the curve.

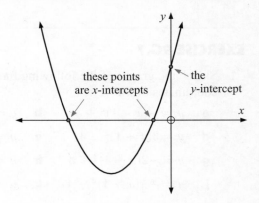

these points are x-intercepts

the y-intercept

INVESTIGATION 8   AXES INTERCEPTS

What to do:

1 For the following quadratic functions, use your **graphing package**
or **graphics calculator** to:

i draw the graph **ii** find the y-intercept **iii** find the x-intercepts (if any exist)

a $y = x^2 - 3x - 4$ **b** $y = -x^2 + 2x + 8$ **c** $y = 2x^2 - 3x$

d $y = -2x^2 + 2x - 3$ **e** $y = (x-1)(x-3)$ **f** $y = -(x+2)(x-3)$

g $y = 3(x+1)(x+4)$ **h** $y = 2(x-2)^2$ **i** $y = -3(x+1)^2$

2 From your observations in question **1**:

a State the y-intercept of a quadratic function in the form $y = ax^2 + bx + c$.

b State the x-intercepts of quadratic function in the form $y = a(x - \alpha)(x - \beta)$.

c What do you notice about the x-intercepts of quadratic functions in the form $y = a(x - \alpha)^2$?

THE y-INTERCEPT

You will have noticed that for a quadratic function of the form $y = ax^2 + bx + c$, the y-intercept is the constant term c.

This is because   any curve cuts the y-axis when $x = 0$.

So, if we substitute $x = 0$ into a function we can find the y-intercept.

For example,   if $y = x^2 - 2x - 3$ and we let $x = 0$

then $y = 0^2 - 2(0) - 3$

$\therefore$ $y = -3$ (the constant term)

EXERCISE 9H.1

1 For the following functions state the y-intercept:

a $y = x^2 + 3x + 3$ **b** $y = x^2 - 5x + 2$ **c** $y = 2x^2 + 7x - 8$

d $y = 3x^2 - x + 1$ **e** $y = -x^2 + 3x + 6$ **f** $y = -2x^2 + 5 - x$

g $y = 6 - x - x^2$ **h** $y = 8 + 2x - 3x^2$ **i** $y = 5x - x^2 - 2$

THE x-INTERCEPTS

You will have noticed that for a quadratic function of the form $y = a(x - \alpha)(x - \beta)$, the x-intercepts are α and β.   **Note:** α and β are the **zeros** of the quadratic function.

This is because   any curve cuts the x-axis when $y = 0$.

So, if we substitute $y = 0$ into the function we get $a(x - \alpha)(x - \beta) = 0$

$\therefore$ $x = \alpha$ or β {by the Null Factor law}

This suggests that x-intercepts are easy to find when the quadratic is in **factorised** form.

Example 24

Find the zeros of:

a $y = 2(x - 3)(x + 2)$ **b** $y = -(x - 4)^2$

a   We let $y = 0$

   $\therefore \quad 2(x - 3)(x + 2) = 0$

   $\therefore \quad x = 3 \quad \text{or} \quad x = -2$

   $\therefore$ the zeros are 3 and -2.

b   We let $y = 0$

   $\therefore \quad -(x - 4)^2 = 0$

   $\therefore \quad x = 4$

   $\therefore$ the zero is 4.

EXERCISE 9H.2

1   For the following functions, find the zeros:

   a $y = (x - 3)(x + 1)$

   b $y = -(x - 2)(x - 4)$

   c $y = 2(x + 3)(x + 2)$

   d $y = -3(x - 4)(x - 5)$

   e $y = 2(x + 3)^2$

   f $y = -5(x - 1)^2$

> If a quadratic function has only one x-intercept then its graph must touch the x-axis.

FACTORISING TO FIND x-INTERCEPTS (ZEROS)

If the quadratic function is given in the form $y = ax^2 + bx + c$ and we wish to find the zeros, we let $y = 0$ and solve for x by **factorising**.

In general:

> for any quadratic function of the form $y = ax^2 + bx + c$, the zeros can be found by solving the equation $ax^2 + bx + c = 0$.

You will recall from **Chapter 8** that quadratic equations may have *two solutions*, *one solution* or *no solutions*.

Consequently, parabolas drawn from quadratic functions can have
- two zeros
- one zero, or
- no zeros.

Example 25

Find the zeros of the quadratic functions:

a $y = x^2 - 6x + 9$ **b** $y = -x^2 - x + 6$

a   When $y = 0$,

   $x^2 - 6x + 9 = 0$

   $\therefore \quad (x - 3)^2 = 0$

   $\therefore \quad x = 3$

   $\therefore$ the zero is 3.

b   When $y = 0$,

   $-x^2 - x + 6 = 0$

   $\therefore \quad x^2 + x - 6 = 0$

   $\therefore \quad (x + 3)(x - 2) = 0$

   $\therefore \quad x = -3 \text{ or } 2$

   $\therefore$ the zeros are -3 and 2.

EXERCISE 9H.3

1 For the following functions find the zeros:

a $y = x^2 - 9$

b $y = 2x^2 - 6$

c $y = x^2 + 7x + 10$

d $y = x^2 + x - 12$

e $y = 4x - x^2$

f $y = -x^2 - 6x - 8$

g $y = -2x^2 - 4x - 2$

h $y = 4x^2 - 24x + 36$

i $y = x^2 - 4x + 1$

j $y = x^2 + 4x - 3$

k $y = x^2 - 6x - 2$

l $y = x^2 + 8x + 11$

I GRAPHS FROM AXES INTERCEPTS

Consider the quadratic function $y = 3(x - 1)^2 + 2$.   From our previous exercises we can easily sketch the graph of this function without the use of technology.

$$y = 3(x - 1)^2 + 2$$

$$a = 3 \qquad h = 1 \qquad k = 2$$

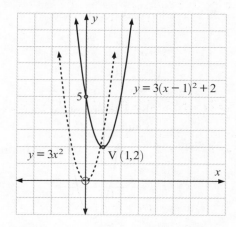

This graph has the same shape as the graph of $y = 3x^2$ and has vertex $(1, 2)$.

However, on expanding:
$$y = 3(x - 1)^2 + 2$$
$$\therefore \quad y = 3(x^2 - 2x + 1) + 2$$
$$\therefore \quad y = 3x^2 - 6x + 3 + 2$$
$$\therefore \quad y = 3x^2 - 6x + 5$$

From this we can see that:

> the graph of a quadratic of the form $y = ax^2 + bx + c$ has the same shape as the graph of $y = ax^2$.

EXERCISE 9I

1   **i**   Use your **graphing package** or **graphics calculator** to graph, on the same set of axes, the following quadratic functions.

     ii   Compare the shapes of the two graphs.

a $y = 2x^2$ and $y = 2x^2 - 3x + 1$

b $y = -2x^2$ and $y = -2x^2 + 5$

c $y = 3x^2$ and $y = 3x^2 - 5x$

d $y = -x^2$ and $y = -x^2 - 6x + 4$

Example 26

Sketch the graph of the following by considering:

   i the value of a **ii** the y-intercept   **iii** the x-intercepts.

a $y = x^2 - 2x - 3$ **b** $f : x \longmapsto -2(x+1)(x-2)$

---

a $y = x^2 - 2x - 3$

  i since $a = 1$ the parabola opens upwards i.e.,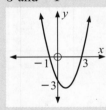

  ii y-intercept occurs when $x = 0$, i.e., $y = -3$
     i.e., y-intercept is -3

  iii x-intercepts occur when $y = 0$

     $\therefore \quad x^2 - 2x - 3 = 0$
     $\therefore \quad (x-3)(x+1) = 0$
     $\therefore \quad x = 3$ or $x = -1$

     i.e., the x-intercepts are 3 and -1

Sketch:

b $y = -2(x+1)(x-2)$

  i Since $a = -2$ the parabola opens downwards i.e.,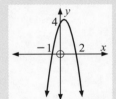

  ii y-intercept occurs when $x = 0$

     $\therefore \quad y = -2(0+1)(0-2)$
           $y = -2 \times 1 \times -2$
           $y = 4$

     i.e., y-intercept is 4

  iii x-intercepts occur when $y = 0$

     $\therefore \quad -2(x+1)(x-2) = 0$
     $\therefore \quad x = -1$ or $x = 2$

     i.e., x-intercepts are -1 and 2

Sketch:

Example 27

Sketch the graph of $y = 2(x-3)^2$ by considering:

a the value of a **b** the y-intercept   **c** the x-intercepts.

$y = 2(x-3)^2$

a Since $a = 2$ the parabola opens upwards i.e.,

b y-intercept occurs when $x = 0$

    $\therefore \quad y = 2(0-3)^2 = 18$

  i.e., y-intercept is 18

c x-intercepts occur when $y = 0$

    $\therefore \quad 2(x-3)^2 = 0$
       $\therefore \quad x = 3$

  i.e., x-intercept is 3 {only one x-intercept $\therefore$ *touches*}

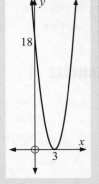

2   Sketch the graphs of the following by considering:

>   **i**   the value of a **ii**   the y-intercept   **iii**   the x-intercepts.

>   **a** $y = x^2 - 4x + 4$ **b** $y = (x - 1)(x + 3)$ **c** $y = 2(x + 2)^2$
>   **d** $y = -(x - 2)(x + 1)$ **e** $y = -3(x + 1)^2$ **f** $y = -3(x - 4)(x - 1)$
>   **g** $y = 2(x + 3)(x + 1)$ **h** $f : x \longmapsto 2x^2 + 3x + 2$ **i** $f : x \longmapsto -2x^2 - 3x + 5$

Recall from **Exercise 9G**:

- the graph of a quadratic function is a **parabola**
- the curve is symmetrical about an **axis of symmetry**
- the curve has a **turning point** or **vertex**.

Example 28

>   **a**   Sketch the graph of $y = 2(x - 2)(x + 4)$ using axis intercepts.
>   **b**   Find the equation of the axis of symmetry and the coordinates of the vertex.

---

a $y = 2(x - 2)(x + 4)$

Since $a = 2$ the parabola opens upwards   i.e.,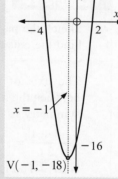

When $x = 0$
$$y = 2 \times -2 \times 4 = -16$$
i.e., y-intercept is -16.

When $y = 0$
$$\therefore \quad 2(x - 2)(x + 4) = 0$$
$$\therefore \quad x = 2 \quad \text{or} \quad x = -4$$
i.e., x-intercepts are 2 and -4.

b   Axis of symmetry is halfway between x-intercepts
$$\therefore \quad \text{axis of symmetry is } x = -1 \qquad \{-1 \text{ is the average of } -4 \text{ and } 2\}$$
when $x = -1, \quad y = 2(-1 - 2)(-1 + 4)$
$$= 2 \times -3 \times 3$$
$$= -18$$
i.e.,   coordinates of vertex are $(-1, -18)$.

---

3   For each of the following find the equation of the axis of symmetry:

a   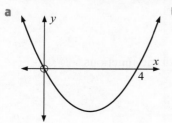   **b**   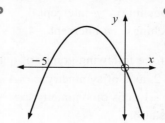   **c**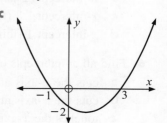

d 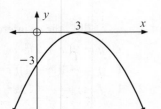   **e** 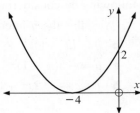   **f**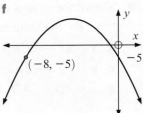

4 For each of the following quadratic functions:

  i   sketch the graph using axes intercepts and hence find
  ii the equation of the axis of symmetry
  iii the coordinates of the vertex.

a $y = x^2 + 4x + 4$ **b** $y = x(x - 4)$ **c** $y = 3(x - 2)^2$

d $y = -(x - 1)(x + 3)$ **e** $y = -2(x - 1)^2$ **f** $y = -3(x + 2)(x - 2)$

g $y = 2(x + 1)(x + 4)$ **h** $y = 2x^2 - 3x - 2$ **i** $y = -2x^2 - x + 3$

Example 29

Sketch the parabola which has x-intercepts -3 and 1, and y-intercept -2.
Find the equation of the axis of symmetry.

The axis of symmetry lies halfway
between the x-intercepts $\therefore$ axis of
symmetry is $x = -1$.

$\{\dfrac{-3 + 1}{2} = -1\}$

Note: The graph must open upwards.
Can you see why?

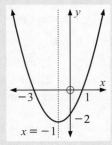

---

5 For each of the following:

  i sketch the parabola
  ii find the equation of the axis of symmetry.

a x-intercepts 3 and -1, y-intercept -4

b x-intercepts 2 and -2, y-intercept 4

c x-intercept -3 (touching), y-intercept 6

d x-intercept 1 (touching), y-intercept -4

6 Find all x-intercepts of the following graphs of quadratic functions:

a cuts the x-axis at 2, axis of symmetry $x = 4$

b cuts the x-axis at -1, axis of symmetry $x = -3$

c touches the x-axis at 3.

7 Consider the quadratic function
$y = ax^2 + bx + c$ whose graph cuts the x-axis at
A and B. Let the equation of the axis of symmetry be $x = h$.   The x-intercepts are an equal
distance (d) from the axis of symmetry.

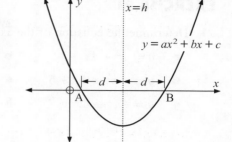

a   Find in terms of h and d the coordinates of
A and B.

b   Substitute the coordinates of A into
$y = ax^2 + bx + c$ to create equation (1).

c   Substitute the coordinates of B into $y = ax^2 + bx + c$ to create equation (2).

d   Use equations (1) and (2) to show that $h = \dfrac{-b}{2a}$.

J   AXIS OF SYMMETRY AND VERTEX

AXIS OF SYMMETRY

As we have seen from the previous exercise:

> the equation of the **axis of symmetry** of $y = ax^2 + bx + c$ is $x = \dfrac{-b}{2a}$.

The problem with the method used to demonstrate this is that not all quadratic functions have
a graph which cuts the x-axis.

To prove this formula we can use an expansion method and compare coefficients.

Proof: Suppose $y = ax^2 + bx + c$ is converted to $y = a(x - h)^2 + k$

i.e., $y = a(x^2 - 2hx + h^2) + k$

i.e., $y = ax^2 - 2ahx + [ah^2 + k]$.

Comparing the coefficients of x we obtain $-2ah = b$

$$\therefore \quad h = \dfrac{-b}{2a}.$$

Example 30

Find the equation of the axis of symmetry of $y = 2x^2 + 3x + 1$.

$y = 2x^2 + 3x + 1$ has $a = 2, \quad b = 3, \quad c = 1$

$\therefore$ axis of symmetry has equation $x = \dfrac{-b}{2a} = \dfrac{-3}{2 \times 2}$

i.e., $x = -\dfrac{3}{4}$

EXERCISE 9J

1 Determine the equation of the axis of symmetry of:

  a $y = x^2 + 4x + 1$ **b** $y = 2x^2 - 6x + 3$ **c** $y = 3x^2 + 4x - 1$

  d $y = -x^2 - 4x + 5$ **e** $y = -2x^2 + 5x + 1$ **f** $y = \frac{1}{2}x^2 - 10x + 2$

  g $y = \frac{1}{3}x^2 + 4x$ **h** $y = 100x - 4x^2$ **i** $y = -\frac{1}{10}x^2 + 30x$

TURNING POINT (OR VERTEX)

The **turning point** (or **vertex**) of any parabola is the point at which the function has a

maximum value (for $a < 0$)   or, a **minimum value** (for $a > 0$) .

As the turning point lies on the axis of symmetry, its x-coordinate will be $x = \dfrac{-b}{2a}$.

The y-coordinate can be found by substituting for x into the function, i.e., $f\left(\dfrac{-b}{2a}\right)$.

Example 31

Determine the coordinates of the vertex of $y = 2x^2 - 8x + 1$.

$y = 2x^2 - 8x + 1$ has $a = 2, \quad b = -8, \quad c = 1$

and so $\dfrac{-b}{2a} = \dfrac{-(-8)}{2 \times 2} = 2$

$\therefore$ equation of axis of symmetry is $x = 2$

and when $x = 2, \quad y = 2(2)^2 - 8(2) + 1$
$= 8 - 16 + 1$
$= -7$

$\therefore$ the vertex has coordinates $(2, -7)$.

---

2 Find the turning point (vertex) for the following quadratic functions:

  a $y = x^2 - 4x + 2$ **b** $y = x^2 + 2x - 3$

  c $f(x) = 2x^2 + 4$ **d** $f(x) = -3x^2 + 1$

  e $y = 2x^2 + 8x - 7$ **f** $y = -x^2 - 4x - 9$

  g $f(x) = 2x^2 + 6x - 1$ **h** $y = 2x^2 - 10x + 3$

  i $f : x \longmapsto -\frac{1}{2}x^2 + x - 5$

  j $f : x \longmapsto -2x^2 + 8x - 2$

The vertex is sometimes called the maximum turning point or the minimum turning point depending on whether the graph is opening downwards or upwards.

Example 32

For the quadratic function $y = -x^2 + 2x + 3$:
- **a** find its axes intercepts
- **b** find the equation of the axis of symmetry
- **c** find the coordinates of the vertex
- **d** sketch the function showing all important features.

a When $x = 0$, $y = 3$

$\therefore$ y-intercept is 3.

When $y = 0$, $-x^2 + 2x + 3 = 0$

$\therefore$ $x^2 - 2x - 3 = 0$

$\therefore$ $(x - 3)(x + 1) = 0$

$\therefore$ $x = 3$ or -1

so, x-intercepts are 3 and -1.

c From **b** when $x = 1$

$$y = -(1)^2 + 2(1) + 3$$
$$= -1 + 2 + 3$$
$$= 4$$

$\therefore$ vertex is $(1, 4)$.

b $a = -1$, $b = 2$, $c = 3$

$\therefore$ $\dfrac{-b}{2a} = \dfrac{-2}{-2} = 1$

$\therefore$ axis of symmetry is $x = 1$.

d

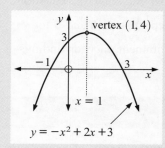

3 For each of the following quadratic functions find :
- **i** the axes intercepts
- **ii** the equation of the axis of symmetry
- **iii** the coordinates of the vertex
- **iv** and hence sketch the graph.

a $y = x^2 - 2x - 8$ **b** $f(x) = x^2 + 3x$ **c** $y = 4x - x^2$

d $y = x^2 + 4x + 4$ **e** $y = x^2 + 3x - 4$ **f** $f(x) = -x^2 + 2x - 1$

g $f(x) = -x^2 - 6x - 8$ **h** $y = -x^2 + 3x - 2$ **i** $y = 2x^2 + 5x - 3$

j $f : x \longmapsto 2x^2 - 5x - 12$ **k** $y = -3x^2 - 4x + 4$ **l** $f : x \longmapsto -\frac{1}{4}x^2 + 5x$

INVESTIGATION 9   SOLVING QUADRATIC EQUATIONS GRAPHICALLY

Many equations of the form $x^2 + bx + c = 0$ cannot be solved by simple factorisation of the LHS.

For example, the LHS of $x^2 - 4x + 1 = 0$ does not have factors with rational numbers.

One possible method of solution is:

- Draw the graph of $y = x^2 - 4x + 1$.
- Now $x^2 - 4x + 1 = 0$ when $y = 0$ and this occurs at the x-intercepts of the graph of $y = x^2 - 4x + 1$.
- Thus the **x-intercepts** are the required **solutions** to $x^2 - 4x + 1 = 0$.

What to do:

1 Use a **graphing package** or your **graphics calculator** to graph $y = x^2 - 4x + 1$.

2 Determine the x-intercepts (zeros) of $y = x^2 - 4x + 1$ correct to 3 decimal places using the built in functions of either your **graphing package** or **graphics calculator**. Hence state the solutions to $x^2 - 4x + 1 = 0$ correct to 2 decimal places.

3 Repeat the procedure for:

     a $x^2 + 4x + 2 = 0$ **b** $x^2 + 6x - 2 = 0$ **c** $2x^2 - 3x - 7 = 0$

     d $x^2 + 2x + 5 = 0$ **e** $-x^2 - 4x - 6 = 0$ **f** $4x^2 - 4x + 1 = 0$

4 Use your **graphing package** or **graphics calculator** to draw the graph of $y = x^2 - x + 4$.

5 Comment on the relationship between the graph of $y = x^2 - x + 4$ and the x-axis. Can you solve the equation $x^2 - x + 4 = 0$? Explain your answer.

6 Repeat steps **4** and **5** for $y = -x^2 + x - 6$.

GRAPHING PACKAGE

7 Comment on your findings.

K   WHERE FUNCTIONS MEET

Consider the graphs of a quadratic function and a linear function on the same set of axes.

Notice that we could have:

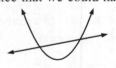   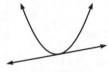   

       cutting   **touching**   **missing**

(2 points of intersection)   (1 point of intersection)   (no points of intersection)

The graphs could meet and the coordinates of the points of intersection of the graphs of the two functions can be found by *solving the two equations simultaneously*.

Example 33

Find the coordinates of the points of intersection of the graphs with equations $y = x^2 - x - 18$ and $y = x - 3$.

$y = x^2 - x - 18$ meets $y = x - 3$ where

$$x^2 - x - 18 = x - 3$$

$$\therefore \quad x^2 - 2x - 15 = 0 \qquad \{\text{RHS} = 0\}$$

$$\therefore \quad (x - 5)(x + 3) = 0 \qquad \{\text{factorising}\}$$

$$\therefore \quad x = 5 \text{ or } -3$$

Substituting into $y = x - 3$, when $x = 5$, $y = 2$ and when $x = -3$, $y = -6$.

$\therefore$ graphs meet at $(5, 2)$ and $(-3, -6)$.

EXERCISE 9K

1 Find the coordinates of the point(s) of intersection of the graphs with equations:

  a $y = x^2 - 2x + 8$ and $y = x + 6$

  b $y = -x^2 + 3x + 9$ and $y = 2x - 3$

  c $y = x^2 - 4x + 3$ and $y = 2x - 6$

  d $y = x^2 - 3x + 2$ and $y = x - 3$

  e $y = -x^2 + 4x - 7$ and $y = 5x - 4$

  f $y = x^2 - 5x + 9$ and $y = 3x - 7$

2 Use a **graphing package** or a **graphics calculator** to find the coordinates of the points of intersection (to two decimal places) of the graphs with equations:

  a $y = x^2 - 3x + 7$ and $y = x + 5$

  b $y = x^2 - 5x + 2$ and $y = x - 7$

  c $y = -x^2 - 2x + 4$ and $y = x + 8$

  d $y = -x^2 + 4x - 2$ and $y = 5x - 6$

GRAPHING PACKAGE

L   QUADRATIC MODELLING

There are many situations in the real world where the relationship between two variables is a quadratic function.

This means that the graph of such a relationship will be either $\smile$ or $\frown$ and the function will have a minimum or maximum value.

> For $y = ax^2 + bx + c$:
>
> - if $a > 0$,   the **minimum** value of y occurs
>   at $x = -\dfrac{b}{2a}$
>
> - if $a < 0$,   the **maximum** value of y occurs
>   at $x = -\dfrac{b}{2a}$.

The process of finding the maximum or minimum value of a function is called **optimisation**.

Optimisation is a very useful tool when looking at such issues as:

- maximising profits
- minimising costs
- maximising heights reached etc.

Example 34

The height H metres, of a rocket t seconds after it is fired vertically upwards is given by $H(t) = 80t - 5t^2$, $t \geqslant 0$.

a How long does it take for the rocket to reach its maximum height?

b What is the maximum height reached by the rocket?

c How long does it take for the rocket to fall back to earth?

a
$$H(t) = 80t - 5t^2$$
$$\therefore \quad H(t) = -5t^2 + 80t \quad \text{where} \quad a = -5 \quad \therefore$$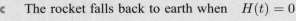

The maximum height reached occurs when $t = \dfrac{-b}{2a}$

$$\text{i.e.,} \quad t = \dfrac{-80}{2(-5)}$$

$$\therefore \quad t = 8$$

i.e., the maximum height is reached after 8 seconds.

b $H(8) = 80 \times 8 - 5 \times 8^2$
$= 640 - 320$
$= 320$

i.e., the maximum height reached is 320 m.

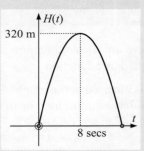

c The rocket falls back to earth when $H(t) = 0$
$$\therefore \quad 0 = 80t - 5t^2$$
$$\therefore \quad 5t^2 - 80t = 0$$
$$\therefore \quad 5t(t - 16) = 0 \qquad \{\text{factorising}\}$$
$$\therefore \quad t = 0 \text{ or } t = 16$$

i.e., the rocket falls back to earth after 16 seconds.

EXERCISE 9L

1 The height H metres, of a ball hit vertically upwards t seconds after it is hit is given by $H(t) = 36t - 2t^2$.

a How long does it take for the ball to reach its maximum height?

b What is the maximum height of the ball?

c How long does it take for the ball to hit the ground?

2 A skateboard manufacturer finds that the cost $\$C$ of making x skateboards per day is given by $C(x) = x^2 - 24x + 244$.

a How many skateboards should be made per day to minimise the cost of production?

b What is the minimum cost?

c What is the cost if no skateboards are made in a day?

3 The hourly profit (P) obtained from operating a fleet of n taxis is given by
$P(n) = 84n - 45 - 2n^2$.

    a What number of taxis gives the maximum hourly profit?

    b What is the maximum hourly profit?

    c How much money is lost per hour if no taxis are on the road?

Example 35

A vegetable gardener has 40 m of fencing to enclose
a rectangular garden plot where one side is an existing
brick wall. If the two equal sides are x m long:

    a   show that the area enclosed is given by
$A = x(40 - 2x)$ m^2

    b   find the dimensions of the vegetable garden of
maximum area.

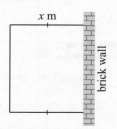

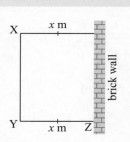

    a   Side XY $= 40 - 2x$ m.

         Now area $=$ length $\times$ width

         $\therefore$ $A = x(40 - 2x)$ m^2.

    b $A = 40x - 2x^2 = -2x^2 + 40x$

        is a quadratic in x, with $a = -2$, $b = 40$, $c = 0$.

        As $a < 0$, shape is

        So, maximum area occurs when $x = \dfrac{-b}{2a}$

                            $= \dfrac{-40}{-4}$

                            $= 10$

      $\therefore$ area is maximised when YZ $= 10$ m
and XY $= 20$ m.

---

4 A rectangular paddock to enclose horses is to be made with one side being a straight water
drain. If 1000 m of fencing is available for the other 3 sides, what are the dimensions
of the paddock if it is to enclose maximum area?

5 1800 m of fencing is available to fence 6
identical pig pens as shown in the diagram
alongside.

    a Explain why $9x + 8y = 1800$.

    b Show that the total area of each pen
is given by $A = -\frac{9}{8}x^2 + 225x$ m^2.

    c If the area enclosed is to be a maximum,
what is the shape of each pen?

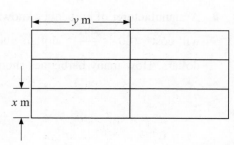

6 If 500 m of fencing is available to make 4 rectangular pens of identical shape, find the dimensions that maximise the area of each pen if the plan is:

a

b

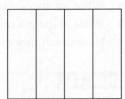

Example 36

A manufacturer of pot-belly stoves has the following situation to consider.

If x are made per week, each one will cost $(50 + \dfrac{400}{x})$ dollars and the total receipts per week for selling them would be $(550x - 2x^2)$ dollars.

How many pot-belly stoves should be made per week in order to maximise profits?

Total profit, $P = $ receipts $-$ costs

$\therefore \quad P = (550x - 2x^2) - \underbrace{(50 + \dfrac{400}{x})x}$

$\qquad\qquad\qquad\qquad\qquad\uparrow\qquad\quad\uparrow$
$\qquad\qquad\qquad\qquad$ cost for one $\quad x$ of them

$\therefore \quad P = 550x - 2x^2 - 50x - 400$

$\therefore \quad P = -2x^2 + 500x - 400$ dollars

which is a quadratic in x, with $a = -2$, $b = 500$, $c = -400$.

Since $a < 0$, shape is

$\therefore \quad P$ is maximised when $x = \dfrac{-b}{2a} = \dfrac{-500}{-4} = 125$

$\therefore \quad$ produce 125 of them per week.

---

7 The total cost of producing x toasters per day is given by $C = (\frac{1}{10}x^2 + 20x + 25)$ dollars, and the selling price of each toaster is $(44 - \frac{1}{5}x)$ dollars. How many toasters should be produced each day in order to maximise the total profit?

8 A manufacturer of barbeques knows that if x of them are made each week then each one will cost $(60 + \dfrac{800}{x})$ dollars and the total receipts per week will be $(1000x - 3x^2)$ dollars. How many barbeques should be made per week for maximum profits?

GRAPHICS CALCULATOR INVESTIGATION   TUNNELS AND TRUCKS

A tunnel is parabolic in shape with dimensions shown:

A truck carrying a wide load is 4.8 m high and 3.9 m wide and needs to pass through the tunnel. Your task is to determine if the truck will fit through the tunnel.

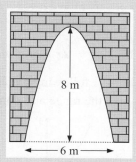

What to do:

1 If a set of axes is fitted to the parabolic tunnel as shown, state the coordinates of points A, B and C.

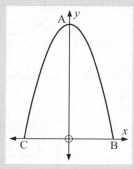

2 Using a **graphics calculator**:
   a enter the x-coordinates of A, B and C into **List 1**
   b enter the y-coordinates of A, B and C into **List 2**.

3 Draw a **scatterplot** of points A, B and C.

4 Set your calculator to display 4 decimal places and determine the equation of the parabolic boundary of the tunnel in the form $y = ax^2 + bx + c$, by fitting a **quadratic model** to the data.

5 Place the end view of the truck on the same set of axes as above.

   What is the equation of the truck's roofline?

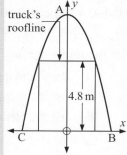

6 You should have found that the equation of the parabolic boundary of the tunnel is $y = -0.8889x^2 + 8$ and the equation of the truck's roofline is $y = 4.8$.

   Graph these equations on the same set of axes. Calculate the **points of intersection** of the graphs of these functions.

7 Using the points of intersection found in **6**, will the truck pass through the tunnel? What is the maximum width of a truck that is 4.8 m high if it is to pass through the tunnel?

8 Investigate the maximum width of a truck that is 3.7 m high if it is to pass through the tunnel.

9 What is the maximum width of a 4.1 m high truck if it is to pass through a parabolic tunnel 6.5 m high and 5 m wide?

REVIEW SET 9A

1 For $f(x) = 5x - 2$ and $g(x) = x^2 - 3x - 15$ find:

   a $f(-2)$ **b** $g(1)$ **c** x if $g(x) = 3$

2 If $f(x) = 2x - x^2$ find: **a** $f(2)$ **b** $f(-3)$ **c** $f(-\frac{1}{2})$

3 For the following graphs determine:

   i the range and domain   **ii** the x and y-intercepts   **iii** whether it is a function.

   a 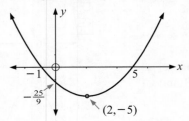   **b**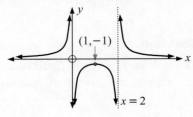

4 For each of the following graphs find the domain and range:

   a 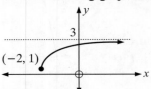   **b**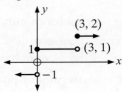

5 If $f(x) = ax + b$ (a and b are constants), find a and b for $f(1) = 7$ and $f(3) = -5$.

6 Find a, b and c if $f(0) = 5$, $f(-2) = 21$ and $f(3) = -4$
for $f(x) = ax^2 + bx + c$.

7 For the following functions of $f : x \longmapsto f(x)$ on $-2 \leqslant x \leqslant 2$ where $x \in Z$:

   i draw a mapping diagram to represent $f(x)$

   ii list the elements of the domain of $f(x)$ using set notation

   iii list the elements of the range of $f(x)$ using set notation.

   a $2x + 5$ **b** $x^2 - x + 2$

REVIEW SET 9B

1 For $f(x) = 2x^2 + x - 2$ and $g(x) = 3 - 2x$ find:

   a $f(-1)$ **b** $g(4)$ **c** x if $f(x) = 4$

2 On separate axes sketch $y = x^2$ and hence sketch:

   a $y = -\frac{1}{2}x^2$ **b** $y = (x + 2)^2 + 5$ **c** $y = -(x - 1)^2 - 3$

3 For $y = -2(x - 1)(x + 3)$ find the:

   a   **i** direction the parabola opens   **ii** y-intercept

      iii x-intercepts   **iv** equation of the axis of symmetry

   b Hence, sketch the graph showing all of the above features.

4 For $y = x^2 - 2x - 15$ find the:

    a   **i** y-intercept   **ii** x-intercepts

         iii   equation of the axis of symmetry   **iv**   coordinates of the vertex

    b   Hence sketch the graph showing all of the above features.

5 Find the coordinates of the point(s) of intersection of the graphs with equations
$y = x^2 - 5x + 10$ and $y = 3x - 6$.

6 A stone was thrown from the top of a cliff 60 metres above sea level. The height H metres, of the stone above sea level t seconds after it was released is given by $H(t) = -5t^2 + 20t + 60$.

    a   Find the time taken for the stone to reach its maximum height.

    b   What is the maximum height above sea level reached by the stone?

    c   How long is it before the stone strikes the water?

7 Consider $f(x) = \dfrac{1}{x^2}.$ **a**   Sketch the graph of this function using technology.

                                         b   State the domain and range of the function.

REVIEW SET 9C

1 Find the zeros of the quadratic function $y = x^2 - 9x + 20$.

2 On separate sets of axes sketch $y = x^2$ and hence sketch:

    a $y = 2x^2$ **b** $y = -(x + 3)^2 + 4$

3 For $y = 3(x - 2)^2$ find the:

    a   **i**   direction the parabola opens   **ii** y-intercept

         iii x-intercepts   **iv**   equation of the axis of symmetry

    b   Hence sketch the graph showing all of the above features.

4 Find the coordinates of the points of intersection of the graphs with equations
$y = -x^2 - x + 1$ and $y = -4x + 5$.

5 The height H metres of a cannonball t seconds after it is fired into the air is given by $H(t) = -4t^2 + 16t + 9$.

    a   Find the time taken for the cannonball to reach its maximum height.

    b   What is the maximum height reached by the cannonball?

    c   How long does it take for the cannonball to fall back to earth?

6 For $y = x^2 - 4x - 1$:

    a   use completing the square to write in the form $y = (x - h)^2 + k$.

    b   Hence, state the coordinates of the vertex.

    c   Find the y-intercept.

    d   Sketch the graph, showing the vertex and y-intercept.

7 A clothing manufacturer finds that the cost $\$C$, of making x leather jackets per day is given by $C(x) = x^2 - 18x + 200$.

    a   What is the cost to the manufacturer if no leather jackets are made in a day?

    b   How many leather jackets should be made per day to minimise the production costs?

    c   What is this minimum cost?

Chapter 10

Numerical trigonometry

Contents:

A   RIGHT ANGLED TRIANGLE TRIGONOMETRY

LABELLING RIGHT ANGLED TRIANGLES

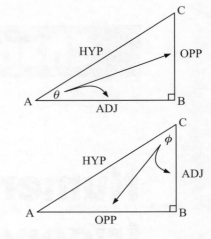

For the given right angled triangle, the **hypotenuse (HYP)** is the side which is opposite the right angle and is the longest side of the triangle.

For the angle marked θ:

- BC is the side **opposite (OPP)** angle θ
- AB is the side **adjacent (ADJ)** angle θ.

Notice that, for the angle marked ϕ:

- AB is the side **opposite (OPP)** angle ϕ
- BC is the side **adjacent (ADJ)** angle ϕ.

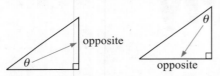
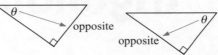

The third side is alongside the angle θ and so is called the **adjacent** side.

Example 1

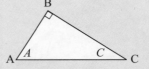

In the diagram alongside, find the:

a hypotenuse	**b** side opposite angle A
c side adjacent to angle A	**d** side opposite angle C
e side adjacent to angle C?	

---

a The hypotenuse is AC.	**b** BC is the side opposite angle A.
c AB is the side adjacent to angle A.	**d** AB is the side opposite angle C.
e BC is the side adjacent to angle C.	

EXERCISE 10A

1 In the diagrams below, name the:

    i hypotenuse
    ii side opposite the angle marked θ
    iii side adjacent to the angle marked θ

a
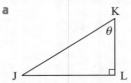

b
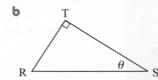

c
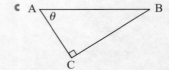

2 The right angled triangle alongside has hypotenuse of length
   a units and other sides of length b units and c units. θ and
   ϕ are the two acute angles. Find the length of the side:

   a opposite θ **b** opposite ϕ
   c adjacent to θ **d** adjacent to ϕ

B   THE TRIGONOMETRIC RATIOS

Consider a right angled triangle which is
similar to $\triangle$ OMP with sides labelled OPP
for opposite, ADJ for adjacent and HYP for
hypotenuse.

We define the three trigonometric ratios,
$\sin\theta$, $\cos\theta$ and $\tan\theta$ in terms of fractions of
sides of right angled triangles. These are:

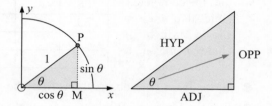

$$\sin \theta = \frac{\text{OPP}}{\text{HYP}}, \qquad \cos \theta = \frac{\text{ADJ}}{\text{HYP}} \qquad \text{and} \qquad \tan \theta = \frac{\text{OPP}}{\text{ADJ}}$$

These three formulae are called the **trigonometric ratios** and are the tools we use for finding
sides and angles of right angled triangles.

FINDING TRIGONOMETRIC RATIOS

> **Example 2**
>
> For the following triangle find:
> **a** $\sin\theta$ **b** $\cos\phi$ **c** $\tan\theta$
>
>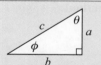
>
> **a** $\sin\theta = \dfrac{\text{OPP}}{\text{HYP}} = \dfrac{b}{c}$ **b** $\cos\phi = \dfrac{\text{ADJ}}{\text{HYP}} = \dfrac{b}{c}$ **c** $\tan\theta = \dfrac{\text{OPP}}{\text{ADJ}} = \dfrac{b}{a}$

EXERCISE 10B.1

1 For each of the following triangles find:

   i $\sin\theta$ **ii** $\cos\theta$ **iii** $\tan\theta$ **iv** $\sin\phi$ **v** $\cos\phi$ **vi** $\tan\phi$

   a

   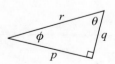

   b

   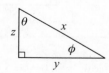

   c

   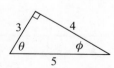

   d

   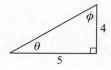

   e

   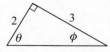

   f

   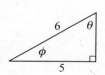

INVESTIGATION 1   COMPLEMENTARY ANGLES

$37°$ and $53°$ are complementary since their sum is $90°$.

$\theta°$ and $(90 - \theta)°$ are therefore **complements** of each other.

WORKSHEET

Your task is to determine if a relationship exists between the sine and cosine of an angle and its complement.

What to do:

1 Use your calcula-
tor to complete a
table like the one
shown which in-
cludes some angles
of your choice.

θ	$\sin \theta$	$\cos \theta$	$90 - \theta$	$\sin (90 - \theta)$	$\cos (90 - \theta)$
17			73		
38					
59					

2 Write down your observations from the tabled values.

3 Use the figure alongside to prove that your observations
above are true for all angles θ, where $0 < \theta < 90$.

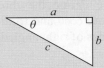

4 Investigate possible connections between $\tan \theta$ and $\tan(90 - \theta)$.

FINDING SIDES

In a right angled triangle, if we are given another angle and a side we can find:

- the third angle using the 'angle sum of a triangle is $180°$'
- the other sides using trigonometry.

The method: *Step 1:*   Redraw the figure and mark on it HYP, OPP, ADJ relative to the
given angle.

Step 2:   For the given angle choose the correct trigonometric ratio which
can be used to set up an equation.

Step 3:   Set up the equation.

Step 4:   Solve to find the unknown.

Example 3

Find the unknown
length in the following
(to 2 dec. places):

a
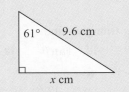

b
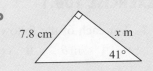

a
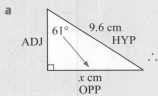

Now $\sin 61° = \dfrac{x}{9.6}$ $\left\{ \dfrac{\text{OPP}}{\text{HYP}} \right\}$

$\therefore$ $\sin 61° \times 9.6 = x$ $\{\times \text{ both sides by } 9.6\}$

$\therefore$ $x \doteqdot 8.40$ $\{\boxed{\sin}\ 61\ \boxed{\times}\ 9.6\ \boxed{\text{ENTER}}\}$

b

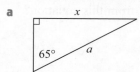

$$\text{Now} \quad \tan 41^o = \frac{7.8}{x} \qquad \left\{ \frac{\text{OPP}}{\text{ADJ}} \right\}$$

$$\therefore \quad x \times \tan 41^o = 7.8 \qquad \{\times \text{ both sides by } x\}$$

$$\therefore \quad x = \frac{7.8}{\tan 41^o} \qquad \{\div \text{ both sides by } \tan 41^o\}$$

$$\therefore \quad x \doteqdot 8.97 \qquad \{7.8 \boxed{\div} \boxed{\text{tan}} \; 41 \; \boxed{\text{ENTER}}\}$$

EXERCISE 10B.2

1 Set up a trigonometric equation connecting the angle and the sides given:

a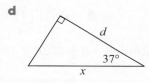

b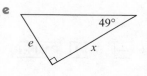

c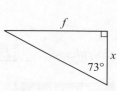

d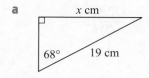

e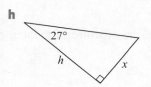

f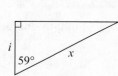

g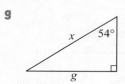

h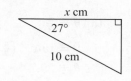

i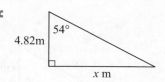

2 Find, to 2 decimal places, the unknown length in:

a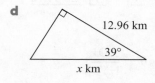

b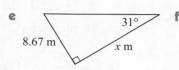

c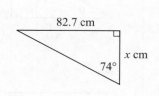

d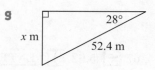

e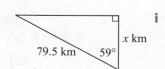

f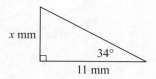

g

h

i

3 Find, to one decimal place, *all* the unknown angles and sides of:

a
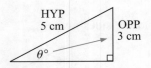

b
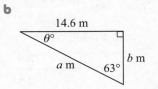

c
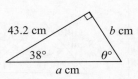

FINDING ANGLES

In the right angled triangle $\sin \theta = \frac{3}{5}$.

How do we find θ from this equation?

We are looking for angle θ with a sine of $\frac{3}{5}$.

If $\sin^{-1}(....)$ reads "the angle with a sine of ....",   we can write $\theta = \arcsin \left(\frac{3}{5}\right)$ or

$$\theta = \sin^{-1} \left(\frac{3}{5}\right).$$

Calculator: $\boxed{\text{2nd}}$ $\boxed{\text{sin}^{-1}}$ 3 $\boxed{\div}$ 5 $\boxed{)}$ $\boxed{\text{ENTER}}$ *Answer:* $\theta \doteqdot 36.9^{o}$

Example 4

Find, to one decimal place, the measure of the angle marked θ in:

a

b

a

$\tan \theta = \frac{4}{7}$ {as $\tan \theta = \dfrac{\text{OPP}}{\text{ADJ}}$}

$\therefore \quad \theta = \tan^{-1} \left(\frac{4}{7}\right)$

$\therefore \quad \theta \doteqdot 29.7$

         { $\boxed{\text{2nd}}$ $\boxed{\text{tan}^{-1}}$ 4 $\boxed{\div}$ 7 $\boxed{)}$ $\boxed{\text{ENTER}}$ }

So, the angle measure is 29.7^{o} .

b

$\cos \theta = \dfrac{2.67}{5.92}$ {as $\cos \theta = \dfrac{\text{ADJ}}{\text{HYP}}$}

$\therefore \quad \theta = \cos^{-1} \left(\dfrac{2.67}{5.92}\right)$

$\therefore \quad \theta \doteqdot 63.2$

         { $\boxed{\text{2nd}}$ $\boxed{\text{cos}^{-1}}$ 2.67 $\boxed{\div}$ 5.92 $\boxed{)}$ $\boxed{\text{ENTER}}$ }

So, the angle measure is 63.2^{o} .

EXERCISE 10B.3

1 Find, to one decimal place, the measure of the angle marked θ in:

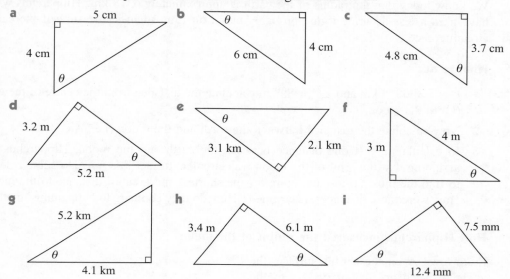

2 Find to 1 decimal place, all the unknown sides and angles of the following:

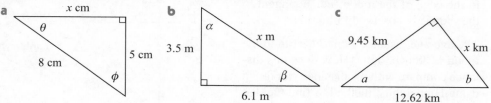

3 Check your answers for x in question **2** using the Pythagorean rule.

4 Find θ using trigonometry in the following. What conclusions can you draw?

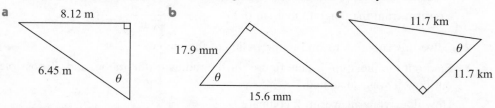

INVESTIGATION 2   HIPPARCHUS AND THE UNIVERSE

How Hipparchus measured the distance to the moon.

Think of A and B as two towns on the earth's equator. The moon is directly overhead town A. From B, the moon is just visible (MB is a tangent to the earth and is therefore perpendicular to BC). Angle C is the difference in longitude between towns A and B.

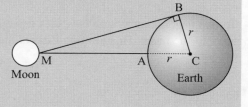

Hipparchus (2nd Century BC) calculated that angle to be approximately 89°.

We know today that the radius of the earth is approximately 6378 km. Hipparchus would have used a less accurate figure probably based on Eratosthenes' measure of the earth's circumference.

What to do:

1. Use $r = 6378$ km and $\angle C = 89^o$ to calculate the distance from the centre of the earth (C) to the moon.

2. Now calculate the distance **between** the earth and the moon (i.e., AM).

3. In calculating just one distance between the earth and the moon, Hipparchus was assuming that the orbit of the moon was circular. In fact it is not. Do some research to find the most up-to-date figure for the shortest and greatest distance to the moon. How were these distances determined? How do they compare to Hipparchus' method?

How Hipparchus measured the radius of the moon.

From town A on the earth's surface, the angle between an imaginary line to the centre of the moon and an imaginary line to the edge of the moon (i.e., a tangent to the moon) is observed to be 0.25^o.

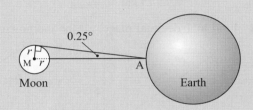

The average distance from the earth to the moon is 384 403 km. (How does this distance compare with the distance you found using Hipparchus' method to find the distance to the moon?)

What to do:

1. Confirm from the diagram that $\sin 0.25^o = \dfrac{r}{r + 384\,403}$.

2. Solve this equation to find r, the radius of the moon.

3. Find out the most up-to-date figure for the radius of the moon, and, if possible, find out how it was calculated.

   How does your answer to **2** compare?

ISOSCELES TRIANGLES

To use trigonometry with isosceles triangles we invariably draw the **perpendicular** from the apex to the base. This altitude **bisects** the base.

Example 5

Find, to 4 s.f. the unknowns in the following diagrams:

a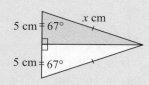

b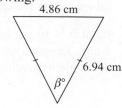

a

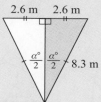

In the shaded right angled triangle

$$\cos 67° = \frac{5}{x}$$

$$\therefore \quad x = \frac{5}{\cos 67°}$$

$$\therefore \quad x \doteqdot 12.8$$

Calc: 5 $\div$ cos 67) **ENTER**

b

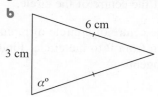

In the shaded right angled triangle

$$\sin\left(\frac{\alpha}{2}\right) = \frac{2.6}{8.3}$$

$$\therefore \quad \frac{\alpha}{2} = \arcsin\left(\frac{2.6}{8.3}\right)$$

$$\therefore \quad \alpha = 2 \times \arcsin\left(\frac{2.6}{8.3}\right) \doteqdot 36.5$$

Calc: 2 $\times$ 2nd sin⁻¹ 2.6 $\div$ 8.3) **ENTER**

EXERCISE 10B.4

1 Find, correct to 4 significant figures, the unknowns in the following:

a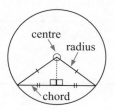

b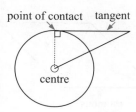

c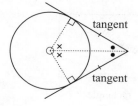

CHORDS AND TANGENTS

Right angled triangles occur in chord and tangent problems.

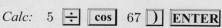

Example 6

A chord of a circle subtends an angle of $112°$ at its centre. Find the length of the chord if the radius of the circle is 6.5 cm.

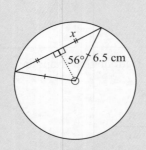

We complete an isosceles triangle and draw the line from the apex to the base.

For the $56°$ angle, HYP = 6.5, OPP = x,

$$\sin 56° = \frac{x}{6.5}$$

$$\therefore \quad 6.5 \times \sin 56° = x$$

$$\therefore \quad x \doteqdot 5.389$$

$$\therefore \quad 2x \doteqdot 10.78$$

$\therefore$ chord is 10.78 cm long.

2   Find the value of the unknown in:

a

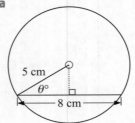

b

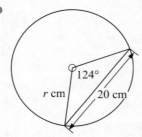

c

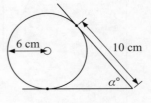

3   A chord of a circle subtends an angle of $89°$ at its centre. Find the length of the chord given that the circle's diameter is 11.4 cm.

4   A chord of a circle is 13.2 cm long and the circle's radius is 9.4 cm. Find the angle subtended by the chord at the centre of the circle.

5   Point P is 10 cm from the centre of a circle of radius 4 cm. Tangents are drawn from P to the circle. Find the angle between the tangents.

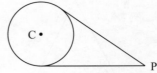

OTHER FIGURES

Sometimes right angled triangles can be found in other geometric figures such as rectangles, rhombi and trapezia.

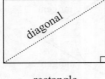

rectangle

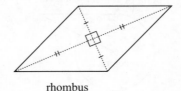

rhombus

trapezium

Often right angled triangle trigonometry can be used in these figures if sufficient information is given.

Example 7

A rhombus has diagonals of length 10 cm and 6 cm respectively.
Find the smaller angle of the rhombus.

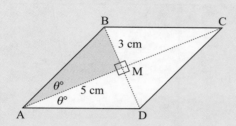

The diagonals bisect each other at right angles, so   AM = 5 cm and BM = 3 cm.

In $\triangle$ABM, θ will be the smallest angle as it is opposite the shortest side.

$$\tan \theta = \tfrac{3}{5}$$
$$\therefore \quad \theta = \tan^{-1}\left(\tfrac{3}{5}\right)$$
$$\text{i.e.,} \quad \theta \doteqdot 30.964$$

But the required angle is 2θ as the diagonals bisect the angles at each vertex,
$\therefore$ angle is $61.93°$.

---

6 A rectangle is 9.2 m by 3.8 m. What angle does its diagonal make with its longer side?

7 The diagonal and the longer side of a rectangle make an angle of $43.2°$. If the longer side is 12.6 cm, find the length of the shorter side.

8 A rhombus has diagonals of length 12 cm and 7 cm respectively. Find the larger angle of the rhombus.

9 The smaller angle of a rhombus measures $21.8°$ and the shorter diagonal is 13.8 cm. Find the lengths of the sides of the rhombus.

Example 8

Find x given:

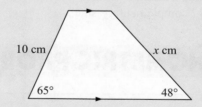

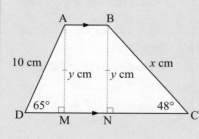

We draw perpendiculars AM and BN to DC creating right angled triangles and rectangle ABNM.

In $\triangle$ADM, $\sin 65° = \dfrac{y}{10}$ and $\therefore$ $y = 10 \sin 65°$

In $\triangle$BCN, $\sin 48° = \dfrac{y}{x} = \dfrac{10 \sin 65°}{x}$

$$\therefore \quad x = \dfrac{10 \sin 65°}{\sin 48°} \doteqdot 12.20$$

10 **a** Find the value of x in:

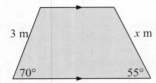

3 m x m

70°   55°

b Find the unknown angle in:

5   6

70° $\alpha°$

11 A stormwater drain is to have the shape as
shown. Determine the angle the left hand
side makes with the bottom of the drain.

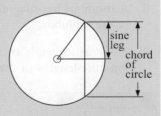

|← 5 m →|

3 m

$\beta°$ 100°

2 m

HISTORICAL NOTE

The origin of the term "sine" is quite fascinating. **Arbyabhata**, a Hindu
mathematician who studied trigonometry in the 5th century AD, called the
sine-leg of the circle diagram "ardha-jya" which means "half-chord".

This was eventually shortened to "jya". Arab scholars later
translated Arbyabhata's work into Arabic and initially phonet-
ically translated "jya" as "jiba" but since this meant nothing
in Arabic they very shortly began writing the word as "jaib"
which has the same letters but means "cove" or "bay".

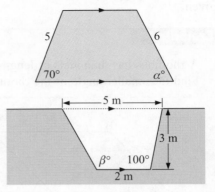

Finally in 1150 an Italian, **Gerardo of Cremona**, translated this
work into Latin and replaced "jaib" with "sinus" which means
"bend" or "curve" but is commonly used in Latin to refer to a
bay or gulf on a coastline. The term "sine" that we use today
comes from this Latin word "sinus". The term "cosine" comes
from the fact that the sine of an angle is equal to the cosine
of its complement. In 1620, **Edmund Gunter** introduced the
abbreviated "co sinus" for "complementary sine".

sine
leg
chord
of
circle

C   TRIGONOMETRIC PROBLEM SOLVING

Example 9

Find the height of a tree which casts a shadow of 12.4 m when the sun makes an
angle of $52°$ to the horizon.

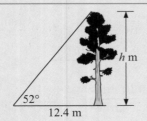

h m

52°

12.4 m

Let h m be the tree's height.

For the $52°$ angle,   OPP $= h$, ADJ $= 12.4$

$$\therefore \quad \tan 52° = \frac{h}{12.4}$$

$$\therefore \quad 12.4 \times \tan 52° = h$$

$$\therefore \quad h \doteqdot 15.87 \quad \therefore \quad \text{height is 15.9 m.}$$

Reminder:

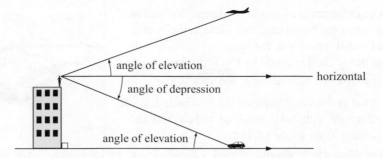

EXERCISE 10C

1 Find the height of a flagpole which casts a shadow of 9.32 m when the sun makes an angle of $63°$ to the horizontal.

2 A hill is inclined at $18°$ to the horizontal. If the base of the hill is at sea level find:
   a my height above sea level if I walk 1.2 km up the hill
   b how far I have walked up the hill if I am 500 metres above sea level.

3 A surveyor standing at A notices two posts B and C on the opposite side of a canal. The posts are 120 m apart. If the angle of sight between the posts is $37°$, how wide is the canal?

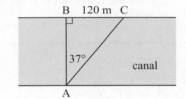

4 A train must climb a constant gradient of 5.5 m for every 200 m of track. Find the angle of incline.

5 Find the angle of elevation to the top of a 56 m high building from point A, which is 113 m from its base. What is the angle of depression from the top of the building to A?

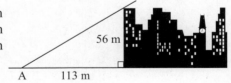

6

120 m

B

sea

The angle of depression from the top of a 120 m high vertical cliff to a boat B is $16°$.

Find how far the boat is from the base of the cliff.

7 Sarah measures the angle of elevation to the top of a redwood in the Montgomery State Reserve as $23.6°$ from a point which is 250 m from its base. Her eye level, where the angle measurement was taken, is 1.5 m above the ground. Assuming the ground to be horizontal, find the height of the tree.

8 Ingrid measures the angle of elevation from a point on level ground to the top of a building 120 metres high to be $32°$. She walks towards the building until the angle of elevation is $45°$. How far does she walk?

9 From a point A, 40 metres from the base of a building B, the angle of elevation to the top of the building C is $51°$, and to the top of the flagpole D on top of the building is $56°$. Find the height of the flagpole.

10 For a circular track of radius r metres, banked at θ degrees to the horizontal, the ideal velocity (the velocity that gives no tendency to sideslip) in metres per second is given by the formula:

$v = \sqrt{gr \tan \theta}$,   where $g = 9.8$ m/s^2.

   a What is the ideal velocity for a vehicle travelling on a circular track of radius 100 m, banked at an angle of $15°$?

   b At what angle should a track of radius 200 m be banked, if it is designed for a vehicle travelling at 20 m/s?

Example 10

A builder designs a roof structure as illustrated. The pitch of the roof is the angle that the roof makes with the horizontal. Find the pitch of the roof.

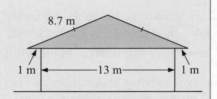

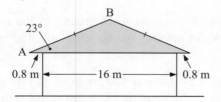

Using the right angled triangle created from the isosceles triangle, for angle θ:

ADJ $= 7.5$,   HYP $= 8.7$

$\therefore$ $\cos \theta = \dfrac{7.5}{8.7}$

$\therefore$ $\theta = \cos^{-1}\left(\dfrac{7.5}{8.7}\right)$

$\therefore$ $\theta \doteqdot 30.450....$

$\therefore$ the pitch is approximately $30\frac{1}{2}°$.

---

11 Find θ, the pitch of the roof.

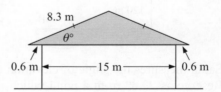

12

If the pitch of the given roof is $23°$, find the length of the timber beam AB.

13 AC is a straight shore line and B is a boat out at sea. Find the shortest distance from the boat to the shore if A and C are 5 km apart.

(**Hint:** the shortest distance from B to AC is the perpendicular distance.)

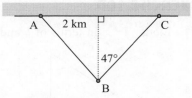

14

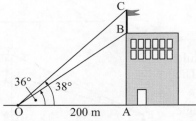

From an observer O, the angles of elevation to the bottom and the top of a flagpole are $36°$ and $38°$ respectively. Find the height of the flagpole.

(**Hint:** Find AB and AC.)

15 The angle of depression from the top of a 150 m high cliff to a boat at sea is $7°$. How much closer to the cliff must the boat move for the angle of depression to become $19°$?

16 A helicopter flies horizontally at 100 kmph. An observer notices that it took 20 seconds for the helicopter to fly from directly overhead to being at an angle of elevation of $60°$. Find the height of the helicopter above the ground.

17 An open right-circular cone has a vertical angle measuring $40°$ and a base radius of 30 cm.

Find the
a height of the cone
b capacity of the cone in litres.
$(V = \frac{1}{3}\pi r^2 h)$

D | CONSTRUCTING TRIGONOMETRIC FORMULAE

Many interesting, but often difficult problems can be solved by first constructing a formula and then using the formula to solve them.

In this exercise we will construct formulae using right angled triangle trigonometry.

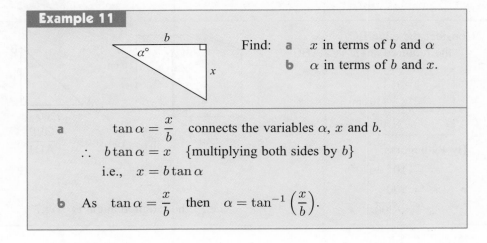

Example 11

Find: **a** x in terms of b and α
b α in terms of b and x.

a $\tan \alpha = \dfrac{x}{b}$ connects the variables α, x and b.

$\therefore \ b \tan \alpha = x$ {multiplying both sides by b}

i.e., $x = b \tan \alpha$

b As $\tan \alpha = \dfrac{x}{b}$ then $\alpha = \tan^{-1}\left(\dfrac{x}{b}\right)$.

EXERCISE 10D

1 Find: **i** x in terms of b and α **ii** α in terms of x and b.

a

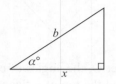

b

c

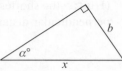

2 A child's eye is 1 m above floor level. The child is viewing a painting AB on a wall as shown. The bottom of the painting is 2 m above floor level.

If the child's eye is x m from the wall and the angle of view of the painting is θ^o, find θ in terms of x.

(**Hint:** $\theta = \angle BEN - \angle AEN$)

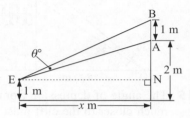

3

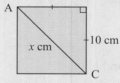

In a rugby game Ray O'Farrell makes a place kick for goal to try and earn extra points. If Ray is 30 m from the goal line, how far is he from the nearer goal post Q and what is the angle of view to the goal posts that Ray faces?

E   3-DIMENSIONAL PROBLEM SOLVING

Right angled triangles occur frequently in 3-dimensional figures. Consequently, trigonometry can be used to find unknown angles and lengths.

Example 12

A cube has sides of length 10 cm. Find the angle between the diagonal AB of the cube and one of the edges at A.

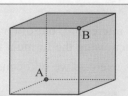

Consider the edge BC.

∴ the required angle is $\angle ABC$.

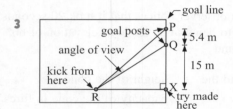

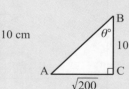

By Pythagoras:

$$x^2 = 10^2 + 10^2$$
$$\therefore \quad x^2 = 200$$
$$\therefore \quad x = \sqrt{200}$$

$\tan \theta = \dfrac{\sqrt{200}}{10}$ {as $\tan \theta = \dfrac{\text{OPP}}{\text{ADJ}}$}

$\therefore \quad \theta = \arctan(\dfrac{\sqrt{200}}{10})$

$\therefore \quad \theta \doteqdot 54.74$

i.e.,   the required angle is 54.7^o.

EXERCISE 10E.1

1 The figure alongside is a cube with sides of length 12 cm. Find:

   a EG   **b** ∠AGE

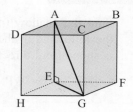

2

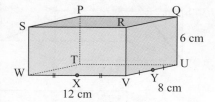

The figure alongside is a rectangular prism with dimensions as shown. X and Y are the midpoints of the edges WV and VU respectively. Find:

   a TX   **b** angle TXP

   c TY   **d** angle PYT

3 In the triangular prism alongside, find:

   a DF

   b angle AFD

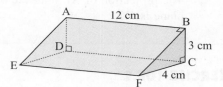

4

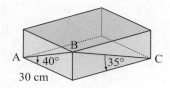

AB and BC are wooden support struts on a crate. Find the total length of wood required to make the two struts.

5 All edges of a square-based pyramid are 10 cm in length. Find the angle between a slant edge and a base diagonal.

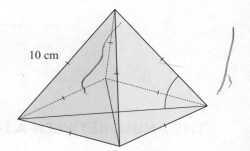

SHADOW LINES (PROJECTIONS)

Consider a wire frame in the shape of a cube as shown in the diagram alongside. Imagine a light source shining down directly on this cube from above.

The shadow cast by wire AG would be EG and this is called the **projection** of AG onto the base plane EFGH. Similarly the projection of BG onto the base plane is FG.

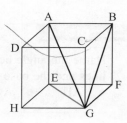

Example 13

Find the shadow (projection) of each of the following in the base plane if a light is shone from directly above the figure:

a UP   **b** WP   **c** VP   **d** XP

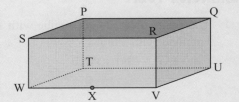

a The projection of UP onto the base plane is UT.

b The projection of WP onto the base plane is WT.

c The projection of VP onto the base plane is VT.

d The projection of XP onto the base plane is XT.

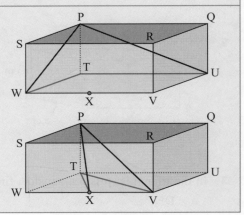

EXERCISE 10E.2

1 Find each of the following projections in the base planes of the given figures:

a

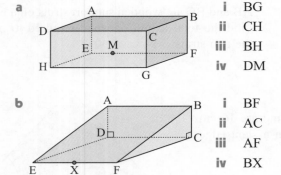

  i BG
  ii CH
  iii BH
  iv DM

c

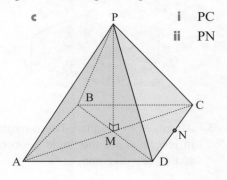

  i PC
  ii PN

b

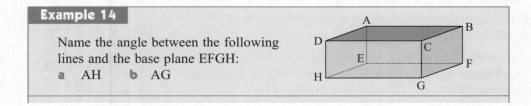

  i BF
  ii AC
  iii AF
  iv BX

THE ANGLE BETWEEN A LINE AND A PLANE

The angle between a line and a plane is the angle between the line and its projection on the plane.

Example 14

Name the angle between the following lines and the base plane EFGH:

a AH   **b** AG

a The projection of AH onto the base plane EFGH is EH
∴   the required angle is ∠AHE.

b The projection of AG onto the base plane EFGH is EG
∴   the required angle is ∠AGE.

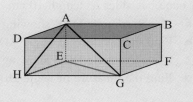

EXERCISE 10E.3

1 For each of the following figures name the angle between the given line and the base plane:

a **i** AF **ii** AG   **b** **i** QY **ii** QW   **c** **i** AS **ii** AY
   iii BH **iv** BX   **iii** PZ **iv** PY

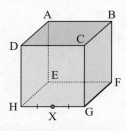

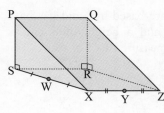

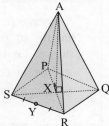

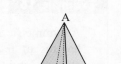

Example 15

Find the angle that:

a PV makes with QV

b SU makes with SQ.

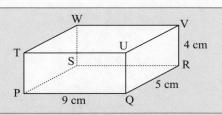

a

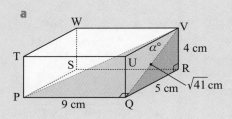

The required angle is ∠PVQ = $\alpha°$ say.
We shade △PQV which is right angled at Q.

Now in △QRV,   QV $= \sqrt{5^2 + 4^2}$
$= \sqrt{41}$ {Pythagoras}

and so, $\tan \alpha = \frac{9}{\sqrt{41}}$ {in △PQV}

∴ $\alpha \doteqdot 54.57$

∴   the required angle is $54.57°$.

b

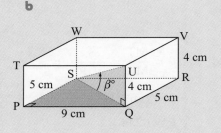

The required angle is marked $\beta°$.
We shade △SQU which is right angled at Q.

Now in △PQS,   SQ $= \sqrt{5^2 + 9^2}$
$= \sqrt{106}$ {Pythagoras}

and so, $\tan \beta = \frac{4}{\sqrt{106}}$

∴ $\beta \doteqdot 21.23$

∴   the required angle is $21.23°$.

2   A cube has sides of length 10 cm.

   a   Sketch triangle DAE showing the right angle.

   b   Find the measure of angle ADE.

   c   Sketch triangle DBH showing the right angle.

   d   Find the measure of angle DBH.

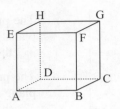

3   M and N are the midpoints of QR and BC respectively.

   a   Draw a sketch of the triangle DMN showing which angle is the right angle.

   b   Find the length of DN.

   c   Find the measure of angle DMN.

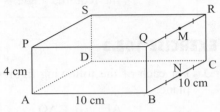

Example 16

A symmetric square-based pyramid has base lengths of 6 cm and a height of 8 cm as shown. Find the measure of:

a   angle TNM

b   angle TRM

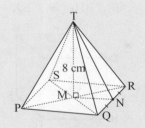

a

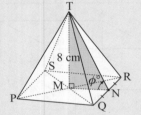   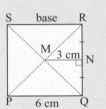   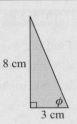

ϕ is the required angle in $\triangle$TNM   and $\tan\phi = \frac{8}{3}$ $\therefore$ $\phi = \tan^{-1}\left(\frac{8}{3}\right)$

$\therefore$ $\phi \doteqdot 69.44$

$\therefore$ angle TNM measures $69.4°$

b

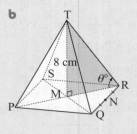   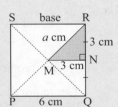

In $\triangle$MNR,

$a = \sqrt{3^2 + 3^2} = \sqrt{18}$

{Pythagoras}

θ is the required angle in $\triangle$MRT   and $\tan\theta = \dfrac{MT}{MR} = \dfrac{8}{\sqrt{18}}$

$\therefore$ $\theta = \tan^{-1}\left(\frac{8}{\sqrt{18}}\right)$

$\therefore$ $\theta \doteqdot 62.06$

$\therefore$ angle TRM measures $62.1°$

4 A square-based pyramid has base lengths
3 cm and slant edges 5 cm as shown.

   a Draw an above view (plan) of its base
   ABCD.

   b Find the length of diagonal AC.

   c Find the measure of angle ECN.

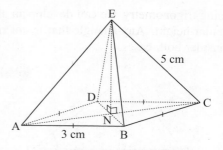

5

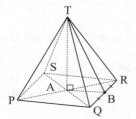

A square-based pyramid has height 10 cm
and base lengths 12 cm.

   a Find the measure of angle ABT.

   b Draw a sketch of triangle AQT
   showing the right angle.

   c Find the measure of angle AQT.

6 The pyramid of Cheops in Egypt has a height of 145 m and has base lengths 230 m.
Find the angle between a sloping edge and the base diagonal.

F   AREAS OF TRIANGLES

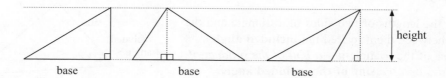

If we know the base and height measurements of a triangle we can calculate the area using
area $= \frac{1}{2}$ base $\times$ height.

However, cases arise where we do not know the height but we can still calculate the area.

These cases are:

- knowing two sides and the angle between
  them (called the **included angle**)

  For example:

  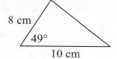

- knowing all three sides

  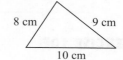

LABELLING TRIANGLES

If triangle ABC has angles of size A^o, B^o, C^o,
the sides opposite these angles are labelled a,
b and c respectively.

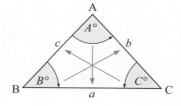

Using trigonometry, we can develop an alternative formula that does not depend on a perpendicular height. Any triangle that is not right angled must be either acute or obtuse. We will consider both cases.

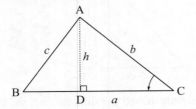
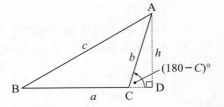

In both triangles a perpendicular is constructed from A to D on BC (extended if necessary).

$$\sin C = \frac{h}{b} \qquad\qquad \sin(180 - C) = \frac{h}{b}$$

$$\therefore \quad h = b\sin C \qquad\qquad\qquad \therefore \quad h = b\sin(180 - C)$$

$$\text{but} \quad \sin(180 - C) = \sin C$$

$$\therefore \quad h = b\sin C$$

So, $\quad$ area $= \frac{1}{2}ah = \frac{1}{2}ab\sin C$.

Using different altitudes we could also show that the area of $\triangle$ABC is given by

$$\boxed{\textbf{Area} = \tfrac{1}{2}bc\sin A = \tfrac{1}{2}ac\sin B = \tfrac{1}{2}ab\sin C.}$$

Summary:

Given the lengths of two sides of a triangle and the angle between them (called the **included angle**), the area of the triangle is   *a half of the product of two sides and the* **sine of the included angle.**

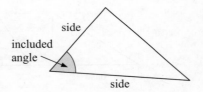

Example 17	
Find the area of triangle ABC:	Area $= \frac{1}{2}ac\sin B$
 11 cm ⟍ A B ∡28° ⟍ C 15 cm	$= \frac{1}{2} \times 15 \times 11 \times \sin 28^o$
	$\doteqdot 38.73$ cm^2

EXERCISE 10F

1   Find the area of:

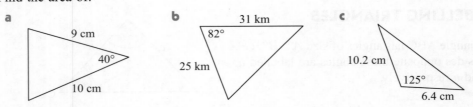

2 If triangle ABC has area 150 cm², find the value of x:

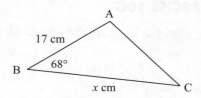

3 A parallelogram has two adjacent sides of length 4 cm and 6 cm respectively. If the included angle measures 52°, find the area of the parallelogram.

4

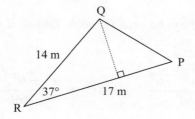

   a Find the area of triangle PQR to 3 decimal places.

   b Hence, find the length of the altitude from Q to RP.

G   THE COSINE RULE

The **cosine rule** involves the sides and angles of a triangle.

In any $\triangle$ABC:

$$a^2 = b^2 + c^2 - 2bc\cos A$$
or
$$b^2 = a^2 + c^2 - 2ac\cos B$$
or
$$c^2 = a^2 + b^2 - 2ab\cos C$$

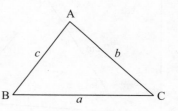

The **cosine rule** can be used to solve triangles given:

- two sides and an included angle
- three sides.

Example 18

Find, correct to 2 decimal places, the length of BC.

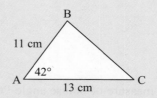

By the cosine rule:

$$BC^2 = 11^2 + 13^2 - 2 \times 11 \times 13 \times \cos 42°$$
$$\therefore \quad BC \doteqdot \sqrt{(11^2 + 13^2 - 2 \times 11 \times 13 \times \cos 42°)}$$
$$\therefore \quad BC \doteqdot 8.801.....$$

$\therefore$ BC is 8.80 cm in length.

EXERCISE 10G

1 Find the length of the remaining side in the given triangle:

a

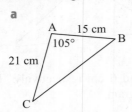

b

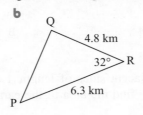

c

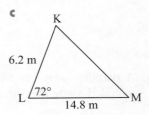

Rearrangement of the original cosine rule formulae can be used for angle finding if we know all three sides. The formulae for finding the angles are:

$$\cos A = \frac{b^2 + c^2 - a^2}{2bc} \qquad \cos B = \frac{c^2 + a^2 - b^2}{2ca} \qquad \cos C = \frac{a^2 + b^2 - c^2}{2ab}$$

Example 19

In triangle ABC, if AB = 7 cm, BC = 8 cm and CA = 5 cm, find the measure of angle BCA.

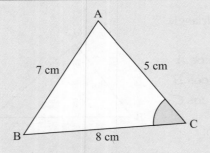

By the cosine rule:

$$\cos C = \frac{(5^2 + 8^2 - 7^2)}{(2 \times 5 \times 8)}$$

$$\therefore \quad C = \cos^{-1}\left(\frac{(5^2 + 8^2 - 7^2)}{(2 \times 5 \times 8)}\right)$$

$$\therefore \quad C = 60$$

So, angle BCA measures 60°.

2 Find the measure of all angles of:

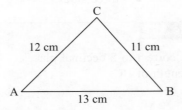

3 Find the measure of obtuse angle PQR.

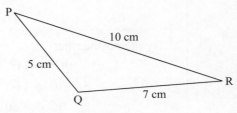

4 Find:

    a the smallest angle of a triangle with sides 11 cm, 13 cm and 17 cm

    b the largest angle of a triangle with sides 4 cm, 7 cm and 9 cm.

THE SINE RULE

The **sine rule** is a set of equations which connects the lengths of the sides of any triangle with the sines of the angles of the triangle. The triangle does not have to be right angled for the sine rule to be used.

THE SINE RULE

> In any triangle ABC with sides a, b and c units in length, and opposite angles A, B and C respectively,
>
> $$\frac{\sin A}{a} = \frac{\sin B}{b} = \frac{\sin C}{c} \quad \text{or} \quad \frac{a}{\sin A} = \frac{b}{\sin B} = \frac{c}{\sin C}$$

Note: The sine rule is used to solve problems involving triangles given either:

- **two angles** and **one side**, or
- **two sides** and a **non-included** angle.

FINDING SIDES

Example 20

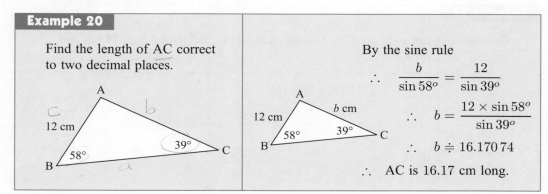

Find the length of AC correct to two decimal places.

By the sine rule

$$\therefore \quad \frac{b}{\sin 58^o} = \frac{12}{\sin 39^o}$$

$$\therefore \quad b = \frac{12 \times \sin 58^o}{\sin 39^o}$$

$$\therefore \quad b \doteqdot 16.170\,74$$

$\therefore$ AC is 16.17 cm long.

EXERCISE 10H.1

1 Find the value of x:

a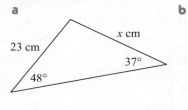

b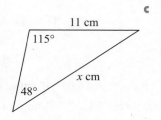

c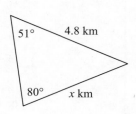

2 In triangle ABC find:

- **a** a if $A = 63^o$, $B = 49^o$ and $b = 18$ cm
- **b** b if $A = 82^o$, $C = 25^o$ and $c = 34$ cm
- **c** c if $B = 21^o$, $C = 48^o$ and $a = 6.4$ cm

FINDING ANGLES

The problem of finding angles using the sine rule is more complicated because there may be two possible answers.

You will need a blank sheet of paper, a ruler, a protractor and a compass for the tasks that follow. In each task you will be required to construct triangles from given information.

Task 1: Draw AB = 10 cm. At A construct an angle of 30°. Using B as centre, draw an arc of a circle of radius 6 cm. Let the arc intersect the ray from A at C. How many different positions may C have and therefore how many different triangles ABC may be constructed?

Task 2: As before, draw AB = 10 cm and construct a 30° angle at A. This time draw an arc of radius 5 cm based on B. How many different triangles are possible?

Task 3: Repeat, but this time draw an arc of radius 3 cm on B. How many different triangles are possible?

Task 4: Repeat with an arc of radius 12 cm from B. How many possible triangles?

Example 21

Find the measure of angle C in triangle ABC if AC is 7 cm, AB is 11 cm and angle B measures 25°.

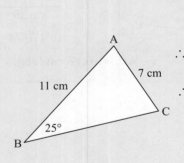

By the sine rule

$$\frac{\sin C}{c} = \frac{\sin B}{b}$$

$$\therefore \quad \frac{\sin C}{11} = \frac{\sin 25^o}{7}$$

$$\therefore \quad \sin C = \frac{11 \times \sin 25^o}{7}$$

$$\therefore \quad C = \sin^{-1}\left(\frac{11 \times \sin 25^o}{7}\right) \quad \text{or its supplement}$$

$$\therefore \quad C \doteqdot 41.61^o \text{ or } 180^o - 41.61^o$$
$$\{\text{as } C \text{ may be obtuse}\}$$

$$\therefore \quad C \doteqdot 41.61^o \text{ or } 138.39^o$$

$\therefore$ C measures 41.61^o if angle C is acute
or C measures 138.39^o if angle C is obtuse.

In this example there is insufficient information to determine the actual shape of the triangle.

Note: Sometimes there is information in the question which enables us to **reject** one of the answers.

Example 22

Find the measure of angle L in triangle KLM given that angle LKM measures 56°, LM = 16.8 m and KM = 13.5 m.

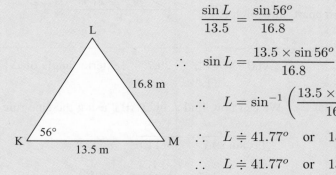

$$\frac{\sin L}{13.5} = \frac{\sin 56^o}{16.8} \qquad \text{\{the sine rule\}}$$

$$\therefore \quad \sin L = \frac{13.5 \times \sin 56^o}{16.8}$$

$$\therefore \quad L = \sin^{-1}\left(\frac{13.5 \times \sin 56^o}{16.8}\right) \text{ or its supplement}$$

$$\therefore \quad L \doteqdot 41.77^o \quad \text{or} \quad 180^o - 41.77^o$$

$$\therefore \quad L \doteqdot 41.77^o \quad \text{or} \quad 138.23^o$$

But reject $L = 138.23^o$ as $138.23^o + 56^o > 180^o$ which is impossible.

$$\therefore \quad \angle L \doteqdot 41.77^o.$$

EXERCISE 10H.2

1 Triangle ABC has $\angle B = 40^o$, $b = 8$ cm and $c = 11$ cm. Show that there are two possible values for angle C and solve the triangle in each case.

2 In triangle ABC, find the measure of:
    a   angle A if $a = 14.6$ cm, $b = 17.4$ cm and $\angle ABC = 65^o$
    b   angle B if $b = 43.8$ cm, $c = 31.4$ cm and $\angle ACB = 43^o$
    c   angle C if $a = 6.5$ km, $c = 4.8$ km and $\angle BAC = 71^o$.

I   USING THE SINE AND COSINE RULES

First decide which rule to use.

If the triangle is right angled then the trigonometric ratios or Pythagoras' Theorem can be used, and for some problems adding an extra line or two to the diagram may result in a right triangle.

However, if you have to choose between the sine and cosine rules, the following checklist may assist you.

Use the **cosine rule** when given	• three sides • two sides and an included angle.
Use the **sine rule** when given	• one side and two angles • two sides and a non-included angle (but beware of the *ambiguous case* which can occur when the smaller of the two given sides is opposite the given angle).

Example 23

The angles of elevation to the top of a mountain are measured from two beacons A and B, at sea.

These angles are as shown on the diagram.

If the beacons are 1473 m apart, how high is the mountain?

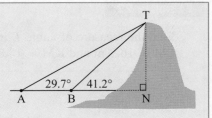

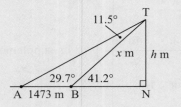

$\angle ATB = 41.2° - 29.7°$ {exterior angle of Δ}
$\quad\quad\ = 11.5°$

We can now find x in $\triangle ABT$ using the sine rule

i.e., $\dfrac{x}{\sin 29.7} = \dfrac{1473}{\sin 11.5}$

$\therefore \quad x = \dfrac{1473}{\sin 11.5} \times \sin 29.7$

$\therefore \quad x \doteqdot 3660.62 \dots$

Now, in $\triangle BNT$,

$$\sin 41.2° = \frac{h}{x}$$

$\therefore \quad \sin 41.2° \times x = h$

$\therefore \quad h = \sin 41.2° \times 3660.62 \dots$

$\therefore \quad h \doteqdot 2411$

So, the mountain is about 2410 m high. (3 sig. figs.)

EXERCISE 10I.1

1 Manny wishes to determine the height of a flag pole. He takes a sighting of the top of the flagpole from point P. He then moves further away from the flagpole by 20 metres to point Q and takes a second sighting. The information is shown in the diagram alongside. How high is the flagpole?

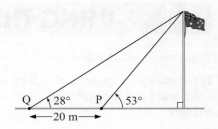

2

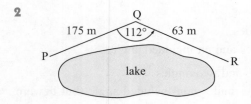

To get from P to R, a park ranger had to walk along a path to Q and then to R as shown.

What is the distance in a straight line from P to R?

3

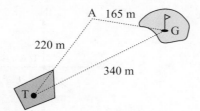

A golfer played his tee shot a distance of 220 m to a point A. He then played a 165 m six iron to the green. If the distance from tee to green is 340 m, determine the number of degrees the golfer was off line with his tee shot.

4 A Communications Tower is constructed on top of a building as shown. Find the height of the tower.

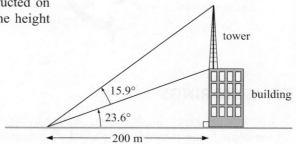

5 A soccer goal is 5 metres wide. When a player is 21 metres from one goal post and 19 metres from the other, he shoots for goal. What is the angle of view of the goals that the player sees?

6 From the foot of a building I have to look upwards at an angle of $22°$ to sight the top of a tree. From the top of the building, 150 metres above ground level, I have to look down at an angle of $50°$ below the horizontal to sight the tree top.

    a How high is the tree?   **b** How far from the building is this tree?

DIRECTION AND BEARINGS

Direction is often quoted as an angle East or West of North or South.

Directions quoted in this manner are called compass bearings.

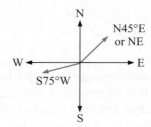

For example,

If the direction of *Town B* from *Town A* is N48°E ("North, 48 degrees, East") then a diagram to depict this direction is given alongside:

If we are considering the direction of *Town B* from *Town A*, the compass points are drawn at Town A. Starting at North, rotate towards East through an angle of $48°$, and the direction line is drawn from Town A to Town B.

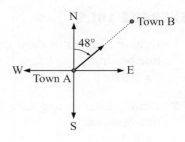

The direction of *Town A* from *Town B* would be
S48°W ("South, 48 degrees, West").

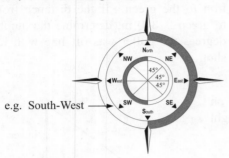

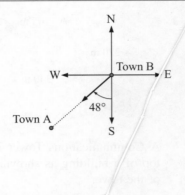

e.g. South-West

TRUE BEARINGS

A true bearing gives the direction, relative to North only,
of the angle being measured in a clockwise direction.

For example, if the direction of Town B from Town A
is N48°E then the **true bearing** of Town B from Town
A is given as 048° (sometimes written as 048°T; the 'T'
standing for True bearing).

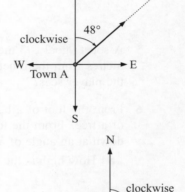

The bearing of Town A from Town B would be 228°.

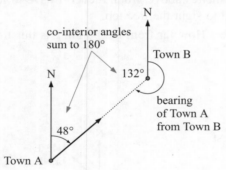

Note: Bearings can be angles from 000° (North) up to 360° (which would also be North).
It is customary to give all bearings as three digits so that there is no doubt that a
digit has been accidentally omitted.

The word 'True' is often omitted when quoting this type of bearing.

EXERCISE 10I.2

1 Draw a diagram to show the direction then convert the following compass bearings to
   true bearings:

   a   S70°E   b   N24°W   c   S38°W

2 Draw a diagram and state the direction in compass bearings for each of the following
   true bearings:

   a   082°   b   176°   c   314°

Example 24

For the diagram given find the bearing of:
a   B from A
b   C from B
c   A from C

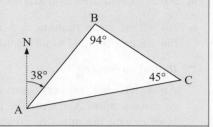

a   In the diagram above, a line showing the direction of North at A is given.

The bearing of B from A will be the angle between this North line at point A and the line joining A to B, measured in a clockwise direction.

This is the angle shown as $38°$ on the diagram hence the bearing of B from A is $038°$.

b   A line showing the direction of North has been drawn at B. The bearing of C from B will be the angle between this line and the line joining B to C.

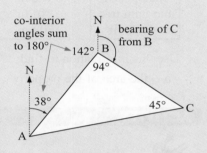

The bearing of C from B
$$= 360° - 94° - 142°$$
$$= 124°$$

c   A North line is drawn at C and the angle is measured from this line to the line joining C to A, in the clockwise direction.

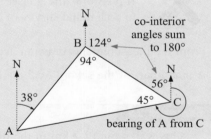

The bearing of A from C
$$= 360° - 45° - 56°$$
$$= 259°$$

3   In the diagram given find the bearing of:
a   B from A   b   B from C
c   A from B   d   A from C

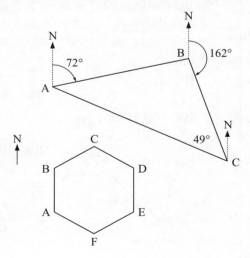

4   ABCDEF is a regular hexagon with the point B being due North of A.

Find the bearing of the point:
a   B from A   b   C from B
c   D from C   d   E from D
e   F from E   f   A from F

5 An aeroplane flies 200 km on a course 137°T. How far east of its starting point is it?

6 An athlete starts running in a direction N63°E and after 45 minutes is 4300 m north of the starting point. Find the athlete's average speed in kmph.

7 **Radial surveying**

Anna has her property surveyed.   The surveyor chooses a point X within the property and using his laser device measures the distances of the boundary corners from X. The surveyor also gives bearings of the corner points from X.

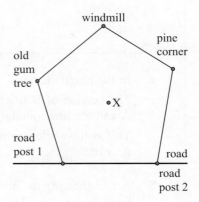

Corner	Distance from X	Bearing from X
Pine corner	368 m	048
road post 2	439 m	132
road post 1	412 m	213
old gum tree	516 m	280
windmill	508 m	343

Use this information to find:

a the area of the property   **b** the perimeter of the property.

ACTIVITY   **RADIAL SURVEYING**

If possible, hire a laser surveying device or obtain the services of a surveyor to do a radial survey of a section of the school grounds.

From the measurements draw an accurate scale diagram and calculate the area enclosed.

A small part of the school oval marked with flags is acceptable.

Example 25

Gus walks for 12.2 km in the direction 123° from camp S and then for 8.9 km in the direction 226° to camp F. Find:
a the distance   **b** the bearing of F from S.

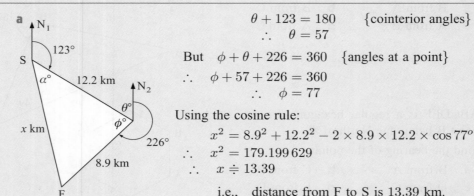

a

$$\theta + 123 = 180 \qquad \{\text{cointerior angles}\}$$
$$\therefore \quad \theta = 57$$

But $\phi + \theta + 226 = 360$ {angles at a point}
$$\therefore \quad \phi + 57 + 226 = 360$$
$$\therefore \quad \phi = 77$$

Using the cosine rule:
$$x^2 = 8.9^2 + 12.2^2 - 2 \times 8.9 \times 12.2 \times \cos 77°$$
$$\therefore \quad x^2 = 179.199\,629$$
$$\therefore \quad x \doteqdot 13.39$$

i.e.,   distance from F to S is 13.39 km.

b We require angle α.

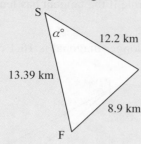

By the cosine rule:

$$\therefore \quad \cos\alpha = \frac{12.2^2 + 13.39^2 - 8.9^2}{2 \times 12.2 \times 13.39}$$

$$\therefore \quad \cos\alpha \doteqdot 0.761\,891......$$

$$\therefore \quad \alpha \doteqdot 40.4^o$$

and $\quad \alpha + 123 \doteqdot 163.4$

$\therefore \quad$ bearing of F from S is 163.4^o.

8 A yacht sails 6 km in a direction 127° and then sails 4 km in a direction 068°. Find the:
 a distance of the yacht from its starting point
 b bearing of the yacht from its starting point.

9 An aircraft takes off from an airport and flies 400 km in a direction of N40°E. It then changes course and flies in a direction S20°E until it is 500 km from the airport.
 a What is the direction of the airport from the aircraft's final position?
 b How far did it fly along the second leg of the journey?

10 A man walks a distance of 8 km in a direction S56°W . He then walks due East for 20 km. What is the distance and bearing of the man from his starting point?

11 Two cyclists depart from the same point. One travels due East at 18 km/h while the other travels North-West at 20 km/h. How long (to the nearest minute) will it be before they are 80 km apart?

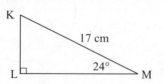

REVIEW SET 10A

1 Find all unknown sides and angles.

2 Metal brackets as shown alongside are attached to a wall so that hanging baskets may be hung from them. Using the dimensions given, find:
 a the length of the diagonal support
 b the angle the diagonal support makes with the wall.

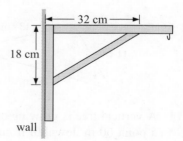

3 Find the angle of elevation to the top of a mountain 2300 m high from a point 5.6 km from its base.

4 A tangent from a point P is drawn to a circle of radius 6.4 cm. Find the angle between the tangent and the line joining P to the centre of the circle if the tangent has length 13.6 cm.

5 The larger angle of a rhombus measures 114° and the longer diagonal is 16.4 cm. Find the lengths of the sides of the rhombus.

6 From the top of a cliff 200 m above sea level the angles of depression to two fishing boats are 6.7^o and 8.2^o respectively. How far apart are the boats?

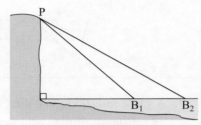

7

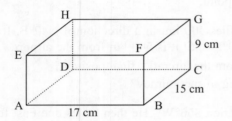

   a Sketch triangle DHE showing which angle is the right angle.

   b Find the measure of angle HDE.

   c Sketch triangle ACG showing which angle is the right angle.

   d Find the measure of angle AGC.

8 Determine the area of:

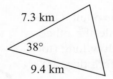

9 Determine the shaded area:

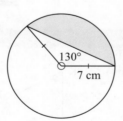

10 Find the value of x:

   a

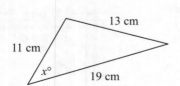

   b

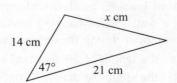

11 A vertical tree is growing on the side of a hill with slope of 10° to the horizontal. From a point 50 m downhill from the tree, the angle of elevation to the top of the tree is 18°. Find the height of the tree.

12 Phuong and Duc are considering buying a block of land and the land agent supplies them with the given accurate sketch. Find the area of the property giving your answer in:

   a m²

   b hectares.

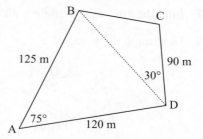

13 A ship sails 156 km on a bearing 147° and then sails 114 km on a bearing 063°. Find:

   a the distance of the ship from its starting point

   b the bearing of the ship from its starting point.

14 A rally car drives at 140 km/h for 45 minutes on a bearing of 032° and then 180 km/h for 40 minutes on a bearing 317°. Find the distance and bearing of the car from its starting point.

REVIEW SET 10B

1 Find the unknown sides and angles for:

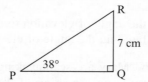

2 When an extension ladder rests against a wall it reaches 4 m up the wall. The ladder is extended a further 0.8 m without moving the foot of the ladder and it now rests against the wall 1 m further up. Find:

   a the length of the extended ladder

   b the increase in the angle that the ladder makes with the ground now that the ladder is extended.

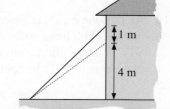

3 A chord of a circle subtends an angle of 114° at its centre. Find the radius of the circle given that the length of the chord is 10.4 cm.

4 An isosceles triangle is drawn with base angles 32° and base 24 cm. Find the base angles of an isosceles triangle with the same base but with double the area.

5 An aeroplane flying at 10 000 m is at an angle of elevation of 36°. If two minutes later, the angle of elevation is 21°, determine the speed of the plane.

6 Find the angle that:

   a DF makes with DB.

   b AG makes with BG

   Hint: In **b** use the cosine rule.

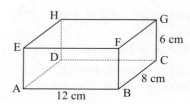

7   Find the area of a triangle with sides 6.4 km and 8.9 km and included angle 63°

8   Determine the value of x:

a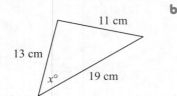

b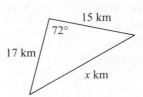

9   Find all unknown sides and angles of $\triangle PQR$ if $\angle P = 42^o$, $PQ = 21$ cm   and
$QR = 17.3$ cm.

10   Find the area of quadrilateral ABCD:

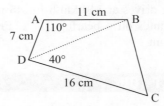

11   From point A, the angle of elevation to the top of a tall building is 20°. On walking 80
m towards the building the angle of elevation is now 23°. How tall is the building?

12   Ahmed, Sue and Mary are sea-kayaking. Ahmed is 430 m from Sue on a bearing of
113° while Mary is on a bearing of 203° and a distance 310 m from Sue. Find the
distance and bearing of Ahmed from Mary.

13   At 12 noon a ship leaves port and travels at 8 km/h along a straight course in the direction
N25^{o}E. At 2 pm a second ship leaves the same port and travels at 9 km/h in the direction
S45^{o}E. The radio communication between the two ships has a range of 75 km. Will the
two ships be able to communicate by radio at 6 pm?

REVIEW SET 10C

1   A yacht sails 8.6 km due east and then 13.2 km south. Find the distance and bearing of
the yacht from its starting point.

2   Determine the height of a tree which casts a shadow of 13.7 m when the sun is at an
angle of 28°.

3   A flagpole 19.6 m high is supported by 3 wires which meet the ground at an angle of
56°. Determine the total length of the three wires.

4   In the given figure $AB = 1$ cm   and
$AC = 3$ cm. Find:

   a   the radius of the circle

   b   the angle subtended by chord BC
   at the centre of the circle.

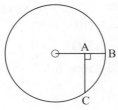

5 In the illustrated roof structure:
   a how long is the timber beam AB
   b at what angle is the beam inclined to the horizontal?

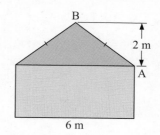

6

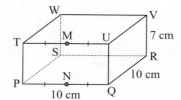

M and N are the midpoints of TU and PQ respectively.
   a Draw a sketch of triangle RMN showing which angle is the right angle.
   b Find the length of RN.
   c Find the measure of angle RMN.

7 ABCD is a square-based pyramid. E, the apex of the pyramid is vertically above M, the point of intersection of AC and BD.

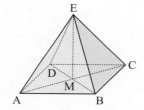

If an Egyptian Pharoah wished to build a square-based pyramid with all edges 200 m, find:

   a how high (to the nearest metre) the pyramid would reach above the desert sands
   b the measure of the angle between a slant edge and a base diagonal.

8 Determine the area of a triangle with sides 11 cm, 9 cm and included angle $65°$.

9 Find the unknown sides and angles:

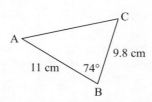

10 An orienteer walks 11.5 km on a bearing of $217°$ and then 8.3 km on a bearing of $271°$. Find the distance and bearing of the orienteer from his starting point.

11 A triangle has an area of 85 cm². If two of the sides are 18 cm and 11 cm, determine the length of the third side.

12 A pilot takes off from an airforce base for a destination 400 km due West of the base. To avoid air turbulence the pilot flies on a bearing of $280°$ for the first 80 km. What course should he now fly to arrive at his destination?

13 The diagram alongside shows a circular entertainment area. It has a paved hexagonal area with plants growing in the garden (shown as the shaded sectors).

If the radius of the circle is 7 metres, find the area of the garden.

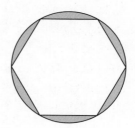

Chapter **11**

Perimeter, area and volume

THE METRIC SYSTEM

The **Metric System** of units is based on the decimal system and was formally established in France in the late 18th century. This system has gradually spread to most countries of the world.

The **Metric System** (International system of units) has three base units which are in frequent use (and four others that are used infrequently). The frequently used ones are:

Base unit	Abbreviation	Used for measuring
metre	m	length
gram	g	mass
second	s	time

Also in common use are:

Unit	Abbreviation	Used for measuring
litre	L	capacity
tonne	t	heavy masses
square metre	m^2	area
cubic metre	m^3	volume
metres per second	m/s	speed
newton	N	force
joule	J	energy
watt	W	power

CONVERSION DIAGRAM FOR METRIC UNITS

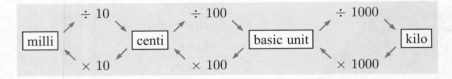

For example, to convert **milli**metres into **centi**metres we divide by 10.

To convert **kilo**grams into grams (**basic** unit) we multiply by 1000.

A   CONVERSION OF UNITS

I use tenths of metres, that is 10 cm lots for all my measurements.

I use millimetres for all my measurements to avoid decimal numbers.

MASS CONVERSIONS

Mass conversions are shown in the following table.

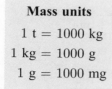

Mass units

$1\ t = 1000\ kg$

$1\ kg = 1000\ g$

$1\ g = 1000\ mg$

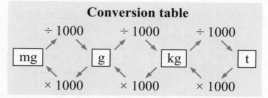

Conversion table

Example 1

Convert: **a** 2.3 kg to grams **b** 8 470 000 g to tonnes

a	Larger to smaller	**b**	Smaller to larger

a Larger to smaller

∴ × by 1000.

  2.3 kg

$= (2.3 \times 1000)$ g

$= 2300$ g

b Smaller to larger

∴ ÷ by 1000 then ÷ by 1000.

  8 470 000 g

$= (8\,470\,000 \div 1000 \div 1000)$ tonnes

$= 8.47$ tonnes

EXERCISE 11A.1

1 Convert:

a 3200 g to kg	**b** 1.87 t to kg	**c** 47 835 mg to kg
d 4653 mg to g	**e** 2.83 t to g	**f** 0.0632 t to g
g 74 682 g to t	**h** 1.7 t to mg	**i** 91 275 g to kg

2 Solve the following problems:

a Find the total mass in kg of 3500 nails each of mass 1.7 grams.

b If each clothes peg has a mass of 6.5 grams, how many pegs are there in 13 kg?

c If a brick has a mass of 1.25 kilograms, how many bricks could a truck with a load limit of 8 tonnes carry?

3 In peppermint flavoured sweets, 0.1 gram of peppermint extract is used per sweet. How many sweets can be made from a drum containing 0.15 t of peppermint extract?

4 A publisher produces a book weighing 856 grams. 6000 of the books are printed and are to be transported from the printer to the nearest port.

a How many tonnes of books are to be sent?

b If the transport costs are $250 per tonne, what will be the cost of sending the books?

LENGTH CONVERSIONS

The following table shows the relationship between various **length units**:

$1\ m = 100\ cm$	$1\ km = 1000\ m$	$1\ cm = 10\ mm$	$1\ mm = \frac{1}{10}\ cm$
$= 1000\ mm$	$= 100\,000\ cm$	$= \frac{1}{100}\ m$	$= \frac{1}{1000}\ m$
$= \frac{1}{1000}\ km$	$= 1\,000\,000\ mm$		

However, we may find it easier to use the following conversion diagram.

CONVERSION DIAGRAM

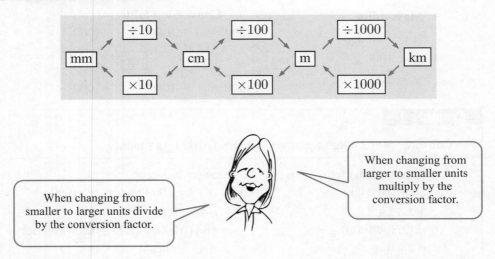

When changing from smaller to larger units divide by the conversion factor.

When changing from larger to smaller units multiply by the conversion factor.

Example 2

Convert: **a**   16.73 m to cm   **b**   48 380 cm to km

a   Larger to smaller   ∴   we × by 100

So,   16.73 m $= (16.73 \times 100)$ cm

               $= 1673$ cm

b   Smaller to larger   ∴   we ÷ by 100 and then ÷ by 1000

∴   48 380 cm $= (48\,380 \div 100 \div 1000)$ km

                    $= 0.4838$ km

EXERCISE 11A.2

1 Convert:

   a   8250 cm to m   **b**   295 mm to cm   **c**   6250 m to km

   d   73.8 m to cm   **e**   24.63 cm to m   **f**   9.761 m to km

2 Convert:

   a   413 cm to mm   **b**   3754 km to m   **c**   4.829 km to cm

   d   26.9 m to mm   **e**   0.47 km to cm   **f**   3.88 km to mm

3   I have 45 coils of fencing wire with each coil being 255 m long. How many kilometres of wire do I have?

4   Jess has 4.75 km of copper wire and needs to cut it into 12.5 cm lengths to be used in electric toasters. How many lengths can he make?

B ▏ PERIMETER

OPENING PROBLEM   BRICK EDGING

 You are asked to quote on the supply and installation of bricks around a lawn. The bricks are expensive and are not returnable. Consequently, you need to accurately calculate how many are needed and what they will cost. You draw a rough sketch of what the house owner wants. You take it back to your office to do the calculations.

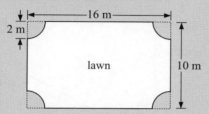

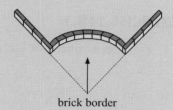

brick border

Questions we need to ask ourselves could be:

- How far is it around the lawn?
- How many bricks will I need?
- What will the cost of the bricks to do the job?
- Set up a spreadsheet to handle the calculations as you are expecting dozens of jobs of this nature in the future.

- What is the length of one brick?
- What is the cost of one brick?

PERIMETER

The **perimeter** of a figure is the measurement of the distance around its boundary.

For a **polygon**, the perimeter is obtained by adding the lengths of all of its sides.

For a **circle**, the perimeter has a special name, the **circumference**.

DEMO

Following is a summary of some **perimeter formulae**:

Shape	Formula	Shape	Formula
square l	$P = 4l$	rectangle w l	$P = 2l + 2w$ or $P = 2(l + w)$
polygon a b c d e	$P = a + b + c + d + e$	circle d r	$C = \pi d$ or $C = 2\pi r$

Example 3

Find the perimeter of: **a**   **b**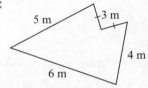

a $P = 13.2 + 2 \times 9.7$ m	**b** $P = 2 \times a + 2 \times (a+4)$ cm
$= 32.6$ m	$= (4a+8)$ cm

EXERCISE 11B.1

1 Find the perimeter of each of the following figures:

a 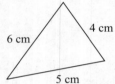   **b** 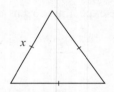   **c**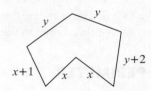

2 Find the perimeter of the following figures:

a 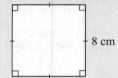   **b** 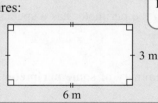   **c**

Example 4

Find the perimeter of the following figures:

> Double the sum of the length and width.

a   8 cm   **b**   3 m
6 m

a $P = 4s$	**b** $P = 2(l+w)$
$\therefore$ $P = 4 \times 8$ cm	$\therefore$ $P = 2(6+3)$ m
i.e., $P = 32$ cm	i.e., $P = 18$ m

3 Use $P = 4s$ for squares and $P = 2(l+w)$ for rectangles to find the perimeters of the following figures:

a 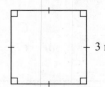   **b** 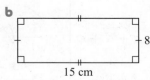   **c**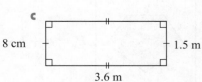

4 A rectangular field 220 metres long and 300 metres wide is to be fenced.

    a Draw and label a diagram of the field.

    b Find the total length of fencing required.

5 A triathlon course has 3 legs:

- a 1.2 km swim
- an 8.3 km bicycle ride
- a 6.3 km run.

Find:

    a the total distance around the track

    b the average speed of a contestant who took 1 hour 10 minutes to complete the course.

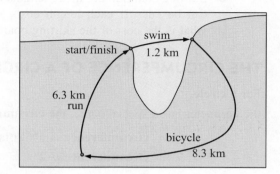

6 A rectangular swimming pool is 50 m long and 20 m wide and has brick paving 3 m wide around it. Find the outer perimeter of the brick paving.

7

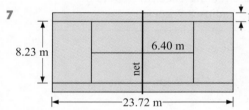

A tennis court has the dimensions shown.

    a What is the perimeter of the court?

    b Find the total length of all the marked lines.

8 Calculate the cost of rewiring the garden marquee illustrated. The wire is 32 cents per metre and 5% extra wire is required to tie the wires into the piping. The wire is sold by the full metre lengths.

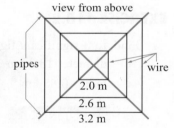

9 A farmer wanted to fence his private garden to keep out wandering cattle. The garden measured 26 m × 40 m. The fence had 5 strands of wire, and posts were placed every 2 metres.

    a Calculate the perimeter.

    b What length of wire was needed?

    c How many posts were needed?

    d Find the total cost of the fence if wire was $0.28 per metre and each post cost $3.95.

10 The old back fence is to be replaced by a wooden paling fence. The fence is to be 1.5 metres high and 51 metres long. Posts are 3 metres apart and there are 2 railings the length of the fence. Use the following table to calculate the cost of the wood for the fence.

Item	Size per unit	Cost per unit
Post	150 mm × 100 mm	$3.85 per post
Railings	75 mm × 50 mm	$2.20 per metre
Palings	150 mm wide	$1.90 per paling

11 A room has the shape and dimensions shown.
Skirting board is laid around the edge of the room.

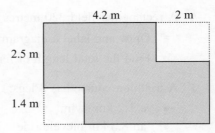

    a How much skirting board would be needed?

    b Skirting comes in 2.4 metre lengths which
can be cut. If each length costs $2.48,
what is the cost of the skirting board?

THE CIRCUMFERENCE OF A CIRCLE

For a **circle**,

the perimeter has a special name, the **circumference**, and we use either

> **circumference = π × diameter** $(C = \pi d)$ or
>
> **circumference = 2 × π × radius** $(C = 2\pi r)$

Example 5

Find, to 3 significant figures, the circumference of a circle of:
  a diameter 13.8 m   **b** radius 3.7 km

---

  a $C = \pi d$ **b** $C = 2\pi r$

          $= \pi \times 13.8$ $= 2 \times \pi \times 3.7$

          $\doteqdot 43.4$ m $\doteqdot 23.2$ km

EXERCISE 11B.2

1 Find, to 3 significant figures, the circumference of a circle of:

    a diameter 13.2 cm   **b** radius 8.6 m   **c** diameter 115 m

    d radius 0.85 km   **e** diameter 7.2 m   **f** radius 235 cm

---

Example 6

Find, to 3 significant figures,
the perimeter of:

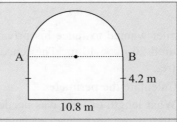

---

    diameter of semi-circle $= d = 10.8$

    $\therefore$ distance from A to B around semi-circle

          $= \frac{1}{2}(\pi d)$

          $= \frac{1}{2} \times \pi \times 10.8$

          $\doteqdot 16.96$ m

    $\therefore$ $P \doteqdot 10.8 + 2 \times 4.2 + 16.96$ m

          $\doteqdot 36.2$ m

2 Find, to 3 significant figures, the perimeter of:

a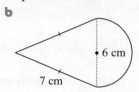

b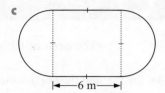

c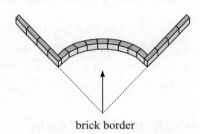

3 How far did I live from school if my bike wheel had a diameter of 32 centimetres and the number of revolutions made by my wheel from home to school was 239? (Consider the relationship between the circumference of the wheel and a revolution of the wheel.)

The **Opening Problem** revisited

4 A lawn is basically rectangular with 2 m radius quarter-circles removed from each of its corners. Bricks of length 220 mm, each costing \$1.70 are used to border the lawn.

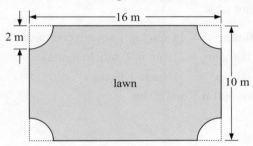

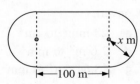

brick border

a Find the perimeter of the lawn.
b How many bricks are necessary to border the lawn?   (Allow 2% for wastage.)
c Find the total cost of the bricks.

> You could use a graphics calculator to solve
> $2 \times \pi \times r = 11.7$

Example 7

Find the radius of a circle with circumference 11.7 m.

$$C = 2\pi r$$
$$\therefore \quad 11.7 = 2 \times \pi \times r$$
$$\therefore \quad 11.7 \doteqdot 6.283 \times r$$
$$\therefore \quad \frac{11.7}{6.283} \doteqdot r \qquad \{\text{dividing both sides by } 6.283\}$$
$$\therefore \quad r \doteqdot 1.86$$

$\therefore$ the radius is approximately 1.86 m.

---

5 Find:
   a the diameter of a circular pond of perimeter 37.38 m
   b the radius of a circular pond of perimeter 32.67 m.

6 An athletics track is to have a perimeter of 400 m and straights of length 100 m.

Find the radius of the 'bend'.

AREA

> The **area** of a closed figure is the number of square units it contains.

As for the length measurements, we will use the **International System of Units**.

The relationships between the various units of area can be obtained using the conversions for length.

For example,   since 1 cm = 10 mm

then   1 cm × 1 cm = 10 mm × 10 mm

i.e., $1 \text{ cm}^2 = 100 \text{ mm}^2$

AREA UNITS

Area conversions are shown in the following table:

$$1 \text{ cm}^2 = 10 \text{ mm} \times 10 \text{ mm} = 100 \text{ mm}^2 \quad \{1 \text{ cm}^2 \text{ is 1 square centimetre}\}$$
$$1 \text{ m}^2 = 100 \text{ cm} \times 100 \text{ cm} = 10\,000 \text{ cm}^2 \quad \{1 \text{ m}^2 \text{ is 1 square metre}\}$$
$$1 \text{ ha} = \quad 100 \text{ m} \times 100 \text{ m} \quad = 10\,000 \text{ m}^2 \quad \{1 \text{ ha is 1 hectare}\}$$
$$1 \text{ km}^2 = 1000 \text{ m} \times 1000 \text{ m} = 1\,000\,000 \text{ m}^2 \text{ or } 100 \text{ ha}$$

CONVERSION DIAGRAM

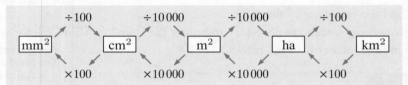

CONVERSION OF UNITS

Example 8

Convert: **a** $650\,000 \text{ m}^2$ into ha   **b** 2.5 m^2 into mm^2

a   Smaller to larger

∴   ÷ by 10 000.

So, $650\,000 \text{ m}^2$

$= (650\,000 \div 10\,000) \text{ ha}$

$= 65 \text{ ha}$

b   Larger to smaller

∴   × by 10 000 and then × by 100

∴ 2.5 m^2

$= (2.5 \times 10\,000 \times 100) \text{ mm}^2$

$= 2\,500\,000 \text{ mm}^2$

EXERCISE 11C.1

1   Convert:

    a 23 mm^2 to cm^2
    b   3.6 ha to m^2
    c 726 cm^2 to m^2

    d 7.6 m^2 to mm^2
    e 8530 m^2 to ha
    f   0.354 ha to cm^2

    g 13.54 cm^2 to mm^2
    h 432 m^2 to cm^2
    i $0.004\,82 \text{ m}^2$ to mm^2

2   **a**   I have purchased a 4.2 ha property.

Council regulations allow me to have 5 free range chickens for every 100 m². How many free range chickens am I allowed to have?

b   Sam purchased 1000 m² of dress material, and needs to cut it into rectangles of area 200 cm² for a patchwork quilt. This can be done with no waste. How many rectangles can Sam cut out?

AREA

> The area within a shape is the number of square units enclosed by the shape.

The formulae for calculating the area of the most commonly occurring shapes are given below:

Shape	Figure	Formula
Rectangle		$\text{Area} = \text{length} \times \text{width}$
Triangle		$\text{Area} = \frac{1}{2} \times \text{base} \times \text{height}$
Parallelogram		$\text{Area} = \text{base} \times \text{height}$
Trapezium		$\text{Area} = \left(\dfrac{a+b}{2}\right) \times h$
Circle		$\text{Area} = \pi r^2$
Sector		$\text{Area} = \left(\dfrac{\theta}{360}\right) \times \pi r^2$

Example 9

Find the area of the rectangles: **a**   **b**

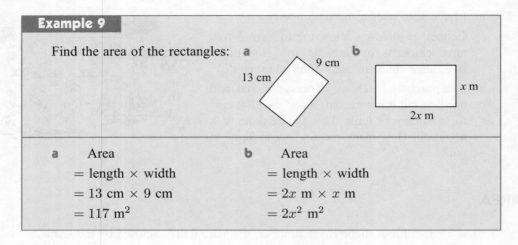

a	Area	**b**	Area
	$= \text{length} \times \text{width}$		$= \text{length} \times \text{width}$
	$= 13 \text{ cm} \times 9 \text{ cm}$		$= 2x \text{ m} \times x \text{ m}$
	$= 117 \text{ m}^2$		$= 2x^2 \text{ m}^2$

EXERCISE 11C.2

1 Find the area of the following rectangles:

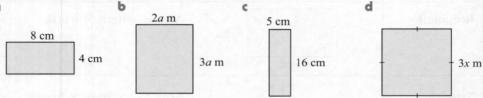

Example 10

Find the area of the
following triangles: **a**   **b**

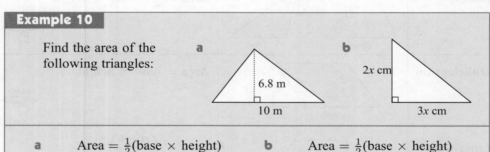

a   Area $= \frac{1}{2}(\text{base} \times \text{height})$ **b**   Area $= \frac{1}{2}(\text{base} \times \text{height})$

$\therefore$ $A = \frac{1}{2}(10 \times 6.8) \text{ m}^2$ $\therefore$ $A = \frac{1}{2}(3x \times 2x) \text{ cm}^2$

$\therefore$ $A = \frac{1}{2} \times 68 \text{ m}^2$ $\therefore$ $A = \frac{1}{2}(6x^2)$

i.e., $A = 34 \text{ m}^2$ i.e., $A = 3x^2 \text{ cm}^2$

2 Find the area of each of the following triangles:

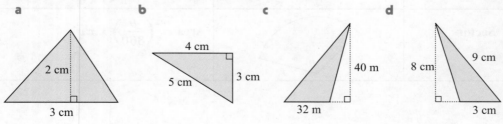

3 Find the area of each of the following triangles:

a

2a cm

a cm

b
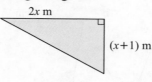
2x m

(x+1) m

c
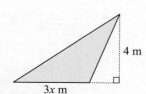
4 m

3x m

Example 11

Find the area of:

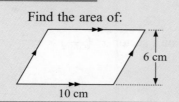
6 cm

10 cm

Area = base × height

∴ $A = 10$ cm × 6 cm

i.e., $A = 60$ cm^2

---

4 Find the area of each of the following parallelograms:

a
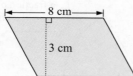
8 cm

3 cm

b
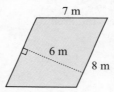
7 m

6 m

8 m

c
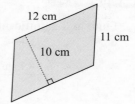
12 cm

11 cm

10 cm

5 Find the area of each of the following parallelograms:

a
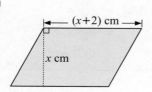
(x+2) cm

x cm

b
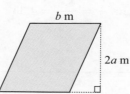
b m

2a m

c
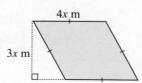
4x m

3x m

Example 12

Find the area of:

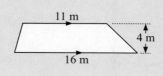
11 m

4 m

16 m

Area = $\left(\dfrac{a+b}{2}\right) \times h$

= $\left(\dfrac{11+16}{2}\right) \times 4$

= 54 m^2

---

6 Find the area of each of the following trapezia:

a
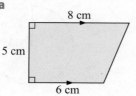
8 cm

5 cm

6 cm

b
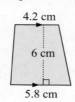
4.2 cm

6 cm

5.8 cm

c
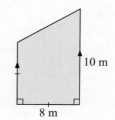
10 m

8 m

7 Find the area of each of the following trapezia:

a
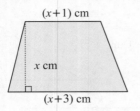
(x+1) cm

x cm

(x+3) cm

b
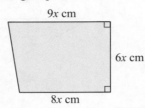
9x cm

6x cm

8x cm

c
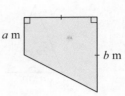
a m

b m

CIRCLES AND SEGMENTS

Area of a circle $= \pi r^2$

r

Area of a sector $= \left(\frac{\theta}{360}\right) \times \pi r^2$

r

$\theta°$

Example 13

Find, to 3 significant figures, the area of:

a

8.7 cm

b

60°

8 cm

> The area of a sector is a fraction of the area of a circle!

a $A = \pi r^2$ where $r = \dfrac{8.7}{2} = 4.35$

$\therefore$ $A = \pi \times (4.35)^2$

$\therefore$ $A \doteqdot 59.4$ cm^2

b   Area $= \dfrac{\theta}{360} \times \pi r^2$

$= \dfrac{60}{360} \times \pi \times 8^2$

$\doteqdot 33.5$ cm^2

EXERCISE 11C.3

1 Find the area of:

a
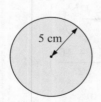
5 cm

b
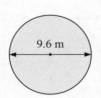
9.6 m

c

2x m

2 Find the area of:

a
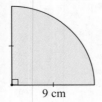
9 cm

b
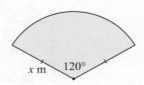
x m   120°

c

4 cm

135°

AREAS OF COMPOSITE SHAPES

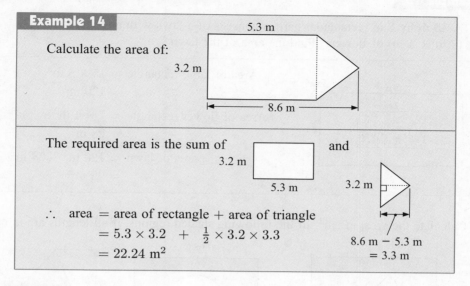

Example 14

Calculate the area of:

5.3 m

3.2 m

8.6 m

The required area is the sum of

3.2 m

5.3 m

and

3.2 m

8.6 m − 5.3 m
= 3.3 m

∴ area = area of rectangle + area of triangle

$= 5.3 \times 3.2 \ + \ \frac{1}{2} \times 3.2 \times 3.3$

$= 22.24 \text{ m}^2$

EXERCISE 11C.4

1 Calculate the area of the following composite shapes (all measurements are in cm).

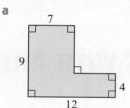

a

7

9

12

4

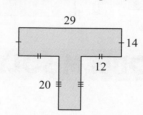

b

29

14

20 12

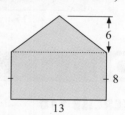

c

6

8

13

Example 15

Find a formula for the area, A, of:

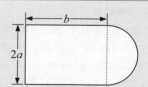

b

$2a$

Area = area of rectangle + area of semi-circle

∴ $A = 2a \times b \ + \ \frac{1}{2} \times \pi \times a^2$ {as the radius of the semi-circle is a units}

∴ $A = \left(2ab + \dfrac{\pi a^2}{2}\right)$ units²

2 Find a formula for the area, A, of the following regions:

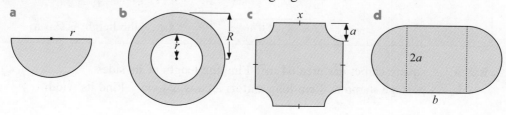

a

r

b

r

R

c

x

a

d

$2a$

b

Example 16

A 15 m by 8 m rectangular garden consists of a lawn with an inner rectangle (7 m × 4 m) of flowers. Find the area of the lawn.

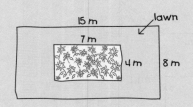

Area of large rectangle = 15 × 8 m²
$$= 120 \text{ m}^2$$

Area of flower rectangle = 7 × 4 m²
$$= 28 \text{ m}^2$$

$\therefore$ area of lawn = 120 m² − 28 m²
$$= 92 \text{ m}^2$$

3 Calculate the area, in cm², of the following shaded regions, if all lengths are in cm.

a

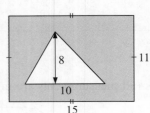

b

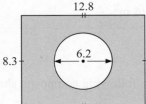

c

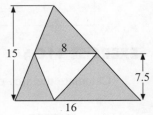

D   PROBLEM SOLVING WITH AREAS

EXERCISE 11D

1 Find the area of:

  a a square station property with sides of length 16.72 km
  b a rectangular housing block which is 42.7 m by 21.3 m
  c a triangular garden patch with base 12.7 m and altitude 10.5 m
  d a circular pond of radius 23.8 m
  e a sector shaped cloth of angle 120° and radius 4.43 m.

Example 17

A triangular advert-
ising sign has base
1.6 m and area 2 m².
Find its height.

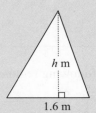

Area of triangle = $\frac{1}{2}(b \times h)$
$$\therefore \quad 2 = \tfrac{1}{2}(1.6 \times h)$$
$$\therefore \quad 1.6 \times h = 4$$
$$\therefore \quad h = 2.5$$

i.e.,   the height is 2.5 m

2 **a** A square carpet has area 64 m². Find the length of its sides.
  b A postage stamp is 3 cm long and its area is 5.4 cm². Find its width.

c The base of a triangular sail is 1.5 m long and the area of the sail is 1.8 m². Find the height of the sail.

3 A wheat paddock is 1.2 km by 0.65 km. Find:

a the area of the paddock in hectares

b the total cost of planting the wheat crop at $79 per hectare.

4

lawn	25 m

30 m

A rectangle of lawn is 30 m × 25 m and has a 1 m wide path around it. Draw a diagram and find the total area of concrete for the path.

5 What is the cost of laying 'instant lawn' (turf) over an 80 m by 120 m rectangular playing field?

The turf comes in $\frac{1}{2}$ m wide strips and costs $15 for 10 m.

6 A cropduster can carry 240 kg of fertiliser at a time. It is necessary to spread 50 kg of fertiliser per hectare. How many flights must be made to fertilise a 1.2 km by 450 m rectangular property?

Example 18

A rectangular lawn is 12 m by 8 m and a 3.2 m square garden bed is dug out of the lawn. The lawn is then reseeded with boxes of seed, where each box costs $8.50 and covers 18 m². Find:

a the area of the lawn

b the number of boxes of seed required

c the total cost of the seed.

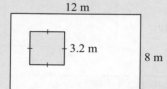

12 m

3.2 m

8 m

a Area of lawn
= area of rectangle − area of square
= $(12 \times 8) - (3.2 \times 3.2)$ m²
= 85.76 m²

b Number of boxes
= $\dfrac{\text{area of lawn}}{\text{area covered by one box}}$
= $\dfrac{85.76 \text{ m}^2}{18 \text{ m}^2}$
$\doteqdot$ 4.76 boxes
∴   require 5 boxes of seeds

c Total cost
= number of boxes × cost of 1 box
= 5 × $8.50
= $42.50

7 A swimming pool has the dimensions as shown and is to be tiled with tiles which are 25 cm by 20 cm.

The tiles cost $21.50 a square metre.

10 m

1 m

5 m

2 m

A

10.05 m

a Find the area of side A.

b Find the total surface area of the 4 walls and bottom of the pool.

c How many tiles are necessary for 1 m² of coverage?

d How many tiles are necessary to complete the task?

e What is the total cost of tiling the pool given that:

• adhesive costs $2.35 per m²   and   • labour costs $22 per m²?

8 A cylindrical tank of base diameter 8 m and height 6 m is to have a non-porous lining on its circular base and curved walls. The lining costs $3.20 per m² on the base and $4.50 per m² on the walls.

a What is the base radius?

b What is the area of the base?

c What is the cost of lining the base?

d Explain why the area of the curved wall is found by using $2\pi r \times h$ where r is the radius and h is the height.

e Find the area of the curved wall.

f What is the cost of the lining of the curved wall?

g What is the total cost of the lining to the nearest $10?

9 Pete's Pizza Bar sells 3 sizes of pizza. The small pizza supreme is 15 cm across and costs $4.20. The medium size is 30 cm across and costs $9.60. The super size has a diameter of 35 cm and costs $14.00. Which of the pizzas gives the best value?
(**Hint:** Calculate the price per square unit.)

10 The diagram shows a room to be painted twice (including ceiling). The door is 0.8 m by 2.2 m and the window is 183 cm by 91 cm. The door also has to be stained on both sides with 2 coats of stain. Use the following table to calculate the total cost of the stain and paint:

2.4 m

4.2 m

3.5 m

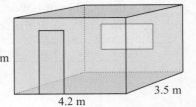

Type of paint	Size	Area covered	Cost per tin
wall paint	4 litres	16 m²	$32.45
	2 litres	8 m²	$20.80
wood stain	2 litres	10 m²	$23.60
(for doors)	1 litre	5 m²	$15.40

INVESTIGATION 1   RECTANGLES OF FIXED PERIMETER

You are to investigate how area changes for a fixed (given) perimeter.

What to do:

1 Examine the following figures and comment on their perimeters and areas:

2 Use a spreadsheet to help you solve this problem:

"Of all the rectangles which have a perimeter of 27.4 units, what is the shape of the one with largest area?"

You may wish to follow these steps:

- Draw a sketch and use x to specify side AB, say.
- Find BC say, in terms of x.
- Write down a formula for finding the area in terms of x.
- Type in and fill down formulae as shown.

	A	B	C	D	E
1	x	other side	area		
2	0	=E$2/2-A2	=A2*B2	perimeter =	27.4
3	=A2+1	=E$2/2-A3	=A3*B3		
4					
5					
6	↓	↓	↓		
7		fill down to row 14			

- Select the largest area in column C and find the value of x which produces this result.
- Refine your answer for x by changing cell A3 to $=$A2$+ 0.1$ and starting A2 with a number close to the value of x.
- Further refine x correct to 2 decimal places.

3 Choose a different fixed perimeter, i.e., change 27.4 to something else and use your spreadsheet to maximise the area for this perimeter.

4 Find out how to use a **graphics calculator** to solve the above problem.

---

11 You are to design a rectangular duck enclosure of area 100 m². Wire netting must be purchased for three of the sides as the fourth side is an existing fence of another duck enclosure.

Naturally you wish to minimise the length of netting required to do the job, and therefore minimise your costs.

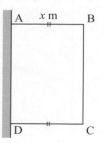

a If AB is x m long, find BC in terms of x.

b Explain why the required length of netting is given by $L = 2x + \dfrac{100}{x}$ m.

c Use technology to find the value of x which will make L a minimum.

d Draw a sketch of the 'best' shape showing the dimensions of the enclosure.

12 An industrial shed is to have a total floor space of 600 m² and is to be divided into three rectangular rooms of equal size. The walls will cost \$60 per metre to build.

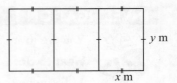

    a Calculate the area of the shed in terms of x and y.

    b Explain why $y = \dfrac{200}{x}$.

    c Show that the total cost of the walls is given by $C = 360x + \dfrac{48\,000}{x}$ dollars.

    d Use technology to find x (nearest cm) which will minimise the cost of the walls.

    e Draw a sketch of the 'best' shape showing all dimensions.

E   SURFACE AREA

SOLIDS WITH PLANE FACES

> The **surface area** of a solid with plane faces is the sum of the areas of the faces.

Note: This means that the surface area is the same as the area of the **net** required to make the figure.

For example,   surface area of 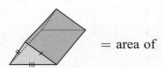 = area of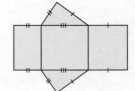

Example 19

Find the surface area of the rectangular prism:

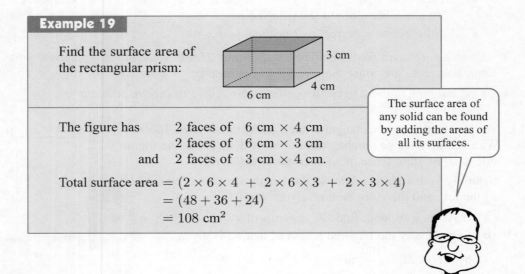

The figure has   2 faces of   6 cm × 4 cm
                  2 faces of   6 cm × 3 cm
        and   2 faces of   3 cm × 4 cm.

$$\begin{aligned}
\text{Total surface area} &= (2 \times 6 \times 4 \ + \ 2 \times 6 \times 3 \ + \ 2 \times 3 \times 4) \\
&= (48 + 36 + 24) \\
&= 108 \text{ cm}^2
\end{aligned}$$

> The surface area of any solid can be found by adding the areas of all its surfaces.

EXERCISE 11E.1

1 Find the total surface area of the following rectangular prisms:

a

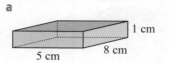

1 cm
8 cm
5 cm

b

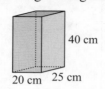

40 cm
20 cm   25 cm

c

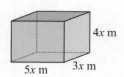

$4x$ m
$5x$ m $3x$ m

2 Find the total surface area of:

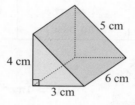

5 cm
4 cm
6 cm
3 cm

3 Draw the following and hence find the surface area:

  a an ice cube with sides 2.5 cm

  b a block of cheese measuring 14 cm by 8 cm by 3 cm

  c a wooden wedge with a 3 cm by 4 cm by 5 cm triangular cross-section and length 8 cm.

4 A shade-house with the dimensions illustrated is to be covered with shade cloth. The cloth costs \$4.75 per square metre.

  a Find the area of each end of the shade-house.

  b Find the total area to be covered with cloth.

  c Find the total cost of the cloth needed given that 5% more than the calculated amount is necessary for wrapping around the piping substructure.

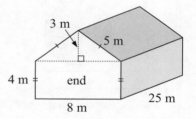

3 m
5 m
4 m   end
8 m   25 m

OBJECTS WITH CURVED SURFACES

There are three objects which have a simple formula for calculating the area of a curved surface. These are   • a cylinder   • a cone   • a sphere.

CYLINDER

Object	Hollow cylinder	Hollow can	Solid cylinder
Figure	hollow h r hollow	hollow h r solid	solid h r solid
Outer surface area	$A = 2\pi rh$ (no ends)	$A = 2\pi rh + \pi r^2$ (one end)	$A = 2\pi rh + 2\pi r^2$ (two ends)

CONE

Object	Hollow cone	Solid cone
Figure		
Outer surface area	$A = \pi r s$ (no base)	$A = \pi r s + \pi r^2$ (solid)

SPHERE

$$A = 4\pi r^2$$

Example 20

Find, to 1 decimal place, the surface area of these solids:

a 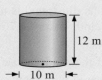   10 cm

b 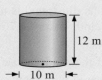   12 m, 10 m

c 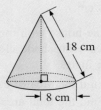   18 cm, 8 cm

a Surface area
$= 4\pi r^2$
$= 4 \times \pi \times 5^2$
$\doteqdot 314.2 \text{ cm}^2$

b Surface area
$= 2\pi r h + 2\pi r^2$
$= 2 \times \pi \times 5 \times 12$
$\quad + 2 \times \pi \times 5^2$
$\doteqdot 534.1 \text{ m}^2$

c Surface area
$= \pi r s + \pi r^2$
$= \pi \times 8 \times 18 + \pi \times 8^2$
$\doteqdot 653.5 \text{ cm}^2$

EXERCISE 11E.2

1 Find the total surface area of these solids:

a 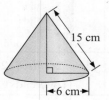   12.6 cm

b 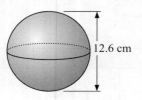   12 cm, 10 cm

c 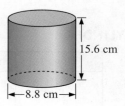   15.6 cm, 8.8 cm

d 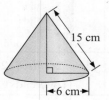   15 cm, 6 cm

e 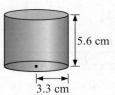   5.6 cm, 3.3 cm

f   8.2 cm

2 Find the total area of surface of the solid hemi-sphere shown.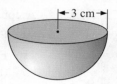

3 Find the total surface area of these solids:
   a a cylinder with height $3x$ cm and base radius x cm
   b a sphere with radius $2x$ cm
   c a cone with radius $2x$ cm and slant height $4x$ cm
   d a cone with radius $3x$ cm and height $4x$ cm

Example 21

Find the outer surface area of:

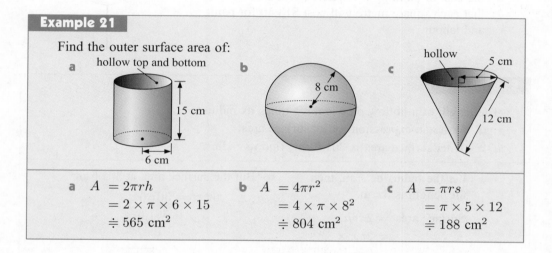

  a $A = 2\pi r h$
      $= 2 \times \pi \times 6 \times 15$
      $\doteqdot 565$ cm^2

  b $A = 4\pi r^2$
      $= 4 \times \pi \times 8^2$
      $\doteqdot 804$ cm^2

  c $A = \pi r s$
      $= \pi \times 5 \times 12$
      $\doteqdot 188$ cm^2

---

4 Find the outer surface area of the following:

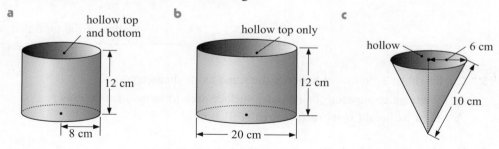

5 A wharf is to have 24 cylindrical concrete pylons, each of diameter 0.6 m and of length 10 m. Each is to be coated with a salt resistant material.
   a Find the total surface area of one pylon.
   b Coating the pylons with the material costs \$45.50 per m^2. Find the cost of coating one pylon.
   c Find the total cost for coating the 24 pylons.

6

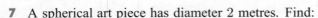

A conical tent has base radius 2 m and height 5 m.

  a Find the slant height s, to 2 decimal places.

  b Find the area of canvas necessary to make the tent (base included).

  c If canvas costs \$18 per m^2, find the cost of canvas needed.

7 A spherical art piece has diameter 2 metres. Find:

  a the surface area of the sphere

  b the cost of painting the sphere (with 3 coats) given that each square metre will cost \$13.50 for paint and labour.

Example 22

The length of a hollow pipe is three times its radius.

  a Write an expression for its surface area.

  b If its surface area is 301.6 m^2, find its radius.

a Let the radius be x m, then the length is $3x$ m.

$$\begin{aligned} \text{Surface area} &= 2\pi rh \\ &= 2\pi x \times 3x \\ &= 6\pi x^2 \text{ m}^2 \end{aligned}$$

b But the surface area is 301.6 m^2

$$\begin{aligned} \therefore \quad 6\pi x^2 &= 301.6 \\ \therefore \quad x^2 &= \frac{301.6}{6\pi} \\ \therefore \quad x &= \sqrt{\frac{301.6}{6\pi}} \quad \{x > 0\} \\ \therefore \quad x &\doteqdot 4 \text{ m} \end{aligned}$$

i.e.,   the radius of the pipe is 4 m.

---

8 The height of a hollow cylinder is the same as its diameter.

  a Write an expression for its outer surface area in terms of its radius r.

  b Find its height if its surface area is 91.6 m^2.

9 The slant height of a hollow cone is three times its radius.

  a Write an expression for its outer surface area in terms of its radius r.

  b Find the slant height if its surface area is 21.2 cm^2.

  c Hence find its height.

F VOLUME

The **volume** of a solid is the amount of space it occupies. It is measured in cubic units.

UNITS OF VOLUME

Volume can be measured in cubic millimetres, cubic centimetres or cubic metres.

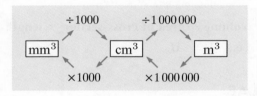

1 cm^3
$= 10 \text{ mm} \times 10 \text{ mm} \times 10 \text{ mm}$
$= 1000 \text{ mm}^3$

1 m^3
$= 100 \text{ cm} \times 100 \text{ cm} \times 100 \text{ cm}$
$= 1\,000\,000 \text{ cm}^3$

CONVERSION DIAGRAM

$$\boxed{mm^3} \quad \xrightarrow{\div 1000} \quad \boxed{cm^3} \quad \xrightarrow{\div 1\,000\,000} \quad \boxed{m^3}$$

$$\boxed{mm^3} \quad \xleftarrow{\times 1000} \quad \boxed{cm^3} \quad \xleftarrow{\times 1\,000\,000} \quad \boxed{m^3}$$

Example 23

Convert the following:
a 5 m^3 to cm^3 **b** $25\,000 \text{ mm}^3$ to cm^3

a 5 m^3
$= (5 \times 100^3) \text{ cm}^3$
$= (5 \times 1\,000\,000) \text{ cm}^3$
$= 5\,000\,000 \text{ cm}^3$

b $25\,000 \text{ mm}^3$
$= (25\,000 \div 10^3) \text{ cm}^3$
$= (25\,000 \div 1000) \text{ cm}^3$
$= 25 \text{ cm}^3$

EXERCISE 11F.1

1   Convert the following:

   a 8.65 cm^3 to mm^3 **b** $86\,000 \text{ mm}^3$ to cm^3 **c** $300\,000 \text{ cm}^3$ to m^3

   d 124 cm^3 to mm^3 **e** 300 mm^3 to cm^3 **f** 3.7 m^3 to cm^3

2   **a** $30\,000$ ingots of lead each with volume 250 cm^3 are required by a battery manufacturer. How many cubic metres of lead does the manufacturer need to purchase?

   b   A manufacturer of lead sinkers for fishing has 0.237 m^3 of lead. If each sinker is 5 cm^3 in volume, how many sinkers can be made?

UNIFORM SOLIDS

For solids of uniform cross-section, cuts made perpendicular to the length are **uniform**, that is, are always the same shape and size.

Examples: A is the cross-sectional area, l is the length of the solid

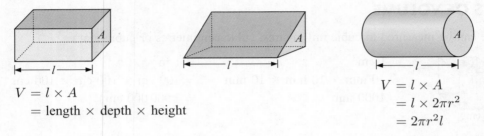

$V = l \times A$
  $= \text{length} \times \text{depth} \times \text{height}$

$V = l \times A$
  $= l \times 2\pi r^2$
  $= 2\pi r^2 l$

For all solids of uniform cross-section,

> **volume = area of cross-section × length**
>
> i.e., $V = Al.$

VOLUME FORMULAE

The following volume formulae are very common and have been considered in previous years:

Object	Figure	Volume
Solids of uniform cross-section	height end ← height → end	**Volume of uniform solid = area of end × length** DEMO
Pyramids and cones	height height h base base (solids that taper to a point)	**Volume of a pyramid or cone** $= \frac{1}{3}(\text{area of base} \times \text{height})$ DEMO
Spheres	r	**Volume of a sphere** $= \frac{4}{3}\pi r^3$

Example 24

Calculate the volume of the uniform solid shown.

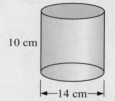

area 11.7 cm^2 5.9 cm

$$\therefore \; V = Al \qquad \text{\{as this is a solid of uniform cross-section\}}$$
$$= \text{area of cross-section} \times \text{length}$$
$$= 11.7 \text{ cm}^2 \times 5.9 \text{ cm}$$
$$\doteqdot 69.0 \text{ cm}^3$$

Example 25

Find the volume of the following:

a
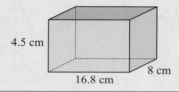

4.5 cm

16.8 cm   8 cm

b

10 cm

|← 14 cm →|

a Volume	= area of end × height	**b** Volume	= area of end × height
	= length × width × height		= $\pi r^2 \times h$
	= $16.8 \times 8 \times 4.5$		= $\pi \times 7^2 \times 10$
	$\doteqdot 605 \text{ cm}^3$		$\doteqdot 1540 \text{ cm}^3$

EXERCISE 11F.2

1 Calculate the volume of:

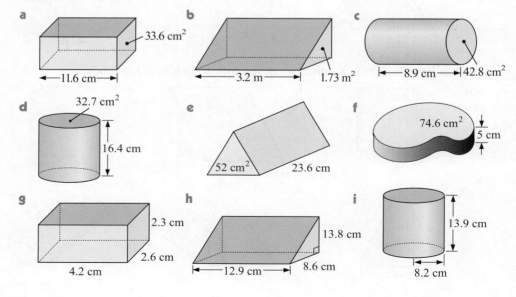

a 33.6 cm^2 ← 11.6 cm →

b ← 3.2 m → 1.73 m^2

c ← 8.9 cm →| 42.8 cm^2

d 32.7 cm^2 16.4 cm

e 52 cm^2 23.6 cm

f 74.6 cm^2 5 cm

g 2.3 cm 2.6 cm 4.2 cm

h 13.8 cm ← 12.9 cm → 8.6 cm

i 13.9 cm 8.2 cm

Example 26

Calculate the volume of the solid shown (to 3 significant figures).

a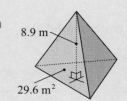
8.9 m
29.6 m²

b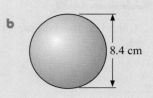
8.4 cm

a The solid tapers to a point

$\therefore$ $V = \frac{1}{3}$ area of base × height

$= \frac{1}{3} \times 29.6 \times 8.9$ m³

$\doteqdot 87.8$ m³

b The sphere has $r = 4.2$ cm

Now $V = \frac{4}{3}\pi r^3$

$\therefore$ $V = \frac{4}{3} \times \pi \times (4.2)^3$

$\doteqdot 310$ cm³

2 Calculate, to 3 significant figures, the volume of:

a
24.9 cm
142.3 cm²

b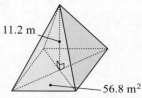
11.2 m
56.8 m²

c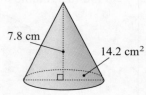
7.8 cm
14.2 cm²

d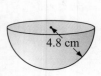
4.8 cm

e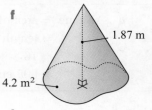
3.7 cm

f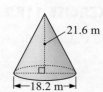
1.87 m
4.2 m²

g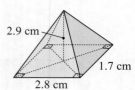
2.9 cm
1.7 cm
2.8 cm

h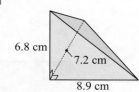
6.8 cm
7.2 cm
8.9 cm

i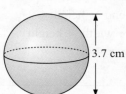
21.6 m
18.2 m

3 Find a formula for the volume, V, of the following illustrated solids:

a
3x
x
2x

b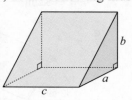
b
a
c

c
4r
2r

d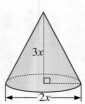
3x
2x

e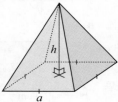
h
a

f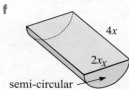
4x
2x
x
semi-circular

Example 27

A box has a square base and its height is 12 cm. If the volume of the box is 867 cm^3, find its length.

Let the length be x cm.

volume $=$ length $\times$ depth $\times$ height

$\therefore \quad 867 = x \times x \times 12$

$\therefore \quad 12x^2 = 867$

$\therefore \quad x^2 = \frac{867}{12}$

$\therefore \quad x = \sqrt{\frac{867}{12}} \quad \{x > 0\}$

$\therefore \quad x \doteq 8.5$

i.e.,   the length of the base is 8.5 cm.

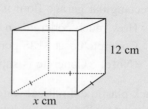

4 **a** Find the length of the side of a cube if its volume is 2.52 cm^3.

    b Find the height of a cylinder with base area 23.8 cm^2 and volume 142.8 cm^3.

5 A concrete path 1 m wide and 10 cm deep is placed around a circular lawn of diameter 20 m.

    a Draw a plan view of the situation.

    b Find the surface area of the concrete.

    c Find the volume of concrete required to lay the path.

6 A swimming pool has dimensions as shown alongside.

    a Find the area of a trapezium-shaped side.

    b Determine the volume of water required to fill the pool.

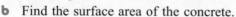

7

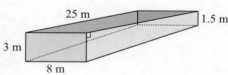
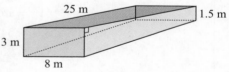

I decided to go camping with friends. My tent has the dimensions as shown. One of my friends is very fussy about health and insists that each person in the tent should have 3 m^3 of space.

Using my friend's condition, how many people could occupy the tent?

> Remember that percentage increase
> $= \dfrac{\text{increase}}{\text{original}} \times 100\%$

8 A circular cake tin has a radius of 20 cm and a height of 7 cm. When cake mix was added to the tin its height was 2 cm. After the cake was cooked it rose to 1.5 cm below the top of the tin.

    a Sketch these two situations.

    b Find the volume of the cake mix.

    c Find the volume of the cooked cake.

    d What was the percentage increase in the volume of the cake when it cooked?

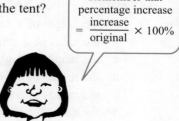

9 The Water Supply department uses huge concrete pipes to drain storm water away.

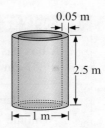

0.05 m

2.5 m

1 m

   a What is the outside radius of a pipe?

   b What is the inside radius of a pipe?

   c Find the volume of concrete necessary to make one pipe.

   (**Hint:** Consider vol. large cylinder − vol. small cylinder.)

10 A rectangular garage floor is 9.2 m by 6.5 m is to be concreted to a depth of 120 mm.

   a How much concrete, in m³, would cover the floor?

   b Concrete costs \$135 per m³. How much would it cost to concrete the floor?

   (**Note:**   Concrete is supplied in multiples of 0.2 m³.)

11 Black plastic cylindrical water pipe is to have an internal diameter of 13 mm and walls of thickness 2 mm. 1000 km of piping is required for a watering project. The piping is made from bulk plastic which weighs 2.3 tonnes per cubic metre. How many tonnes of black plastic are required for the project?

Example 28

Find, to 3 significant figures, the value of r if a cylinder has base radius r cm, height 3.9 cm and has a volume of 54.03 cm³.

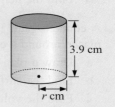

3.9 cm

r cm

$$V = \text{area of cross-section} \times \text{length}$$
$$\therefore \quad \pi \times r^2 \times 3.9 = 54.03$$
$$\therefore \quad r^2 = \frac{54.03}{\pi \times 3.9} \qquad \{\text{dividing both sides by } \pi \times 3.9\}$$
$$\therefore \quad r = \sqrt{\frac{54.03}{\pi \times 3.9}}$$
$$\therefore \quad r \doteqdot 2.10$$

---

12 Find:

   a the height of a rectangular prism of base 5 cm by 3 cm and volume 40 cm³

   b the sides of a cube of butter of volume 34.01 cm³

   c the height of a glass cone of base diameter 24.6 cm and volume 706 cm³

   d the radius of a weather balloon of volume 73.62 m³

   e the radius of a steel cylinder of height 4.6 cm and volume 43.75 cm³

   f the radius of the base of a conical bin of height 6.2 m and volume 203.9 m³

G CAPACITY

The **capacity** of a container is the quantity of fluid it is capable of holding.

Reminder:
- **1 millilitre (mL)** of fluid fills a container of size:

- **1 litre (L)** of fluid fills a container of size 10 cm by 10 cm by 10 cm, i.e., 1 L = 1000 mL

- We say, 1 mL ≡ 1 cm³ and 1000 L = 1 kL ≡ 1 m³.

Below is a **conversion table for capacities**

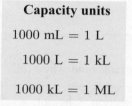

Capacity units
1000 mL = 1 L
1000 L = 1 kL
1000 kL = 1 ML

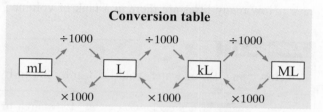

Conversion table

Example 29

Convert: **a** 4.2 L to mL   **b** 36 800 L to kL

a 4.2 L = (4.2 × 1000) mL
= 4200 mL

b 36 800 L = (36 800 ÷ 1000) kL
= 36.8 kL

EXERCISE 11G

1 Convert:
- **a** 3.76 L into mL
- **b** 47 320 L into kL
- **c** 3.5 kL into L
- **d** 0.423 L into mL
- **e** 0.054 kL into mL
- **f** 58 340 mL into kL

2
- **a** A chemist makes up 20 mL bottles of eye drops from a drum of eye drop solution. If the full drum has internal capacity of 0.0275 kL, how many bottles of eye drop solution can the chemist fill?
- **b** 1000 dozen bottles of wine each of capacity 750 mL need to be filled from tanks of capacity 1000 L for export. How many tanks are needed?

VOLUME-CAPACITY CONNECTION

Alongside is a table which shows the connection between volume and capacity units. Remember that the **capacity** of a container is the amount of fluid it can contain.

Volume		Capacity
1 cm³	≡	1 mL
1000 cm³	≡	1 L
1 m³	≡	1 kL
1 m³	≡	1000 L

Example 30

Convert: **a** 9.6 L to cm^3 **b** 3240 L to m^3

a   9.6 L
= (9.6 × 1000) cm^3
= 9600 cm^3

b   3240 L
= (3240 ÷ 1000) m^3
= 3.24 m^3

---

3 Convert:

a 83 kL to m^3 **b** 3200 mL to cm^3 **c** 2300 cm^3 to L

d 7154 m^3 to L   **e** 0.46 kL to m^3 **f** 4.6 kL to cm^3

4 **a** What is the capacity (in mL) of a bottle of volume 25 cm^3?

b Find the volume of a tank (in m^3) if its capacity is 32 kL.

c How many litres are there in a tank of volume 7.32 m^3?

Example 31

How many kL of water would a 3 m by 2.4 m by 1.8 m tank hold when full?

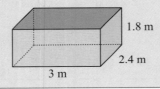

$$V = \text{area of cross-section} \times \text{height}$$
$$= (3 \times 2.4) \times 1.8 \text{ m}^3$$
$$= 12.96 \text{ m}^3$$
$$\therefore \quad \text{capacity is 12.96 kL.}$$

---

5 Find the capacity (in kL) of the following tanks (to 3 significant figures):

a

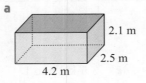

b

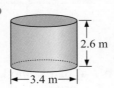

c

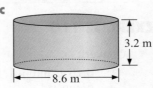

6 A new perfume in a 36 mm (internal) diameter spherical bottle comes onto the market.

a If the bottle is full, calculate its capacity in mL.

b If the bottle costs $25 and the bottle and its contents cost $105, what is the cost of the perfume per bottle?

c How much does the perfume cost per mL?

Example 32

17.3 mm of rain falls on a flat rectangular shed roof of length 10 m and width 6.5 m. All of the water goes into a cylindrical tank of base diameter 4 m. By how much does the water level in the tank rise in mm?

For the roof: Area is in m^2 $\therefore$ we convert 17.3 mm to metres.

$$17.3 \text{ mm} = (17.3 \div 1000) \text{ m} = 0.0173 \text{ m}$$

$\therefore$ volume of water collected by roof $=$ area of roof $\times$ depth

$$= 65 \text{ m}^2 \times 0.0173 \text{ m}$$
$$= 1.1245 \text{ m}^3$$

For the tank: Area of base $= \pi r^2$

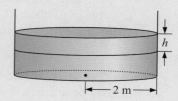

$$= \pi \times 2^2 \text{ m}^2$$
$$\doteqdot 12.566 \text{ m}^2$$

$\therefore$ increased volume $=$ area of base $\times$ height

$$= 12.566 \times h \text{ m}^2$$

So, $12.566 \times h = 1.1245$ {volume in tank = volume from roof}

$\therefore$ $h = \dfrac{1.1245}{12.566}$ {dividing both sides by 12.566}

$\therefore$ $h \doteqdot 0.0895 \text{ m}$ which is close to 9 cm.

---

7 A rectangular tank is 3 m by 2.4 m by 2 m and is full of water. The water is pumped into an empty cylindrical tank of base radius 2 m. How high up the tank will the water level rise?

8 12 mm of rain falls on the roof. All the water goes into a tank of base diameter 4 m.

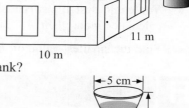

  a Find the volume of water which fell on the roof.

  b How many kL of water enter the tank?

  c By how much will the water level rise in the tank?

9 A conical wine glass has dimensions as shown.

  a How many mL does the glass hold if it is 75% full?

  b If the wine is poured into a cylinder of the same diameter, how high will it rise?

10 The design department of a fish packing company wants to change the size of their cylindrical tins. The original tin is 15 cm high and 7.2 cm in diameter. The new tin is to have approximately the same volume. If the new tin is 10 cm in diameter, how high must it be? Give your answer to the nearest mm.

11 A fleet of trucks all have containers with the shape as illustrated. Wheat is to be transported in these containers, and its level must not exceed a mark 10 cm below the top. How many truck-loads of wheat will be necessary to fill a cylindrical silo of internal diameter 8 m and height 25 m?

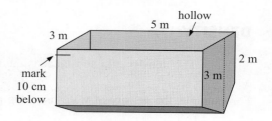

INVESTIGATION 2   MINIMISING MATERIAL

Consider the following problem.

Your boss, Pat, asks you to design a rectangular box-shaped container which is open at the top and is to contain exactly 1 litre of fluid. Pat intends manufacturing millions of these containers and wishes to keep manufacturing costs to a minimum. Consequently, Pat insists that the least amount of material is used. One other condition is conveyed to you: 'The base measurements must be in the ratio 2 : 1'.

SPREADSHEET

Below is one method which could be used to solve the problem.

- The base is to be in the ratio 2 : 1, therefore let dimensions be x cm and $2x$ cm. The height is also unknown, thus, let it be y cm. x and y are called **variables**. As the values of x and y vary, the container changes size. Draw a sketch:

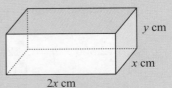

- What exactly are we trying to do? We are attempting to minimise the amount of material used, i.e., make the *surface area as small as possible.*

- Explain why the volume $V = 2x^2y$.

- Explain why $2x^2y = 1000$.

  Explain why $y = \dfrac{500}{x^2}$.

	A	B	C
1	x values	y values	A values
2	1	=500/A2^2	=2*A2^2+6*A2*B2
3	=A2+1	↓	↓
4	↓	fill down	

- Show that the total surface area is $A = 2x^2 + 6xy$.

- You could design a **spreadsheet**, which for given values of x (= 1, 2, 3, 4, ...), calculates y and then A *or*   use a **graphics calculator** to graph A against x.

- Find the smallest value of A and the value of x which produces it.

H   DENSITY

Imagine two containers with volume 1 m^3. Which container would weigh more, if one container was filled with lead and the other container was filled with feathers? Obviously the container of lead would have greater **mass** as it is more compact than the feathers. We say that lead has a higher **density** than feathers.

MASS

The **mass** of an object is the amount of matter it contains.

One gram is the mass of one cubic centimetre of pure water at 4°C.

DENSITY

The **density** of a substance is the mass per unit volume of the substance.

For example, the density of water is   1 g per 1 cm^3 or $\dfrac{1 \text{ g}}{1 \text{ cm}^3}$ or   1 g/cm^3

In general,
$$\text{density} = \frac{\text{mass}}{\text{volume}}.$$

Table of common densities:

Substance	Density (g/cm³)
pine	0.65
paper	0.93
water	1.00
steel	7.82
lead	11.37
uranium	18.97

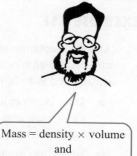

Mass = density × volume
and
Volume = mass ÷ density.

Example 33

Find the density of a metal with mass 25 g and volume 4 cm³.

$$\text{Density} = \frac{\text{mass}}{\text{volume}}$$

$$\therefore \quad \text{density} = \frac{25 \text{ g}}{4 \text{ cm}^3}$$

$$\therefore \quad \text{density} = 6.25 \text{ g/cm}^3$$

EXERCISE 11H

1 Find the density of:
   a a metal with mass 10 g and volume 2 cm³
   b a substance with mass 2 g and volume 1.5 cm³.

2 Find the mass of a 5 cm diameter sphere of a metal which has a density of 8.7 g/cm³.

3 What volume is occupied by a lump of metal with a mass of 3.62 kg and having a density of 11.6 g/cm³?

4 Find the density of:
   a a cube of chocolate with sides 2 cm, and mass of 18 grams
   b a 15 cm by 8 cm by 4 cm rectangular block of metal with mass of 9.6 kg.

5 A gold ingot with dimensions shown has a mass of 12.36 kg. Determine the density of gold.

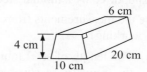

4 cm   6 cm   20 cm   10 cm

6 Determine the depth of a rectangular block of metal with base 7 cm by 5 cm if the block has total mass 1285 g and the metal has density 7.5 g/cm³.

7 If polystyrene spherical beads have an average diameter of 0.6 cm and the density of polystyrene is 0.15 g/cm³, estimate the total number of beads present in a box containing beads with a total mass of 8 kg.

8 Determine the total mass of stone required to build a pyramid which has a square base and all edges have length 200 m. (The density of the stone is 2.25 tonnes per m³.)

HARDER APPLICATIONS

EXERCISE 11I

1 A raised rectangular garden bed is to be built using old bricks. The bricks measure 225 mm × 105 mm × 75 mm and the mortar between them should be 10 mm thick. The garden bed is to have outside dimensions of approximately 2.6 m × 1 m × 0.65 m high.

   a How many bricks are required for the first course? How many courses are required? (You may assume no bricks will be cut. Show all calculations and a sketch may be helpful.)

   b Show the *actual* measurements of the garden bed, inside and outside, on a sketch diagram.

   c How many bricks should be ordered for the project? Explain.

   d What volume of soil will be needed to fill the bed? Explain what volume of soil you would probably order and why.

   e Approximately how much mortar (in litres) would be required for the construction?

2 A feed silo (made from a hemisphere, a cylinder and a cone as shown) is constructed out of sheet steel 3 mm thick.

   a Show why the depth (height) of the cone must be 70 cm and hence find the **slant height** of the conical section.

   b Calculate the surface area of each of the three sections and thus show that the total amount of steel used is about 15.7 square metres.

   c Show that the silo would hold about 5.2 cubic metres of grain when completely full.

   d If grain has a density of 0.68 grams per cubic centimetre and the density of steel is 8 grams per cubic centimetre find the weight of the silo when it is full.

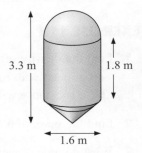

3 When work has to be done on under ground telephone cables, workers often use a small tent over the opening. This shelters them from bad weather and stops passers-by from absentmindedly falling in on top of them. The cover is constructed from canvas and this is supported by a tubular steel frame, as shown below.

Use the measurements from the diagram of the frame to calculate the following (showing all working):

   a The length of steel tubing needed to make the frame (in reasonable units).

   b The amount of room inside the tent (volume) in cubic metres.

   c The amount of canvas (in m²) needed for the cover.

   d How many people do you think could comfortably work in this tent? If there was a sudden downpour of rain, what do you think would be the maximum number of workers who could shelter in the tent? Give reasons for both of your answers.

4 A ready mixed concrete tanker is to be constructed in steel in the form of a cylinder with conical ends as shown alongside.

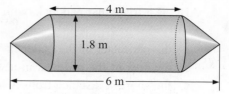

Show all your working in the following questions.

a Calculate the volume held in the tanker by working out the volume of the cylinder and the conical ends.

b How *long* would the tanker be if the ends were hemispheres instead of cones (but the cylindrical section remained the same)? Explain your reasoning.

c How much more or less concrete would fit in the tanker if the ends were hemispheres instead of cones?

d The surface area of the tanker with the conical ends is about 30 m². The tanker with hemispherical ends has surface area of nearly 33 m². Verify one of these figures by your own calculations.

e Using the figures you have from **a** to **d**, discuss the advantages of each design. Overall, which is the better design for the tanker? Give reasons.

REVIEW SET 11A

1 Joanne decides to put a single electric wire along all fences of her cattle breeding property. Find:

a the total length of electric wire needed

b the total cost of the electric wire if it costs $0.42 per metre.

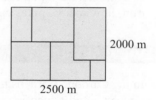

2 Find, correct to 2 decimal places, the circumference of a circle of diameter 16.4 m.

3 Find the area of:

a a rectangular farming property 3.4 km by 2.1 km

b a sector shaped garden bed of angle 120° and radius 5.64 m.

4 Find the volume of:

a

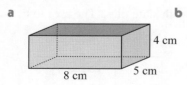

b

c

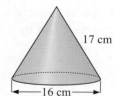

5 Find the surface area of the following solids:

a

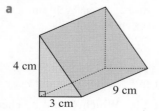

b

6 ABC is a triangular paddock, with its measurements as shown. Find:

    a the perimeter of the paddock

    b the area of the paddock in hectares

    c the shortest distance from A to the fence joining BC.   (**Hint:** use **b**)

7 Water is siphoned from a conical water tower of base diameter 3 m and height 4 m. If the cone was originally full, and the water is transferred to a rectangular tank of base measurements 10 m by 8 m, by how much will the water in the rectangular tank rise?

8 A spherical ball of metal has mass 6483 g and radius 6.2 cm.

    a Determine the density of the metal.

    b A cube of side length 5 cm is made of the same metal. What is its mass?

9 Cans used for canned soup have a base diameter 7 cm and height 10 cm.

    a How many such cans can be filled from a vat containing 2000 litres of soup?

    b Calculate the total surface area of metal required to make the cans in **a**.

10 A manufacturer of spikes has 0.245 kL of molten iron. If each spike contains 15 mL of iron, how many spikes can be made?

REVIEW SET 11B

1 In the Cardone Gardens there is a 4 m diameter circular pond which is surrounded by a 1.5 m wide brick path. A chain fence is placed around the path with posts every 2 m.

    a Find the length of the chains required for the fence. Allow 10% extra for the sag in the chain.

    b How many posts are needed?

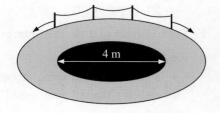

2 Louise wanted to have a new outdoor patio paved with slate tiles. The area to be covered was 4.8 m by 3.6 m. She had to decide between:

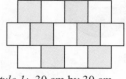

style 1: 30 cm by 30 cm costing $4.55 per tile

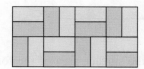

style 2: 40 cm by 20 cm costing $3.95 per tile

    a Calculate the area to be paved.

    b Find the area of each style of slate tile.

    c Find the number of each style of slate tile needed.

    d Compare the total cost of using each style of slate tile. Which would be cheaper?

3 An athletics track is to have a perimeter of 400 m and straights of 78 m.

Find the radius of the circular 'bend'.

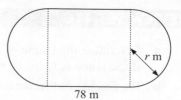

78 m

4 Find the formula for the area A, of:

a

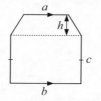

b

2x

5 A toolshed with the dimensions illustrated is to be painted with zinc-alum. Two coats are required with zinc-alum costing $8.25 per litre. Find:

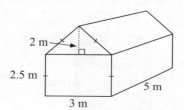

a the area to be painted (including the roof). Reduce your answer by 5% to allow for windows.

b The total cost if the zinc-alum covers 5 m² per litre and must be bought in whole litres.

6 Calculate, correct to 3 significant figures, the volume of:

a

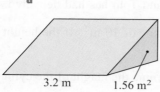

3.2 m   1.56 m²

b

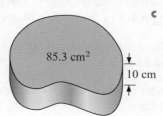

85.3 cm²   10 cm

c

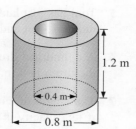

1.2 m   0.4 m   0.8 m

7 500 dozen bottles of wine, each of capacity 750 mL are to be filled from tanks of capacity 1000 L. How many tanks are needed?

8 15.4 mm of rain falls on a rectangular shed roof of length 12 m and width 5.5 m. All of the water goes into a cylindrical tank of base diameter 4.35 m. By how much does the water level in the tank rise in mm?

9 **a** Determine the density of a metal for which a 12 cm diameter sphere has a mass of 4.8 kilograms.

b Calculate the mass of an ingot of gold 12 cm by 4 cm by 3 cm if gold has a density of 10.57 g/cm³.

10 **a** A sphere has surface area 100 cm². Find its
   i radius   **ii** volume.

b A sphere has volume 27 m³. Find its
   i radius   **ii** surface area.

REVIEW SET 11C

1 a Determine the length of fencing around a circular playing field of radius 80 metres if the fence is to be 10 metres from the edge of the field.

   b If a fencing contractor quotes $12.25 per metre of fence, what will his total quotation be?

2 A sector of a circle has radius 12 cm and angle $135°$.

   a What fraction of a whole circle is this sector?

   b Find the perimeter of the sector.

   c Find the area of the sector.

3 Find the outer surface area (to 3 significant figures) of:

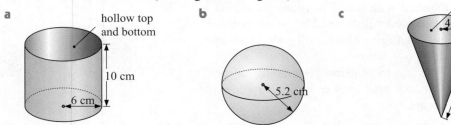

4 A plastic beach ball has a radius of 27 cm. Find its volume.

5 Tom has just had a load of sand delivered. The pile of sand is in the shape of a cone of radius 1.6 m and height 1.2 m. Find the volume of the sand Tom has had delivered.

6 A cylindrical drum for storing industrial waste has a volume of 10 m^3. If the height of the drum is 3 m, find its radius.

7 The height of a cylinder is three times its diameter. Determine the height of the cylinder if its capacity is 100 litres.

8 A cubic metre of brass is melted down and cast into solid door handles with shape as shown.

   a Find the volume of one door handle.

   b How many handles can be cast?

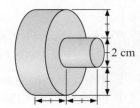

9 A concrete tank has an external diameter of 8 m and an internal height of 5 m. The walls and base of the tank are 20 cm thick.

   a Find the volume of concrete in the base.

   b Find the volume of concrete in the walls.

   c Find the total volume of concrete required.

   d Find the cost of the concrete at $135 per m^3.

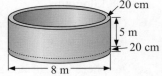

10 Determine the total mass of steel required to construct 450 m of steel piping with internal diameter 1.46 m and external diameter 1.50 m. The density of steel used is 2.36 tonnes per cubic metre.

Chapter 12

Sequences and series

Contents:

A   NUMBER PATTERNS

An important skill in mathematics is to be able to
- **recognise** patterns in sets of numbers,
- **describe** the patterns in words, and
- **continue** the patterns.

A list of numbers where there is a pattern is called a **number sequence**.
The members (numbers) of a sequence are said to be its **terms**.

For example,   3, 7, 11, 15, ... form a number sequence.

The first term is 3,   the second term is 7,   the third term is 11, etc.

We describe this pattern in words:

"The sequence starts at 3 and each term is 4 more than the previous one."

Thus, the fifth term is 19, and the sixth term is 23, etc.

Example 1

Describe the sequence: 14, 17, 20, 23, ... and write down the next two terms.

The sequence starts at 14 and each term is 3 more than the previous term.
The next two terms are 26 and 29.

EXERCISE 12A

1 Write down the first four terms of the sequences described by the following:
   a Start with 4 and add 9 each time.
   b Start with 45 and subtract 6 each time.
   c Start with 2 and multiply by 3 each time.
   d Start with 96 and divide by 2 each time.

2 For each of the following write down a description of the sequence and find the next two terms:

   a 8, 16, 24, 32, ...   **b** 2, 5, 8, 11, ...   **c** 36, 31, 26, 21, ...
   d 96, 89, 82, 75, ...   **e** 1, 4, 16, 64, ...   **f** 2, 6, 18, 54, ...
   g 480, 240, 120, 60, ... **h** 243, 81, 27, 9, ...   **i** 50000, 10000, 2000, 400, ...

3 Find the next two terms of:
   a 95, 91, 87, 83, ...   **b** 5, 20, 80, 320, ...   **c** 45, 54, 63, 72, ...

4 Describe the following number patterns and write down the next three terms:
   a 1, 4, 9, 16, ...   **b** 1, 8, 27, 64, ...   **c** 2, 6, 12, 20, ...
   [**Hint:** In **c** $2 = 1 \times 2$ and $6 = 2 \times 3$.]

5 Find the next two terms of:
   a 1, 16, 81, 256, ...   **b** 1, 1, 2, 3, 5, 8, ... **c** 6, 8, 7, 9, 8, 10, ...
   d 2, 3, 5, 7, 11, ...   **e** 2, 4, 7, 11, ...   **f** 3, 4, 6, 8, 12, ...

SPREADSHEET   "NUMBER PATTERNS"

A spreadsheet is a computer software program that enables you to do calculations, write messages, draw graphs and do "what if" calculations.

This exercise will get you using a spreadsheet to construct and investigate number patterns.

What to do:

1 To form a number pattern with a spreadsheet like "start with 7 and add 4 each time" follow the given steps.

	A	B	C
1			
2			
3			
4			
5			

A spreadsheet consists of a series of rectangles called **cells** and each cell has a position according to the **column** and **row** it is in. Cell **B2** is shaded.

All formulae start with =

Step 1: Open a new spreadsheet.

Step 2: In cell A1,
type the **label** 'Value'

Step 3: In cell A2,
type the **number** 7

Step 4: In cell A3,
type the **formula** $=A2 + 4$

	A	B
1	Value	
2	7	
3	=A2+4	

Step 5: Press **ENTER**. Your spreadsheet should look like this:

	A	B
1	Value	
2	7	
3	11	

Step 6: Highlight cell A3 and position your cursor on the right hand bottom corner until it changes to a **+** . Click the left mouse key and "drag" the cursor down to Row 10. This is called **filling down**.

Filling down copies the formula from A3 to A4 and so on.

	A	B
1	Value	
2	7	
3	11	
4		

Step 7: You should have generated the first nine members of the number sequence as shown:

	A	B
1	Value	
2	7	
3	11	
4	15	
5	19	
6	23	
7	27	
8	31	
9	35	
10	39	

Step 8:   Use the **fill down** process to answer the following questions:

   a   What is the first member of the sequence greater than 100?

   b   Is 409 a member of the sequence?

2   Now that you are familiar with a basic spreadsheet, try to generate the first 20 members of the following number patterns:

   a   Start with 132 and subtract 6 each time. Type:

   Value in B1,   132 in B2,   =B2 − 6 in B3,   then fill down to Row 21.

   b   Start with 3 and multiply by 2 each time. Type:

   Value in C1,   3 in C2,   =C2*2 in C3,   then fill down to Row 21.

   c   Start with 1 000 000 and divide by 5 each time. Type:

   Value in D1,   1 000 000 in D2,   =D2/5 in D3,   then fill down to Row 21.

3   Click on the appropriate icon to learn how to use a **graphics calculator** to generate **sequences** of numbers such as those above.

B   SEQUENCES OF NUMBERS

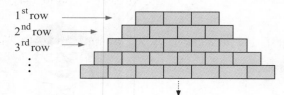

1st row
2nd row
3rd row
⋮

Consider the illustrated tower of bricks. The top row, or first row, has three bricks. The second row has four bricks. The third row has five, etc.

If u_n represents the number of bricks in row n (from the top) then

   $u_1 = 3, \quad u_2 = 4, \quad u_3 = 5, \quad u_4 = 6, \quad ......$

The number pattern: 3, 4, 5, 6, ... is called a **sequence** of numbers.

This sequence can be specified by:

- **Using words**   The top row has three bricks and each successive row under it has one more brick.

- **Using an explicit formula** $u_n = n + 2$ is the **general term** (or **nth term**) formula for $n = 1, 2, 3, 4, 5, ...$ etc.

   Check: $u_1 = 1 + 2 = 3$ ✓
   $u_2 = 2 + 2 = 4$ ✓
   $u_3 = 3 + 2 = 5$ ✓ etc.

Early members of a sequence can be graphed. Each term is represented by a dot.

The dots *must not* be joined.

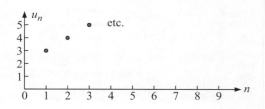

ARITHMETIC SEQUENCES

Number patterns like the one above where we **add** (or **subtract**) the same fixed number to get the next number are called **arithmetic sequences**.

Further examples where arithmetic sequence models apply are:

- Simple interest accumulated amounts at the end of each period.

  For example: on a \$1000 investment at 7% simple interest p.a. (per annum) the value of the investment at the end of successive years is:

  \$1000, \$1070, \$1140, \$1210, \$1280, ...

- The amount still owed to a friend when repaying a personal loan with fixed weekly repayments.

  For example: if repaying \$75 each week to repay a \$1000 personal loan the amounts still owing are: \$1000, \$925, \$850, \$775, ...

GEOMETRIC SEQUENCES

Instead of adding (or subtracting) a fixed number to get the next number in a sequence we sometimes **multiply** (or **divide**) by a fixed number.

When we do this we create **geometric sequences**.

Consider investing \$6000 at a fixed rate of 7% p.a. compound interest over a lengthy period. The initial investment of \$6000 is called the principal.

After 1 year, its value is	\$6000 × 1.07	{to increase by 7% we multiply to 107%}
After 2 years, its value is	($6000 × 1.07) × 1.07	
	$= \$6000 \times (1.07)^2$	
After 3 years, its value is	$\$6000 \times (1.07)^3$, etc.	

The amounts \$6000, \$6000 × 1.07, $\$6000 \times (1.07)^2$, $\$6000 \times (1.07)^3$,   etc.   form a geometric sequence where each term is multiplied by 1.07 which is called the **common ratio**.

Once again we can specify the sequence by:

- **Using words**   The initial value is \$6000 and after each successive year the increase is 7%.

- **Using an explicit formula** $u_n = 6000 \times (1.07)^{n-1}$ for $n = 1, 2, 3, 4, ...$

  Check: $u_1 = 6000 \times (1.07)^0 = 6000$ ✓

  $u_2 = 6000 \times (1.07)^1$ ✓

  $u_3 = 6000 \times (1.07)^2$ ✓   etc.

Notice that u_n is the amount after $n - 1$ years.

Other examples where geometric models occur are:

- Problems involving depreciation.

  For example: The value of a \$12 000 photocopier may decrease by 20% p.a.

  i.e.,   \$12 000,   \$12 000 × 0.8, $\$12\,000 \times (0.8)^2$, ... etc.

- In fractals, as we shall see later in the chapter on page **417**.

OPENING PROBLEM

A circular stadium consists of sections as illustrated. The diagram shows the tiers of concrete steps for the final section, **Section K**. Each seat is to be 0.45 m wide. AB, the arc for the front of the first row is 14.4 m long.

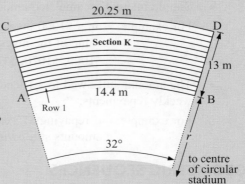

For you to consider:

1 How wide is each concrete step?

2 What is the length of the arc of the back of Row 1, Row 2 , Row 3, etc?

3 How many seats are there in Row 1, Row 2, Row 3, ... Row 13?

4 How many sections are there in the stadium?

5 What is the total seating capacity of the stadium?

6 What is the radius of the 'playing surface'?

To solve problems like the **Opening Problem** and many others, a detailed study of **sequences** and their sums (called **series**) is required.

NUMBER SEQUENCES

A **number sequence** is a set of numbers defined by a rule for positive integers.

Sequences may be defined in one of the following ways:

- by using a formula which represents the **general term** (called the nth **term**)
- by giving a description in words
- by listing the first few terms and assuming that the pattern represented continues indefinitely.

THE GENERAL TERM

u_n, T_n, t_n, A_n, etc. can all be used to represent the **general term** (or nth **term**) of a sequence and are defined for $n = 1, 2, 3, 4, 5, 6, ...$

$\{u_n\}$ represents the sequence that can be generated by using u_n as the nth **term**.

For example, $\{2n + 1\}$ generates the sequence $3, 5, 7, 9, 11, ...$

EXERCISE 12B

1 List the first *five* terms of the sequence:

a $\{2n\}$ **b** $\{2n + 2\}$ **c** $\{2n - 1\}$

d $\{2n - 3\}$ **e** $\{2n + 3\}$ **f** $\{2n + 11\}$

g $\{3n + 1\}$ **h** $\{4n - 3\}$ **i** $\{5n + 4\}$

2 List the first *five* terms of the sequence:

    a $\{2^n\}$ **b** $\{3 \times 2^n\}$ **c** $\{6 \times (\frac{1}{2})^n\}$ **d** $\{(-2)^n\}$

3 List the first *five* terms of the sequence $\{15 - (-2)^n\}$.

C   ARITHMETIC SEQUENCES

An **arithmetic sequence** is a sequence in which each term differs from the previous one by the same fixed number.

For example: 2, 5, 8, 11, 14, ... is arithmetic as $5 - 2 = 8 - 5 = 11 - 8 = 14 - 11$, etc.

Likewise,   31, 27, 23, 19, ... is arithmetic as $27 - 31 = 23 - 27 = 19 - 23$, etc.

ALGEBRAIC DEFINITION

> $\{u_n\}$ is **arithmetic** $\iff u_{n+1} - u_n = d$ for all positive integers n where d is a constant (the **common difference**).

Note:
- $\iff$ is read as 'if and only if'
- If $\{u_n\}$ is arithmetic then $u_{n+1} - u_n$ is a constant *and* if $u_{n+1} - u_n$ is a constant then $\{u_n\}$ is arithmetic.

THE 'ARITHMETIC' NAME

If a, b and c are any consecutive terms of an arithmetic sequence then

$$b - a = c - b \quad \text{\{equating common differences\}}$$
$$\therefore \quad 2b = a + c$$
$$\therefore \quad b = \frac{a + c}{2}$$

i.e.,   middle term = arithmetic mean (average) of terms on each side of it.

      Hence the name *arithmetic sequence*.

THE GENERAL TERM FORMULA

Suppose the first term of an arithmetic sequence is u_1 and the common difference is d.

Then $u_2 = u_1 + d$ $\therefore$ $u_3 = u_1 + 2d$ $\therefore$ $u_4 = u_1 + 3d$ etc.

        then, $u_n = u_1 + \underline{(n - 1)}d$

                The coefficient of d is one less than the subscript.

So,   for an **arithmetic sequence** with **first term** u_1 and **common difference** d the **general term** (or nth **term**) is $u_n = u_1 + (n - 1)d$.

Example 2

Consider the sequence   2, 9, 16, 23, 30, ...

  a   Show that the sequence is arithmetic.

  b   Find the formula for the general term u_n.

  c   Find the 100th term of the sequence.

  d   Is **i** 828 **ii** 2341   a member of the sequence?

a $9 - 2 = 7$ So, assuming that the pattern continues,
       $16 - 9 = 7$ consecutive terms differ by 7
       $23 - 16 = 7$ $\therefore$ the sequence is arithmetic with $u_1 = 2$, $d = 7$.
       $30 - 23 = 7$

b $u_n = u_1 + (n-1)d$ $\therefore$ $u_n = 2 + 7(n-1)$ i.e., $u_n = 7n - 5$

c   If $n = 100$, $u_{100} = 7(100) - 5 = 695$.

d   **i**   Let $u_n = 828$ **ii**   Let $u_n = 2341$
       $\therefore$ $7n - 5 = 828$ $\therefore$ $7n - 5 = 2341$
       $\therefore$ $7n = 833$ $\therefore$ $7n = 2346$
       $\therefore$ $n = 119$ $\therefore$ $n = 335\frac{1}{7}$

   $\therefore$ 828 is a term of the sequence.   which is not possible as n is an
   In fact it is the 119th term.   integer. $\therefore$ 2341 cannot be a term.

EXERCISE 12C

1   Consider the sequence   6, 17, 28, 39, 50, ...

  a   Show that the sequence is arithmetic.   **b**   Find the formula for its general term.

  c   Find its 50th term.   **d**   Is 325 a member?

  e   Is 761 a member?

2   Consider the sequence 87, 83, 79, 75, ...

  a   Show that the sequence is arithmetic.   **b**   Find the formula for the general term.

  c   Find the 40th term.   **d**   Is -143 a member?

3   A sequence is defined by $u_n = 3n - 2$.

  a   Show that the sequence is arithmetic.   **(Hint:**   Find $u_{n+1} - u_n$.)

  b   Find u_1 and d.

  c   Find the 57th term.

  d   What is the least term of the sequence which is greater than 450?

4   A sequence is defined by $u_n = \dfrac{71 - 7n}{2}$.

  a   Show that the sequence is arithmetic.   **b**   Find u_1 and d.   **c**   Find u_{75}.

  d   For what values of n are the terms of the sequence less than -200?

Example 3

Find k given that $3k + 1$, k and -3 are consecutive terms of an arithmetic sequence.

Since the terms are consecutive,

$$k - (3k + 1) = -3 - k \qquad \text{\{equating common differences\}}$$
$$\therefore \quad k - 3k - 1 = -3 - k$$
$$\therefore \quad -2k - 1 = -3 - k$$
$$\therefore \quad -1 + 3 = -k + 2k$$
$$\therefore \quad 2 = k$$

or $\qquad k = \dfrac{(3k + 1) + (-3)}{2}$ {middle term is average of other two}

$$\therefore \quad k = \dfrac{3k - 2}{2} \qquad \text{which when solved gives} \quad k = 2.$$

5 Find k given the consecutive arithmetic terms:

    a $\quad 32, k, 3$ **b** $\quad k, 7, 10$ **c** $\quad k + 1, 2k + 1, 13$

    d $\quad k - 1, 2k + 3, 7 - k$ **e** $\quad k, k^2, k^2 + 6$ **f** $\quad 5, k, k^2 - 8$

Example 4

Find the general term u_n for an arithmetic sequence given that
$u_3 = 8$ and $u_8 = -17$.

$$u_3 = 8 \qquad \therefore \quad u_1 + 2d = 8 \quad ....(1) \qquad \{u_n = u_1 + (n-1)d\}$$
$$u_8 = -17 \quad \therefore \quad u_1 + 7d = -17 \quad ....(2)$$

We now solve (1) and (2) simultaneously

$$-u_1 - 2d = -8$$
$$\underline{\quad u_1 + 7d = -17 \quad}$$
$$\therefore \quad 5d = -25 \qquad \text{\{adding the equations\}}$$
$$\therefore \quad d = -5$$

So in (1) $\qquad u_1 + 2(-5) = 8$ *Check:*
$$\therefore \quad u_1 - 10 = 8 \qquad\qquad\qquad u_3 = 23 - 5(3)$$
$$\therefore \quad u_1 = 18 \qquad\qquad\qquad\quad = 23 - 15$$
$$\text{Now} \quad u_n = u_1 + (n-1)d \qquad\quad = 8 \;\checkmark$$
$$\therefore \quad u_n = 18 - 5(n-1) \qquad u_8 = 23 - 5(8)$$
$$\therefore \quad u_n = 18 - 5n + 5 \qquad\qquad = 23 - 40$$
$$\therefore \quad u_n = 23 - 5n \qquad\qquad\quad = -17 \;\checkmark$$

6 Find the general term u_n for an arithmetic sequence given that:

    a $\quad u_7 = 41$ and $u_{13} = 77$ **b** $\quad u_5 = -2$ and $u_{12} = -12\frac{1}{2}$

c   the seventh term is 1 and the fifteenth term is -39

d   the eleventh and eighth terms are -16 and $-11\frac{1}{2}$ respectively.

Example 5

Insert four numbers between 3 and 12 so that all six numbers are in arithmetic sequence.

If the numbers are $3,\ 3+d,\ 3+2d,\ 3+3d,\ 3+4d,\ 12$

$$\text{then} \quad 3+5d=12$$
$$\therefore \quad 5d=9$$
$$\therefore \quad d=\tfrac{9}{5}=1.8$$

So we have $3,\ 4.8,\ 6.6,\ 8.4,\ 10.2,\ 12.$

---

7   **a**   Insert three numbers between 5 and 10 so that all five numbers are in arithmetic sequence.

b   Insert six numbers between -1 and 32 so that all eight numbers are in arithmetic sequence.

8   Consider the finite arithmetic sequence $36,\ 35\frac{1}{3},\ 34\frac{2}{3},\ ....,\ -30.$

a   Find u_1 and d.   **b**   How many terms does the sequence have?

9   An arithmetic sequence starts $23,\ 36,\ 49,\ 62,\ .....$ What is the first term of the sequence to exceed $100\,000$?

D   GEOMETRIC SEQUENCES

A sequence is **geometric** if each term can be obtained from the previous one by multiplying by the same non-zero constant.

For example: $2,\ 10,\ 50,\ 250,\ ....$ is a geometric sequence as

$$2 \times 5 = 10 \quad \text{and} \quad 10 \times 5 = 50 \quad \text{and} \quad 50 \times 5 = 250.$$

Notice that $\frac{10}{2} = \frac{50}{10} = \frac{250}{50} = 5,$ i.e.,   each term divided by the previous one is constant.

Algebraic definition:

$\{u_n\}$ is **geometric** $\Leftrightarrow$ $\dfrac{u_{n+1}}{u_n} = r$ for all positive integers n

where r is a **constant** (the **common ratio**).

Notice: • $2,\ 10,\ 50,\ 250,\ ....$ is geometric with $r=5$.

• $2,\ -10,\ 50,\ -250,\ ....$ is geometric with $r=-5$.

THE 'GEOMETRIC' NAME

If a, b and c are any consecutive terms of a geometric sequence then

$$\frac{b}{a} = \frac{c}{b} \quad \text{\{equating common ratios\}}$$

$\therefore \quad b^2 = ac \quad$ and so $\quad b = \pm\sqrt{ac} \quad$ where $\sqrt{ac}$ is the **geometric mean** of a and c.

THE GENERAL TERM

Suppose the first term of a geometric sequence is u_1 and the common ratio is r.

Then $\quad u_2 = u_1 r \qquad \therefore \quad u_3 = u_1 r^2 \qquad \therefore \quad u_4 = u_1 r^3 \quad$ etc.

$$\text{then} \quad u_n = u_1 r^{n-1}$$

The power of r is one less than the subscript.

So,

> for a **geometric sequence** with **first term u_1** and **common ratio r**,
> the **general term** (or nth **term**) is $\quad u_n = u_1 r^{n-1}$.

Example 6

For the sequence $\quad 8$, 4, 2, 1, $\frac{1}{2}$, ...

a Show that the sequence is geometric.   **b** Find the general term u_n.

c Hence, find the 12th term as a fraction.

---

a
$$\frac{4}{8} = \tfrac{1}{2}$$

So, assuming the pattern continues, consecutive terms have a common ratio of $\frac{1}{2}$.

$$\frac{2}{4} = \tfrac{1}{2}$$

$\therefore \quad$ the sequence is geometric with $\quad u_1 = 8 \quad$ and $r = \frac{1}{2}$.

$$\frac{1}{2} = \tfrac{1}{2}$$

$$\frac{\frac{1}{2}}{1} = \tfrac{1}{2}$$

b $u_n = u_1 r^{n-1} \qquad \therefore \quad u_n = 8\left(\tfrac{1}{2}\right)^{n-1} \quad$ or $\quad u_n = 2^3 \times (2^{-1})^{n-1}$
$$= 2^3 \times 2^{-n+1}$$
$$= 2^{3+(-n+1)}$$
$$= 2^{4-n}$$

c $u_{12} = 8 \times \left(\tfrac{1}{2}\right)^{11}$ (See chapter **3** for exponent simplification.)

$$= \frac{8}{2^{11}}$$

$$= \tfrac{1}{256}$$

EXERCISE 12D

1 For the geometric sequence with first two terms given, find b and c:

    a   2, 6, b, c, ...   **b**   10, 5, b, c, ...   **c**   12, -6, b, c, ...

2 **a** Show that the sequence   5, 10, 20, 40, ...   is geometric.

    b Find u_n and hence find the 15th term.

3 **a** Show that the sequence   12, -6, 3, -1.5, ...   is geometric.

    b Find u_n and hence find the 13th term (as a fraction).

4 Show that the sequence   8, -6, 4.5, -3.375, ...   is geometric and hence find the 10th term as a decimal.

5 Show that the sequence $8, 4\sqrt{2}, 4, 2\sqrt{2}, ...$ is geometric and hence find, in simplest form, the general term u_n.

Example 7

$k - 1$, $2k$ and $21 - k$ are consecutive terms of a geometric sequence. Find k.

Since the terms are geometric, $\dfrac{2k}{k-1} = \dfrac{21-k}{2k}$ {equating r's}

$$\therefore \quad 4k^2 = (21-k)(k-1)$$
$$\therefore \quad 4k^2 = 21k - 21 - k^2 + k$$
$$\therefore \quad 5k^2 - 22k + 21 = 0$$
$$\therefore \quad (5k-7)(k-3) = 0$$
$$\therefore \quad k = \tfrac{7}{5} \ \text{ or } \ 3$$

Check: If $k = \tfrac{7}{5}$ terms are: $\tfrac{2}{5}, \tfrac{14}{5}, \tfrac{98}{5}$.   ✓   {$r = 7$}

            If $k = 3$ terms are: 2, 6, 18.   ✓   {$r = 3$}

---

6 Find k given that the following are consecutive terms of a geometric sequence:

    a   7, k, 28   **b** k, $3k$, $20 - k$ **c** k, $k + 8$, $9k$

Example 8

A geometric sequence has $u_2 = -6$ and $u_5 = 162$.   Find its general term.

$$u_2 = u_1 r = -6 \quad \ \ \dots (1) \quad \{\text{using } u_n = u_1 r^{n-1} \text{ with } n = 2\}$$
$$\text{and} \quad u_5 = u_1 r^4 = 162 \quad \dots (2)$$

$$\text{So,} \quad \frac{u_1 r^4}{u_1 r} = \frac{162}{-6} \quad \{(2) \div (1)\}$$

$$\therefore \quad r^3 = -27$$
$$\therefore \quad r = \sqrt[3]{-27}$$
$$\therefore \quad r = -3$$

and so in (1) $u_1(-3) = -6$
$$\therefore \quad u_1 = 2$$

Thus $u_n = 2 \times (-3)^{n-1}$.

Note: $(-3)^{n-1} \neq -3^{n-1}$
as we do not know the value of n.
If n is odd, then $(-3)^{n-1} = 3^{n-1}$
If n is even, then $(-3)^{n-1} = -3^{n-1}$

7 Find the general term u_n, of the geometric sequence which has:

a $u_4 = 24$ and $u_7 = 192$ **b** $u_3 = 8$ and $u_6 = -1$

c $u_7 = 24$ and $u_{15} = 384$ **d** $u_3 = 5$ and $u_7 = \frac{5}{4}$

Example 9

Find the first term of the geometric sequence $6, 6\sqrt{2}, 12, 12\sqrt{2}, ...$
which exceeds 1400.

First we find u_n :
Now $u_1 = 6$ and $r = \sqrt{2}$
so as $u_n = u_1 r^{n-1}$ then $u_n = 6 \times (\sqrt{2})^{n-1}$.

Next we need to find n such that $u_n > 1400$.

Using a graphics calculator with $Y_1 = 6 \times (\sqrt{2})^{\wedge}(n-1)$,   we view a
table of values:

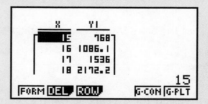

So, the first term to exceed 1400 is u_{17} where $u_{17} = 1536$.

Note: Later we can solve problems like this one using logarithms.

---

8 **a** Find the first term of the sequence $2, 6, 18, 54, ...$ which exceeds $10\,000$.

b Find the first term of the sequence $4, 4\sqrt{3}, 12, 12\sqrt{3}, ...$ which exceeds 4800.

c Find the first term of the sequence $12, 6, 3, 1.5, ...$ which is less than 0.0001.

E   COMPOUND INTEREST

Consider the following: You invest $1000 in the bank.

You leave the money in the bank for 3 years.

You are paid an interest rate of 10% p.a.

The interest is added to your investment each year.

An interest rate of 10% p.a. is paid, *increasing the value* of your investment yearly.

Your percentage increase each year is 10%,   i.e., $100\% + 10\% = 110\%$ of the value at the start of the year, which corresponds to a *multiplier* of 1.1 .

After one year your investment is worth $\$1000 \times 1.1 = \$1100$

After two years it is worth $\$1100 \times 1.1$

$= \$1000 \times 1.1 \times 1.1$

$= \$1000 \times (1.1)^2 = \$1210$

After three years it is worth $\$1210 \times 1.1$

$= \$1000 \times (1.1)^2 \times 1.1$

$= \$1000 \times (1.1)^3$

This suggests that if the money is left in your account for n years it would amount to $\$1000 \times (1.1)^n$.

Note: $u_1 = \$1000$ = initial investment

$u_2 = u_1 \times 1.1$ = amount after 1 year

$u_3 = u_1 \times (1.1)^2$ = amount after 2 years

$u_4 = u_1 \times (1.1)^3$ = amount after 3 years

$\vdots$

$u_{15} = u_1 \times (1.1)^{14}$ = amount after 14 years

$\vdots$

$u_n = u_1 \times (1.1)^{n-1}$ = amount after $(n-1)$ years

$u_{n+1} = u_1 \times (1.1)^n$ = amount after n years

In general, $\boxed{u_{n+1} = u_1 \times r^n}$ is used for compound growth u_1 = initial investment

r = growth multiplier

n = number of years

u_{n+1} = amount after n years

Example 10

$5000 is invested for 4 years at 7% p.a. compound interest.
What will it amount to at the end of this period?

$u_5 = u_1 \times r^4$ is the amount after 4 years

$= 5000 \times (1.07)^4$ {for a 7% increase 100% becomes 107%}

$\doteqdot 6553.98$ {5000 $\boxed{\times}$ 1.07 $\boxed{\wedge}$ 4 $\boxed{\text{ENTER}}$ }

So, it amounts to $6553.98 .

EXERCISE 12E

1 **a** What will an investment of $3000 at 10% p.a. compound interest amount to after 3 years?

   b What part of this is interest?

2 How much compound interest is earned by investing 20 000 euro at 12% p.a. if the investment is over a 4 year period?

3 **a** What will an investment of 30 000 Yen at 10% p.a. compound interest amount to after 4 years?

   b What part of this is interest?

4 How much compound interest is earned by investing $80 000 at 9% p.a., if the investment is over a 3 year period?

Example 11

Dana has $25 000 to invest at 9% p.a. for a 3 year period. What will her investment amount to if the interest is paid into her account:
a each quarter **b** each month?

a $u_1 = 25\,000$

   $n = 3 \times 4$ quarters $= 12$ quarters

   $r = 100\% + \frac{1}{4}$ of $9\% = 102.25\% = 1.0225$

   Now $u_{n+1} = u_1 \times r^n$

   $\phantom{Now\ u_{n+1}} = 25\,000 \times (1.0225)^{12}$

   $\phantom{Now\ u_{n+1}} \doteqdot 32\,651.25$

   So, it amounts to $32 651.25

b $u_1 = 25\,000$

   $n = 3 \times 12$ months $= 36$ months

   $r = 100\% + \frac{1}{12}$ of $9\% = 100.75\% = 1.0075$

   Now $u_{n+1} = u_1 \times r^n$

   $\phantom{Now\ u_{n+1}} = 25\,000 \times (1.0075)^{36}$

   $\phantom{Now\ u_{n+1}} \doteqdot 32\,716.13$

   So, it amounts to $32 716.13

> If interest is paid each quarter, it is paid at the end of each 3 month period or *quarter* year.

---

5 What will an investment of 100 000 Yen amount to after 5 years if it earns 8% p.a. compounded semi-annually?

6 What will an investment of £45 000 amount to after 28 months if it earns 7.5% p.a. compounded quarterly?

Example 12

How much should I invest now if I want the maturing value to be $10 000 in 4 years' time, if I am able to invest at 8.5% p.a. compounded annually?

$u_1 = ?,\quad u_5 = 10\,000,\quad r = 1.085$

$\qquad u_5 = u_1 \times r^4 \qquad$ {using $\ u_{n+1} = u_1 \times r^n$}

$\therefore \quad 10\,000 = u_1 \times (1.085)^4$

$\therefore \quad u_1 = \dfrac{10\,000}{(1.085)^4}$

$\therefore \quad u_1 \doteqdot 7215.74 \qquad$ {10 000 $\boxed{\div}$ 1.085 $\boxed{\wedge}$ 4 $\boxed{\textbf{ENTER}}$}

So, you should invest $7215.74 now.

7 How much money must be invested now if you require $20 000 for a holiday in 4 years' time and the money can be invested at a fixed rate of 7.5% p.a. compounded annually?

8 What initial investment is required to produce a maturing amount of £15 000 in 60 months' time given that a fixed rate of 5.5% p.a. compounded annually is guaranteed?

9 How much should I invest now if I want a maturing amount of 25 000 euro in 3 years' time and the money can be invested at a fixed rate of 8% p.a. compounded quarterly?

10 What initial investment is required to produce a maturing amount of 40 000 Yen in 8 years' time if your money can be invested at 9% p.a., compounded monthly?

F   GROWTH AND DECAY

Problems of growth and decay involve successive multiplications by a constant number and so geometric sequence models are appropriate.

Example 13

The initial population of rabbits on a farm was 50. The population increased by 7% each week.

a How many rabbits were present after:
  i 15 weeks   **ii** 30 weeks?

b How long would it take for the population to reach 500?

We notice that $\quad u_1 = 50\quad$ and $\quad r = 1.07$

$\qquad\qquad u_2 = 50 \times 1.07 =$ the population after 1 week

a **i** $\qquad u_{n+1} = u_1 \times r^n$ $\qquad\qquad$ **ii** and

$\qquad \therefore \quad u_{16} = 50 \times (1.07)^{15}$ $\qquad\qquad\qquad u_{31} = 50 \times (1.07)^{30}$

$\qquad\qquad\qquad \doteqdot 137.95....$ $\qquad\qquad\qquad\qquad\quad \doteqdot 380.61....$

$\qquad\qquad$ i.e.,   138 rabbits $\qquad\qquad\qquad\qquad$ i.e.,   381 rabbits

b $u_{n+1} = u_1 \times (1.07)^n$ after n weeks

So, we need to find when $50 \times (1.07)^n = 500$
$$(1.07)^n = 10$$

Trial and error on your calculator gives
$n \doteqdot 34$ weeks

or using the **Equation Solver** gives $n \doteqdot 34.03$

The solution is: 34.03 weeks.

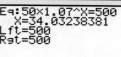

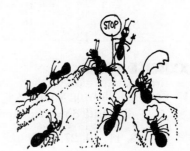

EXERCISE 12F

1 A nest of ants initially consists of 500 ants.
The population is increasing by 12% each week.

   a How many ants will there be after
      i 10 weeks   **ii** 20 weeks?

   b Use technology to find how many weeks it
     will take for the ant population to reach 2000.

2 The animal *Eraticus* is endangered. Since 1995 there has only been one colony remaining and in 1995 the population of the colony was 555. Since then the population has been steadily decreasing at 4.5% per year.
Find:

   a the expected population in the year 2010

   b the year in which we would expect the population to have declined to 50.

3 A herd of 32 deer is to be left unchecked on a large island off the coast of Alaska. It is estimated that the size of the herd will increase each year by about 18%.

   a Estimate the size of the herd after: **i** 5 years   **ii** 10 years.

   b How long will it take for the herd size to reach 5000?

4 A population of rodents was initially 400 but increased by 54% per month.

   a Find the expected population size after 18 months.

   b Find the number of months required for the population to reach 1 000 000.

G   SERIES

A **series** is the addition of the terms of a sequence,

i.e., $u_1 + u_2 + u_3 + \ldots + u_n$ is a series.

The **sum** of a series is the result when all terms of the series are added.

Notation: $S_n = u_1 + u_2 + u_3 + \ldots + u_n$ is the sum of the first n terms.

Example 14

For the sequence   1, 4, 9, 16, 25, ...

a   Write down an expression for S_n.

b   Find S_n for $n = 1, 2, 3, 4$ and 5.

a $S_n = 1^2 + 2^2 + 3^2 + 4^2 + ... + n^2$ {all terms are perfect squares}

b $S_1 = 1$

$S_2 = 1 + 4 = 5$

$S_3 = 1 + 4 + 9 = 14$

$S_4 = 1 + 4 + 9 + 16 = 30$

$S_5 = 1 + 4 + 9 + 16 + 25 = 55$

EXERCISE 12G.1

1   For the following sequences:

      i   write down an expression for S_n **ii**   find S_5.

    a   3, 11, 19, 27, ...   **b**   42, 37, 32, 27, ...   **c** $12, 6, 3, 1\frac{1}{2}, ...$

    d $2, 3, 4\frac{1}{2}, 6\frac{3}{4}, ...$ **e** $1, \frac{1}{2}, \frac{1}{4}, \frac{1}{8}, ...$ **f**   1, 8, 27, 64, ...

ARITHMETIC SERIES

An **arithmetic series** is the addition of successive terms of an arithmetic sequence.

For example: 21, 23, 25, 27, ..., 49   is an arithmetic sequence.

So, $21 + 23 + 25 + 27 + ... + 49$ is an arithmetic series.

SUM OF AN ARITHMETIC SERIES

Recall that if the first term is u_1 and the common difference is d, then the terms are:
$u_1, u_1 + d, u_1 + 2d, u_1 + 3d$, etc.

Suppose that u_n is the last or final term of an arithmetic series.

Then, $S_n = u_1 + (u_1 + d) + (u_1 + 2d) + ... + (u_n - 2d) + (u_n - d) + u_n$

but, $S_n = u_n + (u_n - d) + (u_n - 2d) + ... + (u_1 + 2d) + (u_1 + d) + u_1$ {reversing them}

Adding these two expressions vertically we get

$$2S_n = \underbrace{(u_1 + u_n) + (u_1 + u_n) + (u_1 + u_n) + ... + (u_1 + u_n) + (u_1 + u_n) + (u_1 + u_n)}_{n \text{ of these}}$$

$\therefore$ $2S_n = n(u_1 + u_n)$

i.e., $S_n = \dfrac{n}{2}(u_1 + u_n)$ where $u_n = u_1 + (n-1)d$

so $S_n = \dfrac{n}{2}(u_1 + u_n)$ *or* $S_n = \dfrac{n}{2}(2u_1 + (n-1)d)$

Example 15

Find the sum of $4 + 7 + 10 + 13 + \ldots$ to 50 terms.

The series is arithmetic with $u_1 = 4$, $d = 3$ and $n = 50$.

So, $S_{50} = \frac{50}{2}(2 \times 4 + 49 \times 3)$ {Using $S_n = \frac{n}{2}(2u_1 + (n-1)d)$}

$= 25(8 + 147)$

$= 3875$

EXERCISE 12G.2

1 Find $1 + 5 + 9 + 13 + 17 + 21 + 25$:

   a   by simply adding

   b   using $S_n = \dfrac{n}{2}(u_1 + u_n)$

   c   using $S_n = \dfrac{n}{2}(2u_1 + (n-1)d)$

2 Find the sum of:

   a $3 + 7 + 11 + 15 + \ldots$ to 20 terms

   b $\frac{1}{2} + 3 + 5\frac{1}{2} + 8 + \ldots$ to 50 terms

   c $100 + 93 + 86 + 79 + \ldots$ to 40 terms

   d $50 + 48\frac{1}{2} + 47 + 45\frac{1}{2} + \ldots$ to 80 terms

---

Example 16

Find the sum of $-6 + 1 + 8 + 15 + \ldots + 141$.

The series is arithmetic with $u_1 = -6$, $d = 7$ and $u_n = 141$.

First we need to find n.   Now $u_n = 141$

$\therefore \quad u_1 + (n-1)d = 141$

$\therefore \quad -6 + 7(n-1) = 141$

$\therefore \quad 7(n-1) = 147$

$\therefore \quad n - 1 = 21$

$\therefore \quad n = 22$

Using $S_n = \frac{n}{2}(u_1 + u_n)$ *or* $S_n = \frac{n}{2}(2u_1 + (n-1)d)$

$\therefore \quad S_{22} = \frac{22}{2}(-6 + 141)$ $S_{22} = \frac{22}{2}(2 \times -6 + 21 \times 7)$

$= 11 \times 135$ $= 11 \times 135$

$= 1485$ $= 1485$

3 Find the sum of:

    a $5 + 8 + 11 + 14 + ... + 101$

    b $50 + 49\frac{1}{2} + 49 + 48\frac{1}{2} + ... + (-20)$

    c $8 + 10\frac{1}{2} + 13 + 15\frac{1}{2} + ... + 83$

4 An arithmetic series has seven terms. The first term is 5 and the last term is 53. Find the sum of the series.

5 An arithmetic series has eleven terms. The first term is 6 and the last term is -27. Find the sum of the series.

6 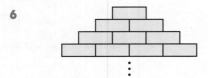 A bricklayer builds a triangular wall with layers of bricks as shown. If the bricklayer uses 171 bricks, how many layers are placed?

7 Each section of a soccer stadium has 44 rows with 22 seats in the first row, 23 in the second row, 24 in the third row, and so on. How many seats are there

    a   in row 44

    b   in a section

    c   in the stadium which has 25 sections?

8 Find the sum of:

    a   the first 50 multiples of 11

    b   the multiples of 7 between 0 and 1000

    c   the integers between 1 and 100 which are not divisible by 3.

9 Prove that the sum of the first n positive integers is $\dfrac{n(n+1)}{2}$,

    i.e.,   show that $1 + 2 + 3 + 4 + ... + n = \dfrac{n(n+1)}{2}$.

10 Consider the series of odd numbers $1 + 3 + 5 + 7 + ...$

    a   What is the nth odd number, that is, u_n?

    b   Prove that "the sum of the first n odd numbers is n^2".

    c   Check your answer to **b** by finding S_1, S_2, S_3 and S_4.

11 Find the first two terms of an arithmetic sequence where the sixth term is 21 and the sum of the first seventeen terms is 0.

12 Three consecutive terms of an arithmetic sequence have a sum of 12 and a product of -80. Find the terms.   (**Hint:** Let the terms be $x - d$, x and $x + d$.)

13 Five consecutive terms of an arithmetic sequence have a sum of 40. The product of the middle and the two end terms is 224. Find the terms of the sequence.

GEOMETRIC SERIES

A **geometric series** is the addition of successive terms of a geometric sequence.

For example,   1, 2, 4, 8, 16, ..., 1024   is a geometric sequence.

So, $1 + 2 + 4 + 8 + 16 + ... + 1024$ is a geometric series.

SUM OF A GEOMETRIC SERIES

Recall that if the first term is u_1 and the common ratio is r, then the terms are:

$$u_1, \ u_1 r, \ u_1 r^2, \ u_1 r^3, \ .... \quad \text{etc.}$$

So, $S_n = u_1 + u_1 r + u_1 r^2 + u_1 r^3 + ... + u_1 r^{n-2} + u_1 r^{n-1}$

$$\underset{u_2}{\uparrow} \quad \underset{u_3}{\uparrow} \quad \underset{u_4}{\uparrow} \quad \underset{u_{n-1}}{\uparrow} \quad \underset{u_n}{\uparrow}$$

and   for $r \neq 1$, $\quad S_n = \dfrac{u_1(r^n - 1)}{r - 1} \quad$ or $\quad S_n = \dfrac{u_1(1 - r^n)}{1 - r}.$

Proof: If $S_n = u_1 + u_1 r + u_1 r^2 + u_1 r^3 + ... + u_1 r^{n-2} + u_1 r^{n-1}$ .... (1)

then $rS_n = (u_1 r + u_1 r^2 + u_1 r^3 + u_1 r^4 + ... + u_1 r^{n-1}) + u_1 r^n$

$\therefore \quad rS_n = (S_n - u_1) + u_1 r^n \qquad \{\text{from (1)}\}$

$\therefore \quad rS_n - S_n = u_1 r^n - u_1$

$\therefore \quad S_n(r - 1) = u_1(r^n - 1)$

$\therefore \quad S_n = \dfrac{u_1(r^n - 1)}{r - 1} \quad$ or $\quad \dfrac{u_1(1 - r^n)}{1 - r} \quad$ p.v. $\quad r \neq 1.$

Example 17

Find the sum of $2 + 6 + 18 + 54 + ...$ to 12 terms.

The series is geometric with $u_1 = 2$, $\ r = 3$ and $\ n = 12$.

So, $S_{12} = \dfrac{2(3^{12} - 1)}{3 - 1} \qquad \left\{ \text{Using } \ S_n = \dfrac{u_1(r^n - 1)}{r - 1} \right\}$

$\qquad = \dfrac{2(3^{12} - 1)}{2}$

$\qquad = 531\,440$

EXERCISE 12G.3

1 Find the sum of the following series:

    a $12 + 6 + 3 + 1.5 + ...$ to 10 terms

    b $\sqrt{7} + 7 + 7\sqrt{7} + 49 + ...$ to 12 terms

    c $6 - 3 + 1\frac{1}{2} - \frac{3}{4} + ...$ to 15 terms

    d $1 - \frac{1}{\sqrt{2}} + \frac{1}{2} - \frac{1}{2\sqrt{2}} + ...$ to 20 terms

Example 17

Find a formula for S_n for $9 - 3 + 1 - \frac{1}{3} + \ldots$ to n terms.

The series is geometric with $u_1 = 9$, $r = -\frac{1}{3}$, "n" $= n$.

So, $S_n = \dfrac{u_1(1 - r^n)}{1 - r}$

$ = \dfrac{9(1 - (-\frac{1}{3})^n)}{\frac{4}{3}}$

$\therefore \quad S_n = \frac{27}{4}(1 - (-\frac{1}{3})^n)$

Note:
This answer cannot be simplified as we do not know if n is odd or even.

2 Find a formula for S_n for:

   a $\sqrt{3} + 3 + 3\sqrt{3} + 9 + \ldots$ to n terms

   b $12 + 6 + 3 + 1\frac{1}{2} + \ldots$ to n terms

   c $0.9 + 0.09 + 0.009 + 0.0009 + \ldots$ to n terms

   d $20 - 10 + 5 - 2\frac{1}{2} + \ldots$ to n terms

3 Each year a sales-person is paid a bonus of \$2000 which is banked into the same account which earns a fixed rate of interest of 6% p.a. with interest being paid annually. The amount at the end of each year in the account is calculated as follows:

$$A_0 = 2000$$
$$A_1 = A_0 \times 1.06 + 2000$$
$$A_2 = A_1 \times 1.06 + 2000 \quad \text{etc.}$$

   a Show that $A_2 = 2000 + 2000 \times 1.06 + 2000 \times (1.06)^2$.

   b Show that $A_3 = 2000[1 + 1.06 + (1.06)^2 + (1.06)^3]$.

   c Hence find the total bank balance after 10 years. (Assume no fees and charges.)

4 Consider $S_n = \frac{1}{2} + \frac{1}{4} + \frac{1}{8} + \frac{1}{16} + \ldots + \dfrac{1}{2^n}$.

   a Find S_1, S_2, S_3, S_4 and S_5 in fractional form.

   b From **a** guess the formula for S_n.

   c Find S_n using $S_n = \dfrac{u_1(1 - r^n)}{1 - r}$.

   d Comment on S_n as n gets very large.

   e What is the relationship between the given diagram and **d**?

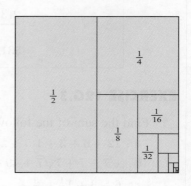

5

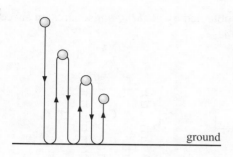

A ball takes 1 second to hit the ground when dropped. It then takes 90% of this time to rebound to its new height and this continues until the ball comes to rest.

a Show that the total time of motion is given by $1 + 2(0.9) + 2(0.9)^2 + 2(0.9)^3 + ...$

b Find S_n for the series in **a**.

c How long does it take for the ball to come to rest?

ground

Note: This diagram is inaccurate as the motion is really up and down on the same spot. It has been separated out to help us visualise what is happening.

INVESTIGATION   **VON KOCH'S SNOWFLAKE CURVE**

C_1 , C_2 , C_3 , C_4 , ...

To draw **Von Koch's Snowflake curve** we

DEMO

- start with an equilateral triangle, C_1
- then divide each side into 3 equal parts _____._____
- then on each middle part draw an equilateral triangle ____/____
- then delete the side of the smaller triangle which lies on C_1. ____/____

The resulting curve is C_2, and C_3, C_4, C_5, ... are found by 'pushing out' equilateral triangles on each edge of the previous curve as we did with C_1 to get C_2.

We get a sequence of special curves C_1, C_2, C_3, C_4,.... and Von Koch's curve is the limiting case, i.e., when n is infinitely large for this sequence.

Your task is to investigate the perimeter and area of Von Koch's curve.

What to do:

1 Suppose C_1 has a perimeter of 3 units. Find the perimeter of C_2, C_3, C_4 and C_5.

(**Hint:** _____._____ becomes ____/____ i.e., 3 parts become 4 parts.)

Remembering that Von Koch's curve is C_n, where n is infinitely large, find the perimeter of Von Koch's curve.

2 Suppose the area of C_1 is 1 unit^2. Explain why the areas of C_2, C_3, C_4 and C_5 are

$A_2 = 1 + \frac{1}{3}$ units^2 $A_3 = 1 + \frac{1}{3}[1 + \frac{4}{9}]$ units^2

$A_4 = 1 + \frac{1}{3}[1 + \frac{4}{9} + (\frac{4}{9})^2]$ units^2 $A_5 = 1 + \frac{1}{3}[1 + \frac{4}{9} + (\frac{4}{9})^2 + (\frac{4}{9})^3]$ units^2.

Use your calculator to find A_n where $n = 1, 2, 3, 4, 5, 6, 7$, etc., giving answers which are as accurate as your calculator permits.

What do you think will be the area within Von Koch's snowflake curve?

3 Similarly, investigate the sequence of curves obtained by *pushing out* squares on successive curves from the middle third of each side,

i.e.,   the curves C_1, C_2, C_3, C_4, etc.

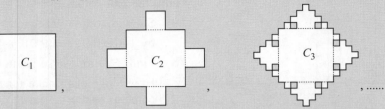

Region contains 8 holes.

REVIEW SET 12A

1 List the first four members of the following sequences defined by:

    a $u_n = 3^{n-2}$ **b** $u_n = \dfrac{3n+2}{n+3}$ **c** $u_n = 2^n - (-3)^n$

2 A sequence is defined by $u_n = 68 - 5n$.

    a Prove that the sequence is arithmetic.

    b Find u_1 and d.

    c Find the 37th term.

    d What is the first term of the sequence less than -200?

3   **a** Show that the sequence   3, 12, 48, 192, ...   is geometric.

    b Find u_n and hence find u_9.

4 Find k if $3k$, $k-2$ and $k+7$ are consecutive terms of an arithmetic sequence.

5 Find the general term of an arithmetic sequence given that $u_7 = 31$ and $u_{15} = -17$. Hence, find the value of u_{34}.

6 A sequence is defined by $u_n = 6(\frac{1}{2})^{n-1}$.

    a Prove that the sequence is geometric.

    b Find u_1 and r.

    c Find the 16th term to 3 significant figures.

7 Show that   28, 23, 18, 13, ...   is arithmetic and hence find u_n and the sum S_n of the first n terms in simplest form.

8 Find k given that 4, k and $k^2 - 1$ are consecutive geometric terms.

REVIEW SET 12B

1 **a** Determine the number of terms in the sequence $24, 23\frac{1}{4}, 22\frac{1}{2}, ..., -36$.

   b Find the value of u_{35} for the sequence in **a**.

   c Find the sum of the terms of the sequence in **a**.

2 Insert six numbers between 23 and 9 so that all eight numbers are in arithmetic sequence.

3 Find the formula for u_n, the general term of:

   a $86, 83, 80, 77, ...$ **b** $\frac{3}{4}, 1, \frac{7}{6}, \frac{9}{7}, ...$ **c** $100, 90, 81, 72.9, ...$

   [**Note:** One of these sequences is neither arithmetic nor geometric.]

4 Find the sum of:

   a $3 + 9 + 15 + 21 + ...$ to 23 terms   **b** $24 + 12 + 6 + 3 + ...$ to 12 terms.

5 Find the first term of the sequence $5, 10, 20, 40, ...$ which exceeds $10\,000$.

6 What will an investment of 6000 euro at 7% p.a. compound interest amount to after 5 years if the interest is compounded:

   a   annually   **b**   quarterly   **c**   monthly?

7 Determine the general term of a geometric sequence given that its sixth term is $\frac{16}{3}$ and its tenth term is $\frac{256}{3}$.

8 Olaf is employed as a sales consultant with a starting salary of 50000 euro. He is promised a 3% pay rise each year of his employment.

   a   How much does Olaf earn in his 5th year of employment?

   b   How many years must Olaf work before his salary reaches 60 000 euro?

REVIEW SET 12C

1 A geometric sequence has $u_6 = 24$ and $u_{11} = 768$. Determine the general term of the sequence and hence find:

   a u_{17} **b**   the sum of the first 15 terms.

2 How many terms of the series $11 + 16 + 21 + 26 + ...$ are needed to exceed a sum of 450?

3 Find the first term of the sequence $24, 8, \frac{8}{3}, \frac{8}{9}, ...$ which is less than 0.001.

4 **a** Determine the number of terms in the sequence $128, 64, 32, 16, ..., \frac{1}{512}$.

   b Find the sum of these terms.

5 Find k, given that $k, \ k + 9, \ 16k$ are consecutive terms of a geometric sequence.

6 $12\,500$ is invested in an account which pays 8.25% p.a. compounded. Find the value of the investment after 5 years if the interest is compounded:

   a   half-yearly   **b**   monthly.

7 How much should be invested at a fixed rate of 9% p.a. compounded interest if you wish it to amount to $20 000 after 4 years with interest paid monthly?

8 In 1998 there were 3000 koalas on Koala Island. Since then, the population of koalas on the island has increased by 5% each year.

    a How many koalas were on the island in 2001?

    b In what year will the population first exceed 5000?

Chapter 13

Financial mathematics

Contents:

FOREIGN EXCHANGE

When you are in another country you must use the currency of that country. This means that you must exchange your money for the equivalent amount of local currency (money). The equivalent amount is found using **exchange rates**.

Exchange rates show the relationship between the values of currencies. They are published daily in newspapers, in bank windows and on the internet at various bank and travel agency sites.

Exchange rates change constantly. This table is an example of a typical newspaper **currency exchange table**.

It shows how much $1 Australian is worth in some other currencies.

The exchange rates in any country are usually given as the amount of foreign currency equal to one unit of the local currency.

CURRENCIES					
currency exchange rates quoted by the Commonwealth Bank on Jul 13					
	Bank buys	Bank sells		Bank buys	Bank sells
US dollar	0.7303	0.7216	Malta lira	0.2534	0.2443
Europe euro	0.5933	0.5769	NZdollar	1.1087	1.0826
UK pound	0.3935	0.3851	Norway kroner	5.0118	4.8797
Canada dollar	0.9665	0.9432	Pakistan rupee	n/a	n/a
China renminbi	n/a	n/a	PNG kina	n/a	2.0261
Denmark kroner	4.4034	4.2874	Philippines peso	n/a	38.538
Fiji dollar	n/a	1.2359	Singapore dollar	1.2462	1.209
Fr Pacific franc	71.51	67.56	S Africa rand	4.5082	4.2725
Hong Kong dollar	5.7386	5.5579	Sri Lanka rupee	76.13	70.06
India rupee	33.67	32.068	Sweden krona	5.4512	5.3076
Indonesia rupiah	n/a	n/a	Switzerland franc	0.8998	0.8761
Japan yen	79.45	77.35	Thailand bant	30.09	27.56
Malaysia ringgit	n/a	n/a	Vanuatu vatu	n/a	79.55

For example, suppose you are from Europe with 800 euro and you travel to India. You need to know how many Indian rupee there are to 1 euro. From an Indian exchange table similar to this one, we might find that 1 euro = 55.376 rupee.

So, your 800 euro will buy 800×55.376 rupee $\div 44\,300$ rupee.

Example 1

Given that 1 South African rand = $0.2754 Singapore, find how many Singapore dollars you could buy for 2500 South African rand.

1 South African rand = $0.2754 Singapore

$\therefore$ 2500 South African rand = $2500 \times \$0.2754$ Singapore
= $688.50 Singapore

Notice that both **buying and selling rates** at the bank are included in the table. These two quantities differ as the bank makes a profit on all money exchanges.

EXERCISE 13A.1

1 If $1 US buys 0.5417 UK pounds, find how many UK pounds could be bought for:
    a $560 US   **b** $980 US   **c** $2179 US

2 If $1 Australian buys 4.908 Norwegian kroner, how many kroner could be bought for:
    a $275 Australian   **b** $675 Australian   **c** $3825 Australian?

3 If 1 Japanese Yen buys 0.06872 Swedish krona, how many krona could be bought for:
    a 6000 Yen   **b** 12 500 Yen   **c** 36 800 Yen?

4 If $1 Canadian buys $5.706 Hong Kong, find how many Hong Kong dollars could be bought for:
    a $235 Canadian   **b** $1250 Canadian   **c** $8500 Canadian

5 If 1 euro buys 50.631 Thailand baht, find how many baht could be bought for:

    a   55 euro   **b**   644 euro   **c**   2695 euro

BUYING AND SELLING INTERNATIONAL CURRENCY

Selling

- How much foreign currency will you receive by selling other currency?
  Use the **selling exchange rate** and the formula:

> **Foreign currency bought = other currency sold × selling exchange rate**

Example 2

Given that $1 Australian = 0.4032 UK pounds (when selling), convert $500 Australian into United Kingdom pounds.

$\therefore$ UK currency bought $= 500 \times 0.4032$ pounds
$= 201.60$ pounds

- How much will it cost you in your currency if you have to purchase foreign currency?
  You are selling your currency, so use the selling exchange rate and the formula:

$$\textbf{Cost in currency you have} = \frac{\textbf{foreign currency bought}}{\textbf{selling exchange rate}}$$

Example 3

What does it cost in New Zealand dollars to buy $2000 US, if $1 NZ = $0.6328 US?

$\therefore$ cost in $ NZ $= \dfrac{2000}{0.6328} \doteqdot \$3160.55$ US

Buying

- You have **currency from another country** and want to change it to **your country's currency**. You are **buying** your currency. Use the **buying exchange rate** and:

$$\textbf{Your currency bought} = \frac{\textbf{foreign currency sold}}{\textbf{buying exchange rate}}$$

Example 4

If you have $1000 Hong Kong and exchange it for Phillipine pesos, how many pesos will you receive if 1 peso = $0.1454 HK?

Pesos bought $= \dfrac{1000}{0.1454} \doteqdot 6880$ pesos

EXERCISE 13A.2

1 Convert 680 euro into New Zealand dollars given that $1 NZ = 0.5258 euro (selling).

2 What does it cost in Canadian dollars to buy £700, given that $1 Canadian = £0.4073.

3 If you have $490 US and exchange it for Danish kroner, how many kroner will you receive? 1 Danish kroner = $0.16645 US.

4 Convert $2540 Singapore into Indian rupees if 27.014 rupee = $1 Singapore (selling).

5 What does it cost in Malaysian ringgits to buy 560 Brazilian real given that 1 ringgit = 0.33058 Brazilian real.

6 Convert 1650 Swiss francs into euro given that 1 euro = 1.5268 Swiss francs (selling).

Conversion graphs are line graphs which enable us to convert from one quantity to another.

Example 5

The graph alongside shows the relationship between Australian dollars and English pounds on a particular day. Find:

a the number of dollars in 250 pounds

b the number of pounds in 480 dollars

c whether a person with $360 could afford to buy an item valued at 200 pounds.

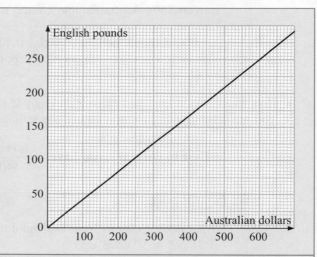

a 250 pounds is equivalent to $600.

b $480 is equivalent to 200 pounds.

c $360 is equivalent to 150 pounds.
∴ cannot afford to buy the item.

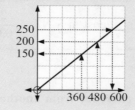

7 Use the currency conversion graph of **Example 5** to estimate:
  a the number of dollars in   **i** 130 pounds   **ii** 240 pounds
  b the number of pounds in   **i** $400   **ii** $560

ACTIVITY   CURRENCY TRENDS

Over a period of a month collect from the daily newspaper or from the internet, the currency conversions which compare your currency to the currency of another country. Graph your results updating the graph each day. You could use   www.x-rates.com/calculator.html .

COMMISSION ON CURRENCY EXCHANGE

When any currency trader (such as a bank) exchanges currency for a customer a commission is paid by the customer for this service. The commission could vary from $\frac{1}{2}$% to 3%.

The commission could be calculated using   •   a fixed percentage   •   the buy/sell values.

Example 6

A bank changes US dollars to other currency at a fixed commission of 1.5%.
Max wishes to convert $200 US to baht where $1 US buys 40.23 Thai baht.
a   What commission is charged?   b   What does the customer receive?

a   Commission
    $= \$200$ US $\times 1.5\%$
    $= \$200$ US $\times 0.015$
    $= \$3$ US

b   ∴   customer receives
    197×40.23 baht
    $\doteqdot 7925$ baht

Example 7

A currency exchange service exchanges 1 euro for Japanese Yen using: 'buy at 135.69, sell at 132.08'. Cedric wishes to exchange 800 euro for Yen.
a   How many Yen will he receive?
b   If the Yen in a is converted immediately back to euro, how many euro are bought?
c   What is the resultant commission on the double transaction?

a   Cedric receives

$800 \times 132.08 \doteqdot 105\,700$ Yen
(using the selling rate as the
bank is selling currency)

b   Cedric receives

$\dfrac{105\,700}{135.69} \doteqdot 779$ euro

(using the buying rate as the
bank is buying currency)

c   The resultant commission is $800 - 779 = 21$ euro.

EXERCISE 13A.3

1   A bank exchanges UK pounds for a commission of 1.5%.
     i   What commission is charged?
     ii   What does the customer receive for these transactions?
   a   Converting 500 UK pounds to US dollars where £1 UK buys $1.8734 US.
   b   Converting 350 UK pounds to euro where £1 UK buys $.5071 euro.
   c   Converting £1200 UK to New Zealand dollars where £1 UK buys $2.8424 NZ.

2   A bank exchanges US dollars for a commission of 1.8%.
     i   What commission is charged?
     ii   What does the customer receive for these transactions?
   a   Converting $400 US to £ UK if $1 US buys £0.533 79

   b Converting $700 US to $ Australian if $1 US buys $1.3728 Australian

   c Converting $1300 US to euro if $1 US buys 0.804 41 euro.

3 A currency exchange service exchanges 1 Mexican peso for Thai baht using: 'buy at 3.584, sell at 3.4807'. Sergio wishes to exchange 400 pesos for Thai baht.

   a How many baht will he receive?

   b If he immediately changes the baht back to pesos how many will he get?

   c What is the resultant commission for the double transaction?

4 A currency exchange service exchanges 1 South African rand to Indian rupees using: 'buy at 7.8086, sell at 7.5641'. Jonte wishes to exchange 375 rand for rupees.

   a How much will he receive?

   b If he immediately changes the rupees back to rand how many will he get?

   c What is the resultant commission for the double transaction?

TRAVELLERS CHEQUES

When travelling overseas some people carry their money as **travellers cheques**. They are more convenient than carrying large amounts of cash. They provide protection in case of accidental loss or theft. If necessary travellers cheques may be quickly replaced.

Travellers cheques are usually purchased from a bank before you leave your country. You should take the currency of the country you are visiting or a widely acceptable currency like US dollars. Usually banks who provide travellers cheques charge 1% of the value of the cheques when they are issued.

So,   **cost of travellers cheques** $= \dfrac{\textbf{amount of foreign currency}}{\textbf{selling exchange rate}} \times \textbf{101\%}$

Note: It is also possible to buy foreign currency using a credit card that is accepted internationally, such as Visa or Mastercard. Currency can be purchased using your credit cards at banks and automatic teller machines (ATMs) in most countries.

Example 8

If you want to buy 2000 UK pounds worth of travellers cheques, what will it cost in Australian dollars, if $1 Australian = 0.4032 pounds?

$$\text{cost} = \frac{2000}{0.4032} \times 1.01 = \$5009.90 \text{ Australian}$$

EXERCISE 13A.4

1 Calculate the cost of purchasing travellers cheques in:

   a US dollars, if you have $4200 Singapore and 1 US dollar = $1.7058 Singapore

   b Norwegian kroner, if you have 327 euro and 1 Norwegian kroner = 0.11740 euro

   c New Zealand dollars, if you have $16 000 Canadian and $1 NZ = $0.86141 Can.

   d Thailand baht, if you have 300 English pounds and 1 baht = £0.013177

   e Japanese Yen, if you have $1540 NZ and 1 Yen = $0.014034 NZ

B | SIMPLE INTEREST

Under this method, interest is calculated on the full amount borrowed or lent for the entire period of the loan or investment.

For example,   if $2000 is borrowed at 8% p.a. for 3 years, the interest payable for 1 year is 8% of $2000 = $2000 × 0.08

So, for 3 years it would be ($2000 × 0.08) × 3

From examples like this one we construct the **simple interest formula**.

Simple interest is often called flat rate interest.

This is $I = C \times r \times n$ where:

I, is the **$ amount of interest**
C, is the **principal** (amount borrowed)
r, is the **simple interest per annum as a decimal**
n, is the **time (or length) of the loan**, and is always expressed in terms of **years**.

Example 9

Calculate the simple interest on a loan of $8000 at a rate of 7% p.a. over 18 months.

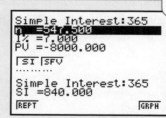

7% means 7 out of 100 i.e., 7 ÷ 100 = 0.07

$C = 8000$, $r = 0.07$, $n = \frac{18}{12} = 1.5$

Now $I = C \times r \times n$
so, $I = 8000 \times 0.07 \times 1.5$
$\therefore$ $I = 840$

i.e.,   simple interest is $840.

EXERCISE 13B.1

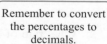

1   Calculate the simple interest on a loan of:

a $4000 at a rate of 6% p.a. over 2 years

b $9500 at a rate of 8.7% p.a. over an 18 month period

c $20 000 at a rate of 6.5% p.a. over a 5 year 4 month period

d   6000 Yen at a rate of 9.3% p.a. over a 263 day period.

Remember to convert the percentages to decimals.
Hint: Shift the decimal point two places to the left!

2   Which loan charges less interest:

• $15 000 at a rate of 7% p.a. simple interest for 4 years, or

• $15 000 at a rate of 6.55% p.a. simple interest for $4\frac{1}{2}$ years?

We can also use the same formula to find the other three variables C, r, and n in the equation.

Example 10

How much is borrowed if a rate of 6.5% p.a. simple interest results in an interest charge of $3900 after 5 years?

You could use
$$C = \dfrac{I}{r \times n}$$

$I = 3900, \qquad r = 0.065, \qquad n = 5$

Now $\quad I = C \times r \times n$

So, $\quad 3900 = C \times 0.065 \times 5$

Thus, $\quad C = 12\,000$

i.e., $12\,000 was borrowed.

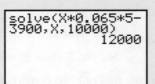

```
solve(X*0.065*5-
3900,X,10000)
            12000
```

3 How much is borrowed if:

   a a rate of 6.5% p.a. simple interest results in a charge of $910 after 5 years

   b a rate of 9% p.a. simple interest results in a charge of £7560 after 4 years?

4 An investor wants to earn $3500 in 5 months. How much would he need to invest given that the current simple interest rates are $6\frac{3}{4}\%$?

You could use
$$r = \dfrac{I}{C \times n}$$

Example 11

If you wanted to earn $6000 in interest on a 4 year loan of $18\,000, what rate of simple interest would you need to charge?

$I = 6000, \qquad C = 18\,000, \qquad n = 4$

Now $\quad I = C \times r \times n$

So $\quad 6000 = 18\,000 \times \frac{r}{100} \times 4$

Thus, $\quad r = 0.083333$

$\qquad = 0.083333 \times 100\%$

$\qquad = 8.3333\%$

i.e., the simple interest rate is $8\frac{1}{3}\%$ p.a.

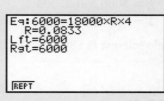

```
Eq:6000=18000×R×4
  R=0.0833
Lft=6000
Rgt=6000

REPT
```

5 What rate of simple interest is charged if you want to earn:

   a $800 after 4 years on $6000   **b** $936 after 18 months on $7800?

6 What rate of simple interest would need to be charged on a loan of $28\,000, if you wanted to earn $3970 in interest over 2 years?

Convert the decimals to percentages by multiplying by 100, i.e., shift the decimal point two places to the right!

7 A student wants to buy a car costing $7500 in 15 month's time. She has already saved $6000 and deposits this in an account that pays simple interest. What rate of simple interest must the account pay to enable the student to reach her target?

Example 12

How long would it take to earn interest of $4760 on a loan of $16 000 if a rate of 8.5% p.a. simple interest is charged?

$I = 4760$, $C = 16\,000$, $r = 0.085$

Now $I = C \times r \times n$

So, $4760 = 16\,000 \times 0.085 \times n$

Thus, $n = 3.5$

i.e.,   it would take $3\frac{1}{2}$ years to earn $4760 interest.

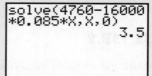

```
solve(4760-16000
*0.085*X,X,0)
              3.5
```

8 How long would it take to earn simple interest of:

  a $5625 on a loan of $30 000 at a rate of 7.5% p.a.

  b 9334 euro on a loan of 35 900 euro at a rate of 4% p.a.?

9 If you deposited $9000 in an investment account that paid simple interest at a rate of 5.75% p.a., how long would it take to earn $3105 in interest?

CALCULATING REPAYMENTS

Whenever money is borrowed, the amount borrowed (or **principal**) must be repaid along with the interest charges applicable to that loan. In addition, it must be repaid within the loan period. The full amount of the repayment can be made either by:

- making **one payment** on a set date at the conclusion of the loan period, or
- making numerous (usually equal) **periodic payments** over the loan period.

When the loan is repaid by making equal periodic payments over the loan period, the amount of each periodic payment is calculated by dividing the total to be repaid by the number of payments to be made,

$$\textbf{Periodic repayment} = \frac{\textbf{principal + interest}}{\textbf{number of repayments}} \quad \text{i.e.,} \quad R_P = \frac{C + I}{N}$$

Example 13

Calculate the monthly repayments on a loan of $7000 at $8\frac{1}{2}$% p.a. simple interest over 4 years.

Step 1: **Calculate interest**

$C = 7000$ Now $I = C \times r \times n$

$r = 0.085$ $I = 7000 \times 0.085 \times 4$

$n = 4$ $\therefore$ $I = 2380$ i.e., interest is $2380.

Step 2: **Calculate repayments**

$C = 7000$ Now $R_P = \dfrac{C + I}{N} = \dfrac{7000 + 2380}{48}$

$I = 2380$

$N = 4 \times 12 = 48$ months $\therefore \quad R_P = 195.42$

i.e., monthly repayments of \$195.42 are needed.

EXERCISE 13B.2

1 Calculate the monthly repayments on a loan of \$6800 at $8\frac{1}{2}\%$ p.a. simple interest over $2\frac{1}{2}$ years.

2 If a loan of 10 000 baht at a simple interest rate of $5\frac{3}{4}\%$ p.a. for 10 years is to be repaid each half year, what should be the size of each repayment?

3 A young couple obtain a loan from friends for \$15 000 for 36 months at a simple interest rate of $4\frac{1}{2}\%$ p.a.. Calculate the quarterly repayments they must make on this loan.

4 Justine arranges a loan of \$8000 from her parents and repays \$230 per month for $3\frac{1}{2}$ years. How much interest does she pay on the loan?

5 Eric approaches two friends for a loan and receives the following offers:
Rachel can lend \$12 000 at $5\frac{1}{4}\%$ p.a. simple interest repayable monthly for $3\frac{1}{2}$ years.
Lesley can lend \$12 000 at $4\frac{3}{4}\%$ p.a. simple interest repayable monthly for $4\frac{1}{2}$ years.
Eric can only afford a maximum repayment of \$300 per month. Which loan should he accept?

© Jim Russell - General Features

C   COMPOUND INTEREST

Compound interest is a method of calculating interest in which the *interest is added to the principal each period* so that the principal continues to grow throughout the life of the loan or investment (unlike simple interest where the principal remains constant throughout the period of the loan or investment).

Thus, the interest generated in one period then earns interest itself in the next period.

Example 14

Calculate the interest paid on a deposit of $6000 at 8% p.a. compounded annually for 3 years.

Compounded annually means interest is added to the principal annually so we need to calculate interest each year. We will use the simple interest formula and set out our calculations in table form.

Year	Principal (1)	Interest ($= C \times r \times n$) (2)	Balance (1) + (2)
1	$6000.00	$6000.00 $\times$ 0.08 $\times$ 1 = $480.00	$6480.00
2	$6480.00	$6480.00 $\times$ 0.08 $\times$ 1 = $518.40	$6998.40
3	$6998.40	$6998.40 $\times$ 0.08 $\times$ 1 = $559.87	$7558.27

Thus, the $6000.00 grows to $7558.27 after 3 years,

i.e., $7558.27 − $6000.00 = $1558.27   is interest.

Notice that the amount of interest paid increases from year to year. ($480.00, $518.40, $559.87).   Can you see why?

Note: **Total interest earned = final balance − principal.**

EXERCISE 13C.1

1 Find the final value of a compound interest investment of:
 a $4500 after 3 years at 6% p.a. with interest calculated annually
 b $6000 after 4 years at 5% p.a. with interest calculated annually
 c £8200 after 2 years at 6.5% p.a. with interest calculated annually.

2 Find the total interest earned for the following compound interest investments:
 a 950 euro after 2 years at 5.8% p.a. with interest calculated annually
 b $3850 after 3 years at 9.25% p.a. with interest calculated annually
 c $17 500 after 4 years at 8.4% p.a. with interest calculated annually.

3 Mei Ling invests $15 000 into an account which pays 8% p.a. compounded annually. Find:
 a the value of her account after 2 years
 b the total interest earned after 2 years.

4 Brit places 8000 kroner in a fixed term investment account which pays 6.5% p.a. compounded annually.
 a How much will she have in her account after 3 years?
 b What interest has she earned over this period?

DIFFERENT COMPOUNDING PERIODS

Interest can be compounded more than once per year.

Commonly, interest can be compounded:

- half-yearly   (two times per year)
- quarterly   (four times per year)
- monthly   (12 times per year)
- daily!   (365 or 366 times a year)

Example 15

Calculate the final balance of a $10 000 investment at 6% p.a. where interest is compounded quarterly for one year.

We need to calculate the interest generated each quarter.

Quarter	Principal (1)	Interest ($= C \times r \times n$) (2)	Balance (1) + (2)
1	$10 000.00	$10 000.00 $\times 0.06 \times \frac{1}{4} =$ $150.00	$10 150.00
2	$10 150.00	$10 150.00 $\times 0.06 \times \frac{1}{4} =$ $152.25	$10 302.25
3	$10 302.25	$10 302.25 $\times 0.06 \times \frac{1}{4} =$ $154.53	$10 456.78
4	$10 456.78	$10 456.78 $\times 0.06 \times \frac{1}{4} =$ $156.85	$10 613.63

Thus, the final balance would be $10 613.63

EXERCISE 13C.2

1 Mac places $8500 in a fixed deposit account that pays interest at the rate of 6% p.a. compounded quarterly. How much will Mac have in his account after 1 year?

2 Michaela invests her savings of $24 000 in an account that pays 5% p.a. compounded monthly. How much interest will she earn in 3 months?

3 Compare the interest paid on an investment of $45 000 at 8.5% p.a. over 2 years if the interest is:

　a simple interest　　　b compounded $\frac{1}{2}$ yearly　　　c compounded quarterly.

COMPOUND INTEREST FORMULAE

$$A = C \times \left(1 + \tfrac{r}{100}\right)^n \quad \text{and} \quad I = C \times \left(1 + \tfrac{r}{100}\right)^n - C \quad \text{where}$$

A is the **future value** (or **final balance**)

C is the **present value** or **principal** (amount originally invested)

r is the **interest rate per compound period**

n is the **number of periods** (i.e., the number of times the interest is compounded).

Example 16

> Rework **Example 15**   i.e., calculate the final balance of a $10 000 investment at 6% p.a. where interest is compounded quarterly for one year.

$$C = 10\,000, \quad r = \frac{6\%}{4} = 1.5\% \text{ (per quarter)}, \quad n = 1 \times 4 = 4$$

Now $A = C \times (1 + \frac{r}{100})^n$

So, $A = 10\,000 \times (1 + 0.015)^4$

∴ $A = 10\,613.64$

i.e.,   the final balance is $10\,613.64

Notice that 1.5% per quarter is paid for 4 quarters.

Example 17

> How much interest is earned if $8800 is placed in an account that pays $4\frac{1}{2}\%$ p.a. compounded monthly for $3\frac{1}{2}$ years?

$C = 8800$

$r = \dfrac{4.5}{12} = 0.375$

$n = 12 \times 3\frac{1}{2} = 42$

Now $I = C \times (1 + \frac{r}{100})^n - C$

So, $I = 8800 \times (1 + 0.00375)^{42} - 8800$

∴ $I = \$1498.08$

EXERCISE 13C.3

1   Ali places £9000 in a savings account that pays 8% p.a. compounded quarterly. How much will she have in the account after 5 years?

2   How much interest would be earned on a deposit of $2500 at 5% p.a. compounded half yearly for 4 years?

3   Compare the interest earned on 35 000 Yen left for 3 years in an account paying $4\frac{1}{2}\%$ p.a. when the interest is:
   a   simple interest
   b   compounded annually
   c   compounded half yearly
   d   compounded quarterly
   e   compounded monthly.

4   Jai recently inherited $92 000 and must invest it for 10 years before he spends any of it. The two banks in his town offer the following terms:
   Bank A: $5\frac{1}{2}\%$ p.a. compounded yearly.
   Bank B: $5\frac{1}{4}\%$ p.a. compounded monthly.
   Which bank offers Jai the greater interest on his inheritance?

5   Mimi has 28 000 euro to invest and can place it in an account that pays 8% p.a. simple interest or one that pays $7\frac{1}{2}\%$ p.a. compounded monthly. Which account will earn her more interest over a 4 year period and how much more will it be?

INVESTIGATION 1   DOUBLING TIME

Many investors pose the question: "How long will it take to double my money?"

Most graphics calculators have an in-built **finance program** that can be used to investigate financial scenarios. For example, the TI-83 has a TVM Solver.

TVM stands for **time value of money** and the **TVM Solver** can be used to find any of the variables below given the other variables.

On this screen:

- N represents the **number of time periods**
- $I\%$ represents the **interest rate per year**
- PV represents the **present value** of the investment
- PMT represents the **payment each time period**
- FV represents the **future value** of the investment
- P/Y is the **number of payments per year**
- C/Y is the **number of compounding periods per year**
- PMT : END BEGIN   lets you choose between the payments at the end of a time period or at the *beginning* of a time period. Most interest payments are made at the *end* of the time periods.

Consider an investment of $5000 at 7.2% p.a. compounded annually for 10 years.

To investigate this using the TVM solver on the calculator, set up the TVM screen as shown:

Note: All money being invested is considered as outgoings and is entered as a negative value. ⟶

There are no payments into the account during the term of the investment, so PMT is set to 0.

```
N=10
I%=7.2
PV=-5000
PMT=0
FV=■
P/Y=1
C/Y=1
PMT:END BEGIN
```

Highlight FV and press **ALPHA** **SOLVE** to find the future value.

The investment amounts to $10 021.16, ⟶
i.e.,   the $5000 *doubles* in value.

```
N=10
I%=7.2
PV=-5000
PMT=0
■FV=10021.15681
P/Y=1
C/Y=1
PMT:END BEGIN
```

What to do:

1 Use the in-built finance program on a **graphics calculator** to calculate the amount the following investments grow to if interest is compounded **annually**:

   a $10 000 invested at 8% p.a. for 9 years

   b $10 000 invested at 6% p.a. for 12 years

   c $10 000 invested at 4% p.a. for 18 years.

2 You should notice that each investment approximately *doubles* in value. Can you see a pattern involving the interest rate and time of the investment?   [**Hint:**   Multiply the rate by time (in years).]

3 Write a rough rule that would tell an investor:

   a how long they need to invest their money at a given annual compound rate for it to double in value

   b the annual compound rate they need to invest at for a given time period for it to double in value.

4 Use your rule to approximate the time needed for a $6000 investment to double in value at the following annual compound rates:

   a 2% p.a.   **b** 5% p.a.   **c** 10% p.a.   **d** 18% p.a.

Check your approximations by using a **graphics calculator**.

5 Use your rule to approximate the annual compound interest rate required for a $50 000 investment to double in value in:

   a 20 years   **b** 10 years   **c** 5 years   **d** 2 years.

Check your approximations by using a **graphics calculator**.

Example 18

Holly invests $15 000 in an account that pays 4.25% p.a. compounded monthly. How much is her investment worth after 5 years?

To answer this using the TVM function on the calculator, first set up the TVM screen. **Note:** The initial investment is considered as an outgoing and is entered as a negative value.

For **TI-83**:    For **Casio**: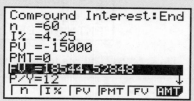

Holly has $18 544.53 after 5 years.

EXERCISE 13C.4

Use a graphics calculator to answer the following questions.

1 If I deposit $6000 in a bank account that pays 5% p.a. compounded daily, how much will I have in my account after 2 years?

2 When my child was born I deposited 2000 in a bank account paying 4% p.a. compounded half-yearly. How much would my child receive on her 18th birthday?

3 Calculate the compound interest earned on an investment of $13 000 for 4 years if the interest rate is 7% p.a. compounded quarterly.

4 Investigate the difference a change in interest rate makes by considering 5000 rupees invested for 3 years when the annual interest rate compounded monthly is:

   a 2%   **b** 4%   **c** 8%

Compare the interest earned in each case and write a brief statement explaining the results.

FINDING THE PRESENT VALUE

The **present value** (principal) is the amount of money that we invest now, at a given compound rate over a given time interval, so that it will amount to a given future value.

Example 19

How much does Halena need to deposit into an account to collect $50 000 at the end of 3 years if the account is paying 5.2% p.a. compounded quarterly?

Formula solution:

Given $A = 50\,000$

$$r = \frac{5.2}{4} = 1.3$$

$$n = 3 \times 4 = 12$$

Using $A = C \times (1 + \frac{r}{100})^n$

$\therefore \quad 50\,000 = C \times (1 + \frac{1.3}{100})^{12}$

$\therefore \quad C = 42\,820.99$ {using solver}

i.e., $42\,821 needs to be deposited.

Graphics Calculator Solution:

To answer this using the TVM function on the calculator, set up the TVM screen as shown: **Note:**   There are $3 \times 4 = 12$ quarter periods.

For **TI-83**:

For **Casio**

Thus, $42 821 needs to be deposited.

EXERCISE 13C.5

1   Calculate the amount you would need to invest now in order to accumulate 25 000 Yen in 5 years time, if the interest rate is 4.5% p.a. compounded monthly.

2   You have set your sights on buying a car costing $23 000 in two years time. Your bank account pays 5% p.a. compounded semi-annually. How much would you need to deposit now in order to be able to buy your car in two years time?

3   You have just won the lottery and decide to invest the money. Your accountant advises you to deposit your winnings in an account that pays 5% p.a. compounded daily. Your accountant states that after two years your winnings have grown to $88 413.07. How much did you win in the lottery?

4   Thirty years ago your father purchased some land in the country which is now worth 225 000 baht. If inflation over that period averaged 3.5% p.a. (and assuming the increase in value is due to inflation only) what was the original cost of the property?

5   Before leaving for overseas on a three year trip to India, I deposit a sum of money in an account that pays 6% p.a. compounded quarterly. When I return from the trip, the amount in my account stands at $9564.95. How much interest has been added since I have been away?

FINDING THE TIME PERIOD

Often we wish to know **how long** we must invest money for it to grow to a specified sum in the future.

Example 20

For how long must Magnus invest $4000 at 6.45% p.a. compounded half-yearly if it is to amount to $10 000?

Formula solution:

Given $A = 10\,000$

$C = 4000$

$r = \dfrac{6.45}{2} = 3.225$

Using $A = C \times (1 + \frac{r}{100})^n$

$\therefore \quad 10\,000 = 4000 \times (1 + \frac{3.225}{100})^n$

$\therefore \quad 10\,000 = 4000 \times (1.03225)^n$

$\therefore \quad n = 28.9 \quad$ {using solver}

Thus, 29 half-years are required, i.e., 14.5 years.

Graphics calculator:

To answer this using the TVM function on the calculator, set up the TVM screen as shown. We then need to find the number of periods required.

For **TI-83**: 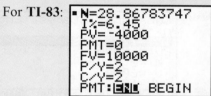   For **Casio**: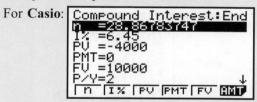

$n = 28.9$, thus, 29 half-years are required, i.e., 14.5 years.

Note: We find the number of compounding periods and need to convert to the time units required. The appropriate time periods in the above example are years.

EXERCISE 13C.6

1 Your parents give you $8000 to buy a car but the car you want costs $9200. You deposit this money in an account that pays 6% p.a. compounded monthly. How long will it be before you have enough money to buy the car you want?

2 A couple inherit $40 000 and deposit it in an account that pays $4\frac{1}{2}$% p.a. compounded quarterly. They withdraw the money as soon as they have over $45 000. How long did they keep the money in that account?

3 A business deposits £80 000 in an account that pays $5\frac{1}{4}$% p.a. compounded monthly. How long will it take before they double their money?

4 An investor deposits $12 000 in an account paying 5% p.a. compounded daily. How long will it take the investor to earn $5000 in interest?

FINDING THE ANNUAL RATE OF INCREASE

When comparing two investments where the present and final values are known over a particular time period, we can calculate the **annual rate of increase**.

Example 21

If Iman deposits $5000 in an account that compounds interest monthly and 2.5 years later the account totals $6000, what annual rate of interest was paid?

Formula solution:

Given $A = 6000$
$C = 5000$
$n = 2.5 \times 12 = 30$

Using $A = C(1 + \frac{r}{100})^n$

$\therefore$ $6000 = 5000 \times (1 + \frac{r}{100})^{30}$

$\therefore$ $r = 0.609\,589$ per month

$\therefore$ $r = 0.609\,589 \times 12$ per year

$\therefore$ $r \doteqdot 7.32\%$ per year

i.e.,   7.32% p.a. is required.

Graphics calculator:

To answer this using the TVM function on the calculator, set up the TVM screen as shown.   **Note:** $2.5 \times 12 = 30$ months

For **TI-83**:    For **Casio**: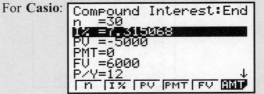

We find the annual interest rate required.

Thus, 7.32% p.a. is required.

EXERCISE 13C.7

1 An investor purchases coins for $10\,000 and hopes to sell them 3 years later for $15\,000. What would be the annual increase in value of the coins over this period, assuming uniform annual increases?

2 If I deposit $5000 in an account that compounds interest monthly and $3\frac{1}{2}$ years later the account totals $6165, what annual rate of interest did the account pay?

3 A young couple invest their savings of 9000 Yen in an account where the interest is compounded annually. Three years later the account balance is 10\,493.22 Yen. What interest rate was being paid?

4 An investor purchased a parcel of shares for $15\,000 and sold them 4 years later for $24\,500. He also purchased a house for $105\,000 and sold it 7 years later for $198\,000. Which investment had the greater average annual percentage increase in value?

FIXED TERM DEPOSITS

As the name suggests, deposits can be locked away for a fixed time period (from one month to ten years) at a fixed interest rate.

The interest is calculated on the daily balance and can be paid monthly, quarterly, half-yearly or annually.

The interest can be compounded so that the principal increases during the fixed term. However, the interest can also be paid out; many retirees live off interest that fixed term deposits generate.

Generally, the rate offered increases if the money is locked away for a longer period of time. We will consider scenarios where the interest is compounded as an application of **compound interest**.

Below is a typical schedule of rates offered by a financial institution:

$5000 to $25 000 and $25 000 to $100 000

Term (months)	Interest at Maturity		Monthly Interest		Quarterly Interest		Half Yearly Interest	
	$5k - $25k	$25k - $100k	$5k - $25k	$25k - $100k	$5k - $25k	$25k - $100k	$5k - $25k	$25k - $100k
1	3.80%	4.50%						
2	4.15%	4.75%						
3	5.20%	5.20%						
4	5.40%	6.00%						
5*	5.45%*	6.25%*						
6	5.45%	5.50%					5.45%	5.50%
7 - 8*	6.10%*	6.35%*					6.05%*	6.30%*
9 - 11	5.65%	5.90%					5.65%	5.90%
12 - 17*	6.30%*	6.40%*	6.15%*	6.20%*	6.15%*	6.25%*	6.20%*	6.30%*
18 - 23	6.10%	6.40%	5.95%	6.20%	5.95%	6.25%	6.00%	6.30%
24 - 29	6.20%	6.40%	6.05%	6.20%	6.05%	6.25%	6.10%	6.30%
30 - 35	6.20%	6.40%	6.05%	6.20%	6.05%	6.25%	6.10%	6.30%
36 - 41	6.20%	6.40%	6.05%	6.20%	6.05%	6.25%	6.10%	6.30%
42 - 47	6.20%	6.40%	6.05%	6.20%	6.05%	6.25%	6.10%	6.30%
48 - 59	6.20%	6.40%	6.05%	6.20%	6.05%	6.25%	6.10%	6.30%
60 - 120	6.20%	6.40%	6.05%	6.20%	6.05%	6.25%	6.10%	6.30%

* indicates special offers

Please note: All interest rates are per annum. For terms of 12 months or more, interest must be paid at least annually.

© Jim Russell - General-Features

Example 22

For the financial institution whose rates are listed above, compare the interest offered if $45 000 is deposited for 15 months and the interest is compounded:

a   monthly   **b**   quarterly.

a $45 000 deposited for 15 months with the interest compounded monthly receives 6.2% p.a. Using a graphics calculator, we solve for FV.

For **TI-83**:    For **Casio**: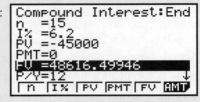

Interest $= \$48\,616.50 - \$45\,000 = \$3616.50$

b $45 000 deposited for 15 months with the interest compounded quarterly receives 6.25% p.a. Using a graphics calculator, we solve for FV.

For **TI-83**: 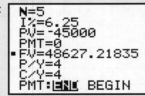   For **Casio**: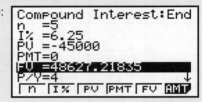

Interest $= \$48\,627.22 - \$45\,000 = \$3627.22$

Thus, the quarterly option is more by $10.72.

Example 23

$10 000 is invested in a fixed term deposit for 24 months, with interest paid monthly. Find the effective after tax return if the investor's tax rate is 48.5 cents in the dollar.

$10 000 deposited for 24 months with the interest compounded monthly receives 6.05% p.a.

{from the Fixed Term Deposit Rate table}

Using a graphics calculator, we solve for FV.

For **TI-83**:
```
N=24
I%=6.05
PV=-10000
PMT=0
•FV=11282.81863
P/Y=12
C/Y=12
PMT:END BEGIN
```

Interest $= \$11\,282.82 - \$10\,000$

$\qquad\quad = \$1282.82$

Tax $= 48.5\%$ of $1282.82

$\qquad = \$622.17$

For **Casio**:
```
Compound Interest:End
n  =24
I% =6.05
PV =-10000
PMT=0
FV =11282.81863
P/Y=12
 n  I% PV PMT FV AMT
```

After tax return $= \$1282.82 - \$622.17$

$\qquad\qquad\qquad = \$660.65$ (which is an average of 3.30% p.a.)

EXERCISE 13C.8

In the exercises below refer to the Fixed Term deposit rates on page **439**:

1   Compare the interest offered if \$18 000 is deposited for 18 months and the interest is compounded: **a**   monthly   **b**   half-yearly.

2   Calculate the effective after tax return for the investments in **1** if the investor's tax rate is: **a**   48.5 cents in the dollar   **b**   31.5 cents in the dollar.

3   Derk wins \$50 000 and decides to deposit it in a fixed term deposit for one year.
   a   Which option would you advise Derk to invest in?
   b   How much interest would he earn?
   c   What is the effective after tax return if Derk's tax rate is 44.5%?

THE EFFECTIVE INTEREST RATE ON AN INVESTMENT

Because interest rates are applied in different ways, comparing them can be misleading.

Consider \$10 000 invested at 6% compounded annually.

After 1 year,

$$C = \$10\,000 \times (1 + \tfrac{0.06}{1})^1$$
$$= \$10\,000 \times (1.06)^1$$
$$= \$10\,600$$

Now consider \$10 000 invested at 5.85% compounded monthly.

After 1 year,

$$C = \$10\,000 \times (1 + \tfrac{0.0585}{12})^{12}$$
$$= \$10\,000 \times (1.004875)^{12}$$
$$= \$10\,600.94$$

Hence, 5.85% p.a. compounded monthly is equivalent to 6% p.a. compounded annually.

We say 5.85% p.a. compounded monthly is a **nominal rate** (the named rate) and it is equivalent to an **effective rate** of 6% p.a. compounded on an annual basis.

> The **effective rate** is the **equivalent annualised rate** (i.e., the **interest rate compounded annually**).

Conversion of Nominal (Compound) Rates to an effective rate:

$$r = (1 + i)^c - 1 \quad \text{where} \quad$$
   r is the **effective rate**
   i is the **rate per compound interest period**
   c is the **number of compound periods per annum**

Example 24

Which is the better rate offered:
   4% p.a. compounded monthly   or   4.2% p.a. compounded quarterly?

Given $i = \dfrac{0.04}{12} = 0.003\,33 \ldots$

   $c = 12$

$$r = (1 + i)^c - 1$$
$$= (1.003\,33 \ldots)^{12} - 1$$
$$= 0.040\,74 \ldots$$

i.e., effective rate is 4.07% p.a.

Given $i = \dfrac{0.042}{4} = 0.0105$

$c = 4$

$$\begin{aligned} r &= (1+i)^c - 1 \\ &= (1.0105)^4 - 1 \\ &= 0.04266 \ .... \end{aligned}$$

i.e.,   effective rate is 4.27% p.a.

Thus, the better rate for an investment is 4.2% p.a. compounded quarterly.

CALCULATOR CONVERSION OF NOMINAL RATE TO EFFECTIVE RATE

A graphics calculator can be used to do the calculations above. There is an effective rate conversion function in the finance program.

The syntax is   Eff(nominal rate, number of compound periods p.a.)

For example,   Eff(4, 12) = 4.07   and   Eff(4.2, 4) = 4.27

Note: If **effective rate** is given, use **annual compounds**.

EXERCISE 13C.9

1   Which is the better rate offered:
    A: 5.4% p.a. compounded half-yearly   or   B: 5.3% p.a. compounded quarterly?

2   Which is the better rate offered:
    P: 7.6% p.a. compounded monthly   or   Q: 7.75% p.a. compounded half-yearly?

3   Find the effective rate of interest on an investment where the nominal rate is:
    a   4.95% p.a. compounded annually   b   4.9% p.a. compounded monthly
    c   Which rate would you invest in?

4   Find the effective rate of interest on an investment where the nominal rate is:
    a   7.75% p.a. compounded daily   b   7.95% p.a. compounded half-yearly
    c   Which rate would you invest in?

5   A bank offers an investment rate of 6.8% p.a. They claim the account will effectively yield 7.02% p.a. How many times is the interest compounded p.a.?

6   Assume you have $40 000 to invest in a fixed term deposit for one year. Consult the rates on page **439** and calculate the effective rate of interest if it is paid:
    a   monthly
    b   quarterly
    c   half-yearly
    d   at maturity.
    e   Which option would you take and how much interest would you receive?

CONVERSION OF A SIMPLE INTEREST RATE TO AN EFFECTIVE RATE

An investment offering a simple interest rate over a number of years can also be converted to an equivalent rate compounded annually.

Example 25

Calculate the effective interest rate for a simple interest rate of 6.5% p.a. applied for 5 years.

Consider $100 invested.   Interest $= C \times r \times n = \$100 \times 0.065 = \$32.50$

Thus, the $100 grows to $\$100 + \$32.50 = \$132.50$

If the effective rate was $r\%$ p.a. then using $C \times (1 + \dfrac{r}{100})^n = A$

$$\therefore \quad 100 \times (1 + \frac{r}{100})^5 = 132.50$$

$$\therefore \quad r = 5.79 \quad \text{\{solver\}}$$

i.e.,   the effective interest rate is $\doteqdot 5.79\%$ p.a.

EXERCISE 13C.10

1 Calculate the effective interest rate for a simple interest rate of:

    a   7.5% p.a. over 5 years
       b   7.5% p.a. over 10 years

    c   7.5% p.a. over 15 years
       d   7.5% p.a. over 20 years

2 Calculate the effective interest rate for a simple interest rate of:

    a   8.4% p.a. over 4 years
       b   4.9% p.a. over 7 years

    c   6.25% p.a. over 10 years
       d   3.75% p.a. over 3 years

3 Five year government bonds offer a coupon rate of 7.25% p.a. Calculate the effective rate offered on this investment.

D   DEPRECIATION

When a car, piece of machinery, office furniture etc. is used it will lose value over time due to wear-and-tear, obsolescence and other factors. We say that it **depreciates** (decreases) in value.

> **Depreciation** is the loss in value of an item over time.

Where such things are essential for a business to earn an income, the Taxation Office may allow that business to claim this depreciation as a **tax deduction**.

Many items are depreciated by a constant percentage for each year of their useful life. They are said to be depreciated on their **reduced balance**.

The following table shows how a computer costing $8500 depreciates over 3 years at 30% each year. The depreciated value is also called the **book value**.

Age (years)	Depreciation	Book Value
0		$8500.00
1	30% of $8500.00 = $2550.00	$8500.00 − $2550.00 = $5950.00
2	30% of $5950.00 = $1785.00	$5950.00 − $1785.00 = $4165.00
3	30% of $4165.00 = $1249.50	$4165.00 − $1249.50 = $2915.50

Notice that the annual depreciation decreases each year as it is calculated on the reduced balance of the item.

We can also use the concept of chain percentage decreases to calculate the value of the computer after 3 years.

Each year, the computer is only $70\% = (100\% - 30\%)$ of its previous value.

$$\begin{aligned}\therefore \quad \text{value after 3 years} &= \$8500 \times 0.7 \times 0.7 \times 0.7 \\ &= \$8500 \times (0.7)^3 \\ &= \$2915.50\end{aligned}$$

When calculating depreciation, the **annual multiplier** is $(1 - \frac{r}{100})$
where r is the annual depreciation rate percentage.

The compound interest calculating formula can again be used to calculate the value of an asset that depreciates by a constant percentage each year.

However this time r is **negative**.

> The **depreciation formula** is:
>
> $$A = C \times (1 + \tfrac{r}{100})^n \quad \text{where}$$
>
> A is the **future value** after n time periods
> C is the **original purchase price**
> r is the **depreciation rate per period and r is negative**
> n is the **number of periods**.

Depreciation is an example of exponential decay!

EXERCISE 13D

1   **a**   Copy and complete the following table to find the value of a deep fryer purchased by a fish and chip shop for $15 000 and depreciated by 15% each year for 3 years:

Age (years)	Depreciation	Book Value
0		$15 000
1	15% of $15 000 = $2250	
2		
3		

b   Calculate how much depreciation can be claimed as a tax deduction by the shop in:
   i   Year 1   **ii**   Year 2   **iii**   Year 3

Example 26

A photocopier was purchased for $12 500 and depreciated at 15% each year.
a   Find its value after five years.   **b**   By how much did it depreciate?

a   Now $A = C \times (1 + \frac{r}{100})^n$ where $C = 12\,500, \ r = -15, \ n = 5$
$$\begin{aligned}&= 12\,500 \times (1 - 0.15)^5 \\ &= 12\,500 \times (0.85)^5 \\ &\doteqdot 5546.32 \qquad \text{i.e., after 5 years the value is } \$5546.32\end{aligned}$$

b   Depreciation $= \$12\,500 - \$5546.32 = \$6953.68$

2 **a** Find the future value of a truck which is purchased for \$225 000 if it depreciates at 25% p.a. for 5 years.

  b By how much did it depreciate?

3 **a** If I buy a car for \$32 400 and keep it for 3 years, what will its value be at the end of that period given that its annual depreciation rate is 20%?

  b By how much did it depreciate?

Example 27

A vending machine bought for \$15 000 is sold 3 years later for \$9540. Calculate its annual rate of depreciation.

$A = 9540$, $C = 15\,000$, $n = 3$

$$A = C(1 + \tfrac{r}{100})^n$$

$\therefore \quad 9540 = 15\,000 \times (1 + \tfrac{r}{100})^3$

$\therefore \quad r \doteqdot 0.140025$ $\qquad$ {using solver}

i.e., the annual rate of depreciation is 14.0% p.a.

```
Eq:A=C(1+R÷100)^N
  R=-14.00252396
Lft=9540
Rgt=9540

REPT
```

---

4 A printing press costing £250 000 was sold 4 years later for £80 000. At what yearly rate did it depreciate in value?

5 A 4-wheel-drive vehicle was purchased for \$45 000 and sold for \$28 500 after 2 years and 3 months. Find its annual rate of depreciation.

6 The Taxation Office allows industrial vehicles to be depreciated at $7\tfrac{1}{2}$% each 6 months.

  a What would be the value in 2 years' time of vehicles currently worth \$240 000?

  b By how much have they depreciated?

E   PERSONAL LOANS

A common way to borrow money to finance purchases such as cars, boats, renovations, overseas holidays, education expenses or share portfolios is to take out a personal loan. Banks, credit unions and finance companies will all offer personal loans with differing terms, conditions and interest rates.

Personal loans are usually **short term** (6 months to 7 years) and can be either **secured** or **unsecured**. Car loans are usually **secured**. This means the car acts as **security** in the event that the borrower defaults on the payments. The bank has the right to sell the car and take the money owed to them. A loan to pay for an overseas holiday may be **unsecured**. Such loans will usually charge higher interest rates than **secured** loans.

Interest is calculated on the **reducing balance** of the loan so that the interest reduces as the loan is repaid. Borrowers can usually choose between **fixed** or **variable** interest rates.

Fixed rate loans have fixed repayments for the entire loan period. This may appeal to people on a tight budget.

Variable rate loans have interest rates that fluctuate with economic changes and thus repayments may vary. However, higher repayments than the minimum are allowed if you want to try to pay the loan off sooner.

INTEREST

Interest will be the biggest cost involved in repaying a personal loan. The borrower will be given an indication of the regular repayment amount (for example, per month) based on the loan amount, time of the loan and the interest rate charged.

A table of monthly repayments follows and is based on borrowing $1000.

Loan term (months)	*Table of Monthly Repayments per $1000*						
	Annual interest rate						
	10.0%	10.5%	11.0%	11.5%	12.0%	12.5%	13.0%
12	87.9159	88.1486	88.6151	88.8488	89.0829	89.3173	89.3173
18	60.0571	60.2876	60.5185	60.7500	60.9820	61.2146	61.4476
24	46.1449	46.3760	46.6078	46.8403	47.0735	47.3073	47.5418
30	37.8114	38.0443	38.2781	38.5127	38.7481	38.9844	39.2215
36	32.2672	32.5024	32.7387	32.9760	33.2143	33.4536	33.6940
42	28.3168	28.5547	28.7939	29.0342	29.2756	29.5183	29.7621
48	25.3626	25.6034	25.8455	26.0890	26.3338	26.5800	26.8275
54	23.0724	23.0724	23.5615	23.8083	24.0566	24.3064	24.5577
60	21.2470	21.4939	21.7424	21.9926	22.2444	22.4979	22.7531

Note: The resulting monthly instalments are rounded off to the **next 10 cents**.
For example, $485.51 becomes $485.60

Example 28

Francine takes out a personal loan for $16 500 to buy a car. She negotiates a term of 4 years at 11.5% p.a. interest. Calculate the:
a   monthly repayments   **b**   total repayments   **c**   interest charged.

a   From the table, the monthly repayments on each $1000 for 4 years
(48 months) at 11.5% p.a. = $26.0890

$\therefore$ repayments on $16 500   = $26.0890 × 16.5   {16.5 lots of $1000}
= $430.4685
$\doteqdot$ $430.50   {next 10 cents}

i.e., $430.50 per month.

b   Total repayments   = monthly repayment × number of months
= $430.50 × 48
= $20 664

i.e., $20 664 is repaid in total.

c   Interest   = total repayments − amount borrowed
= $20 664 − $16 500
= $4164   i.e., $4164 is paid in interest.

The in-built finance program on a **graphics calculator** can also be used to calculate the monthly repayments on a loan.

For example, the TVM Solver on a TI-83 will calculate the monthly repayments for **Example 28** if the following information is entered:

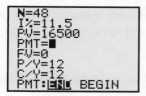

Highlight *PMT* and press **ALPHA** **SOLVE** to find the monthly repayment.

Round the monthly repayment of $430.47 off to the **next 10 cents** i.e., $430.50.

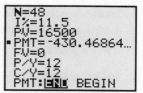

Example 29

Show the progress of Francine's loan from **Example 28** for the first 3 months (assume March, April, May) and state the outstanding balance.

With a reducing balance loan, interest is calculated on the balance, added on and the repayment deducted,

   i.e.,   outstanding balance = previous balance + interest − repayment

A table is a convenient way of showing the progress:

Mth.	Opening balance	Days	Interest	Repmt.	Closing bal.
Mar	$16 500.00	31	$16 500.00 \times 0.115 \times \frac{31}{365}$ $= 161.16	$430.50	$16 500 +$161.16 −$430.50 = $16 230.66
Apr	$16 230.66	30	$16 230.66 \times 0.115 \times \frac{30}{365}$ $= 153.41	$430.50	$16 230.66 +$153.41 −$430.50 = $15 953.57
May	$15 953.57	31	$15 953.57 \times 0.115 \times \frac{31}{365}$ $= 155.82	$430.50	$15 953.57 +$155.82 −$430.50 = $15 678.89

Thus, the outstanding balance after 3 months is $15 678.89

EXERCISE 13E

1 Raphael takes out a personal loan for $12 000 to go overseas. He will repay it over 5 years at 12% p.a. Calculate the:

   a monthly repayments   **b** total repayments   **c** interest charged

   d outstanding balance of the loan after 2 months. (Assume one month is $\frac{1}{12}$ year.)

2 Jay and Penni need $9500 to fund house renovations. They take out a personal loan over 3 years at 10.5% p.a. Calculate the:

   a monthly repayments   **b** total repayments   **c** interest charged

   d outstanding balance of the loan after 3 months. (Assume one month is $\frac{1}{12}$ year.)

3 Binh-vu needs 15 000 Yen to buy a boat. His bank offers him a personal loan at 12.5% p.a. Calculate the total interest he will pay if he repays it over:

   a 2 years   **b** 5 years.

4 Becky wants to borrow $25 000 to purchase some shares. Calculate the total interest charged for the following options:

   a Balance Bank: offer 11.5% over 5 years

   b Cash Credit Union: offer 12.5% over 3 years.

What would you recommend for Becky?

INVESTIGATION 2   BUYING A CAR

Use the skills and knowledge you have built up in this chapter to **investigate** the following scenario. You could possibly set up a **spreadsheet** to help you.

Alma is considering buying a new car (she has an old one that still gets her around). She wants to spend around $20 000. She has $5000 in her savings and a spare $500 per month to either save or use on repayments.

What to do:

1 Assume she banks with 'Balance Bank' and invests the $5000 in a 6 month Term Deposit Account. She also opens a Cash Management Account and saves $500 per month into it. Investigate how long it will take her to have around $20 000 saved. Detail your assumptions and calculations, including interest earned. (You may or may not want to consider tax.)

2 Assume she wants to buy a $20 000 car now. She has two choices:

   • **Personal Loan:**   available at 11% p.a. over 3 years

  or • **Paying on terms:**   the dealer will accept a 10% deposit and $420 per month over 5 years.

Investigate the costs involved in each option. Detail any assumptions you make (like the size of the personal loan) and calculate total costs and interest paid. Which option would you recommend and why?

3 Investigate a combination of **saving** and **borrowing**, i.e., could Alma save for 6 months or a year and then borrow the money? Detail any assumptions you make and set out all calculations.

4 Based on your investigations to **1**, **2** and **3**, how would **you** advise Alma to buy the car? Detail your assumptions and calculations.

THE EFFECT OF INFLATION

The **Consumer Price Index (CPI)** measures the increase in price of a general 'basket' of goods and services over time and is an accepted method of measuring inflation. Inflation effectively erodes the purchasing power of money as a fixed amount of money will not be able to purchase the same amount of goods and services over time as prices increase.

For example, you may have $20 000 ready to purchase a new car. If you delay buying the car now, a similar new car may cost $23 000 in a few years. Your $20 000 has lost some of its purchasing power due to the inflationary effect. Of course, the $20 000 could be invested at a rate greater than that of inflation to counter this.

Many investors take the effects of inflation into account when they assess the returns received from investments. The **real rate of return** takes into account the effect of inflation.

Example 30

$10 000 is deposited in a fixed term account for 3 years with interest of 5.4% p.a. compounded monthly. Inflation over the period averages 2.5% p.a.

a Calculate the value of the investment after three years.

b What is the value of the $10 000 indexed for inflation?

c What is the real increase in value of the investment?

d Calculate the real average annual percentage increase in the investment.

---

a Using a graphics calculator, we solve for FV.

$N = 36$, $I\% = 5.4$, $PV = -10\,000$, $PMT = 0$, $P/Y = 12$, $C/Y = 12$

The investment is worth $11 754.33 after three years.

```
N=36
I%=5.4
PV=-10000
PMT=0
•FV=11754.32999
P/Y=12
C/Y=12
PMT:END BEGIN
```

b Inflation increases at 2.5% p.a. on a compound basis.

$$\begin{aligned}
\text{Indexed value} &= \$10\,000 \times 1.025 \times 1.025 \times 1.025 \\
&= \$10\,000 \times (1.025)^3 \\
&= \$10\,768.91
\end{aligned}$$

c Real increase in value of the investment $= \$11\,754.33 - \$10\,768.91 = \$985.42$

d Using a graphics calculator we find the real average annual percentage increase in the investment, by solving for I%.

$N = 3$, $PV = -10\,768.91$, $PMT = 0$, $FV = 11\,754.33$, $P/Y = 1$, $C/Y = 1$

After inflation there is effectively a 2.96% p.a. increase in the investment.

EXERCISE 13F.1

1 Casey deposited $50 000 in a fixed term account for five years with interest of 5.7% p.a. compounded quarterly. Inflation over the period averages 2.3% p.a.

a Find the value of her investment after five years.

b Calculate the value of the $50 000 indexed for inflation.

c Find the real increase in value of her investment.

d What is the real average annual percentage increase in her investment?

2 Gino invested $20 000 in a fixed term deposit for three years with interest of 3.85% p.a. compounded monthly. Inflation over the period averages 3.4% p.a.

    a What is the value of Gino's investment after three years?

    b Index the $50 000 for inflation.

    c Calculate the real increase in value of Gino's investment.

    d Find the real average annual percentage increase in the investment.

3 Jordan leaves $5000 in an account paying 4.15% p.a. compounded annually for 2 years. Inflation runs at 3.5% p.a. in year 1 and 5.2% p.a. in year 2. Has the real value of the $5000 increased or decreased?

4 Hoang requires $1000 per week to maintain his lifestyle. Assuming inflation increases at an average rate of 3% p.a., how much will Hoang require per week if he wishes to maintain his current lifestyle in: **a**   10 years   **b**   20 years   **c**   30 years?

DISCOUNTING VALUES BY INFLATION

A loaf of bread is a regular purchase for most families. Imagine a loaf of bread costs $2.30 and that the cost of the loaf has risen by the inflation rate in the last 20 years.

If the average annual inflation rate over this period had been 3.8%, what would a loaf of bread have cost 20 years ago?

We let the cost of a loaf of bread 20 years ago be $x.

Thus, $x \times (1.038)^{20} = \$2.30$

$$\therefore \quad x = \frac{\$2.30}{(1.038)^{20}} = \$1.09$$

i.e., the loaf of bread may have cost around $1.09 twenty years ago.

Notice that when we want to discount a value by the inflation rate we *divide*.

Example 31

In 2001, $4000 was invested in a term deposit for 5 years at 5.3% p.a. interest compounded monthly. Inflation over the same period averaged 3.6% p.a.

    a Calculate the amount in the account after 5 years.

    b What is the value of the deposit in 2001 dollars?

    a Using a graphics calculator,

        N = 60,   I% = 5.3,   PV = −4000,   PMT = 0,   P/Y = 12,   C/Y = 12

        There is $5210.68 in the account in 2006.

    b Value the deposit in 2001 dollars $= \dfrac{\$5210.68}{(1.036)^5} = \$4366.12$

EXERCISE 13F.2

1 Mandy invests $15 000 in 2002 in a term deposit for three years at 6.15% p.a. interest compounded quarterly. Inflation over the three year period averages 4.3% p.a.

    a Calculate the amount in the account after three years.

    b What is the value of the deposit in 2002 dollars?

2 Suppose Francis deposits 8000 in a term deposit account at the start of 2003 and receives 4.8% p.a. interest compounded monthly.

    a What amount is in the account after ten years?

    b What is the value of the investment in 2003 dollars if inflation is expected to rise by an average of: **i** 2.5% p.a.   **ii** 3.5% p.a.   **iii** 4.5% p.a.?

3 Christianne invests $25 000 in five year bonds paying 6.25% p.a. in 2003.

    a How much interest will she receive over the five years?

    b Christianne receives her initial investment back at maturity. What is the value of her capital invested in 2003 dollars if inflation has averaged:

        i 3.5% p.a.   **ii** 5.5% p.a.?

4 At the start of 2003, Jon leaves $3000 in a savings account paying 0.5% p.a. interest compounded annually whilst he travels overseas for 4 years.

    a How much will there be in Jon's account when he returns from overseas in 2007?

    b If inflation has averaged 4.2% p.a. over the four year period, what is the value of the account in 2003 dollars?

    c What interest rate did Jon need to invest at to make a real return on his money?

REVIEW SET 13A

1 If 1 UK pound buys 1.846 $US, find how many $US could be bought for £650.

2 If you have 2100 Norwegian kroner and you exchange it for euro, how many euro will you receive if 1 euro = 8.447 Norwegian kroner?

3 Josie places $9600 in an account paying $5\frac{3}{4}\%$ simple interest. How long will it take the account to earn $3000 in interest?

4 Mary borrowed $13 000 from Wally and over 3 years repaid $15 250. What simple interest rate was Mary being charged?

5 Jali deposits her lottery winnings into an account that pays 5.5% p.a. simple interest. After 18 months she has earned $32 600 in interest. How much did she win in lottery?

6 Sven sells his second house and deposits the proceeds of $87 000 in a term deposit account for nine months. The account pays $9\frac{3}{4}\%$ p.a. compounded monthly. How much interest will he earn over this period?

7 Val receives a $285 000 superannuation payment when she retires. She checks a number of banks and finds the following rates being offered:

        Bank A: $6\frac{1}{2}\%$ p.a. simple interest

        Bank B: 6% p.a. compounded quarterly

        Bank C: $5\frac{3}{4}\%$ p.a. compounded monthly.

  Compare the interest that would be received over a ten year period from these banks. In which bank should Val deposit her superannuation?

8 **a** Find the future value of a tractor which is purchased for $48 000 if it depreciates at 15% p.a. for 5 years.

    b By how much did it depreciate?

9 Marcel deposited $25 000 in a fixed term account for 3 years with interest of 5.4% p.a. compounded quarterly. Inflation over the period averages 2.1% p.a.

    a Find the value of his investment after 3 years.

    b Calculate the value of the $25 000 indexed for inflation.

    c Find the real increase in value of his investment.

    d What is the real average annual percentage increase in his investment?

REVIEW SET 13B

1 A bank exchanges 1200 Swiss francs to euro for a commission of 1.8%.

    a What commission is charged?

    b What does the customer receive for the transaction if 1 Swiss franc $= 0.6807$ euro.

2 Pieter has $28 000 to invest and places it in an account that pays $4\frac{1}{2}$% p.a. interest compounded monthly. After 18 months he withdraws all of the money from that account and places it in another account that pays $4\frac{3}{4}$% p.a. interest compounded quarterly. After a further 15 months he withdraws the full amount from the account. How much does he withdraw at this time?

3 Before leaving for overseas on a three year trip to India, I leave a sum of money in an account that pays 6% p.a. compounded semi-annually. When I return from the trip, the amount in my account stands at $5970.26. How much interest has been added since I've been away?

4 Alberto takes out a personal loan for $23 000 at 12.5% p.a. over 5 years to buy a car. Find: **a** his monthly repayments **b** the total cost of the car **c** the interest charged.

5 Jana is about to purchase a car for $18 500 and is considering the following options:

    Option 1:   Buy on terms from a car dealer. The terms are 10% deposit and $117.90 per week for 48 months.

    Option 2:   Borrow the full amount from the bank and repay the loan at $445.10 per month for 5 years.

    a Calculate the total cost of the car for each option.

    b Calculate the amount of interest charged for each option.

    Which option would you recommend for Jana and why?

6 Ian obtains a personal loan of £25 000 for renovating his home. The loan is to be repaid monthly over 5 years and the interest rate is 8.5% p.a.

    a What monthly repayments will Ian have to make?

    b How much interest will Ian pay?

    c What will be the outstanding balance on the loan after 3 months? (Assume each month is $\frac{1}{12}$ year.)

7 Retief invests $10 000 in 2004 in a term deposit for 3 years at 6.45% p.a. interest compounded quarterly. Inflation over the three year period averages 3.1% p.a.

    a Calculate the amount in the account after 3 years.

    b What is the value of the deposit in 2004 dollars?

Chapter 14

Probability

Contents:

In the study of chance, we need a mathematical method to describe the likelihood of an event happening. We do this by carefully assigning a number which lies between 0 and 1 (inclusive).

> An event which has a 0% chance of happening (i.e., is impossible) is assigned a probability of 0.
>
> An event which has a 100% chance of happening (i.e., is certain) is assigned a probability of 1.
>
> All other events can then be assigned a probability between 0 and 1.

The number line below shows how we could interpret different probabilities:

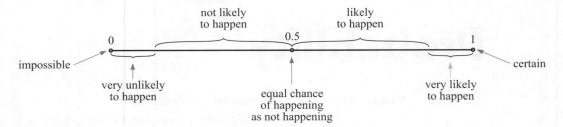

The assigning of probabilities is usually based on either:

- observing the results of an experiment (experimental probability), or
- using arguments of symmetry (theoretical probability).

Probability theory is the study of the *chance* (or likelihood) of events happening.

The study of the theory of chance has vitally important applications in physical and biological sciences, economics, politics, sport, life insurance, quality control, production planning in industry and a host of other areas.

HISTORICAL NOTE

The development of modern probability theory began in 1653 when gambler Chevalier de Mere contacted mathematician **Blaise Pascal** with a problem on how to divide the stakes when a gambling game is interrupted during play. Pascal involved **Pierre de Fermat**, a lawyer and amateur mathematician, and together they solved the problem. While think-

Blaise Pascal   *Pierre de Fermat*

ing about it they laid the foundations upon which the laws of probability were formed.

In the late 17th century, English mathematicians compiled and analysed mortality tables. These tables showed how many people died at different ages. From these tables they could estimate the probability that a person would be alive at a future date. This led to the establishment of the first life-insurance company in 1699.

A EXPERIMENTAL PROBABILITY

In experiments involving chance we agree to use appropriate language to accurately describe what we are doing and the results we are obtaining.

- The number of **trials** is the total number of times the experiment is repeated.
- The **outcomes** are the different results possible for one trial of the experiment.
- The **frequency** of a particular outcome is the number of times that this outcome is observed.
- The **relative frequency** of an outcome is the frequency of that outcome expressed as a fraction or percentage of the total number of trials.

When a small plastic cone was tossed into the air 279 times it fell on its *side* 183 times and on its *base* 96 times.

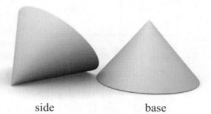

The relative frequencies of *side* and *base* are
$\frac{183}{279} \doteqdot 0.656$ and $\frac{96}{279} \doteqdot 0.344$ respectively.

side   base

In the absence of any further data we say that the relative frequency of each event is our best estimate of the probability of each event occurring.

That is,   Experimental probability = relative frequency.

We write   Experimental P(side) = 0.656,   Experimental P(base) = 0.344

INVESTIGATION 1   TOSSING DRAWING PINS

If a drawing pin finishes we say it has finished on its *back*

and if we say it has finished on its *side*.

If two drawing pins are tossed simultaneously the possible results are:

two backs   *back and side*   *two sides*

What to do:

1 Obtain two drawing pins of the same shape and size. Toss the pair 80 times and record the outcomes in a table.

2 Obtain relative frequencies (experimental probabilities) for each of the three events.

3 Pool your results with four other people and so obtain experimental probabilities from 400 tosses.   **Note:** The others must have pins from the same batch, i.e., the same shape.

4 Which gives the more reliable estimates, your results or the groups'? Why?

5 Keep your results as they may be useful later in this chapter.

Note: In some cases, such as in the investigation above, experimentation is the only way of obtaining probabilities.

EXERCISE 14A

1 When a batch of 145 paper clips were dropped onto 6 cm by 6 cm squared paper it was observed that 113 fell completely inside squares and 32 finished up on the grid lines. Find, to 2 decimal places, the estimated probability of a clip falling:

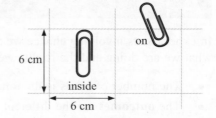

    a inside a square   **b** on a line.

2

Length	Frequency
0 - 19	17
20 - 39	38
40 - 59	19
60+	4

Jose surveyed the length of TV commercials (in seconds). Find to 3 decimal places the estimated probability that a randomly chosen TV commercial will last:

    a 20 to 39 seconds   **b** more than a minute

    c between 20 and 59 seconds (inclusive)

3 Betul keeps records of the number of phone calls she receives over a period of consecutive days.

    a For how many days did the survey last?

    b Estimate Betul's chance of receiving:

        i no phone calls on one day

        ii 5 or more phone calls on a day

        iii less than 3 phone calls on a day.

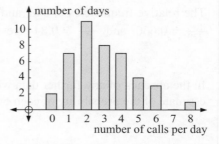

4 Pat does a lot of travelling in her car and she keeps records on how often she fills her car with petrol. The table alongside shows the frequencies of the *number of days between refills*. Estimate the likelihood that:

    a there is a four day gap between refills

    b there is at least a four day gap between refills.

Days between refills	Frequency
1	37
2	81
3	48
4	17
5	6
6	1

B   CHANCE INVESTIGATIONS

In experimental probability, the **relative frequency** of an outcome gives us an **estimate** for the **probability** of that outcome.

In general, the greater the number of trials, the more we can rely on our estimate of the probability.

The most commonly used equipment for experimental probability and games of chance is described below:

COINS

When a **coin** is tossed there are two possible sides that could show upwards; the *head* (usually the head of a monarch, president or leader) or the *tail* (the other side of the coin). We would expect a head (H) and a tail (T) to have equal chance of occurring, i.e., we expect each to occur 50% of the time.

So,   the probability of obtaining a head is $\frac{1}{2}$, and

the probability of obtaining a tail is $\frac{1}{2}$.

The table below shows actual experimental results obtained for tossing a coin:

Number of tosses	H	T	%H	%T
10	7	3	70.0	30.0
100	56	44	56.0	44.0
1000	491	509	49.1	50.9

each is nearly 50%

These experimental results support our expectations and suggest the general rule:

"The more times we repeat an experiment, like tossing a coin, the closer the results will be to the theoretical results we expect to obtain".

When an experiment is repeated many times and the experimental probability is found, we often call this the **long run probability**.

DICE

[**Note:**   **Dice** is the plural of **die.**]

The most commonly used dice are small cubes with the numbers 1, 2, 3, 4, 5 and 6 marked on them by way of dots.

The numbers on the face of a cubic die are arranged such that the sum of each pair of opposite faces is seven.

SPINNERS

A simple **spinner** consists of a regular polygon (or sometimes a circle with equal sectors) with a toothpick or match through its centre.

Alongside is a square spinner showing a result of 1 since it has come to rest on the side marked 1.

A circular spinner such as the one shown alongside may be used instead of a die, providing all angles are of $60°$. The result shown is 2, since the pointer came to rest on the sector marked 2.

What to do:

1 Toss *one coin* 40 times. Record the number of heads resulting in a table.

Result	Tally	Frequency	Relative frequency
1 head			
0 head			

2 Toss *two coins* 60 times. Record the number of heads resulting in a table.

Result	Tally	Frequency	Relative frequency
2 heads			
1 head			
0 head			

3 Toss *three coins* 80 times. Record the number of heads resulting in a table.

Result	Tally	Frequency	Relative frequency
3 heads			
2 heads			
1 head			
0 head			

4 Share your results to **1**, **2** and **3** with several others. Comment on any similarities and differences.

5 Pool your results and find new relative frequencies for tossing one coin, two coins, tossing three coins.

6 Click on the icon to examine a coin tossing simulation.

COIN TOSSING

In ⟨Number of coins⟩ type 1 . In ⟨Number of flips⟩ type 10 000 .
Click ⟨Start⟩ and click ⟨Start⟩ 9 more times, each time recording the % Frequency for each possible result. Comment on these results. Do your results agree with what you expected?

7 Repeat **6** but this time with *two coins* and then repeat **6** but this time with *three coins*.

From the previous investigation you should have observed that there are roughly twice as many 'one head' results as there are 'no heads' or 'two heads'. The explanation for this is best seen using two different coins where you could get:

two heads   one head   one head   no heads

This shows that we should expect two heads : one head : no heads to be 1 : 2 : 1. However, due to chance, there will be variations from this when we look at experimental results.

INVESTIGATION 3   DICE ROLLING EXPERIMENTS

You will need: At least one normal six-sided die with numbers 1 to 6 on its faces. Several dice would be useful to speed up the experimentation.

WORKSHEET

What to do:

1 Examine a die. List the possible outcomes for the uppermost face when the die is rolled.

2 Consider the possible outcomes when the die is rolled 60 times.

Copy and complete the following table of your **expected results**:

Outcomes	Expected frequency	Expected relative frequency
⋮		

3 Roll the die 60 times and record the result on the uppermost face in a table like the one following:

Outcome	Tally	Frequency	Relative frequency
1			
2			
⋮			
6			
Total		60	

4 Pool as much data as you can with other students.
  • Look at similarities and differences from one set to another.
  • Look at the overall pooled data added into one table.

5 How close to your expectation were your results?

6 Use the die rolling simulation from the computer package on the CD to roll the die 10 000 times and repeat this 10 times. On each occasion, record your results in a table like that in **3**. Do your results further confirm your expected results?

7 These are the different possible results when a pair of dice is rolled.

The illustration given shows that when two dice are rolled there are 36 possible outcomes. Of these, $\{1, 3\}$, $\{2, 2\}$ and $\{3, 1\}$ give a sum of 4.

Using the illustration above, copy and complete the following table of expected (theoretical) results:

Sum	2	3	4	5	⋯	12
Fraction of total			$\frac{3}{36}$			
Fraction as decimal			0.083			

8 If a pair of dice is rolled 360 times, how many of each result (2, 3, 4, ..., 12) would you expect to get? Extend your table of **7** by adding another row and write your **expected frequencies** within it.

9 Toss two dice 360 times and record in a table the *sum of the two numbers* for each toss.

Sum	Tally	Frequency	Rel. Frequency
2			
3			
4			
⋮			
12			
Total		360	1

10 Pool as much data as you can with other students and find the overall relative frequency of each *sum*.

WORKSHEET   SIMULATION

11 Use the two dice simulation from the computer package on the CD to roll the pair of dice 10000 times. Repeat this 10 times and on each occasion record your results in a table like that of **9**. Are your results consistent with your expectations?

C | ESTIMATING PROBABILITIES FROM DATA

Statistical information can be used to calculate probabilities in many situations.

Example 1

Short-Term Visitors to Australia

Main reason for journey	April 2003 '000	May 2003 '000	June 2003 '000
Convention/conference	8.3	14.8	8.8
Business	27.2	33.9	32.0
Visiting friends/relatives	77.5	52.7	59.9
Holiday	159.3	119.3	156.5
Employment	4.2	4.3	5.5
Education	9.8	7.9	12.5
Other	35.2	28.0	33.2
Total	321.5	260.9	308.3

Source: Australian Bureau of Statistics

The table shows the number of short-term visitors coming to Australia in the period April - June 2003 and the main reason for their visit.

a What is the probability that a person who arrived in June was here on holiday?

b What is the probability that a person coming to Australia arrived in May?

c Lars arrived in Australia in April, May or June 2003. He came to visit his brother. What is the probability that he arrived in April?

---

a $P(\text{on holiday in June}) = \dfrac{156.5}{308.3}$ ←— number on holiday in June
←— total number for June

$\phantom{P(\text{on holiday in June})} \doteqdot 0.508$

b   There were $321.5 + 260.9 + 308.3 = 890.7$ thousand short-term visitors during the three months.

$\therefore \ P(\text{arrived in May}) = \dfrac{260.9}{890.7}$

$\phantom{\therefore \ P(\text{arrived in May})} \doteqdot 0.293$

c $77.5 + 52.7 + 59.9 = 190.1$ thousand people came to Australia to visit friends or relatives during the period.

$\therefore \ P(\text{arrived in April}) = \dfrac{77.5}{190.1}$

$\phantom{\therefore \ P(\text{arrived in April})} \doteqdot 0.408$

EXERCISE 14C

1 The table shows data from a survey conducted at five schools on the rate of smoking amongst 15 year old students.

School	No. of 15 year olds		No. of smokers	
	Male	Female	Male	Female
A	45	51	10	11
B	36	42	9	6
C	52	49	13	13
D	28	33	9	10
E	40	39	7	4
Total	201	214	48	44

  a What is the probability that a randomly chosen female 15 year old student at school **C** is a smoker?

  b What is the probability that a randomly chosen 15 year old student at school **E** is a smoker?

  c If a 15 year old is chosen at random from the five schools, what is the probability that he or she is a smoker?

2 The given table shows complaints received by the Telecommunications Ombudsman concerning internet services over a four year period.

Reason	1998/99	1999/00	2000/01	2001/02
Access	585	1127	2545	-
Billing	1822	2102	3136	3582
Contracts	242	440	719	836
Credit control	3	44	118	136
Customer Service	12	282	1181	1940
Disconnection	n/a	n/a	n/a	248
Faults	86	79	0	2384
Privacy	93	86	57	60
Provision	172	122	209	311
Total	3015	4282	7965	9497

  a What is the probability that a complaint received in 2000/01 is about customer service?

  b What is the probability that a complaint received at any time during the 4 year period related to billing?

  c What is the probability that a complaint received in 2001/02 did *not* relate to either billing or faults?

3 The table provides data on average daily maximum temperatures in Auburn during summer.

Summer Temperatures in Auburn

	Month		
(Assume that there are 28 days in February)	Dec	Jan	Feb
Mean no. days max. $\geqslant 40°$ C	0.3	1.2	0.7
Mean no. days max. $\geqslant 35°$ C	3.0	5.8	5.3
Mean no. days max. $\geqslant 30°$ C	9.4	12.3	12.6

  a Find the probability that on a February day in Auburn the maximum temperature will:

   i be 35°C or higher
   ii not exceed 30°C

  b Find the probability that on any summer day in Auburn the temperature will be 30°C or higher.

  c It is a 40°C summer's day in Auburn. What is the probability that the month is January?

D   SAMPLE SPACE

A **sample space** is the set of all possible outcomes of an experiment.

There are a variety of ways of representing or illustrating sample spaces.

LISTING OUTCOMES

Example 2

List the sample space of possible outcomes for:
a   tossing a coin   b   rolling a die.

a   When a coin is tossed, there   b   When a die is rolled, there are 6
are two possible outcomes.   possible outcomes.
∴   sample space = {H, T}   ∴   sample space = {1, 2, 3, 4, 5, 6}

2-DIMENSIONAL GRIDS

When an experiment involves more than one operation we can still use listing to illustrate the sample space. However, a grid can often be a better way of achieving this.

Example 3

Illustrate the possible outcomes when 2 coins are tossed by using a 2-dimensional grid.

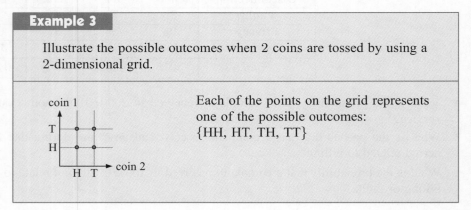

Each of the points on the grid represents one of the possible outcomes:
{HH, HT, TH, TT}

TREE DIAGRAMS

The sample space in **Example 3** could also be represented by a tree diagram. The advantage of tree diagrams is that they can be used when more than two operations are involved.

Example 4

Illustrate, using a tree diagram, the possible outcomes when:
a   tossing two coins
b   drawing two marbles from a bag containing a number of red, green and yellow marbles.

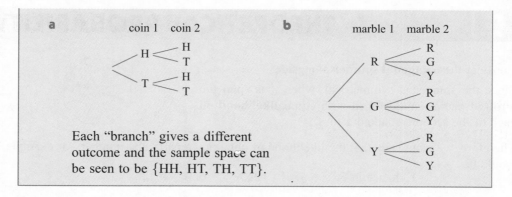

Each "branch" gives a different outcome and the sample space can be seen to be {HH, HT, TH, TT}.

Example 5

John plays Peter at tennis and the first to win two sets wins the match. Illustrate the sample space using a tree diagram.

If J means "John wins the set" and P means "Peter wins the set" then the tree diagram display is:

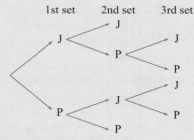

Note: The sample space is   S = {JJ, JPJ, JPP, PJJ, PJP, PP}.

EXERCISE 14D

1   List the sample space for the following:
   a   twirling a square spinner labelled A, B, C, D
   b   the sexes of a 2-child family
   c   the order in which 4 blocks A, B, C and D can be lined up
   d   the 8 different 3-child families.

2   Illustrate on a 2-dimensional grid the sample space for:
   a   rolling a die and tossing a coin simultaneously
   b   rolling two dice
   c   rolling a die and spinning a spinner with sides A, B, C, D
   d   twirling two square spinners; one labelled A, B, C, D and the other 1, 2, 3, 4.

3   Illustrate on a tree diagram the sample space for:
   a   tossing a 5-cent and 10-cent coin simultaneously
   b   tossing a coin and twirling an equilateral triangular spinner labelled A, B and C
   c   twirling two equilateral triangular spinners labelled 1, 2 and 3 and X, Y and Z
   d   drawing two tickets from a hat containing a number of pink, blue and white tickets.

E | THEORETICAL PROBABILITY

Consider the **octagonal spinner** alongside.

Since the spinner is symmetrical, when it is spun the arrowed marker could finish with **equal likelihood** on each of the sections marked 1 to 8.

Therefore, we would say that the likelihood of obtaining a particular number, for example, 4, would be

$$1 \text{ chance in } 8, \quad \tfrac{1}{8}, \quad 12\tfrac{1}{2}\% \quad \text{or} \quad 0.125$$

This is a **mathematical** (or **theoretical**) probability and is based on what we theoretically expect to occur.

> The **theoretical probability** of a particular event is a measure of the chance of that event occurring in any trial of the experiment.

If we are interested in the event of getting a result of *6 or more* from one spin of the octagonal spinner, there are three favourable results (6, 7 or 8) out of the eight possible results, and each of these is equally likely to occur.

We read $\tfrac{3}{8}$ as '3 chances in 8'.

So, the probability of a result of 6 or more is $\tfrac{3}{8}$,

i.e., P(6 or more) $= \tfrac{3}{8}$

In general, for an event E containing **equally likely** possible results:

$$P(E) = \frac{\text{the number of members of the event E}}{\text{the total number of possible outcomes}}.$$

Example 6

A ticket is *randomly selected* from a basket containing 3 green, 4 yellow and 5 blue tickets. Determine the probability of getting:

a   a green ticket   **b**   a green or yellow ticket
c   an orange ticket   **d**   a green, yellow or blue ticket

The sample space is   {G, G, G, Y, Y, Y, Y, B, B, B, B, B}
which has $3 + 4 + 5 = 12$ outcomes.

a   P(green)

$= \tfrac{3}{12}$

$= \tfrac{1}{4}$

b   P(a green or a yellow)

$= \tfrac{3+4}{12}$

$= \tfrac{7}{12}$

c   P(orange)

$= \tfrac{0}{12}$

$= 0$

d   P(green, yellow or blue)

$= \tfrac{3+4+5}{12}$

$= 1$

In **Example 6** notice that in **c** an orange result cannot occur and the calculated probability is 0, which fits the fact that it has *no chance* of occurring.

Also notice in **d**, a green, yellow or blue result is certain to occur. It is 100% likely which is perfectly described using a 1.

The two events of   *no chance of occurring* with probability 0 and
                    certain to occur with probability 1 are two extremes.

Consequently, for any event E, $0 \leqslant P(E) \leqslant 1$.

COMPLEMENTARY EVENTS

Example 7

An ordinary 6-sided die is rolled once. Determine the chance of:

a   getting a 6   **b**   not getting a 6
c   getting a 1 or 2   **d**   not getting a 1 or 2

The sample space of possible outcomes is   {1, 2, 3, 4, 5, 6}

a   P(6)   **b**   P(not getting a 6)
       $= \frac{1}{6}$ = P(1, 2, 3, 4 or 5)
                                                $= \frac{5}{6}$

c   P(1 or 2)   **d**   P(not getting a 1 or 2)
       $= \frac{2}{6}$ = P(3, 4, 5, or 6)
                                                $= \frac{4}{6}$

In **Example 7**, did you notice that   P(6) + P(not getting a 6) = 1   and that
                    P(1 or 2) + P(not getting a 1 or 2) = 1?

This is no surprise as *getting a 6* and *not getting a 6* are **complementary events** where one of them **must occur**.

NOTATION

If E is an event, then E′ is the **complementary event** of E.

So, $P(E) + P(E') = 1$

A useful rearrangement is: P(E **not** occurring) = 1 − P(E occurring)

EXERCISE 14E

1   A marble is randomly selected from a box containing 5 green, 3 red and 7 blue marbles. Determine the probability that the marble is:

a   red   **b**   green   **c**   blue
d   not red   **e**   neither green nor blue   **f**   green or red

2   A carton of a dozen eggs contains eight brown eggs.
The rest are white.

   a   How many white eggs are there in the carton?

   b   What is the probability that an egg selected at
random is: **i**   brown   **ii**   white?

3   In a class of 32 students, eight have one first name, nineteen have two first names and
five have three first names. A student is selected at random. Determine the probability
that the student has:

   a   no first name   **b**   one first name   **c**   two first names   **d**   three first names.

4   A dart board has 36 sectors, labelled 1 to 36.
Determine the probability that a dart thrown
at the board hits:

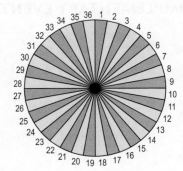

   a   a multiple of 4

   b   a number between 6 and 9 inclusive

   c   a number greater than 20

   d   9

   e   a multiple of 13

   f   an odd number that is a multiple of 3.

5   What is the probability that a randomly chosen person has his/her next birthday on:

   a   a Tuesday   **b**   a week-end   **c**   in July   **d**   in January or February?

6   List the six different orders in which Antti, Kai and Neda may sit in a row. If the three
of them sit randomly in a row, determine the probability that:

   a   Antti sits in the middle   **b**   Antti sits at the left end

   c   Antti sits at the right end   **d**   Kai and Neda are seated together

7   **a**   List the 8 possible 3-child families, according to the gender of the children. E.g.,
GGB means *"the first is a girl, the second is a girl, and the third is a boy"*.

   b   Assuming that each of these is equally likely to occur, determine the probability
that a randomly selected 3-child family consists of:

     i   all boys   **ii**   all girls   **iii**   boy, then girl, then girl

     iv   two girls and a boy   **v**   a girl for the eldest   **vi**   at least one boy.

8   **a**   List, in systematic order, the 24 different orders in which four people A, B, C and
D may sit in a row.

   b   Hence, determine the probability that when the four people sit at random in a row:

     i   A sits on one end   **ii**   B sits on one of the two middle seats

     iii   A and B are seated together

     iv   A, B and C are seated together, not necessarily in that order.

9   List the possible outcomes when four coins are tossed simultaneously. Hence determine
the probability of getting:

   a   all heads   **b**   two heads and 2 tails   **c**   more tails than heads

   d   at least one tail   **e**   exactly one head.

INVESTIGATION 4 — A PROBABILITY EXPERIMENT

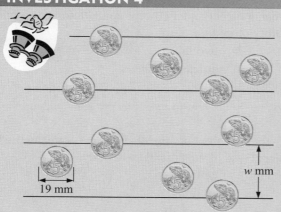

5-cent coins are 19 mm in diameter. When a 5-cent coin is dropped onto a sheet of paper containing many parallel lines which are more than 19 mm apart, the coin may or may not finish on a line.

If a coin finishes on a line you lose the game. If it does not land on a line you win the game.

Your task is to determine the probability of winning the game for various values of w mm, the distance between the lines.

What to do:

1 For $w = 20$, toss a 5-cent coin 100 times onto the lined paper and hence estimate the chance of winning.

2 Repeat for $w = 25, 30, 45, 50, 55, 60$.

3 Construct a graph of **P(winning)** against w using your experimental results of **1** and **2** above:

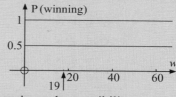

4 Can this graph be obtained using theoretical calculations rather than performing experiments? Investigate the possibility.

F   USING GRIDS TO FIND PROBABILITIES

Two dimensional grids give us excellent visual displays of sample spaces. From these we can count favourable outcomes and so calculate probabilities.

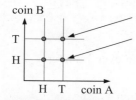

This point represents 'a tail from coin A' and 'a tail from coin B'.
This point represents 'a tail from coin A' and 'a head from coin B'.
There are four members of the sample space.

Example 8

Use a two-dimensional grid to illustrate the sample space for tossing a coin and rolling a die simultaneously. From this grid determine the probability of:

a tossing a head   b getting a tail and a 5   c getting a tail or a 5

There are 12 members in the sample space.

a P(head) $= \frac{6}{12} = \frac{1}{2}$ b P(tail and a '5') $= \frac{1}{12}$

c P(tail or a '5') $= \frac{7}{12}$ {the enclosed points}

EXERCISE 14F

1   Draw the grid of the sample space when a 5-cent and a 10-cent coin are tossed simultaneously. Hence determine the probability of getting:

    a   two heads   **b**   two tails

    c   exactly one head   **d**   at least one head

Example 9

Two square spinners, each with 1, 2, 3 and 4 on their edges, are twirled simultaneously. Draw a two-dimensional grid of the possible outcomes.

Use your grid to determine the probability of getting:

    a   a 3 with each spinner   **b**   a 3 and a 1

    c   an even result for each spinner

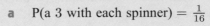

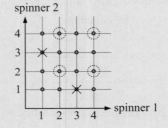

The sample space has 16 members.

a   P(a 3 with each spinner) $= \frac{1}{16}$

b   P(a 3 and a 1) $= \frac{2}{16}$ {crossed points}
                   $= \frac{1}{8}$

c   P(an even result for each spinner)
      $= \frac{4}{16}$ {circled points}
      $= \frac{1}{4}$

---

2   A coin and a pentagonal spinner with sectors 1, 2, 3, 4 and 5 are rolled and spun respectively.

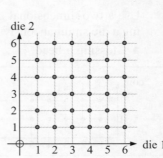

    a   Draw a grid to illustrate the sample space of possible outcomes.

    b   How many outcomes are possible?

    c   Use your grid to determine the chance of getting:

        i   a tail and a 3   **ii**   a head and an even number

        iii   an odd number   **iv**   a head or a 5

3   A pair of dice is rolled. The 36 different possible 'pair of dice' results are illustrated below on a 2-dimensional grid.

Use the 2-dimensional grid of the 36 possible outcomes to determine the probability of getting:

    a   two 3's   **b**   a 5 and a 6

    c   a 5 or a 6   **d**   at least one 6

    e   exactly one 6   **f**   no sixes

    g   a sum of 7   **h**   a sum greater than 8

    i   a sum of 7 or 11   **j**   a sum of no more than 8.

DISCUSSION

Read and discuss:

Three children have been experimenting with a coin, tossing it in the air and recording the outcomes. They have done this 10 times and have recorded 10 tails. Before the next toss they make the following statements:

Jackson: "It's got to be a head next time!"

Sally: "No, it always has an equal chance of being a head or a tail. The coin cannot remember what the outcomes have been."

Amy: "Actually , I think it will probably be a tail again, because I think the coin must be biased - it might be weighted somehow so that it is more likely to give a tail."

G   COMPOUND EVENTS

Consider the following problem:

Box X contains 2 blue and 2 green balls and Box Y contains 3 red and 1 white ball. A ball is randomly selected from each of the boxes. Determine the probability of getting "a blue ball from X and a red ball from Y."

By illustrating the sample space on a two-dimensional grid as shown alongside, it can be seen that as 6 of the 16 possibilities are blue from X and red from Y and each outcome is equally likely,

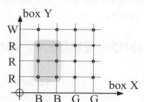

$P(\text{blue from } X \textbf{ and } \text{red from } Y) = \frac{6}{16}$

The question arises, "Is there a quicker, easier way to find this probability?"

INVESTIGATION 5   PROBABILITIES OF COMPOUND EVENTS

The purpose of this investigation is to find, if possible, a rule for finding P(A *and* B) for events A and B.

A coin is tossed and at the same time, a die is rolled. The result of the coin toss will be called outcome A, and likewise for the die, outcome B.

What to do:

a Copy and complete; given that P(A) and P(B) are theoretical and P(A and B) is experimental from 200 trials.

P(A *and* B)	P(A)	P(B)
P(a head and a 4) =		
P(a head and an odd number) =		
P(a tail and a number larger than 1) =		
P(a tail and a number less than 3) =		

b What is the connection between P(A and B) and P(A), P(B)?

INVESTIGATION 6   REVISITING DRAWING PINS

 Since we cannot find by theoretical argument the probability that a drawing pin will land on its back ⊥ , the question arises for tossing two drawing pins, does

$$P(\text{back } and \text{ back}) = P(\text{back}) \times P(\text{back})?$$

What to do:

1   From **Investigation 1** on page **455**, what is your estimate of P(back *and* back)?

2   **a**   Count the number of drawing pins in a full packet. They must be identical to each other and the same ones that you used in **Investigation 1**.

     b   Drop the whole packet onto a solid surface and count the number of *backs* and *sides*. Repeat this many times. Pool results with others and then estimate P(back).

3   Find   P(back) × P(back)   using **2 b**.

4   Is   P(back *and* back) ≑ P(back) × P(back)?

From **Investigation 5 and 6**, it seems that:

If A and B are two events, where the occurrence of one of them does not affect the occurrence of the other, then

$$P(A \textbf{ and } B) = P(A) \times P(B).$$

Before we can formulate a rule, we need to distinguish between **independent** and **dependent** events.

INDEPENDENT EVENTS

> **Independent events** are events where the occurrence of one of the events **does not** affect the occurrence of the other event.

Consider again the example on the previous page. Suppose we happen to choose a blue ball from box X. This in no way affects the outcome when we choose a ball from box Y. The two events "a blue ball from X" and "a red ball from Y" are **independent events**.

In general: If A and B are **independent events** then $P(A \textbf{ and } B) = P(A) \times P(B)$.

This rule can be extended for any number of independent events.

For example: If A, B and C are all **independent events**, then
                $P(A \textbf{ and } B \textbf{ and } C) = P(A) \times P(B) \times P(C)$.

Example 10

A coin and a die are tossed simultaneously. Determine the probability of getting a head and a 3 without using a grid.

$$
\begin{aligned}
P(\text{head and } 3) &= P(H) \times P(3) \qquad \{\text{as events are clearly physically independent}\} \\
&= \tfrac{1}{2} \times \tfrac{1}{6} \\
&= \tfrac{1}{12}
\end{aligned}
$$

EXERCISE 14G.1

1 At a mountain village in New Guinea it rains on average 6 days a week. Determine the probability that it rains on:

    a   any one day   **b**   two successive days   **c**   three successive days.

2 A coin is tossed 3 times. Determine the probability of getting the following sequences of results:

    a   head, then head, then head   **b**   tail, then head, then tail

3 A school has two photocopiers. On any one day, machine A has an 8% chance of malfunctioning and machine B has a 12% chance of malfunctioning.

    Determine the probability that on any one day both machines will:

    a   malfunction   **b**   work effectively.

4 A couple decide that they want 4 children, none of whom will be adopted. They will be disappointed if the children are not born in the order boy, girl, boy, girl. Determine the probability that:

    a   they will be happy with the order of arrival

    b   they will be unhappy with the order of arrival.

5 Two marksmen fire at a target simultaneously. Jiri hits the target 70% of the time and Benita hits it 80% of the time. Determine the probability that:

    a   they both hit the target

    b   they both miss the target

    c   Jiri hits it but Benita misses

    d   Benita hits it but Jiri misses.

6

An archer always hits a circular target with each arrow shot, and hits the bullseye 2 out of every 5 shots on average. If 3 arrows are shot at the target, determine the probability that the bullseye is hit:

    a   every time

    b   the first two times, but not on the third shot

    c   on no occasion.

DEPENDENT EVENTS

Suppose a hat contains 5 red and 3 blue tickets. One ticket is randomly chosen, its colour is noted and it is thrown in a bin. A second ticket is randomly selected. What is the chance that it is red?

If the first ticket was red,   P(second is red) $= \frac{4}{7}$ ⟵ 4 reds remaining ⟵ 7 to choose from

If the first ticket was blue,   P(second is red) $= \frac{5}{7}$ ⟵ 5 reds remaining ⟵ 7 to choose from

So, the probability of the second ticket being red **depends** on what colour the first ticket was. Here we have **dependent events**.

Two or more events are **dependent** if they are **not independent**.

Dependent events are events where the occurrence of one of the events *does affect* the occurrence of the other event.

For compound events which are dependent, a similar product rule applies as to that for independent events:

If A and B are dependent events then

P(A then B) = P(A) × P(B given that A has occurred).

Example 11

A box contains 4 red and 2 yellow tickets. Two tickets are randomly selected, one by one from the box, *without* replacement. Find the probability that:

a both are red   **b** the first is red and the second is yellow.

a   P(both red)

= P(first selected is red *and* second is red)

= P(first selected is red) × P(second is red given that the first is red)

= $\frac{4}{6} \times \frac{3}{5}$ ← 3 reds remain out of a total of 5 after a red first draw

= $\frac{2}{5}$ 4 reds out of a total of 6 tickets

b   P(first is red *and* second is yellow)

= P(first is red) × P(second is yellow given that the first is red)

= $\frac{4}{6} \times \frac{2}{5}$ ← 2 yellows remain out of a total of 5 after a red first draw

= $\frac{4}{15}$ 4 reds out of a total of 6 tickets

Example 12

A hat contains tickets with numbers 1, 2, 3, ..., 19, 20 printed on them. If 3 tickets are drawn from the hat, without replacement, determine the probability that all are prime numbers.

> In each fraction the bottom number is the total from which the selection is made and the top number is *"how many of the particular event we want"*.

There are 20 numbers of which 8 are primes:

  {2, 3, 5, 7, 11, 13, 17, 19} are primes.

∴   P(3 primes)

= P(1st drawn is prime *and* 2nd is prime *and* 3rd is prime)

= $\frac{8}{20} \times \frac{7}{19} \times \frac{6}{18}$

    8 primes out of 20 numbers
    7 primes out of 19 numbers after a successful first draw
≑ 0.049 12   6 primes out of 18 numbers after two successful draws

EXERCISE 14G.2

1 A box contains 7 red and 3 green balls. Two balls are randomly selected from the box without replacement. Determine the probability that:

Drawing two balls simultaneously is the same as selecting one ball after another with no replacement.

    a both are red

    b the first is green and the second is red

    c a green and a red are obtained.

2 A bin contains 12 identically shaped chocolates of which 8 are strawberry-creams. If 3 chocolates are selected at random from the bin, determine the probability that:

    a they are all strawberry-creams   **b** none of them are strawberry-creams.

3 A lottery has 100 tickets which are placed in a barrel. Three tickets are drawn at random from the barrel to decide 3 prizes. If John has 3 tickets in the lottery, determine his probability of winning:

    a first prize   **b** first and second prize

    c all 3 prizes   **d** none of the prizes

4 A hat contains 7 names of players in a tennis squad including the captain and the vice captain. If a team of 3 is chosen at random by drawing the names from the hat, determine the probability that it does not:

    a contain the captain   **b** contain the captain or the vice captain.

H   USING TREE DIAGRAMS

Tree diagrams can be used to illustrate sample spaces provided that the alternatives are not too numerous. Once the sample space is illustrated, the tree diagram can be used for determining probabilities.

Consider two archers:

Li with probability $\frac{3}{4}$ of hitting a target and Yuka with probability $\frac{4}{5}$.

They both shoot simultaneously.

The tree diagram for this information is:

			outcome	probability

H = hit   M = miss   Li's results   Yuka's results

$$\begin{array}{ccc}
 & \text{H} & \text{H and H} & \frac{3}{4} \times \frac{4}{5} = \frac{12}{20} \\
 & \text{M} & \text{H and M} & \frac{3}{4} \times \frac{1}{5} = \frac{3}{20} \\
 & \text{H} & \text{M and H} & \frac{1}{4} \times \frac{4}{5} = \frac{4}{20} \\
 & \text{M} & \text{M and M} & \frac{1}{4} \times \frac{1}{5} = \frac{1}{20}
\end{array}$$

total   1

Notice that:

- The probabilities for hitting and missing are marked on the branches.
- There are *four* alternative paths and each branch shows a particular outcome.
- All outcomes are represented and the probabilities are obtained by **multiplying**.

Example 13

A box of chocolates contains equal numbers of milk and dark chocolates. It has 15 hard centred milk chocolates, 5 soft centred milk chocolates, 10 hard centred dark chocolates and 10 soft centred dark chocolates.

a Copy and complete the diagram for randomly selecting a chocolate from a full box.

b What is the probability of getting:
  i a soft centred dark chocolate
  ii a hard centred chocolate?

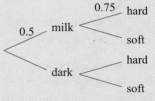

a

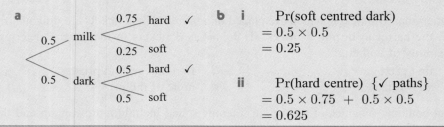

b **i** Pr(soft centred dark)
  $= 0.5 \times 0.5$
  $= 0.25$

ii Pr(hard centre) {✓ paths}
  $= 0.5 \times 0.75 + 0.5 \times 0.5$
  $= 0.625$

EXERCISE 14H

1 Jason takes the car to school two days a week and the other days he rides his bike. If he has the car the chance that he is late is 10% but if he rides it is 30%.

  a Copy and complete the tree diagram.
  b What is the probability that on a randomly selected day Jason was:
    i riding and not late   **ii** late?

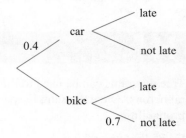

2 At 20 November 2003, 4% of prison inmates were female. 15% of female inmates were under 20 while 10% of males were under 20.

  a Copy and complete the tree diagram.
  b Find the probability that an inmate was:
    i male and under 20   **ii** under 20.
  c 13% of the females under 20 were aged 16. What is the probability that a female inmate was under 20 but not aged 16?

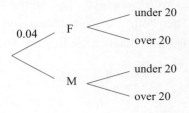

3 Of those students playing musical instruments 60% were female. 20% of the females and 30% of the males play the violin.

  a Copy and complete the tree diagram.
  b What is the probability that a randomly selected student:
    i is male and does not play the violin
    ii plays the violin?

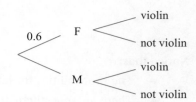

4 Some students at the local high school were asked
if they had a takeaway meal last week. 55% of the
students were male. 50% of the females and 60% of
the males had had a takeaway meal.

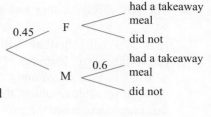

a Copy and complete the tree diagram.

b What is the probability that a randomly selected
student:

   i   was female and had a takeaway meal

   ii did not have a takeaway meal?

5 **a** Copy and complete this tree diagram
about people in the armed forces.

b What is the probability that a member
of the armed forces:

   i   is an officer

   ii   is not an officer in the navy

   iii is not an army or air force officer?

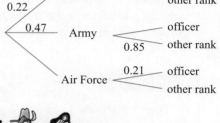

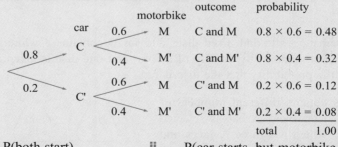

Example 14

Carl is not having much luck lately. His car will only start 80% of the time
and his motorbike will only start 60% of the time.

a Draw a tree diagram to illustrate this situation.

b Use the tree diagram to determine the chance that:

   i both will start   **ii** Carl has no choice but to use his car.

a C = car starts M = motorbike starts

		outcome	probability
car	0.6 → M	C and M	$0.8 \times 0.6 = 0.48$
C 0.8	0.4 → M'	C and M'	$0.8 \times 0.4 = 0.32$
0.2 C'	0.6 → M	C' and M	$0.2 \times 0.6 = 0.12$
	0.4 → M'	C' and M'	$0.2 \times 0.4 = 0.08$
		total	1.00

b **i**   P(both start)   **ii**   P(car starts, but motorbike does not)

       = P(C and M)   = P(C and M$'$)

       = 0.8×0.6 = 0.8×0.4

       = 0.48 = 0.32

6 The probability of rain tomorrow is estimated to be $\frac{1}{5}$.

If it does rain, Mudlark will start favourite with probability $\frac{1}{2}$ of winning. If it is fine he has a 1 in 20 chance of winning. Display the sample space of possible results of the horse race on a tree diagram. Determine the probability that Mudlark will win tomorrow.

7 Machine A makes 40% of the bottles produced at a factory. Machine B makes the rest. Machine A spoils 5% of its product, while Machine B spoils only 2%. Determine the probability that the next bottle inspected at this factory is spoiled.

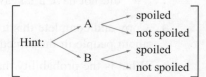

Example 15

Bag A contains 3 red and 2 yellow tickets. Bag B contains 1 red and 4 yellow tickets. A bag is randomly selected by tossing a coin and one ticket is removed from it. Determine the probability that it is yellow.

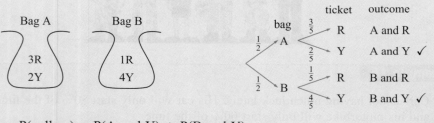

$$P(\text{yellow}) = P(A \text{ and } Y) + P(B \text{ and } Y)$$
$$= \tfrac{1}{2} \times \tfrac{2}{5} + \tfrac{1}{2} \times \tfrac{4}{5} \qquad \{\text{branches marked with a } \checkmark\}$$
$$= \tfrac{6}{10}$$
$$= \tfrac{3}{5}$$

---

8 Jar A contains 2 white and 3 red discs and Jar B contains 3 white and 1 red disc. A jar is chosen at random (by the flip of a coin) and one disc is taken at random from it. Determine the probability that the disc is red.

9 Three bags contain different numbers of blue and red marbles. A bag is selected using a die which has three A faces and two B faces and one C face.

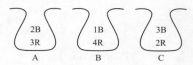

One marble is then selected randomly from the bag. Determine the probability that it is:

  a   blue   **b**   red.

SAMPLING WITH AND WITHOUT REPLACEMENT

SAMPLING

Sampling is the process of selecting an object from a large group of objects and inspecting it, noting some feature(s). The object is then either **put back** (sampling **with replacement**) or **put to one side** (sampling **without replacement**).

Sometimes the inspection process makes it impossible to return the object to the large group.

Such processes include:

- Is the chocolate hard- or soft-centred? Bite it or squeeze it to see.
- Does the egg contain one or two yolks? Break it open and see.
- Is the object correctly made? Pull it apart to see.

This sampling process is used to maintain Quality Control in industrial processes.

Consider a box containing 3 red, 2 blue and 1 yellow marble. Suppose we wish to sample two marbles:

- by **replacement** of the first before the second is drawn
- by **not replacing** the first before the second is drawn.

Examine how the tree diagrams differ:

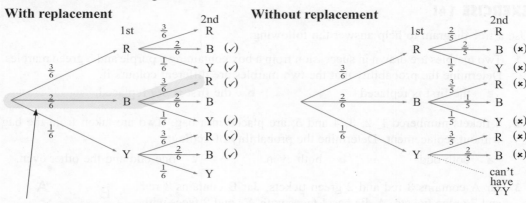

This branch represents Blue with the 1st draw and Red with the second draw and this is written BR.

Notice that:

- with replacement
  $P(\text{two reds}) = \frac{3}{6} \times \frac{3}{6} = \frac{1}{4}$

- without replacement
  $P(\text{two reds}) = \frac{3}{6} \times \frac{2}{5} = \frac{1}{5}$

Example 16

For the example of the box containing 3 red, 2 blue and 1 yellow marble find the probability of getting two different colours:

a if replacement occurs   b if replacement does not occur.

a   P (two different colours)

= P (RB or RY or BR or BY or YR or YB)   {ticked ones}

$= \frac{3}{6} \times \frac{2}{6} + \frac{3}{6} \times \frac{1}{6} + \frac{2}{6} \times \frac{3}{6} + \frac{2}{6} \times \frac{1}{6} + \frac{1}{6} \times \frac{3}{6} + \frac{1}{6} \times \frac{2}{6}$

$= \frac{22}{36}$

$= \frac{11}{18}$

b   P (two different colours)

= P(RB or RY or BR or BY or YR or YB)   {crossed ones}

$= \frac{3}{6} \times \frac{2}{5} + \frac{3}{6} \times \frac{1}{5} + \frac{2}{6} \times \frac{3}{5} + \frac{2}{6} \times \frac{1}{5} + \frac{1}{6} \times \frac{3}{5} + \frac{1}{6} \times \frac{2}{5}$

$= \frac{22}{30}$

$= \frac{11}{15}$

Note: When using tree diagrams to assist in solving probability questions, the following rules should be used:

- The probability for each branch is calculated by **multiplying** the probabilities along that path.
- If two or more branch paths meet the description of the compound event, the probability of each path is found and then they are **added**.

EXERCISE 14I

Use a tree diagram to help answer the following:

1   Two marbles are drawn in succession from a box containing 2 purple and 5 green marbles. Determine the probability that the two marbles are different colours if:

    a   the first is replaced   **b**   the first is *not* replaced.

2   5 tickets numbered 1, 2, 3, 4 and 5, are placed in a bag. Two are taken from the bag without replacement. Determine the probability of getting:

    a   both odd   **b**   both even   **c**   one odd and the other even.

3   Jar A contains 3 red and 2 green tickets. Jar B contains 3 red and 7 green tickets. A die has 4 faces with A's and 2 faces with B's, and when rolled it is used to select either jar A or jar B.

When a jar has been selected, two tickets are randomly selected without replacement from it. Determine the probability that:

    a   both are green   **b**   they are different in colour.

4   Marie has a bag of sweets which are all identical in shape. The bag contains 6 orange drops and 4 lemon drops. She selects one sweet at random, eats it and then takes another, also at random. Determine the probability that:

    a   both sweets were orange drops   **b**   both sweets were lemon drops

    c   the first was an orange drop and the second was a lemon drop

    d   the first was a lemon drop and the second was an orange drop

Add your answers to **a**, **b**, **c** and **d**. Explain why the answer must be 1.

Example 17

A bag contains 5 red and 3 blue marbles. Two marbles are drawn simultaneously from the bag. Determine the probability that at least one is red.

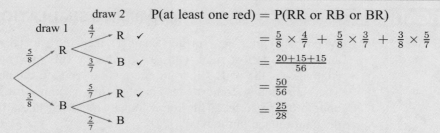

P(at least one red) = P(RR or RB or BR)

$$= \frac{5}{8} \times \frac{4}{7} + \frac{5}{8} \times \frac{3}{7} + \frac{3}{8} \times \frac{5}{7}$$

$$= \frac{20+15+15}{56}$$

$$= \frac{50}{56}$$

$$= \frac{25}{28}$$

Note: *Alternatively,*   P(at least one red) = 1 − P(no reds)   {complementary events}

$$= 1 - P(BB), \quad \text{etc}$$

5 A cook selects an egg at random from a carton containing 6 ordinary eggs and 3 double-yolk eggs. She cracks the egg into a bowl and sees whether it has two yolks or not. She then selects another egg at random from the carton and checks it.

Let S represent "a single yolk egg" and D represent "a double yolk egg".

   a Draw a tree diagram to illustrate this sampling process.

   b What is the probability that both eggs had two yolks?

   c What is the probability that both eggs had only one yolk?

6

Petra selects a chocolate at random from a box containing 10 hard-centred and 15 soft-centred chocolates. She bites it to see whether it is hard-centred or not. She then selects another chocolate at random from the box and checks it. (She eats both chocolates.)

Let H represent "a hard-centred chocolate" and S represent "a soft-centred chocolate".

   a Draw a tree diagram to illustrate this sampling process.

   b What is the probability that both chocolates had hard centres?

   c What is the probability that both chocolates had soft centres?

7 A bag contains four red and two blue marbles. Three marbles are selected simultaneously. Determine the probablity that:

   a all are red   **b** only two are red   **c** at least two are red.

8 Bag A contains 3 red and 2 white marbles. Bag B contains 4 red and 3 white marbles. One marble is randomly selected from A and its colour noted. If it is red, 2 reds are added to B. If it is white, 2 whites are added to B. A marble is then selected from B.

What are the chances that the marble selected from B is white?

9 A man holds two tickets in a 100-ticket lottery in which there are two winning tickets. If no replacement occurs, determine the probability that he will win:

    a   both prizes   **b**   neither prize   **c**   at least one prize.

INVESTIGATION 7   SAMPLING SIMULATION

When balls enter the 'sorting' chamber they hit a metal rod and may go left or right with *equal chance*. This movement continues as the balls fall from one level of rods to the next. The balls finally come to rest in collection chambers at the bottom of the sorter. This sorter looks very much like a tree diagram rotated through $90°$.

Click on the icon to open the simulation. Notice that we can use the sliding bar to alter the probabilities of balls going to the left or right at each rod.

What to do:

1 To simulate the results of tossing *two coins*, set the bar to 50%

and the sorter to show

SIMULATION

Run the simulation 200 times and repeat this four more times. Record each set of results.

2 A bag contains 7 blue and 3 red marbles and *two marbles* are randomly selected from it, the first being *replaced* before the second is drawn.

The sorter should show 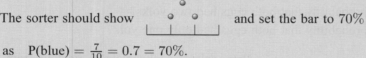 and set the bar to 70%

as $P(\text{blue}) = \frac{7}{10} = 0.7 = 70\%$.

Run the simulation a large number of times and use the results to estimate the probabilities of getting: **a** two blues   **b** one blue   **c** no blues.

3 The tree diagram representation of the marble selection in **2** is:

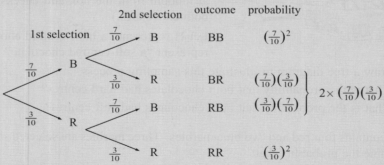

   a The tree diagram gives us expected, theoretical probabilities for the different outcomes. Do they agree with the experimental results obtained in **2**?

   b Write down the algebraic expansion of $(a + b)^2$.

   c Substitute $a = \frac{7}{10}$ and $b = \frac{3}{10}$ in the $(a + b)^2$ expansion. What do you notice?

4 From the bag of 7 blue and 3 red marbles, *three* marbles are randomly selected *with replacement*.

Set the sorter to ⬤ and the bar to 70%.

Run the simulation a large number of times to obtain experimental estimates of the probabilities of getting:

   a three blues   **b** two blues   **c** one blue   **d** no blues.

5 a Use a tree diagram showing 1st selection, 2nd selection and 3rd selection to find theoretical probabilities of getting the results of **4**.

   b Show that $(a + b)^3 = a^3 + 3a^2b + 3ab^2 + b^3$ and use this expansion with $a = \frac{7}{10}$ and $b = \frac{3}{10}$ to also check the results of **4** and **5a**.

J ▌ PROBABILITIES FROM VENN DIAGRAMS

Example 18

The Venn diagram alongside represents a sample space, S, of all children in a class. Each dot represents a student. The event, A, shows all those students with blue eyes. Determine the probability that a randomly selected child:

   a has blue eyes   **b** does not have blue eyes.

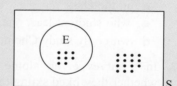

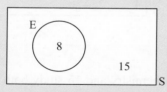

$n(S) = 23, \quad n(A) = 8$

   a P(blue eyes) $= \dfrac{n(A)}{n(S)} = \dfrac{8}{23}$

   b P(not blue eyes) $= \dfrac{15}{23}$

        {as 15 of the 23 are not in A}

  or   P(not blue) $= 1 -$ P(blue eyes)   {complementary events}

                 $= 1 - \frac{8}{23}$

                 $= \frac{15}{23}$

EXERCISE 14J

1 On separate Venn diagrams, using two events A and B that intersect, shade the region representing:

   a in A   **b** in B   **c** in both A and B

   d in A or B   **e** in B but not in A   **f** in exactly one of A or B

Example 19

If the Venn diagram alongside illustrates the number of people in a sporting club who play tennis (T) and hockey (H), determine the number of people:

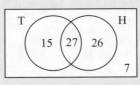

a   in the club
b   who play hockey
c   who play both sports
d   who play neither sport
e   who play at least one sport

a   Number in the club
$= 15 + 27 + 26 + 7 = 75$

b   Number who play hockey
$= 27 + 26 = 53$

c   Number who play both sports $= 27$

d   Number who play neither sport
$= 7$

e   Number who play at least one sport
$= 15 + 27 + 26 = 68$

2   The Venn diagram alongside illustrates the number of students in a particular class who study Chemistry (C) and History (H). Determine the number of students:

a   in the class   **b**   who study both subjects
c   who study at least one of the subjects
d   who only study Chemistry.

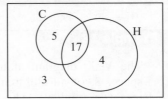

3   In a survey at an alpine resort, people were asked whether they liked skiing (S) or snowboarding (B). Use the Venn diagram to determine the number of people:

a   in the survey   **b**   who liked both activities
c   who liked neither activity
d   who liked exactly one of the activities.

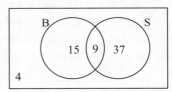

Example 20

In a class of 30 students, 19 study Physics, 17 study Chemistry and 15 study both of these subjects. Display this information on a Venn diagram and hence determine the probability that a randomly selected class member studies:

a   both subjects   **b**   at least one of the subjects
c   Physics, but not Chemistry   **d**   exactly one of the subjects
e   neither subject   **f**   Chemistry if it is known that the student studies Physics

Let P represent the event of 'studying Physics', and C represent the event of 'studying Chemistry'.

Now
$$a + b = 19 \quad \{\text{as 19 study Physics}\}$$
$$b + c = 17 \quad \{\text{as 17 study Chemistry}\}$$
$$b = 15 \quad \{\text{as 15 study both}\}$$
$$a + b + c + d = 30 \quad \{\text{as there are 30 in the class}\}$$
$$\therefore \quad b = 15, \quad a = 4, \quad c = 2, \quad d = 9.$$

a P(studies both)

$= \frac{15}{30}$ or $\frac{1}{2}$

b P(studies at least one subject)

$= \frac{4+15+2}{30}$

$= \frac{21}{30}$

$= \frac{7}{10}$

c P(studies P, but not C)

$= \frac{4}{30}$

$= \frac{2}{15}$

d P(studies exactly one)

$= \frac{4+2}{30}$

$= \frac{6}{30}$

$= \frac{1}{5}$

e P(studies neither)

$= \frac{9}{30}$

$= \frac{3}{10}$

f P(studies C if it is known studies P)

$= \frac{15}{15+4}$

$= \frac{15}{19}$

4 In a class of 40 students, 19 play tennis, 20 play netball and 8 play neither of these sports. A student is randomly chosen from the class. Determine the probability that the student:

- **a** plays tennis
- **b** does not play netball
- **c** plays at least one of the sports
- **d** plays one and only one of the sports
- **e** plays netball, but not tennis
- **f** plays tennis knowing he/she plays netball.

5 50 married men were asked whether they gave their wife flowers or chocolates for their last birthday. The results were: 31 gave chocolates, 12 gave flowers and 5 gave both chocolates and flowers. If one of the married men was chosen at random, determine the probability that he gave his wife:

- **a** chocolates or flowers
- **b** chocolates but not flowers
- **c** neither chocolates nor flowers
- **d** flowers if it is known that he did not give her chocolates.

6 The medical records for a class of 30 children showed whether they had previously had measles or mumps. The records showed 24 had had measles, 12 had had measles and mumps, and 26 had had measles or mumps. If one child from the class is selected randomly from the group, determine the probability that he/she has had:

- **a** mumps
- **b** mumps but not measles
- **c** neither mumps nor measles
- **d** measles if it is known that the child has had mumps.

7

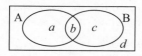

From the Venn diagram $P(A) = \dfrac{a+b}{a+b+c+d}$

a Use the Venn diagram to find:

  i P(B)   **ii** P(A and B)   **iii** P(A or B)   **iv** P(A) + P(B) − P(A and B)

b What is the connection between P(A or B) and P(A) + P(B) − P(A and B)?

8

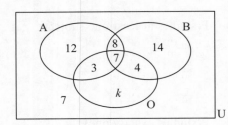

In the Venn diagram, U is the set of all members of a gymnastic club.

The members indicate their liking for apples (A), bananas (B) and oranges (O). There are 60 members in the club.

a Find the value of k.

b If a randomly chosen member is asked about their preferences for this fruit, what is the probability that the member likes:

   i   only bananas   **ii**   bananas and oranges

   iii none of these fruit   **iv** at least one of these fruits

   v   all of the fruits   **vi** apples and bananas, but not oranges

   vii oranges or bananas   **vii** exactly one of the three varieties of fruit

K   LAWS OF PROBABILITY

THE ADDITION LAW

From question **7** of the previous exercise we showed that

$$\text{for two events A and B,}\quad P(A \cup B) = P(A) + P(B) - P(A \cap B).$$

This is known as the **addition law of probability**, and can be written as

$$P(\textbf{either } A \textbf{ or } B) = P(A) + P(B) - P(\textbf{both } A \textbf{ and } B)$$

Example 21

If $P(A) = 0.6$, $P(A \cup B) = 0.7$ and $P(A \cap B) = 0.3$, find $P(B)$.

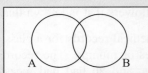

Since $P(A \cup B) = P(A) + P(B) - P(A \cap B)$,

then $0.7 = 0.6 + P(B) - 0.3$

$\therefore P(B) = 0.4$

or   Using a Venn diagram with the probabilities on it,

$$a + 0.3 = 0.6 \quad \therefore \quad a = 0.3$$

$$a + b + 0.3 = 0.7$$

$$\therefore \quad a + b = 0.4$$

$$\therefore \quad 0.3 + b = 0.4$$

$$\therefore \quad b = 0.1$$

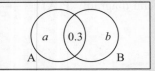

$$\therefore \quad P(B) = 0.3 + b = 0.4$$

MUTUALLY EXCLUSIVE EVENTS

> If A and B are **mutually exclusive** events then $P(A \cap B) = 0$
> and so the addition law becomes $P(A \cup B) = P(A) + P(B)$.

Example 22

A box of chocolates contains 6 with hard centres (H) and 12 with soft centres (S).
a   Are the events H and S mutually exclusive?
b   Find   **i** $P(H)$ **ii** $P(S)$ **iii** $P(H \cap S)$ **iv** $P(H \cup S)$.

---

a   Chocolates cannot have both a hard and a soft centre.
$\therefore$ H and S are mutually exclusive.

b   **i** $P(H)$ **ii** $P(S)$ **iii** $P(H \cap S)$ **iv** $P(H \cup S)$
$= \frac{6}{18}$ $= \frac{12}{18}$ $= 0$ $= \frac{18}{18}$
$= \frac{1}{3}$ $= \frac{2}{3}$ $= 1$

CONDITIONAL PROBABILITY WITH VENN DIAGRAMS

Suppose we have two events A and B, then

> A $|$ B is used to represent that 'A occurs knowing that B has occurred'.

Example 23

In a class of 25 students, 14 like Pizza and 16 like iced coffee. One student likes
neither and 6 students like both. One student is randomly selected from the class.
What is the probability that the student:
a   likes Pizza   **b**   likes Pizza given that he/she likes iced coffee?

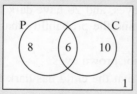

The Venn diagram of the situation is shown.

a $P(\text{Pizza}) = \frac{14}{25}$ {of the 25 students 14 like Pizza}

b $P(\text{Pizza/iced coffee}) = \frac{6}{16}$
{of the 16 who like iced coffee, 6 like Pizza}

If A and B are events then   **Proof:** $P(A \mid B)$

$$P(A \mid B) = \frac{P(A \cap B)}{P(B)}.$$

$$= \frac{b}{b+c} \quad \text{\{Venn diagram\}}$$

$$= \frac{b/(a+b+c+d)}{(b+c)/(a+b+c+d)}$$

$$= \frac{P(A \cap B)}{P(B)}$$

It follows that $P(A \cap B) = P(A \mid B)P(B)$ or $P(A \cap B) = P(B \mid A)P(A)$

Example 24

In a class of 40, 34 like bananas, 22 like pineapples and 2 dislike both fruits.
If a student is randomly selected, find the probability that the student:

a likes both fruits   **b** likes at least one fruit
c likes bananas given that he/she likes pineapples
d dislikes pineapples given that he/she likes bananas.

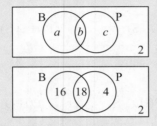

B represents students who like bananas
P represents students who like pineapples

We are given that $a + b = 34$
$$b + c = 22$$
$$a + b + c = 38$$

$\therefore \quad c = 38 - 34 \quad$ and so $\quad b = 18$
$\quad\quad = 4 \quad\quad\quad$ and $\quad a = 16$

a   P(likes both)   **b**   P(likes at least one)   **c**   P(B | P)   **d**   P(P′ | B)

$= \frac{18}{40}$ $= \frac{38}{40}$ $= \frac{18}{22}$ $= \frac{16}{34}$

$= \frac{9}{20}$ $= \frac{19}{20}$ $= \frac{9}{11}$ $= \frac{8}{17}$

EXERCISE 14K

1 In a group of 50 students, 40 study Mathematics, 32 study Physics and each student studies at least one of these subjects.

    a From a Venn diagram find how many students study both subjects.
    b If a student from this group is randomly selected, find the probability that he/she:
        i studies Mathematics but not Physics
       ii studies Physics given that he/she studies Mathematics.

2 In a class of 40 students, 23 have dark hair, 18 have brown eyes, and 26 have dark hair, brown eyes or both. A child is selected at random. Determine the probability that the child has:

    a dark hair and brown eyes   **b** neither dark hair nor brown eyes
    c dark hair but not brown eyes   **d** brown eyes given that the child has dark hair.

3 50 students go bushwalking. 23 get sunburnt, 22 get bitten by ants and 5 are both sunburnt and bitten by ants. Determine the probability that a randomly selected student:

    a escaped being bitten   **b** was either bitten or sunburnt
    c was neither bitten nor sunburnt
    d was bitten, given that the student was sunburnt
    e was sunburnt, given that the student was not bitten.

4 30 students sit for an examination in both French and English. 25 pass French, 24 pass English and 3 fail both. Determine the probability that a student who:

    a passed French, also passed English   **b** failed English, passed in French.

5 400 families were surveyed. It was found that 90% had a TV set and 60% had a computer. Every family had at least one of these items. If one of these families is randomly selected, find the probability it has a TV set given that it has a computer.

6 In a certain town 3 newspapers are published. 20% of the population read A, 16% read B, 14% read C, 8% read A and B, 5% read A and C, 4% read B and C and 2% read all 3 newspapers. A person is selected at random.

Determine the probability that the person reads:

  a none of the papers   **b** at least one of the papers

  c exactly one of the papers   **d** either A or B

  e A, given that the person reads at least one paper

  f C, given that the person reads either A or B or both.

7 In a printing business there are three printing machines. Machine P operates 84% of the time, Q operates 84% of the time and R 87% of the time. P and Q operate 75% of the time, Q and R 77% of the time and P and R 76%. All three are in operation 73% of the time. Determine the probability that:

  a P or Q is operating   **b** none is operating

  c exactly one is operating   **d** P and Q, but not R, are operating.

CONDITIONAL PROBABILITY WITH TREE DIAGRAMS

Example 25

> Bin A contains 3 red and 2 white tickets. Bin B contains 4 red and 1 white. A die with 4 faces marked A and two faces marked B is rolled and used to select bin A or B. A ticket is then selected from this bin. Determine the probability that:
>
> **a** the ticket is red   **b** the ticket was chosen from B given it is red.

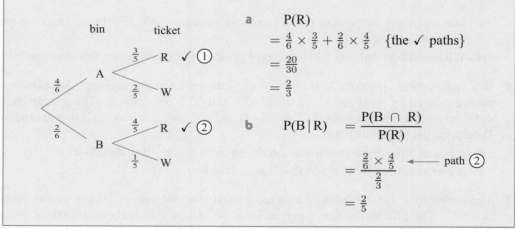

a $P(R)$

$ = \frac{4}{6} \times \frac{3}{5} + \frac{2}{6} \times \frac{4}{5}$ {the ✓ paths}

$ = \frac{20}{30}$

$ = \frac{2}{3}$

b $P(B \,|\, R) = \dfrac{P(B \cap R)}{P(R)}$

$ = \dfrac{\frac{2}{6} \times \frac{4}{5}}{\frac{2}{3}}$ ⟵ path ②

$ = \frac{2}{5}$

8 Urn A contains 2 red and 3 blue marbles, and urn B contains 4 red and 1 blue marble. Peter selects an urn by tossing a coin, and takes a marble from that urn.

  a Determine the probability that it is red.

  b Given that the marble is red, what is the probability it came from B?

9 The probability that Greta's mother takes her shopping is $\frac{2}{5}$. When Greta goes shopping with her mother she gets an icecream 70% of the time. When Greta does not go shopping with her mother she gets an icecream 30% of the time.

Determine the probability that

   a Greta's mother buys her an icecream when shopping.

   b Greta went shopping with her mother, given that her mother buys her an icecream.

10 A sociologist examined the criminal justice system. Following exhaustive interviews which included the use of lie detector test results, she published her findings. Her results were given on a tree diagram.

   G ≡ guilty   C ≡ convicted

   G′ ≡ not guilty   C′ ≡ not convicted

   a What percentage of people were correctly judged?

   b What is the probability of convicting a person given he/she is guilty?

   c What is the probability of acquitting a person given he/she is innocent?

   d Which of the answers to **b** and **c** would you prefer to be the higher?

   e What is the probability that a randomly selected person on trial will be convicted?

   f What is the probability that a randomly selected person on trial is guilty given that he/she is not convicted?

11 Two printing machines X and Y are used so that X produces 60% of the finished print run and Y produces 40% of it. 97% of the printing from X is perfect whereas 95% of the printing from Y is perfect.

   a What is the probability that a piece of printing is perfect given that it was printed on X?

   b What is the probability that a piece of printing is perfect given that it was printed on Y?

   c Determine the probability that a randomly selected piece of printing from the machines is perfect.

   d What is the probability that a piece of printing is from X given that it is perfect?

12 In a factory three operators A, B and C work shifts on a certain machine. The number of parts produced by A, B and C are 25%, 40% and 35% respectively. The percentage of defectives produced is 1% for A, 1.5% for B and 3% for C. A part is chosen at random. Determine the probability that:

   a it is defective given that it was produced by B   **b** it is defective

   c it was produced by B given that it was defective.

13 The probability that a randomly selected person over the age of 50 has bowel cancer is 0.007. The probability that a person (over 50) reacts positively to a medical test to detect bowel cancer is 0.973 if the person has it and 0.017 if the person does not have it. A person (over 50) is selected at random. What is the probability that this person:

   a reacts postively to the test

   b has bowel cancer given that he/she

      i reacts positively to the test   **ii** reacts negatively to the test?

L INDEPENDENT EVENTS REVISITED

A and B are **independent events** if the occurrence (or non-occurrence) of one event does not affect the occurrence of the other,

$$\text{i.e.,} \quad P(A \mid B) = P(A) \quad \text{and} \quad P(B \mid A) = P(B).$$

So, as $P(A \cap B) = P(A \mid B) P(B)$,

A and B are **independent events** $\Leftrightarrow$ $P(A \cap B) = P(A) P(B)$.

Example 26

When two coins are tossed, A is the event of getting 2 heads. When a die is rolled, B is the event of getting a 5 or 6. Prove that A and B are independent events.

$P(A) = \frac{1}{4}$ and $P(B) = \frac{2}{6}$. Therefore, $P(A) P(B) = \frac{1}{4} \times \frac{2}{6} = \frac{1}{12}$

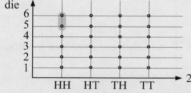

$P(A \cap B)$
$= P(2 \text{ heads } \textbf{and} \text{ a 5 or a 6})$
$= \frac{2}{24}$
$= \frac{1}{12}$

So, as $P(A \cap B) = P(A) P(B)$, the events A and B are independent.

Example 27

$P(A) = \frac{1}{2}$, $P(B) = \frac{1}{3}$ and $P(A \cup B) = p$. Find p if:

a A and B are mutually exclusive
b A and B are independent.

a If A and B are mutually exclusive, $A \cap B = \phi$
and so $P(A \cap B) = 0$
But $P(A \cup B) = P(A) + P(B) - P(A \cap B)$
$\therefore \quad p = \frac{1}{2} + \frac{1}{3} - 0$
i.e., $p = \frac{5}{6}$

b If A and B are independent, $P(A \cap B) = P(A) P(B)$
$= \frac{1}{2} \times \frac{1}{3}$
$= \frac{1}{6}$

$\therefore \quad P(A \cup B) = \frac{1}{2} + \frac{1}{3} - \frac{1}{6}$
i.e., $p = \frac{2}{3}$

EXERCISE 14L

1 If P(R) = 0.4, P(S) = 0.5 and P(R ∪ S) = 0.7, are R and S independent events?

2 If P(A) = $\frac{2}{5}$, P(B) = $\frac{1}{3}$ and P(A ∪ B) = $\frac{1}{2}$, find:

   a P(A ∩ B)   **b** P(B | A)   **c** P(A | B)

Are A and B independent events?

3 If P(X) = 0.5, P(Y) = 0.7, and X and Y are independent events, determine the probability of the occurrence of:

   a both X and Y   **b** X or Y   **c** neither X nor Y

   d X but not Y   **e** X given that Y occurs.

4 The probability that A, B and C solve a particular problem is $\frac{3}{5}$, $\frac{2}{3}$, $\frac{1}{2}$, respectively.

If they all try, determine the probability that the group solves the problem.

5 **a** Determine the probability of getting at least one six when an ordinary die is rolled 3 times.

   b If a die is rolled n times, find the smallest n such that
P(at least one 6 in n throws) > 99%.

6 A and B are independent events. Prove that A and B′ are also independent events.

REVIEW SET 14A

1 Donald keeps records of the number of clients he telephones over a consecutive period of days.

   a For how many days did the survey last?

   b Estimate Donald's chances of telephoning:

      i no clients on a day

      ii four or more clients on a day

      iii less than three clients on a day.

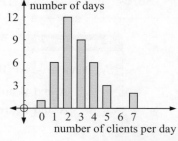

2 David conducted a survey to determine the ages of people walking through a shopping mall. The results are shown in the table alongside. Find, correct to 3 decimal places, the estimated probability that the next person David meets in the shopping mall is:

Age	Frequency
0 - 19	19
20 - 39	45
40 - 59	37
60+	11

   a between 20 and 39 years of age

   b less than 40 years of age   **c** at least 20 years of age.

3 A farmer fences his rectangular property, as shown alongside, into 9 rectangular paddocks:

If a paddock is selected at random, what is the probability that

   a it has no fences on the boundary of the property

   b it has one fence on the boundary of the property

   c it has two fences on the boundary of the property?

4 When a box of drawing pins dropped onto the floor it was observed that 47 landed on their backs and 31 landed on their sides. Find, correct to 2 decimal places, the estimated probability of a drawing pin landing: **a**   on its back   **b**   on its side.

back   side

5 A saw mill receives logs of various lengths from a plantation. The length of a log is important in being able to produce timber of the length required. The following data indicates the lengths of the latest 100 logs received.

Length	Frequency
8 - 8.9	3
9 - 9.9	4
10 - 10.9	14
11 - 11.9	12
12 - 12.9	18
13 - 13.9	20
14 - 14.9	14
15 - 15.9	7
16 - 16.9	8

a   What is the probability of a log being less than 11 metres long arriving at the saw mill?

b   What is the probability of a log being longer than 15 metres arriving at the saw mill?

c   In the next batch of 50 logs, how many would be expected to be between 11 m and 15 m long?

6 At peak hour railway crossings are closed 30% of the time. If you have to drive through three railway crossings during peak hour, what are the chances you will have to stop at least once?

7 In a golf match, Annette has 70% chance of hitting the green when using a nine iron and Kari has 90% chance when using the same club. If, at a particular hole, they both elect to use a nine iron to play to the green, determine the probability that:

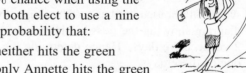

a   both hit the green   **b**   neither hits the green

c   at least one hits the green   **d**   only Annette hits the green

8 Jar A contains 3 white and 2 red marbles. Jar B contains 6 white and 4 red marbles. A jar is selected at random and then two marbles are selected without replacement. Determine the probability that:

a   both marbles are white   **b**   two red marbles are picked from Jar A.

9 Given that $n(U) = 60$:

a   Find a.

b   If A represents all students who like athletics, B represents all students who like basketball and T represents all students who like tennis, find the probability that a randomly chosen student:

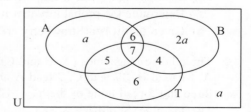

i   likes athletics

ii   dislikes basketball

iii   likes basketball and tennis

iv   likes athletics or tennis

v   likes all three sports

vi   likes exactly one sport

vii   likes athletics and tennis but not basketball.

REVIEW SET 14B

1 List the different orderings in which 4 people A, B, C and D could line up. If they line up at random, determine the probability that:

    a A is next to C   **b** there is exactly one person between A and C.

2 A coin is tossed and a square spinner, labelled A, B, C, D, is twirled. Determine the probability of obtaining:

    a a head and consonant   **b** a tail and C   **c** a tail or a vowel.

3 A class contains 25 students. 13 play tennis, 14 play volleyball and 1 plays neither of these two sports. A student is randomly selected from the class. Determine the probability that the student:

    a plays both tennis and volleyball   **b** plays at least one of these two sports.

    c plays volleyball given that he/she does not play tennis.

4 Niklas and Rolf play tennis and the winner is the first to win two sets. Niklas has a 40% chance of beating Rolf in any set. Draw a tree diagram showing the possible outcomes and hence determine the probability that Niklas will win the match.

5 The probability that a man will be alive in 25 years is $\frac{3}{5}$, and the probability that his wife will be alive is $\frac{2}{3}$. Determine the probability that in 25 years:

    a both will be alive   **b** at least one will be alive

    c only the wife will be alive.

6 If I buy 4 tickets in a 500 ticket lottery, determine the probability that I win:

    a the first 3 prizes   **b** at least one of the first 3 prizes.

7 A school photocopier has a 95% chance of working on any particular day. Draw a tree diagram showing the possibilities for the working or otherwise of the photocopier over 2 consecutive days. Hence determine the likelihood that it will be working on at least one of these days.

8 Box A contains 7 Strawberry creams and 3 Turkish delights,

Box B contains 6 Strawberry creams and 4 Turkish delights,

Box C contains 5 Strawberry creams and 5 Turkish delights.

A bin is chosen using a die with 3 faces marked A, 2 marked B and one marked C and one chocolate is chosen at random from the box.

    a Determine the probability that it is a Strawberry cream

    b Given that it is a Strawberry cream, find the probability it came from box B.

9 Three horse magazines (A, B and C) are read by people in an equestrian club. 35 read A, 31 read B, 35 read C, 7 read A and B, 11 read B and C, 8 read A and C, 5 read all three and 2 read none of them.

    a Draw a Venn diagram and show on it the given information.

    b How many people are in the equestrian club?

    c If a member is chosen at random, what is the probability that he/she reads:

        i A only   **ii** all of them   **iii** B or C

        iv A and B, but not C   **v** exactly one of them   **vi** at least one of them?

REVIEW SET 14C

1 Systematically list the sexes of 4-child families. Hence determine the probability that a randomly selected 4-child family consists of two children of each sex.

2 A pair of dice is rolled. Graph the sample space of all possible outcomes. Hence determine the probability that:

    a a sum of 7 or 11 results   **b** a sum of at least 8 results.

3 In a group of 40 students, 22 study Economics, 25 study Law and 3 study neither of these subjects. Determine the probability that a randomly chosen student studies:

    a both Economics and Law   **b** at least one of these subjects
    c Economics given that he/she studies Law.

4 An integer lies between 0 and 100. Determine the likelihood that is divisible by 6 or 8.

5 A bag contains 3 red, 4 yellow and 5 blue marbles. Two marbles are randomly selected from the bag (without replacement). What is the probability that:

    a both are blue   **b** both are the same colour
    c at least one is red   **d** exactly one is yellow?

6 What is meant by: **a** independent events   **b** disjoint events?

7 On any one day it could rain with 25% chance and be windy with 36% chance. Draw a tree diagram showing the possibilities with regard to wind and rain on a particular day. Hence determine the probability that on a particular day there will be:

    a rain and wind   **b** rain or wind.

8 A, B and C have 10%, 20% and 30% chance of independently solving a certain maths problem. If they all try independently of one another, what is the probability that this group will solve the problem?

9 Jon goes cycling on three mornings of each week (at random). When he goes cycling he has eggs for breakfast 70% of the time. When he does not go cycling he has eggs for breakfast 25% of the time. Determine the probability that he

    a has eggs for breakfast   **b** goes cycling given that he has eggs for breakfast.

Chapter **15**

Logic

Contents:

INTRODUCTION

Mathematical logic deals with converting worded statements to symbols and applying rules of deduction to them.

Mathematical logic was originally suggested by **G W Leibnitz** 1646-1716). George Boole (1815-1864) introduced the symbolism which provided the tools for the analysis. Others who contributed to this field of mathematical endeavour were: Bertrand Russell, Augustus DeMorgan, David Hilbert, John Venn, Guiseppe Peano and Gottlob Freje.

Logical reasoning is essential in Mathematics.

Good mathematical argument requires basic **definitions** and **axioms**. These are simple statements that we accept without proof.

These mark the starting point of any reasoning (logic) we perform.

The remaining part of logic is built around definite, clearly stated rules.

 # PROPOSITIONS

Propositions are statements which may be true or false.

Note: Comments, opinions and questions are not propositions.

Propositions may be **indeterminate**. For example, 'your grandfather's height is 177 cm' would not have the same answer (true or false) for all people.

Example 1

Which of the following statements are propositions? If they are propositions, are they true, false or indeterminate?

a $20 \div 4 = 80$ **b** $25 \times 8 = 200$ **c**   Mathematics is important.

a   This is a proposition and is false as $20 \div 4 = 5$.

b   This is a proposition and is true.

c   This is an indeterminate comment which has no common answer for all people who study it,   i.e.,   it is not a proposition.

EXERCISE 15A.1

1   Which of the following statements are propositions? If they are propositions, are they true, false or indeterminate?

     a $11 - 5 = 7$ **b** $12 \in \{\text{odd numbers}\}$

     c $\frac{3}{4} \in Q$ **d** $2 \notin Q$

     e   A hexagon has 6 sides.   **f** $37 \in \{\text{prime numbers}\}$

     g   How tall are you?   **h**   All squares are rectangles.

     i   Is it snowing?   **j**   A rectangle is not a parallelogram.

     k   Your brother is 13.   **l**   Do you like dramatic movies?

m   Joan sings loudly.

o   Alternate angles are equal.

n   The angle sum of a quadrilateral is 360°.

p   Parallel lines eventually meet.

NOTATION USED FOR PROPOSITIONS

We represent propositions by letters such as p, q and r.

For example, p: It always rains on Tuesdays.

q: $37 + 9 = 46$

r: x is an even number

NEGATION

The **negation** of a proposition is its opposite.

If p represents a proposition, then $\neg p$ represents its negation. $\neg p$ is read *not p*.

So, p: It is raining. $\neg p$: It is not raining.

Truth values are T for true and F for false.

Every proposition must take a truth value of T or F.

This table shows that $\neg p$ $\begin{cases} \text{is false when } p \text{ is true} \\ \text{is true when } p \text{ is false.} \end{cases}$

p: It is raining.	$\neg p$: It is not raining.	$\neg(\neg p)$
T	F	T
F	T	F

Notice that p and $\neg(\neg p)$ are equivalent statements.

2 For each of the following: **i**   write down the negation

ii   indicate if the statement or its negation is true

a p: All rectangles are parallelograms.

b m: $\sqrt{5}$ is an irrational number.   **c** r: 7 is a rational number.

d q: $23 - 14 = 12$ **e** r: $52 \div 4 = 13$

f s: The difference between any two odd numbers is even.

g t: The product of consecutive integers is always even.

h u: All obtuse angles are equal.

i p: All trapeziums are parallelograms.

j q: If a triangle has two equal angles it is isosceles.

3 What is the negation of these statements for $x \in R$?

a $x < 5$ **b** $x \geqslant 3$ **c** $y < 8$ **d** $y \leqslant 10$

4 Two propositions r and s are given:

i   Is s the negation of r?

ii   If s is not the negative of r, give a reason why this is so.

a r: Bic scored more than 60%. s: Bic scored less than 60%.

b r: Jon is at soccer practice. s: Jon is at music practice.

    c r: The problem is easy to solve. s: The problem is hard to solve.

5 In question **4**, if s was obtained from r using its opposite, will this method give the correct negation?

6 Where possible, try to avoid using the word *not* in writing the negations of:

    a p: the football team won its game   **b** q: she has less than two sisters

    c r: this suburb has a swimming pool   **d** s: this garage is in a mess

The wording of a negation may depend on the domain of the variable.

Example 2

Find the negation of:

a x is a dog for $x \in \{\text{dogs, cats}\}$

b $x \geqslant 2$ for $x \in N$ **c** $x \geqslant 2$ for $x \in Z$

---

a x is a cat   **b** $x = \{0, 1\}$

c $x < 2$ for $x \in Z$ or $x = \{..... -3, -2, -1, 0, 1\}$

---

7 Find the negation of:

    a $x \geqslant 5$ for $x \in Z^+$

    b x is a cow for $x \in$ $\{\text{horses, sheep, cows, goats, deer}\}$

    c $x \geqslant 0$ for $x \in Z$

    d x is a male student for $x \in \{\text{students}\}$

    e x is a female student for $x \in \{\text{females}\}$

NEGATION AND VENN DIAGRAMS

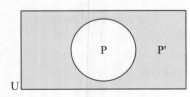

If U is the universal **set** and P is a set within U, then P$'$, called the **complement** of P, consists of all elements of U which are not in P.

Notice that $P \cap P' = \varnothing$ and $P \cup P' = U$.

We can also use **Venn diagrams** to represent a proposition and its negation. If p is a proposition and $\neg p$ is its negation, we can show this as:

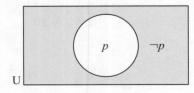

Notice that:

- P is the **truth set** of p

  $\uparrow$

  proposition

- P being the truth set of p means that P is the set whose elements have the characteristics of p.

Example 3

For $U = \{x \mid 0 < x < 10, \quad x \in N\}$ and proposition $p: x$ is a prime number, find the truth sets of p and $\neg p$.

If P is the truth set of p then $P = \{2, 3, 5, 7\}$.

The truth set of $\neg p$ is $P' = \{1, 4, 6, 8, 9\}$

The Venn diagram representation is:

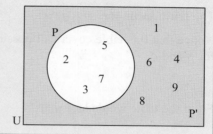

EXERCISE 15A.2

1 **a** Find the truth sets of these statements:

    i x is a multiple of 3, x is between 20 and 30

    ii x is an even number, $1 < x \leqslant 10$.

    iii x is a factor of 42

  b Draw Venn diagrams to represent **a i** and **a ii**.

2 If $U = \{$students in Year 11$\}$, $M = \{$students who study music$\}$, $O = \{$students who play in the orchestra$\}$,   draw a Venn diagram to represent the statements:

  a All music students play in the school orchestra.

  b None of the orchestral students study music.

  c No-one in the orchestra does not study music.

3 **a** **i** Represent $U = \{x \mid 5 < x < 15, \quad x \in N\}$ and $p: \ x < 9$
        on a Venn diagram.

    ii List the truth set of $\neg p$.

  b **i** Represent $U = \{x \mid x < 10, \quad x \in N\}$ and $p: \ x$ is a multiple of 2
        on a Venn diagram.

    ii List the truth set of $\neg p$.

B COMPOUND PROPOSITIONS

Compound propositions are statements which are formed using connectives such as **and** and **or**.

For example,   •   For breakfast Pepe will have cereal **or** toast.

        p: Pepe will have cereal q: Pepe will have toast

    •   Sally ran fast **and** then she threw the javelin.

        p: Sally ran fast q: Sally threw the javelin

CONJUNCTION

When two propositions are joined using the word **and**, the new proposition is the **conjunction** of the original propositions.

Notation   If p and q are propositions, $p \wedge q$ is used to denote their conjunction.

The **truth table** for $p \wedge q$ is:

Notice that this table covers all possible cases for p and q.

p	q	$p \wedge q$
T	T	T
T	F	F
F	T	F
F	F	F

From the table we deduce that:

A conjunction is true only when both original propositions are true.

For example, if p: Eli had coffee for lunch

q: Eli had a pie for lunch

then $p \wedge q$:   Eli had coffee and a pie for lunch

We see that $p \wedge q$ is true only when p and q are true.

If either of p or q is not true, or both p and q are not true, then $p \wedge q$ is not true.

THE VENN DIAGRAM FOR CONJUNCTION

If p: x is a multiple of 3, $x < 20$ and q: x is a multiple of 4, $x < 20$
     then $p \wedge q$: x is a multiple of 3 and 4, $x < 20$.

The Venn diagram representation of $p \wedge q$ is:

P is the truth set of p,   i.e.,   has the property p

Q is the truth set of q,   i.e.,   has the property q

Note: If $x \in$ P   and $x \in$ Q   then $x \in$ P $\cap$ Q.

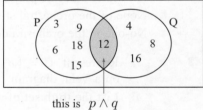

this is $p \wedge q$

DISJUNCTION

When two propositions are joined by the word **or**, the new proposition is the **disjunction** of the original propositions.

Notation   If p and q are propositions:

• $p \vee q$ is used to denote their **inclusive disjunction**

• $p \veebar q$ is used to denote their **exclusive disjunction**.

The **inclusive disjunction** is true when **one or both** propositions are true.

The **exclusive disjunction** is true when **only one of** the propositions is true.

'**or**' is usually given in the **inclusive** sense.

If we wish to use it exclusively we say 'p or q, but not both' or 'exactly one of p or q'.

Inclusive disjunction $(p \vee q)$	Exclusive disjunction $(p \veebar q)$
p or q means p or q or both p and q.	p or q means p or q both not both.

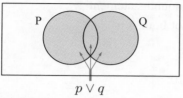

$p \vee q$

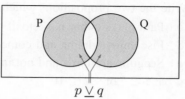

$p \veebar q$

Truth table :

p	q	$p \vee q$
T	T	T
T	F	T
F	T	T
F	F	F

Examples :

p :   Kay's eyes are blue.

q :   Kay's hair is brown.

Kay's eyes are blue and her hair is brown is a possibility.

So, $p \vee q$, the inclusive disjunction would be used.

Truth table :

p	q	$p \veebar q$
T	T	F
T	F	T
F	T	T
F	F	F

Examples :

p :   Sam travels to London all the way by road.

q :   Sam travels to London all the way by rail.

Notice that travelling by road excludes the possibility of travelling by rail, i.e., travelling by road and rail is impossible.

So, $p \veebar q$, the exclusive disjunction would be used.

EXERCISE 15B

1 If p : the shirt is new, and q : the shirt is clean, write down the meaning of:

   a $\neg p$ **b** $\neg q$ **c** $p \vee q$ **d** $p \wedge q$

   e $p \wedge \neg q$ **f** $\neg p \wedge q$ **g** $\neg(\neg p)$ **h** $\neg(\neg q)$

2 If r : Kelly is a good driver, and s : Kelly has a good car, write down in symbolic form:

   a Kelly is not a good driver   **b** Kelly does not have a good car

   c Kelly is a good driver and has a good car

   d Kelly does not have a good car and is not a good driver

   e Kelly has a good car or Kelly is a good driver.

3 If x : Sergio would like to go swimming tomorrow, and
    y : Sergio would like to go bowling tomorrow:

   a Sergio would not like to go swimming tomorrow

   b Sergio would like to go swimming and bowling tomorrow

   c Sergio would like to go swimming or bowling tomorrow

d Sergio would not like to go both swimming and bowling tomorrow

e Sergio would not like to go swimming and go bowling tomorrow

f Sergio would like either to go swimming or go bowling, but not both, tomorrow.

4 Write the two propositions which make the compound statement:

a Pam plays either basketball or tennis, but not both.

b Fisc enjoys drama and comedy.

c Seagulls are birds and not mammals.

d $x > 2$ or $x \leqslant 10$ **e** $2 \leqslant x < 10$.

Example 4

Find the truth sets for p and q and $p \wedge q$ when:

$p:$ x is even and less than 13, $q:$ x is a multiple of 4, less than 20

Let P represent the truth set of p and Q represent the truth set of q.

Then,   P $= \{2, 4, 6, 8, 10, 12\}$ and   Q $= \{4, 8, 12, 16\}$

and the truth set of $p \wedge q$ is $\{4, 8, 12\}$ which is P $\cap$ Q.

---

5 List the truth sets of p, q and $p \wedge q$ for:

a $p:$ x is odd and less than 10, $q:$ x is prime and less than 12

b $p:$ x is a number ending in 3, $x < 20$, $q:$ x is a square number < 50.

6 Find the truth sets of p, q and $p \vee q$ for:

a $p:$ x is a factor of 20, $q:$ x is a factor of 21

b $p:$ x is a multiple of 7, $x < 40$; $q:$ x ends in 0, $0 \leqslant x < 40$.

7 For these statements, determine if their conjunction is true or false. Give a reason.

a $x:$ 23 is an odd number, $y:$ $y = 2x + 1$ is the equation of a line

b $x:$ an emu can fly, $y:$ all birds can fly

c $x:$ the seventh term of $\{1, 2, 4, 8, 16, \ldots\}$ is 64,

$y:$ the 10th term of $\{80, 70, 60, 50, \ldots\}$ is -20.

8 For these statements, determine if their inclusive disjunction is true or false. Give a reason.

a $m:$ 23 is a prime number, $n:$ The graph of $y = x^2$ is a parabola.

b $m:$ 87 is a prime number, $n:$ If $y = \sin \theta$ then $y > 1$.

9 For these statements, determine if their exclusive disjunction is true or false. Give a reason.

a $p:$ 42 is a multiple of 4, $q:$ 42 is a multiple of 3

b $a:$ Paris is the capital of France, $b:$ Rome is the capital of Italy.

10 **a** What can be deduced about x if:

i $x \geqslant 7$ is false   **ii** $3 \leqslant x \leqslant 6$ is false?

b What can be deduced about statements p and q given that:

i $p \vee q$ is true   **ii** $p \wedge q$ is false?

TRUTH TABLES AND LOGICAL EQUIVALENCE

Previous truth tables for negation, conjunction and disjunction can be summarised in one table. This is:

Table 1

p	q	Negation $\neg p$	Conjunction $p \wedge q$	Inclusive disjunction $p \vee q$	Exclusive disjunction $p \underline{\vee} q$
T	T	F	T	T	F
T	F	F	F	T	T
F	T	T	F	T	T
F	F	T	F	F	F

These truth tables can be taken as definitions of these statements.

LOGICAL EQUIVALENCE

If two compound propositions have the same T/F column we say that the two propositions are **logically equivalent** (logically the same).

Example 5

Show that $\neg(p \wedge q)$ and $\neg p \vee \neg q$ are logically equivalent.

The truth table for $\neg(p \wedge q)$ is:

p	q	$p \wedge q$	$\neg(p \wedge q)$
T	T	T	F
T	F	F	T
F	T	F	T
F	F	F	T

The truth table for $\neg p \vee \neg q$ is:

As the last columns of T/Fs are identical then:
$\neg(p \wedge q)$ and $\neg p \vee \neg q$ are logically equivalent,

i.e., $\neg(p \wedge q) = \neg p \vee \neg q$

p	q	$\neg p$	$\neg q$	$\neg p \vee \neg q$
T	T	F	F	F
T	F	F	T	T
F	T	T	F	T
F	F	T	T	T

The results $\neg(p \wedge q) = \neg p \vee \neg q$

and $\neg(p \vee q) = \neg p \wedge \neg q$ are called **deMorgan properties**.

These properties can be shown to be correct by using truth tables as in **Example 5** or by using a Venn diagram.

Showing that $(P \cap Q)' = P' \cup Q'$ establishes that $\neg(p \wedge q) = \neg p \vee \neg q$.

EXERCISE 15C

1   By using truth tables as in **Example 5** establish that $\neg (p \vee q) = \neg p \wedge \neg q$.

2   Use a Venn diagram to show that $(P \cap Q)' = P' \cup Q'$.

> **Hint:**   Shade $P \cap Q$ and then $(P \cap Q)'$ on one Venn diagram.
>
> Shade P', Q' and then $P' \cup Q'$ on another.   Does the shading match?

Example 6

Use a deMorgan property to find the negation of $3 < x < 7$.

$3 < x < 7$ means that $x > 3$ **and** $x < 7$.

$$\neg (3 < x < 7) = \neg (x > 3 \wedge x < 7)$$
$$= \neg (x > 3) \vee \neg (x < 7) \qquad \text{\{DeMorgan's property\}}$$
$$\text{i.e.,} \quad x \leqslant 3 \quad \text{or} \quad x \geqslant 7$$

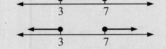

represents $3 < x < 7$

is its negation.

---

3   Use deMorgan's properties to find the negation of:

a $x = -4$ or $x = 5$ **b** $5 \leqslant x \leqslant 9$

c $x < -1$ or $x > 7$ **d**   the number ends in 2 or 8

e   the enclosure is an ellipse or a circle

f   use the cosine rule or the sine rule

g   lines PQ and RS are parallel and equal in length

h $\triangle$KLM is a right angled isosceles triangle.

4   Use truth tables to establish the following logical equivalences:

a $\neg(\neg p) = p$ **b** $p \wedge p = p$ **c** $p \vee (\neg p \wedge q) = p \vee q$

5   **a**   Construct a truth table for $(\neg p \wedge q) \vee (p \wedge \neg q)$.

b   The truth table summary at the start of this section, (Table 1), will help identify what $(\neg p \wedge q) \vee (p \wedge \neg q)$ is logically equivalent to.

TAUTOLOGIES

A **tautology** is a compound statement which is **true for all possibilities** in the truth table.

Example 7

		p	$\neg q$	$p \vee \neg q$
Show that $p \vee \neg p$ is a tautology.	The truth table is :	T	F	T
		F	T	T

A **logical contradiction** is a compound statement which is **false for all possibilities** in the truth table.

6 Identify which of the following compound statements are tautologies, which are logical contradictions, and which are neither:

    a $\neg p \wedge p$ **b** $p \wedge p$ **c** $p \wedge p \wedge p$ **d** $(\neg p \wedge \neg q) \vee (p \vee q)$

7 Why are these true statements?

    a Any two tantologies are logically equivalent.

    b Any two logical contradictions are logically equivalent.

8 What can be said concerning:

    a the negation of a logical contradiction

    b the negation of a tautology

    c the disjunction of a tautology and any other statement r?

D   TRUTH TABLES FOR THREE PROPOSITIONS

When three propositions are considered for combining, we usually denote them p, q and r.

All possible combinations of the truth values for p, q and r are shown in the table alongside.

Notice the systematic listing of these possibilities.

p	q	r
T	T	T
T	T	F
T	F	T
T	F	F
F	T	T
F	T	F
F	F	T
F	F	F

Example 8

Construct a truth table for the compound proposition $(p \vee q) \wedge r$.					

p	q	r	$p \vee q$	$(p \vee q) \wedge r$
T	T	T	T	T
T	T	F	T	F
T	F	T	T	T
T	F	F	T	F
F	T	T	T	T
F	T	F	T	F
F	F	T	F	F
F	F	F	F	F

EXERCISE 15D

1 Construct truth tables for these compound statements:

    a $\neg p \vee (q \wedge r)$ **b** $(p \vee \neg q) \wedge r$ **c** $(p \vee q) \vee (p \wedge \neg r)$

2 a Use truth tables to establish that $(p \wedge q) \wedge r = p \wedge (q \wedge r)$.
This result shows that conjunction ($\wedge$) is associative.

b Establish the truth or otherwise of $(p \vee q) \vee r = p \vee (q \vee r)$.

3 Use Venn diagrams to illustrate the truth sets of p, q and r for:

a $(p \wedge q) \wedge r$ **b** $p \wedge (q \wedge r)$

4

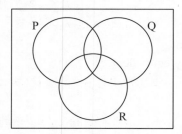

P is the truth set of statement p.

Q is the truth set of statement q.

R is the truth set of statement r.

a Shade the truth set for $p \vee (q \wedge r)$ and $(p \vee q) \wedge (p \vee r)$ on different Venn diagrams.

b What can be deduced from **a**?

5 Is it true that $p \wedge (q \vee r) = (p \wedge q) \vee (p \wedge r)$?
You will need to consider a Venn diagram answer.

E   IMPLICATION

If a compound statement can be formed using an "If ....., then ......" means of connection then the statement is an **implication**.

Consider p: Kato has a TV set   and q: Kato can watch TV.

If Kato has a TV set, then Kato can watch TV is called an **implicative statement**.

Using symbols, we write "if p, then q" as $p \Rightarrow q$ which is read as p *implies q*.

THE TRUTH TABLE FOR IMPLICATION

Consider the implicative statement $x > 3 \Rightarrow x^2 > 9$ which we know is true.

By substituting various values for x we notice that three combinations of truth values occur (TT, FT and FF).

For example,

x	x^2	$x > 3$	$x^2 > 9$
1	1	F	F
4	16	T	T
-5	25	F	T

The fourth possibility (TF) cannot occur, i.e., it is not possible for the first statement to be true and the second false.

The truth table for **implication**, i.e., $p \Rightarrow q$ is:

p	q	$p \Rightarrow q$	
T	T	T	**Table 2**
T	F	F	
F	T	T	
F	F	T	

p is called the **antecedent** and q is called the **consequant**.

Thus: The truth value of the implication ($\Rightarrow$) is only false with p is true and q is false,   i.e.,   the antecedent is true and the consequant is false.

Another interpretation is $p \Rightarrow q$ means that q is true whenever p is true.

EQUIVALENCE

Two statements are **equivalent** if one implies the other and vice versa.

Equivalence is denoted by the symbol $\Leftrightarrow$.

So, for two statements, p and q, $p \Leftrightarrow q$ is the conjunction of the two implications $p \Rightarrow q$, and $q \Rightarrow p$, i.e., $(p \Rightarrow q) \wedge (q \Rightarrow p)$.

The phrase *if and only if* (denoted 'iff') is also used when talking about equivalence.

Example: For p: 'I will pass the exam.'   and q: 'The exam is easy.'

$p \Rightarrow q$:   *If* I pass the exam, *then* the exam is easy.

$q \Rightarrow p$:   *If* the exam is easy, *then* I will pass it.

$p \Leftrightarrow q$:   I will pass the exam *if and only if* the exam is easy.

Notice that:

p	q	$p \Rightarrow q$	$q \Rightarrow p$	$(p \Rightarrow q) \wedge (q \Rightarrow p)$
T	T	T	T	T
T	F	F	T	F
F	T	T	F	F
F	F	T	T	T

So, the truth table for **equivalence** $p \Leftrightarrow q$ is:

p	q	$p \Leftrightarrow q$
T	T	T
T	F	F
F	T	F
F	F	T

EXERCISE 15E

1   In the following implicative statements, state the antecedent and the consequant.

a   If an integer ends in 0 or 5 then it is divisible by 5.

b   If the temperature is low enough then the lake will freeze.

c   If $2x + 5 = 17$, then $2x = 12$.

d   If you jump all 8 hurdles then you may win the race.

e   If $x^2 + 5 = 9$, then $x = \pm 2$.

2   For statements p and q: i   write down the conditional statement $p \Rightarrow q$

ii   state whether the conditional statement is true or false.

a p: $17 > 10$ q: $17 + 3 > 10 + 3$

b p: $x > 4$ q: $x^2 > 16$

c p: $x^2 > 9$ q: $x > 3$

d p: $7 + 5 > 9$ q: $\dfrac{7 + 5}{3} > 4$

3 Suppose p represents 'it is raining' and q represents 'there are puddles forming'. Write the following statements in symbols:

   a If it is raining then puddles are forming.

   b If puddles are forming then it is raining.

   c Puddles are not forming.

   d It is not raining.

   e If it is not raining, then puddles are not forming.

   f If it is raining, then puddles are not forming.

   g If there are no puddles, then it is raining.

4 Construct truth tables for:

   a $p \Rightarrow \neg q$ **b** $\neg q \Rightarrow \neg p$ **c** $(p \wedge q) \Rightarrow p$ **d** $q \wedge (p \Rightarrow q)$

   e $p \Leftrightarrow \neg q$ **f** $(p \Leftrightarrow q) \wedge \neg p$ **g** $p \Rightarrow (p \wedge \neg q)$ **h** $(p \Rightarrow q) \Rightarrow \neg p$

5 **a** By examining truth tables, show that $p \Leftrightarrow q$ and $(p \wedge q) \vee (\neg p \wedge \neg q)$ are logically equivalent.

   b Give a mathematical example for p and q which shows that the equivalence in **a** is justified.

6 Which of these forms are logically equivalent to the negation of $q \Rightarrow p$?

   a $p \Rightarrow q$ **b** $\neg q \Rightarrow p$ **c** $q \Rightarrow \neg p$ **d** $\neg p \Rightarrow \neg q$

7 If a logical form has a truth value which is always F then it is called a **logical contradiction**. (If the truth values are always T it is called a **tautology**.) Which of the following are logical contradictions and which are tautologies? (They may be neither.)

   a $p \Rightarrow (\neg p \wedge q)$ **b** $p \wedge q \Rightarrow p \vee q$ **c** $(p \Rightarrow \neg q) \vee (\neg p \Rightarrow q)$

F   CONVERSE, INVERSE AND CONTRAPOSITIVE

THE CONVERSE

> The **converse** of the statement $p \Rightarrow q$ is the statement $q \Rightarrow p$.

Notice that the equivalence $p \Leftrightarrow q$ is the conjunction of $p \Rightarrow q$ and its converse $(q \Rightarrow p)$.

Example 9

For p: the triangle is isosceles, and q: two angles of the triangle are equal, state $p \Rightarrow q$ and its converse $q \Rightarrow p$.

$p \Rightarrow q$ is If the triangle is isosceles, then two of its angles are equal.

$q \Rightarrow p$ is If two angles of the triangle are equal, then the triangle is isosceles.

Example 10

What is the converse of   "if $x > 10$, then $x > 4$"?

p: $x > 10$, and q: $x > 4$.   The statement is $p \Rightarrow q$.

The converse is $q \Rightarrow p$, and is   'If $x > 4$ then $x > 10$'.

Notice that the converse is false.

THE INVERSE

The **inverse** statement of $p \Rightarrow q$ is the statement $\neg p \Rightarrow \neg q$.

The inverse has truth table:

p	q	$\neg p$	$\neg q$	$\neg p \Rightarrow \neg q$
T	T	F	F	T
T	F	F	T	T
F	T	T	F	F
F	F	T	T	T

Notice that this is the same truth table as $q \Rightarrow p$.   (See the table on page **507**.)

So,   the converse and inverse of an implication are logically equivalent.

THE CONTRAPOSITIVE

The **contrapositive** of the statement $p \Rightarrow q$ is the statement $\neg q \Rightarrow \neg p$.

The **contrapositive** has truth table:

p	q	$\neg q$	$\neg p$	$\neg q \Rightarrow \neg p$
T	T	F	F	T
T	F	T	F	F
F	T	F	T	T
F	F	T	T	T

Notice that the truth table for $\neg q \Rightarrow \neg p$ is the same as that for $p \Rightarrow q$,
(See **Table 2** on page **506**)

i.e.,   the implication and its contrapositive are logically equivalent.

EXERCISE 15F

1   Write the converse and inverse for:
   a   If $5x - 2 = 13$,   then $x = 3$.
   b   If two triangles are similar, then they are equiangular.
   c   If $2x^2 = 12$,   then $x = \pm\sqrt{6}$.
   d   If a figure is a parallelogram, then its opposite sides are equal in length.
   e   If a triangle is equilateral, then its three sides are equal in length.

2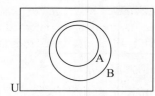

a   Which of these implications are true?

   i $a \in A \Rightarrow a \in B$ ii $b \in B \Rightarrow b \in A$

   iii $c \in B \Rightarrow c \notin A$ iv $c \notin B \Rightarrow c \notin A$

b   Which if **ii**, **iii** or **iv** is the contrapositive of **i**?

3   Write down the contrapositives of these statements:

a   All rose bushes have thorns.

b   No umpire makes correct decisions all the time.

c   No good soccer player has poor kicking skills.

d   Liquids always take the shape of the container in which they are placed.

e   If a person is fair and clever then the person is a doctor.

4   a   State the contrapositive of: "All high school students study Mathematics."

b   What can be deduced (if anything) about these statements?

   i   Keong, who is a high school student

   ii   Tamra, who does not study Mathematics

   iii   Eli, who studies English and Mathematics.

5   Write down the contrapositive of:

a x is divisible by 3 $\Rightarrow$ x^2 is divisible by 9

b x is a number ending in 2 $\Rightarrow$ x is even

c   PQRS is a rectangle $\Rightarrow$ PQ $\parallel$ SR   and   PS $\parallel$ QR

d   KLM is an equilateral triangle $\Rightarrow$ $\angle$KML measures 60°.

6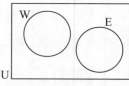

W represents all weak students and E represents all Year 11 students.

a   Copy and complete:

   i   No weak students are....

   ii   No Year 11 students are....

b   Copy and complete:

   i   If $x \in W$ then   ....

   ii   If $x \in E$ then   ....

c   What is the relationship between the implications in **b**?

G   VALID ARGUMENTS

An **argument** is made up of premises (propositions) that lead to a conclusion. An argument is usually indicated by the words 'therefore' or 'hence'.

Example 11

What conclusion results from: "Emus cannot fly. Jane is an Emu"?

Conclusion is: Jane cannot fly.

The argument above can be written as

Let p: Emus cannot fly
        q: Jane is an emu

p
$\underline{q}$ premise
q conclusion

Everything above the line is the premise. Below the line is the conclusion.

Arguments can have their validity tested by analysing the premises and conclusions. A mathematical argument assumes that the premise is always true. Logic is not concerned with whether the premises are true or false, but rather with what can be validly inferred from them.

In logical form, the argument above becomes $p \wedge q \Rightarrow q$.

The truth table is :

p	$\wedge$	q	$\Rightarrow$	q
T	T	T	T	T
T	F	F	T	F
F	F	T	T	T
F	F	F	T	F

The premise is represented by the truth values of the '$\wedge$' on the left side. The argument is represented by the truth values of '$\Rightarrow$'.

By showing that the logical form is a tautology we are demonstrating that the argument is valid.

Example 12

p: x is a positive integer
q: $x + 5 < 2$.

Is this a valid argument?

$$p \Rightarrow q$$
$$\underline{\qquad q \qquad}$$
$$p$$

In logical form $(p \Rightarrow q) \wedge q \Rightarrow p$.   The truth table is constructed as follows.

Step 1: Write down the known truth values.

$p \Rightarrow q$	$\wedge$	q	$\Rightarrow$	p
T		T		T
F		F		T
T		T		F
T		F		F

Step 2: Complete the premise by finding the truth values of the conjunction.

$p \Rightarrow q$	$\wedge$	q	$\Rightarrow$	p
T	T	T		T
F	F	F		T
T	T	T		F
T	F	F		F

Step 3: • Use the values for the premise and conclusion to find the truth values for the implication.

• The logical form is not a tautology. The argument is invalid.

$p \Rightarrow q$	$\wedge$	q	$\Rightarrow$	p
T	T	T	T	T
F	F	F	T	T
T	T	T	F	F
T	F	F	T	F

• What values of p and q lead to a false conclusion? Solve the expression q and state the value of x. Explain why this value makes this argument invalid.

EXERCISE 15G

1 Write the following arguments in symbolic form:

a $p \Rightarrow q$
   $\dfrac{\neg q}{\neg p}$

b $p \vee q$
   $\dfrac{\neg p}{q}$

c $\dfrac{p \vee q}{p}$

d $p \Rightarrow q$
   $\dfrac{\neg p}{\neg q}$

e $p \Rightarrow q$
   $\dfrac{q \Rightarrow p}{p}$

2 Construct truth tables for each part in question **1**. Which of the arguments are valid?

3 p : x is a prime number q : x is an odd number

Write the following in symbolic form:

   a x is an odd number and x is a prime number

   b x is not a prime number

   c If x is an odd number and a prime number, then x is a prime number.

   d If x is an odd number and a prime number, and x is not a prime number, then x is not an odd number.

   Construct a truth table for part **d**. Is this a valid argument?

4 p : It is raining q : I wear my raincoat.

"I am not wearing my raincoat, $\therefore$ it is not raining".

Test the validity of this argument. Give an example that makes this statement invalid.

5 Determine the validity of the following arguments written in logical form:

   a $(p \wedge q) \Rightarrow p$

   b $(p \Rightarrow q) \wedge \neg q \Rightarrow p$

   c $(p \Rightarrow q) \wedge (q \Rightarrow p) \Rightarrow (p \Leftrightarrow q)$

   d $(p \wedge \neg q) \Rightarrow (\neg p \vee q)$

6 Show that the following are invalid arguments:

   a $(p \Rightarrow q) \wedge q \Rightarrow p$

   b $(q \Rightarrow p) \wedge (p \Rightarrow q) \Rightarrow p$.

INVESTIGATION   **SYLLOGISMS**

A **syllogism** is an argument consisting of three lines, the first two of which are true. The third line is supposed to be the logical conclusion from the first two lines.

Example 1:

If I had wings like a seagull I could fly.
I have got wings like a seagull.
Therefore, I can fly.

Example 2:

If I had wings like a seagull I could fly.
I can fly.
Therefore, I have wings like a seagull.

The arguments in these examples can be written as:

   Example 1: $p \Rightarrow q$
      $\dfrac{p}{q}$

   Example 2: $p \Rightarrow q$
      $\dfrac{q}{p}$

Their validity can be tested by truth tables. The first example is a valid argument, the second is invalid. Check these for yourself.

Example 3: All cows have four legs.
Bimbo is not a cow.
Hence, Bimbo does not have four legs.

The syllogism is comprised of two propositions.

p: X is a cow
q: X has four legs

Write the argument in logical form and show that it is invalid.

What to do

1 Test the validity of the following syllogisms.

a All prime numbers greater than two are odd. 15 is odd.
Hence, 15 is a prime number.

b All mathematicians are clever. Jules is not clever.
Therefore, Jules is not a mathematician.

c All rabbits eat grass. Peter is a rabbit.
Therefore, Peter eats grass.

2 Give the third line of the following syllogisms to reach a correct conclusion.

a All cats have fur.
Jason is a cat.
Therefore, ...

b Students who waste time fail.
Takuma wastes time.
Hence, ...

c All emus cannot fly.
Fred cannot fly.
Therefore, ...

H ARGUMENTS WITH THREE PROPOSITIONS

Example 13

The chain rule is an example of logical reasoning:

p: $2x + 5 = 13$
q: $2x = 8$
r: $x = 4$

The argument is written as

$$p \Rightarrow q$$
$$\dfrac{q \Rightarrow r}{p \Rightarrow r}$$

In logical form $(p \Rightarrow q) \land (q \Rightarrow r) \Rightarrow (p \Rightarrow r)$ with truth table:

p	q	r
T	T	T
T	T	F
T	F	T
T	F	F
F	T	T
F	T	F
F	F	T
F	F	F

$(p \Rightarrow q)$	$\land$	$(q \Rightarrow r)$	$\Rightarrow$	$(p \Rightarrow r)$
T	T	T	T	T
T	F	F	T	F
F	F	T	T	T
F	F	T	T	F
T	T	T	T	T
T	F	F	T	T
T	T	T	T	T
T	T	T	T	T

* The logical form of the rule **is** a tautology. Hence the argument is valid.

Thus, if $2x + 5 = 13$ then $x = 4$.

EXERCISE 15H

1 Given the propositions: p: It is sunny q: I am warm r: I feel happy

Write the following in words.

a $(p \land q) \Rightarrow r$ b $p \land \neg q \Rightarrow \neg r$ c $q \land r \Rightarrow p$

2 Which, if any, of the following are valid?

a $p \Rightarrow q$
$\dfrac{q \Rightarrow r}{(q \land r)}$

b $(p \land q) \lor r \Rightarrow (p \lor r)$ c $((p \land q) \Rightarrow r) \land p \Rightarrow r$

3 Consider the argument $p \Rightarrow q$
$\dfrac{q \Rightarrow r}{p \Rightarrow \neg r}$ What values of p, q and r lead to an invalid argument?

4 If I do not like a subject, I do not work hard. If I do not work hard I fail. I passed, therefore I must like the subject.

a Identify the propositions p, q and r.   b Write the above argument in logical form.

c Is the conclusion a result of valid reasoning?

REVIEW SET 15A

1 Which of the following are propositions? State whether they are T, F or indeterminate.

a Sheep have four legs.   b Do cows have four legs?

c Alicia is good at Mathematics.   d I think my favourite team will win.

e Vicki is very clever.   f There are 7 days in a week.

g Put your shoes on.   h All cows are brown.

i $a^2 + b^2 = c^2$

j The opposite sides of a parallelogram are equal.

2 Given p: x is an even number, q: x is divisible by 3.   Write the following in words:

a $\neg p$ b $p \lor q$ c $p \veebar q$ d $p \Rightarrow q$
e $\neg p \land q$ f $\neg p \veebar q$ g $p \Rightarrow \neg q$ h $\neg p \Rightarrow \neg q$

3 Let p: x is a prime number, q: x is a multiple of 7.

Write the following in symbolic language:

a If x is a prime number then x is a multiple of 7.

b x is not a prime number.

c x is a multiple of 7 and not a prime number.

d x is either a prime number of a multiple of 7 but not both.

e x is neither a prime number nor a multiple of 7.

In each case, write down a number that satisfies the statement:

4 Write the implication, the inverse, converse and contrapositive of the following propositions in both words and symbols.

a p: I love swimming.   b p: I like food.
q: I live near the sea. q: I eat a lot.

5 Represent the following on Venn diagrams:

   a $p \veebar q$ **b** $\neg(p \vee q)$ **c** $\neg p \wedge q$

   d $\neg p$ **e** $\neg p \vee q$ **f** $\neg(p \wedge q \wedge r)$

6 p: x is a factor of 12 q: x is an odd number < 10

List the truth sets of p, q, $p \wedge q$, $p \vee q$.

7 Which of the following implications are true?

   a If x is an even number then it is divisible by 2.

   b If a triangle is right angled then it has at least one obtuse angle.

   c If a coin is tossed then it falls as heads or tails.

   d If the base angles of a triangle are equal then the triangle is isosceles.

   e If London is the capital of England then all birds have wings.

REVIEW SET 15B

1 Draw a valid conclusion from the following:

   a If I study hard then I will pass this course. I study hard.

   b If it is raining, then the sky is cloudy. The sky is blue.

   c If the bus is late, I will be late for school. The bus is late.

   d If the sum of digits is divisible by 3 then the number is divisible by 3. The sum of digits is 14.

2 Write negations for the following: Avoid the word 'not' if possible.

   a Eddy is good at football.   **b** The maths class includes more than 10 boys.

   c The writing is illegible.   **d** Ali owns a new car.

3 Write the following statements as implications:

   a All birds have two legs.   **b** Snakes are not mammals.

   c No rectangle has five sides.   **d** This equation has no real solutions.

4 'Positive' and 'negative' are defined as follows:

$$x \text{ is positive} \iff x > 0 \qquad x \text{ is negative} \iff x < 0$$

   a Is zero positive or negative?

   b What is the negation of 'x is negative' when $x \in \{\text{rational numbers}\}$?

5 Write the following in symbolic language:

   a 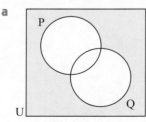   **b** 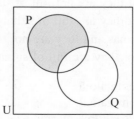   **c**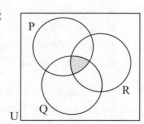

6 Which of the following pairs are logically equivalent?

   a $p \Rightarrow q$ and $\neg q \Rightarrow \neg p$ **b** $\neg(p \wedge q)$ and $\neg p \vee \neg q$

   c $p \Leftrightarrow q$ and $(p \wedge q) \wedge \neg q$ **d** $\neg p \Rightarrow \neg q$ and $q \Rightarrow p$

7 Construct truth tables for the following and state whether the statements are tautologies, contradictions or neither:

a $(p \Rightarrow q) \wedge q \Rightarrow p$

b $(p \wedge q) \wedge \neg (p \vee q)$

c $\neg p \Leftrightarrow q$

d $(p \vee \neg q) \Rightarrow q$

e $(\neg p \vee q) \Rightarrow r$

f $p \wedge \neg q$

REVIEW SET 15C

1 Given $p:$ x is a multiple of s, $18 < x < 30$
  $q:$ x is a factor of 24,
  $r:$ x is an even number, $18 < x < 30$

a List the truth sets of P, Q, R, $p \wedge q \wedge r$

b List the elements of: i $p \wedge q$ ii $q \wedge r$ iii $p \wedge r$

2 Use truth tables to determine the validity of the following arguments:

a $p \Rightarrow q$
  $\underline{\neg p}$
  $\neg q$

b $p \vee q$
  $\underline{\neg q}$
  $\neg p$

c $p \Rightarrow q$
  $\underline{q \Rightarrow r}$
  $r \vee q$

3 Express the following in logical form. Determine whether or not the argument is valid.

a If the sun is shining I will wear my shorts. The sun is shining $\therefore$ I wear shorts.

b All teachers work hard. Marty is not a teacher. Therefore Marty does not work hard.

c If Fred is a dog he has fur. If Fred has fur he has a cold nose.
Fred is a dog $\therefore$ Fred has a cold nose.

4 Write down, in words, the inverse, converse and contrapositive for the implication:
"The diagonals of a rhombus are equal." Which of the three are correct statements?

5 *And* and *or* are used with inequalities. For example: $x \geqslant 5$ means $x > 5$ or $x = 5$ and $6 < x < 9$ means $x > 6$ and $x < 9$.

a Is $x \geqslant 5$ true or false when x is i 2 ii 7? On a number line, illustrate $x \geqslant 5$ by the union of two sets.

b Is $6 < x < 9$ true or false when x is i 2 ii 7? On a number line, illustrate $6 < x < 9$ by the intersection of two sets.

6 Write each of the following using logic symbols.

Let $p:$ cakes are sweet $q:$ cakes are full of sultanas.

a If cakes are not sweet they are not full of sultanas.

b If cakes are not sweet they are full of sultanas.

c Cakes are full of sultanas and they are not sweet.

d Cakes are not sweet or they are full of sultanas.

7 $p:$ The plane leaves from gate 5.
  $q:$ The plane leaves from gate 2.
  $r:$ The plane does not leave this morning.

a Write the following logic statement in words: $p \Rightarrow (\neg r \wedge \neg q)$

b Write in symbols: The plane leaves this morning if and only if it leaves from gate 2 or from gate 5.

Chapter 16

Exponential and trigonometric functions

Contents:

INTRODUCTION TO EXPONENTIAL FUNCTIONS

Consider a population of 100 mice which is growing under plague conditions.

If the mouse population doubles each week we can construct a **table** to show the population number (P) after t weeks.

t (weeks)	0	1	2	3	4	......
P	100	200	400	800	1600	......

We can also represent this information graphically as shown below:

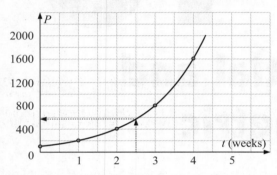

If we use a smooth curve to join the points, we could predict the mouse population when $t = 2.5$ weeks!

Can we find a relationship between P and t?   Examine the following:

t	P values
0	$100 = 100 \times 2^0$
1	$200 = 100 \times 2^1$
2	$400 = 100 \times 2^2$
3	$800 = 100 \times 2^3$
4	$1600 = 100 \times 2^4$

So, the relationship which connects P and t is $P = 100 \times 2^t$.

This is an **exponential relationship** and the graph above is an **exponential graph**.

t is an **index** or **exponent**.

Notice that we could write $P(t) = 100 \times 2^t$ as P is a **function** of t.

In the table above we have evaluated $P(t)$ for various values of t.

A | EVALUATING EXPONENTIAL FUNCTIONS

Recall that if we are asked to find $f(a)$, we find the value of the **function** when $x = a$.

Example 1

For the function $f(x) = 2^x - 3$,   find: **a** $f(0)$ **b** $f(2)$ **c** $f(-2)$

a $f(0) = 2^0 - 3$
$ = 1 - 3$
$ = -2$

b $f(2) = 2^2 - 3$
$ = 4 - 3$
$ = 1$

c $f(-2) = 2^{-2} - 3$
$ = \frac{1}{4} - 3$
$ = -2\frac{3}{4}$

EXERCISE 16A

1 If $f(x) = 2^x + 3$ find the value of:

   a $f(0)$ **b** $f(1)$ **c** $f(2)$ **d** $f(-1)$ **e** $f(-2)$

2 If $f(x) = 5 \times 2^x$ find the value of:

   a $f(0)$ **b** $f(1)$ **c** $f(2)$ **d** $f(-1)$ **e** $f(-2)$

3 If $g(x) = 2^{x+1}$ find the value of:

   a $g(0)$ **b** $g(1)$ **c** $g(3)$ **d** $g(-1)$ **e** $g(-3)$

4 If $h(x) = 2^{-x}$ find the value of:

   a $h(0)$ **b** $h(2)$ **c** $h(4)$ **d** $h(-2)$ **e** $h(-4)$

5 If $P(x) = 6 \times 3^{-x}$ find the value of:

   a $P(0)$ **b** $P(1)$ **c** $P(-1)$ **d** $P(2)$ **e** $P(-2)$

B   GRAPHING SIMPLE EXPONENTIAL FUNCTIONS

We will consider the simplest exponential function $y = 2^x$.

A **table of values** can be constructed for $y = 2^x$.

x	-3	-2	-1	0	1	2	3
y	$\frac{1}{8}$	$\frac{1}{4}$	$\frac{1}{2}$	1	2	4	8

If $x = -3$,   If $x = 0$,   If $x = 3$,

$y = 2^{-3} = \frac{1}{8}$ $y = 2^0 = 1$ $y = 2^3 = 8$

Graph of $y = 2^x$:

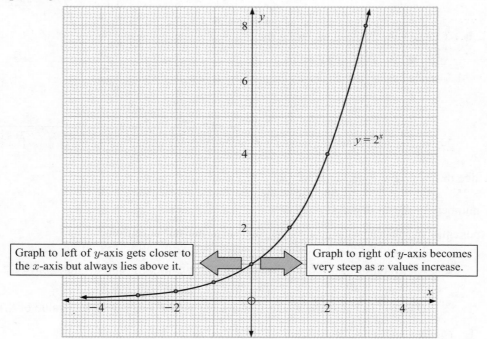

Graph to left of y-axis gets closer to the x-axis but always lies above it.

Graph to right of y-axis becomes very steep as x values increase.

HORIZONTAL ASYMPTOTE

We say that the graph of $y = 2^x$ is **asymptotic** to the x-axis, or the x-axis is a **horizontal asymptote** for the graph of $y = 2^x$.

All exponential graphs have a horizontal asymptote, but not necessarily the x-axis.

EXERCISE 16B

1 Alongside is the graph of $y = 2^x$.
Use the graph to estimate, to one decimal place, the value of:

  a $2^{0.5}$ (find the y-value when $x = 0.5$)

  b $2^{1.6}$

  c $2^{2.1}$

  d   Check your estimations using the ∧ key of your calculator.

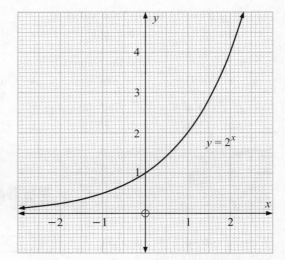

2 Alongside is the graph of $y = 3^x$.

  a   Use the graph to estimate, to one decimal place, the value of:

      i 3^0 **ii** 3^1

      iii $3^{0.5}$ **iv** $3^{1.2}$

  b   Check your estimations using the ∧ key of your calculator.

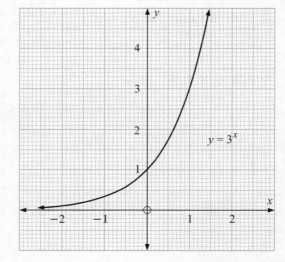

Notice that we have used the graphs of $y = 2^x$ and $y = 3^x$ to find the value of numbers raised to **decimal** powers.

Exponential graphs can also be used to solve **exponential equations**.

3   **a**   Use the graph of $y = 2^x$ in question **1** to estimate x to one decimal place if:

      i $2^x = 4$ **ii** $2^x = 3$ **iii** $2^x = 1$ **iv** $2^x = 0.8$

      (**Hint:**   In **i** find the x value of a point on the graph with a y-value of 4.)

  b   Check your estimations using the ∧ key on your calculator.

4 **a** Use the graph of $y = 3^x$ in question **2** to estimate x to one decimal place if:

  i $3^x = 3$ **ii** $3^x = 4$ **iii** $3^x = 1$ **iv** $3^x = 0.4$

  b Check your estimations using your calculator.

INVESTIGATION 1   EXPONENTIAL GRAPHS

An exponential function is one of the form $y = k \times a^{\lambda x} + c$, where a, c, k and λ are real constants and $a > 0$. A simple example of an exponential function is $f(x) = 3^x$.

This investigation is intended to analyse the influence of the parameters a, c, k and λ on the graph of $y = k \times a^{\lambda x} + c$.

What to do:

1 **a** On the same set of axes, use a graphing package or graphics calculator to graph the following functions: $y = 2^x$, $y = 3^x$, $y = 10^x$, and $y = (1.3)^x$.

  b The functions in **a** are all members of the family $y = a^x$ where $a > 1$.
    i What effect does changing a have on the shape of the graph?
    ii What is the y-intercept of each graph?
    iii What is the horizontal asymptote of each graph?

2 **a** On the same set of axes, use a graphing package or graphics calculator to graph the following functions: $y = \left(\frac{1}{2}\right)^x$, $y = \left(\frac{1}{5}\right)^x$, and $y = (0.875)^x$.

  b The functions in **a** are all members of the family $y = a^x$ where $0 < a < 1$.
    i What effect does changing a have on the shape of the graph?
    ii What is the y-intercept of each graph?
    iii What is the horizontal asymptote of each graph?

3 On the same set of axes, use a graphing package or graphics calculator to graph the following functions: $y = 2^x$, $y = 2^x + 1$, and $y = 2^x - 2$.
These functions are all members of the family $y = 2^x + c$ where $c \in R$.

  a What effect does changing c have on the position of the graph?
  b What effect does changing c have on the shape of the graph?
  c What is the horizontal asymptote of each graph?
  d What is the horizontal asymptote of $y = 2^x + c$?

4 **a** On the same set of axes, use a graphing package or graphics calculator to graph the following two sets of functions:
    i $y = 2^x$, $y = 3 \times 2^x$, and $y = \frac{1}{2} \times 2^x$.
    ii $y = -2^x$, $y = -3 \times 2^x$ and $y = -\frac{1}{2} \times 2^x$

  b The functions in **a** are all members of the family $y = k \times 2^x$ where $k \in R$.
    i Comment on the effect when $k > 0$ and when $k < 0$.
    ii What is the equation of the horizontal asymptote of each graph?

5 **a** On the same set of axes, use a graphing package or graphics calculator to graph the following functions:
    i $y = 2^x$, $y = 2^{2x}$, and $y = 2^{4x}$ **ii** $y = 2^x$, $y = 2^{\frac{1}{2}x}$, and $y = 2^{\frac{1}{3}x}$

  b The functions in **a** are all members of the family $y = 2^{\lambda x}$ where $\lambda \in R$.
Comment on the effect when: **i** $\lambda > 1$ **ii** $0 < \lambda < 1$.

6 Summarise your findings. Discuss how a, c, k and λ affect the graph of $y = k \times a^{\lambda x} + c$.

From your investigation you should have discovered that:

for the general exponential function $y = k \times a^{\lambda x} + c$

► a and λ control how steeply the graph increases or decreases

► c controls vertical position and $y = c$ is the equation of the horizontal asymptote.

►   •   if $k > 0$, $a > 1$ •   if $k > 0$, $0 < a < 1$

           i.e.,   increasing   i.e.,   decreasing

        •   if $k < 0$, $a > 1$ •   If $k < 0$, $0 < a < 1$

           i.e.,   decreasing   i.e.,   increasing

HORIZONTAL ASYMPTOTES

From the previous investigation we noted that for the general exponential function $y = k \times a^{\lambda x} + c$, $y = c$ is the **horizontal asymptote**.

> All exponential graphs are similar in shape and have a horizontal asymptote.

We can actually obtain reasonably accurate sketch graphs of exponential functions using

- the horizontal asymptote
- the y-intercept
- two other points, say when $x = 2$, $x = -2$

Example 2

Sketch the graph of $y = 2^{-x} - 3$.

For $y = 2^{-x} - 3$
the horizontal asymptote is $y = -3$

when $x = 0$, $y = 2^0 - 3$
                 $= 1 - 3$
                 $= -2$

$\therefore$ the y-intercept is -2

when $x = 2$, $y = 2^{-2} - 3$
                 $= \frac{1}{4} - 3$
                 $= -2\frac{3}{4}$

when $x = -2$, $y = 2^2 - 3$
                  $= 4 - 3$
                  $= 1$

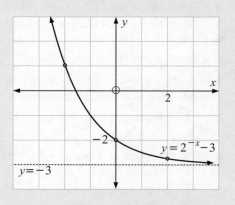

---

5   Sketch the graphs of:

    a $y = 2^x + 1$ **b** $y = 2 - 2^x$ **c** $y = 2^{-x} + 3$ **d** $y = 3 - 2^{-x}$

INVESTIGATION 2   **SOLVING EXPONENTIAL EQUATIONS**

Consider the exponential equation $2^x = 14$.

We notice that for $2^x = 14$, x must lie between
3 and 4 (as $2^3 = 8$ and $2^4 = 16$).

A **graphics calculator** can be used to solve this equation by drawing the graphs of
$y = 2^x$ and $y = 14$ and finding the **point of intersection**.

What to do:

1 Draw the **graph** of $Y_1 = 2$^X   using your graphics calculator.

2 Use **trace** to estimate x when $y = 14$.

3 Draw the **graph** of $Y_2 = 14$ on the same set of axes as $Y_1 = 2$^X.

4 Check the estimation in **2** by finding the coordinates of the **point of intersection** of
the graphs.

5 Use your graphics calculator and the method described above to solve for x, correct
to 3 decimal places:

  a $2^x = 3$ **b** $2^x = 7$ **c** $2^x = 34$

  d $2^x = 100$ **e** $3^x = 12$ **f** $3^x = 41$

(You may have to change the *viewing window*.)

C | EXPONENTIAL GROWTH

In this exercise we will examine situations where quantities are increasing (growth) or decreasing (decay) exponentially.

Populations of animals, people, bacteria, etc usually grow in an exponential way whereas
radioactive substances and items that depreciate usually decay exponentially.

BIOLOGICAL GROWTH

Consider a population of 100 mice which under favourable conditions is increasing by 20%
each week. To increase a quantity by 20%, we multiply it by 120% or 1.2.

So, if P_n is the population after n weeks, then

$P_0 = 100$ {the *original* population}
$P_1 = P_0 \times 1.2 = 100 \times 1.2$
$P_2 = P_1 \times 1.2 = 100 \times (1.2)^2$
$P_3 = P_2 \times 1.2 = 100 \times (1.2)^3$

$\vdots$ etc.

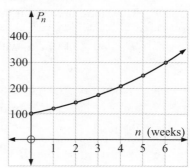

and from this pattern we see that $P_n = 100 \times (1.2)^n$.

Alternatively:

This is an example of a *geometric sequence* and we could have found the rule to generate it.
Clearly $r = 1.2$ and so as $P_n = P_0 r^n$,   then $P_n = 100 \times (1.2)^n$ for $n = 0, 1, 2, 3, ...$

FINANCIAL GROWTH

A further example showing compounding growth could be that of an investment of $5000 at 6% p.a. for a period of n months where this interest is paid monthly.

6% p.a. means a monthly growth of $\dfrac{6\%}{12} = 0.5\%$.

So, each month our investment increases by 0.5%.

This means that our investment increases to 100.5% of what it was the previous month.

As $100.5\% = 1.005$ we have:

$$
\begin{aligned}
A_0 &= 5000 \quad \text{\{the \textit{original} amount\}} \\
A_1 &= 5000 \times 1.005 \\
A_2 &= A_1 \times 1.005 = 5000 \times (1.005)^2 \\
A_3 &= A_2 \times 1.005 = 5000 \times (1.005)^3
\end{aligned}
$$

$\vdots$ etc.

And so, in general $A_n = 5000 \times (1.005)^n$

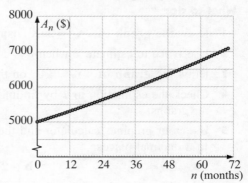

Alternatively:

This is once again an example of a *geometric sequence* with $A_0 = 5000$ and $r = 1.005$.
Consequently, $A_n = A_0 r^n$.   So, $A_n = 5000 \times (1.005)^n$ for $n = 0, 1, 2, 3, ...$

Note: For growth to occur, $r > 1$.

Example 3

The population size of rabbits on a farm is given, approximately, by $R = 50 \times (1.07)^n$ where n is the number of weeks after the rabbit farm was established.

a   What was the original rabbit population?
b   How many rabbits were present after 15 weeks?
c   How many rabbits were present after 30 weeks?
d   Sketch the graph of R against n ($n \geqslant 0$).
e   How long it would it take for the population to reach 500?

$R = 50 \times (1.07)^n$ where R is the population size and
n is the number of weeks from the start.

a   When $n = 0$, $R = 50 \times (1.07)^0$
$$= 50 \times 1$$
$$= 50 \qquad \text{i.e., } 50 \text{ rabbits originally.}$$

b   When $n = 15$, $R = 50 \times (1.07)^{15}$
$$\doteqdot 137.95 \qquad \text{i.e., } 138 \text{ rabbits.}$$

c   When $n = 30$, $R = 50 \times (1.07)^{30}$
$$\doteqdot 380.61 \qquad \text{i.e., } 381 \text{ rabbits.}$$

d

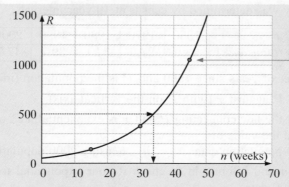

when $n = 45$

$R = 50 \times (1.07)^{45}$

$\doteq 1131$

e From the graph, the approximate number of weeks to reach 500 rabbits is 34.

This solution can also be found using the solver facility of a calculator: *Answer:* $n \doteq 34.0$

or

by finding the intersection of $y = 50 \times (1.07)^x$ and $y = 500$ (as shown)

EXERCISE 16C   You are encouraged to use technology wherever possible.

1 The population of a nest of ants, n weeks after it is established, is given by $P = 500 \times (1.12)^n$.

   a How many ants were originally in the nest?

   b How many ants were in the nest after: **i**   10 weeks   **ii**   20 weeks?

   c Sketch the graph of P against n for $n \geqslant 0$.

   d Use your graph or technology to find how many weeks it takes for the ant population to reach 2000.

2 The weight of bacteria in a culture, t hours after it has been established, is given by the formula

$$W(t) = 20 \times (1.007)^t \text{ grams.}$$

   a Find the original weight of bacteria in the culture.

   b Find the weight of the bacteria after 24 hours.

   c Sketch the graph of $W(t)$ against t for $t \geqslant 0$.

   d Use your graph or technology to find how long it takes for the weight to reach 100 grams.

3 The population of wasps in a nest, n days after it was discovered, is given by $P = 250 \times (1.06)^n$.

   a How many wasps were in the nest originally?

   b Find the number of wasps after   **i**   25 days   **ii**   8 weeks.

   c Sketch the graph of P against n using the above.

   d How long does it take for the population to double? (**Hint:** Use your graph or technology.)

4   The population of a city was determined by census at 10 year intervals.

Year	1960	1970	1980	1990	2000
Population (thousands)	23.0	27.6	33.1	39.7	47.7

Suppose x is the number of years since 1960 and P is the population (in thousands).

a Draw the graph of P against x (with P on the vertical axis).

b It is suspected that the law connecting P and x is exponential, i.e., of the form
$P = a \times b^x$.   Find the value of:
    i a, using 1960 population data   **ii** b, using 2000 population data.

c For the values of a and b found in **b**, check if your exponential formula fits the 1970, 1980 and 1990 data.

d Use your formula to predict the city's population in the year   **i** 2010 **ii** 2050.

e Use a **graphics calculator** to draw a **scatterplot** of the data graphed in **a**.

f Hence, find the exponential law connecting P and x. How does this law compare to the law you found in **b**?

g Use **f** to check your predictions in **d**.

D   EXPONENTIAL DECAY

Decay problems occur when the size of a variable decreases over time. Many quantities which decay over time are examples of **exponential decay**.

Example 4

When a CD player is switched off, the current dies away according to the formula
$I(t) = 24 \times (0.25)^t$ amps, where t is the time in seconds.

a   Find $I(t)$ when $t = 0$, 1, 2 and 3.

b   What current flowed in the CD player at the instant when it was switched off?

c   Plot the graph of $I(t)$ against t $(t \geqslant 0)$ using the information above.

d   Use your graph and/or technology to find how long it takes for the current to reach 4 amps.

a $I(t) = 24 \times (0.25)^t$ amps

$I(0)$
$= 24 \times (0.25)^0$
$= 24$ amps

$I(1)$
$= 24 \times (0.25)^1$
$= 6$ amps

$I(2)$
$= 24 \times (0.25)^2$
$= 1.5$ amps

$I(3)$
$= 24 \times (0.25)^3$
$= 0.375$ amps

b   When $t = 0$, $I(0) = 24$
$\therefore$ 24 amps of current flowed.

c
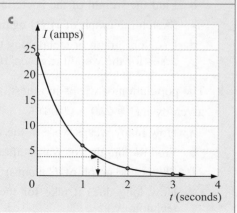

> **d**   From the graph above, the approximate time to reach 4 amps is 1.3 seconds.
>
>   Using a calculator, the solution to 3 sig. figs. is $\doteqdot 1.29$ seconds.

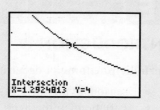

EXERCISE 16D

1   When a liquid in a container is placed in a refrigerator, its temperature (in °C) is given by $T = 100 \times (0.933)^t$, t being the time in minutes. Find:

   a   the initial temperature of the liquid

   b   the temperature after   **i**   10 min   **ii**   20 min   **iii**   30 min.

   c   Draw the graph of T against t for $t \geqslant 0$, using the information above.

   d   Use your graph or a **graphics calculator** to estimate the number of minutes taken for the liquid to reach   **i**   40°C   **ii**   10°C.

2   The weight of a radioactive substance t years after being set aside is given by
$W(t) = 250 \times (0.998)^t$ grams.

   a   How much radioactive substance was put aside?

   b   Determine the weight of the substance after:

      i   400 years   **ii**   800 years   **iii**   1200 years.

   c   Sketch the graph of W against t for $t \geqslant 0$, using the above information.

   d   Use your graph or **graphics calculator** to find how long it takes for the substance to decay to 125 grams.

3   The marsupial *Eraticus* is endangered. There is only one colony remaining and survey details of its numbers have been determined at 5 year intervals:

Year	1975	1980	1985	1990	1995	2000
Number	255	204	163	131	104	84

Let n be the time since 1975 and P be the population size.

   a   Graph P against n (with P on the vertical axis).

   b   It is suspected that the law connecting P and n is of the form $P = a \times b^n$ where a and b are constants. Find:

      i   the value of a, using the 1975 data

      ii   the value of b, using the 2000 population data.

   c   Check if the data for 1980, 1985, 1990 and 1995 fit the law.

   d   In what year do you expect the population size to reduce to 50?

   e   Use a **graphics calculator** to draw a **scatterplot** of the data graphed in **a**.

   f   Hence find the exponential law connecting P and n.
How does this law compare to the law you found in **b**?

   g   Use **f** to check your prediction in **d**.

E   # PERIODIC FUNCTIONS

INTRODUCTION

Periodic phenomena occur in the physical world in:

- seasonal variations in our climate
- variations in the average maximum and minimum monthly temperatures at a place
- the number of daylight hours at a place
- variations in the depth of water in a harbour due to tidal movement
- the phases of the moon   etc.

Periodic phenomena also occur in the living world in animal populations.

These phenomena illustrate variable behaviour which is repeated over time. This repetition may be called periodic, oscillatory or cyclic in different situations.

In this topic we will consider various data sets which display periodic behaviour.

OPENING PROBLEM

      A Ferris wheel rotates at a constant speed. The wheel's radius is 10 m and the bottom of the wheel is 2 m above ground level.
From a point in front of the wheel Andrew is watching a green light on the perimeter of the wheel. Andrew notices that the green light moves in a circle. He then considers how high the light is above ground level at two second intervals and draws a scatterplot of his results.

- What would his scatterplot look like?
- Could a known function be used to model the data?
- How could this function be used to find the light's position at any point in time?
- How could this function be used to find the time when the light is at a maximum (or minimum) height?
- What part of the function would indicate the time interval over which one complete cycle occurs?

Click on the icon to visit a simulation of the Ferris wheel.

You are to view the light on the Ferris wheel:

- from a position in front of the wheel
- from a side-on position   **DEMO**
- from above the wheel.

Now observe the graph of height above (or below) the wheel's axis over time as the wheel rotates at a constant rate.

OBSERVING PERIODIC BEHAVIOUR

Consider the table below which shows the mean monthly maximum temperature (^{o}C) for Cape town.

Month	Jan	Feb	Mar	Apr	May	Jun	Jul	Aug	Sep	Oct	Nov	Dec
Temp	28	27	$25\frac{1}{2}$	22	$18\frac{1}{2}$	16	15	16	18	21	24	26

If this data is graphed using a scatterplot, assigning January = 1, February = 2 etc., for the 12 months of the year, the graph shown is obtained.

(**Note:** The points are not joined as interpolation has no meaning here.)

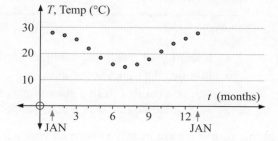

The temperature shows a variation from an average of 28°C in January through a range of values across the months and the cycle will repeat itself for the next 12 months.

It is worthwhile noting that later we will be able to establish a function which approximately fits this set of points.

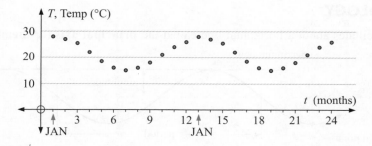

Graphs which have this basic shape where the cycle is repeated over and over are called **sine waves**.

GATHERING PERIODIC DATA

* Maximum and minimum monthly temperatures are obtained from an internet site.
* Tidal details can be obtained from daily newspapers.

ACTIVITY   **BICYCLE DATA**

On a flat surface such as a tennis court mark a chalk line with equal intervals of 20 cm. On a tyre of a bicycle wheel mark a white spot using correcting fluid. Start with the spot at the bottom of the tyre on the first marked interval. Wheel the bike until the bottom of the tyre is on the second marked interval. Use a metre rule to measure the height of the spot above the ground.

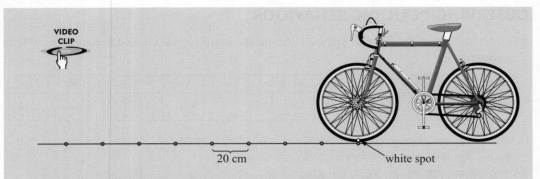

20 cm   white spot

a Record your result and continue until you have 20 or more data values.
b Plot this data on a set of axes.
c Are you entitled to fit a smooth curve through these points or should they be left as discrete points? Keep your results for future analysis.

In this course we are mainly concerned with periodic phenomena which show a wave pattern when graphed.

the wave

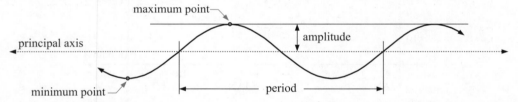

principal axis

TERMINOLOGY

The wave oscillates about a horizontal line called the **principal axis** (or **mean line**).

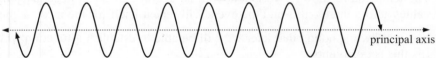

maximum point

principal axis

amplitude

minimum point

period

A **maximum point** occurs at the top of a crest and a **minimum point** at the bottom of a trough.

The **amplitude** is the distance between a maximum (or minimum) point and the principal axis.

PERIODICITY

A **periodic function** is one which repeats itself over and over in a horizontal direction.

The **period** of a periodic function is the length of one repetition or cycle.

If $f(x)$ is a periodic function with period p then $f(x + p) = f(x)$ for all x and p is the smallest positive value for this to be true.

Notice that $f(x + 12) = f(x)$ for all x on the temperature curve.
This means that $f(0) = f(12) = f(24)$,   etc.
$f(1) = f(13) = f(25)$,   etc.
$f(2.35) = f(14.35) = f(26.35)$,   etc.

EXERCISE 16E

1 For each set of data below, draw a scatterplot and decide whether or not the data exhibits approximately periodic behaviour.

a

x	0	1	2	3	4	5	6	7	8	9	10	11	12
y	0	1	1.4	1	0	−1	−1.4	−1	0	1	1.4	1	0

b

x	0	1	2	3	4
y	4	1	0	1	4

c

x	0	0.5	1.0	1.5	2.0	2.5	3.0	3.5
y	0	1.9	3.5	4.5	4.7	4.3	3.4	2.4

d

x	0	2	3	4	5	6	7	8	9	10	12
y	0	4.7	3.4	1.7	2.1	5.2	8.9	10.9	10.2	8.4	10.4

2 The following tabled values show actual bicycle wheel data as determined by the method described earlier.

Distance travelled (cm)	0	20	40	60	80	100	120	140	160
Height above ground (cm)	0	6	23	42	57	64	59	43	23
Distance travelled (cm)	180	200	220	240	260	280	300	320	340
Height above ground (cm)	7	1	5	27	40	55	63	60	44
Distance travelled (cm)	360	380	400						
Height above ground (cm)	24	9	3						

a Plot the graph of height against distance.

b Is the data periodic, and if so find estimates of:
  - **i** the equation of the principal axis
  - **ii** the maximum value
  - **iii** the period
  - **iv** the amplitude

c Is it reasonable to fit a curve to this data, or should we leave it as discrete points?

3 Which of these graphs show periodic behaviour?

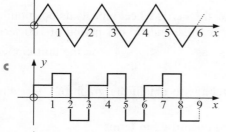

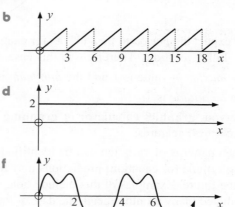

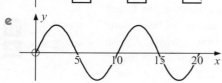

4

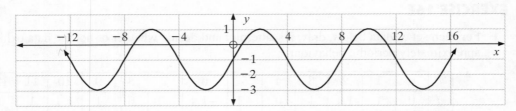

For the given periodic function:

  a state its amplitude   **b** state its period.
  c State the coordinates of its first maximum point for positive x values.
  d What is the distance between successive maxima?
  e What is the equation of its principal axis?

F   SINE FUNCTIONS

Returning to the Ferris wheel we will examine the graph obtained when plotting
the height of the light above or below
the principal axis against the time in seconds. We do this for a wheel of radius
10 m which takes 100 seconds for one
full revolution.

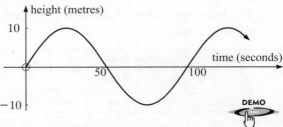

We observe that the amplitude is 10 and the period is 100 seconds.

The family of sine curves can have different amplitudes and different periods. We will
examine such families in this section.

THE BASIC SINE CURVE

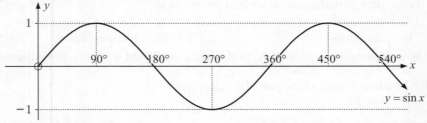

We expect the *period* to be $360°$, as for example, the Ferris wheel repeats its positioning after
one full revolution. Therefore, in this case $360°$ corresponds to our period of 100 seconds.

The *maximum* value is 1 and the *minimum* is -1 as $-1 \leqslant y \leqslant 1$ on the unit circle.

The *amplitude* is 1.

Use your **graphics calculator** or **graphing package** to obtain the graph of $y = \sin x$ to
check these features.

When patterns of variation can be identified and quantified in terms
of a formula (or equation) predictions may be made about behaviour
in the future. Examples of this include tidal movement which can be
predicted many months ahead, the date of the full moon in the future
for setting Good Friday etc.

INVESTIGATION 3   THE FAMILY $y = A\sin x$

If using a graphics calculator, make sure that the mode is set in **degrees** and that your viewing window is appropriate.

GRAPHING PACKAGE

What to do:

1 Use technology to graph on the same set of axes:

   a $y = \sin x$ and $y = 2\sin x$ **b** $y = \sin x$ and $y = 0.5\sin x$

   c $y = \sin x$ and $y = -\sin x$ $(A = -1)$

2 For each of $y = \sin x$, $y = 2\sin x$, $y = 0.5\sin x$, $y = -\sin x$ record the maximum and minimum values and state the period and amplitude. If using a calculator use the built in functions to find the maximum and minimum values.

3 How does A affect the function $y = A\sin x$?

4 State the amplitude of: **a** $y = 3\sin x$ **b** $y = \sqrt{7}\sin x$ **c** $y = -2\sin x$

INVESTIGATION 4   THE FAMILY $y = \sin Bx$, $B > 0$

What to do:

1 Use technology to graph on the same set of axes:

   a $y = \sin x$ and $y = \sin 2x$ **b** $y = \sin x$ and $\sin(\frac{1}{2}x)$

2 For each of $y = \sin x$, $y = \sin 2x$, $y = \sin(\frac{1}{2}x)$ record the maximum and minimum values and state the period and amplitude.

3 How does B affect the function $y = \sin Bx$?

GRAPHING PACKAGE

4 State the period of:

   a $y = \sin 3x$ **b** $y = \sin(\frac{1}{3}x)$ **c** $y = \sin(1.2x)$ **d** $y = \sin Bx$

From the previous investigations you should have observed that:

- In $y = A\sin x$, A affects the amplitude and the amplitude is $|A|$.

If $A = 2$, vertical distances from the x-axis to $y = \sin x$ are doubled to get the graph of $y = 2\sin x$.

This is a **vertical dilation**.

If $A = \frac{1}{2}$, vertical distances are halved.

The graph of $y = -\sin x$ is the graph of $y = \sin x$ reflected in the x-axis.

The graph of $y = -2\sin x$ is the graph of $y = \sin x$ vertically dilated with factor 2 and then reflected in the x-axis.

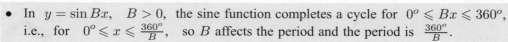

- In $y = \sin Bx$, $B > 0$, the sine function completes a cycle for $0° \leqslant Bx \leqslant 360°$, i.e., for $0° \leqslant x \leqslant \frac{360°}{B}$, so B affects the period and the period is $\frac{360°}{B}$.

If $B = 2$, the period is halved.

If $B = \frac{1}{2}$, the period is doubled.

Here we have a **horizontal dilation** of factor $\frac{1}{B}$.

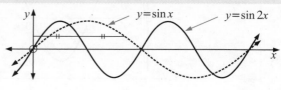

Example 5

Without using technology sketch the graphs of:

a $y = 2 \sin x$ **b** $y = -2 \sin x$ for $0° \leqslant x \leqslant 360°$.

a The amplitude is 2, and the period is $360°$.

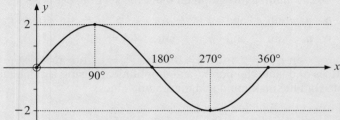

We place the 5 points as shown and fit the sine wave to them.

b The amplitude is 2, the period is $360°$, and it is the reflection of $y = 2 \sin x$ in the x-axis.

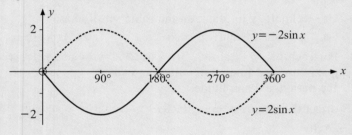

EXERCISE 16F.1

1 Without using technology draw the graphs of the following for $0° \leqslant x \leqslant 360°$:

a $y = 3 \sin x$ **b** $y = -3 \sin x$ **c** $y = \frac{3}{2} \sin x$ **d** $y = -\frac{3}{2} \sin x$

Example 6

Without using technology sketch the graph of $y = \sin 2x$, $0° \leqslant x \leqslant 360°$.

The period is $\frac{360°}{2} = 180°$.

So, for example, the maximum values are $180°$ apart.

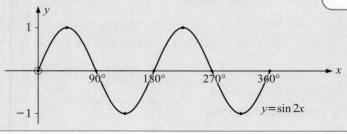

As $\sin 2x$ has half the period of $\sin x$, the first maximum is at $45°$ not $90°$.

2 Without using technology sketch the graphs of the following for $0° \leqslant x \leqslant 540°$:

    a $y = \sin 3x$ **b** $y = \sin \left(\frac{x}{2}\right)$

3 State the period of:

    a $y = \sin 4x$ **b** $y = \sin \left(\frac{x}{3}\right)$ **c** $y = \sin(0.6x)$

4 Find B given that the function $y = \sin Bx$, $B > 0$ has period:

    a $900°$ **b** $120°$ **c** $2160°$

INVESTIGATION 5   THE FAMILY $y = \sin x + C$

What to do:

GRAPHING PACKAGE

1 Use technology to graph on the same set of axes:

    a $y = \sin x$ and $y = \sin x + 3$ **b** $y = \sin x$ and $y = \sin x - 2$

2 For each of $y = \sin x$, $y = \sin x + 3$ and $y = \sin x - 2$ record the maximum and minimum values and state the period and amplitude.

3 What transformation moves $y = \sin x$ to $y = \sin x + C$?

From **Investigation 5** we observe that:

- $y = \sin x + C$ is a **vertical translation** of $y = \sin x$ through C units.

EXERCISE 16F.2

1 Find a formula in the form $y = A \sin(Bx) + C$ which produces the following graphs:

a

b

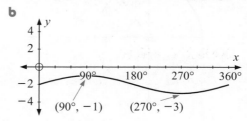

c

d

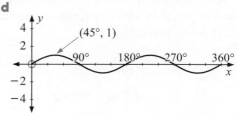

e

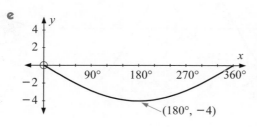

f

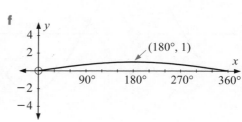

g

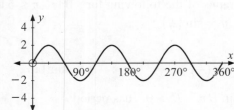

h

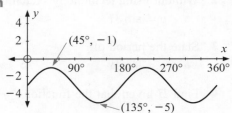

2 Without using technology, sketch the graphs of the following for $0° \leqslant x \leqslant 360°$:

a $y = \sin x + 1$	**b** $y = \sin x - 2$	**c** $y = 1 - \sin x$	
d $y = 2\sin x - 1$	**e** $y = \sin(3x) + 1$	**f** $y = 1 - \sin(2x)$	**GRAPH PAPER**

Click on the icon to obtain printable graph paper for the above question.

G   COSINE FUNCTIONS

We return to the Ferris wheel to see the cosine function being generated.

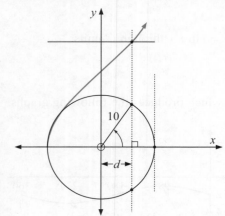

Click on the icon to inspect a simulation of the view from above the wheel.

The graph being generated over time is a **cosine function**.

This is no surprise as $\cos\theta = \dfrac{d}{10}$

i.e., $d = 10\cos\theta$.

DEMO

Now view the relationship between the sine and cosine functions. Notice that the functions are identical in shape, but the cosine function is $90°$ units left of the sine function under a horizontal translation.

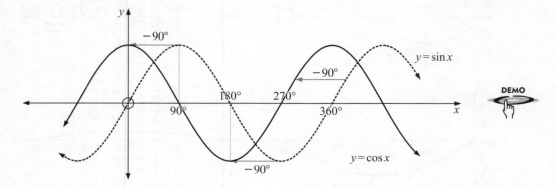

EXERCISE 16G

1   Given the graph of $y = \cos x$,
sketch the graphs of:

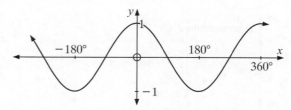

   a $y = \cos x + 2$ **b** $y = \cos x - 1$ **c** $y = \frac{2}{3} \cos x$

   d $y = \frac{3}{2} \cos x$ **e** $y = -\cos x$ **f** $y = \cos (2x)$

   g $y = \cos \left(\frac{x}{2}\right)$ **h** $y = 3 \cos (2x)$

2   Without graphing them, state the periods of:

   a $y = \cos (3x)$ **b** $y = \cos \left(\frac{x}{3}\right)$ **c** $y = \cos \left(\frac{x}{2}\right)$

3   The general cosine function is $y = A \cos (Bx) + C$.
State the geometrical significance of A, B and C.

4   For the following graphs, find the cosine function representing them:

  a

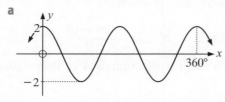

  b

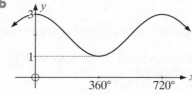

  c

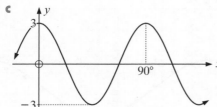

  d

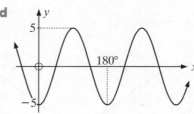

H | MODELLING USING SINE AND COSINE FUNCTIONS

Sine and cosine functions can be useful for modelling certain biological and physical phenomena in nature which are periodic (or roughly so).

MEAN MONTHLY TEMPERATURE

The mean monthly maximum temperature (°C) for Adelaide is as shown in the given table

Month	Jan	Feb	Mar	Apr	May	Jun	Jul	Aug	Sep	Oct	Nov	Dec
Temp	28.0	27.0	24.7	21.5	18.2	15.6	15.1	15.8	18.2	21.5	24.8	27.1

and the graph over a two year period is follows:

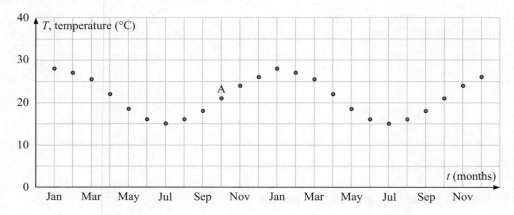

We will attempt to model this data to $y = A \cos Bx + C$ i.e., $T = A \cos Bt + C$.

Now the period is 12 months, so $\dfrac{360^o}{B} = 12$ months and $\therefore$ $B = 30^o$ per month.

The amplitude $= \dfrac{\text{max.} - \text{min.}}{2} \doteqdot \dfrac{28 - 15}{2} \doteqdot 6.5,$ so $A = 6.5$.

The principal axis is midway between max. and min. $\therefore$ $D = \dfrac{28 + 15}{2} = 21.5$.

The model is therefore $T = 6.5 \cos(30t) + 21.5$ and is superimposed on the original data in the graph that follows.

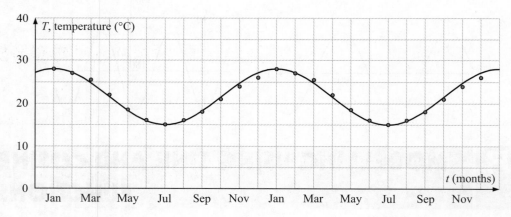

EXERCISE 16H

1   The following data gives the mean monthly minimum temperature (^{o}C) for Adelaide.

Month	Jan	Feb	Mar	Apr	May	Jun	July	Aug	Sept	Oct	Nov	Dec
Temp	16	$14\frac{1}{2}$	12	10	$7\frac{1}{2}$	7	$7\frac{1}{2}$	$8\frac{1}{2}$	$10\frac{1}{2}$	$12\frac{1}{2}$	14	15

Fit a cosine model to this temperature data. The temperature T, is to be a function of time.

2   The depth of water, D metres, at a port entrance was recorded every hour for 12 hours. The data collected is shown below:

Time (t hours)	0	1	2	3	4	5	6	7	8	9	10	11	12
Distance (D m)	4.5	5.25	5.8	6	5.8	5.25	4.5	3.75	3.2	3	3.2	3.75	4.5

a   Estimate the   **i** amplitude   **ii** period   **iii** vertical translation.

b   Hence, find a sine model which connects depth (D) and time (t).

3   Revisit the **Opening Problem** on page **528**.

The wheel takes 100 seconds to complete one revolution.

Find the cosine model which gives the height of the light above the ground at any point in time.   (**Hint**: A could be positive or negative)

Assume at time $t = 0$, the light is at its lowest point.

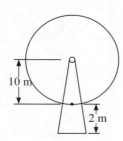

I   EQUATIONS INVOLVING SINE & COSINE

Linear equations such as $2x + 3 = 11$ have exactly one solution.

Quadratic equations, i.e., equations of the form $ax^2 + bx + c = 0$, $a \neq 0$ have at most two real solutions.

Trigonometric equations generally have infinitely many solutions unless a restrictive domain such as $0^o \leqslant x \leqslant 360^o$ is given.   We will examine solving sine equations using:
- preprepared graphs   • technology   • algebraic methods.

For the Ferris Wheel **Opening Problem** the model is $H = -10\cos(3.6t) + 12$.

We can easily check this by substituting
$t = 0, 25, 50, 75$

$H(0) = -10\cos(0) + 12 = -10 + 12 = 2$ ✓

$H(25) = -10\cos(90) + 12 = 12$ ✓

$H(50) = -10\cos(180) + 12 = 22$ ✓

etc.

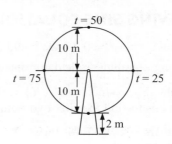

However, we may be interested in the times when the light is 16 m above the ground, which means that we need to solve the equation $-10\cos(3.6t) + 12 = 16$ which is of course a **cosine equation**.

GRAPHICAL SOLUTION OF SINE EQUATIONS

Sometimes simple sine graphs on grid paper are available and estimates of solutions can be obtained. To solve $\sin x = 0.3$, we observe where the horizontal line $y = 0.3$ meets the graph of $y = \sin x$.

EXERCISE 16I

1

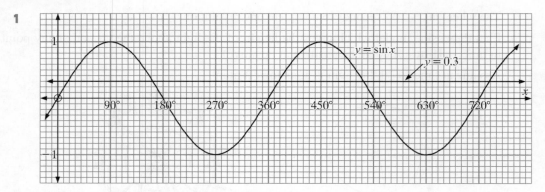

Use the graph of $y = \sin x$ to find approximate solutions of:

a $\sin x = 0.3$ for $0^o \leqslant x \leqslant 720^o$ **b** $\sin x = -0.4$ for $180^o \leqslant x \leqslant 360^o$

2

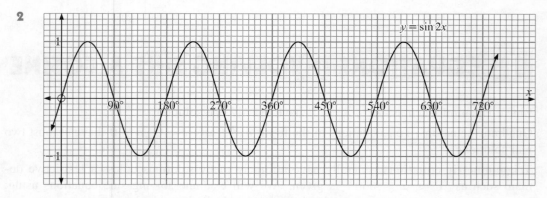

Use the graph of $y = \sin 2x$ to find approximate solutions of:

a $\sin 2x = 0.7$ **b** $\sin 2x = -0.3$ for $0^o \leqslant x \leqslant 540^o$

SOLVING SINE EQUATIONS USING TECHNOLOGY

To solve $\sin x = 0.3$ we could use either a **graphing package** or **graphics calculator**.

If using a graphics calculator make sure the **mode** is set to **degrees**.

Graph $Y_1 = \sin X$ and $Y_2 = 0.3$. Use the built in functions to find the first two points of intersection.

These are $X = 17.46$ and $X = 162.5$ for $0^o \leqslant x \leqslant 360^o$.

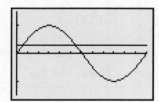
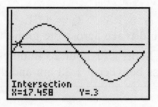

3   Use technology to solve for $0° \leqslant x \leqslant 360°$.

  a $\sin x = 0.414$ **b** $\sin x = -0.673$ **c** $\sin x = 1.289$

  d $\sin(2x) = 0.162$ **e** $\sin\left(\frac{x}{2}\right) = 0.606$ **f** $\sin\left(\frac{2x}{3}\right) = 0.9367$

J | USING SINE AND COSINE MODELS

Example 7

The height $h(t)$ metres of the tide above mean sea level on January 24th at Outer Harbour is modelled approximately by $h(t) = 3\sin(30t)$ where t is the number of hours after midnight.

a   Graph $y = h(t)$ for $0 \leqslant t \leqslant 24$.

b   When was high tide and what was the maximum height?

c   What was the height at 2 pm?

d   If a ship can cross the harbour provided the tide is at least 2 m above mean sea level, when is crossing possible on January 24?

a $h(t) = 3\sin(30t)$ has period $= \frac{360}{30} = 12$ hours   and $h(0) = 0$

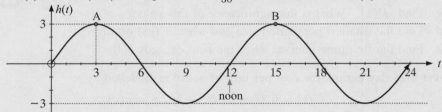

b   High tide is at 3 am and 3 pm, and maximum height is 3 m above the mean as seen at points A and B.

c   At 2 pm, $t = 14$ and $h(14) = 3\sin(30 \times 14) \doteqdot 2.60$

So the tide is 2.6 m above the mean.

d

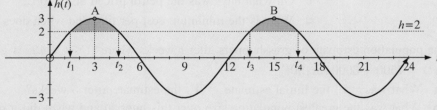

We need to solve $h(t) = 2$ i.e., $3\sin(30t) = 2$.

Using a graphics calculator with $Y_1 = 3\sin(30X)$ and $Y_2 = 2$

we obtain $t_1 = 1.39$, $t_2 = 4.61$, $t_3 = 13.39$, $t_4 = 16.61$
or   you could **trace** across the graph to find these values.

Now   1.39 hours = 1 hour 23 minutes, etc.

$\therefore$ can cross between   1:23 am and 4:37 am   or   1:23 pm and 4:37 pm.

EXERCISE 16J

1   The model for the height of a light on a Ferris Wheel
    is $H(t) = 20 - 19\sin(120t)$,   where H is the height in
    metres above the ground, t is in minutes.

a   Where is the light at time $t = 0$?

b   At what time was the light at its lowest in the first
    revolution of the wheel?

c   How long does the wheel take to complete one
    revolution?

d   Sketch the graph of the $H(t)$ function over one
    revolution.

2   The population of water buffalo is given by
    $P(t) = 400 + 250\sin(90t)$ where t is the number of
    years since the first estimate was made.

a   What was the initial estimate?

b   What was the population size after:

    i   6 months   ii   two years?

c   Find $P(1)$.   What is the significance of this value?

d   Find the smallest population size and when it first occurs.

e   Find the first time interval when the herd exceeds 500.

3   Over a 28 day period, the cost per litre of petrol is modelled by

$$C(t) = 6.8\cos(22.5t) + 107.8 \text{ cents/L.}$$

a   True or false?

    i   "The cost/litre oscillates about 107.8 cents
       with maximum price $1.17."

    ii   "Every 14 days, the cycle repeats itself."

b   What is the cost at day 7?

c   On what days was the petrol priced at $1.10/L?

d   What is the minimum cost per litre and when does it occur?

4   The population estimate of grasshoppers after t weeks where $0 \leqslant t \leqslant 12$ is given by
    $P(t) = 7500 + 3000\sin(90t)$.

a   What was: i   the initial estimate   ii   the estimate after 5 weeks?

b   What was the greatest population size over this interval and when did it occur?

c   When is the population   i   9000   ii   6000?

d   During what time interval(s) does the population size exceed 10 000?

REVIEW SET 16A

1 If $f(x) = 3 \times 2^x$, find the value of: **a** $f(0)$ **b** $f(3)$ **c** $f(-2)$

2 On the same set of axes draw the graphs of   **a** $y = 2^x$ **b** $y = 2^x - 4$, stating the y-intercepts and the equations of the horizontal asymptotes.

3 The temperature of a liquid t minutes after it was heated is given by $T = 80 \times (0.913)^t$ °C. Find:
   a the initial temperature of the liquid
   b the temperature after   **i** $t = 12$ **ii** $t = 24$ **ii** $t = 36$ minutes.
   c Draw the graph of T against t, $t \geqslant 0$, using the above or technology.
   d Hence, find the time taken for the temperature to reach $25°C$.

4 If $g(x) = 2^{x-1}$, find the value of   **a** $g(0)$ **b** $g(-2)$ **c** $g(3)$

5 On the same set of axes draw graphs of   **a** $y = 2^x$ **b** $y = 2^{-x}$ stating the y-intercept and the equation of the horizontal asymptote.

6 A mob of kangaroos P, t years after an initial count, is given by $P = 1000 \times (1.26)^t$.
   a Find the original number of kangaroos.
   b Find the number of kangaroos when   **i** $t = 6$ **ii** $t = 12$ **iii** $t = 18$ years.
   c Sketch the graph of P against t, $(t \geqslant 0)$, using the above or technology.
   d Hence, determine the time taken for the population to reach 5000 kangaroos.

REVIEW SET 16B

1 If $f(x) = 3^x - 2$, find the value of   **a** $f(0)$ **b** $f(2)$ **c** $f(-2)$

2 The weight of a radioactive substance after t years is given by $W = 1500 \times (0.993)^t$ grams.
   a Find the original amount of radioactive material.
   b Find the amount of radioactive material remaining after   **i** 400 years   **ii** 800 years.
   c Sketch the graph of W against t, $t \geqslant 0$, using the above or technology.
   d Hence, find the time taken for the weight to reduce to 100 grams.

3 On the same set of axes draw the graphs of:
   a $y = 2^x$ and $y = 2^x + 2$ **b** $y = 2^x$ and $y = 2^{x+2}$

4 Sketch the graphs of   **a** $y = 2^{-x} - 5$ **b** $y = 4 - 2^x$.

5 A population of seals t years after a colony was established is given by $P_t = 40 \times 2^{0.3t}$, $t \geqslant 0$.   Find:
   a the initial size of the population
   b the population after   **i** 5 years   **ii** 10 years   **iii** 20 years.
   c Sketch the graph of P_t against t using only your results from **a** and **b**.
   d Use technology to graph $Y_1 = 40 \times 2^{0.3X}$ and check your answers to **a**, **b** and **c**.

REVIEW SET 16C

1 Without using technology draw the graph of $y = 4\sin x$ for $0° \leqslant x \leqslant 360°$.

2 Without using technology draw the graph of $y = \sin(3x)$ for $0° \leqslant x \leqslant 360°$.

3 State the period of: **a** $y = 4\sin\left(\frac{x}{3}\right)$ **b** $y = -2\sin(4x)$

4 Use the graph of $y = \sin x$ on page **540** to find to 1 decimal place the solutions of:

   a $\sin x = 0.4$ for $0° \leqslant x \leqslant 360°$ **b** $\sin x = -0.2$ for $180° \leqslant x \leqslant 540°$

5 The table below gives the mean monthly maximum temperature ($°C$) for Northland.

Month	Jan	Feb	Mar	Apr	May	Jun	Jul	Aug	Sept	Oct	Nov	Dec
Temp	24.8	28.3	30.9	31.8	30.9	28.3	24.8	21.3	18.6	17.7	18.6	21.2

A sine function of the form $T \doteqdot A\sin(Bt) + C$ is used to model the data. Find estimates of constants A, B and C without using technology. Use Jan $\equiv 1$, Feb $\equiv 2$, etc.

6 Use technology to solve for $0° \leqslant x \leqslant 400°$:

   a $\sin x = 0.382$ **b** $\sin\left(\frac{x}{2}\right) = 1 - \cos x$

7 The population estimate, in thousands, of a species of water beetle where $0 \leqslant t \leqslant 8$ and t is the number of weeks after the initial population estimate was made, is given by $P(t) = 5 + 2\sin(60t)$.

   a Find the initial population. **b** Find the smallest and largest population sizes.
   c Find time interval(s) where the population size exceeds 6500.

REVIEW SET 16D

1 On the same set of axes, sketch the graphs of $y = \cos x$ and $y = \cos x - 3$.

2 On the same set of axes, sketch the graphs of $y = \cos x$ and $y = 3\cos(2x)$.

3 For the following graphs, find the cosine function representing them:

a
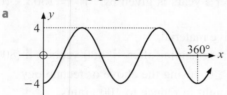

b
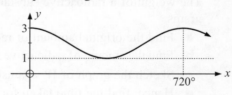

4 Use the graph of $y = \sin(2x)$ on page **540** to find approximate solutions for:

   a $\sin(2x) = 0.8$ **b** $\sin(2x) = -0.2$ for $0° \leqslant x \leqslant 360°$

5 Use technology to solve $\cos x = 0.4379$ for $0° \leqslant x \leqslant 540°$.

6 In an industrial city, the amount of pollution in the air becomes greater during the working week when factories are operating, and lessens over the weekend.

The amount of pollutants in a cubic metre of air is given by $P(t) = 40 + 12\sin\left(\frac{360t}{7}\right)$ where t is the number of days after midnight on Saturday night.

   a What was the minimum level of pollution?
   b At what time during the week does this minimum level occur?

7 Use technology to solve:

   a $4\cos(2x) + 1 = 0$ for $0° \leqslant x \leqslant 200°$ **b** $\cos(2x) = \sin x$, $0° \leqslant x \leqslant 360°$.

Chapter 17

More functions

In this chapter we will review some familiar functions and investigate some new ones.

These are the functions we will consider:

Name	General form
Polynomial:	
• Linear	$f(x) = ax + b, \quad a \neq 0$
• Quadratic	$f(x) = ax^2 + bx + c, \quad a \neq 0$
• Cubic	$f(x) = ax^3 + bx^2 + cx + d, \quad a \neq 0$
• Quartic	$f(x) = ax^4 + bx^3 + cx^2 + dx + e, \quad a \neq 0$
Reciprocal:	$f(x) = \dfrac{k}{x}, \quad x \neq 0$
Exponential:	$f(x) = a^x, \quad a > 0, \quad a \neq 1$
Trigonometric:	$f(x) = \begin{cases} a\sin bx + c \\ a\cos bx + c, \quad a, b, c \in Q \end{cases}$

Note: $f(x) = ax + b$ and $f: \; x \mapsto ax + b$ are two different notations for the same function type.

Although the above functions have different graphs, they do have some similar features.

The main features we are interested in are:

- the axes intercepts (where the graph cuts the x and y-axes)
- gradients
- turning points (maxima and minima)
- values of x where the function does not exist
- the presence of asymptotes (lines or curves that the graph approaches).

TERMINOLOGY

Graphs have many different features. It is important to use the correct descriptive terminology when talking about graphical features. Here are a few worth noting:

•
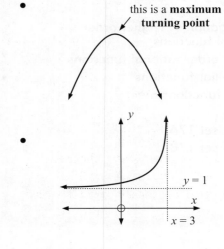

this is a **maximum turning point**

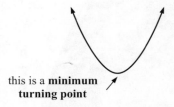

this is a **minimum turning point**

•

This curve has a **horizontal asymptote** ($y = 1$) and a **vertical asymptote** ($x = 3$).

As we choose larger positive y values and larger negative x values, the graph gets closer and closer to the asymptotes.

POLYNOMIAL FUNCTIONS

The **degree (order)** of a polynomial depends upon the highest power present within the polynomial.

For example, $f(x) = ax^2 + bx + c, \ a \neq 0$ is a degree 2 polynomial (quadratic) as it has no powers higher than 2.

A degree 3 polynomial has general form $f(x) = ax^3 + bx^2 + cx + d, \ a \neq 0$ and contains no higher powers than 3.

Hence, the common names used to describe particular families of polynomials such as **linear**, **quadratic**, **cubic** and **quartic** can be re-named more systematically to degree 1, degree 2, degree 3 and degree 4 respectively.

This system also allows higher order polynomials to be named more easily, i.e., degree 7 must have general form $f(x) = ax^7 + bx^6 + cx^5 + dx^4 + ex^3 + fx^2 + gx + h, \ a \neq 0$

In previous chapters, we have looked at degree 1 (linear) and degree 2 (quadratic) polynomials. While you should be able to find the important characteristics of each of these polynomials algebraically (i.e., axes intercepts, gradient, vertex etc.), the focus of this chapter is on the use of technology to sketch and analyse more complex and unfamiliar functions.

A   CUBIC POLYNOMIALS

A **cubic polynomial** (degree 3) has form $f(x) = ax^3 + bx^2 + cx + d$
where $a \neq 0$,   and a, b, c and d are constants.

OPENING PROBLEM

A 40 cm by 30 cm sheet of tinplate is to be used to make a cake tin.

Squares are cut from its corners and the metal is then folded upwards along the dashed lines.

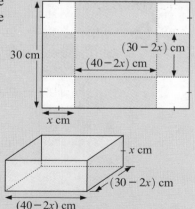

Edges are fixed together to form the open rectangular tin.

The depth of the tin is x cm and its length and width are $(40 - 2x)$ cm and $(30 - 2x)$ cm respectively.

Consequently the capacity of the cake tin V, is given by

$$V(x) = x(40 - 2x)(30 - 2x) \text{ cm}^3$$
$$\text{or} \quad V(x) = 4x(20 - x)(15 - x) \text{ cm}^3$$

If expanded, $V(x) = 4x(300 - 35x + x^2)$

i.e., $V(x) = 4x^3 - 140x^2 + 1200x$

For you to consider:
- How does the capacity of the tin change as x changes?
- What are the restrictions on the x values, if any?
- What sized squares must be cut out for the cake tin to have maximum capacity?

Reminder:

For the function $V(x) = 4x^3 - 140x^2 + 1200x$; since the highest power of x is 3 we have a **cubic polynomial**.

$\quad V(x) = 4x(20 - x)(15 - x)$ is the **factored form** of the given cubic

$\quad V(x) = 4x^3 - 140x^2 + 1200x$ is its **expanded** form.

Graphing cubics by hand is tedious, so we will use technology to assist us.

HISTORICAL NOTE

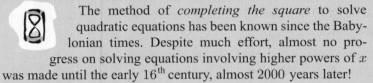

The method of *completing the square* to solve quadratic equations has been known since the Babylonian times. Despite much effort, almost no progress on solving equations involving higher powers of x was made until the early 16th century, almost 2000 years later!

In 1500, **dal Ferro** solved equations of the type $x^3 + mx = n$, for example, $x^3 - 5x = 7$.

In 1530, **Niccolo Fontana**, known as Tartaglia, solved equations of the type $x^3 + mx^2 = n$.

A general method of solving all *third degree equations* (of the form $ax^3 + bx^2 + cx + d = 0$) was found by **Leonhard Euler** in 1732. The general solution to fourth degree equations (of the form $ax^4 + bx^3 + cx^2 + dx + e = 0$) was found by **Lodovico Ferrari** soon after. One night in 1732, a Frenchman, **Evariste Galois**, scribbled a brief outline of a proof that no general solution for fifth degree (or higher degree) equations exists. The next day, he was engaged in a duel over a romantic involvement. Galois was killed in that duel. He was 21 years old at the time.

INVESTIGATION 1   GRAPHING SOME FAMILIES OF CUBICS

This investigation is best done using technology.

GRAPHING PACKAGE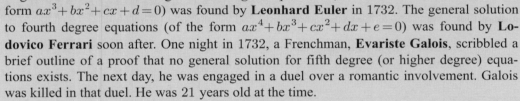

What to do:

1 a Use technology to assist you to draw sketch graphs of $y = f(x)$ if:

   i $f(x) = x^3$, $\ f(x) = 2x^3$, $\ f(x) = \frac{1}{2}x^3$, $\ f(x) = \frac{1}{3}x^3$

   ii $f(x) = x^3$ and $f(x) = -x^3$

   iii $f(x) = -x^3$, $\ f(x) = -2x^3$, $\ f(x) = -\frac{1}{2}x^3$

b What is the geometrical significance of a in $f(x) = ax^3$?
You should comment on the sign of a and the size of a.

2 a Use technology to assist you to draw sketch graphs of:

   i $f(x) = x^3$, $\ f(x) = x^3 + 2$, $\ f(x) = x^3 - 3$

   ii $f(x) = x^3$, $\ f(x) = (x + 2)^3$, $\ f(x) = (x - 3)^3$

   iii $f(x) = (x - 1)^3 + 2$, $\ f(x) = (x + 2)^3 + 1$

b What is the geometrical significance of:

- k in the family of cubics of the form $f(x) = x^3 + k$
- h in the family of cubics of the form $f(x) = (x - h)^3$
- h and k in the family of cubics of the form $f(x) = (x - h)^3 + k$?

From the previous investigation you should have discovered that:

- for $a > 0$,   the graph's shape is   For $a < 0$ it is   ↘ .

- $f(x) = (x - h)^3 + k$ is the translation of $f(x) = x^3$ through h units horizontally

  k units vertically .

GRAPHING BASIC CUBIC POLYNOMIALS

The basic cubic function is
$y = x^3$ with graph:

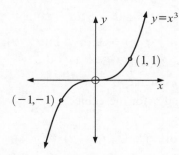

So, $y = (x - h)^2 + k$ has the
same shape, but is translated
h units horizontally and
k units vertically.

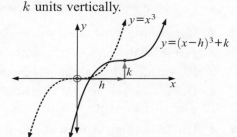

EXERCISE 17A.1

1 For each of the following:
   i   use technology to observe the function
   ii draw a sketch graph for $y = f(x)$ without plotting points from a table of values
   iii find the y-intercept and include it on your sketch graph.

   a $f(x) = x^3 + 2$

   b $f(x) = -x^3 + 2$

   c $f(x) = \frac{1}{2}x^3 + 3$

   d $f(x) = 2x^3 - 1$

   e $f(x) = (x - 1)^3$

   f $f(x) = (x + 2)^3$

   g $f(x) = -(x + 1)^3$

   h $f(x) = (x - 1)^3 + 2$

   i $f(x) = -(x - 2)^3 - 5$

Example 1	
Write $y = 2(x - 1)^3 + 4$ in general form.	$y = 2(x - 1)^3 + 4$ $\therefore \quad y = 2(x - 1)(x - 1)^2 + 4$ $\therefore \quad y = 2(x - 1)(x^2 - 2x + 1) + 4$ $\therefore \quad y = 2[x^3 - 2x^2 + x - x^2 + 2x - 1] + 4$ $\therefore \quad y = 2[x^3 - 3x^2 + 3x - 1] + 4$ $\therefore \quad y = 2x^3 - 6x^2 + 6x - 2 + 4$ i.e., $y = 2x^3 - 6x^2 + 6x + 4$

2 Write in general form:

   a $y = 3(x + 1)^3 - 2$

   b $y = \frac{1}{2}(x + 2)^3 + 2$

   c $y = -(x + 1)^3 + 5$

INVESTIGATION 2   CUBICS IN LINEAR FACTORED FORM

Cubics are observed in the form $f(x) = a(x - \alpha)(x - \beta)(x - \gamma)$, $a \neq 0$.

We will consider graphs of this form and also $f(x) = a(x - \alpha)^2(x - \beta)$
where one of the linear factors is repeated.

Once again technology should be used.

GRAPHING PACKAGE

What to do:

1 a Use technology to assist you to sketch graphs of $f(x) = (x - 1)(x + 1)(x + 3)$,
$f(x) = 2(x - 1)(x + 1)(x + 3)$, $f(x) = \frac{1}{2}(x - 1)(x + 1)(x + 3)$ and
$f(x) = -2(x - 1)(x + 1)(x + 3)$.

   b For each graph in **a** state the x-intercepts and the y-intercept.

   c Discuss the geometrical significance of a in $f(x) = a(x - \alpha)(x - \beta)(x - \gamma)$.

2 a Use technology to assist you to sketch graphs of $f(x) = 2x(x + 1)(x - 2)$,
$f(x) = 2(x + 3)(x - 1)(x - 2)$ and $f(x) = 2x(x + 2)(x - 1)$.

   b For each graph in **a** state the x-intercepts and y-intercept.

   c Discuss the geometrical significance of α, β and γ for the cubic
$f(x) = a(x - \alpha)(x - \beta)(x - \gamma)$.

3 a Use technology to assist you to sketch graphs of $f(x) = (x - 2)^2(x + 1)$,
$f(x) = (x + 1)^2(x - 3)$, $f(x) = 2(x - 3)^2(x - 1)$ and $f(x) = -2(x + 1)(x - 2)^2$.

   b For each graph in **a**, state the x-intercepts and the y-intercept.

   c Discuss the geometrical significance of α and β for the cubic
$f(x) = a(x - \alpha)^2(x - \beta)$.

From the previous investigation you should have discovered that:

- if $a > 0$, shape is ⁀ or ⁀ if $a < 0$ shape is ⁀ or ⁀

- for a cubic in the form $y = a(x - \alpha)(x - \beta)(x - \gamma)$ the graph has x-intercepts
  α, β and γ and the graph crosses over or **cuts** the x-axis at these points

- for a cubic in the form $y = a(x - \alpha)^2(x - \beta)$ the graph **touches** the x-axis
  at α and **cuts** it at β.

EXERCISE 17A.2

1 For each of the following:

   i   use technology to observe the function

   ii draw a sketch graph for $y = f(x)$ without plotting points from a table of
          values

   iii find the y-intercept and include it on your sketch graph.

   a $y = (x + 1)(x - 2)(x - 3)$ **b** $y = -2(x + 1)(x - 2)(x - \frac{1}{2})$

   c $y = \frac{1}{2}x(x - 4)(x + 3)$ **d** $y = 2x^2(x - 3)$

   e $y = -\frac{1}{4}(x - 2)^2(x + 1)$ **f** $y = -3(x + 1)^2(x - \frac{2}{3})$

THE ZEROS OF A POLYNOMIAL

The **zeros** of any polynomial are the values of x which make y have a value of zero. These are clearly the x-intercepts of the graph of the polynomial.

So, the zeros of $y = a(x - \alpha)(x - \beta)(x - \gamma)$ are α, β and γ.

MAXIMUM AND MINIMUM TURNING POINTS

The illustration shows typical maximum and minimum turning points.

The coordinates of these points are easily found using technology.

Click on the icon to obtain the **graphing package** and/or instructions when using a **graphics calculator**.

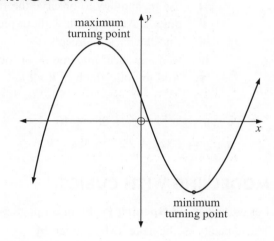

GRAPHING CUBICS FROM GENERAL FORM

What are the zeros of a cubic in expanded form $y = ax^3 + bx^2 + cx + d$ $(a \neq 0)$?

While algebraic methods can be used to factorise a cubic polynomial, such techniques are beyond the scope of this course. However, technology can be used to find zeros of cubics and higher order polynomials. Using technology, it is also possible to find other significant features of a function such as local maximum and local minimum values.

Consider $y = 3x^3 - 14x^2 + 5x + 2$. Using a graphics calculator we graph the cubic.

The zeros are found to be $0.666\,666$, $-0.236\,068$ and $4.236\,068$.

Note: It is often necessary to alter the view window. For this graph these settings were used.

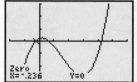 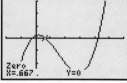 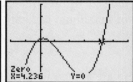 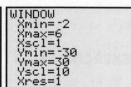

Rounding the zeros to 3 significant figures we have 0.667, -0.236 and 4.24.

We can also find the local maximum and local minimum values:

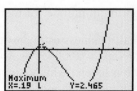 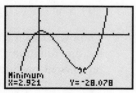

The local maximum was found to be $(0.190, 2.47)$ and the local minimum was $(2.92, -28.1)$ (correct to 3 s.f.).

The y-intercept can also be found using technology or algebraically by substituting $x = 0$ into our equation. For this function, the y-intercept is $y = 2$.

EXERCISE 17A.3

1 For each of the following:
    i use technology to observe the function
    ii draw a sketch graph without plotting actual points
    iii find the axes intercepts
    iv find any local maximums or local minimums.
    v Add to your graph in **ii** all key features found in **iii** and **iv**.

a $y = -2x^3 + 8x^2 + 6x - 36$ **b** $y = x^3 + 5x^2 + 2x - 8$

c $y = 3x^3 - 7x^2 - 28x + 10$ **d** $y = -3x^3 + 9x^2 + 27x + 15$

e $y = -3x^3 - 24x^2 - 48x$ **f** $y = -x^3 + \frac{7}{2}x^2 + \frac{7}{2}x - 6$

MODELLING WITH CUBICS

Let us revisit the **Opening Problem** on page **547**.

The capacity of the cake dish is given by

$$V = x(40 - 2x)(30 - 2x)$$
$$= 4x(20 - x)(15 - x)$$
i.e., $V = 4x^3 - 140x^2 + 1200$ cm^3

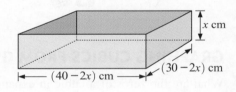
x cm
$(40 - 2x)$ cm
$(30 - 2x)$ cm

In this problem there are restrictions on the values that x may have.

Clearly $x > 0$. But also $30 - 2x > 0$ means that $x < 15$.

So, x is restricted by $0 < x < 15$.

We can use technology to graph the cubic on the interval $0 < x < 15$.

We can then determine the value of x in this interval so that V is a maximum.

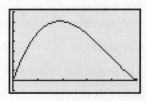

This value of x tells us the size of the squares to be cut out to produce the cake tin of maximum capacity. Find it.

GRAPHING
PACKAGE

EXERCISE 17A.4

1 In a coal mine, there are x men in each shift at the coalface, and the output of coal is given by $T = \frac{1}{30}x^2(20 - x)$ tonnes of coal.

    a What are the restrictions on the values of x?

    b What is the optimum number of men per shift? Use technology to assist.

2 The owner of a small business estimates that the profit from producing x items is given by the function $P(x) = 0.003x^3 - 1.5x^2 + 200x - 1000$. This function is based on current production levels, which cannot exceed 250 items due to limited space and resources. How many items should be produced to maximise the profit?

3 A scientist working for Crash Test Barriers, Inc. is trying to design a crash test barrier whose ideal characteristics are shown graphically below. The independent variable is the time after impact, measured in milliseconds. The dependent variable is the distance that the barrier has been depressed because of the impact, measured in millimetres.

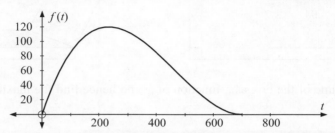

a The equation for this graph is of the form $f(t) = kt(t - a)^2$. From the graph, what is the value of a? What does it represent?

b If the ideal crash barrier is depressed by 85 mm after 100 milliseconds, find the value of k, and hence find the equation of the graph given.

c What is the maximum amount of depression, and when does it occur?

4 W. F. Weeks and W. J. Campbell in the Journal of Glaciology used a cubic model for towing a flat iceberg from Amery Ice Shelf in Antarctica to Australia. The equation is $V = 7.75 - 0.35x + 0.0625x^2 - 0.0208x^3$, where V is the number of cubic kilometres of ice remaining and $x + 3$ is the number of thousand kilometres travelled by the iceberg, $0 \leqslant x \leqslant 4$. The volume of ice changed very little in the first 3000 km of the tow, as it was still in Antarctic conditions.

a What was the initial volume of ice?

b What was the volume of ice remaining after the iceberg had been towed 4000 km (that is, when $x = 1$)?

c If the total length of the journey was 7000 km, what percentage of the original volume of ice still remained?

d How many kilometres was the iceberg towed before it lost 10% of its volume?

5 In the last year (starting 1st January), the volume of water (in megalitres) in a particular reservoir after t months could be described by the model $V(t) = -t^3 + 30t^2 - 131t + 250$.

The reservoir authority rules that if the volume falls below 100 ML, irrigation is prohibited. During which months, if any, was irrigation prohibited in the last twelve months?

Include in your answer a neat sketch of any graphs you may have used.

6 A closed box (like a pizza box) is to be formed from a sheet of cardboard 64 cm by 40 cm by cutting equal squares, of side x cm, from two corners of the short side, and two equal rectangles of width x cm from the other two corners and folding along the dotted lines as shown in the diagram.

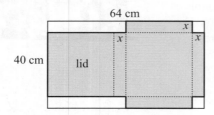

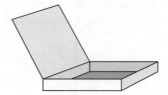

Find the volume of the box as a function of x and hence find the maximum volume.

B   QUARTIC POLYNOMIALS

Quartic polynomials (degree 4) are of the form $f(x) = ax^4 + bx^3 + cx^2 + dx + e$ where a, b, c, d and e are constants, $a \neq 0$.

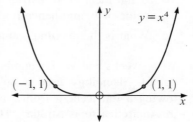

The basic or simplest quartic polynomial is $y = x^4$ and its graph is shown alongside.

Notice that it opens in the **positive direction** and is **never negative**.

INVESTIGATION 3   GRAPHING QUARTICS

Once again this investigation is best done using technology.

GRAPHING
PACKAGE

What to do:

1 a Use technology to assist you to draw sketch graphs of:
  i $f(x) = x^4$, $f(x) = 2x^4$ and $f(x) = \frac{1}{2}x^4$ **ii** $f(x) = -x^4$, $f(x) = -\frac{1}{3}x^4$

  b What is the geometrical significance of the sign of a and the size of a?

2 a Use technology to assist you to draw sketch graphs of:
$$f(x) = (x+3)(x+1)(x-2)(x-3), \quad f(x) = (x-2)^2(x+1)(x-3)$$
$$f(x) = (x-1)^3(x+2), \quad \text{and} \quad f(x) = (x-2)^4$$

  b Copy and complete:
  • a factor $(x - \alpha)$ or $(x - \alpha)^3$ indicates that the graph ...... the x-axis at α
  • a factor $(x - \alpha)^2$ or $(x - \alpha)^4$ indicates that the graph ...... the x-axis at α.

  c There is a difference in the geometrical cutting of the x-axis due to $(x - \alpha)$ compared with $(x - \alpha)^3$. What is it?

From the previous investigation you should have discovered that:

> ▶ For $y = ax^4$, if $a > 0$ the graph opens upwards
>   if $a < 0$ the graph opens downwards.
>
> ▶ The size of a controls the width of the quartic graph.
>
> ▶ If a quartic is fully factored into linear factors, for:
>
>   • a **single factor** $(x - \alpha)$, the graph **cuts** the x-axis at α
>
> e.g.
>
>   • a **squared factor** $(x - \alpha)^2$, the graph **touches** the x-axis at α
>
> e.g.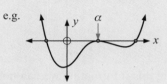
>
>   • a **cubed factor** $(x - \alpha)^3$, the graph **cuts** the x-axis at α, but is 'flat' at α
>
> e.g.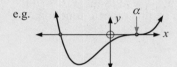
>
>   • a **quadruple factor** $(x - \alpha)^4$, the graph **touches** the x-axis but is 'flat' at that point.
>
> e.g.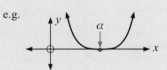

EXERCISE 17B.1

1 For each of the following:

  i use technology to observe the function

  ii draw a sketch graph without plotting actual points

  iii find the axes intercepts

  iv find any local maximums or local minimums

  v Add to your graph in **ii** all key features found in **iii** and **iv**.

a $y = (x + 4)(x + 1)(x - 2)(x - 3)$

b $y = -x(x + 2)(x - 1)(x - 4)$

c $y = 2(x - 1)^2(x + 2)(2x - 5)$

d $y = -(x - 4)(x + 3)(2x - 1)^2$

e $y = 3(x + 1)^2(x - 2)^2$

f $y = -2(x - 1)^2(x + 2)^2$

g $y = 4(x - 1)^3(2x + 3)$

h $y = -x^3(2x - 3)$

i $y = 3(x - 1)^4$

j $y = -2(x + 2)^4$

GRAPHING QUARTICS FROM GENERAL FORM

Using technology, we can find all important features of a degree 4 polynomial.

Consider the function $y = x^4 + 3x^3 + x^2 + x + 1$.

Using a graphics calculator, we find two zeros at $x \doteq -2.72$ and $x \doteq -0.70$ and a minimum value (no maximum) of $y \doteq -5.03$ when $x \doteq -2.07$.

For our function, the y-intercept is $y = 1$ occuring when $x = 0$.

We can also find the y-intercept using technology.

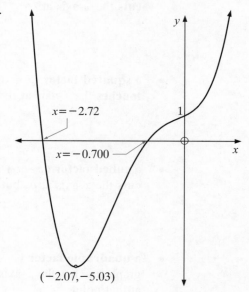

Using this information we can now sketch a graph showing all of the important features of this quartic function.

$(-2.07, -5.03)$

EXERCISE 17B.2

1   For each of the following:
   i   use technology to observe the function
   ii   draw a sketch graph without plotting actual points
   iii find the axes intercepts
   iv   find any local maximums or local minimums
   v   Add to your graph in ii all key features found in iii and iv.

   a $y = 2x^4 - 8x^3 + 7x + 2$ b $y = -2x^4 + 2x^3 + x^2 + x + 2$
   c $y = -\frac{1}{2}x^4 + 2x^2 + 2$ d $y = 2x^4 + 6x^3 + x + 2$
   e $y = \frac{4}{5}x^4 + 5x^3 + 5x^2 + 2$

2   Find the maximum value of:
   a $y = -x^4 + 2x^3 + 5x^2 + x + 2$ on $0 \leqslant x \leqslant 4$
   b $y = -2x^4 + 5x^2 + x + 2$ on the interval:
      i $-2 \leqslant x \leqslant 2$ ii $-2 \leqslant x \leqslant 0$ iii $0 \leqslant x \leqslant 2$.

3   Find the minimum value of $f(x)$ on $-3 \leqslant x \leqslant 0$ if $f(x) = x^4 + 3x^3 + x^2 + x + 4$.

C | THE RECTANGULAR HYPERBOLA

A contract painter estimates that it will take him 36 days to paint the walls of a hostel if he works alone. How long would it take to paint the walls if two people are available and work at the same rate? What about the time for three people, four people, etc.?

We can tabulate and graph the time (T) against the number of people painting (n).

n	1	2	3	4	6	9	12	18	36
T	36	18	12	9	6	4	3	2	1

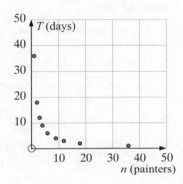

Note:
- The time taken T depends on the number of painters n. So, n is the independent variable.
- It is not sensible to join these points with a curve. Why?
- Did you notice that $nT = 36$ for all tabled values?

Now consider the situation of manufacturing envelopes with area 36 cm^2.

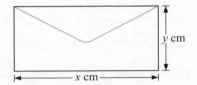

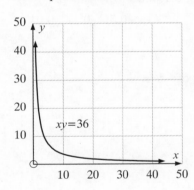

If each envelope is x cm by y cm, then $xy = 36$.

A table of values identical to those for the painter problem could be formed and the graph drawn.

However, on this occasion we have a continuous graph, not discrete points, as any positive value of x is possible.

RECTANGULAR HYPERBOLAE

Firstly, consider $xy = 36$ where x can take any value; positive, zero or negative.

What would the graph look like?

Of course, $xy = 36$ can be written as $y = \dfrac{36}{x}$ and graphs of these functions are called **rectangular hyperbolae** or reciprocal functions.

DISCUSSION

Whether in the form $xy = 36$ or $y = \dfrac{36}{x}$ we can see that neither x nor y can be zero.

INVESTIGATION 4   THE FAMILY OF CURVES $y = \dfrac{A}{x}$

The use of a **graphing package** or
graphics calculator is recommended.

GRAPHING
PACKAGE

What to do:

1 On the same set of axes draw the graphs $y = \dfrac{1}{x}$, $y = \dfrac{2}{x}$, $y = \dfrac{4}{x}$.

2 The functions in **1** are members of the family $y = \dfrac{A}{x}$ where A is 1, 4 or 8.
Describe the effect changes in the A values have on the graphs for $A > 0$.

3 Repeat **1** for $y = \dfrac{-1}{x}$, $y = \dfrac{-2}{x}$, $y = \dfrac{-4}{x}$.

4 Comment on shape changes in **3**.

5 Explain why no graph exists when $x = 0$.

6 Consider the function $y = \dfrac{10}{x}$.

    a Find y values for $x = 1000, 100, 10, 5, 2, 1, 0.5, 0.1, 0.01$

    b Find y values for $x = -1000, -100, -10, -5, -2, -1, -0.5, -0.1, -0.01$

    c Draw a sketch graph of $y = \dfrac{10}{x}$.

    d Without calculating new values, sketch the graph of $y = \dfrac{-10}{x}$.

From the above investigation you should have discovered that:

- All graphs of relations of the form $y = \dfrac{A}{x}$

  ▶ have lines of symmetry $y = x$ and $y = -x$

  ▶ are point symmetric about the origin O.   (This means that under
     a $180°$ rotation about O, they map onto themselves.)

- All graphs of relations of the form $y = \dfrac{A}{x}$ are **asymptotic** to the x

  and y-axes. This means that the further we move away from the origin,
  the closer the graph approaches these lines without ever reaching them.

-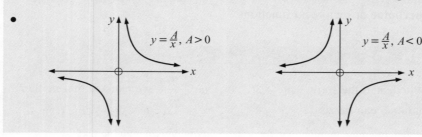

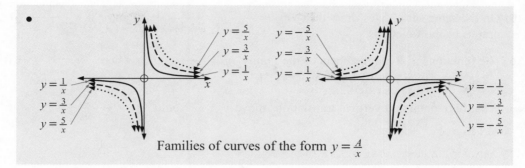

Families of curves of the form $y = \frac{4}{x}$

EXERCISE 17C.1

1 On the same set of axes draw graphs of $y = \dfrac{8}{x}$ and $y = -\dfrac{8}{x}$.

2 Draw the graph of $y^2 = \dfrac{16}{x^2}$.

3 Determine the equations of the following rectangular hyperbolae:

a
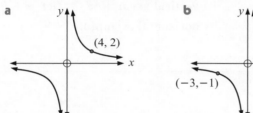
(4, 2)

b
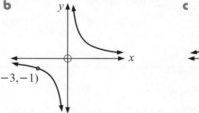
(−3, −1)

c
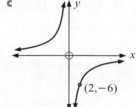
(2, −6)

4 Find the coordinates of $xy = 4$ which are closest to the origin (0, 0).

INVESTIGATION 5   FUNCTIONS OF THE FORM $y = \dfrac{A}{x-h} + k$

The use of a **graphing package** or
graphics calculator is recommended.

GRAPHING
PACKAGE

What to do:

1 On the same set of axes draw the graphs of the functions

$$y = \frac{4}{x}, \quad y = \frac{4}{x-2} \quad \text{and} \quad y = \frac{4}{x+3}.$$

2 The functions in **1** are members of the family $y = \dfrac{4}{x-h}$ where h is 0, 2 or −3.

What effect has changing values of h on the members of this family?

3 On the same set of axes draw the graphs of the functions

$$y = \frac{4}{x}, \quad y = \frac{4}{x} + 3 \quad \text{and} \quad y = \frac{4}{x} - 2.$$

4 The functions in **3** are members of the family $y = \dfrac{4}{x} + k$ where k is 0, 3 or −2.

5 On the same set of axes draw the graphs of the functions $\quad y = \dfrac{4}{x}, \quad y = \dfrac{4}{x-2} + 3 \quad$ and $\quad y = \dfrac{4}{x+1} - 5.$

6 The functions in **5** are members of the family $\quad y = \dfrac{4}{x-h} + k \quad$ where $\quad h = 0, \ k = 0$
or $\ h = 2, \ k = 3 \ $ or $\ h = -1, \ k = -5.$

What horizontal and vertical translations move $y = \dfrac{4}{x} \quad$ onto $\quad y = \dfrac{4}{x-h} + k$?

FUNCTIONS OF THE FORM $\ y = \dfrac{A}{x-h} + k$

From the investigation you should have discovered that:

- $y = \dfrac{A}{x}$ moves to $y = \dfrac{A}{x-h} + k$ under a translation of $\quad h$ units horizontally
and k units vertically.

- For $\ y = \dfrac{A}{x-h} + k$, the line $\ x = h$ is its **vertical asymptote** (let $x - h = 0$)
the line $\ y = k$ is its **horizontal asymptote**.

Example 2

For the function $\quad y = \dfrac{6}{x-2} + 4$

  i What is the vertical asymptote?   **ii** What is the horizontal asymptote?
  iii What is the x-intercept?   **iv** What is the y-intercept?
  v Using only **i** to **iv** sketch the graph of the function. Check using technology.

a $y = \dfrac{6}{x-2} + 4$

  i Letting $x - 2 = 0$, the VA
     is $x = 2$

  ii The HA is $y = 4$

  iii When $y = 0$, $\qquad \dfrac{6}{x-2} = -4$

   $\therefore \quad -4(x - 2) = 6$
   $\therefore \qquad -4x + 8 = 6$
   $\therefore \qquad\quad -4x = -2$
   $\therefore \qquad\qquad x = \tfrac{1}{2}$

   $\therefore$ the x-intercept is $\tfrac{1}{2}$

  iv When $x = 0$, $y = \dfrac{6}{-2} + 4 = 1$

   $\therefore$ the y-intercept is 1

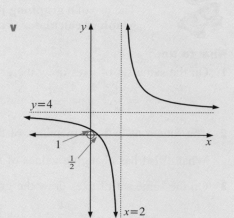

EXERCISE 17C.2

1   For the following functions:

   i   what is the vertical asymptote   **ii**   what is the horizontal asymptote

   iii   what is the y-intercept   **iv**   what is the x-intercept?

   v   Using technology to sketch the graph and mark features **i** to **iv** on it.

   a $y = \dfrac{3}{x-1} + 2$ **b** $y = \dfrac{4}{x-3} + 1$ **c** $y = \dfrac{1}{x+2} + 3$

   d $y = \dfrac{8}{x+2} + 1$ **e** $y = \dfrac{6}{x+2} - 4$ **f** $y = \dfrac{2}{x+4} - 1$

2   **a**   Solve the following equations:

   i $\dfrac{2x+1}{3x+4} = \dfrac{2}{3}$ **ii** $\dfrac{5x-1}{2x+4} = \dfrac{5}{2}$ **iii** $\dfrac{3x-1}{-4x+3} = -\dfrac{3}{4}$

   b   What have you discovered about the solution of the equation $\dfrac{ax+b}{cx+d} = \dfrac{a}{c}$?

   c   Graph   **i** $y = \dfrac{2x+1}{3x+4}$ **ii** $y = \dfrac{5x-1}{2x+4}$ **iii** $y = \dfrac{3x-1}{-4x+3}$

   d   From **b** and **c**, state the equation of the horizontal asymptote of $y = \dfrac{ax+b}{cx+d}$.

3   For the following functions:

   i   what is the vertical asymptote   **ii**   what is the horizontal asymptote

   iii   what is the y-intercept   **iv**   what is the x-intercept?

   v   Using technology to sketch the graph and mark features **i** to **iv** on it.

   a $y = \dfrac{2x+3}{x-1}$ **b** $y = \dfrac{3x+1}{x+2}$ **c** $y = \dfrac{x-6}{x+3}$

   d $y = \dfrac{-2x+3}{x-2}$ **e** $y = \dfrac{4x+2}{3x-4}$ **f** $y = \dfrac{-5x+1}{2x-1}$

APPLICATIONS OF RECTANGULAR HYPERBOLAE

In the following exercise we will examine models and then use technology to assist us.

Example 3

An experimental breeding colony of gorillas is set up and the size of the colony at time t years $(0 \leqslant t \leqslant 8)$ is given by $N = 20 - \dfrac{400}{t-10}$.

a   What was the original size of the colony?

b   What is the size of the colony after   **i**   3 years   **ii**   8 years?

c   How long would it take for the colony to reach a size of 100?

d   Draw the graph of N against t, $0 \leqslant t \leqslant 8$,   using only **a**, **b** and **c**.

$$N = 20 - \frac{400}{t - 10}$$

a   When $t = 0$, $N = 20 - \frac{400}{-10}$
$$= 20 + 40$$
$$= 60 \text{ gorillas.}$$

b   **i**   When $t = 3$,
$$N = 20 - \frac{400}{-7}$$
$$\doteqdot 77 \text{ gorillas}$$

   ii   When $t = 8$,
$$N = 20 - \frac{400}{-2}$$
$$= 200 \text{ gorillas}$$

c   When $N = 100$,
$$100 = 20 - \frac{400}{t - 10}$$
$$\therefore \quad 80 = \frac{-400}{t - 10}$$
$$\therefore \quad t - 10 = -5$$
$$\therefore \quad t = 5$$
   i.e.,   it would take 5 years.

d
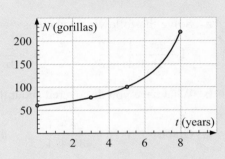

EXERCISE 17C.3

1   In order to remove noxious weeds from his property a farmer sprays a weedicide over it. The chemical is slow acting and the number of weeds per hectare remaining after t days is given by $N = 20 + \dfrac{100}{t + 2}$ weeds per hectare.

   a   How many weeds per hectare were alive before spraying the weedicide?
   b   How many weeds were alive after 8 days?
   c   How long did it take for the number of weeds still alive to reach 40 per hectare?
   d   Sketch the graph of N against t, using **a**, **b** and **c** and your calculator.
   e   Is the chemical spraying program going to eradicate all weeds?

2   The current remaining in an electrical circuit t seconds after it is switched off, is given by $I = \dfrac{1000}{t + 4}$ amps.

   a   What current flowed in the circuit at the instant before it was switched off?
   b   Find the current after   **i**   6 seconds   **ii**   20 seconds.
   c   How long would it take for the current to reach 10 amps?
   d   Graph I against t, using only **a**, **b** and **c**.
   e   Use a calculator to graph the function.
       Use **trace** to check your answers to **a**, **b** and **c**.

3 A helicopter on its helipad climbs vertically to a height h metres above the pad so that

at time t minutes, $h(t) = 2500\left(1 - \dfrac{1}{t+1}\right)$.

   a Check that $h(0) = 0$.

   b Find the height reached after   **i** 9 minutes   **ii** 20 minutes.

   c Use technology to help you sketch the graph of h against t, $(t \geqslant 0)$ and check your answers to **a** and **b**.

   d Is there an altitude beyond which the helicopter does not go? Explain.

4 On a wet day a cyclist was travelling at a constant speed along a bitumen road.

He braked suddenly and his speed afterwards was given by $S = \dfrac{20}{t+1} - 4$ m/s, for $t \geqslant 0$.

   a How fast was he travelling at the instant when the brakes were applied?

   b How long did it take for him to reach a speed of 10 m/s?

   c How long did it take for him to become stationary?

   d Using technology to assist, graph S against t for the braking interval.

   e The inequality $t \geqslant 0$ should be replaced with a more appropriate one. What should it be?

INVESTIGATION 6   HIGHER ORDER RATIONAL FUNCTIONS

The **rational functions** we will examine are of the form

$y = \dfrac{f(x)}{g(x)}$ where $f(x)$ and $g(x)$ could be linear or quadratic.

We will examine some of these functions using suitable technology.

What to do:

1 Firstly we examine rational functions of the form $y = \dfrac{\text{linear}}{\text{quadratic}}$.

   a Examine the graphs of $y = \dfrac{x+2}{(x-2)^2}$, $y = \dfrac{x+2}{(x-1)(x-3)}$ and $y = \dfrac{x+2}{x^2+4}$.

   b Draw a fully labelled sketch of each graph.

   c List similarities and differences between the graphs and explain what causes the differences.

2 Finally, examine $y = \dfrac{\text{quadratic}}{\text{quadratic}}$ rational functions.

   a Examine the graphs of

$$y = \dfrac{(x-1)(x-3)}{x(x+2)}, \quad y = \dfrac{2(x-1)(x-3)}{x^2+2} \quad \text{and} \quad y = \dfrac{3x^2+2}{x^2+1}.$$

   b Draw fully labelled sketches of these graphs.

   c List similarities and differences between the graphs and explain what causes the differences.

   d What is the horizontal asymptote of $y = \dfrac{ax^2+bx+c}{ex^2+fx+g}$?

D | HIGHER ORDER RATIONAL FUNCTIONS

EXERCISE 17D

1 For each of the following:

    i use technology to observe the function

    ii draw a sketch graph without plotting actual points

    iii find the axes intercepts

    iv find any local maximums or local minimums

    v Add to your graph in **ii** all key features.

a $y = \dfrac{x-2}{x^2-1}$ **b** $y = \dfrac{x+2}{x^2-2x}$ **c** $y = \dfrac{x+3}{(x-1)(x-4)}$

d $y = \dfrac{x(x-2)}{x^2-1}$ **e** $y = \dfrac{-x^2+2x}{x^2-3x+2}$ **f** $y = \dfrac{(x-1)(x+4)}{(x+1)(x-4)}$

E | UNFAMILIAR FUNCTIONS

So far, we have grouped particular functions into categories such as polynomial, exponential, trigonometric and reciprocal. As such, we have looked at the individual characteristics of each type of function.

For example, trigonometric functions have a periodic nature which is easily recognised whereas exponential functions are either constantly increasing or constantly decreasing as x increases.

While all of these function types are important to this course, they are only a small subset of the infinite set of all functions. It would be impossible to formally investigate all function types. As a consequence, we rely on the use of technology to help explore unfamiliar functions types.

For example, a quadratic function $y = x^2$ has a distinct parabolic curve and an exponential function $y = 2^x$ has its own distinct shape as shown alongside:

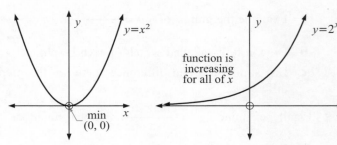

Having explored each of these in some detail within this book, you can probably sketch such graphs without the use of a graphics calculator. However, if we multiply these functions together, it may not be so easy to sketch the graph of $y = x^2 \times 2^x$ without the use of technology.

In fact, the function $y = x^2 \times 2^x$ has graph:

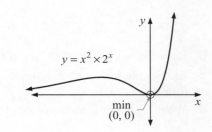

Notice, that some of the properties of the original functions are displayed by the new composite function.

Example 4

Consider the function $f : x \mapsto \dfrac{2^x}{x}$.   Using technology to assist:
- **a**   find any turning points
- **b**   determine the equation of any asymptotes
- **c**   find any axes intercepts
- **d**   specify the domain and range of the function
- **e**   sketch a graph showing all important features.

Using technology, we obtain a graph:
on $-5 \leqslant x \leqslant 5$ and $-2 \leqslant y \leqslant 4$.

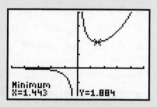

- **a**   There is a local minimum at $(1.44, \ 1.88)$ (3 s.f.)
- **b**   The graph is asymptotic about $y = 0$ and $x = 0$.
  (**Note:**   This is confirmed using **table** mode or by altering the view window.)
- **c**   There are no axes intercepts.
- **d**   Domain is $\{x : \ x \neq 0\}$ Range is $\{y : \ y < 0 \ \text{or} \ y > 1.88\}$
- **e**   Sketching:

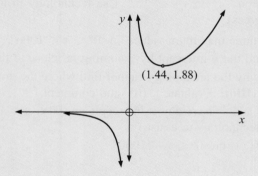

EXERCISE 17E

1   Consider the function $f : x \mapsto \sin(2^x)$ on $0^\circ \leqslant x \leqslant 8.5^\circ$.
Use technology to assist (making sure your calculator is in degrees).

- **a**   Determine the value of any turning points (maximum or minimum).
- **b**   Determine the x-intercepts on $0^\circ \leqslant x \leqslant 8.5^\circ$.
- **c**   Specify the range on $0^\circ \leqslant x \leqslant 8.5^\circ$.
- **d**   Comment on the period of $y = \sin(2^x)$ compared to $y = \sin(2x)$.
- **e**   Sketch a graph of $y = \sin(2^x)$ on $0^\circ \leqslant x \leqslant 8.5^\circ$.

2 Consider the function $f : x \mapsto 2^{\cos x}$ on $-30^o \leqslant x \leqslant 390^o$.
(**Note:** Make sure your calculator is in degrees.)

a Determine the value of any turning points.

b Determine the y-intercept.

c Specify the range on $-30^o \leqslant x \leqslant 390^o$.

d Hence determine the **i** period **ii** amplitude.

e Sketch a graph of $y = 2^{\cos x}$ on $-30^o < x < 390^o$.

3 Consider the function $f : x \mapsto x^5 + 4x^4 + 5x^3 + 2x^2$ on the domain $-2.5 \leqslant x \leqslant 1$.

a Show that $x^2(x + 1)^2(x + 2) = x^5 + 4x^4 + 5x^3 + 2x^2$
(**Hint:** Expand the left-hand side.)

b Find the x-intercepts.

c Find the y-intercept.

d Determine the position of any turning points on the domain given.

e Sketch a graph of $y = f(x)$ on $-2.5 \leqslant x \leqslant 1$ showing *all* important points.

4 Consider the function $f : x \mapsto 3x \times 2^x$ on $-6 \leqslant x \leqslant 1$. Using technology, find:

a the y-intercept

b the minimum value of $f(x)$ on $-6 \leqslant x \leqslant 1$

c the equation of the horizontal asymptote

d $f(-6)$ and $f(1)$.

e Sketch a graph of $y = f(x)$ on $-6 \leqslant x \leqslant 1$ showing *all* of the features calculated above.

5 Consider the function $f : x \mapsto x^{\sin x}$. Use technology to assist (making sure your calculator is in degrees).

a Find the first three maximum values on $0^o \leqslant x \leqslant 1000^o$.

b **i** Try to find the y-intercept by 'zooming-in' close to the y-axis.

ii Explain why the technology cannot find where the graph *appears* to touch the y-axis. (**Hint:** Evaluate $f(0)$ and comment.)

c Does $f(x) = x^{\sin x}$ have any x-intercepts on $0^o \leqslant x \leqslant 1000^o$?
(**Hint:** Use technology to assist.)

d Sketch $y = f(x)$ on $0^o \leqslant x \leqslant 1000^o$.

F   # WHERE FUNCTIONS MEET

Recall the graphs of a quadratic function and a linear function on the same set of axes.
Notice that we could have:

cutting
(2 points of intersection)

touching
(1 point of intersection)

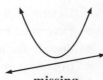

missing
(no points of intersection)

The graphs could meet and the coordinates of the points of intersection of the graphs of the two functions can be found by *solving the two equations simultaneously* or by using technology.

Example 5

Find the coordinates of the points of intersection of the graphs with equations $y = x^2 - x - 18$ and $y = x - 3$.

$y = x^2 - x - 18$ meets $y = x - 3$ where

$$x^2 - x - 18 = x - 3$$
$$\therefore \quad x^2 - 2x - 15 = 0 \qquad \{RHS = 0\}$$
$$\therefore \quad (x - 5)(x + 3) = 0 \qquad \{\text{factorising}\}$$
$$\therefore \quad x = 5 \text{ or } -3$$

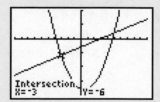

Substituting into $y = x - 3$, when $x = 5$, $y = 2$ and when $x = -3$, $y = -6$.

$\therefore$ graphs meet at $(5, 2)$ and $(-3, -6)$.

EXERCISE 17F.1

1 Find the coordinates of the point(s) of intersection of the graphs with equations:

    a $y = x^2 - 2x + 8$ and $y = x + 6$

    b $y = -x^2 + 3x + 9$ and $y = 2x - 3$

    c $y = x^2 - 4x + 3$ and $y = 2x - 6$

    d $y = x^2 - 3x + 2$ and $y = x - 3$

2 Use technology to find the coordinates of the points of intersection (to two decimal places) of the graphs with equations:

    a $y = x^2 - 3x + 7$ and $y = x + 5$

    b $y = x^2 - 5x + 2$ and $y = x - 7$

    c $y = -x^2 - 2x + 4$ and $y = x + 8$

    d $y = -x^2 + 4x - 2$ and $y = 5x - 6$

Now consider the graph of a linear and a rectangular hyperbola. We could have:

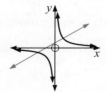 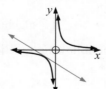 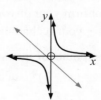

Again, if the graphs intersect, we can use two methods to obtain the intersection points: algebraically or using technology.

Example 6

Find where the line $y = x - 3$ meets $y = \dfrac{4}{x}$.

$y = x - 3$ meets $y = \dfrac{4}{x}$ when

$$x - 3 = \frac{4}{x}$$

$\therefore \quad x(x - 3) = 4$

$\therefore \quad x^2 - 3x = 4$

$\therefore \quad x^2 - 3x - 4 = 0$

$\therefore \quad (x - 4)(x + 1) = 0$

$\therefore \quad x = 4 \text{ or } x = -1$

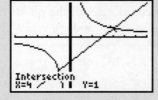

Thus $y = x - 3$ meets $y = \dfrac{4}{x}$ at $(4, 1)$ and $(-1, -4)$.

3 Find without technology where the line:

   a $y = x + 4$ meeets $y = -\dfrac{3}{x}$

   b $y = x - 5$ meets $y = \dfrac{6}{x}$

   c $y = 2x - 1$ meets $y = \dfrac{10}{x}$

   d $y = -2x + 2$ meets $y = \dfrac{4}{x}$

4 Use technology to find where the line:

   a $y = 2x - 3$ meets $y = \dfrac{4}{x}$

   b $y = 4x + 3$ meets $y = -\dfrac{10}{x}$

   c $y = 3x - 2$ meets $y = \dfrac{6}{x}$

   d $y = 3x + 5$ meets $y = \dfrac{12}{x}$

   e $3x - 5y = 10$ meets $y = \dfrac{5}{x}$

   f $3x + 2y = 6$ meets $y = \dfrac{10}{x}$

5 a Use your calculator to draw, on the same set of axes, the graphs of:

   A $y = \dfrac{10}{x}$

   B $y = -2x - 9$

  b Is the line a tangent to the curve? At first glance it certainly appears to be.

  c Confirm your answer by using:

   i your technology

   ii algebraic methods as in **Example 6.**

6 Find the points of intersection of the following functions using a graphics calculator:

   a $y = (x + 1)^2$ and $y = -x^2 + x + 4$

   b $y = 6(x - \dfrac{1}{x})$ and $y = 5$

   c $y = 6x - 1$ and $y = \dfrac{10}{x - 2}$

In order to find the points of intersection between higher order polynomials and other functions, the use of algebra becomes increasingly difficult. Hence, we will focus on the use of technology to explore such situations.

For example, to find the intersection between $y = x + 2$ and $y = 2^x$ algebraically, we must use a logarithmic approach, which is beyond the scope of this course. Using technology however, we can obtain a clear picture of these two graphs drawn on the same axes (as shown below).

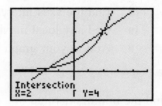

Hence, the two solutions $(-1.69, 0.310)$ and $(2, 4)$ can be obtained (correct to 3 s.f.).

In the following exercises, use technology to find intersection points.

EXERCISE 17F.2

1 Using the 'graph' mode of your graphics calculator, find the intersection(s) between the following sets of functions:

  a $y = 5^{-x}$ and $y = 5x + 10$

  b $y = 3^x$ and $y = 2x + 1$

  c $y = 2^{-x}$ and $y = x^2 - 2x - 3$

  d $y = -3^x$ and $y = -x^4$

  e $y = 5^x$ and $y = \dfrac{1}{x}$

  f $y = 4^x$ and $y = \dfrac{x}{2x - 1}$

2 **a** Using the 'solver' mode of your calculator, find the solution to $x^3 = \dfrac{2}{x}$.

  b Using the 'graph' mode of your calculator, find the intersection points between $y = x^3$ and $y = \dfrac{2}{x}$.

  c Comment on the advantages and disadvantages of using 'graph' mode compared to 'solver' mode.

3 Find the intersection point(s) between:

  a $y = x^2 \times 2^x$ and $y = -(x - 3)(x + 2)$

  b $y = \dfrac{3^x}{x}$ and $y = 5 - x$

  c $y = \dfrac{x + 1}{(x - 1)(x + 3)}$ and $y = 2^x$

  d $y = 2x - 2^x$ and $y = x - 3$

  e $y = \sin x$ and $y = \cos 2x$ on $0^\circ \leqslant x \leqslant 360^\circ$

REVIEW SET 17A

1 For each of the following:

  i use technology to observe the function

  ii draw a sketch graph for $y = f(x)$ without plotting points from a table of values

  iii find the y-intercept and include it on your sketch graph.

  a $y = x^3 + 3$

  b $y = -2x^2(x + 4)$

2 Write in general form: $y = 2(x - 1)^3 - 5$.

3 For each of the following:
>   **i**   use technology to observe the function
>   **ii** draw a sketch graph without plotting actual points
>   **iii** find the axes intercepts
>   **iv** find any local maximums or local minimums.
>   **v**   Add to your graph in **ii** all key features.

     a $y = (x-2)^2(x+1)(x-3)$ **b** $y = 3x^3 - x^2 + 2x - 1$

4 Find the minimum value of $y = x^4 - x^3 - 3x^2 + x + 2$ on the interval:

     a $-2 \leqslant x \leqslant 2$ **b** $-1 \leqslant x \leqslant 1$ **c** $0 \leqslant x \leqslant 2$

5 For the function $y = \dfrac{x-1}{(x+3)(x-2)}$:

     a use technology to observe the function
     b draw a sketch graph without plotting actual points
     c find the axes intercepts
     d find any local maximums or local minimums.
     e Add to your graph in **b** all key features.

REVIEW SET 17B

1 For each of the following:
>   **i**   use technology to observe the function
>   **ii** draw a sketch graph without plotting actual points
>   **iii** find the axes intercepts
>   **iv** find any local maximums or local minimums.
>   **v**   Add to your graph in **ii** all key features.

     a $y = -(x-1)(x+1)(x-3)^2$ **b** $y = x^4 - 4x^3 + 6x^2 - 4x + 1$

2 Find the minimum value of $y = x^4 - x^3 + 2x^2 + 3x - 1$ on $-2 \leqslant x \leqslant 2$.

3 Use technology to find the coordinates of the points of intersection (to two decimal places) of the graphs with equations $y = x^2 - 5x + 6$ and $y = 2x - 1$.

4 For the function $y = \dfrac{2}{x+1} - 3$:

     a What is the vertical asymptote?   **b** What is the horizontal asymptote?
     c What is the x-intercept?   **d** What is the x-intercept?
     e Using only **a** to **b** sketch the graph of the function. Check using technology.

5 Consider the function $f : x \to x \times 2^x$ on $-4 \leqslant x \leqslant 1$. Using technology, find:
     a the y-intercept
     b the minimum value of $f(x)$ on $-4 \leqslant x \leqslant 1$
     c the equation of the horizontal asymptote
     d $f(-4)$ and $f(1)$.
     e Sketch a graph of $y = f(x)$ on $-4 \leqslant x \leqslant 1$ showing *all* of the features calculated above.

Chapter 18

Two variable statistics

TWO VARIABLE ANALYSIS

Often a statistician will want to know how two variables are **associated** or **related**.

To find such a relationship the first step is to construct and observe a **scatterplot**. On the horizontal axis we put the **independent variable** and on the vertical axis the **dependent variable**. A typical scatterplot could look like this:

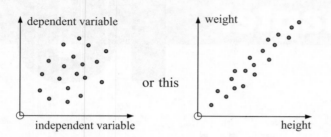

or this

The weight of a person is usually dependent on their height.

OPENING PROBLEM

The *height* and *weight* of a squad of hockey players is to be investigated. The raw data is given:

Player	Height	Weight	Player	Height	Weight	Player	Height	Weight
A	183	85	H	167	67	O	167	72
B	170	74	I	169	74	P	171	69
C	174	76	J	163	67	Q	170	76
D	168	69	K	161	69	R	174	71
E	167	68	L	172	74	S	162	71
F	177	74	M	160	64	T	187	84
G	162	62	N	160	62	U	175	77

Things to think about:

- Are the variables categorical or quantitative?
- What is the dependent variable?
- What would the scatterplot look like? Are the points close to being linear?
- Does an increase in the independent variable generally cause an increase or a decrease in the dependent variable?
- Is there a way of indicating the strength of a linear connection for the variables?
- How can we find the equation of 'the line of best fit' and how can we use it?

The **scatterplot** for the **Opening Problem** is drawn alongside. *Height* is the independent variable and is represented on the horizontal axis.

Notice that in general, as the *height* increases so does the *weight*.

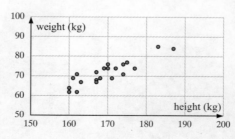

CORRELATION

> **Correlation** refers to the relationship or association between two variables.

In looking at the correlation between two variables we should follow these steps:

Step 1:   Look at the scatterplot for any **pattern**.

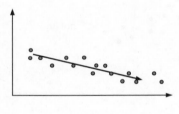

For a generally *upward* trend we say that the correlation is **positive**, and in this case an increase in the independent variable means that the dependent variable generally increases.

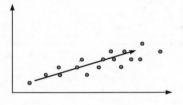

For a generally *downward* trend we say that the correlation is **negative**, and in this case an increase in the independent variable means that the dependent variable generally decreases.

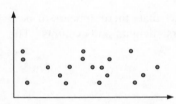

For *randomly scattered* points (with no upward or downward trend) there is usually **no correlation**.

Step 2:   Look at the spread of points to make a judgement about the **strength** of the correlation. For **positive relationships** we would classify the following scatterplots as:

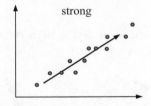
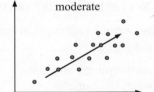
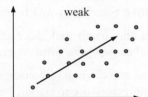

Similarly there are strength classifications for **negative relationships**:

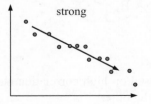
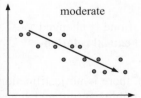
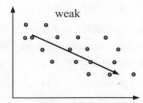

Step 3:   Look at the pattern of points to see whether or not it is **linear**.

These points appear to be roughly linear.   These points do not appear to be linear.

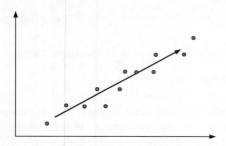

Step 4:   Look for and investigate any **outliers**. These appear as isolated points away from the main body of data.

Outliers should be investigated as sometimes they are mistakes made in recording the data or plotting it.

Genuine extraordinary data should be included.

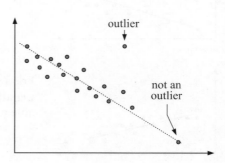

Looking at the scatterplot for the **Opening Problem** we can say that 'there appears to be a not very strong positive correlation between the hockey players' heights and weights. The relationship appears to be linear with no possible outliers'.

CAUSATION

Correlation between two variables does not necessarily mean that one variable causes the other. Consider the following:

The *arm length* and *running speed* of a sample of young children were measured and a strong, positive correlation was found to exist between the variables.

Does this mean that short arms cause a reduction in running speed or that a high running speed causes your arms to grow long?

These are obviously nonsense assumptions and the strong positive correlation between the variables is attributed to the fact that both *arm length* and *running speed* are closely related to a third variable, *age*. *Arm length* increases with age as does *running speed* (up to a certain age).

When variables are related so that if one is changed the other changes then we can say a **causal relationship** exists between the variables.

In cases where this is *not* apparent, there is no justification, based on high correlation alone, to conclude that changes in one variable cause the changes in the other.

EXERCISE 18A

1 For each of the scatterplots below state:

   i whether there is positive, negative or no association between the variables

   ii whether the relationship between the variables appears to be linear or not

   iii the strength of the association (zero, weak, moderate or strong).

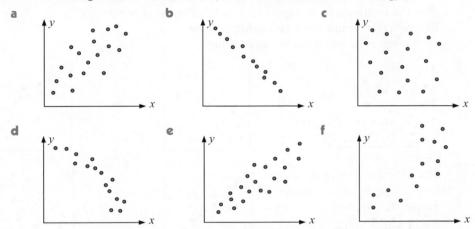

2 Copy and complete the following:

   a If the variables x and y are positively associated then as x increases y ....

   b If there is negative correlation between the variables m and n then as m increases, n ....

   c If there is no association between two variables then the points on the scatterplot appear to be ....

3 The results of a group of students for a Maths test and a Science test are compared:

Student	A	B	C	D	E	F	G	H	I	J
Maths test	64	67	69	70	73	74	77	82	84	85
Science test	68	73	68	75	78	73	77	84	86	89

Construct a scatterplot for the data. (Make the scale on both axes from 60 to 90.)

4 The scores awarded by two judges at a diving competition are shown in the table.

Competitor	P	Q	R	S	T	U	V	W	X	Y
Judge A	5	6.5	8	9	4	2.5	7	5	6	3
Judge B	6	7	8.5	9	5	4	7.5	5	7	4.5

   a Construct a scatterplot for this data with Judge A's scores on the horizontal axis and Judge B's scores on the vertical axis.

   b Copy and complete the following comments on the scatterplot:

   There appears to be ...., ... correlation between Judge A's scores and Judge B's scores. This means that as Judge A's scores increase, Judge B's scores ....

5 a What is meant by the independent and dependent variables?

   b Give another name for each of the variables in **a**.

   c When graphing, which variable is placed on the horizontal axis?

6 For the following scatterplots comment on:

   i the existence of any *pattern* (positive, negative or no association)

   ii the relationship *strength* (zero, weak, moderate or strong)

   iii whether or not the relationship is linear

   iv whether or not there are any outliers.

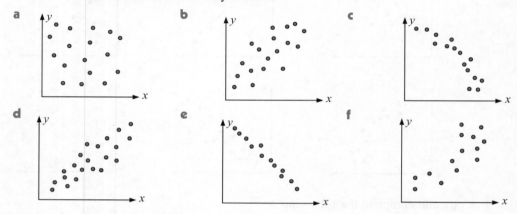

7 What is meant by causation?

LINE OF BEST FIT

> **Regression** is the method of fitting a line to a set of data and then finding the equation of the line.

The line is often called the **model**.

The **regression line** is often called '**the line of best fit**' and can be used to predict a value of the dependent variable for a given value of the independent variable. There are several ways to fit a straight line to a data set. We will examine two of them:

 • The line of best fit '**by eye**'.

 • The '**least squares**' regression line (**linear regression**).

THE 'BY EYE' METHOD: (Hockey player data)

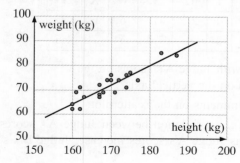

Your guess is as good as mine!

'By eye' we draw a line which best fits with about equal numbers of points on either side (but not necessarily so). The average distances away from the line should balance.

We now select two points on this line and use $\dfrac{y - b}{x - a} = m$ to find the equation.

In our hockey players data these are (160, 64) and (190, 88).

Now $m \doteqdot \dfrac{88 - 64}{190 - 160}$ So, the equation is $\dfrac{y - 64}{x - 160} \doteqdot 0.8$

$\doteqdot \frac{24}{30}$ $\therefore$ $y - 64 \doteqdot 0.8(x - 160)$

$\doteqdot 0.8$ $\therefore$ $y - 64 \doteqdot 0.8x - 128$

i.e., $y \doteqdot 0.8x - 64$

The problem with this method is that the answer will vary from one person to another. Selecting the line and choosing two points on it can be very difficult.

B   MEASURING CORRELATION

When dealing with linear association we can use the concept known as **correlation** to measure the strength and direction of association.

The **correlation coefficient** (r) lies between -1 and 1.

POSITIVE CORRELATION

> An association between two variables is described as a **positive correlation** if an increase in one variable results in an increase in the other in an approximately linear manner.

The strength of the association is best measured with the **correlation coefficient (r)** that ranges between 0 and 1 for positive correlation.

An r value of 0 suggests that there is no linear association present (or **no correlation**).

An r value of 1 suggests that there is a perfect linear association present (or **perfect positive correlation**).

The correlation between the *height* and the *weight* of people is positive and lies between 0 and $+1$. It is not an example of perfect positive correlation because, for example, not all short people are of light weight. However, taller people are generally heavier than shorter people.

The r values in between 0 and 1 represent varying degrees of linearity.

Scatter diagrams for positive correlation:

The scales on each of the four graphs are the same.

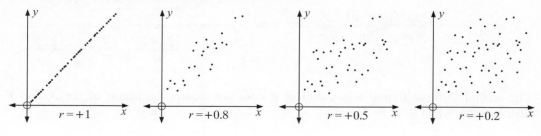

NEGATIVE CORRELATION

An association between two variables is described as a **negative correlation** if an increase in one variable results in a decrease in the other in an approximately linear manner.

The strength of the association is best measured with the **correlation coefficient** (r) that ranges between 0 and -1 for negative correlation.

An r value of -1 suggests that there is a perfect linear association present (or **perfect negative correlation**).

Scatter diagrams for negative correlation:

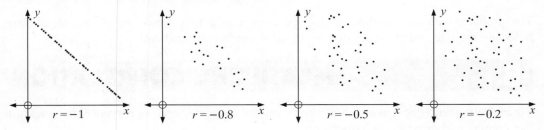

PEARSON'S CORRELATION COEFFICIENT

Pearson's correlation coefficient, for finding the degree of linearity between two random variables X and Y, given n ordered pairs (x_1, y_1), (x_2, y_2), (x_3, y_3), ..., (x_n, y_n) of data is:

$$r = \frac{s_{xy}}{s_x s_y} \quad \text{where} \quad s_{xy} = \frac{\sum(x - \overline{x})(y - \overline{y})}{n} \quad \text{or} \quad = \sum xy - \frac{(\sum x)(\sum y)}{n}$$

$$s_x = \sqrt{\frac{\sum(x - \overline{x})^2}{n}} \quad \text{or} \quad = \sqrt{\sum x^2 - \frac{(\sum x)^2}{n}}$$

$$s_y = \sqrt{\frac{\sum(y - \overline{y})^2}{n}} \quad \text{or} \quad = \sqrt{\sum y^2 - \frac{(\sum y)^2}{n}}$$

wrong

s_{xy} is called the **covariance** of X and Y

s_x is the **standard deviation** of X

s_y is the **standard deviation** of Y

So,

$$r = \frac{\sum(x - \overline{x})(y - \overline{y})}{\sqrt{\sum(x - \overline{x})^2}\sqrt{\sum(y - \overline{y})^2}} \quad \text{or} \quad r = \frac{\sum xy - \frac{(\sum x)(\sum y)}{n}}{\sqrt{\sum x^2 - \frac{(\sum x)^2}{n}}\sqrt{\sum y^2 - \frac{(\sum y)^2}{n}}}$$

correct

The second of these formulae is useful as it does not require the means of the X and Y distributions, $\overline{x}$ and $\overline{y}$, to be found.

Example 1

A chemical fertiliser company wishes to determine the extent of correlation between '*quantity of compound X used*' and '*lawn growth*' per day. Find the Pearson's correlation coefficient between the two variables.

Lawn	Compound X (g)	Lawn growth (mm)
A	1	3
B	2	3
C	4	6
D	5	8

x	y	xy	x^2	y^2
1	3	3	1	9
2	3	6	4	9
4	6	24	16	36
5	8	40	25	64
totals: 12	20	73	46	118

$n = 4$ (our pairs of data values)

$$\sum x = 12, \quad \sum y = 20, \quad \sum xy = 73$$

$$\sum x^2 = 46, \quad \sum y^2 = 118$$

$$r = \frac{73 - \dfrac{12 \times 20}{4}}{\sqrt{46 - \dfrac{12^2}{4}} \sqrt{118 - \dfrac{20^2}{4}}}$$

$$\therefore \quad r = \frac{13}{\sqrt{10}\sqrt{18}}$$

$$\therefore \quad r \doteqdot 0.969$$

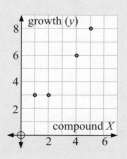

So,
- a positive r means that as x (the mass of chemical compound) increases, then so does y (the lawn growth)
- r close to 1 indicates a very strong positive correlation.

Example 2

In attempting to find if there is any association between *average speed* in the metropolitan area and *age of drivers*, a device was fitted to cars of drivers of different ages.

The results are shown in the scatterplot. The r-value for this association is $+0.027$. Describe the association.

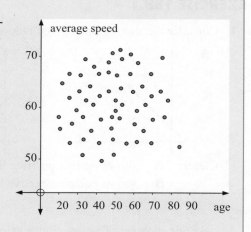

As r is close to zero, there is no correlation between the two variables.

Example 3

Wydox have been trying out a new chemical to control the number of lawn beetles in the soil. Determine the extent of the correlation between the *quantity of chemical used* and the *number of surviving lawn beetles* per square metre of lawn.

Lawn	Amount of chemical (g)	Number of surviving lawn beetles
A	2	11
B	5	6
C	6	4
D	3	6
E	9	3

x	y	xy	x^2	y^2
2	11	22	4	121
5	6	30	25	36
6	4	24	36	16
3	6	18	9	36
9	3	27	81	9
totals 25	30	121	155	218

There are five data points $\therefore$ $n = 5$.

$$\sum x = 25, \quad \sum y = 30, \quad \sum xy = 121$$

$$\sum x^2 = 155, \quad \sum y^2 = 218$$

$$\therefore \quad r = \frac{121 - \dfrac{25 \times 30}{5}}{\sqrt{155 - \dfrac{25^2}{5}} \sqrt{218 - \dfrac{30^2}{5}}}$$

$$\doteqdot -0.859$$

Clearly we have a moderate negative association between the *amount of chemical used* and the *number of surviving lawn beetles*. Generally, the more chemical used, the fewer beetles survive.

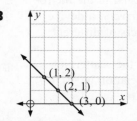

EXERCISE 18B.1

1 Consider the three graphs given below:

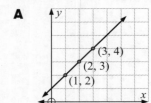

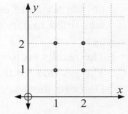

Clearly **A** shows perfect positive linear correlation   **C** shows no correlation.
     B shows perfect negative linear correlation

a For each set of points find r using $r = \dfrac{\sum xy - \dfrac{(\sum x)(\sum y)}{n}}{\sqrt{\sum x^2 - \dfrac{(\sum x)^2}{n}} \sqrt{\sum y^2 - \dfrac{(\sum y)^2}{n}}}$

b Comment on the value of r for each graph.

2 Find Pearson's correlation coefficient for random variables X and Y where:

  a $s_x = 14.7$, $s_y = 19.2$,   and $s_{xy} = 136.8$

  b the standard deviation of the X distribution is 8.71, the standard deviation of the Y distribution is 13.23 and the covariance of X and Y is -9.26

  c $\sum x = 65$, $\sum y = 141$, $\sum xy = 1165$, $\sum x^2 = 505$, $\sum y^2 = 2745$, $n = 11$

3 The scores awarded by two judges at a diving competition are shown in the table.

Competitor	P	Q	R	S	T	U	V	W
Judge A	7	6	8.5	8	5	5	6	7.5
Judge B	6	5	9	8	4.5	3.5	6.5	7

  a Construct a scatterplot for this data with Judge A's scores on the horizontal axis and Judge B's scores on the vertical axis.

  b Copy and complete the following comments on the scatterplot:

  There appears to be ...., ... correlation between Judge A's scores and Judge B's scores. This means that as Judge A's scores increase, Judge B's scores ....

  c Calculate and interpret Pearson's correlation coefficient.

From this point onwards we will use **technology** to find r.

THE COEFFICIENT OF DETERMINATION (r^2)

To help describe the strength of association we calculate the **coefficient of determination** (r^2). This is simply the square of the correlation coefficient (r) and as such the direction of association is eliminated.

Many texts vary on the advice they give. We suggest the rule of thumb given alongside when describing the strength of linear association.

value	strength of association
$r^2 = 0$	no correlation
$0 < r^2 < 0.25$	very weak correlation
$0.25 \leqslant r^2 < 0.50$	weak correlation
$0.50 \leqslant r^2 < 0.75$	moderate correlation
$0.75 \leqslant r^2 < 0.90$	strong correlation
$0.90 \leqslant r^2 < 1$	very strong correlation
$r^2 = 1$	perfect correlation

USING TECHNOLOGY FOR THE CORRELATION COEFFICIENT

We will calculate r for the data set:

x	1	2	3	4	5	6	7
y	5	8	10	13	16	18	20

CALCULATING r USING A GRAPHICS CALCULATOR

Enter the data and find r, r^2 and hence determine the strength of the correlation. Click on the calculator icon of your choice to find detailed instructions.

CALCULATING r USING A STATISTICS PACKAGE

Enter the data and simply read off r, r^2 and the degree of strength of the correlation. Click on the icon to find an easy to use two variable analysis computer package.

STATISTICS
PACKAGE

CALCULATING r USING MS EXCEL

Enter the data in columns A and B as shown and type into cell D2 say,

=CORREL(A1:A7,B1:B7)

to get $r \doteqdot 0.998$

The formula in cell D3 will calculate r^2.

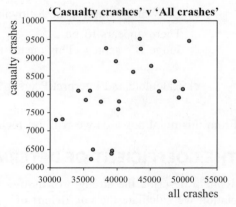

	A	B	C	D
				=CORREL(A1:A7,B1:B7)
1	1	5		
2	2	8	r	0.997726
3	3	10	r^2	=D2^2
4	4	13		
5	5	16		
6	6	18		
7	7	20		

EXERCISE 18B.2

1   The scatterplot alongside shows the association between the number of car crashes in which a casualty occurred and total number of car crashes in each year from 1972 to 1994. Given that the r value is 0.49:

a   find r^2

b   describe the association between these variables.

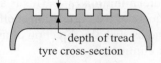

'Casualty crashes' v 'All crashes'

2   In an investigation to examine the association between the *tread depth* (y mm) and the *number of kilometres travelled* (x thousand), a sample of 8 tyres of the same brand was taken and the results are given below.

depth of tread
tyre cross-section

Kilometres (x thousand)	14	17	24	34	35	37	38	39
Tread depth (y mm)	5.7	6.5	4.0	3.0	1.9	2.7	1.9	2.3

a   Draw a scatterplot of the data.   b   Calculate r and r^2 for the tabled data.
c   Describe the association between *tread depth* and the *number of kilometres travelled* for this brand of tyre.

3   A restauranteur believes that during March the number of people wanting dinner (y) is related to the temperature at noon (x°C). Over a period of a fortnight the number of diners and the noon temperature were recorded.

Temperature (x°C)	23	25	28	30	30	27	25	28	32	31	33	29	27
Number of diners (y)	57	64	62	75	69	58	61	78	80	67	84	73	76

a   Draw a scatterplot of the data.   b   Calculate r and r^2 for the data.
c   Describe the association between *number of diners* and *noon temperature* for the restaurant in question.

4 Tomatoes are sprayed with a pesticide-fertiliser mix. The figures below give the *yield of tomatoes* per bush for various *spray concentrations*.

Spray concentration $(x$ mL/L$)$	3	5	6	8	9	11
Yield of tomatoes per bush (y)	67	90	103	120	124	150

a Draw a scatterplot for this data.   **b** Determine the r and r^2 values.

c Describe the association between *yield* and *spray concentration*.

5 It has long been thought that frosty conditions are necessary to 'set' the fruit of cherries and apples. The following data shows *annual cherry yield* and *number of frosts* data for a cherry growing farm over a 7 year period.

Number of frosts, (x)	27	23	7	37	32	14	16
Cherry yield $(y$ tonnes$)$	5.6	4.8	3.1	7.2	6.1	3.7	3.8

a Draw a scatterplot for this data.   **b** Determine the r and r^2 value.

c Describe the association between *cherry yield* and the *number of frosts*.

6 In 2002 a business advertised starting salaries for recently graduated university students depending on whether they held a Bachelor's degree or a PhD, as shown alongside.

Field	Bachelor's degree (x)	PhD (y)
Chemical engineer	38 250	48 750
Computer coder	41 750	68 270
Electrical engineer	38 250	56 750
Sociologist	32 750	38 300
Applied mathematician	43 000	72 600
Accountant	38 550	46 000

a Draw a scatterplot of the data.   **b** Determine r and r^2.

c Describe the association between *starting salaries for Bachelor's degrees* and *starting salaries for PhD's*.

7 World War II saw a peak in the production of aeroplanes. One specific type of plane that was made was the *fighter plane*. It was used in aerial combat and also to shoot at enemy on the ground. The table below contains information on the *maximum speed* and *maximum altitude obtainable* (ceiling) for nineteen fighter planes. Maximum speed is given in km/h $\div$ 1000.   Ceiling is given in m $\div$ 1000.

max. speed	ceiling	max. speed	ceiling	max. speed	ceiling
0.46	8.84	0.68	10.66	0.67	12.49
0.42	10.06	0.72	11.27	0.57	10.66
0.53	10.97	0.71	12.64	0.44	10.51
0.53	9.906	0.66	11.12	0.67	11.58
0.49	9.448	0.78	12.80	0.70	11.73
0.53	10.36	0.73	11.88	0.52	10.36
0.68	11.73				

a Draw a scatterplot for this data.   **b** Determine the r and r^2 value.

c Describe the association between *maximum speed* and *ceiling*.

C   LEAST SQUARES REGRESSION

Let us revisit the **Opening Problem** again. We know that there is quite a strong positive correlation between the *height* and the *weight* of the players.

Consequently, we should be able to find a linear equation which '*best fits*' the data.

This **line of best fit** could be found *by eye*. However, different people will use different lines. So, how do we find mathematically, the line of best fit?

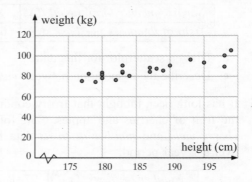

RESIDUALS

A **residual** is the vertical distance between a data point and the possible line of best fit.

That is:

A **residual** is a value of $y - \widehat{y}$ where y is an observed value and $\widehat{y}$ is on the possible line of best fit above or below y.

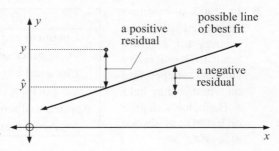

So, there are positive and negative residuals.

LEAST SQUARES REGRESSION LINE FOR y ON x

Statisticians invented a method where the best line results. The process is **minimisation of the sum of the squares of the residuals**.

Suppose the line of best fit is $y = mx + c$.

$$r_1^2 = (y_1 - \widehat{y_1})^2$$
$$r_2^2 = (y_2 - \widehat{y_2})^2$$
$$r_3^2 = (y_3 - \widehat{y_3})^2$$
$$\text{etc.}$$

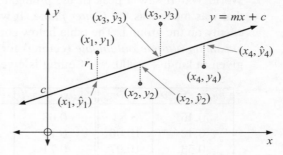

So, we need to minimise
$$S = r_1^2 + r_2^2 + r_3^2 + \ldots\ldots + r_n^2.$$

DEMO

Click on the icon to experiment with finding the 'line of best fit' by minimising the sum of the squares of the residuals.

Write down the function that you find which minimises the sum of the squares of the residuals.

LEAST SQUARES FORMULAE

You can probably imagine the time-consuming work needed to find m and c, especially with 20 or so points (not just three as in the above working).

In fact a formula exists for finding the **least squares regression line** for y on x. It is:

$$y - \overline{y} = \frac{s_{xy}}{s_x^2}(x - \overline{x})$$

Example 4

Use the formulae for calculating m and c for the line of best fit through $(1, 3)$, $(3, 5)$ and $(5, 6)$.

x	y	xy	x^2
1	3	3	1
3	5	15	9
5	6	30	25
$\sum$ 9	14	48	35

So, $\sum x = 9$, $\sum y = 14$, $\sum xy = 48$,

$\sum x^2 = 35$, $n = 3$

$s_{xy} = \sum xy - \dfrac{(\sum x)(\sum y)}{n} = 48 - \dfrac{9 \times 14}{3} = 6$

$s_x^2 = \sum x^2 - \dfrac{(\sum x)^2}{n} = 35 - \dfrac{9^2}{3} = 8$

$\overline{x} = \dfrac{\sum x}{n} = \dfrac{9}{3} = 3$ and $\overline{y} = \dfrac{\sum y}{n} = \dfrac{14}{3}$

So, using $y - \overline{y} = \dfrac{s_{xy}}{s_x^2}(x - \overline{x})$ we get $y - \dfrac{14}{3} = \dfrac{6}{8}(x - 3)$

$y - 4.67 \doteqdot 0.75x - 2.25$

$y \doteqdot 0.75x - 2.25 + 4.67$

$y \doteqdot 0.75x + 2.42$

From this point onwards we will use technology to find the least squares regression line. We can find the **least squares regression** line using:

- a computer package
- a graphics calculator
- a computer spreadsheet

To do this consider the tabled data:

x	1	2	3	4	5	6	7
y	5	8	10	13	16	18	20

USING A STATISTICS PACKAGE

The package is very easy to use. Click on the icon.
Enter the data. Examine all features that the package produces.

STATISTICS
PACKAGE

USING A GRAPHICS CALCULATOR

Enter the data in two **lists** and use your calculator to find the equation of the **regression line**.

USING A COMPUTER SPREADSHEET

The plotting of points, fitting the line of best fit and finding its equation can be easily determined using a spreadsheet such as Microsoft Excel®. Following is a step-by-step procedure for determining the line of best fit.

Step 1:   Enter the data into the cells.

Step 2:   Highlight (blacken) the cells containing the data by clicking the LH mouse button on A1 and dragging through to B7.

You should now see:

Step 3:   Click on , the chart wizard icon.

Click on XY (Scatter)   Then NEXT

Click on FINISH

You should now have a graph showing the 7 points.

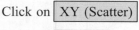

Step 4:   Place the arrow on one of the points and click the RH mouse button once.

Select Add Trendline and click on this with the LH mouse button.

Select Linear   OK   and the line of best fit should be added to your graph.

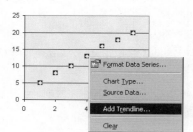

Step 5:   Place the arrow on the trend line and click with RH mouse button.

Select Format Trendline , Options then

select Display equation on chart

and Display R-squared value on chart   OK .

You should now have
$y = 2.5357x + 2.7143$
and $R^2 = 0.9955$ added to your graph.

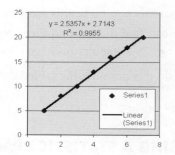

INTERPOLATION / EXTRAPOLATION

The two variables in the following scatterplot are the mass of a platypus (independent variable plotted on the x-axis) and the length of the same platypus (dependent variable plotted on the y-axis) for 14 different animals.

The data was collected in an experiment to discover if there was a relationship between the length and mass of these animals.

Even though the correlation in this case is only moderate, a line of best fit has been drawn to enable predictions to be made.

If we use the equation of the least squares line to predict length values for mass values **in between** the smallest and largest mass values that were supplied from the experiment, we say we are **interpolating** (in between the poles).

If we predict length values for mass values **outside** the smallest and largest mass values that were supplied from the experiment we say we are **extrapolating** (outside the poles).

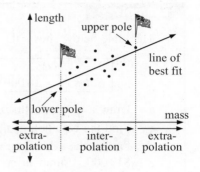

The accuracy of an interpolation depends on how linear the original data was. This can be gauged by determining the correlation coefficient and ensuring that the data is randomly scattered around the line of best fit.

The accuracy of an extrapolation depends not only on 'how linear' the original data was, but also on the assumption that the linear trend will continue past the poles.

The validity of this assumption depends greatly on the situation under investigation.

CARE MUST BE TAKEN WHEN EXTRAPOLATING

The performance of a light spring is under consideration. An attempt is being made to find the connection between the extension (y cm) of the spring and the mass in the basket (x grams).

A typical graph for this experiment is:

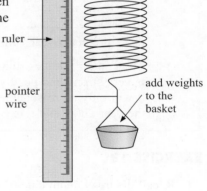

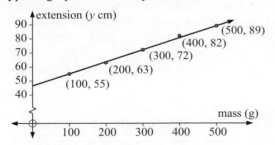

There is a very high positive correlation between the variables, and the line of best fit is determined to be $y \doteqdot 0.087x + 46.1$ cm.

However, it would be dangerous to predict that for a mass of 800 grams the extension would be $0.087 \times 800 + 46.1 = 115.7$ cm because we may have exceeded the elastic limit of the spring somewhere between $x = 500$ grams and $x = 800$ grams, meaning that the spring becomes permanently stretched more than predicted by the graph.

A further example could be the world record for the long jump prior to the Mexico City Olympic Games of 1968. A steady regular increase in the World record over the previous 30 years had been recorded. However, due to the high altitude and a perfect jump, the USA competitor Bob Beamon, shattered the record by a huge amount, not in keeping with previous increases.

Example 5

The table below shows the sales for Hancock's Electronics established in late 1998.

Year	1999	2000	2001	2002	2003	2004
Sales ($ × 10 000)	5	9	14	18	21	27

a Draw a graph to illustrate this data.   **b** Find r^2.

c Find the equation of the line of best fit using the linear regression formula.

d Predict the sales figures for year 2006, giving your answer to the nearest $10 000. Comment on the reasonableness of this prediction.

a Let t be the time in years from 1998 and S be the sales in $10 000's, i.e.,

t	S
1	5
2	9
3	14
4	18
5	21
6	27

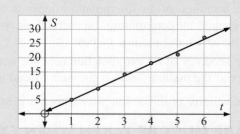

b Using technology, $r^2 = 0.9941$.

c The line of best fit is $S = 4.286t + 0.667$.

d In 2006, $x = 8$ ∴ $S \doteqdot 4.286 \times 8 + 0.667 \doteqdot 35$

i.e., predicted year 2006 sales would be $350 000.

The r and r^2 values suggest that the linear relationship between sales and year is very strong and positive. However, since this prediction is an extrapolation, it will only be reasonable if the trend evident from 1999 to 2006 continues to the year 2006, and this may or may not occur.

EXERCISE 18C

1 Recall the tread depth data of car tyres after travelling thousands of kilometres:

kilometres (X thousand)	14	17	24	34	35	37	38	39
tread depth (Y mm)	5.7	6.5	4.0	3.0	1.9	2.7	1.9	2.3

a Which is the dependent variable?

b Find the equation of the least squares regression line.

c On a scatterplot graph the least squares regression line.

d Use the equation of the line of best fit to estimate the tread depth of a new tyre.

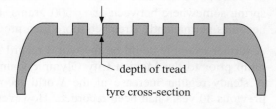

depth of tread

tyre cross-section

e If a tread depth of 2 mm or more is considered to be essential for safe driving, estimate the distance a tyre of this brand should last.

2 Recall the restauranteur's data for the number of diners in March and the temperature at noon.

Temperature $(X°C)$	23	25	28	30	30	27	25	28	32	31	33	29	27
Number of diners (Y)	57	64	62	75	69	58	61	78	80	67	84	73	76

    a What is the independent variable?

    b Find the covariance of X and Y, denoted $Cov(X, Y)$.

    c Find the equation of the least squares regression line.

    d On a scatterplot graph the least squares regression line.

    e How accurate would the interpolation using the regression line be? Why?

3 Recall the spray on tomatoes data:

Spray concentration $(X\ mL/L)$	3	5	6	8	9	11
Yield of tomatoes per bush (Y)	67	90	103	120	124	150

    a Define the role of each variable and produce an appropriate scatterplot.

    b Use the method of least squares to determine the equation of the line of best fit.

    c Give an interpretation for the slope and vertical intercept of this line.

    d Use the equation of the least squares line to predict the yield if the spray concentration was 7 mL/L. Comment on the reasonableness of this prediction.

    e If a 50 mL/L spray concentration was used, would this ensure a large tomato yield? Explain.

4 Recall the frost on cherry data:

Number of frosts, (X)	27	23	7	37	32	14	16
Cherry yield $(Y\ tonnes)$	5.6	4.8	3.1	7.2	6.1	3.7	3.8

    a Define the role of each variable and produce an appropriate scatterplot.

    b Use the method of least squares to determine the equation of the line of best fit.

    c Give an interpretation for the slope and vertical intercept of this line.

    d Use the equation of the least squares line to predict the cherry yield if 29 frosts were recorded. Comment on the reasonableness of this prediction.

    e Use the equation of the least squares line to predict the cherry yield if 1 frost was recorded. Comment on the reasonableness of this prediction.

5 The rate of a chemical reaction in a certain plant depends on the number of frost-free days experienced by the plant over a year which, in turn, depends on altitude. The higher the altitude, the greater the chance of frost. The following table shows the rate of the chemical reaction R, as a function of the number of frost-free days, n.

Frost-free days (n)	75	100	125	150	175	200
Rate of reaction (R)	44.6	42.1	39.4	37.0	34.1	31.2

    a Produce a scatterplot for the data of R against n.

    b Find a linear model which best fits the data. State the value of r^2.

    c Estimate the rate of the chemical reaction when the number of frost free days is:

       i 90   **ii** 215.

    d Complete: "The higher the altitude, the ...... the rate of reaction."

6 The yield (Y kg) of pumpkins on a farm depends on the quantity of fertiliser (X g/m²).

The following table shows corresponding X and Y values.

X	4	13	20	26	30	35	50
Y	1.8	2.9	3.8	4.2	4.7	5.7	4.4

   a Draw a scatterplot of the data and identify an outlier.

   b Calculate the correlation coefficient:

      i with the outlier included   **ii** without the outlier.

   c Calculate the equation of the least squares regression line:

      i with the outlier included   **ii** without the outlier.

   d If you wish to estimate the yield when 15 g/m² are used, which regression line from **c** should be used?

   e Can you explain what may have caused the outlier?

7 Find the least squares regression line for y on x if:

   a $\overline{x} = 6.12$, $\overline{y} = 5.94$, $s_{xy} = -4.28$, $s_x = 2.32$

   b $\overline{x} = 21.6$, $\overline{y} = 45.9$, $s_{xy} = 12.28$, $s_x = 8.77$

8 $n = 6$, $\sum x = 61$, $\sum y = 89$, $\sum xy = 1108$, $\sum(x - \overline{x})^2 = 138$ and $\sum(y - \overline{y})^2 = 284$

   a Find **i** the mean of X **ii** the mean of Y.

   b Find **i** the standard deviation of X **ii** the standard deviation of Y.

   c Find the covariance of X and Y.

   d Find the least squares regression model for y on x.

INVESTIGATION   **SPEARMAN'S RANK ORDER**
                  CORRELATION COEFFICIENT

Suppose we wish to test the degree of agreement between two wine tasting judges at a vintage festival, or between two judges at a diving competition.

Spearman's rank order correlation coefficient can be used for this purpose.

His formula is:

$$t = 1 - \frac{6 \sum d^2}{n(n^2 - 1)}$$

where t is **Spearman's rank order correlation coefficient**
     d is the **difference in the ranking**
     n is the **number of rankings**

Note: $\sum d^2$ is the **sum of the squares of the differences**.

What to do:

1 Amy and Lee are two wine judges. They are considering six red wines: A, B, C, D, E and F. They taste each wine and put them in order of enjoyment from 1 (best) to 6 (worst), and the results of their judging is shown in the table which follows:

Wine	A	B	C	D	E	F
Amy's order	3	1	6	2	4	5
Lee's order	6	5	2	1	3	4

Notice that for wine A, $d = 6 - 3 = 3$
    and for wine C, $d = 6 - 2 = 4$

   a Find Spearman's rank order correlation coefficient for the wine tasting data.

   b Comment on the degree of agreement between their rankings of the wine.

   c What is the significance of the sign of t?

2 Amy and Lee then taste six white wines and their rankings were:

Wine	A	B	C	D	E	F
Amy's order	1	2	4	3	5	6
Lee's order	2	1	3	4	6	5

   a Find Spearman's rank order correlation coefficient for the wine tasting data.

   b Comment on the degree of agreement between their rankings of the wine.

   c What is the significance of the sign of t?

3 Find t for: **a** perfect agreement   **b** completely opposite order.

4 Construct some examples of your own for the following cases:

   a t being close to $+1$ **b** t being close to -1 **c** t being close to 0

   d t being positive   **e** t being negative

   As a consequence of your investigation comment on these five categories.

5 Arrange some competitions of your own choosing and test for rank agreement between the views of two friends. Record all data and show all calculations. You could examine preferences in food tasting, sports watched on TV, etc.

D   THE χ^2 TEST OF INDEPENDENCE

The χ^2 (**chi–squared**) test is the test we use to find if two **classifications** (or **factors**) from the same sample are **independent**, i.e., if the occurrence of one of them does not affect the occurrence of the other.

Examples of two classifications could include:

- *income* and *voting intentions*
- *gender* and *money earning capacity*
- *school year groups* and *canteen improvements*

The χ^2 test examines the difference between the **observed** and **expected** values and

$$\chi^2_{calc} = \sum \frac{(f_o - f_e)^2}{f_e} \qquad \text{where} \quad f_o \text{ is an observed frequency} \\ \text{and} \quad f_e \text{ is an expected frequency.}$$

Small differences between observed and expected frequencies are an indication of the independence between the two classifications.

CALCULATING χ^2

This table shows the results of a sample of 400 randomly selected adults classified according to *gender* and *regular exercise*.

This is a 2×2 **contingency table**.

	Regular exercise	No regular exercise	sum
Male	112	104	216
Female	96	88	184
sum	208	192	400

The question is: "Are *regular exercise* and *gender* independent"?

Consider a general 2×2 contingency table for classifications R and S.

Notice that for independence,

$$P(R_1 \cap S_1) = P(R_1) \, P(S_1) = \frac{w}{n} \times \frac{y}{n}$$

	S_1	S_2	sum
R_1	a	b	w
R_2	c	d	x
sum	y	z	n

$\therefore$ the expected value of $P(R_1 \cap S_1)$

$$= n \times P(R_1 \cap S_1)$$
$$= n \times \frac{w}{n} \times \frac{y}{n}$$
$$= \frac{wy}{n}$$

$n = w + x = y + z$

Likewise, the expected value of $P(R_1 \cap S_2) = \dfrac{wz}{n}$,   etc

So,   the **expected value table** is:

	S_1	S_2	sum
R_1	$\dfrac{wy}{n}$	$\dfrac{wz}{n}$	w
R_2	$\dfrac{xy}{n}$	$\dfrac{xz}{n}$	x
sum	y	z	n

Using this result, the **expected table** for the *regular exercise* and *gender* data is:

	Regular exercise	No regular exercise	sum
Male	$\frac{216 \times 208}{400} \doteqdot 112.3$	$\frac{216 \times 192}{400} \doteqdot 103.7$	216
Female	$\frac{184 \times 208}{400} \doteqdot 95.7$	$\frac{184 \times 192}{400} \doteqdot 88.3$	184
sum	208	192	400

and the χ^2 calculation is:

f_o	f_e	$f_o - f_e$	$(f_o - f_e)^2$	$\dfrac{(f_o - f_e)^2}{f_e}$
112	112.3	-0.3	0.09	0.000801
104	103.7	0.3	0.09	0.000868
96	95.7	0.3	0.09	0.000940
88	88.3	-0.3	0.09	0.001019
			Total	$\doteqdot 0.00363$

So, $\chi^2_{calc} \doteqdot 0.00363$

Since χ^2_{calc} is very small, there is a very close agreement between observed and expected values. This indicates that *regular exercise* and *gender* are independent factors.

Note: If observed and expected values differ considerably, the numerators of each fraction added are large and so χ^2_{calc} would be large.

The question now arises: In a problem like the one considered above, how large would χ^2_{calc} need to be in order for us to conclude that the two factors are not independent?

EXERCISE 18D.1

1 Find χ^2_{calc} for the following contingency tables:

a   Factor M

	M_1	M_2	
N_1	31	22	53
N_2	20	27	47
	51	49	100

b   Factor S

	S_1	S_2	
R_1	28	17	
R_2	52	41	

c   Factor A

	A_1	A_2	
B_1	24	11	
B_2	16	18	
B_3	25	12	

d   Factor T

	T_1	T_2	T_3	T_4	
D_1	31	22	21	16	
D_2	23	19	22	13	

2 Now use a calculator to check your answers to question **1**.

Note: For the *regular exercise - gender*
example a screen dump is:

The χ^2_{calc} answer above was $\doteqdot 0.00363$
which is not accurate to 3 significant
figures due to the rounding errors in the
expected values.

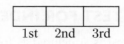

DEGREES OF FREEDOM

The χ^2 distribution is dependent on the number of **degrees of freedom (df)** where

$$\textbf{df} = (r - 1)(c - 1) \quad \text{for a contingency table which is } r \times c \text{ in size.}$$

Our original *regular exercise* and *gender* contingency table is 2×2.
So, df $= (2 - 1) \times (2 - 1) = 1$.
For a 4×3 contingency table, df $= (4 - 1)(3 - 1) = 6$.

A 'rule of thumb' explanation of degrees of freedom

Consider placing the numbers 5, 6 and 8 into the table.

For the first position any one of the three numbers
could be used, i.e., we have freedom to choose.

1st	2nd	3rd

For the second position we have freedom to choose from the remaining two numbers. However
for the remaining position there is no freedom of choice as the remaining number must go
into the third position. So we have 2 degrees of freedom (of choice), which is $3 - 1$.

Now see if you can show that in placing   4 6 5   into   there are
                                          1 2 8
$(3 - 1) \times (3 - 1)$ degrees of freedom.   3 9 7

3 Find the degrees of freedom (df) for the contingency tables of question **1**.

4 Find the df of:

Factor K

	K_1	K_2	K_3	K_4	K_5	K_6	K_7
L_1	2	5	7	3	1	4	9
L_2	6	1	3	8	2	1	7
L_3	4	2	2	5	1	6	5
L_4	3	4	2	4	3	2	4

TABLE OF CRITICAL VALUES

The χ^2 distribution graph is:

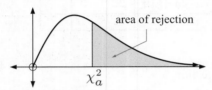

area of rejection

χ_a^2

The values 0.10, 0.05, 0.01, i.e., 10%, 5%, 1% are called **significance levels** and these are the ones which are commonly used. Tables of χ^2 values exist for up to 100 degrees of freedom and for other significance levels than those given alongside.

Degrees of freedom (df)	Area right of table value		
	0.10	0.05	0.01
1	2.71	3.84	6.63
2	4.61	5.99	9.21
3	6.25	7.81	11.34
4	7.78	9.49	13.28
5	9.24	11.07	15.09
6	10.64	12.59	16.81
7	12.02	14.07	18.48
8	13.36	15.51	20.09
9	14.68	16.92	21.67
10	15.99	18.31	23.21

Notice that, at a 5% significance level, with df $= 1$, $\chi_{0.05}^2 = 3.84$.

This means that at a 5% significance level, the departure between observed and expected is too great if $\chi_{calc}^2 > 3.84$.

Likewise, at a 1% significance level, with df $= 7$, the departure between observed and expected is too great if $\chi_{calc}^2 > 18.48$.   {also shaded on the table}

Note: For large values of n, χ^2 has an approximate chi-squared distribution with $(r-1)(c-1)$ degrees of freedom.

Generally, n is sufficiently large if all the expected values are 5 or more.

FORMAL TEST FOR INDEPENDENCE

The formal test is structured as follows:

Step 1:   We state H_0 called the **null hypothesis**. This is a statement that the two classifications being considered are independent.

We state H_1 called the **alternative hypothesis**. This is a statement that the two classifications being considered are not independent

Step 2:   Calculate df according to   df $= (r-1)(c-1)$.

Step 3:   We quote the **significance level** required,   i.e.,   10%, 5% or 1%.

Step 4:   We state the **rejection inequality** $\chi^2_{calc} > k$ where k is obtained from the **table of critical values**.

Step 5:   From the contingency table, find χ^2_{calc} using $\chi^2_{calc} = \sum \dfrac{(f_o - f_e)^2}{f_e}$.

Step 6:   We either accept H_0 or reject H_0, depending on the rejection inequality result.

Step 7:   If operating at a 5% level, we could also use p-**values** to help us with our decision making.   If $p > 0.05$,   we accept H_0.

If $p < 0.05$,   we reject H_0.

Returning to our original *regular exercise/gender* example:

Step 1: H_0 is *regular exercise* and *gender* are independent.

H_1 is *regular exercise* and *gender* are not independent.

Step 2:   df $= (2 - 1)(2 - 1) = 1$

Step 3:   Significance level is 5%.

Step 4:   We reject H_0 if $\chi^2_{calc} > 3.84$.

Step 5: $\chi^2_{calc} \doteqdot 0.00413$

Step 6:   As $\chi^2_{calc} < 3.84$, we accept H_0 in favour of H_1,

i.e., that *regular exercise* and *gender* are independent classifications.

Step 7: $p \doteqdot 0.949$ which is > 0.05, providing further evidence to accept H_0.

```
χ²-Test
 χ²=.00413
 p=.94877
 df=1.00000
```

Example 6

A survey was given to randomly chosen high school students from years 9 to 12 on possible changes to the school's canteen. The contingency table shows the results.

At a 5% level, test whether there is a significant difference between the proportion of students wanting a change in the canteen across the four year groups.

	Year group			
	9	10	11	12
change	7	9	13	14
no change	14	12	9	7

H_0 is *year group* and *change* are independent   (no significant departure).

H_1 is *year group* and *change* are not independent.

df $= (4 - 1)(2 - 1) = 3$ The significance level is 5% or 0.05 .

We reject H_0 if $\chi^2_{calc} > 7.81$.   {from critical values table}

The 2×4 contingency table is:

	Year group				
	9	10	11	12	sum
C	7	9	13	14	43
C′	14	12	9	7	42
sum	21	21	22	21	85

The expected frequency table is:

	Year group			
	9	10	11	12
C	10.6	10.6	11.1	10.6
C′	10.4	10.4	10.9	10.4

f_o	f_e	$f_o - f_e$	$(f_o - f_e)^2$	$\dfrac{(f_o - f_e)^2}{f_e}$
7	10.6	−3.6	12.96	1.223
9	10.6	−1.6	2.56	0.242
13	11.1	1.9	3.61	0.325
14	10.6	3.4	11.56	1.091
14	10.4	3.6	12.96	1.246
12	10.4	1.6	2.56	0.246
9	10.9	−1.9	3.61	0.331
7	10.4	−3.4	11.56	1.112
			Total	5.816

$\therefore \quad \chi^2_{calc} \doteqdot 5.82$

which is not > 7.81

Consequently, we accept H_0, that there is no significant difference between the proportions across the year groups.

Note: Using a calculator we obtain:

Notice the small error in the above table due to rounding.

```
X²-Test
X²=5.812
P=.121
df=3.000
```

EXERCISE 18D.2

1   A random sample of people is taken to find if there is a relationship between *smoking marijuana as a teenager* and *suffering schizophrenia within the next 15 years*. The results are given in the table below:

	Schizophrenic	Non-Schizophrenic
Smoker	58	73
Non-smoker	269	624

Test at a 5% level whether there is a relationship between *smoking marijuana as a teenager* and *suffering schizophrenia within the next 15 years*.

2   Examine the following contingency tables for the independence of classifications P and Q. Use a χ^2 test **i** at a 5% level of significance **ii** at a 10% level of significance.

a

	Q_1	Q_2
P_1	11	17
P_2	21	23
P_3	28	19
P_4	17	28

b

	Q_1	Q_2	Q_3	Q_4
P_1	6	11	14	18
P_2	9	12	21	17
P_3	13	24	16	10

3   The table shows the way in which a randomly chosen group intend to vote in the next election.

Test at a 5% level whether there is any association between the *age of a voter* and *the party they wish to vote for*.

	Age of voter		
	18 to 35	36 to 59	60+
Party A	85	95	131
Party B	168	197	173

4 The following table shows the results of a random sample where *annual income* and *cigarette smoking* are being compared.

	Income level			
	low	*average*	*high*	*very high*
Smoker	82	167	74	31
Non-smoker	212	668	428	168

Test at a 10% level whether lower income people are more likely to be cigarette smokers.

5 This contingency table shows the responses of a randomly chosen sample of 50+ year olds to a survey dealing with peoples *weight* and whether they *have diabetes*.

	Weight			
	light	*medium*	*heavy*	*obese*
Diabetic	11	19	21	38
Non-diabetic	79	68	74	53

Test at a 1% level whether there is a link between *weight* and *suffering diabetes*.

6 The following table is a result of a major investigation considering the two factors of *intelligence level (IQ)* and *cigarette smoking*.

	Intelligence level			
	low	*average*	*high*	*very high*
Non smoker	283	486	226	38
Medium level smoker	123	201	58	18
Heavy smoker	100	147	64	8

Test at a 1% level whether there is a link between *intelligence level (IQ)* and *cigarette smoking*.

REVIEW SET 18A

1 Thomas rode for an hour each day for eleven days and recorded the number of kilometres ridden against the temperature that day.

Temp *t* (°C)	32.9	33.9	35.2	37.1	38.9	30.3	32.5	31.7	35.7	36.3	34.7
d km ridden	26.5	26.7	24.4	19.8	18.5	32.6	28.7	29.4	23.8	21.2	29.7

  a Using technology, construct a scatterplot of the data.
  b Find and interpret Pearson's correlation coefficient.
  c Calculate the equation of the least squares line. How hot must it get before Thomas does not ride at all?

2 The contingency table below shows the results of motor vehicle accidents in relation to whether or not the traveller was wearing a seat belt.

	Serious injury	*Permanent disablement*	*Death*
Wearing a belt	189	104	58
Not wearing a belt	83	67	46

Find χ^2 and test at a 1% level that the wearing of a seat belt and injury or death are independent factors.

3 A drinks vendor varies the price of Supa-fizz on a daily basis, and records the number of sales of the drink (shown below).

Price (p)	$2.50	$1.90	$1.60	$2.10	$2.20	$1.40	$1.70	$1.85
Sales (s)	389	450	448	386	381	458	597	431

a Produce a scatterplot of the data. Do there appear to be any outliers? If so, should they be included in the analysis?

b Calculate the least squares regression line. Could it give an accurate prediction of sales if Super-fizz was priced at 50 cents?

4 Eight identical flower beds (petunias) were watered a varying number of times each week, and the number of flowers each bed produced is recorded in the table below:

Number of waterings (n)	0	1	2	3	4	5	6	7
Flowers produced (f)	18	52	86	123	158	191	228	250

a Which is the independent variable?

b Calculate the equation of the least squares line.

c On a scatterplot of the data, plot the least squares line.

d Violet has two beds of petunias. One she waters five times a fortnight ($2\frac{1}{2}$ times a week), the other ten times a week.

   i How many flowers can she expect from each bed?

   ii Which is the more reliable estimate?

5 Examine the following contingency tables for the independence of classifications P and Q.

Use a χ^2 test **a** at a 5% level of significance

             b at a 1% level of significance.

	Q_1	Q_2	Q_3	Q_4
P_1	19	23	27	39
P_2	11	20	27	35
P_3	26	39	21	30

REVIEW SET 18B

1 The following table gives peptic ulcer rates per 100 of population for differing family incomes in the year 1998.

Income (I thousand $)	10	15	20	25	30	40	50	60	80
Peptic ulcer rate R	8.3	7.7	6.9	7.3	5.9	4.7	3.6	2.6	1.2

a Define the role of each variable and produce an appropriate graph.

b Use the method of least squares determination to find the equation of the line of best fit.

c Give an interpretation of the slope and y-intercept of the line.

d Use the equation of the least squares line to predict the peptic ulcer rate for families with $45 000 incomes.

e What is the x-intercept of this line? Do you think predictions for incomes greater than this will be accurate?

  f Later it is realised that one of the figures was written incorrectly.

    i Which is it likely to be? Why?

    ii Repeat **b** and **d** having removed the incorrect data.

2 The table shows the responses to a survey as to whether the city speed limit should be increased.

Test at a 5% level whether there is any association between the *age of a driver* and *increasing the speed limit*.

	Age of driver		
	18 to 30	31 to 54	55+
Increase	234	169	134
No-increase	156	191	233

3 The following table is a result of a major investigation considering the two factors of *intelligence level* and *business success*

	Intelligence level			
	low	*average*	*high*	*very high*
No success	35	30	41	25
Low success	28	41	26	29
Success	35	24	41	56
High success	52	38	63	72

Test at a 1% level whether there is a link between *intelligence level (IQ)* and *business success*.

4 Safety authorities advise drivers to travel 3 seconds behind the car in front of them as this provides the driver with a greater chance of avoiding a collision if the car in front has to brake quickly or is itself involved in an accident. A test was carried out to find out how long it would take a driver to bring a car to rest from the time a red light was flashed. (This is called *stopping time*, which includes reaction time and braking time.) The following results are for one driver in the same car under the same test conditions.

Speed (v km/h)	10	20	30	40	50	60	70	80	90
Stopping time (t secs)	1.23	1.54	1.88	2.20	2.52	2.83	3.15	3.45	3.83

  a Produce a scatterplot of the data and indicate its most likely model type.

  b Find the linear model which best fits the data. Give evidence as to why you have chosen this model.

  c Use the model to find the stopping time for a speed of:

    i 55 km/h   **ii** 110 km/h

  d What is the interpretation of the vertical intercept?

  e Why does this simple rule apply at all speeds, with a good safety margin?

5 Two supervillains, Silent Predator and the Furry Reaper terrorise Metropolis by abducting fair maidens (most of whom happen to be journalists). Superman believes that they are collaborating, alternatively abducting fair maidens so as not to compete with each other for ransom money. He plots their abduction rate below (in dozens of maidens).

Silent Predator (p)	4	6	5	9	3	5	8	11	3	7	7	4
Furry Reaper (r)	13	10	11	8	11	9	6	6	12	7	10	8

a Plot the data on a scatterplot, and find the least squares regression line (put Silent Predator on the x-axis).

b Is their any evidence for Superman's suspicions? (Calculate the r and r^2 and describe the strength of Silent Predator and Furry Reaper's relationship.)

c What is the estimated number of the Furry Reaper's abductions given that Silent Predator's were 6 dozen?

d Why is the model inappropriate when the Furry Reaper abducts more than 20 dozen maidens?

e Calculate the p- and r-intercepts. What do these values represent?

f If Superman is faced with a choice of capturing one supervillian but not the other, which should he choose? (**Hint:** Use **e**.)

Chapter 19

Introductory differential calculus

Contents:

RATE OF CHANGE

Often we judge sporting performances by using rates. For example:

- Sir Donald Bradman's test cricket batting rate at Test level was 99.94 *runs per innings*
- Matthew's ice hockey scoring rate was 2.29 *goals per game*
- Jodie's typing speed is 63 *words per minute* with an error rate of 2.3 *errors per page*.

We see that:

> A **rate** is a comparison between two quantities of different kinds.

FINDING RATES

Rates are invariably used to compare performance.

Example 1

Josef typed 213 words in 3 minutes and made 6 errors, whereas Marie typed 260 words in 5 minutes and made 7 errors. Compare their performances using rates.

Josef's typing rate $= \dfrac{213 \text{ words}}{3 \text{ minutes}}$ Marie's typing rate $= \dfrac{260 \text{ words}}{4 \text{ minutes}}$

$\qquad\qquad\quad = 71 \text{ words/min}$ $\qquad\qquad\qquad\quad = 65 \text{ words/minute}$

Josef's error rate $= \dfrac{6 \text{ errors}}{213 \text{ words}}$ Marie's error rate $= \dfrac{7 \text{ errors}}{260 \text{ words}}$

$\qquad\qquad\quad \doteq 0.0282 \text{ errors/word}$ $\qquad\qquad\quad\doteq 0.0269 \text{ errors/word}$

$\therefore$ Josef typed at a faster rate but Marie typed with greater accuracy.

One rate we use frequently is speed. **Speed** is the rate of distance travelled per unit of time and is usually measured in km/h or m/s.

EXERCISE 19A.1

1   Karsten's pulse rate was measured at 67 beats/minute.

   a   Explain exactly what this rate means.

   b   How many heart beats would Karsten expect to have each hour?

2   Jana typed a 14 page document and made eight errors. If an average page of typing has 380 words, find Jana's error rate in:

   a   errors/word   **b**   errors/100 words.

3   Niko worked 12 hours for $148.20 whereas Marita worked 13 hours for $157.95. Who worked for the better hourly rate of pay?

4 New tyres have a tread depth of 8 mm. After driving for 32 178 km the tread depth was reduced to 2.3 mm. What was the wearing rate of the tyres in:

    a mm per km travelled   **b** mm per 10 000 km travelled?

5 We left Adelaide at 11.43 am and travelled to Victor Harbor, a distance of 71 km. We arrived there are 12.39 pm. What was our average speed in:

    a km/h   **b** m/s?

AVERAGE RATE OF CHANGE

Consider a trip from Adelaide to Melbourne. The following table gives places along the way, distances travelled and time taken.

Place	Time taken (min)	Distance travelled (km)
Adelaide tollgate	0	0
Tailem Bend	63	98
Bordertown	157	237
Nhill	204	324
Horsham	261	431
Ararat	317	527
Midland H/W Junction	386	616
Melbourne	534	729

We plot the *distance travelled* against the *time taken* to obtain a graph of the situation. Even though there would be variable speed between each place we will join points with straight line segments.

We can find the average speed between any two places.

For example, the average speed from Bordertown to Nhill is:

$$\frac{\text{distance travelled}}{\text{time taken}}$$

$$= \frac{324 - 237 \text{ km}}{204 - 157 \text{ min}}$$

$$= \frac{87 \text{ km}}{\frac{47}{60} \text{ h}}$$

$$\doteq 111 \text{ km/h}$$

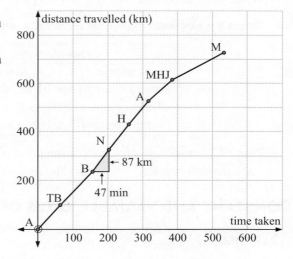

We notice that the average speed is the $\dfrac{y\text{-step}}{x\text{-step}}$ on the graph.

So, the average speed is the **gradient of the line segment** joining the two points which means that the faster the trip between two places, the greater the gradient of the graph.

If $s(t)$ is the distance travelled function then the average speed over the time interval from $t = t_1$ to $t = t_2$ is given by:

$$\text{Average speed} = \frac{s(t_2) - s(t_1)}{t_2 - t_1}.$$

EXERCISE 19A.2

1 For the Adelaide to Melbourne data, find the average speed from:

    a   Tailem Bend to Nhill   **b**   Horsham to Melbourne.

2 Paul walks to the newsagent to get a paper each morning. The travel graph alongside shows Paul's distance from home. Use the graph to answer the following questions.

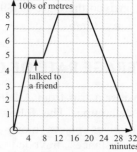

    a   How far is the newsagent from Paul's house?

    b   What is the gradient of the line segment for the first 4 minutes of the walk?

    c   What was Paul's average walking speed for the first 4 minutes (m/min)?

    d   What is the physical representation of the gradient in this problem?

    e   How many minutes did Paul stay at the newsagent's store?

    f   What was Paul's average speed on the return journey?

    g   What total distance did Paul walk?

3 During December Mount Bold reservoir was losing water at a constant rate due to usage and evaporation. On December 12 the estimated water in the reservoir was 53.8 million kL and on December 23 the estimate was 48.2 million kL. No water entered the reservoir. What was the average rate of water loss during this period?

4 Stefan's water consumption invoices for a one year period show:

Period	Consumption
First quarter (Jan 01 to Mar 31)	106.8 kL
Second quarter (Apr 01 to Jun 30)	79.4 kL
Third quarter (Jul 01 to Sep 30)	81.8 kL
Fourth quarter (Oct 01 to Dec 31)	115.8 kL

Find the average rate of water consumption per day for:

    a   the first quarter   **b**   the first six months   **c**   the whole year.

CONSTANT AND VARIABLE RATES OF CHANGE

Let us visit the water filling demonstration. Water flows from a tap at a constant rate into vessels of various shapes. We see the water level change over time. A corresponding *height* against *time* graph shows what is happening to the height as time increases. Click on the icon to investigate the rate of change in height over time for the different shaped vessels given.

DEMO

DISCUSSION

    • In which vessel was there a constant rate of change in height over time?

    • What is the nature of the graph when there is a constant rate of change?

    • What graphical features indicate   ◆ an increasing rate of change

                                        ◆ a decreasing rate of change?

AVERAGE RATES FROM CURVED GRAPHS

Consider the following example where we are to find average rates over a given time interval.

Example 2

The number of mice in a colony was recorded on a weekly basis.

a Estimate the average rate of increase in population for:
  i the period from week 3 to week 6
  ii the seven week period.

b What is the overall trend with regard to population increase over this period?

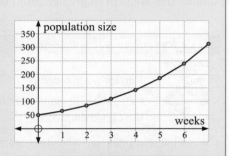

a **i** population rate

$$= \frac{\text{increase in population}}{\text{increase in time}}$$

$$= \frac{(240 - 110) \text{ mice}}{(6 - 3) \text{ weeks}}$$

$$\doteqdot 43 \quad \text{mice/week}$$

ii population rate

$$= \frac{(315 - 50) \text{ mice}}{(7 - 0) \text{ weeks}}$$

$$\doteqdot 38 \quad \text{mice/week}$$

b The population rate is increasing over the seven week period as shown by the increasing y-steps on the graph for equal x-steps.

You should notice that:

> The **average rate of change** between two points on the graph is the **gradient of the chord** (or **secant**) connecting these two points.

EXERCISE 19A.3

1 For the travel graph given alongside find estimates of the average speed:

a in the first 4 seconds
b in the last 4 seconds
c in the 8 second interval.

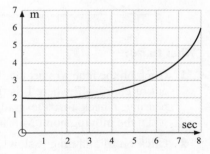

2 The number of lawn beetles per m^2 surviving in a lawn for various doses of poison is given in the graph alongside.

a Estimate the rate of beetle decrease when:
  i the dose increases from 0 to 10 g
  ii the dose increases from 4 to 8 g.

b Describe the effect on the rate of beetle decline as the dose goes from 0 to 14 g.

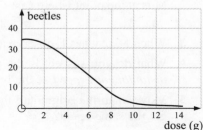

INSTANTANEOUS RATES OF CHANGE

A moving object such as a motor car, an aeroplane or a runner has variable speed. At a particular instant in time, the speed of the object is called its **instantaneous speed**. To examine this concept in greater detail consider the following investigation.

INVESTIGATION 1   INSTANTANEOUS SPEED

Earlier we noticed that:

"The average rate of change between two points on a graph is the gradient of the chord connecting them."

But, what happens if these two points are extremely close together or in fact coincide?

To discover what will happen consider the following problem:

A ball bearing is dropped from the top of a tall building. The distance fallen after t seconds is recorded and the following graph of distance against time is obtained.

The question is,   "What is the speed of the ball bearing at $t = 2$ seconds?

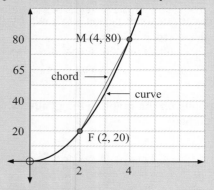

Notice that the average speed in the time interval $2 \leqslant t \leqslant 4$ is

$$= \frac{\text{distance travelled}}{\text{time taken}}$$

$$= \frac{(80 - 20) \text{ m}}{(4 - 2) \text{ s}}$$

$$= \frac{60}{2} \text{ m/s}$$

$$= 30 \text{ m/s}$$

DEMO

What to do:

1 Click on the icon to start the demonstration. The gradient box gives the starting gradient of chord MF where F is the fixed point at which we require the speed of the ball bearing. Check that the gradient of MF is that which is given in the gradient box when the moving point M is at (4, 80).

2 Click on M and drag it slowly towards F. Write down the gradient of the chord when M is at the point where t is:

   a   3   b   2.5   c   2.1   d   2.01

3 When M reaches F observe and record what happens. Why is this so?

4 What do you suspect is the speed of the ball bearing at $t = 2$? This speed is the *instantaneous speed* of the ball bearing at this instant.

5 Move M to the origin and then move it towards F from the other direction. Do you get the same result?

From the investigation you should have discovered that:

> The **instantaneous rate of change** (speed in this case) at a particular point is given by the **gradient of the tangent** to the graph at that point.

VARIABLE RATES OF CHANGE

THE GRAPHICAL METHOD

Consider a cyclist who is stationary at an intersection. The graph alongside shows how the cyclist accelerates away from the intersection.

Notice that the average speed over the first 8 seconds is $\dfrac{100 \text{ m}}{8 \text{ sec}} = 12.5$ m/s.

Notice also that the cyclist's early speed is quite small, but is increasing as time goes by.

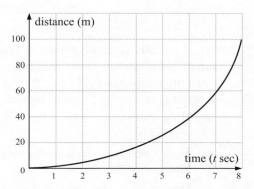

To find the instantaneous speed at any time instant, for example, $t = 4$ we simply draw the tangent at that point and find its gradient.

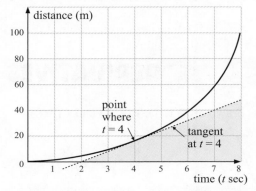

Notice that the tangent passes through $(2, 0)$ and $(7, 40)$

$\therefore$ instantaneous speed at $t = 4$

$= $ gradient of tangent

$= \dfrac{(40 - 0) \text{ m}}{(7 - 2) \text{ s}}$

$= \frac{40}{5}$ m/s

$= 8$ m/s

At any point in time we can use this method to find the speed of the cyclist at that instant.

EXERCISE 19A.4

1   For each of the following graphs find the approximate rate of change at the point shown by the arrow. Make sure your answer contains the correct units.

a

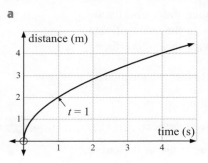

b

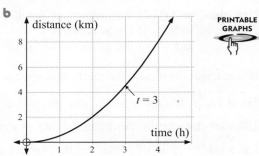

c

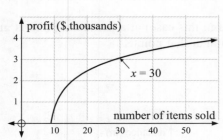

d

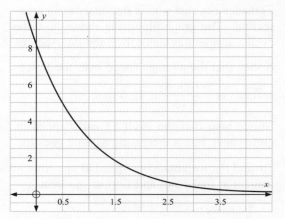

2 Water is leaking from a tank. The amount of water left in the tank (measured in thousands of litres) after x hours is given in the graph alongside.

   a How much water was in the tank originally?

   b How much water was in the tank after 1 hour?

   c How quickly was the tank losing water initially?

   d How quickly was the tank losing water after 1 hour?

B | DERIVATIVES

Following are graphs of $y = x^2$ with tangents drawn at A(1, 1), B(2, 4) and C(3, 9).

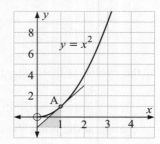
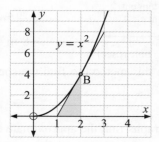
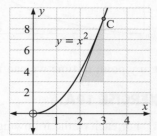

Note that at A(1, 1) the gradient is 2, which is twice the x-coordinate

             at B(2, 4) the gradient is 4, which is twice the x-coordinate

             at C(3, 9) the gradient is 6, which is twice the x-coordinate.

If this trend continues, the gradient is always double the x-coordinate.

So, if (x, y) lies on $y = x^2$, the gradient at this point is $m = 2x$.

Click on the icon to check the validity of the above statement.

Click on the point of contact of the tangent and drag it along the curve.

DEMO

Notation: If y is given in terms of x, the gradient of the tangent at any point is represented

by $\dfrac{dy}{dx}$.   So, for example, if $y = x^2$ then $\dfrac{dy}{dx} = 2x$.

We say '*dee y by dee x*' when we write $\dfrac{dy}{dx}$.

DISCUSSION

What is $\dfrac{dy}{dx}$ for $y = x$, $y = 3x$, $y = -2x$, $y = mx$, $y = mx + c$?

THE DERIVATIVE OF x^n

From the discussion you should have deduced that:

- if $y = x$, then $\dfrac{dy}{dx} = 1$ •   if $y = mx + c$, then $\dfrac{dy}{dx} = m$.

From the introduction, we concluded that: if $y = x^2$, then $\dfrac{dy}{dx} = 2x$.

What about $y = x^3$, $y = x^4$, $y = x^0$, $y = x^{-1}$, $y = x^{-2}$?

What is $\dfrac{dy}{dx}$ in each case, and is there a general rule for finding it?

All of these simple functions have form $y = x^n$.

Can we find $\dfrac{dy}{dx}$ for the general function $y = x^n$?

GRAPHS OF $y = x^n$

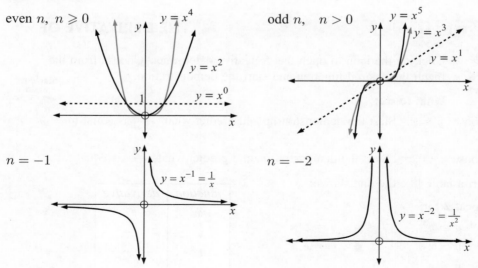

Click on the icon to the graphing package to graph families of power functions (of form $y = x^n$).

GRAPHING
PACKAGE

INVESTIGATION 2   THE DERIVATIVE OF x^3

PRINTABLE GRAPHS

Click on the icon to print the graph of $y = x^3$ on grid paper.

What to do:

1 Carefully draw tangents to the curve at $x = 0, \pm 1, \pm 2, \pm 3$.

2 Use the grid paper to find the approximate gradients of these tangents.

3 Copy and complete:

x	-3	-2	-1	0	1	2	3
$\dfrac{dy}{dx}$							

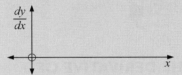

4 Plot the points from the table on axes as shown.

5 What do you observe from the graph?

From the investigation you probably guessed that if

$$y = x^3 \quad \text{then} \quad \frac{dy}{dx} = kx^2 \quad \text{for some constant } k.$$

In the following investigation we will attempt to predict the gradient curve for functions of the form $y = x^n$. The family of curves on previous page may prove useful to help with identification.

DEMO

You will notice that the tangent is the hypotenuse of a right angled triangle and the x-step is 1 unit.

Consequently, the gradient is the y-step and this y-step is translated to the graph below it to help to form the gradient function.

tangent m

1 unit

INVESTIGATION 3   THE DERIVATIVE OF x^n

Click on the icon to open the derivative determiner. Choose from the menu the desired function and start the demonstration.

DERIVATIVE DEMO

What to do:

1 Choose $y = x$. Start the demonstration and predict what $\dfrac{dy}{dx}$ is equal to.

2 Choose $y = x^2$. Start the demonstration and predict what $\dfrac{dy}{dx}$ is equal to.

3 Start a table like the one shown:

4 Repeat **2** for:

  a $y = x^3$ **b** $y = x^4$

  c $y = x^5$ **d** $y = x^{-1}$

  e $y = x^{-2}$

function	derivative
x	
x^2	
x^3	kx^2
$\vdots$	

5 Predict the form of $\dfrac{dy}{dx}$ when $y = x^n$.

You should have discovered from the previous investigation that:

if $y = x^n$ then $\dfrac{dy}{dx} = kx^{n-1}$ for some constant k.

The question now arises: "How do we find what k is equal to?"

To answer this question we will consider the principles of differential calculus.

C ■ THE IDEA OF A LIMIT

We now investigate the slopes of chords (secants) from a fixed point on a curve over successively smaller intervals.

Note:
  • A chord (secant) of a curve is a straight line segment which joins any two points on the curve.

  • A tangent is a straight line which touches a curve at a point.

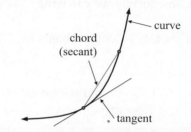

INVESTIGATION 4   THE GRADIENT OF A TANGENT

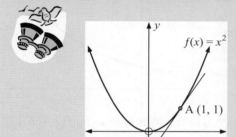

Given a curve, how can we find the gradient of a tangent at any point on it?

For example, the point A(1, 1) lies on the curve $y = x^2$. What is the gradient of the tangent at A?

DEMO

What to do:

1

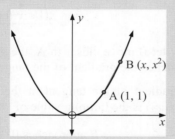

Suppose B lies on $f(x) = x^2$ and B has coordinates (x, x^2).

  a Show that the chord AB has gradient
  $$\frac{f(x) - f(1)}{x - 1} \quad \text{or} \quad \frac{x^2 - 1}{x - 1}.$$

x	Point B	gradient of AB
5	(5, 25)	6
2		
1.5		
1.1		
1.01		
1.001		

  b Copy and complete:

2 Comment on the gradient of AB as x gets closer to 1.

3 Repeat the process as x gets closer to 1, but from the left of A.

4 Click on the icon to view a demonstration of the process.

5 What do you suspect is the gradient of the tangent at A?

The previous investigation shows us that as x approaches 1, the gradient of the chord approaches the gradient of the tangent at $x = 1$.

Notation: We use a horizontal arrow, $\rightarrow$, to represent the word '*approaches*' or the phrase '*tends towards*'.

So, $x \rightarrow 1$ is read as 'x approaches 1'   or   'x tends to 1'.

In the investigation we noticed that the gradient of AB approached a limiting value of 2 as x approached 1, from either side of 1.

Consequently we can write,   as $x \rightarrow 1$, $\dfrac{x^2 - 1}{x - 1} \rightarrow 2$.

This idea is written simply as $\displaystyle \lim_{x \to 1}$ $\dfrac{x^2 - 1}{x - 1}$ $=$ 2

and is read as: the limit as x of $\dfrac{x^2 - 1}{x - 1}$ is 2
approaches 1

D   THE DERIVATIVE FUNCTION

For a non-linear function with equation $y = f(x)$, slopes of tangents at various points continually change.

Our task is to determine a **gradient function** so that when we replace x by a, say, we will be able to find the slope of the tangent at $x = a$.

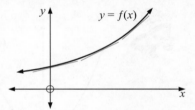

Consider a general function $y = f(x)$ where A is $(x,\ f(x))$ and B is $(x + h,\ f(x + h))$.

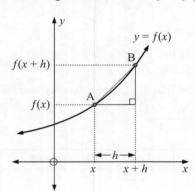

The chord AB has gradient $= \dfrac{f(x + h) - f(x)}{x + h - x}$

$= \dfrac{f(x + h) - f(x)}{h}$.

If we now let B move closer to A, the gradient of AB approaches the gradient of the tangent at A.

So, the gradient of the tangent at the variable point $(x,\ f(x))$ is the limiting value of

$\dfrac{f(x + h) - f(x)}{h}$ as h approaches 0.

Since this gradient contains the variable x it is called a **gradient function.**

DERIVATIVE FUNCTION

The **gradient function**, also known as the **derived function**, or **derivative function** or simply the **derivative** is defined as

$$f'(x) = \lim_{h \to 0} \frac{f(x + h) - f(x)}{h}.$$

[**Note:** $\displaystyle\lim_{h \to 0} \frac{f(x+h) - f(x)}{h}$ is the shorthand way of writing

"the limiting value of $\dfrac{f(x+h) - f(x)}{h}$ as h gets as close as we like to zero."]

Example 3

Find, from first principles, the gradient function of $f(x) = x^2$.

If $f(x) = x^2$, $\begin{aligned}[t] f'(x) &= \lim_{h \to 0} \frac{f(x+h) - f(x)}{h} \\[2mm] &= \lim_{h \to 0} \frac{(x+h)^2 - x^2}{h} \\[2mm] &= \lim_{h \to 0} \frac{x^2 + 2hx + h^2 - x^2}{h} \\[2mm] &= \lim_{h \to 0} \frac{{}^1\!\!\!\not h(2x + h)}{{}_1\!\not h} \qquad \{\text{as } h \neq 0\} \\[2mm] &= 2x \end{aligned}$

Example 4

Find, from first principles, $f'(x)$ if $f(x) = \dfrac{1}{x}$.

If $f(x) = \dfrac{1}{x}$, $\begin{aligned}[t] f'(x) &= \lim_{h \to 0} \frac{f(x+h) - f(x)}{h} \\[3mm] &= \lim_{h \to 0} \left[\frac{\dfrac{1}{x+h} - \dfrac{1}{x}}{h} \right] \times \frac{(x+h)x}{(x+h)x} \\[3mm] &= \lim_{h \to 0} \frac{x - (x+h)}{hx(x+h)} \\[3mm] &= \lim_{h \to 0} \frac{{}^{-1}\!\!\not h}{{}_1\!hx(x+h)} \qquad \{\text{as } h \neq 0\} \\[3mm] &= -\frac{1}{x^2} \qquad \{\text{as } h \to 0, \ x + h \to x\} \end{aligned}$

EXERCISE 19D

1 Find, from first principles, the gradient function of $f(x)$ where $f(x)$ is:

   a x **b**   5   **c** x^3 **d** x^4

[**Note:**   you will need to use $(a+b)^3 = a^3 + 3a^2 b + 3ab^2 + b^3$
$(a+b)^4 = a^4 + 4a^3 b + 6a^2 b^2 + 4ab^3 + b^4$]

2 Find, from first principles, $f'(x)$ given that $f(x)$ is:

    a $2x + 5$ **b** $x^2 - 3x$ **c** $2x^2 + 5x + 4$

3 Find, from first principles, the derivative of $f(x)$ when $f(x)$ is:

    a $\dfrac{1}{x^2}$ **b** $\dfrac{3}{x^3}$

4 Using the results of derivatives in this exercise, copy and complete:

Function	Derivative (in form kx^n)
x	
x^2	$2x = 2x^1$
x^3	
x^4	
x^{-1}	
x^{-2}	
x^{-3}	

Use your table to predict a formula for $f'(x)$ given that $f(x) = x^n$ where n is rational.

E   SIMPLE RULES OF DIFFERENTIATION

Differentiation is the process of finding the derivative (i.e., gradient function).

Notation: If we are given a function $f(x)$ then $f'(x)$ represents the derivative function. However, if we are given y in terms of x then y' or $\dfrac{dy}{dx}$ are commonly used to represent the derivative.

Note:
- $\dfrac{dy}{dx}$ reads "dee y by dee x", or " the derivative of y with respect to x".

- $\dfrac{dy}{dx}$ is **not a fraction**.

- $\dfrac{d(.....)}{dx}$ reads "the derivative of (....) with respect to x.

From question **4** of the previous exercise you should have discovered that if $f(x) = x^n$ then $f'(x) = nx^{n-1}$.

Are there other rules like this one which can be used to differentiate more complicated functions without having to resort to the tedious limit method? In the following investigation we may discover some additional rules.

INVESTIGATION 5   SIMPLE RULES OF DIFFERENTIATION

In this investigation we attempt to differentiate functions of the form cx^n where c is a constant, and functions which are a sum (or difference) of terms of the form cx^n.

What to do:

1 Find, from first principles, the derivatives of: **a** $4x^2$ **b** $2x^3$

2 Compare your results with the derivatives of x^2 and x^3 obtained earlier.
Copy and complete: "If $f(x) = cx^n$, then $f'(x) = \ldots$."

3 Use first principles to find $f'(x)$ for:
a $f(x) = x^2 + 3x$ **b** $f(x) = x^3 - 2x^2$

4 Use **3** to copy and complete: "If $f(x) = u(x) + v(x)$ then $f'(x) = \ldots$."

You should have discovered the following rules for differentiating functions.

Rules

$f(x)$	$f'(x)$	Name of rule
c (a constant)	0	**differentiating a constant**
x^n	nx^{n-1}	**differentiating x^n**
$c\,u(x)$	$c\,u'(x)$	**constant times a function**
$u(x) + v(x)$	$u'(x) + v'(x)$	**sum rule**

Using the rules we have now developed we can differentiate sums of powers of x.

For example, if $f(x) = 3x^4 + 2x^3 - 5x^2 + 7x + 6$ then

$$f'(x) = 3(4x^3) + 2(3x^2) - 5(2x) + 7(1) + 0$$
$$= 12x^3 + 6x^2 - 10x + 7$$

Example 5

Find $f'(x)$ for $f(x)$ equal to: **a** $5x^3 + 6x^2 - 3x + 2$ **b** $7x - \dfrac{4}{x} + \dfrac{3}{x^3}$

c $\dfrac{x^2 + 4x - 5}{x}$

---

a $f(x) = 5x^3 + 6x^2 - 3x + 2$

$\therefore f'(x) = 5(3x^2) + 6(2x) - 3(1)$
$= 15x^2 + 12x - 3$

b $f(x) = 7x - \dfrac{4}{x} + \dfrac{3}{x^3}$

$= 7x - 4x^{-1} + 3x^{-3}$

$\therefore f'(x) = 7(1) - 4(-1x^{-2}) + 3(-3x^{-4})$
$= 7 + 4x^{-2} - 9x^{-4}$
$= 7 + \dfrac{4}{x^2} - \dfrac{9}{x^4}$

c $f(x) = \dfrac{x^2 + 4x - 5}{x}$

$= \dfrac{x^2}{x} + \dfrac{4x}{x} - \dfrac{5}{x}$

$= x + 4 - 5x^{-1}$

$\therefore f'(x) = 1 + 5x^{-2}$

EXERCISE 19E

1 Find $f'(x)$ given that $f(x)$ is:

    a x^3

    b $2x^3$

    c $7x^2$

    d $x^2 + x$

    e $4 - 2x^2$

    f $x^2 + 3x - 5$

    g $x^3 + 3x^2 + 4x - 1$

    h $5x^4 - 6x^2$

    i $3 - 6x^{-1}$

    j $\dfrac{2x - 3}{x^2}$

    k $\dfrac{x^3 + 5}{x}$

    l $\dfrac{x^3 + x - 3}{x}$

Example 6

Find the gradient function of $f(x) = x^2 - \dfrac{4}{x}$ and hence find the gradient of the tangent to the function at the point where $x = 2$.

$$f(x) = x^2 - \frac{4}{x} \qquad\qquad \therefore \quad f'(x) = 2x - 4(-1x^{-2})$$
$$= x^2 - 4x^{-1} \qquad\qquad\qquad = 2x + 4x^{-2}$$
$$\qquad\qquad\qquad\qquad\qquad = 2x + \frac{4}{x^2}$$

If $x = 2$ in the gradient function we will get the gradient of the tangent at $x = 2$.
So, as $f'(2) = 4 + 1 = 5$,   the tangent has gradient of 5.

---

2 Find the gradient of the tangent to:

    a $y = x^2$ at $x = 2$

    b $y = \dfrac{8}{x^2}$ at $x = 9$

    c $y = 2x^2 - 3x + 7$ at $x = -1$

    d $y = 2x - 5x^{-1}$ at $x = 2$

    e $y = \dfrac{x^2 - 4}{x^2}$ at $x = 4$

    f $y = \dfrac{x^3 - 4x - 8}{x^2}$ at $x = -1$

Example 7

If $y = 3x^2 - 4x$,   find $\dfrac{dy}{dx}$ and interpret its meaning.

As $y = 3x^2 - 4x$, $\dfrac{dy}{dx} = 6x - 4$.

$\dfrac{dy}{dx}$ is
- the gradient function or derivative of $y = 3x^2 - 4x$ from which the gradient at any point can be found
- the instantaneous rate of change in y as x changes.

---

3 **a** If $y = 4x - \dfrac{3}{x}$,   find $\dfrac{dy}{dx}$ and interpret its meaning.

b The position of a car moving along a straight road is given by $S = 2t^2 + 4t$ metres where t is the time in seconds. Find $\dfrac{dS}{dt}$ and interpret its meaning.

c The cost of producing and selling x toasters each week is given by $C = 1785 + 3x + 0.002x^2$ dollars. Find $\dfrac{dC}{dx}$ and interpret its meaning.

F   TANGENTS TO CURVES

TANGENTS

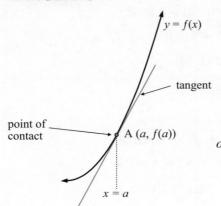

Consider a curve $y = f(x)$.

If A is the point with x-coordinate a, then the gradient of the tangent at this point is $f'(a)$.

The equation of the tangent is

$$\frac{y - f(a)}{x - a} = f'(a) \qquad \text{\{equating gradients\}}$$

$or \quad y - f(a) = f'(a)(x - a)$

Note: •   If a tangent touches $y = f(x)$ at $(a,\ b)$ then it has equation

$$\frac{y - b}{x - a} = f'(a) \quad or \quad y - b = f'(a)(x - a).$$

   •   Vertical and horizontal lines have equations $x = k$ and $y = c$ respectively.

Example 8

Find the equation of the tangent to $f(x) = x^2 + 1$ at the point where $x = 1$.

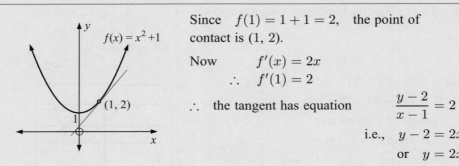

Since $f(1) = 1 + 1 = 2$, the point of contact is $(1, 2)$.

Now $\quad f'(x) = 2x$

$\therefore \quad f'(1) = 2$

$\therefore$ the tangent has equation $\dfrac{y - 2}{x - 1} = 2$

i.e., $\quad y - 2 = 2x - 2$

or $\quad y = 2x$

Example 9

Find the coordinates of any points on the curve with equation
$f(x) = x^3 + 3x^2 - 9x + 5$ where the tangent is horizontal.

$$f(x) = x^3 + 3x^2 - 9x + 5$$
$$f'(x) = 3x^2 + 6x - 9$$

Now tangents are horizontal when they have zero gradient   i.e.,   when $f'(x) = 0$.

i.e., $3x^2 + 6x - 9 = 0$

i.e., $x^2 + 2x - 3 = 0$

i.e., $(x + 3)(x - 1) = 0$ $\therefore$ $x = -3$ or 1

When $x = -3$, $f(-3) = (-3)^3 + 3(-3)^2 - 9(-3) + 5 = 32$

When $x = 1$, $f(1) = 1^3 + 3 \times 1^2 - 9 \times 1 + 5 = 0$

Tangents are horizontal at the points $(-3, 32)$ and $(1, 0)$.

EXERCISE 19F

1 Find the equation of the tangent to $y = f(x)$ for $f(x)$ equal to:

   a $x^3 + 2x$ at $x = 0$ **b** $2x^2 + 5x + 3$ at $x = -2$

   c $x + 2x^{-1}$ at $x = 2$ **d** $\dfrac{x^2 + 4}{x}$ at $x = -1$.

2 Find the coordinates of the point(s) on:

   a $f(x) = x^2 + 3x + 5$ where the tangent is horizontal

   b $f(x) = x^3 + x^2 - 1$ where the tangent has gradient 1

   c $f(x) = x^3 - 3x + 1$ where the tangent has gradient 9

   d $f(x) = ax^2 + bx + c$ where the tangent has zero gradient.

Example 10

Find the equations of any horizontal tangents to $y = x^3 - 12x + 2$.

Let $f(x) = x^3 - 12x + 2$ $\therefore$ $f'(x) = 3x^2 - 12$

But $f'(x) = 0$ for horizontal tangents and so $3x^2 - 12 = 0$

$\therefore$ $3(x^2 - 4) = 0$

$\therefore$ $3(x + 2)(x - 2) = 0$

$\therefore$ $x = -2$ or 2

Now $f(2) = 8 - 24 + 2 = -14$ and
$f(-2) = -8 + 24 + 2 = 18$

i.e.,   points of contact are:

$(2, -14)$ and $(-2, 18)$

$\therefore$ tangents are $y = -14$ and $y = 18$.

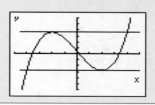

3 **a** Find the equations of the horizontal tangents to $y = 2x^3 + 3x^2 - 12x + 1$.

  b Find k if the tangent to $y = 2x^3 + kx^2 - 3$ at the point where $x = 2$ has gradient 4.

  c Find the equation of the tangents to $y = 1 - 3x + 12x^2 - 8x^3$ which are parallel to the tangent at $(1, 2)$.

4 Find a if:

  a $f(x) = ax^2 + 6x - 3$ has a tangent with gradient 2 at the point where $x = -1$

  b $f(x) = \dfrac{a}{x}$ has a tangent with gradient 3 at the point where $x = 2$.

5 The tangent to the curve $y = x^2 + ax + b$,   where a and b are constants, is $2x + y = 6$ at the point where $x = 1$.   Find the values of a and b.

G   THE SECOND DERIVATIVE

> The **second derivative** of a function $f(x)$ is the derivative of $f'(x)$,
> i.e., **the derivative of the first derivative**.
>
> **Notation:**   We use $f''(x)$,   or y'' or $\dfrac{d^2y}{dx^2}$ to represent the second derivative.

Note that: • $\dfrac{d^2y}{dx^2} = \dfrac{d}{dx}\left(\dfrac{dy}{dx}\right)$ • $\dfrac{d^2y}{dx^2}$ reads "*dee two y by dee x squared*".

THE SECOND DERIVATIVE IN CONTEXT

Michael rides up a hill and down the other side to his friend's house. The dots on the graph show Michael's position at various times t.

$t = 0$ $t = 5$ $t = 15$ $t = 17$ $t = 19$

Michael's place $t = 10$ friend's house

The distance travelled by Michael from his place is given at various times in the following table:

Time of ride (t min)	0	2.5	5	7.5	10	12.5	15	17	19
Distance travelled (s m)	0	498	782	908	989	1096	1350	1792	2500

A cubic model seems to fit this data well with coefficient of determination $r^2 = 0.9992$. The model is $s \doteq 1.18t^3 - 30.47t^2 + 284.52t - 16.08$ metres.

Notice that the model gives $s(0) = -16.08$ m   whereas the actual data gives $s(0) = 0$. This sort of problem often occurs when modelling from data.

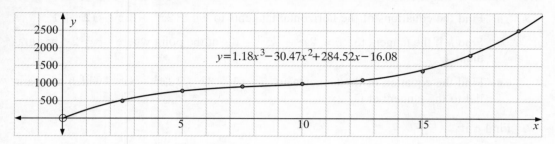

$y = 1.18x^3 - 30.47x^2 + 284.52x - 16.08$

Now $\dfrac{ds}{dt} \doteqdot 3.54t^2 - 60.94t + 284.52$ metres/minute is the instantaneous rate of change in displacement per unit of time, i.e., instantaneous velocity.

The instantaneous rate of change in velocity at any point in time is the acceleration of the moving object and so,

$$\dfrac{d}{dt}\left(\dfrac{ds}{dt}\right) = \dfrac{d^2 s}{dt^2} \quad \text{is the instantaneous acceleration,}$$

i.e., $\dfrac{d^2 s}{dt^2} = 7.08t - 60.94$ metres/minute per minute.

Notice that, when $t = 12$, $s \doteqdot 1050$ m, $\dfrac{ds}{dt} \doteqdot 63$ metres/minute

and $\dfrac{d^2 s}{dt^2} \doteqdot 24$ metres/minute/minute

Example 11

Find $f''(x)$ given that	Now $f(x) = x^3 - 3x^{-1}$
$f(x) = x^3 - \dfrac{3}{x}$.	$\therefore \quad f'(x) = 3x^2 + 3x^{-2}$
	and $f''(x) = 6x - 6x^{-3}$
	$= 6x - \dfrac{6}{x^3}$

EXERCISE 19G

1 Find $f''(x)$ given that:

    a $f(x) = 3x^2 - 6x + 2$ **b** $f(x) = 2x^3 - 3x^2 - x + 5$ **c** $f(x) = 2x^{-2} - 3x^{-1}$

2 Find $\dfrac{d^2 y}{dx^2}$ given that: **a** $y = x - x^3$ **b** $y = x^2 - 5x^{-2}$ **c** $y = \dfrac{4 - x}{x}$

H   CHANGING SHAPE

Recall that $f'(x)$ or $\dfrac{dy}{dx}$ is the **gradient function** of a curve.

The derivative of a function is another function which enables us to find the gradient of a tangent to the curve at any point on it.

For example, if $f(x) = x^2$ then

$$f'(x) = 2x$$

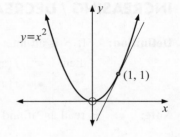

Substituting $x = \frac{1}{4}, \frac{1}{2}, 1$ and 4 gives:

$$f'\left(\tfrac{1}{4}\right) = \tfrac{1}{2}, \quad f'\left(\tfrac{1}{2}\right) = 1, \quad f'(1) = 2, \quad f'(4) = 8$$

i.e., the gradients are $\frac{1}{2}$, 1, 2 and 8 respectively.

Notice also that a tangent to the graph at any point, provided that $x > 0$, has a positive gradient.

This fact is also observed from $f'(x) = 2x$ as $2x$ is never negative for $x > 0$.

MONOTONICITY

Many functions are **increasing** *for all* x whereas others are **decreasing** *for all* x.

For example,

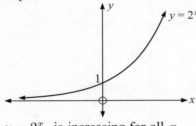

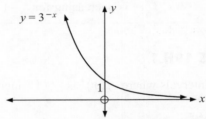

$y = 2^x$ is increasing for all x. $y = 3^{-x}$ is decreasing for all x.

Notice that:

- for an increasing function an increase in x produces an increase in y
- for a decreasing function an increase in x produces a decrease in y.

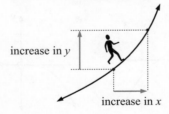

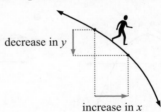

The majority of other functions have intervals where the function is increasing and intervals where it is decreasing.

For example:

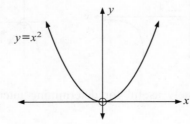

$y = x^2$ decreasing for $x \leqslant 0$ and increasing for $x \geqslant 0$.

Note: $x \leqslant 0$ is an interval of x values. So is $x \geqslant 0$.

INCREASING / DECREASING INTERVALS

Definition: If S is an interval of real numbers and $f(x)$ is defined for all x in S,
then:
- $f(x)$ is **increasing** on S $\Leftrightarrow$ $f'(x) > 0$ for all x in S,
- $f(x)$ is **decreasing** on S $\Leftrightarrow$ $f'(x) < 0$ for all x in S.

Note: $\Leftrightarrow$ is read as 'if and only if'

Example 12

Find intervals where $f(x)$ is:
a   increasing
b   decreasing.

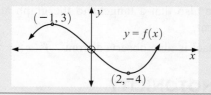

a $f(x)$ is increasing for $x < -1$ and for $x > 2$.
{since tangents have gradients $\geqslant 0$ on these intervals}

b $f(x)$ is decreasing for $-1 < x < 2$.

EXERCISE 19H.1

1 Find intervals where $f(x)$ is   **i** increasing   **ii** decreasing:

a
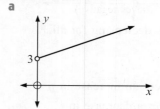

b
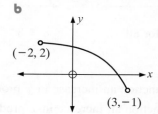

c
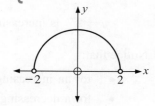

d
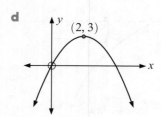

e
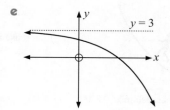

f
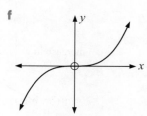

g
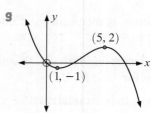

h
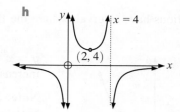

i
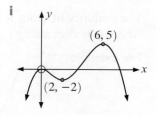

Sign diagrams for the derivative are extremely useful for determining intervals where a function is increasing/decreasing.

Consider the following examples:

- $f(x) = x^2$

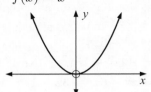

$f'(x) = 2x$

which has sign diagram

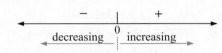

So $f(x) = x^2$ is   decreasing for $x < 0$
and   increasing for $x > 0$.

- $f(x) = -x^2$

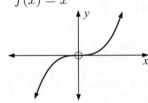

$f'(x) = -2x$

which has sign diagram

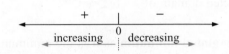

- $f(x) = x^3$

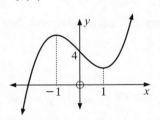

$f'(x) = 3x^2$

which has sign diagram

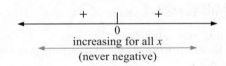

- $f(x) = x^3 - 3x + 4$

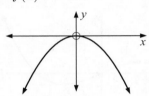

$f'(x) \begin{aligned} &= 3x^2 - 3 \\ &= 3(x^2 - 1) \\ &= 3(x+1)(x-1) \end{aligned}$

which has sign diagram

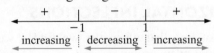

Example 13

Find the intervals where the following functions are increasing/deceasing:
$f(x) = -x^3 + 3x^2 + 5$

$$f(x) = -x^3 + 3x^2 + 5$$
$$\therefore \quad f'(x) = -3x^2 + 6x$$
$$\therefore \quad f'(x) = -3x(x - 2)$$

which has sign diagram

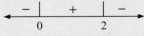

Y1=-X^3+3X²+5

X=2   Y=9

So, $f(x)$ is decreasing for $x < 0$ and for $x > 2$ and is increasing for $0 < x < 2$.

EXERCISE 19H.2

1 Find intervals where $f(x)$ is increasing/decreasing:

    a $f(x) = x^2$ **b** $f(x) = -x^3$

    c $f(x) = 2x^2 + 3x - 4$ **d** $f(x) = x^3 - 6x^2$

    e $f(x) = -2x^3 + 4x$ **f** $f(x) = -4x^3 + 15x^2 + 18x + 3$

    g $f(x) = 2x^3 + 9x^2 + 6x - 7$ **h** $f(x) = x^3 - 6x^2 + 3x - 1$

STATIONARY POINTS

Consider the following graph which has a restricted domain of $-5 \leqslant x \leqslant 6$.

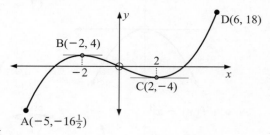

A is a **global minimum** as it is the minimum value of y and occurs at an endpoint of the domain.

B is a **local maximum** as it is a turning point where the curve has shape ⌢ and $f'(x) = 0$ at that point.

C is a **local minimum** as it is a turning point where the curve has shape ⌣ and $f'(x) = 0$ at that point.

D is a **global maximum** as it is the maximum value of y and occurs at the endpoint of the domain.

Note: For local maxima/minima, tangents at these points are **horizontal** and thus have a gradient of 0, i.e., $f'(x) = 0$.

HORIZONTAL INFLECTIONS

It is not true that whenever we find a value of x where $f'(x) = 0$ we have a local maximum or minimum.

For example, $f(x) = x^3$ has $f'(x) = 3x^2$
                and $f'(x) = 0$ when $x = 0$.

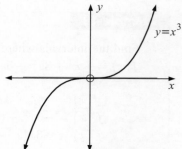

Notice that the x-axis is a tangent to the curve which actually crosses over the curve at O(0, 0).

This tangent is horizontal and O(0, 0) is not a local maximum or minimum.

It is called a **horizontal inflection** (or **inflexion**).

STATIONARY POINTS

A **stationary point** is a point where $f'(x) = 0$. It could be a local maximum, a local minimum or a horizontal inflection.

Consider the following graph:

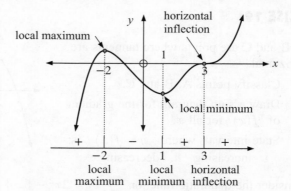

Its **gradient sign diagram** is:

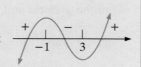

Summary:

Stationary point	Sign diagram of $f'(x)$ near $x = a$	Shape of curve near $x = a$
local maximum	$\overset{+}{\longleftarrow}\underset{a}{\mid}\overset{-}{\longrightarrow}$	$x = a$
local minimum	$\overset{-}{\longleftarrow}\underset{a}{\mid}\overset{+}{\longrightarrow}$	$x = a$
horizontal inflection	$\overset{+}{\longleftarrow}\underset{a}{\mid}\overset{+}{\longrightarrow}$ or $\overset{-}{\longleftarrow}\underset{a}{\mid}\overset{-}{\longrightarrow}$	or $x = a$ $\quad$ $x = a$

Example 14

Find and classify all stationary points of $f(x) = x^3 - 3x^2 - 9x + 5$.

$f(x) = x^3 - 3x^2 - 9x + 5$

$\therefore \quad f'(x) = 3x^2 - 6x - 9$

$\qquad = 3(x^2 - 2x - 3)$

$\qquad = 3(x - 3)(x + 1)$, which has sign diagram: $\overset{+}{\longleftarrow}\underset{-1}{\mid}\overset{-}{}\underset{3}{\mid}\overset{+}{\longrightarrow}$

So, we have a local maximum at $x = -1$ and a local minimum at $x = 3$.

$f(-1) = (-1)^3 - 3(-1)^2 - 9(-1) + 5 = 10$

$\qquad\qquad\qquad \therefore \quad$ local maximum at $(-1, 10)$

$f(3) = 3^3 - 3 \times 3^2 - 9 \times 3 + 5 = -22$

$\qquad\qquad\qquad \therefore \quad$ local minimum at $(3, -22)$

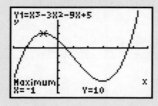

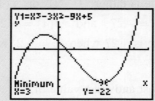

EXERCISE 19I

1   A, B and C are points where tangents are horizontal.

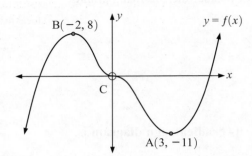

    a   Classify points A, B and C.

    b   Draw a sign diagram for the gradient of $f(x)$ for all x.

    c   State intervals where $y = f(x)$ is:

       i increasing **ii** decreasing.

2   Consider the quadratic function $f(x) = 2x^2 - 5x + 1$

    a   Use quadratic theory to find the equation of the axis of symmetry

    b   Find $f'(x)$ and hence find x when $f'(x) = 0$.   Explain!

3   For each of the following functions, find and classify the stationary points and hence sketch the function showing all important features.

    a $f(x) = x^2 - 2$ **b** $f(x) = x^3 + 1$

    c $f(x) = x^3 - 3x + 2$ **d** $f(x) = x^4 - 2x^2$

    e $f(x) = x^3 - 6x^2 + 12x + 1$ **f** $f(x) = 4x - x^3$

4   At what value of x does the quadratic function, $f(x) = ax^2 + bx + c$, $a \neq 0$, have a stationary point? Under what conditions is the stationary point a local maximum or a local minimum?

5 $f(x) = x^3 + ax + b$ has a stationary point at $(-2, 3)$.

    a   Find the values of a and b.

    b   Find the position and nature of all stationary points.

Example 15

Find the greatest and least value of $x^3 - 6x^2 + 5$ on the interval $-2 \leqslant x \leqslant 5$.

First we graph $y = x^3 - 6x^2 + 5$ on $[-2, 5]$.

The greatest value is clearly when $\dfrac{dy}{dx} = 0$

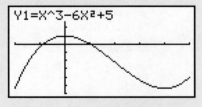

Now $\dfrac{dy}{dx} = 3x^2 - 12x$

$\phantom{Now \dfrac{dy}{dx}} = 3x(x - 4)$

$\phantom{Now \dfrac{dy}{dx}} = 0$ when $x = 0$ or 4

So, the greatest value is $f(0) = 5$ when $x = 0$.

The least value is either $f(-2)$ or $f(4)$, whichever is smaller.

Now $f(-2) = -27$ and $f(4) = -27$

$\therefore$ least value is -27 when $x = -2$ and $x = 4$.

---

6   Find the greatest and least value of:

    a $x^3 - 12x - 2$ for $-3 \leqslant x \leqslant 5$ **b** $4 - 3x^2 + x^3$ for $-2 \leqslant x \leqslant 3$

7 A manufacturing company makes door hinges. The cost function for making x hinges per hour is $C(x) = 0.0007x^3 - 0.1796x^2 + 14.663x + 160$ dollars where $50 \leqslant x \leqslant 150$.

The condition $50 \leqslant x \leqslant 150$ applies as the company has a standing order filled by producing 50 each hour, but knows that production of more than 150 an hour is useless as they will not sell.

Find the minimum and maximum hourly costs and the production levels when each occurs.

One application of differential calculus is the finding of equations of tangents to curves.

There are many other uses, but in this course we consider only:

- **tangents to curves**
- **curve properties**   (monotonicity and change of shape)
- **rates of change**
- **optimisation**   (maxima and minima, local and global)

J   RATES OF CHANGE

Earlier we discovered that:

$$\frac{dy}{dx} \text{ gives the \textbf{rate of change in } } y \text{ \textbf{with respect to} } x.$$

Note:
- if $\dfrac{dy}{dx} > 0$ then y increases as x increases.

- if $\dfrac{dy}{dx} < 0$ then y decreases as x increases.

Notice that:

- If a spherical balloon has volume V and radius r then $\dfrac{dV}{dr}$ is the rate of change in volume with respect to radius.

- If a cricket ball is thrown vertically upwards and its height above the ground is given by h then $\dfrac{dh}{dt}$ is the rate of change in height with respect to time t.

  This will be the velocity of the ball at any time t.

  If h is in metres and t in seconds then $\dfrac{dh}{dt}$ is in $\dfrac{\text{metres}}{\text{second}}$ i.e., m/s.

EXERCISE 19J

1 Find:

a $\dfrac{dM}{dt}$ if $M = t^3 - 3t^2 + 1$

b $\dfrac{dR}{dt}$ if $R = (2t + 1)^2$

c $\dfrac{dT}{dr}$ if $T = r^2 - \dfrac{100}{r}$

d $\dfrac{dA}{dh}$ if $A = 2\pi h + \frac{1}{4}h^2$

2 **a** If A is measured in cm^2 and t is measured in seconds, what are the units for $\dfrac{dA}{dt}$?

   b If V is measured in m^3 and t is measured in minutes, what are the units for $\dfrac{dV}{dt}$?

   c If C is measured in dollars and x is the number of items produced, what are the units for $\dfrac{dC}{dx}$?

Example 16

The volume of air V m^3 in a hot air balloon after t minutes is given by
$V = 2t^3 - 3t^2 + 10t + 2$ where t is the time in minutes, $0 \leqslant t \leqslant 8$. Find:

a the initial volume   **b** the volume at $t = 8$ **c** $\dfrac{dV}{dt}$

d the rate of increase in volume at $t = 4$ min

a when $t = 0$, $V(0) = 2$ m^3

b when $t = 8$, $V(8) = 2 \times 8^3 - 3 \times 8^2 + 10 \times 8 + 2$
    $= 914$ m^3

c $\dfrac{dV}{dt} = 6t^2 - 6t + 10$ m^3/min

d when $t = 4$, $\dfrac{dV}{dt} = 6 \times 4^2 - 6 \times 4 + 10$
      $= 82$ m^3/min

Since $\dfrac{dV}{dt} > 0$, V is increasing.

So, V is increasing at 82 m^3/min at $t = 4$.

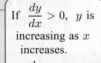

If $\dfrac{dy}{dx} > 0$, y is increasing as x increases.

If $\dfrac{dy}{dx} < 0$, y is decreasing as x increases.

---

3 Growth of bacteria in a dish is given by $B(t) = 0.3t^2 + 30t + 150$ where t is in days, and $0 \leqslant t \leqslant 10$.

   a Find $B'(t)$ and state its meaning.

   b Find $B'(3)$ and state its meaning.

   c Explain why $B(t)$ is increasing over the first 10 days.

4 The height of a palm tree, grown in ideal conditions, is given by $H = 20 - \dfrac{18}{t}$ metres

   for $t \geqslant 1$, where t is the number of years after the tree was planted from an established potted juvenile tree.

   a How high was the palm after 1 year?

   b Find the height of the palm at $t = 2$, $t = 3$, $t = 5$, $t = 10$, $t = 50$.

   c Find $\dfrac{dH}{dt}$. What are the units for $\dfrac{dH}{dt}$?

   d At what rate is the tree growing at $t = 1$, 3 and 10 years?

   e Explain why $\dfrac{dH}{dt} > 0$ for all $t \geqslant 1$. What does this mean in terms of the tree's growth?

5 The resistance to the flow of electricity in a certain metal is given by

$R = 20 + \frac{1}{10}T + \frac{1}{200}T^2$ where T is the temperature (in °C) of the metal.

  a Find the resistance R, at temperatures of 0°C, 20°C and 40°C.

  b Find the rate of change in the resistance at any temperature T.

  c For what values of T does the resistance increase as the temperature increases?

6 The total cost of running a train is given by $C(v) = 200v + \dfrac{10\,000}{v}$ dollars where v is the average speed of the train in km/h.

  a Find the total cost of running the train at: **i**   20 km/h   **ii**   40 km/h.

  b Show that $C'(v) = 200 - \dfrac{10\,000}{v^2}$ and state the units $C'(v)$ is measured in.

  c Find the rate of change in the cost of running the train at speeds of:
  i   10 km/h   **ii**   30 km/h.

7

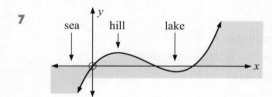

Alongside is a land and sea profile where the x-axis is sea level and y-values give the height of the land or sea bed above (or below) sea level and

$$y = \tfrac{1}{10}x(x-2)(x-3) \text{ km.}$$

  a Find where the lake is located relative to the shore line of the sea.

  b Show that $y = \tfrac{1}{10}(x^3 - 5x^2 + 6x)$ km.

  c Find $\dfrac{dy}{dx}$ and interpret its value when $x = \tfrac{1}{2}$ and when $x = 1\tfrac{1}{2}$ km.

COST MODELS

The cost of manufacturing x items has a **cost function** associated with it. Suppose this cost function is $C(x)$ dollars.

$\dfrac{dC}{dx}$ or $C'(x)$ is the **instantaneous rate of change in cost** with respect to the number of items made.   This is $\dfrac{dC}{dx}$ and is known by economists as the **marginal cost** function.

Thus the marginal cost is approximately the additional cost of producing one more item, i.e., $x + 1$ items instead of x of them.

Most often cost functions are polynomial models.

For example, the cost of producing x items per day may be given by

$$C(x) = \underline{0.000\,13x^3} + \underline{0.002x^2} + \underline{5x} + \underline{2200}$$

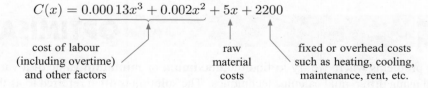

cost of labour	raw	fixed or overhead costs
(including overtime)	material	such as heating, cooling,
and other factors	costs	maintenance, rent, etc.

Example 17

For the cost model $C(x) = 0.000\,13x^3 + 0.002x^2 + 5x + 2200$:

a   find the marginal cost function

b   find the marginal cost when 150 are produced. Interpret this result.

a   The marginal cost function is $C'(x) = 0.000\,39x^2 + 0.004x + 5$

b $C'(150) = \$14.38$ and is the rate at which the costs are increasing with respect to the production level x. It gives an estimated cost for making the 151st shirt.

8 Seablue make jeans and the cost model for making x of them each day is
$$C(x) = 0.0003x^3 + 0.02x^2 + 4x + 2250 \text{ dollars.}$$

  a Find the marginal cost function $C'(x)$.

  b Find $C'(220)$.   What does it estimate?

9 The cost function for producing x items each day is
$$C(x) = 0.000\,072x^3 - 0.000\,61x^2 + 0.19x + 893 \text{ dollars.}$$

  a Find $C'(x)$ and explain what it represents.

  b Find $C'(300)$ and explain what it estimates.

  c Find the actual cost of producing the 301st item.

The **profit** $P(x)$ in making and selling x items is given by

$$P(x) = R(x) - C(x) \quad \text{where} \quad R(x) \text{ is the } \textbf{revenue function} \text{ and}$$
$$C(x) \text{ is the } \textbf{cost function}.$$

$\dfrac{dP}{dx}$ or $P'(x)$ is called the **marginal profit function**.

---

10 The profit made in selling x items is given by $P(x) = 5x - 2000 - \dfrac{x^2}{10\,000}$ dollars.

  a Graph $P(x)$ using technology and determine sales levels which product a profit.

  b Find $P'(x)$ and hence find x when the profit is increasing.

11 The cost of producing x items is given by $C(x) = 0.002x^3 + 0.04x^2 + 10x + 3000$ dollars.   If each item sells for $30, find:

  a the revenue function, $R(x)$ **b** the profit function, $P(x)$

  c the marginal profit function

  d the marginal profit when $x = 50$.   Explain the significance of this result.

K   OPTIMISATION

Many problems where we try to find the **maximum** or **minimum** value of a variable can be solved using differential calculus techniques. The solution is often referred to as the **optimum** solution.

Consider the following problem.

An industrial shed is to have a total floor space of 600 m² and is to be divided into 3 rectangular rooms of equal size. The walls, internal and external, will cost \$60 per metre to build.

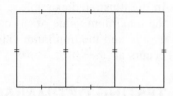

What dimensions should the shed have to minimise the cost of the walls? We let one room be x m by y m as shown.

The total length of wall material is L where $L = 6x + 4y$ metres.

However we do know that the total area is 600 m²,

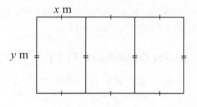

$$\therefore \quad 3x \times y = 600$$

and so $y = \dfrac{200}{x}$ **Note:** $x > 0$ and $y > 0$

Knowing this relationship enables us to write L in terms of one variable (x in this case),

$$\text{i.e.,} \quad L = 6x + 4\left(\frac{200}{x}\right) \text{ m}, \qquad \text{i.e.,} \quad L = \left(6x + \frac{800}{x}\right) \text{ m}$$

and at \$60/metre, the total cost is $C(x) = 60\left(6x + \dfrac{800}{x}\right)$ dollars.

When graphed we have:

Clearly, $C(x)$ is a minimum when $C'(x) = 0$.

Now $C(x) = 360x + 48\,000x^{-1}$

$\therefore \quad C'(x) = 360 - 48\,000x^{-2}$

$\therefore \quad C'(x) = 0$ when $360 = \dfrac{48\,000}{x^2}$

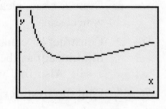

$$\text{i.e.,} \quad x^2 = \frac{48\,000}{360} \doteqdot 133.333$$

$$\text{i.e.,} \quad x \doteqdot 11.547$$

Now when $x \doteqdot 11.547$, $y \doteqdot \dfrac{200}{11.547} \doteqdot 17.321$

and the minimum cost is $C(11.547) \doteqdot 8313.84$ dollars.

So, the floor design is:

and the minimum cost is **\$8313.84**.

WARNING

The maximum/minimum value does not always occur when the first derivative is zero.

It is essential to also examine the values of the function at the end point(s) of the domain for global maxima/minima, i.e., given $a \leqslant x \leqslant b$, you should also consider $f(a)$ and $f(b)$.

Example:

In the illustrated example,
the maximum value of y occurs at
$x = b$ and the minimum value of y
occurs at $x = p$.

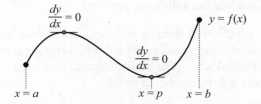

TESTING OPTIMAL SOLUTIONS

If one is trying to optimise a function $f(x)$ and we find values of x such that $f'(x) = 0$, how do we know whether we have a maximum or a minimum solution? The following are acceptable evidence.

SIGN DIAGRAM TEST

If near to $x = a$ where $f'(a) = 0$ the sign diagram is:

- $\dfrac{+\;\;\;-}{a}$ we have a **local maximum**
- $\dfrac{-\;\;\;+}{a}$ we have a **local minimum**.

GRAPHICAL TEST

If we have a graph of $y = f(x)$ showing ⌢ we have a **local maximum** and ⌣ we have a **local minimum**.

OPTIMISATION PROBLEM SOLVING METHOD

The following steps should be followed:

Step 1: Draw a large, clear diagram of the situation. Sometimes more than one diagram is necessary.

Step 2: Construct an equation with the variable to be **optimised** (**maximised** or **minimised**) as the subject of the formula in terms of **one** convenient **variable**, x say. Also find what restrictions there may be on x.

Step 3: Find the **first derivative** and find the value(s) of x when it is **zero**.

Step 4: If there is a restricted domain such as $a \leqslant x \leqslant b$, the maximum/minimum value of the function may occur either when the derivative is zero or at $x = a$ or at $x = b$. Show by the **sign diagram test** that you have a maximum or a minimum situation.

Example 18

Two numbers have a sum of 10. What is the maximum value of their product?

Let one number be x, therefore the other is $10 - x$,

and their product, $P = x(10 - x)$

i.e., $P = 10x - x^2$

$\therefore \quad \dfrac{dP}{dx} = 10 - 2x$ which has sign diagram:

$\therefore$ maximum value of P occurs when $x = 5$ and $P_{\max} = 5 \times 5 = 25$.

EXERCISE 19K

1 The cost of making x tennis racquets each day is given by

$C(x) = x^2 - 20x + 120$ dollars per racquet.

How many racquets should be made per day to minimise the cost per racquet?

2 When a stone is thrown vertically upwards its height above the ground is given by

$h(t) = 49t - 9.8t^2$ metres. Find the maximum height reached.

3 A small business which employs x workers earns a profit $P(x)$ dollars given by

$P(x) = -x^3 + 300x + 1000$. How many workers should be employed to maximise the profit?

Example 19

Find the most economical shape (minimum surface area) for a box with a square base, vertical sides and an open top, given that it must contain 4 litres.

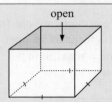

open

Step 1: 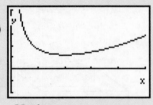 Let the base lengths be x cm and the depth be y cm. Now the volume

y cm

x cm

x cm

$$V = \text{length} \times \text{width} \times \text{depth}$$

$$\therefore \quad V = x^2 y$$

$$\therefore \quad 4000 = x^2 y \ \ .... \ (1) \quad \{\text{as 1 litre} \equiv 1000 \text{ cm}^3\}$$

Step 2: Now total surface area,

$$A = \text{area of base} + 4 \,(\text{area of one side})$$

$$\therefore \quad A(x) = x^2 + 4xy$$

$$\therefore \quad A(x) = x^2 + 4x \left(\frac{4000}{x^2}\right) \qquad \{\text{using (1)}\}$$

$$\therefore \quad A(x) = x^2 + 16\,000x^{-1}$$

Notice:
$x > 0$ as x is a length.

Step 3: Thus $A'(x) = 2x - 16\,000x^{-2}$

and $A'(x) = 0$ when $2x = \dfrac{16\,000}{x^2}$

$$\therefore \quad x = 20 \quad \{\text{using technology}\}$$

Step 4: **Sign diagram test**

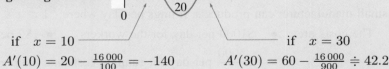

if $x = 10$

$A'(10) = 20 - \frac{16\,000}{100} = -140$

if $x = 30$

$A'(30) = 60 - \frac{16\,000}{900} \doteqdot 42.2$

This test establishes that minimum material is used to make the container when

$x = 20$, and $y = \dfrac{4000}{20^2} = 10$,

i.e.,

10 cm

20 cm

20 cm

is the shape.

4 An open rectangular box has square base and a fixed outer
surface area of 108 cm^2.

a Show that $x^2 + 4xy = 108$.

b From **a**, show that $y = \dfrac{108 - x^2}{4x}$.

c Find a formula for the capacity C, of the container in terms of x only.

d Find $\dfrac{dC}{dx}$ and hence find x when $\dfrac{dC}{dx} = 0$.

e What size must the base be for maximum volume?

Example 20

Square corners are cut from a piece of 12 cm by 12 cm tinplate which is then bent
into the form of an open dish. What size squares should be removed if the volume
is to be a maximum?

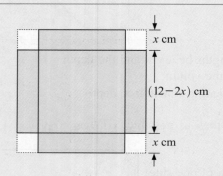

Suppose we cut out x cm by x cm squares.

Volume = length by width by depth

$$\therefore \quad V = (12 - 2x)^2 \times x$$
$$\therefore \quad V = (144 - 48x + 4x^2) \times x$$
$$\therefore \quad V = 144x - 48x^2 + 4x^3$$

$$\therefore \quad V'(x) = 144 - 96x + 12x^2$$
$$= 12(x^2 - 8x + 12)$$
$$= 12(x - 2)(x - 6)$$

which has sign diagram

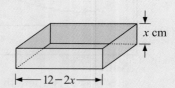

Note: $0 \leqslant x \leqslant 6$

$\therefore$ maximum volume occurs when $x = 2$ cm

$\therefore$ cut out 2 cm squares.

---

5 Square corners are cut from a piece of cardboard 40 cm by 25 cm and the edges are
turned up to form an open box. What size squares must be removed to make a box of
maximum volume?

6 A small manufacturer can produce x fittings per day where $1 \leqslant x \leqslant 10\,000$.

The costs are: • $1000 per day for the workers   • $2 per day per fitting

• $\$\dfrac{5000}{x}$ per day for running costs and maintenance.

How many fittings should be produced daily to minimise costs?

7 A duck farmer wishes to build a rectangular enclosure of area 100 m^2. The farmer must
purchase wire netting for three of the sides as the fourth side is an existing fence of
another duck yard. Naturally the farmer wishes to minimise the length (and therefore

the cost) of the fencing required to complete the job.

   a If the shorter sides are of length x m, show that the required length of wire netting to be purchased is $L = 2x + \dfrac{100}{x}$.

   b Use **technology** to help you sketch the graph of $y = 2x + \dfrac{100}{x}$.

   c Find the minimum value of L and the corresponding value of x when this occurs.

   d Sketch the optimum situation with its dimensions.

8 Radioactive waste is to be disposed of in fully enclosed lead boxes of inner volume 200 cm³. The base of the box has dimensions in the ratio $2 : 1$.

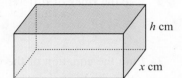

   a What is the inner length of the box?

   b Explain why $x^2h = 100$.

   c Explain why the inner surface area of the box is given by $A(x) = 4x^2 + \dfrac{600}{x}$ cm².

   d Use technology to help sketch the graph of $y = 4x^2 + \dfrac{600}{x}$.

   e Find the minimum inner surface area of the box and the corresponding value of x.

   f Draw a sketch of the optimum box shape with dimensions shown.

9 Consider the manufacture of 1 L capacity tin cans where the cost of the metal used to manufacture them is to be minimised. This means that the surface area is to be as small as possible but still must hold a litre.

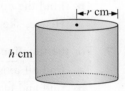

   a Explain why the height h, is given by $h = \dfrac{1000}{\pi r^2}$ cm.

   b Show that the total surface area A, is given by $A = 2\pi r^2 + \dfrac{2000}{r}$ cm².

   c Use technology to help you sketch the graph of A against r.

   d Find the value of r which makes A as small as possible.

   e Draw a sketch of the dimensions of the can of smallest surface area.

10 A beam is to be cut from a log of diameter 1 m, and is to have a rectangular cross-section.

The strength of the beam S is given by $S = kwd^2$, where w is the width, d the depth and k is a positive constant.

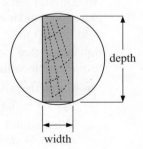

   a Show that $d^2 = 1 - w^2$ using the rule of Pythagoras.

   b Find the dimensions of the strongest beam that can be cut.

11

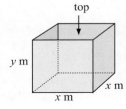

Special boxes are constructed from lead. Each box is to have an internal capacity of one cubic metre and the base is to be square. The cost per square metre of lining each of the 4 sides is twice the cost of lining the base.

   a Show that $y = \dfrac{1}{x^2}$.

b If the base costs $25 per m^2 to line, show that the total cost of lining the box is
$C(x) = 25(x^2 + 8x^{-1})$ dollars.

c What are the dimensions of the box costing least to line?

REVIEW SET 19A

1 Find, using the limit method, the gradient of the tangent to $y = \dfrac{4}{x}$ at $(-2, -2)$.

2 Use the rules of differentiation to find $f'(x)$ for $f(x)$ equal to:

    a $7x^3$ **b** $3x^2 - x^3$ **c** $(2x - 3)^2$ **d** $\dfrac{7x^3 + 2x^4}{x^2}$

3 Find the equation of the tangent to $y = -2x^2$ at the point where $x = -1$.

4 Find $f''(x)$ for: **a** $f(x) = 3x^2 - \dfrac{1}{x}$ **b** $f(x) = (x + 4)^2$

5 $y = 2x$ is a tangent to the curve $y = x^3 + ax + b$ at $x = 1$. Find a and b.

6 Consider the function $f(x) = x^3 - 3x$.

    a Determine the y-intercept.

    b Find the coordinates of any local maxima, local minima or horizontal inflections.

    c Hence, sketch the graph of the function.

7 For the function $f(x) = x^3 - 4x^2 + 4x$:

    a find all axis intercepts

    b find and classify all stationary points

    c sketch the graph of $y = f(x)$ showing features from **a** and **b**.

8 An open box is made by cutting squares out of the corners of a 36 cm by 36 cm square sheet of tinplate. What is the size of the squares to be removed if the volume is to be a maximum?

9 An athletics track consists of two 'straights' of length l m and two semicircular ends of radius x m. The perimeter of the track is to be 400 m.

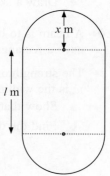

    a Show that $l = 200 - \pi x$, and hence write down the possible values that x may have.

    b Show that the area of the rectangle inside the track is given by $A = 400x - 2\pi x^2$.

    c What values of l and x produce the largest area of the rectangle inside the track?

REVIEW SET 19B

1 Find, from the basic definition, the derivative function of $f(x) = x^3 - 2x$.

2 Find $\dfrac{dy}{dx}$ for: **a** $y = 3x^2 - x^4$ **b** $y = \dfrac{x^3 - x}{x^2}$

3 Find the equation of the tangent to $y = x^3 - 3x + 5$ at the point where $x = 2$.

4 Find all points on the curve $y = 2x + x^{-1}$ which have a tangent parallel to the x-axis.

5 Find the maximum and minimum values of $x^3 - 3x^2 + 5$ for $-1 \leqslant x \leqslant 4$.

6 For the function $f(x) = 2x^3 - 3x^2 - 36x + 7$:
 a find and classify all stationary points
 b find intervals where the function is increasing and decreasing
 c sketch the graph of $y = f(x)$, showing all important features.

7 A rectangular gutter is formed by bending a 24 cm wide sheet of metal as illustrated. Find the dimensions of the gutter so as to maximise the water carried.

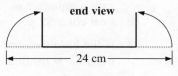

8 A 200 m fence is to be placed around a lawn which has the shape of a rectangle with a semi-circle on one of its sides and dimensions as shown.

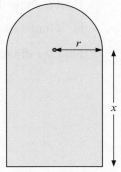

 a Find the expression for the perimeter of the lawn in terms of r and x.
 b Find x in terms of r.
 c Show that the area of the lawn, A, can be written as $A = 200r - r^2 \left(2 + \frac{\pi}{2}\right)$.
 d Find the dimensions of the lawn such that its area is a maximum.

REVIEW SET 19C

1 Find, from first principles, the derivative of $f(x) = x^2 + 2x$.

2 Use the rules of differentiation to find $f'(x)$ for $f(x)$ equal to:

 a $2x^{-3} + x^{-4}$ **b** $\dfrac{1}{x} - \dfrac{4}{x^2}$ **c** $(x - \dfrac{1}{x})^2$ **d** $(4x)^{-1}$

3 Find the equation of the tangent to $y = \dfrac{12}{x^2}$ at the point $(1, 12)$.

4 $f(x) = x^3 + Ax + B$ has a stationary point at $(-2, 3)$.
 a Find A and B.
 b Find the nature of all stationary points.

5 For the function $f(x) = x^3 + x^2 + 2x - 4$:
 a state the y-intercept
 b find the x-intercept(s)
 c find and classify any stationary points
 d on a sketch of the cubic, show the features found in **a**, **b** and **c**.

6 The cost per hour of running a barge up the Rhein is given by $C(v) = 10v + \dfrac{90}{v}$ dollars, where v is the average speed of the barge.

   a Find the cost of running the barge for:

     i two hours at 15 km/h.   **ii** 5 hours at 24 km/h.

   b Find the rate of change in the cost of running the barge at speeds of:

     i 10 km/h   **ii** 6 km/h.

   c At what speed will the cost be a minimum?

7 A manufacturer of open steel boxes has to make one with a square base and a volume of 1 m^3. The steel costs \$2 per square metre.

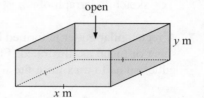

   a If the base measures x m by x m and the height is y m, find y in terms of x.

   b Hence, show that the total cost of the steel is $C(x) = 2x^2 + \dfrac{8}{x}$ dollars.

   c Find the dimensions of the box costing the manufacturer least to make.

ANSWERS

EXERCISE 1A

1 a $5 \in D$ **b** $6 \notin G$ **c** $d \notin \{a, e, i, o, u\}$
  d $\{2, 5\} \subset \{1, 2, 3, 4, 5, 6\}$
  e $\{3, 8, 6\} \not\subset \{1, 2, 3, 4, 5, 6\}$
2 a i $\{9\}$ **ii** $\{5, 6, 7, 8, 9, 10, 11, 12, 13\}$
  b i $\{ \ \}$ or $\varnothing$ **ii** $\{1, 2, 3, 4, 5, 6, 7, 8\}$
  c i $\{1, 3, 5, 7\}$ **ii** $\{1, 2, 3, 4, 5, 6, 7, 8, 9\}$
3 a i disjoint **ii** not disjoint **b** true

EXERCISE 1B

1 a $8 = \frac{8}{1}$ $\therefore$ rational **b** $-11 = \frac{-11}{1}$ $\therefore$ rational
2 $\frac{4}{0}$ is undefined (of form $\frac{p}{q}$, but $q = 0$)
3 a true **b** true **c** true **d** true **e** false **f** true
  g true **h** false
4 a $0.8 = \frac{4}{5}$ **b** $0.75 = \frac{3}{4}$ **c** $0.45 = \frac{9}{20}$
  d $0.215 = \frac{43}{200}$ **e** $0.864 = \frac{108}{125}$
5 a true **b** true **c** false **d** true
6 a $0.444\,444 \dots = \frac{4}{9}$ **b** $0.212\,121 \dots = \frac{7}{33}$
  c $0.325\,325\,325 \dots = \frac{325}{999}$

EXERCISE 1C

1 a 20 **b** 37 **c** 55 **d** 156 **2 a** 458 **b** 4379
3 a \$66\,370, \$130\,515 **b** \$2743 **c** 62 kg
4 a 544 **b** 23 **c** 120
5 a 55 **b** 317 **c** 100\,000 **d** 90
6 a \$5304 **b** \$17\,316 **c** 17

EXERCISE 1D

1 a $2^2 \times 3^3$ **b** 2×5^2 **c** $2 \times 3^3 \times 5$ **d** $5^2 \times 7^2$
  e $2^2 \times 5^3 \times 7$ **f** $3^2 \times 7^2 \times 11^2$
2 a 30 **b** 20 **c** 56 **d** 270 **e** 396 **f** 24\,200
3 a 69\,984 **b** 222\,264 **c** 1\,054\,152 **d** 2\,156\,000
  e 1\,711\,125 **f** 21\,600\,000

EXERCISE 1E

1 a 1, 3, 9 **b** 1, 2, 3, 4, 6, 12 **c** $12 = 2 \times 6$
  d 3×4 or 1×12
2 a 1, 2, 5, 10 **b** 1, 2, 3, 6, 9, 18
  c 1, 2, 3, 5, 6, 10, 15, 30 **d** 1, 5, 7, 35
  e 1, 2, 4, 11, 22, 44 **f** 1, 2, 4, 7, 8, 14, 28, 56
  g 1, 2, 5, 10, 25, 50
  h 1, 2, 3, 4, 6, 7, 12, 14, 21, 28, 42, 84
3 a 4 **b** 5 **c** 7 **d** 20 **e** 8 **f** 44 **g** 18 **h** 12 **i** 4
4 a 6 **b** 9 **c** 9 **d** 24 **e** 22 **f** 25 **g** 45 **h** 13
5 a 8, 10, 12 **b** 17, 19, 21, 23, 25
6 a 2 and 8 **b** 1 and 19, 3 and 17, 5 and 15, 7 and 13
7 a even **b** even **c** even **d** odd **e** odd **f** odd
  g even **h** odd **i** even
8 a 2, 3, 5, 7, 11, 13, 17, 19, 23, 29 **b** Yes, 2
9 a other factors of 5485 are 5, 1097
  b other factors of 8230 are 2, 5, 10, 823, 1646, 4115
  c other factors of 7882 are 2, 7, 14, 563, 1126, 3941
  d other factors of 999 are 3, 9, 27, 37, 111, 333

EXERCISE 1F

1 a 3, 6, 9, 12, 15, 18 **b** 8, 16, 24, 32, 40, 48
  c 12, 24, 36, 48, 60, 72 **2 a** 30 **b** 40
3 a 1, 2, ③, ④, 5, ⑥, 7, ⑧,
  ⑨, 10, 11, ⑫, 13, 14, ⑮, ⑯,
  17, ⑱, 19, ⑳, ㉑, 22, 23, ㉔,
  25, 26, ㉗, ㉘, 29, ㉚

  c 12, 24
4 a 30, 60, 90, 120, 150 **b** 45, 90, 135 **c** 60, 120
  d 60, 120
5 a 6 **b** 12 **c** 40 **d** 60 **e** 24 **f** 24 **g** 36 **h** 180
6 a 204 **b** 495 **7** 36 m **8** 36 seconds

EXERCISE 1G

1 a 5 **b** 11 **c** 11 **d** 7 **e** 99 **f** 21 **g** 2
  h 28 **i** 36 **j** 13 **k** 16 **l** 3
2 a 30 **b** 18 **c** 1 **d** 50 **e** 26 **f** 18 **g** 36
  h 3 **i** 30 **j** 16 **k** 6 **l** 13 **m** 28 **n** 40 **o** 18
3 a 90 **b** 28 **c** 54 **d** 31 **e** 9 **f** 168
4 a 3 **b** 2 **c** 3 **d** 2 **e** 8 **f** 3
5 a 48 **b** 54 **c** 17 **d** 3 **e** 37 **f** 11
6 a 12 **b** -12 **c** 38 **d** 2 **e** -12 **f** -6
  g 30 **h** 3 **i** -11 **j** 4 **k** -1 **l** -2
7 a 45 **b** 6 **c** 14 **d** 10 **e** 10 **f** 1 **g** 4
  h 25 **i** 1 **j** 18 **k** 2 **l** 1

REVIEW SET 1A

1 a i $\{5, 7\}$ **ii** $\{2, 3, 5, 7, 9\}$ **b** no **c** no
2 1200 **3** 166 **4** 1000 times **5** 74 **6** 57 kg
7 \$2685 **8** 2, 4, 8, 16, 32, 64 **9** 60
10 $420 = 2^2 \times 3 \times 5 \times 7$ **11** 13
12 1, 2, 3, 6, 9, 18, 27, 54, 81, 162 **13** 21
14 36, 42, 48 **15** 24 **16** 308 **17** 21

REVIEW SET 1B

1 a $0.75 = \frac{3}{4}$
  b $\sqrt{7} = 2.645\,751 \dots$, i.e., it is a non-terminating decimal number and can therefore not be written in the form $\frac{p}{q}$ for any $p, q \in Z$.
2 true **3 a** 32 **b** 355 **c** 3900 **d** 532 **e** 23
4 a 18 **b** 75 **c** 37 **d** 22
5 a 1, 2, 3, 4, 6, 7, 12, 14, 21, 28, 42, 84 **b** 23, 29
  c $124 = 2^2 \times 31$ **d** 7
6 a $3^2 \times 5^3$ **b** 144 **7** 90 **8** 90 sweets
9 a Z^+ **b** $\varnothing$

EXERCISE 2A.1

1 a 6 h 23 min **b** 7 h 11 min **c** 5 h 58 min
  d 4 h 36 min **e** 5 h 29 min **f** 26 h 20 min
2 1 h 38 min
3 a 6.16 pm **b** 6.15 am **c** 4.50 pm **d** 1.30 pm
  e 3.15 am on the next day **f** 7.35 pm
4 5.10 pm

(top right column)

10 a $2 \times 2 \times 2 \times 3$ **b** $2 \times 2 \times 7$ **c** $3 \times 3 \times 7$
  d $2 \times 2 \times 2 \times 3 \times 3$ **e** $2 \times 2 \times 2 \times 17$
11 a 3 **b** 11 **c** 15
12 a 3 **b** 8 **c** 6 **d** 14 **e** 6 **f** 8 **g** 5 **h** 21

EXERCISE 2A.2

1 a 3 pm **b** 8 pm **c** 9 pm **d** 7 am

2 a 2 am Monday **b** 5 am Monday **c** 9 am Monday
d 12 midnight Monday

3 a 6 pm Tuesday **b** 3 pm Tuesday **c** 9 am Wednesday
d 1 am Wednesday

4 a 3.15 pm Sunday **b** 8.15 am Sunday
c 10.15 pm Saturday **d** 7.15 pm Saturday

5 1 am the following day **6** 5 am

7 a 0200 the next day **b** 2330 the previous day

EXERCISE 2B

1 a i 5°C **ii** 93°C **b i** 68°F **ii** 122°F

2 a -17.8°C **b** 37.8°C **c** -6.7°C

3 a 158°F **b** 14°F **c** 752°F

4 a i 4.4°C **ii** 93.3°C **b i** 68°F **ii** 122°F

6 a -26.7°C **b** 77°F **7** -40°F $= -40^\circ$C

EXERCISE 2C

1 a 7.92 pounds **b** 8.10 km **c** 6.35 cm **d** 59.1 inches
e 0.227 kg **f** 60.96 cm

2 113.5 grams **3** 111.1 miles/hour

4 a 64 inches **b** 163 cm
c 17.1 hands means 17 hands 1 inch, i.e., $17\frac{1}{4}$ hands
d 160 cm

5 a i 177.8 cm **ii** 185.4 cm
b i 6 feet 2 inches **ii** 6 feet 7 inches

6 a 13 pounds 3 ounces **b** 2.38 kg

7 a 69.9 kg **b** 11.9 m

8 a $\frac{1}{8}$ inch $\doteq$ 3.2 mm, $\frac{3}{16}$ inch $\doteq$ 4.8 mm, $\frac{5}{32}$ inch $\doteq$ 4.0 mm
b $\frac{5}{32}$ inch bit

EXERCISE 2D

1 c and **d** are not in scientific notation

2 $1000 = 10^3$
$100 = 10^2$
$10 = 10^1$
$1 = 10^0$
$0.1 = 10^{-1}$
$0.01 = 10^{-2}$
$0.001 = 10^{-3}$
$0.0001 = 10^{-4}$

3 a 82 000 **b** 36 **c** 8.7
d 490 **e** 550 000 **f** 1.91
g 6 100 000 000 **h** 2700

4 a 0.0078 **b** 3.6 **c** 0.055
d 0.48 **e** 0.000 029 **f** 4.63
g 0.376 **h** 0.002 02

5 a 6 800 000 000 people
b 0.000 11 m **c** 0.000 98 mm
d 140 000 light years

6 a 2.0×10^{-3} mm **b** 1.38×10^{11} m **c** 8.3×10^6 $^\circ$C
d 3×10^9 cells

7 a 2.78×10^{10} L **b** 27.8 gigalitres (or 2.78×10 GL)

8 a 1.0×10^{-9} seconds

EXERCISE 2E.1

1 a 80 **b** 80 **c** 300 **d** 640 **e** 3990 **f** 1650
g 9800 **h** 1020

2 a 100 **b** 500 **c** 900 **d** 1000 **e** 5400 **f** 4800
g 13 100 **h** 44 000

3 a 1000 **b** 6000 **c** 10 000 **d** 4000 **e** 65 000
f 123 000 **g** 435 000 **h** 571 000

4 a $15 000 **b** 470 kg **c** $600 **d** 5700 km
e 367 000 L **f** 46 000 people **g** 1 000 000 people
h $6 500 000 **i** 32 000 hectares **j** 36 000 000 beats
k $1 000 000 000

EXERCISE 2E.2

1 a 3.5 **b** 5.36 **c** 7.2 **d** 15.23 **e** 9.025 **f** 12.6
g 0.44 **h** 9.28 **i** 0.01

2 a 499.32 **b** 228.84 **c** 9.11 **d** 31.75 **e** 26.67
f 0.88 **g** 7.41 **h** 5.93 **i** 0.48

EXERCISE 2E.3

1 a 570 **b** 16 000 **c** 71 **d** 3.0 **e** 0.72 **f** 50
g 3.0 **h** 1800 **i** 0.041 **j** 46 000

2 a 43 600 **b** 10 100 **c** 0.667 **d** 0.0368 **e** 0.319
f 0.720 **g** 0.636 **h** 0.0637 **i** 19.0 **j** 257 000

3 a 28.04 **b** 0.005 362 **c** 23 680 **d** 42 370 000
e 0.038 79 **f** 0.006 378 **g** 0.000 899 9 **h** 43.08

4 a 2 **b** 2 or 3 (we do not know if the 0 is significant)
c 2 **d** 3 **e** 4 **f** 3 **g** 3 **h** 3 **i** 5 **j** 5

5 a 42 **b** 6.24 **c** 0.046 **d** 0.25 **e** 440 **f** 2100
g 31 000 **h** 10.3 **i** 1 **j** 1.0 **k** 264 000 **l** 0.037 64

6 a i 30 000 people **ii** 26 000 people **b** 26 000 people

EXERCISE 2E.4

1 a 4.5×10^7 **b** 3.8×10^{-4} **c** 2.1×10^5 **d** 4.0×10^{-3}
e 6.1×10^3 **f** 1.6×10^{-6} **g** 3.9×10^4 **h** 6.7×10^{-2}

2 a 45 000 000 **b** 0.000 38 **c** 210 000 **d** 0.004
e 6100 **f** 0.000 001 6 **g** 39 000 **h** 0.067

3 a 4×10^{-10} **b** 8.75×10^{12} **c** 8.84×10^{-3}
d 4.29×10^{14} **e** 8.75×10^{-16} **f** 7.18×10^{10}

4 a 1.40×10^5 km **b** 6.72×10^5 km **c** 2.45×10^8 km

5 a 17 sheets **b** 167 sheets **c** 8334 sheets

6 a 2×10^4 **b** 6.72×10^5 **c** 3.51×10^7

7 a 6.78×10^{16} **b** 1.22×10^{20} **c** 1.82×10^{-7}
d 7.59×10^{-13} **e** 1.56×10^{17} **f** 1.45×10^{-14}

8 a 60 seconds **b** 60 minutes **c** 24 hours
d 3.1536×10^7 **e** 3.00×10^8 m

EXERCISE 2E.5

1 a 4.00×10^{13} **b** 2.59×10^7 **c** 7.08×10^{-9}
d 4.87×10^{-11} **e** 8.01×10^6 **f** 3.55×10^{-9}

2 a i 2.88×10^8 km **ii** 7.01×10^{10} km
b 2.5×10^8 km/h **c** 0.9 seconds **d** 3.94 times
e C is approx. 32.9 times heavier

EXERCISE 2F.1

1 a 83.8 km/h **b** 83.3 km/h **2** 751.0 km/h **3** 30 m/s
4 a 321.4 km **b** 32.7 km **5** 7 h 25 min 16 sec

EXERCISE 2F.2

1 a $2.76 per kg **b** 14.5 km per litre **c** 27 L/minute
d $9.50/h **e** 4 degrees per hour **f** 5.5 kg/m^2
g 10.45 cents/kilowatt hour **h** 55 words/minute

2 a $466.80 **b** $13.34 per hour

3 B, 11.65 bag per hectare **4** Jo, 26.27 goals/match

5 a 1067 litres per day **b i** $62.40 **ii** 69.33 cents per day

6 a 10 minutes **b** 4 minutes 17 seconds

7 $219.85 **8 a** 13°C **b** 1.16°C per hour

9 a 20 cents per minute **b** 4 kL per second
c 108 litres per hour **d** $2730 per kg
e 52 560 deaths per year   {using 1 year = 365 days}

EXERCISE 2G

1 a 28 371 kilowatt hours **b** 40 688 kilowatt hours

2 a 94 units **b** 122 units **c** 38 units **d** 46 units
e 24 units **f** 56 units

3 a $\frac{1}{5}$ **b** $\frac{3}{5}$ **c** $\frac{9}{10}$ **d** $\frac{3}{10}$

4 a 70 km/h **b** 25 km/h **c** 76 km/h

5 a 4500 rpm **b** 2750 rpm **c** 2500 rpm **d** 4250 rpm

6 a 38.5°C **b** 36.8°C

EXERCISE 2H

1 a $\pm\frac{1}{2}$ mm **b** $\pm\frac{1}{2}$ kg **c** $\pm\frac{1}{2}$ cm **d** ± 50 mL

2 a 154.5 cm to 155.5 cm **b** 209.5 cm to 210.5 cm
c 188.5 cm to 189.5 cm

3 a 52.5 kg to 53.5 kg **b** 94.5 kg to 95.5 kg
c 78.5 kg to 79.5 kg

4 a 26.5 mm to 27.5 mm **b** 38.25 cm to 38.35 cm
c 4.75 m to 4.85 m **d** 1.45 kg to 1.55 kg
e 24.5 g to 25.5 g **f** 3.745 kg to 3.755 kg

5 a 6.4 m **b** 6.0 m **c** 10 cm graduations

6 246 cm and 250 cm **7** between 788 cm and 792 cm

8 a 55.25 cm^2 **b** 41.25 cm^2

9 between 1058.25 cm^2 and 1126.25 cm^2

10 between 31.875 cm^2 and 40.375 cm^2

11 between 144.4 cm^3 and 248.6 cm^3

12 between 1502.1 cm^3 and 1546.7 cm^3

13 between 922.5 cm^3 and 1473.0 cm^3

14 between 332.2 cm^3 and 347.7 cm^3

EXERCISE 2I

1 a i \$2460 **ii** 0.180% **b i** −467 people **ii** 1.48%
c i \$1890 **ii** 0.413% **d i** 189 cars **ii** 6.72%

2 a i 1.238 kg **ii** 19.8% **b i** 2.4 m **ii** 2.46%
c i 3.8 L **ii** 16.0% **d i** 22 hours **ii** 30.6%

3 a 100 m^2 **b** 99.91 m^2 **c** 0.09 m^2 **d** 0.090%

4 a 3254.224 cm^3 **b** 3240 cm^3 **c** −14.224 cm^3 **d** 0.437%

5 a 65.25 km/h **b** 4.75 km/h **c** 7.28%

REVIEW SET 2A

1 6 hours 17 minutes **2 a** 3.35 am next day **b** 9.50 am

3 9 pm **4** 1188 pounds **5** 95°F

6 a 1 392 000 km **b** 0.000 008 4 m

7 a 8.64×10^5 km **b** 3.16×10^8 km

8 a i 6.4 **ii** 6.38 **b i** 0.05 **ii** 0.047

9 a 25.28 km/h **b** 5.583 hours (or 5 hours 35 minutes)

10 a ± 0.5 cm **b** 35.5 cm to 36.5 cm
c between 1260.25 cm^2 and 1332.25 cm^2

11 a i −\$590 **ii** 22.8% **b i** −0.109 cm **ii** 0.417%

REVIEW SET 2B

1 165 minutes **2** 11 hours 4 minutes

3 2 am the next day **4** 28.9°C **5** 180.34 cm

6 a 460 000 000 000 **b** 1.9 **c** 0.0032

7 a 1.276×10^7 m **b** 4.2×10^{-7} cm **8** 313 sheets

9 a i 59.4 **ii** 59.40 **b i** 0.01 **ii** 0.0084

10 a \$508.75 **b** \$14.54 per hour **11** \$230.80

12 a −2.3 m **b** 6.71% **13** between 48 cm and 52 cm

EXERCISE 3A

1 a finite **b** infinite **c** infinite **d** infinite

2 a i The set of all x such that x is an integer between −1
and 7, including −1.
ii A = {−1, 0, 1, 2, 3, 4, 5, 6} **iii** $n(A) = 8$
iv no, A is finite

b i The set of all x such that x is a natural number between
−2 and 8.
ii A = {0, 1, 2, 3, 4, 5, 6, 7} **iii** $n(A) = 8$
iv no, A is finite

c i The set of all x such that x is a real number between 0
and 1, including both 0 and 1.
ii A = {all real numbers between 0 and 1}
iii $n(A) = \infty$ **iv** yes, A is infinite

d i The set of all x such that x is a rational number between
5 and 6, including both 5 and 6.
ii not possible **iii** $n(A)$ is infinite **iv** yes, A is infinite

3 a $\{x \mid x \in Z, \ -100 < x < 100\}$
b $\{x \mid x \in R, \ x > 1000\}$ **c** $\{x \mid x \in Q, \ 2 \leqslant x \leqslant 3\}$

4 a i $\varnothing, \{a\}, \{b\}, \{c\}, \{a, b\}, \{a, c\}, \{b, c\}, \{a, b, c\}$;
8 of them
ii $\varnothing, \{a\}, \{b\}, \{c\}, \{d\}, \{a, b\}, \{a, c\}, \{a, d\}, \{b, c\},$
$\{b, d\}, \{c, d\}, \{a, b, c\}, \{a, b, d\}, \{a, c, d\},$
$\{b, c, d\}, \{a, b, c, d\}$; 16 of them

b If a set has n elements it has 2^n subsets.

5 a yes **b** no **c** yes **d** yes **e** no **f** no

EXERCISE 3B

1 a C′ = {all consonants} **b** C′ = {all positive integers}
c C′ = $\{x \mid x \in Z, \ x \geqslant -4\}$
d C′ = $\{x \mid x \in Q, \ 2 < x < 8\}$

2 a A = {2, 3, 4, 5, 6, 7} **b** A′ = {0, 1, 8}
c B = {5, 6, 7, 8} **d** B′ = {0, 1, 2, 3, 4}
e A ∩ B = {5, 6, 7} **f** A ∪ B = {2, 3, 4, 5, 6, 7, 8}
g A ∩ B′ = {2, 3, 4}

3 a $n(P') = 9$ **b** $n(Q) = 11$ **4 a** false **b** true

5 a B′ = {1, 2, 10, 11, 12} **b** C′ = {1, 2, 3, 4, 12}
c A′ = {1, 8, 9, 10, 11, 12} **d** A ∩ B = {3, 4, 5, 6, 7}
e $(A \cap B)' = \{1, 2, 8, 9, 10, 11, 12\}$
f A′ ∩ C = {8, 9, 10, 11}
g B′ ∩ C = {1, 2, 5, 6, 7, 8, 9, 10, 11, 12}
h $(A \cup C) \cap B' = \{2, 10, 11\}$

6 a P = {2, 3, 5, 7, 11, 13, 17, 19, 23}
b P ∩ Q = {2, 5, 11}
c P ∪ Q = {2, 3, 4, 5, 7, 11, 12, 13, 15, 17, 19, 23}
d $n(P \cup Q) = 12$; $n(P) + n(Q) - n(P \cap Q) = 9 + 6 - 3 = 12$

7 a P = {3, 6, 9, 12, 15, 18, 21, 24, 27}
b P ∩ Q = {12, 24}
c P ∪ Q = {3, 4, 6, 8, 9, 12, 15, 16, 18, 20, 21, 24, 27, 28}
d $n(P \cup Q) = 14$; $n(P) + n(Q) - n(P \cap Q) = 9 + 7 - 2 = 14$

8 a M = {32, 36, 40, 44, 48, 52, 56}, N = {36, 42, 48, 54}
b M ∪ N = {32, 36, 40, 42, 44, 48, 52, 54, 56}
c M ∩ N = {36, 48}
d $n(M \cup N) = 11$; $n(M) + n(N) - n(M \cap N) = 7 + 4 - 2 = 9$

9 a R = {−2, −1, 0, 1, 2, 3, 4}; S = {0, 1, 2, 3, 4, 5, 6}
b R ∩ S = {0, 1, 2, 3, 4}
c R ∪ S = {−2, −1, 0, 1, 2, 3, 4, 5, 6}
d $n(R \cup S) = 9$, $n(R) + n(S) - n(R \cap S) = 7 + 7 - 5 = 9$

10 a C = {−4, −3, −2, −1}
D = {−7, −6, −5, −4, −3, −2, −1, 0}
b C ∩ D = {−4, −3, −2, −1} (= C)
c C ∪ D = {−7, −6, −5, −4, −3, −2, −1, 0} (= D)
d $n(C \cup D) = 8$; $n(C) + n(D) - n(C \cap D) = 4 + 8 - 4 = 8$

11 a A = {4, 8, 12, 16, 20, 24, 28, 32, 36}

B = {6, 12, 18, 24, 30, 36};   C = {12, 24, 36}

b i A ∩ B = {12, 24, 36} **ii** B ∩ C = {12, 24, 36}

 iii A ∩ C = {12, 24, 36} **iv** A ∩ B ∩ C = {12, 24, 36}

c A ∪ B ∪ C = {4, 6, 8, 12, 16, 18, 20, 24, 28, 30, 32, 36}

d $n(A ∪ B ∪ C) = 12$;

 $n(A) + n(B) + n(C) − n(A ∩ B) − n(B ∩ C) −$
 $n(A ∩ C) + n(A ∩ B ∩ C) = 9 + 6 + 3 − 3 − 3 − 3 + 3 = 12$

12 a A = {6, 12, 18, 24, 30}; B = {1, 2, 3, 5, 6, 10, 15, 30};
   C = {2, 3, 5, 7, 11, 13, 17, 19, 23, 29}

b i A ∩ B = {6, 30} **ii** B ∩ C = {2, 3, 5}

 iii A ∩ C = ∅ **iv** A ∩ B ∩ C = ∅

c A ∪ B ∪ C = {1, 2, 3, 5, 6, 7, 10, 11, 12, 13, 15, 17,
   18, 19, 23, 24, 29, 30}

d $n(A ∪ B ∪ C) = 18$;
   $n(A) + n(B) + n(C) − n(A ∩ B) − n(B ∩ C) −$
   $n(A ∩ C) + n(A ∩ B ∩ C) = 5 + 8 + 10 − 2 − 3 − 0 + 0 = 18$

e A ∩ B = {e, h} **f** A ∪ B = {b, d, e, f, h, i, j}

g (A ∪ B)' = {a, c, g, k}

h A' ∪ B' = {a, b, c, d, f, g, i, j, k}

6 a $n(B) = b + c$ **b** $n(A') = c + d$ **c** $n(A ∩ B) = b$

d $n(A ∪ B) = a + b + c$ **e** $n((A ∩ B)') = a + c + d$

f $n((A ∪ B)') = d$

7 a i $n(Q) = 2a + 4$ **ii** $n(P ∪ Q) = 4a + 4$

 iii $n(Q') = 3a − 5$ **iv** $n(U) = 5a − 1$

b i $a = 6$

 ii $n(U) = 31$ gives $a = \frac{32}{5}$ but $a ∈ Z^+$ ∴ no solution

8 a i A = {a, b, c, d, h, j} **ii** B = {a, c, d, e, f, g, k}

 iii C = {a, b, e, f, i, l} **iv** A ∩ B = {a, c, d}

 v A ∪ B = {a, b, c, d, e, f, g, h, j, k}

 vi B ∩ C = {a, e, f} **vii** A ∩ B ∩ C = {a}

 viii A ∪ B ∪ C = {a, b, c, d, e, f, g, h, i, j, k, l}

b i 12 **ii** 12

EXERCISE 3C

1 a 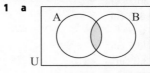 **b**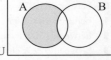

c 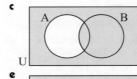 **d**

2 a A = {1, 3, 5, 7, 9}; B = {2, 3, 5, 7}

b A ∩ B = {3, 5, 7}; A ∪ B = {1, 2, 3, 5, 7, 9}

c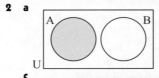

3 a A = {4, 8, 12, 16, 20, 24, 28}; B = {6, 12, 18, 24}

b A ∩ B = {12, 24}
   A ∪ B = {4, 6, 8, 12, 16, 18, 20, 24, 28}

c

4 a R = {2, 3, 5, 7, 11, 13, 17, 19, 23, 29}
   S = {4, 6, 8, 9, 10, 12, 14, 15, 16, 18, 20, 21, 22, 24,
     25, 26, 27, 28}

b R ∩ S = ∅
   R ∪ S = {2, 3, 4, 5, 6, 7, 8, 9, 10, 11, 12, 13, 14, 15,
   16, 17, 18, 19, 20, 21, 22, 23, 24, 25, 26, 27, 28, 29}
     = {x | x ∈ Z⁺, 1 < x < 30}

c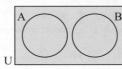

5 a A = {b, d, e, h} **b** B = {e, f, h, i, j}

c A' = {a, c, f, g, i, j, k} **d** B' = {a, b, c, d, g, k}

EXERCISE 3D

1 a 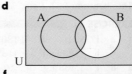 **b**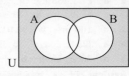

c 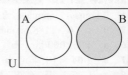 **d**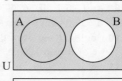

e 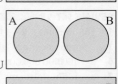 **f**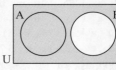

2 a **b**

c **d**

e **f**

g **h**

i

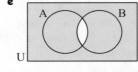
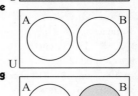
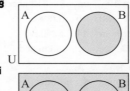

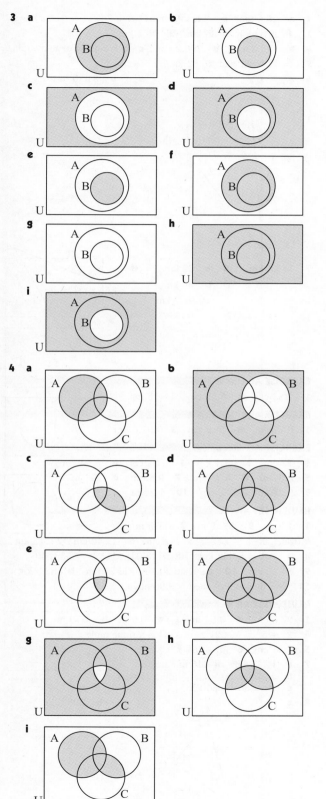

3 a ... **b** ... **c** ... **d** ... **e** ... **f** ... **g** ... **h** ... **i**

4 a ... **b** ... **c** ... **d** ... **e** ... **f** ... **g** ... **h** ... **i**

EXERCISE 3E

1 a 7 **b** 14 **c** 14 **d** 7 **e** 5 **f** 9
2 a 5 **b** 6 **c** 17 **d** 8 **e** 3 **f** 2
3 a 15 **b** 4 **4 a** 18 **b** 6 **5 a** 7 **b** 23
6 a

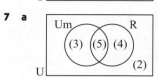

where LH represents long hair and Br represents brown

b i 9 **ii** 6 **iii** 3

7 a

Um ... R

where Um represents taking umbrella and R represents rain

b i 4 days **ii** 2 days

8 13 members **9** 20 people
10 $n(A) + n(B) + n(A \cap B) = a + b + b + c - b$
$$= a + b + c$$
$$= n(A \cup B)$$

11 a 16 **b** 33 **c** 14 **d** 7
12 a 29 **b** 6 **c** 1 **d** 11
13 a 3 **b** 5 **c** 5 **d** 21 **14 a** 3 **b** 6 **c** 9

REVIEW SET 3A

1 a $P = \{2, 3, 4, 5, 6, 7, 8\}$ **b** $n(P) = 7$
2 a i $\{x \mid x \in Z, \ x > 3\}$ **ii** $\{x \mid x \in Q, \ -5 \leqslant x \leqslant 5\}$
   b i infinite **ii** infinite
3 a yes, $Q \subset P$ **b** yes, $\{\} \subset Q$
4 a $A = \{4, 8, 12\}$ **b** $B = \{2, 3, 5, 7, 11, 13\}$
   c $B' = \{0, 1, 4, 6, 8, 9, 10, 12, 14, 15\}$
   d $A \cup B = \{2, 3, 4, 5, 7, 8, 11, 12, 13\}$
   e $A \cap B' = \{4, 8, 12\}$
5 a i $A = \{p, q, r, s\}$ **ii** $B = \{s, t, v, w\}$
   iii $A \cap B = \{s\}$ **iv** $(A \cup B)' = \{a, e, i, o, u\}$
   v $A \cup B' = \{a, e, i, o, p, q, r, s, u\}$
   vi $A' \cap B = \{t, v, w\}$
   b i $n(A \cup B) = 7$ **ii** $n(A \cap B') = 3$
6 a

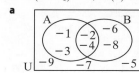

where CE represents company executive and S represents wearing suit

b i 3 **ii** 2

7 5 students **8 a** 20 **b** 4 **c** 10

REVIEW SET 3B

1 a The set of all x such that x is a negative number between -4 and 0.
   b $A = \{-3, -2, -1\}$ **c** A is finite
2 yes, $T \subset S$
3 a $P = \{2, 4, 6, 8, 10, 12, 14, 16, 18\}$; $Q = \{4, 8, 12, 16\}$
   b $P \cap Q = \{4, 8, 12, 16\}$
   c $P \cup Q = \{2, 4, 6, 8, 10, 12, 14, 16, 18\}$ **d** yes, $Q \subset P$
   e $n(P \cup Q) = 9$; $n(P) + n(Q) - n(P \cap Q) = 9 + 4 - 4 = 9$
4 a

b **i** R = {2, 4, 6, 8, 10, 12, 14, 16, 18}; S = {1, 4, 9, 16}
   ii $n(S) = 4$
   iii S′ = {2, 3, 5, 6, 7, 8, 10, 11, 12, 13, 14, 15, 17, 18, 19, 20}
   iv no, since not all members of S are in R

5 a 8 **b** 9

6 a

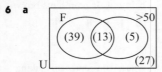

where F represents female and > 50 represents more than 50 years old
  b i 27 **ii** 66

7 11 children **8 a** 16 **b** 4 **c** 36

EXERCISE 4A

1 a $\sqrt{34}$ cm **b** $\sqrt{20}$ cm **c** $\sqrt{72}$ cm **d** $\sqrt{233}$ km
  e $\sqrt{392}$ m **f** $\sqrt{250}$ km **g** $\sqrt{\frac{5}{4}}$ cm **h** 21.5 cm
  i 8.38 cm

2 a 8 cm **b** $\sqrt{15}$ cm **c** $\sqrt{24}$ km **d** 2.53 km
  e $\sqrt{3}$ m **f** 8.98 cm

3 a $x = \sqrt{7}$ **b** $x = \sqrt{2}$ **c** $x = \sqrt{7}$ **d** $x = \sqrt{5}$
  e $x = \sqrt{5}$ **f** $x = \sqrt{8}$

4 a $x = \sqrt{\frac{10}{9}}$ **b** $x = \sqrt{\frac{5}{4}}$ **c** $x = \sqrt{\frac{7}{4}}$ **d** $x = \sqrt{\frac{13}{16}}$
  e $x = 1$ **f** $x = \frac{1}{2}$

5 a $x = \sqrt{48}$ **b** $x = \sqrt{13}$ **c** $x = \sqrt{3}$ **d** $x = \sqrt{7}$
  e $x = \sqrt{3}$ **f** $x = 1$

6 a $y = \sqrt{2},\ x = \sqrt{3}$ **b** $y = \sqrt{21},\ x = \sqrt{5}$
  c $x = \sqrt{21},\ y = \sqrt{22}$

7 a $x = \sqrt{2}$ **b** $x = \sqrt{7} - 1$ **8** $AC = \sqrt{\frac{97}{4}}$ m

9 a $\sqrt{10}$ cm **b** $\sqrt{27}$ m **c** $\sqrt{32}$ m

EXERCISE 4B

1 7.62 cm **2** 4.47 cm × 8.94 cm **3 a** 75.9 cm **b** 270 cm²

4 9.80 cm **5** 5.66 cm **6** 14.4 cm **7 a** 10.4 cm **b** 62.4 cm²

8 a 7.42 cm **b** 22.2 cm²

9 a 173.21 m **b** 17 321 m² **c** $73 612

EXERCISE 4C

1 **b**, **e**, **f** are right angled

2 all are right angled **a** ∠BAC **b** ∠ABC **c** ∠ACB

3 The sum of the squares of two adjacent sides in a rectangular section must equal the square of its diagonal *or* the diagonals must be equal in length.

EXERCISE 4D

1 a $x \doteqdot 0.663$ **b** $x \doteqdot 4.34$ **c** $x \doteqdot 2.23$ **2** 68 cm

3 4.66 m **4 a** 73.9 m **b** 30.9 m **5 a** 8.62 m **b** 7.74 m

6 a 5.97 m **b** 5.13 m **7** $\doteqdot 236.8$ m

8 a 43.97 km to A, 61.55 km to B **b** $247 974 **9** 72 m

10 50.3 m **11 a** 68.3 cm **b** 546.4 cm **c** 13.1 kg

12 a 7.07 m **b** 108.3 m

13 a 3.05 m **b** 15.9 m **c** 79.7 m

14 a 1103.1 m **b** 4274.3 m **15** Yes **16** Yes **17** Yes

EXERCISE 4E

1 a

  b 87.1 km

2 a 26.6 km **b** 2 hrs 40 mins

3 a B 32 km, C 36 km **b** 48.2 km

4 a 113.3 min **b** 106.2 min **c** It is quicker to go by train.

5 13.42 km/h and 26.83 km/h

6 a

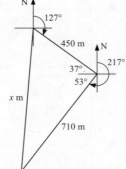

  b 840.6 m

7 180.6 km **8 a** 72 km **d**
  b 56 km
  c 91.2 km

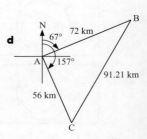

EXERCISE 4F.1

1 2.83 cm **2** 5.66 cm **3** 16 cm

EXERCISE 4F.2

1 8.77 cm **2** 11.2 cm **3** 801.6 km

EXERCISE 4G

1 12 cm **2** 11.7 cm **3** 8.24 cm **4** 3.46 cm

5 8.37 m **6** 5.38 cm **7** 9.89 m **8** 141 m

9 9.38 cm × 9.38 cm **10** 1.22 m

REVIEW SET 4A

1 1.2 m **2** 452.5 m **3** 7.79 m

4 no, $4^2 + 5^2 \neq 8^2$ **5** 42 cm **6** 11.40 cm **7** 13 km

8 a AR = 8.062 km, BR = 13.038 km **b** $538 050

9 84.15 cm **10 a** Kate 20 km, Ric 26 km **b** 32.8 km

11 No, diagonal of shed is 9.90 m

REVIEW SET 4B

1 27.5 cm **2** 3.53 m **3 a** 0.9 m **b** 1.79 m

4 $2^2 + 5^2 = (\sqrt{29})^2$, ∠ABC is a right angle

5 7.94 cm **6** 42.4 m

7 a i 75 km **ii** 45 minutes
  b i 53.4 km **ii** 53 mins 25 secs
  c the sealed road PXQ

8 a

  b 598.5 km

9 53.4 m

REVIEW SET 4C

1 1078 m **2** 6.40 cm **3 a** 7.99 m **b** 11.30 m^2
4 77.9 m **5 a** 11.3 m **b** 45.3 m^2 **6** yes
7 a

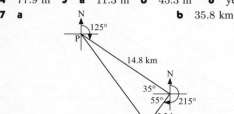

b 35.8 km

8 17.3 cm **9** 8.60 m

EXERCISE 5A

1 a quant. discrete **b** quant. continuous **c** categorical
d quant. discrete **e** quant.e discrete **f** categorical
g quant. discrete **h** quant. continuous
i quant. discrete **j** quant. continuous **k** categorical
l categorical **m** categorical **n** quant. continuous
o quant. discrete **p** categorical **q** quant. discrete
r categorical **s** categorical **t** quant. continuous
u quant. discrete **2** Answers will vary

EXERCISE 5B

1 a continuous **b** continuous **c** discrete **d** continuous
e continuous **f** discrete **g** discrete **h** continuous
2 a number of pets
b Discrete since you can't have part of a pet.
c

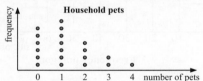

d positively skewed, no outliers **e** 30% **f** 15%
3 a the number of phone calls made in a day
b You can only make whole phone calls. **c** 10%
d 20% **e** ...... two calls per day
f positively skewed with an outlier
g Data value 11 is an outlier.
4 a 50 households **b** 15 households **c** 36%
d positively skewed, no outliers
5 a number of matches in a box **b** discrete
c

No. matches	Tally	Freq.
47	│	1
48	卌	5
49	卌 卌	10
50	卌 卌 卌 卌 │││	23
51	卌 卌	10
52	卌 ││││	9
53	││	2

d

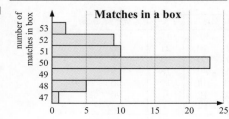

e 38.3%
6 a

No. peas	Tally	Freq.
3	││││	4
4	卌 卌 │││	13
5	卌 卌 │	11
6	卌 卌 卌 卌 卌 │││	28
7	卌 卌 卌 卌 卌 卌 卌 卌 卌 │││	48
8	卌 卌 卌 卌 卌 ││	27
9	卌 卌 ││││	14
10	││││	4
11		0
12		0
13	│	1
Total		150

b

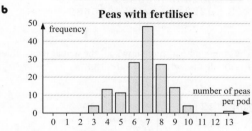

c Yes, data value 13 is an outlier.
d On average the number of peas is higher in the "with fertiliser" group. The mode has increased from 6 to 7.
e Yes, assuming the fertiliser is not too expensive and the peas are as big as they were previously.

EXERCISE 5C

1 a

Test Score	Tally	Freq.
0 - 9		0
10 - 19		0
20 - 29	│	1
30 - 39	││	2
40 - 49	│││	3
50 - 59	卌 ││││	9
60 - 69	卌 卌 │││	13
70 - 79	卌 │││	8
80 - 89	卌 卌	10
90 - 100	││││	4
Total		50

b 28%
c 12%

d More students had a test score in the interval 60 - 69 than in any other interval.
e

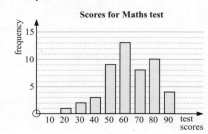

2 a

Stem	Leaf
2	7 5 9 7 4 1
3	4 6 4 3 0 5
4	2 8 6 0
5	7 1 0 8

b

Stem	Leaf
2	1 4 5 7 7 9
3	0 3 4 4 5 6
4	0 2 6 8
5	0 1 7 8

5 | 8 represents 58

3 a 2 **b** 62 **c** 16 **d** 5 **e** 17.1% **f** positively skewed

4 a

Stem	Leaf
0	9
1	8
2	5 8 8 4 9 8 6 2 0 5 0 8 5
3	5 3 3 4 8 2 4 9 5 5 0 5 6 2 3 3 4 4 6
4	2 4 9 1 6 3 5 8 0 2
5	0

5 | 0   represents 50

b

Stem	Leaf
0	9
1	8
2	0 0 2 4 5 5 5 6 8 8 8 8 9
3	0 2 2 3 3 3 3 4 4 4 4 5 5 5 5 6 6 8 9
4	0 1 2 2 3 4 5 6 8 9
5	0

c The stem-and-leaf plot shows all the actual data values.
d i 50 **ii** 9 **e** 24.4% **f** 13.3%

5

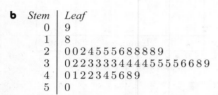

	11C		11P
9 5	0	6 7
8 7 5 4 4 3 2 2 1 0	1	1 2 4 4 5 5 6 8 9
7 7 6 1 1 0	2	2 3 4 5 9 9
7 2 0	3	0

EXERCISE 5D

1 a

Weight (kg)	Tally	Frequency
50 -	||||	4
60 -	∦ ||	7
70 -	∦ ∦ ∦	15
80 -	∦ ∦ |	11
90 -	|||	3
Total		40

b 15
c 26
d 72.5%

2 a 29 **b** 22.2% **3 a** 20 **b** 58.3% **c i** 1218 **ii** 512

EXERCISE 5E

1

Value	Freq	Cum Freq	Rel Freq	Cum Rel Freq
32	3	3	0.0375	0.0375
35	7	10	0.0875	0.125
36	8	18	0.1	0.225
39	11	29	0.1375	0.3625
41	15	44	0.1875	0.55
44	12	56	0.15	0.70
45	8	64	0.1	0.80
46	7	71	0.0875	0.8875
49	5	76	0.0625	0.95
50	4	80	0.05	1.00
Total	80	—	1	—

2 b i 15 **ii** 16.7% **iii** 73.3%
c

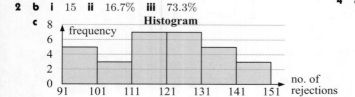

Histogram

2 (right column)

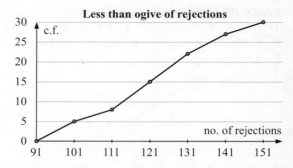

Less than ogive of rejections

3 a

Class	Freq	Cum Freq
1 - 10	8	8
11 - 20	11	19
21 - 30	14	33
31 - 40	17	50
41 - 50	20	70
51 - 60	9	79
61 - 70	7	86
71 - 80	4	90

b

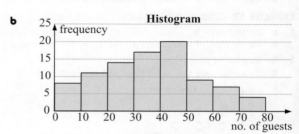

Histogram

Frequency polygon

c

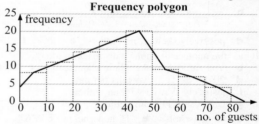

Less than ogive

i 33 nights **ii** 58 nights
d No, less than 30 guests 35% of the time, ∴ did not break even.

4 a

Class	Freq	Cum Freq
1 - 20	15	15
21 - 40	21	36
41 - 60	24	60
61 - 80	18	78
81 - 100	12	90
101 - 120	6	96
121 - 140	4	100

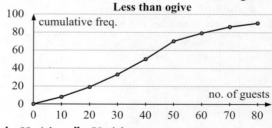

b

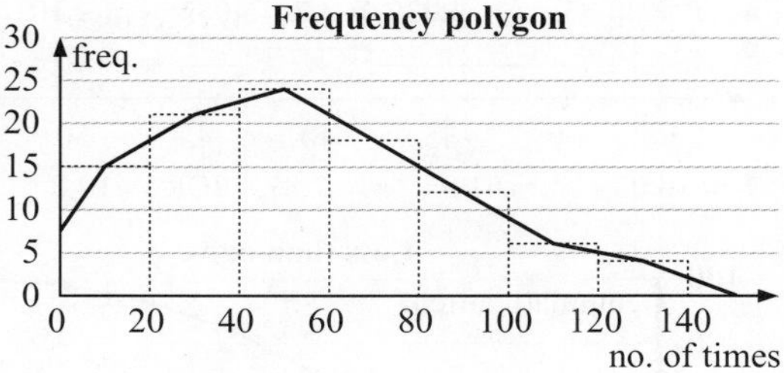
Frequency polygon
freq.
30
25
20
15
10
5
0
0 20 40 60 80 100 120 140
no. of times

c

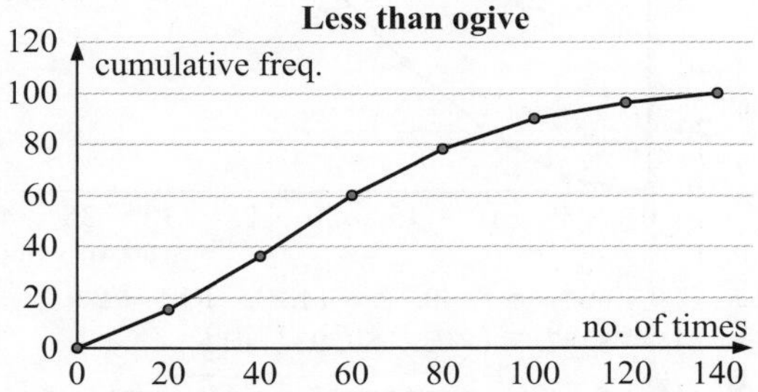
Less than ogive
cumulative freq.
120
100
80
60
40
20
0
0 20 40 60 80 100 120 140
no. of times

i 79% **ii** 30 times

5 **a**

Class	*Tally*	*Freq*	*Cum Freq*
0 - 1.99	||||	4	4
2 - 3.99	~~||||~~ ~~||||~~ ||	12	16
4 - 5.99	~~||||~~ ~~||||~~ ~~||||~~	15	31
6 - 7.99	~~||||~~ ~~||||~~	10	41
8 - 9.99	~~||||~~ |	6	47
10 - 11.99	|||	3	50

b

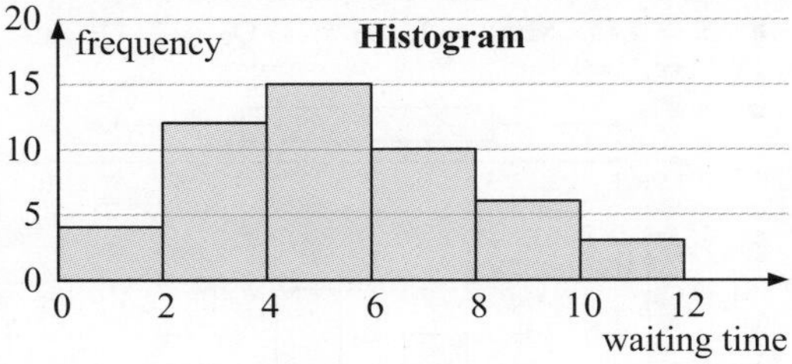
Histogram
frequency
20
15
10
5
0
0 2 4 6 8 10 12
waiting time

c

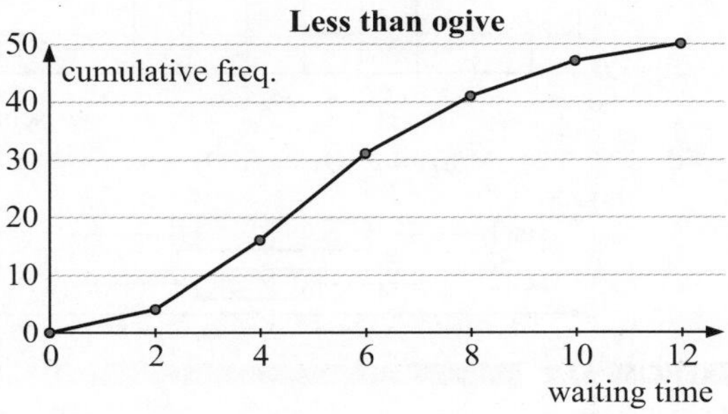
Less than ogive
cumulative freq.
50
40
30
20
10
0
0 2 4 6 8 10 12
waiting time

i 16 customers **ii** 47%

iii In 8 minutes, $\frac{41}{50} \times 100\% = 82\%$ are served which is short of its goal.

EXERCISE 5F.1

1 Team A 91.25, Team B 91.75, $\therefore$ Team B

2 **a** 49 **b** 144 and 147 (bi-modal) **c** 25

3 **a** 29 **b** 107 **c** 149.5

4 **a** 0 **b** 1.7 **c** 1.5 **5** $\doteqdot$ 81.2 mm

6 **a** mean = \$163 770, median = \$147 200 (differ by \$16 570)
b **i** mean selling price **ii** median selling price

7 $x = 15$ **8** 17.25 goals per game **9** $\doteqdot$ 17.7 **10** 10.1

11 6 and 12 **12** **a** $a = 5$, $b = 13$ **b** the mode does not exist

EXERCISE 5F.2

1 **a** mean = 59.45, mode = 60, median = 59

b Both the mean and median number of nails per pack are under the store's claim of 60 nails per pack.

c The mean, as it takes into account all of the data and there are no extreme values.

2 mean $\doteqdot$ 34.6, mode = 35, median = 35

3 **a** $\doteqdot$ 70.9 g **b** $\doteqdot$ 210.1 g **c** 139.25 g

EXERCISE 5F.3

1 **a** $\overline{x} \doteqdot 13.5$ **b** $\overline{x} \doteqdot 50.5$ **2** 31.7

3 **a** $\doteqdot$ 13.6 goals **b** **i** 13.5 goals **ii** 13.6 goals

c The approximations are about the same. **4** $\doteqdot$ 495 mm

5 **a** 70 **b** $\doteqdot$ 411 000 litres, i.e., $\doteqdot$ 411 kL **c** $\doteqdot$ 5870 L

6 **a** 125 people **b** $\doteqdot$ 119 people **c** $\frac{3}{25}$ **d** 137 marks

7 **a** 95 **b** 59.6 kg **c** 25 **d** 36.8% **e** $\frac{9}{19}$ or 47.4%

EXERCISE 5F.4

1 **a** 2 **b** 8 **2** 1 error

3 **a**

Length (x cm)	*Frequency*	*C. frequency*
$24 \leqslant x < 27$	1	1
$27 \leqslant x < 30$	2	3
$30 \leqslant x < 33$	5	8
$33 \leqslant x < 36$	10	18
$36 \leqslant x < 39$	9	27
$39 \leqslant x < 42$	2	29
$42 \leqslant x < 45$	1	30

b

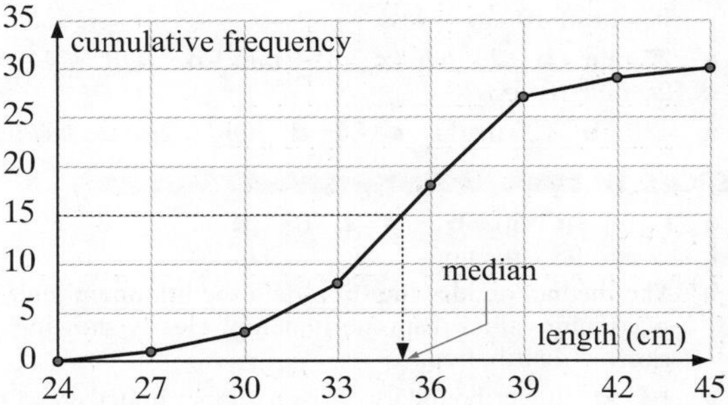
cumulative frequency
35
30
25
20
15
10
5
0
median
length (cm)
24 27 30 33 36 39 42 45

c median $\doteqdot$ 35 cm

d actual median = 34.5, i.e., a good approximation

4

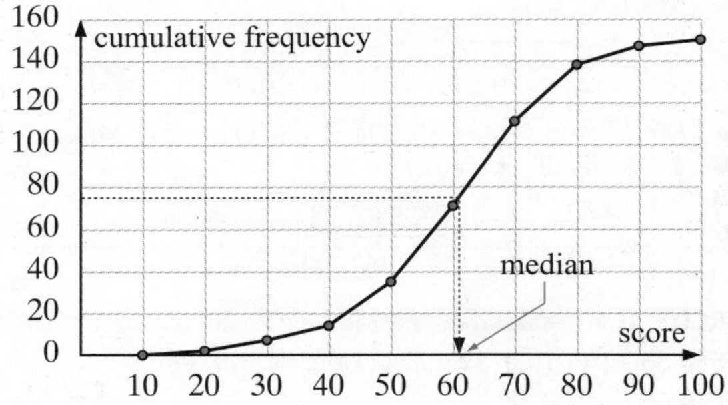
cumulative frequency
160
140
120
100
80
60
40
20
0
median
score
10 20 30 40 50 60 70 80 90 100

a $\doteqdot 61$ students **b** $\doteqdot 91$ students **c** $\doteqdot 76$ students
d 24 (or 25) students **e** 76 marks

5

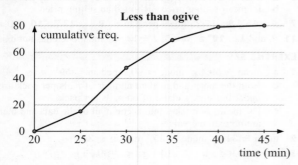

a 29 **b** 33 **c** 26.5
6 a 26 years **b** 41.5% **c i** 0.556 **ii** 0.0287

EXERCISE 5G

1 a i 14 **ii** 4 **iii** 10 **iv** 13
  b i 52 **ii** 45.5 **iii** 6.5 **iv** 21
  c i 31.5 **ii** 19.5 **iii** 12 **iv** 38
  d i 3 **ii** 1 **iii** 2 **iv** 5
2 a median = 2.05 minutes, Q_3 = 3.15 minutes,
       Q_1 = 1.1 minutes
  b range = 5.6 minutes, IQR = 2.05 minutes
  c i "50% of the waiting times were greater than 2.05
       minutes."
    ii "75% of the waiting times were less than 3.15 minutes."
    iii "The minimum waiting time was 0 minutes and the
       maximum waiting time was 5.6 minutes. The waiting
       times were spread over 5.6 minutes."
3 a 6.0 **b** 10.1 **c** 8.15 **d** 7.5 **e** 8.9 **f** 4.1 **g** 1.4
4 a 10 **b** $\doteqdot 28.3\%$ **c** 7 cm **d** IQR $\doteqdot$ 2.6 cm
  e 10 cm, which means that 90% of the seedlings have a
       height of 10 cm or less.
5 a 27 min **b** 29 min **c** $31\frac{1}{2}$ min **d** IQR $\doteqdot 4\frac{1}{2}$ min
  e 28 min 10 sec
6 a 480 **b** 120 marks **c** 84 **d** IQR $\doteqdot$ 28 **e** 107 marks

EXERCISE 5H

1 a i 98 **ii** 30 **b** 73 **c** 68 **d** 75% **e** 22
  f 82 and 98 **g** no
  h The median divides the 'box' and the minimum and
       maximum values disproportionately, clearly showing a
       skewed distribution.
2 a 12 **b** lower boundary = 13.5, upper boundary = 61.5
  c 13.2 and 65 would be outliers
3 a median = 6, Q_1 = 5, Q_3 = 8 **b** 3
  c lower boundary = 0.5, upper boundary = 12.5
  d yes, 13 **e**

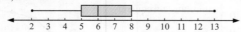

4 a Min_x = 33, Q_1 = 35, Q_2 = 36, Q_3 = 37, Max_x = 40
  b i 7 **ii** 2 **c** no
  d

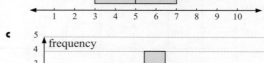

EXERCISE 5I.1

1 $\overline{x}$ = 55, s = 10.9 **2** $\overline{x}$ = 1.69, s = 0.365
3 $\overline{x}$ = 45, s = 3.28

4 a $\overline{x}$ = 40.35, s = 4.23 **b** $\overline{x}$ = 40.6, s = 4.10 **c** 5
  d

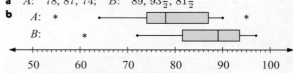

5 $\overline{x}$ = 16.95, s^2 = 61.95, s = 7.87, IQR = 10.6

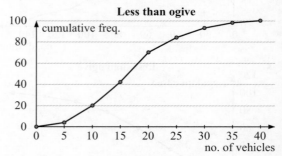

6 $\overline{x}$ = 26.7, s^2 = 202.49, s = 14.23, IQR = 22
7 a A: 78, 87, 74; B: 89, $93\frac{1}{2}$, $81\frac{1}{2}$
  b

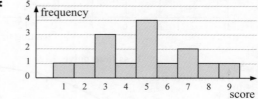

  c B as the distribution is further right than for A.
     Also the median is much higher.

EXERCISE 5I.2

1 Machine B (7.69%)
2 Greater variability in the past 6 months (9.66% to 10.0%)
   may indicate a problem with the machine. **3** males (8.08%)

EXERCISE 5J.1

1 a $\overline{x} \doteqdot 4.87$, Min_x = 1, Q_1 = 3, Q_2 = 5, Q_3 = 7,
     Max_x = 9
  b

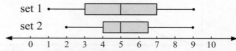

  c

frequency / score (bar chart)

  d $\overline{x} \doteqdot 5.24$, Min_x = 2, Q_1 = 4, Q_2 = 5, Q_3 = 6.5,
     Max_x = 9

set 1 / set 2 (box plots)

EXERCISE 5J.2

1 a discrete
  c

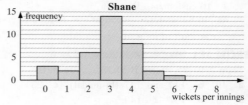

Shane

frequency / wickets per innings

Brett

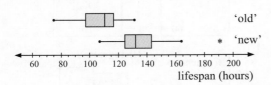

d There are no significant outliers.

e Shane's distribution is reasonably symmetrical.
Brett's distribution is positively skewed.

f Shane has a higher mean ($\doteqdot$ 2.89 wickets) compared with Brett ($\doteqdot$ 2.67 wickets). Shane has a higher median (3 wickets) compared with Brett (2.5 wickets). Shane's modal number of wickets is 3 (14 times) compared with Brett, who has a bi-modal distribution of 2 and 3 (7 times each).

g Shane's range is 6 wickets, compared with Brett's range of 8 wickets. Shane's IQR is 2 wickets, compared with Brett's IQR of 3 wickets. Brett's wicket taking shows greater spread or variability.

h

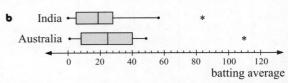

j Generally, Shane takes more wickets than Brett and is a more consistent bowler.

2 a continuous

c For the 'New type' globes, 191 hours could be considered an outlier. However, it could be a genuine piece of data, so we will include it in the analysis.

d

	Old type	New type
Mean	107	134
Median	110.5	132
Range	56	84
IQR	19	18.5

The mean and median are $\doteqdot$ 25% and $\doteqdot$ 19% higher for the 'new type' of globe compared with the 'old type'.
The range is higher for the 'new type' of globe (but has been affected by the 191 hours).
The IQR for each type of globe is almost the same.

e

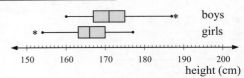

f For the 'old type' of globe, the data is bunched to the right of the median, hence the distribution is negatively skewed. For the 'new type' of globe, the data is bunched to the left of the median, hence the distribution is positively skewed.

g The manufacturer's claim, that the 'new type' of globe has a 20% longer life than the 'old type' seems to be backed up by the 25% higher mean life and 19.5% higher median life.

EXERCISE 5K

1 a

	Year 10	Year 12
Min_x	5	8
Q_1	7.5	10
Median	10	14
Q_3	12	16
Max_x	16	17
b i Range	11	9
ii IQR	4.5	6

2 a i Class B   **ii** Class B   **iii** Class B
b i 52   **ii** 15   **c i** 75%   **ii** 50%
d i almost symmetrical   **ii** positively skewed
e "The students in class A generally scored higher marks. The marks in class B were more varied."

3 a

	Australia	India
Min_x	1	0
Q_1	8	5
Median	24.8	18.84
Q_3	40.1	27.92
Max_x	109.8	83.83
Outliers	109.8	83.83
c Mean	28.97	22.98
Range	108.8	83.83
IQR	32.1	22.915

b

India
Australia
batting average

Australia's measures of centre are higher than for India. The spread of scores for Australia are also higher than for India.

d No, the outliers are genuine data and should be included in the analysis.

4 a

	Boys	Girls
Min_x	160	152
Q_1	167	163
Median	171	166
Q_3	175	170
Max_x	188	177
Outliers	188	152

boys
girls
height (cm)

b The distributions show that in general boys are taller than the girls and are more varied in their heights.

REVIEW SET 5A

1 a discrete numerical   **b** continuous numerical
c categorical   **d** categorical   **e** categorical
f continuous numerical   **g** continuous numerical
h discrete numerical   **i** discrete numerical

2 a

Stem	Leaf
0	7
1	8
2	7 9
3	2 2 5 6 8 9 9
4	0 3 4 4 5 6 6 7 8

1 | 8   means 18 marks

b 45% **c** 20% **d** 90% **e** negatively skewed

3 a Heights can take any value from 170 cm to 205 cm.

b

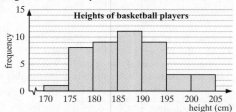

Heights of basketball players

c The modal class is (185-< 190) cm as this occurred the most frequently.

d slightly positively skewed

4 15

5 a highest = 97.5 m, lowest = 64.6 m

b use groups 60 - , 65 - , 70 - , etc.

c

A frequency distribution table for distances thrown by Kapil		
distance (m)	tally	freq. (f)
60 -			1
65 -	|||	3
70 -	﷽	5
75 -	||	2
80 -	﷽ |||	8
85 -	﷽		6
90 -	|||	3
95 < 100	||	2
	Total	30

d i / ii

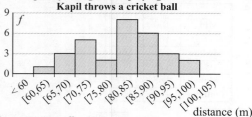

Frequency histogram displaying the distance Kapil throws a cricket ball

distance (m)

e i $\doteqdot$ 81.1 m **ii** 83.0 m

6 a

	A	B
Min	11	11.2
Q_1	11.6	12
Median	12	12.6
Q_3	12.6	13.2
Max	13	13.8

b i

Range	2	2.6
ii IQR	1	1.2

c i The members of squad A generally ran faster times.

  ii The times in squad B were more varied.

7 a $\bar{x} \doteqdot 121.55$, $s \doteqdot 8.32$ **b** $\bar{x} \doteqdot 7.01$, $s \doteqdot 1.05$

8 $a = 5$

9 a

	Girls	Boys
outliers	none	none
shape	approx. symm.	approx. symm.
centre (median)	36.3 sec	34.9 sec
spread (range)	7.7 sec	4.9 sec

b For both the girls' and boys' distributions of times there are no outliers and they are approximately symmetrical. The median swim times for boys is 1.4 seconds lower than for girls but the range of the girls' swim times is 2.8 seconds higher than for boys. The analysis supports the conjecture that boys generally swim faster than girls with less spread of times.

REVIEW SET 5B

1 a continuous numerical **b** categorical **c** categorical

  d continuous numerical **e** discrete numerical

  f discrete numerical **g** categorical

2 a Diameter of bacteria colonies

0	4 8 9
1	3 5 5 7
2	1 1 5 6 8 8
3	0 1 2 3 4 5 5 6 6 7 7 9
4	0 1 2 7 9

leaf unit: 0.1 cm

b i 3.15 cm **ii** 4.5 cm

c The distribution is slightly negatively skewed.

3 a a histogram **b** continuous **c** (55 - < 60) kg

d

Weights (kg)	Freq.	Midpoints	Product
40 -	4	42.5	170
45 -	12	47.5	570
50 -	13	52.5	682.5
55 -	25	57.5	1437.5
60 -	17	62.5	1062.6
65 - < 70	9	67.5	607.5
Total	80		4530

e $\doteqdot 56.63$ kg

4 a 70 **b** 14% **c** $\doteqdot 46\%$

d / e

Weight (kg)	Freq.	Cum. Freq.
48 -	3	3
50 -	7	10
52 -	12	22
54 -	21	43
56 -	24	67
58 - < 60	3	70

f

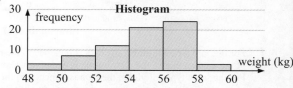

Histogram

g modal class is 56 - **h** no, it is negatively skewed

5 20.7

6 a $a = 8$ and $b = 9$ *or* $a = 9$ and $b = 7$ **b** 7.5

7 a min = 4, Q_1 = 13.4, med = 15.75, Q_3 = 22.4, max = 30

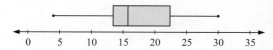

b min = 11, $Q_1 = 19$, med = 25, $Q_3 = 37$, max = 49

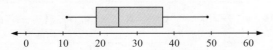

8 a i 101.5 **ii** 98 **iii** 105.5 **b** 7.5 **c** $\overline{x} = 100.2$, $s \doteq 7.78$
9 a i 3.75 **ii** $\doteq 0.24$ **iii** 0.06
  b i $\doteq 4.86$ **ii** $\doteq 4.38$ **iii** $\doteq 19.14$

10 a

	Brand X		Brand Y
	8 4 1	87	
	9 8 6 5 5 2	88	
	9 8 8 7 6 5 5 4 4 3	89	1 2 3 4 4 5 6 8
	8 7 5 4 4 3 1 1	90	0 1 1 1 3 3 3 4 4 6 7 7 9 9
	6 3 0	91	0 3 3 7
		92	1 4 7 8 Left unit: 1 peanut

b

	Brand X	Brand Y
outliers	none	none
shape	approx. symm.	slightly pos. skewed
centre (median)	896.5	903.5
spread (range)	45	37

c The distributions of both Brand X and Brand Y show no outliers. Brand X's distribution is approximately symmetrical but Brand Y's is slightly positively skewed. Brand Y's distribution has a higher median number of peanuts per jar by 7. Brand X jars show a greater spread of peanuts per jar than Brand Y by 8.

EXERCISE 6A

1 a 20 **b** -8 **c** -24 **d** -24 **e** 16 **f** 23
  g -14 **h** 14
2 a $1\frac{1}{2}$ **b** -5 **c** -3 **d** 3
3 a 1 **b** -8 **c** 13 **d** 1 **e** -9 **f** -27 **g** 36 **h** 18
4 a -1 **b** 1 **c** $\doteq 2.24$ **d** 4 **e** $\doteq 3.32$ **f** 5
  g 0 **h** undefined

EXERCISE 6B

1 a $x = -4$ **b** $x = 7$ **c** $x = 11$ **d** $x = -6$
  e $x = 2$ **f** $x = -2$ **g** $x = 3\frac{1}{2}$ **h** $x = 2\frac{1}{2}$
2 a $x = 32$ **b** $x = -10$ **c** $x = -6$ **d** $x = -9$
  e $x = -11$ **f** $x = 10$ **g** $x = 8$ **h** $x = 11$
3 a $x = -\frac{5}{7}$ **b** $x = 0$ **c** $x = -9$ **d** $x = 1$
  e $x = -1\frac{2}{3}$ **f** $x = -\frac{1}{9}$
4 a $x = 9$ **b** $x = \frac{7}{5}$ **c** $x = \frac{11}{8}$ **d** $x = -2$ **e** $x = \frac{2}{5}$
  f $x = -10$ **g** $x = 8$ **h** $x = 0$ **i** $x = \frac{3}{4}$ **j** $x = -\frac{13}{9}$
5 a $x = 1$ **b** $x = 2$ **c** $x = \frac{22}{9}$ **d** $x = -\frac{9}{5}$
  e $x = -\frac{2}{3}$ **f** $x = -\frac{19}{6}$
6 a True for all values of x. **b** No solution.
  c Case **a** represents the intersection of two coincident lines. Case **b** represents two parallel lines.

EXERCISE 6C.1

1 a $x = \frac{6}{7}$ **b** $x = \frac{18}{5}$ **c** $x = 5$ **d** $x = -\frac{11}{2}$ **e** $x = \frac{15}{11}$
  f no solution **g** $x = \frac{8}{9}$ **h** $x = \frac{41}{19}$ **i** $x = -\frac{4}{15}$
2 a $x = \frac{25}{3}$ **b** $x = 21$ **c** $x = \frac{21}{4}$ **d** $x = \frac{6}{5}$
  e $x = \frac{9}{10}$ **f** $x = -4$ **g** $x = -\frac{15}{11}$
  h No solution, $32x = 21x$ gives $x = 0$ but you cannot divide by 0.

3 a $x = -\frac{5}{2}$ **b** $x = -\frac{2}{11}$ **c** no solution **d** no solution
  e $x = \frac{9}{8}$ **f** $x = -\frac{9}{4}$ **g** $x = \frac{31}{26}$ **h** no solution **i** $x = -\frac{1}{6}$

EXERCISE 6C.2

1 a $x \doteq 1.56$ **b** $x = 262$ **c** $x \doteq 1.093$ **d** $x \doteq -25.97$
2 a $w = 55$ **b** $w = 92.5$
3 a $4.44°C$ **b** $-17.8°C$ **c** $93.3°C$
4 a 110 m **b** 190 m
5 a $x = 9.5$ **b** $x \doteq 6.56$ **c** $x \doteq 10.5$ **d** $x \doteq 37.3$

EXERCISE 6D

1 $\frac{19}{7}$ **2** 6 **3** $4\frac{1}{2}$ **4** 11 **5** $\frac{5}{3}$ **6** 14 years
7 9 years **8** 16 2-cent stamps
9 13 one-cent coins, 26 two-cent coins, 11 five-cent coins
10 7 2-cent, 21 5-cent, 7 10-cent stamps
11 950 \$6 tickets, 2850 \$10 tickets, 3350 \$15 tickets

EXERCISE 6E

1 a 35.8 cm **b** 79.6 cm **c** 15.9 m **2 a** 44.1 m **b** 129 m
3 a 80 km/h **b** 260 km **c** 7 h 52 min
4 a 98.5 cm^2 **b** 8.0 m
5 a 7916.8 cm^3 **b** 1.59 cm **c** 0.399 mm
6 a 11.3 km **b** 71.0 m **7 a** 598.3 cm^2 **b** 28.2 cm

EXERCISE 6F

1 a $y = \dfrac{4-x}{2}$ **b** $y = \dfrac{7-2x}{6}$ **c** $y = \dfrac{11-3x}{4}$

  d $y = \dfrac{8-5x}{4}$ **e** $y = \dfrac{20-7x}{2}$ **f** $y = \dfrac{38-11x}{15}$

2 a $y = \dfrac{x-4}{2}$ **b** $y = \dfrac{2x-7}{6}$ **c** $y = \dfrac{3x+12}{4}$

  d $y = \dfrac{4x-18}{5}$ **e** $y = \dfrac{7x-42}{6}$ **f** $y = \dfrac{12x+44}{13}$

3 a $x = b - a$ **b** $x = \dfrac{b}{a}$ **c** $x = \dfrac{d-a}{2}$ **d** $x = t - c$

  e $x = \dfrac{d-3y}{7}$ **f** $x = \dfrac{c-by}{a}$ **g** $x = \dfrac{c+y}{m}$

  h $x = \dfrac{c-p}{2}$ **i** $x = \dfrac{a-t}{3}$ **j** $x = \dfrac{n-5}{k}$

  k $x = \dfrac{a-n}{b}$ **l** $x = \dfrac{a-p}{n}$

4 a $x = ab$ **b** $x = \dfrac{a}{d}$ **c** $x = \dfrac{2}{p}$ **d** $x = 2n$

  e $x = \dfrac{b}{5}$ **f** $x = \pm\sqrt{mn}$

EXERCISE 6G.1

1 a $x = 2$, $y = 3$ **b** $x = 6$, $y = 10$ **c** $x = 2$, $y = -6$
  d $x = 4$, $y = 2\frac{1}{2}$ **e** $x = 0$, $y = -4$ **f** $x = 2$, $y = 3$
2 a $x = 3$, $y = -2$ **b** $x = \frac{2}{5}$, $y = \frac{11}{5}$ **c** $x = 3$, $y = 4$
  d $x = -1$, $y = -5$ **e** $x = -5$, $y = 1$ **f** $x = 2$, $y = -4$
3 a obtain $1 = 4$ **b** no solution
4 a obtain $2 = 2$ **b** an infinite number of solutions

EXERCISE 6G.2

1 a $6x = 6$ **b** $-y = 8$ **c** $5x = 7$ **d** $-6x = -30$
  e $8y = 4$ **f** $-2y = -16$
2 a $x = 2$, $y = -1$ **b** $x = -2$, $y = 5$ **c** $x = 3$, $y = 2$
  d $x = -2$, $y = -1$ **e** $x = 5$, $y = -3$ **f** $x = 4$, $y = -3$

EXERCISE 6G.3

1 **a** $9x+12y=6$ **b** $-2x+8y=-14$ **c** $25x-5y=-15$
d $-21x-9y=12$ **e** $8x+20y=-4$ **f** $-3x+y=1$

2 **a** $x=3,\ y=2$ **b** $x=8,\ y=7$ **c** $x=-2,\ y=3$
d $x=2,\ y=1$ **e** $x=3,\ y=2$ **f** $x=2,\ y=-5$
g $x=5,\ y=2$ **h** $x=4,\ y=1$ **i** $x=19,\ y=-17$

EXERCISE 6G.4

1 **a** $x=6,\ y=10$ **b** $x=2,\ y=3$ **c** $x=3.8,\ y=-1.2$
d $x \doteqdot 2.14, y \doteqdot 2.71$ **e** $x=-2.5,\ y=-8.5$ **f** $x=3,\ y=0$

EXERCISE 6H

1 $16\frac{1}{2}$ and $30\frac{1}{2}$ **2** 16 and 12 **3** 17 and 68

4 pencils, 28 cents; biros, 54 cents

5 toffees, 15 cents; chocolates, 21 cents

6 16 50-cent coins, 27 \$1 coins

7 Amy, \$12.60; Michelle, \$16.80

8 24, 250 g packs; 34, 400 g packs

9 $a=3,\ b=5$ **10** length 11 cm, width 5 cm

EXERCISE 6I

1 **a** $2^1=2,\ 2^2=4,\ 2^3=8,\ 2^4=16,\ 2^5=32,\ 2^6=64$
b $3^1=3,\ 3^2=9,\ 3^3=27,\ 3^4=81$
c $5^1=5,\ 5^2=25,\ 5^3=125,\ 5^4=625$
d $7^1=7,\ 7^2=49,\ 7^3=343$

EXERCISE 6J

1 **a** -1 **b** 1 **c** 1 **d** -1 **e** 1 **f** -1 **g** -1
h -8 **i** -8 **j** 8 **k** -25 **l** 125

2 **a** 512 **b** -3125 **c** -243 **d** $16\,807$ **e** 512
f 6561 **g** -6561 **h** $5.117\,264\,691$ **i** $-0.764\,479\,956$
j $-20.361\,584\,96$

3 **a** $\overline{0.142857}$ **b** $\overline{0.142857}$ **c** $0.\overline{1}$ **d** $0.\overline{1}$
e $0.015\,625$ **f** $0.015\,625$ **g** 1 **h** 1

4 3 **5** 7

EXERCISE 6K

1 **a** 7^5 **b** 5^7 **c** a^9 **d** a^5 **e** b^{13} **f** a^{3+n}
g b^{7+m} **h** m^9

2 **a** 5^7 **b** 11^4 **c** 7^3 **d** a^4 **e** b^3 **f** p^{5-m}
g y^{a-5} **h** b^{2x-1}

3 **a** 3^8 **b** 5^{15} **c** 2^{28} **d** a^{10} **e** p^{20} **f** b^{5n}
g x^{3y} **h** a^{10x}

4 **a** 2^3 **b** 5^2 **c** 3^3 **d** 2^6 **e** 3^4 **f** 3^{a+2} **g** 5^{t-1}
h 3^{3n} **i** 2^{4-x} **j** 3^2 **k** 5^{4x-4} **l** 2^2 **m** 2^{y-2x}
n 2^{2y-3x} **o** 3^{2x} **p** 2^3

5 **a** a^3b^3 **b** a^4c^4 **c** b^5c^5 **d** $a^3b^3c^3$ **e** $16a^4$ **f** $25b^2$
g $81n^4$ **h** $8b^3c^3$ **i** $64a^3b^3$ **j** $\dfrac{a^3}{b^3}$ **k** $\dfrac{m^4}{n^4}$ **l** $\dfrac{32c^5}{d^5}$

6 **a** $8b^{12}$ **b** $\dfrac{9}{x^4y^2}$ **c** $25a^8b^2$ **d** $\dfrac{m^{12}}{16n^8}$ **e** $\dfrac{27a^9}{b^{15}}$
f $32m^{15}n^{10}$ **g** $\dfrac{16a^8}{b^4}$ **h** $125x^6y^9$

7 **a** a^2 **b** $8b^5$ **c** m^3n **d** $7a^5$ **e** $4ab^2$ **f** $\dfrac{9}{2}m^3$
g $40h^5k^3$ **h** $\dfrac{1}{m}$ **i** p^3

8 **a** 1 **b** $\frac{1}{3}$ **c** $\frac{1}{6}$ **d** 1 **e** 4 **f** $\frac{1}{4}$ **g** 8 **h** $\frac{1}{8}$
i 25 **j** $\frac{1}{25}$ **k** 100 **l** $\frac{1}{100}$

9 **a** 1 **b** 1 **c** 3 **d** 1 **e** 2 **f** 1 **g** $\frac{1}{25}$ **h** $\frac{1}{32}$

i 3 **j** $\frac{5}{2}$ **k** $\frac{3}{4}$ **l** 12 **m** $2\frac{1}{4}$ **n** $\frac{4}{5}$ **o** $\frac{8}{7}$ **p** $\frac{7}{2}$

10 **a** $\dfrac{1}{2a}$ **b** $\dfrac{2}{a}$ **c** $\dfrac{3}{b}$ **d** $\dfrac{1}{3b}$ **e** $\dfrac{b^2}{4}$ **f** $\dfrac{1}{4b^2}$ **g** $\dfrac{1}{9n^2}$

h $\dfrac{n^2}{3}$ **i** $\dfrac{a}{b}$ **j** $\dfrac{1}{ab}$ **k** $\dfrac{a}{b^2}$ **l** $\dfrac{1}{a^2b^2}$

11 **a** 3^{-1} **b** 2^{-1} **c** 5^{-1} **d** 2^{-2} **e** 3^{-3} **f** 5^{-2}
g 2^{-3x} **h** 2^{-4y} **i** 3^{-4a} **j** 3^{-2} **k** 5^{-2} **l** 5^{-3}

12 25 days **13** 63 sums

14 **a** $5^3=21+23+25+27+29$
b $7^3=43+45+47+49+51+53+55$
c $12^3=133+135+137+139+141+143$
$\qquad +145+147+149+151+153+155$

15 **a** $1+3x^{-1}$ **b** $5x^{-2}-x^{-1}$ **c** $x^{-2}+2x^{-3}$
d $x+5x^{-1}$ **e** $x+1-2x^{-1}$ **f** $x-3x^{-1}+5x^{-2}$
g $5x^{-1}-1-x$ **h** $16x^{-2}-3x^{-1}+x$
i $5x^2-3+x^{-1}+6x^{-2}$

EXERCISE 6L

1 **a** $x=1$ **b** $x=2$ **c** $x=3$ **d** $x=0$ **e** $x=-1$
f $x=-1$ **g** $x=-3$ **h** $x=2$ **i** $x=0$
j $x=-4$ **k** $x=5$ **l** $x=1$

2 **a** $x=2\frac{1}{2}$ **b** $x=-\frac{2}{3}$ **c** $x=-\frac{1}{2}$ **d** $x=-\frac{1}{2}$
e $x=-1\frac{1}{2}$ **f** $x=-\frac{1}{2}$ **g** $x=-\frac{1}{3}$ **h** $x=\frac{5}{3}$
i $x=\frac{1}{4}$ **j** $x=\frac{7}{2}$ **k** $x=-2$ **l** $x=-4$
m $x=0$ **n** $x=\frac{5}{2}$ **o** $x=-2$ **p** $x=-6$

3 **a** $x \doteqdot 4.32$ **b** $x \doteqdot 6.64$ **c** $x \doteqdot 12.3$ **d** $x \doteqdot 16.3$
e $x \doteqdot 3.10$ **f** $x \doteqdot -0.353$ **g** $x \doteqdot 4.04$ **h** $x \doteqdot 36.8$
i $x \doteqdot 4.95$ **j** $x \doteqdot 6.39$ **k** $x \doteqdot 2.46$ **l** $x \doteqdot 9.88$

REVIEW SET 6A

1 -3 **2** **a** $x=\frac{1}{5}$ **b** $x=\frac{11}{13}$ **3** $x=\dfrac{7y+25}{3}$

4 **a** $x=-1,\ y=-7$ **b** $x=2,\ y=-1$

5 1.720 sec **6** **a** $-\frac{5}{7}$ **b** 9 of 5 kg, 11 of 2 kg

7 **a** -1 **b** 27 **c** $\frac{2}{3}$ **8** **a** a^6b^7 **b** $\dfrac{2}{3x}$ **c** $\dfrac{y^2}{5}$

9 **a** $\dfrac{1}{b^3}$ **b** $\dfrac{1}{ab}$ **c** $\dfrac{a}{b}$ **10** **a** $x=-5$ **b** $x \doteqdot -1.79$

11 **a** 2.28 **b** 0.517 **c** 3.16

REVIEW SET 6B

1 -1 **2** **a** $x=3\frac{1}{2}$ **b** $x=1\frac{1}{2}$

3 **a** $y=\dfrac{3x-5}{8}$ **b** $y=\dfrac{4}{a}$

4 **a** $x=4,\ y=-2$ **b** $x=2,\ y=-1$ **5** 7226 cm^3

6 $-3\frac{6}{7}$ **7** small buses 23, large buses 34

8 **a** 8 **b** $-\frac{4}{5}$ **9** **a** a^{21} **b** p^4q^6 **c** $\dfrac{4b}{a^3}$

10 **a** $\dfrac{1}{x^5}$ **b** $\dfrac{2}{a^2b^2}$ **c** $\dfrac{2a}{b^2}$ **11** **a** $x=4$ **b** $x \doteqdot -4.57$

12 **a** 2.52 **b** 0.224 **c** 3.11

REVIEW SET 6C

1 -2 **2** **a** $x \doteqdot 2.09$ **b** $x=\frac{7}{3}$

3 **a** $y=\dfrac{4x-10}{3}$ **b** $y=\dfrac{c}{m}$ **4** $x=8,\ y=-13$

5 $-2\frac{1}{2}$ **6** 32 of 2 L cartons, 15 of 600 mL bottles

7 **a** 2^{n+2} **b** $-\frac{6}{7}$ **c** $3\frac{3}{8}$ **d** $\dfrac{4}{a^2b^4}$ **e** 2^{2x}

8 a 5^0 **b** $5^{\frac{3}{2}}$ **c** $5^{-\frac{1}{4}}$ **d** 5^{2a+6}

9 a -4 **b** $\dfrac{1}{4a^6}$ **c** $-\dfrac{b^3}{27}$ **10 a** 5.62 **b** 0.347 **c** 2.09

11 a $x = -3$ **b** $x \div 19.5$

EXERCISE 7A.1

1 a $\sqrt{5}$ units **b** 5 units **c** $\sqrt{10}$ units **d** 3 units
e $\sqrt{13}$ units **f** $\sqrt{34}$ units **g** $\sqrt{20}$ units **h** 5 units

EXERCISE 7A.2

1 a $\sqrt{2}$ units **b** 6 units **c** 5 units **d** 4 units
e 5 units **f** $\sqrt{20}$ units **g** $\sqrt{5}$ units **h** $\sqrt{20}$ units
2 a isosceles (AB = AC) **b** scalene
c isosceles (AB = BC) **d** equilateral **e** equilateral
f isosceles (AC = BC), equilateral if $a = 1 \pm b\sqrt{3}$
3 a $\angle$BAC **b** $\angle$ABC **c** $\angle$BAC **d** $\angle$BAC
4 a $b = 0$ or 8 **b** $b = 0$ or 4 **c** $b = \pm 3$ **d** $b = \frac{8}{3}$

EXERCISE 7B.1

1 a $\frac{1}{2}$ **b** 0 **c** $-\frac{3}{2}$ **d** $\frac{1}{3}$ **e** 1 **f** -4 **g** undefined **h** $-\frac{1}{5}$
2 a

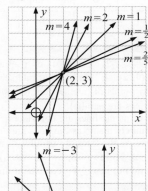

EXERCISE 7B.2

1 a 1 **b** $-\frac{1}{2}$ **c** 3 **d** 0 **e** undefined **f** $\frac{2}{5}$
g $-\frac{1}{5}$ **h** $\frac{1}{3}$
2

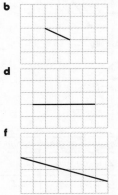

3

EXERCISE 7B.3

1 a -3 **b** $-\frac{5}{3}$ **c** $-\frac{1}{2}$ **d** $-\frac{1}{5}$ **e** $\frac{4}{3}$ **f** $\frac{2}{5}$ **g** $\frac{1}{4}$ **h** 1
2 c, d and **f** are perpendicular
3 a $a = 11$ **b** $a = 0$ **c** $a = 9$ **4 a** $t = -1$ **b** $t = 3\frac{1}{4}$
5 a $t = 2\frac{2}{3}$ **b** $t = -6$ **c** $t = 13\frac{1}{2}$ **d** $t = 3\frac{2}{7}$
6 a $a = 4$ **b i** $\frac{1}{2}$ **ii** -2
c gradient of AP $\times$ gradient of BP $= -1$, i.e., AP $\perp$ BP
$\therefore \angle$APB $= 90^o$
7 a collinear **b** collinear **c** not collinear **d** collinear
8 a $c = 3$ **b** $c = -5$

EXERCISE 7C

1 a $8\frac{1}{3}$ **b** $8\frac{1}{3}$ m/s **c** constant speed as gradient is constant
2 a 83.6 km/h **b i** 90 km/h **ii** 110 km/h **c** B to C
3 a retainer of $80 paid **b** gradient is 15, paid $15 per hour
c i $170 **ii** $350 **d** $25 per hour
4 a A has gradient 12, B has gradient 9
b Gradient is number of kilometres travelled per litre of petrol. **c** $26.11
5 a $2 flag-fall (a fee regardless of distance travelled)
b AB has gradient 1.2, BC has gradient 1, these values give the charge per km.
c Gradient is 1.1, average fare is $1.10 per km (excluding flag-fall).

EXERCISE 7D

1 a $(-1, 3)$ **b** $(-\frac{1}{2}, -2)$ **c** $(1, 1)$ **d** $(2, \frac{1}{2})$
e $(1, -1\frac{1}{2})$ **f** $(-3, 1\frac{1}{2})$ **g** $(-3\frac{1}{2}, \frac{1}{2})$ **h** $(-1, \frac{1}{2})$
2 a $(4, 4)$ **b** $(2, 0)$ **c** $(1\frac{1}{2}, 2\frac{1}{2})$ **d** $(0, 6)$
e $(1, -\frac{1}{2})$ **f** $(1, 1)$ **g** $(1\frac{1}{2}, 2)$ **h** $(-3\frac{1}{2}, 2)$
3 a B$(2, -4)$ **b** B$(3, -2)$ **c** B$(-2, 5)$ **d** B$(1, 6)$
e B$(-3, 1)$ **f** B$(-5, 1)$
5 a P$(-11, 9)$ **b** P$(12, 3)$ **6** C$(2, -5)$ **7** P$(11, -3)$
8 C$(0, 2)$ **9** $\sqrt{\frac{65}{2}}$ units
10 a D$(9, -1)$ **b** R$(3, 1)$ **c** X$(2, -1)$
11 a i P$(\frac{9}{2}, 4)$ **ii** Q$(6, 1)$ **iii** R$(\frac{9}{2}, -3)$ **iv** S$(0, 0)$
b i $-\frac{2}{3}$ **ii** $\frac{8}{3}$ **iii** $-\frac{2}{3}$ **iv** $\frac{8}{3}$
c PQRS is a parallelogram

EXERCISE 7E

1 a $y = -4$ **b** $x = 5$ **c** $x = -1$ **d** $y = 2$
e $y = 0$ **f** $x = 0$
2 $(4\frac{3}{4}, 8)$ **3 a** $(2\frac{1}{3}, 7)$ **b** 8.028 km **c** yes (6.162 km)
4 $(13.41, 8)$ or $(-5.41, 8)$

EXERCISE 7F

1 a $y = 4x - 13$ **b** $y = -3x - 5$ **c** $y = -5x + 32$
d $y = \frac{1}{2}x + \frac{7}{2}$ **e** $y = -\frac{1}{3}x + \frac{8}{3}$ **f** $y = 6$
2 a $2x - 3y = -11$ **b** $3x - 5y = -23$ **c** $x + 3y = 5$
d $2x + 7y = -2$ **e** $4x - y = -11$ **f** $2x + y = 7$
3 a $5x - 2y = 4$ **b** $2x + y = 3$ **c** $y = -2$
d $x + 5y = 2$ **e** $x - 6y = 11$ **f** $2x + 3y = -11$
4 a $x - 3y = -3$ **b** $5x - y = 1$ **c** $x - y = 3$
d $4x - 5y = 10$ **e** $x - 2y = -1$ **f** $2x + 3y = -5$
5 a 3 **b** -2 **c** 0 **d** undefined **e** $\frac{2}{3}$ **f** -3
g $\frac{2}{7}$ **h** $-\frac{2}{7}$ **i** $\frac{3}{4}$

4 a $\frac{2}{3}$ **b** -4 **c** 1 **d** 0 **e** $-\frac{2}{5}$ **f** undefined

6 a yes **b** no **c** yes

7 a $k = 1$ **b** $k = -11$ **8 a** $a = 2$ **b** $a = 7$

9 a i $(3, 5\frac{1}{2})$ **ii** $(4\frac{1}{5}, 4)$
  b state the possible x-value restriction, i.e., $1 < x < 5$
  c no

10 a $y = 2x + 7$ **b** $y = 4x - 6$ **c** $y = -3x - 1$
  d $y = -\frac{1}{2}x + 2$ **e** $y = 8$ **f** $x = 2$

11 a $y = x + 1$ **b** $y = \frac{4}{3}x - 1$ **c** $y = -\frac{2}{3}x + 2$
  d $y = \frac{2}{3}x + \frac{13}{3}$ **e** $y = -2x - 2$ **f** $y = -\frac{3}{7}x - \frac{9}{7}$

12 a $M = \frac{1}{3}t + 2$ **b** $N = \frac{1}{2}x - 1$ **c** $G = -\frac{1}{2}s + 2$
  d $H = -\frac{5}{4}g + 2$ **e** $F = \frac{1}{10}x + 1$ **f** $P = -\frac{1}{6}t - 2$

EXERCISE 7G.1

1 a $y = \frac{1}{2}x + 2$

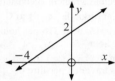

b $y = 2x + 1$

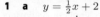

c $y = -x + 3$

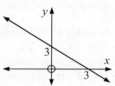

d $y = -3x + 2$

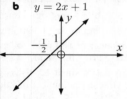

e $y = -\frac{1}{2}x$

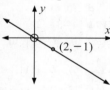

f $y = -2x - 2$

g $y = \frac{3}{2}x$

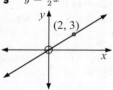

h $y = \frac{2}{3}x + 2$

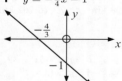

i $y = -\frac{3}{4}x - 1$

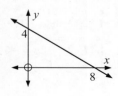

2 a $x + 2y = 8$

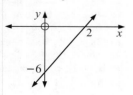

b $3x - y = 6$

c $2x - 3y = 6$

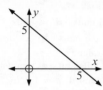

d $4x + 3y = 12$

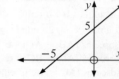

e $x + y = 5$

f $x - y = -5$

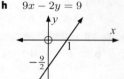

g $2x - y = -4$

h $9x - 2y = 9$

i $3x + 4y = -15$

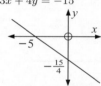

3

Equation of line	Slope	x-int.	y-int.
a	$2x - 3y = 6$	$\frac{2}{3}$	3	-2
b	$4x + 5y = 20$	$-\frac{4}{5}$	5	4
c	$y = -2x + 5$	-2	$\frac{5}{2}$	5
d	$x = 8$	undef.	8	no y-int.
e	$y = 5$	0	no x-int.	5
f	$x + y = 11$	-1	11	11

EXERCISE 7G.2

1 a $(2, -1)$ **b** $(-1, -2)$ **c** $(3, 0)$ **d** $(-3, 6)$ **e** $(\frac{2}{3}, 7)$
  f $(-3, 2)$ **g** $(4, 2)$ **h** no point of intersection
  i infinitely many points of intersection (lines coincident)

2 a none, as the lines are parallel
  b infinitely many, as the lines are coincident
  c If $k = 5$, infinitely many, as the lines are coincident;
    if $k \neq 5$, none, as the lines are parallel.

3 a $(4, 2)$ **b** $(-2, 3)$ **c** $(-3, 6)$ **d** $(4, 0)$
  e parallel lines do not meet **f** coincident lines

EXERCISE 7H

1 a $x - y = 4$ **b** $2x - y = -6$ **c** $12x - 10y = -35$
  d $y = 1$

2 $2x - 3y = -5$ **3** $(3, 3)$

4 a P(5, 3), Q(4, 0), R(2, 2)
  b i $x - y = 2$ **ii** $3x + y = 12$ **iii** $x + 3y = 8$
  c X$(3\frac{1}{2}, 1\frac{1}{2})$ **d** yes

e The perpendicular bisectors meet at a point.

f X is the centre of the circle which could be drawn through A, B and C.

REVIEW SET 7A

1 **a** 5 units **b** $-\frac{4}{3}$ **c** (1, 4)

2 x-intercept is 2, y-intercept is -5, gradient is $\frac{5}{2}$

3 $2x + y = 6$ **4** $a = 7\frac{1}{5}$ **5** $c = -5$

6 **a** $y = -3x + 4$ **b** $x + 2y = 5$ **7** (8, 7) **8** 5 units

9 **a** $4x - 3y = 19$ **b** $4y - x = 6$

10 **a** **i** $y = 2$ **ii** $x = 0$ **iii** $4x - 3y = -18$
b **i** $(-3, 2)$ **ii** $(0, 6)$

11 C(3, 0) **12** $k = 4 \pm 2\sqrt{6}$

REVIEW SET 7B

1 $(5, 1\frac{1}{2})$ **2** **a** $y = -3x + 5$ **b** $y = -\frac{3}{2}x + \frac{5}{2}$

3 **a** $(0, 7)$, $(\frac{14}{3}, 0)$ **b** $(0, -4)$, $(\frac{12}{5}, 0)$ **4** yes

5 **a** $3x + 5y = 12$ **b** $7x + 2y = 32$

6 **a** $k = \frac{3}{4}$ **b** $k = -12$ **7** $(-1, 4\frac{1}{2})$ **8** $(\frac{11}{4}, 5)$

9 **a** PQ = PR = $\sqrt{20}$ units **b** (4, 4)
c gradient of PM = $-\frac{1}{3}$, gradient of QR = 3
d

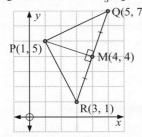

10 **a** gradient of MN = 0 = gradient of AC
b MN = 3 units, AC = 6 units

REVIEW SET 7C

1 **a** $2x - 3y = 17$ **b** $x - 3y = 11$ **2** **a** $k = -\frac{21}{5}$ **b** $k = \frac{15}{7}$

3 **a** $T_1(0, 2 + 2\sqrt{2})$, $T_2(0, 2 - 2\sqrt{2})$
b $2\sqrt{2}x - y = -2\sqrt{2} - 2$

4 **a**

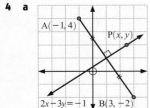

b $2x - 3y = -1$

5 isosceles (KL = LM), $\angle$KLM = 90°

6 K(10, 15) **7** $5x - 3y = 4$ **8** $x - 2y = -9$

9 **a** **i** gradient of AB = $-\frac{1}{4}$ = gradient of DC
ii gradient of BC = $\frac{3}{5}$ = gradient of AD
b a parallelogram
c AB = $\sqrt{68}$ units = DC, BC = $\sqrt{306}$ units = AD
d **i** $(\frac{5}{2}, \frac{5}{2})$ **ii** $(\frac{5}{2}, \frac{5}{2})$ **e** diagonals bisect each other

10 **a** AB = BC = CD = DA = 5 units **b** (2, 1) and (2, 1)
c gradient of AC = -2, gradient of BD = $\frac{1}{2}$

EXERCISE 8A.1

1 **a** $6x$ **b** $20x$ **c** $-14x$ **d** $-6x$ **e** $2x^2$ **f** $6x^2$
g $-2x^2$ **h** $-12x$ **i** $2x^2$ **j** $-3x^3$ **k** $2x^3$
l $-6d^2$ **m** a^2 **n** $4a^2$ **o** $2a^4$ **p** $-3a^3$

2 **a** $3x + 6$ **b** $10 - 2x$ **c** $-x - 2$ **d** $x - 3$
e $-2x - 8$ **f** $3 - 6x$ **g** $x^2 + 3x$ **h** $2x^2 - 10x$
i $a^2 + ab$ **j** $ab - a^2$ **k** $2x^2 - x$ **l** $2x^3 - 2x^2 - 4x$

3 **a** $5x - 2$ **b** $3a - 2b$ **c** $a + 2b$ **d** $15 - 3y$
e $-6y - 10$ **f** $15x - 8$ **g** $a + 5b$ **h** $x^2 + 6x - 6$
i $x^2 + 2x + 6$ **j** $2x^2 - x$ **k** $-2x^2 + 2x$ **l** $x^2 - y^2$
m $5 - 3x$ **n** $8x - 8$ **o** $6x^2 - 22x$

EXERCISE 8A.2

1 **a** $A_1 = ac$ **b** $A_2 = ad$ **c** $A_3 = bc$ **d** $A_4 = bd$
e $A = ac + ad + bc + bd$
$ac + ad + bc + bd = (a + b)(c + d)$

2 **a** $x^2 + 10x + 21$ **b** $x^2 + x - 20$ **c** $x^2 + 3x - 18$
d $x^2 - 4$ **e** $x^2 - 5x - 24$ **f** $6x^2 + 11x + 4$
g $1 + 2x - 8x^2$ **h** $12 + 5x - 2x^2$ **i** $6x^2 - x - 2$
j $25 - 10x - 3x^2$ **k** $7 + 27x - 4x^2$ **l** $25x^2 + 20x + 4$

3 **a** $x^2 - 4$ **b** $a^2 - 25$ **c** $16 - x^2$ **d** $4x^2 - 1$
e $25a^2 - 9$ **f** $16 - 9a^2$

4 **a** $x^2 + 6x + 9$ **b** $x^2 - 4x + 4$ **c** $9x^2 - 12x + 4$
d $1 - 6x + 9x^2$ **e** $9 - 24x + 16x^2$ **f** $25x^2 - 10xy + y^2$

EXERCISE 8A.3

1 **a** $x^2 - 4$ **b** $x^2 - 4$ **c** $4 - x^2$ **d** $4 - x^2$ **e** $x^2 - 1$
f $1 - x^2$ **g** $x^2 - 49$ **h** $c^2 - 64$ **i** $d^2 - 25$
j $x^2 - y^2$ **k** $16 - d^2$ **l** $25 - e^2$

2 **a** $4x^2 - 1$ **b** $9x^2 - 4$ **c** $16y^2 - 25$ **d** $4y^2 - 25$
e $9x^2 - 1$ **f** $1 - 9x^2$ **g** $4 - 25y^2$ **h** $9 - 16a^2$
i $16 - 9a^2$

3 **a** $4a^2 - b^2$ **b** $a^2 - 4b^2$ **c** $16x^2 - y^2$ **d** $16x^2 - 25y^2$
e $4x^2 - 9y^2$ **f** $49x^2 - 4y^2$

EXERCISE 8A.4

1 **a** $A_1 = a^2$ **b** $A_2 = ab$ **c** $A_3 = ab$ **d** $A_4 = b^2$
e $A = a^2 + 2ab + b^2$, $(a + b)^2 = a^2 + 2ab + b^2$

2 **a** $x^2 + 10x + 25$ **b** $x^2 + 8x + 16$ **c** $x^2 + 14x + 49$
d $a^2 + 4a + 4$ **e** $9 + 6c + c^2$ **f** $25 + 10x + x^2$

3 **a** $x^2 - 6x + 9$ **b** $x^2 - 4x + 4$ **c** $y^2 - 16y + 64$
d $a^2 - 14a + 49$ **e** $25 - 10x + x^2$ **f** $16 - 8y + y^2$

4 **a** $9x^2 + 24x + 16$ **b** $4a^2 - 12a + 9$ **c** $9y^2 + 6y + 1$
d $4x^2 - 20x + 25$ **e** $9y^2 - 30y + 25$ **f** $49 + 28a + 4a^2$
g $1 + 10x + 25x^2$ **h** $49 - 42y + 9y^2$ **i** $9 + 24a + 16a^2$

5 **a** $x^4 + 4x^2 + 4$ **b** $y^4 - 6y^2 + 9$ **c** $9a^4 + 24a^2 + 16$
d $1 - 4x^2 + 4x^4$ **e** $x^4 + 2x^2y^2 + y^4$ **f** $x^4 - 2a^2x^2 + a^4$

6 **a** $-x^2 - 3x - 8$ **b** $x^2 + x + 2$ **c** $2x^2 + 6x + 5$
d $-6x - 13$ **e** $11 - 13x + 3x^2$ **f** $10x^2 - 7x - 5$
g $3x^2 - 2x - 10$ **h** $3x^2 - x - 10$ **i** $2x^2 + 2x + 5$
j $-6x - 3$

EXERCISE 8B

1 **a** $x^3 + 4x^2 + 5x + 6$ **b** $x^3 + 5x^2 + 2x - 8$
c $x^3 + 3x^2 + 3x + 2$ **d** $x^3 + 4x^2 - 6x - 5$
e $2x^3 + 3x^2 + 9x + 4$ **f** $3x^3 - 5x^2 - 7x + 6$
g $2x^3 + 3x^2 + 4$ **h** $6x^3 - 5x^2 + 5x - 2$

2 **a** $x^3 + 3x^2 + 3x + 1$ **b** $x^3 + 9x^2 + 27x + 27$
c $x^3 - 3x^2 + 3x - 1$ **d** $x^3 - 9x^2 + 27x - 27$
e $8x^3 + 12x^2 + 6x + 1$ **f** $27x^3 - 54x^2 + 36x - 8$

3 **a** $x^3 + 5x^2 + 6x$ **b** $x^3 - 3x^2 - 4x$ **c** $x^3 - 5x^2 + 6x$
d $2x^3 + 8x^2 + 6x$ **e** $-2x^3 + 10x^2 - 8x$ **f** $x^3 + x^2 - 6x$
g $6x - 9x^2 - 6x^3$ **h** $x - x^2 - 6x^3$

4 **a** $x^3 + 6x^2 + 11x + 6$ **b** $x^3 + x^2 - 10x + 8$

c $x^3 - 8x^2 + 19x - 12$ **d** $2x^3 + x^2 - 5x + 2$
e $3x^3 + 14x^2 + 17x + 6$ **f** $4x^3 + 16x^2 - x - 4$
g $-3x^3 + 7x^2 - 4$ **h** $-3x^3 + 10x^2 - x - 6$

5 a 4 **b** 6 **c** 6 **d** 9 **e** 8 **f** 12 **g** 8 **h** 12

EXERCISE 8C.1

1 a $x(3x+5)$ **b** $x(2x-7)$ **c** $3x(x+2)$ **d** $4x(x-2)$
e $x(2x-9)$ **f** $3x(x+5)$ **g** $4x(1+2x)$ **h** $5x(1-2x)$
i $4x(3-x)$

2 a $(x+2)(x-3)$ **b** $(x-1)(x-4)$ **c** $(x+1)(x+3)$
d $(x-2)(x+1)$ **e** $(x+3)(x+4)$ **f** $(x+4)(x+6)$
g $(x-3)(x-4)$ **h** $(x+4)(x+2)$ **i** $(x-4)(x-9)$

EXERCISE 8C.2

1 a $(x+2)(x-2)$ **b** $(2+x)(2-x)$ **c** $(x+9)(x-9)$
d $(5+x)(5-x)$ **e** $(2x+1)(2x-1)$ **f** $(3x+4)(3x-4)$
g $(2x+3)(2x-3)$ **h** $(6+7x)(6-7x)$

2 a $3(x+3)(x-3)$ **b** $-2(x+2)(x-2)$
c $3(x+5)(x-5)$ **d** $-5(x+1)(x-1)$
e $2(2x+3)(2x-3)$ **f** $-3(3x+5)(3x-5)$

3 a $(x+\sqrt{3})(x-\sqrt{3})$ **b** no linear factors
c $(x+\sqrt{15})(x-\sqrt{15})$ **d** $3(x+\sqrt{5})(x-\sqrt{5})$
e $(x+1+\sqrt{6})(x+1-\sqrt{6})$ **f** no linear factors
g $(x-2+\sqrt{7})(x-2-\sqrt{7})$
h $(x+3+\sqrt{17})(x+3-\sqrt{17})$ **i** no linear factors

4 a $(x+3)(x-1)$ **b** $4(x+2)(x-1)$ **c** $(x-5)(x+3)$
d $3(x+1)(3-x)$ **e** $(3x+2)(x-2)$ **f** $(2x+3)(4x-3)$
g $(3x-1)(x+3)$ **h** $8x(x-1)$ **i** $-3(4x+3)$

EXERCISE 8C.3

1 a $(x+3)^2$ **b** $(x+4)^2$ **c** $(x-3)^2$ **d** $(x-4)^2$
e $(x+1)^2$ **f** $(x-5)^2$ **g** $(y+9)^2$ **h** $(m-10)^2$
i $(t+6)^2$

2 a $(3x+1)^2$ **b** $(2x-1)^2$ **c** $(3x+2)^2$ **d** $(5x-1)^2$
e $(4x+3)^2$ **f** $(5x-2)^2$ **g** $(x-1)^2$ **h** $2(x+2)^2$
i $3(x+5)^2$

EXERCISE 8C.4

1 a $(x+2)(x+1)$ **b** $(x+3)(x+2)$ **c** $(x-3)(x+2)$
d $(x+5)(x-2)$ **e** $(x+7)(x-3)$ **f** $(x+4)^2$
g $(x-7)^2$ **h** $(x+7)(x-4)$ **i** $(x+5)(x+2)$
j $(x-8)(x-3)$ **k** $(x+11)(x+4)$ **l** $(x+7)(x-6)$
m $(x-8)(x+7)$ **n** $(x-9)^2$ **o** $(x-8)(x+4)$

2 a $2(x-4)(x+1)$ **b** $3(x+4)(x-1)$ **c** $5(x+3)(x-1)$
d $4(x+5)(x-4)$ **e** $2(x-5)(x+3)$ **f** $3(x+7)(x-3)$
g $2(x-5)(x+4)$ **h** $3(x-2)^2$ **i** $7(x+4)(x-1)$
j $5(x+5)(x-2)$ **k** $2(x+7)(x-3)$ **l** $(x-8)(x+4)$

EXERCISE 8C.5

1 a $3x(x+3)$ **b** $(2x+1)(2x-1)$ **c** $5(x+\sqrt{3})(x-\sqrt{3})$
d $x(3-5x)$ **e** $(x+8)(x-5)$ **f** $2(x+4)(x-4)$
g does not factorise **h** $(x+5)^2$ **i** $(x-3)(x+2)$
j $(x-13)(x-3)$ **k** $(x-12)(x+5)$ **l** $(x-4)(x+2)$
m $(x+5)(x+6)$ **n** $(x+8)(x-2)$ **o** $(x-8)(x+3)$
p $3(x+6)(x-4)$ **q** $4(x-5)(x+3)$ **r** $3(x-11)(x-3)$

EXERCISE 8D

1 a $(2x+3)(x+1)$ **b** $(2x+5)(x+1)$ **c** $(7x+2)(x+1)$
d $(3x+4)(x+1)$ **e** $(3x+1)(x+4)$ **f** $(3x+2)(x+2)$
g $(4x+1)(2x+3)$ **h** $(7x+1)(3x+2)$ **i** $(3x+1)(2x+1)$
j $(6x+1)(x+3)$ **k** $(5x+1)(2x+3)$ **l** $(7x+1)(2x+5)$

2 a $(2x+1)(x-5)$ **b** $(3x-1)(x+2)$ **c** $(3x+1)(x-2)$
d $(2x-1)(x+2)$ **e** $(2x+5)(x-1)$ **f** $(5x+1)(x-3)$
g $(5x-3)(x-1)$ **h** $(11x+2)(x-1)$ **i** $(3x+2)(x-3)$
j $(2x+3)(x-3)$ **k** $(3x-2)(x-5)$ **l** $(5x+2)(x-3)$
m $(3x-2)(x+4)$ **n** $(2x-1)(x+9)$ **o** $(2x-3)(x+6)$
p $(2x-3)(x+7)$ **q** $(5x+2)(3x-1)$ **r** $(21x+1)(x-3)$
s $(3x-2)^2$ **t** $(4x-5)(3x+8)$ **u** $(8x-3)(2x+5)$

EXERCISE 8E.1

1 a $x = \pm 4$ **b** $x = \pm 3$ **c** $x = \pm 3$ **d** $x = \pm\sqrt{6}$
e no solution **f** $x = 0$ **g** $x = \pm 3$ **h** no solution
i $x = \pm\sqrt{3}$

2 a $x = 5$ or -1 **b** $x = 1$ or -9 **c** no solution
d $x = 4 \pm \sqrt{2}$ **e** no solution **f** $x = -2$
g $x = -2\frac{1}{2}$ **h** $x = 0$ or $\frac{4}{3}$ **i** $x = \dfrac{1 \pm \sqrt{24}}{2}$

EXERCISE 8E.2

1 a $x = 0$ or -3 **b** $x = 0$ or 5 **c** $x = 1$ or 3
d $x = 0$ or 2 **e** $x = 0$ or $-\frac{1}{2}$ **f** $x = -2$ or $\frac{1}{2}$
g $x = -\frac{1}{2}$ or $\frac{1}{2}$ **h** $x = -2$ or 7 **i** $x = 5$ or $-\frac{2}{3}$
j $x = 0$ **k** $x = 3$ **l** $x = \frac{1}{2}$

EXERCISE 8E.3

1 a $x = 0$ or -2 **b** $x = 0$ or $-\frac{5}{2}$ **c** $x = 0$ or $\frac{3}{4}$
d $x = 0$ or $\frac{5}{4}$ **e** $x = 0$ or 3 **f** $x = 0$ or $\frac{1}{2}$

2 a $x = -7$ or -2 **b** $x = -5$ or -6 **c** $x = -5$ or 3
d $x = -4$ or 3 **e** $x = 3$ or 2 **f** $x = 2$
g $x = 3$ or -2 **h** $x = 12$ or -5 **i** $x = 10$ or -7
j $x = -5$ or 2 **k** $x = 3$ or 4 **l** $x = 12$ or -3

3 a $x = \frac{1}{2}$ or 2 **b** $x = -3$ or $\frac{1}{3}$ **c** $x = -4$ or $-\frac{5}{3}$
d $x = \frac{1}{2}$ or -3 **e** $x = \frac{1}{2}$ or 5 **f** $x = -1$ or $-\frac{5}{2}$
g $x = -\frac{1}{3}$ or -4 **h** $x = -\frac{2}{5}$ or 3 **i** $x = \frac{1}{2}$ or -9
j $x = 1$ or $-\frac{5}{2}$ **k** $x = \frac{4}{3}$ or -2 **l** $x = \frac{3}{2}$ or -6

4 a $x = \frac{1}{3}$ or $-\frac{5}{2}$ **b** $x = \frac{2}{3}$ or $-\frac{1}{2}$ **c** $x = -\frac{1}{2}$ or $-\frac{1}{3}$
d $x = -\frac{1}{21}$ or 3 **e** $x = \frac{2}{5}$ or $-\frac{1}{2}$ **f** $x = -\frac{3}{10}$ or 1

5 a $x = -4$ or -3 **b** $x = -3$ or 1 **c** $x = \pm 3$
d $x = -1$ or $\frac{2}{3}$ **e** $x = -\frac{1}{2}$ **f** $x = \frac{5}{2}$ or 4

6 a $x = \pm\sqrt{6}$ **b** $x = \pm\sqrt{8}$ **c** $x = \pm\sqrt{10}$
d $x = 4$ or -3 **e** $x = -1$ or -5 **f** $x = 2$ or -1
g $x = \frac{1}{2}$ or -1 **h** $x = 1$ or $-\frac{1}{3}$ **i** $x = -1$ or 4

EXERCISE 8F

1 a i 1 **ii** $(x+1)^2 = 6$ **b i** 1 **ii** $(x-1)^2 = -6$
c i 9 **ii** $(x+3)^2 = 11$ **d i** 9 **ii** $(x-3)^2 = 6$
e i 25 **ii** $(x+5)^2 = 26$ **f i** 16 **ii** $(x-4)^2 = 21$
g i 36 **ii** $(x+6)^2 = 49$ **h i** $\frac{25}{4}$ **ii** $(x+\frac{5}{2})^2 = 4\frac{1}{4}$
i i $\frac{49}{4}$ **ii** $(x-\frac{7}{2})^2 = 16\frac{1}{4}$

2 a $x = 2 \pm \sqrt{3}$ **b** $x = 1 \pm \sqrt{3}$ **c** $x = 2 \pm \sqrt{7}$
d $x = -1 \pm \sqrt{2}$ **e** no solution **f** $x = -2 \pm \sqrt{3}$

g $x = -3 \pm \sqrt{6}$ **h** no solution **i** $x = -4 \pm \sqrt{2}$

3 a $x = -1$ or -2 **b** $x = 2 \pm \sqrt{12}$ **c** $x = 2$ or 3

d $x = \dfrac{-1 \pm \sqrt{5}}{2}$ **e** $x = \dfrac{-3 \pm \sqrt{13}}{2}$ **f** $x = \dfrac{-5 \pm \sqrt{33}}{2}$

4 a -0.551 or -5.45 **b** no solutions **c** -2.59 or -5.41
d 0.225 or -2.22 **e** 3.29 or 0.709 **f** 1.63 or 0.368

EXERCISE 8G

1 8 or -9 **2** 8 or -12 **3** 8 or -3 **4** 15 and -6
5 -2 and 3 or 2 and -3 **6** width is 7 cm
7 altitude is 9 cm **8** 8 m $\times$ 12 m
9 Either 5 m from wall and 14 m wide
    or 7 m from wall and 10 m wide.
10 a $x = 5$ **b** $x = 6$ **11** 8 cm, 15 cm, 17 cm
12 15 rows **13** BC is 5 cm or 16 cm **14** 3 cm $\times$ 3 cm
15 c $1\frac{1}{2}$ metres wide

REVIEW SET 8A

1 a $2x - 10$ **b** $7 - 7x$
2 a $3x^2 + 4x - 4$ **b** $4x^2 + 4x + 1$ **c** $9x^2 - 1$
3 a $(x-7)(x+3)$ **b** $(2x+5)(2x-5)$ **c** $3(x-6)(x+4)$
  d $(3x+2)(2x-1)$
4 a $x = 3$ or 8 **b** $x = -\frac{2}{5}$ or $\frac{3}{2}$
5 $(x+3)^2 = -2$ has no solution **6** 13 cm $\times$ 20 cm
7 a $-x^2 - 8x - 4$ **b** $3x^3 + 4x^2 + 17x - 14$

REVIEW SET 8B

1 a $-x$ **b** $2x^2 - 5x$
2 a $2x^2 + x - 15$ **b** $9x^2 - 24x + 16$ **c** $9x^2 - 4$
3 a $(x-11)(x+3)$ **b** $(x+8)(x-4)$ **c** $(x-5)^2$
4 a $x = -8$ or 3 **b** $x = \pm 3$ **c** $x = -\frac{3}{4}$ or $\frac{1}{2}$
5 $x = 1 \pm \sqrt{101}$ **6** 9 cm $\times$ 12 cm **7** 5 or 2
8 a $5x^2 - 5x - 5$ **b** $2x^3 - 3x^2 + 4x + 3$

REVIEW SET 8C

1 a $-14x + 35$ **b** $7x - 18$
2 a $5x^2 + 14x - 3$ **b** $9x^2 - 6x + 1$ **c** $1 - 25x^2$
3 a $(x-9)(x+2)$ **b** $(2x-7)(2x+7)$ **c** $2(x-10)(x+3)$
  d $(5x+2)(x+2)$
4 a $x = 2$ or 3 **b** $x = 4$ **c** $x = -\frac{7}{3}$ or 3
5 $x = 7 \pm \sqrt{42}$ **6** The number is $2 + \sqrt{3}$ or $2 - \sqrt{3}$.
7 8 cm, 15 cm, 17 cm
8 a $-x^2 - 3x - 15$ **b** $2x^3 + 3x^2 + x + 15$

EXERCISE 9A

1 a, d, e **2** a, b, c, e, h **3** No, e.g., $x = 1$
4 $y = \pm\sqrt{9 - x^2}$, i.e., a vertical line cuts it more than once.

EXERCISE 9B

1 a Domain $\{x : \ x \geqslant -1\}$, Range $\{y : \ y \leqslant 3\}$
b Domain $\{x : \ -1 < x \leqslant 5\}$, Range $\{y : \ 1 < y \leqslant 3\}$
c Domain $\{x : \ x \neq 2\}$, Range $\{y : \ y \neq -1\}$
d Domain $\{x : \ x \in R\}$, Range $\{y : \ 0 < y \leqslant 2\}$
e Domain $\{x : \ x \in R\}$, Range $\{y : \ y \geqslant -1\}$
f Domain $\{x : \ x \in R\}$, Range $\{y : \ y \leqslant \frac{25}{4}\}$
g Domain $\{x : \ x \geqslant -4\}$, Range $\{y : \ y \geqslant -3\}$
h Domain $\{x : \ x \in R\}$, Range $\{y : \ y > -2\}$
i Domain $\{x : \ x \neq \pm 2\}$, Range $\{y : \ y \leqslant -1$ or $y > 0\}$

2 a Domain $\{x : \ x \geqslant 0\}$, Range $\{y : \ y \geqslant 0\}$
b Domain $\{x : \ x \neq 0\}$, Range $\{y : \ y > 0\}$
c Domain $\{x : \ x \leqslant 4\}$, Range $\{y : \ y \geqslant 0\}$
d Domain $\{x : \ x \in R\}$, Range $\{y : \ y \geqslant -2\frac{1}{4}\}$
e Domain $\{x : \ x \in R\}$, Range $\{y : \ y \leqslant 2\frac{1}{12}\}$
f Domain $\{x : \ x \neq 0\}$, Range $\{y : \ y \leqslant -2$ or $y \geqslant 2\}$
g Domain $\{x : \ x \neq 2\}$, Range $\{y : \ y \neq 1\}$
h Domain $\{x : \ x \in R\}$, Range $\{y : \ y \in R\}$
i Domain $\{x : x \neq -1$ or $2\}$, Range $\{y : y \leqslant \frac{1}{3}$ or $y \geqslant 3\}$
j Domain $\{x : \ x \neq 0\}$, Range $\{y : y \geqslant 2\}$
k Domain $\{x : \ x \neq 0\}$, Range $\{y : \ y \leqslant -2$ or $y \geqslant 2\}$
l Domain $\{x : \ x \in R\}$, Range $\{y : \ y \geqslant -8\}$

EXERCISE 9C

1 a 2 **b** 8 **c** -1 **d** -13 **e** 1
2 a -3 **b** 3 **c** 3 **d** -3 **e** $7\frac{1}{2}$
3 a 2 **b** 2 **c** -16 **d** -68 **e** $\frac{17}{4}$
4 a $7 - 3a$ **b** $7 + 3a$ **c** $-3a - 2$ **d** $10 - 3b$ **e** $1 - 3x$
5 a $2x^2 + 19x + 43$ **b** $2x^2 - 11x + 13$ **c** $2x^2 - 3x - 1$
  d $2x^4 + 3x^2 - 1$ **e** $2x^4 - x^2 - 2$
6 a i $-\frac{7}{2}$ **ii** $-\frac{3}{4}$ **iii** $-\frac{4}{9}$ **b** $x = 4$ **c** $\dfrac{2x + 7}{x - 2}$ **d** $x = \frac{9}{5}$

7 f is the function which converts x into $f(x)$ whereas $f(x)$ is the value of the function at any value of x.
8 $f(a)f(b) = 2^a 2^b = 2^{a+b}$ {index law} and $f(a + b) = 2^{a+b}$
9 a $x + 3$ **b** $4 + h$
10 a \$6210, value after 4 years **b** $t = 4.5$, the time for the photocopier to reach a value of \$5780. **c** \$9650
11

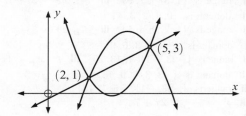

12 $f(x) = -2x + 5$ **13** $a = 3, \ b = -2$
14 $T(x) = 3x^2 - x - 4$

EXERCISE 9D

1 a i

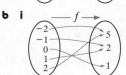

  ii domain is $\{-2, -1, 0, 1, 2\}$
  iii range is $\{-7, -4, -1, 2, 5\}$

b i

  ii domain is $\{-2, -1, 0, 1, 2\}$
  iii range is $\{1, 2, 5\}$

c i

  ii domain is $\{-2, -1, 0, 1, 2\}$
  iii range is $\{-5, -1, 3, 7, 11\}$

d i

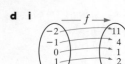

ii domain is $\{-2, -1, 0, 1, 2\}$
iii range is $\{1, 2, 4, 7, 11\}$

e i

ii domain is $\{-2, -1, 0, 1, 2\}$
iii range is $\{-8, -1, 0, 1, 8\}$

f i

ii domain is $\{-2, -1, 0, 1, 2\}$
iii range is $\{\frac{1}{9}, \frac{1}{3}, 1, 3, 9\}$

g i

ii domain is $\{-2, -1, 0, 1, 2\}$
iii range is $\{\frac{1}{5}, \frac{1}{4}, \frac{1}{3}, \frac{1}{2}, 1\}$

h i

ii domain is $\{-2, -1, 1, 2\}$
iii range is $\{-2, -\frac{1}{2}, \frac{5}{2}, 4\}$

2 a i domain is $\{-2, -1, 0, 1, 2\}$
ii range is $\{1, 3, 5, 7, 9\}$　**iii** $f : x \longmapsto 2x + 5$
b i domain is $\{0, 1, 2, 3, 4\}$
ii range is $\{1, 2, 3, 4, 5\}$　**iii** $f : x \longmapsto 5 - x$
c i domain is $\{-3, 0, 3, 6, 9\}$
ii range is $\{-23, -11, 1, 13, 25\}$　**iii** $f : x \longmapsto 4x - 11$
d i domain is $\{-3, 3, 6, 9, 15\}$
ii range is $\{-23, -11, -5, 1, 13\}$
iii $f : x \longmapsto -2x + 7$

3 a i

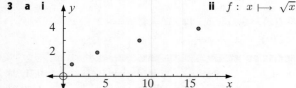

ii $f : x \longmapsto \sqrt{x}$

b i

ii $f : x \longmapsto x^2$

c i

ii $f : x \longmapsto 2^x$

d i

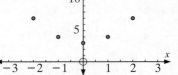

ii $f : x \longmapsto x^2 + 3$

EXERCISE 9E

1 a \$170　**b** \$320　**c** \$720
2 a 100°C　**b** 0°C　**c** 40°C　**d** 190°C
3 a $V(0) = \$25\,000$, $V(0)$ is the initial value of the car
b $V(3) = \$16\,000$, $V(3)$ is the value of the car after 3 years
c $t = 5$, where 5 years is the time taken for the value of the car to decrease to \$10 000
4 $f(x) = 4x - 1$
5 a

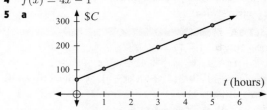

b $C(t) = 45t + 60$　**c** \$352.50
6 a

t	0	1	2	3	4	5
V	265	254	243	232	221	210

b

c $V(t) = 265 - 11t$　**d i** 100 litres　**ii** $\doteqdot 24.1$ min
7 a $C(n) = 158 + 365n$　**b** \$1526.75　**c** $\doteqdot 13\,266$ km
8 a $W(s) = 250 + 50s$　**b** \$1925　**c** \$11 600
9 a

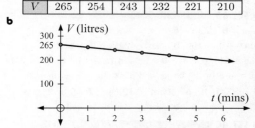

b $n = 5$　**c** $n > 5$　**d** 55
10 a $R(n) = 7n$, $C(n) = 300 + 2.5n$
b

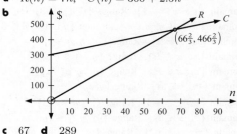

c 67　**d** 289

11 a $C(n) = 6000 + 3.25n$, $R(n) = 9.5n$

b

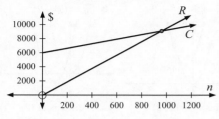

c 960 **d** 2560

12 a $C(n) = 2100 + 28n$, $R(n) = 70n$

b

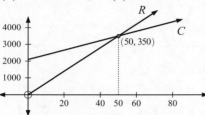

50 carburettors required to break even

c $P(n) = 42n - 2100$

d i \$3150 **ii** at least 81 carburettors

EXERCISE 9F

1 **a**, **c**, **d**, **e**

2 a $y = 20$ **b** $y = 27$ **c** $y = -4$ **d** $y = 37$

3 a 3 **b** -5 **c** 8

4 a no **b** no **c** yes **d** no

5 a $x = -3$ **b** $x = -3$ or -2 **c** $x = 1$ or 4 **d** no solution

6 a $x = 0$ or $\frac{2}{3}$ **b** $x = 3$ or -2 **c** $x = \frac{1}{2}$ or -7 **d** $x = 3$

7 a i 25 m **ii** 25 m **iii** 45 m

b i 2 secs and 4 secs **ii** 0 secs and 6 secs

c once going up and once coming down

8 a i $-\$30$ **ii** \$105 **b** 6 or 58 cakes

EXERCISE 9G.1

1 a

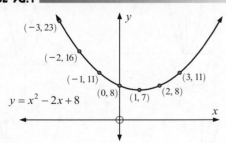

$y = x^2 - 2x + 8$

b

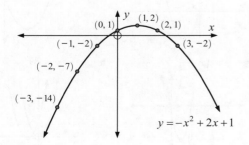

$y = -x^2 + 2x + 1$

c

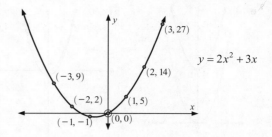

$y = 2x^2 + 3x$

d

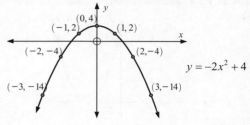

$y = -2x^2 + 4$

e

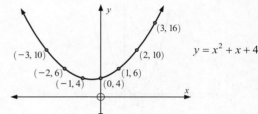

$y = x^2 + x + 4$

f

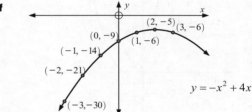

$y = -x^2 + 4x - 9$

EXERCISE 9G.2

1 a

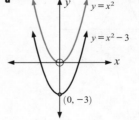

b

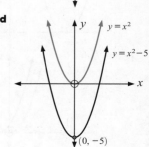

c

d

e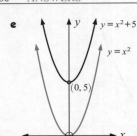

f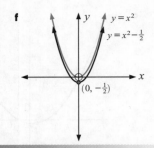

EXERCISE 9G.3

1 a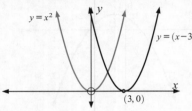

b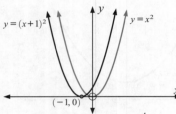

c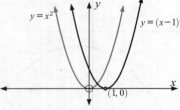

d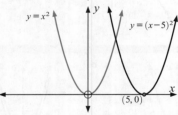

e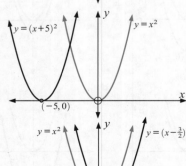

f

EXERCISE 9G.4

1 a $y = (x-1)^2 + 3$

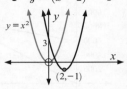

vertex is at $(1, 3)$

b $y = (x-2)^2 - 1$

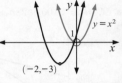

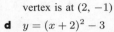

vertex is at $(2, -1)$

c $y = (x+1)^2 + 4$

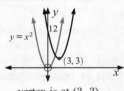

vertex is at $(-1, 4)$

d $y = (x+2)^2 - 3$

vertex is at $(-2, -3)$

e $y = (x+3)^2 - 2$

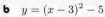

vertex is at $(-3, -2)$

f $y = (x-3)^2 + 3$

vertex is at $(3, 3)$

EXERCISE 9G.5

1 a $y = (x-1)^2 + 2$

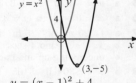

b $y = (x-3)^2 - 5$

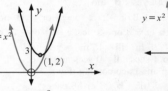

c $y = (x+2)^2 - 6$

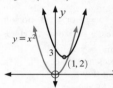

d $y = (x-1)^2 + 4$

e $y = (x-2)^2 - 4$

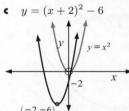

f $y = (x+\frac{3}{2})^2 - \frac{9}{4}$

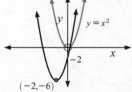

g $y = (x+\frac{5}{2})^2 - \frac{33}{4}$

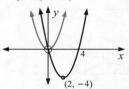

h $y = (x-\frac{3}{2})^2 - \frac{1}{4}$

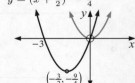
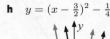

i $y = (x - \frac{5}{2})^2 - \frac{21}{4}$

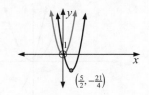

$(\frac{5}{2}, -\frac{21}{4})$

3 **a** $y = (x - 2)^2 + 3$ **b** $y = (x + 3)^2 - 6$
   c $y = (x + 2)^2 + 1$ **d** $y = (x + 1)^2 - 5$
   e $y = (x - \frac{3}{2})^2 - \frac{5}{4}$ **f** $y = (x - \frac{9}{2})^2 - \frac{101}{4}$

EXERCISE 9G.6

1 a

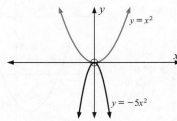

$y = x^2$

$y = 5x^2$

i $y = 5x^2$ is 'thinner' than $y = x^2$
ii graph opens upwards

b

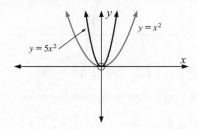

$y = x^2$

$y = -5x^2$

i $y = -5x^2$ is 'thinner' than $y = x^2$
ii graph opens downwards

c

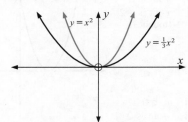

$y = x^2$

$y = \frac{1}{3}x^2$

i $y = \frac{1}{3}x^2$ is 'wider' than $y = x^2$
ii graph opens upwards

d

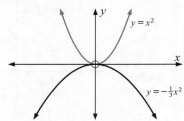

$y = x^2$

$y = -\frac{1}{3}x^2$

i $y = -\frac{1}{3}x^2$ is 'wider' than $y = x^2$
ii graph opens downwards

e

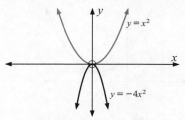

$y = x^2$

$y = -4x^2$

i $y = -4x^2$ is 'thinner' than $y = x^2$
ii graph opens downwards

f

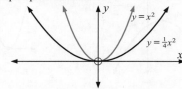

$y = x^2$

$y = \frac{1}{4}x^2$

i $y = \frac{1}{4}x^2$ is 'wider' than $y = x^2$
ii graph opens upwards

EXERCISE 9G.7

1 a

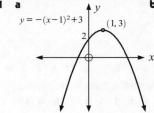

$y = -(x - 1)^2 + 3$ $(1, 3)$

b

$y = 2x^2 + 4$ $(0, 4)$

c

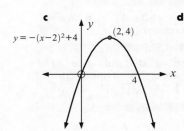

$y = -(x - 2)^2 + 4$ $(2, 4)$

d

$y = 3(x + 1)^2 - 4$ $(-1, -4)$

e

$y = \frac{1}{2}(x + 3)^2$ $4\frac{1}{2}$ $(-3, 0)$

f

$(-3, 1)$ $-3\frac{1}{2}$ $y = -\frac{1}{2}(x + 3)^2 + 1$

g

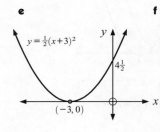

$y = -2(x + 4)^2 + 3$ $(-4, 3)$

h

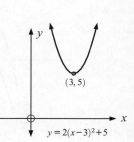

$(3, 5)$ $y = 2(x - 3)^2 + 5$

i

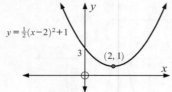

$y = \frac{1}{2}(x-2)^2 + 1$

3

$(2, 1)$

j

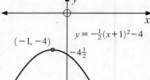

$y = -\frac{1}{2}(x+1)^2 - 4$

$(-1, -4)$

$-4\frac{1}{2}$

k

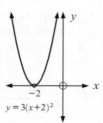

-2

$y = 3(x+2)^2$

l

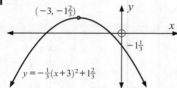

$(-3, -1\frac{2}{3})$

$-1\frac{1}{3}$

$y = -\frac{1}{3}(x+3)^2 + 1\frac{2}{3}$

3 a G **b** A **c** E **d** B **e** I **f** C **g** D **h** F **i** H

1 a 3 **b** 2 **c** -8 **d** 1 **e** 6 **f** 5 **g** 6 **h** 8 **i** -2

1 a 3 and -1 **b** 2 and 4 **c** -3 and -2 **d** 4 and 5
e -3 (touching) **f** 1 (touching)

1 a ± 3 **b** $\pm\sqrt{3}$ **c** -5 and -2 **d** 3 and -4 **e** 0 and 4
f -4 and -2 **g** -1 (touching) **h** 3 (touching)
i $2 \pm \sqrt{3}$ **j** $-2 \pm \sqrt{7}$ **k** $3 \pm \sqrt{11}$ **l** $-4 \pm \sqrt{5}$

1 a i

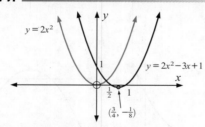

$y = 2x^2$

1

$y = 2x^2 - 3x + 1$

$\frac{1}{2}$ 1

$\left(\frac{3}{4}, -\frac{1}{8}\right)$

ii same shape, both open upwards

b i

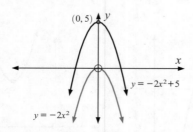

$(0, 5)$

$y = -2x^2 + 5$

$y = -2x^2$

ii same shape, both open downwards

c i

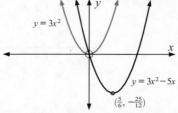

$y = 3x^2$

$y = 3x^2 - 5x$

$\left(\frac{5}{6}, -\frac{25}{12}\right)$

ii same shape, both open upwards

d i

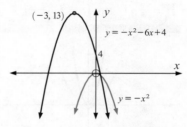

$(-3, 13)$

$y = -x^2 - 6x + 4$

4

$y = -x^2$

ii same shape, both open downwards

2 a i $a > 0$
   ii y-int is 4
   iii x-int is 2
       (touches)

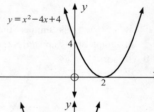

$y = x^2 - 4x + 4$

4

2

b i $a > 0$
   ii y-int is -3
   iii x-int's are
       -3 and 1

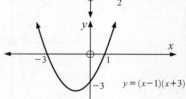

-3 1

-3

$y = (x-1)(x+3)$

c i $a > 0$
   ii y-int is 8
   iii x-int is -2
       (touches)

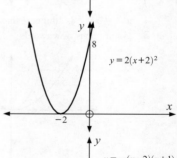

8

$y = 2(x+2)^2$

-2

d i $a < 0$
   ii y-int is 2
   iii x-int's are
       -1 and 2

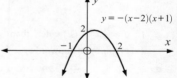

$y = -(x-2)(x+1)$

2

-1 2

e i $a < 0$
   ii y-int is -3
   iii x-int is -1
       (touches)

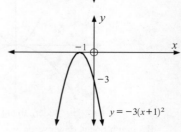

-1

-3

$y = -3(x+1)^2$

f **i** $a < 0$
ii y-int is -12
iii x-int's are 1 and 4

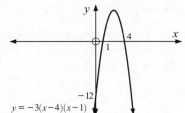

$y = -3(x-4)(x-1)$

g **i** $a > 0$
ii y-int is 6
iii x-int's are -3 and -1

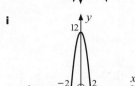

$y = 2(x+3)(x+1)$

h **i** $a > 0$
ii y-int is 2
iii no x-int

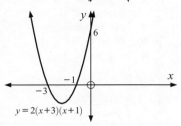

$y = 2x^2 + 3x + 2$

i **i** $a < 0$
ii y-int is 5
iii x-int's are $-\frac{5}{2}$ and 1

$y = -2x^2 - 3x + 5$

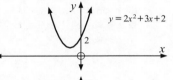

3 **a** $x = 2$ **b** $x = -\frac{5}{2}$ **c** $x = 1$ **d** $x = 3$
e $x = -4$ **f** $x = -4$

4 **a** **i**

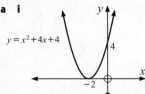

$y = x^2 + 4x + 4$

ii $x = -2$
iii $(-2, 0)$

b **i**

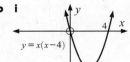

$y = x(x-4)$

ii $x = 2$
iii $(2, -4)$

c **i**

$y = 3(x-2)^2$

ii $x = 2$
iii $(2, 0)$

d **i**

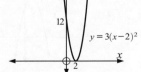

$y = -(x-1)(x+3)$

ii $x = -1$
iii $(-1, 4)$

e **i**

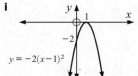

$y = -2(x-1)^2$

ii $x = 1$
iii $(1, 0)$

f **i**

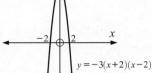

$y = -3(x+2)(x-2)$

ii $x = 0$
iii $(0, 12)$

g **i**

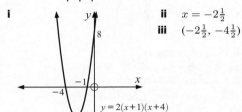

$y = 2(x+1)(x+4)$

ii $x = -2\frac{1}{2}$
iii $(-2\frac{1}{2}, -4\frac{1}{2})$

h **i**

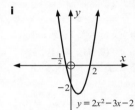

$y = 2x^2 - 3x - 2$

ii $x = \frac{3}{4}$
iii $(\frac{3}{4}, -\frac{25}{8})$

i **i**

$y = -2x^2 - x + 3$

ii $x = -\frac{1}{4}$
iii $(-\frac{1}{4}, \frac{25}{8})$

5 **a** **i**

ii axis of symmetry $x = 1$

b **i**

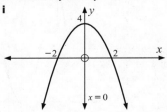

ii axis of symmetry $x = 0$

c i

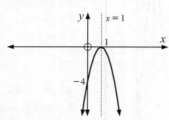

ii axis of symmetry $x = -3$

d i

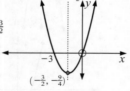

ii axis of symmetry $x = 1$

6 a 2 and 6   **b** -1 and -5 **c** 3 (touching)

7 a $A(h - d, 0)$ $B(h + d, 0)$
 b $ah^2 - 2adh + ad^2 + bh - bd + c = 0$
 c $ah^2 + 2adh + ad^2 + bh + bd + c = 0$

EXERCISE 9J

1 a $x = -2$ **b** $x = \frac{3}{2}$ **c** $x = -\frac{2}{3}$ **d** $x = -2$ **e** $x = \frac{5}{4}$
 f $x = 10$ **g** $x = -6$ **h** $x = \frac{25}{2}$ **i** $x = 150$

2 a $(2, -2)$ **b** $(-1, -4)$ **c** $(0, 4)$ **d** $(0, 1)$
 e $(-2, -15)$ **f** $(-2, -5)$ **g** $(-\frac{3}{2}, -\frac{11}{2})$ **h** $(\frac{5}{2}, -\frac{19}{2})$
 i $(1, -\frac{9}{2})$ **j** $(2, 6)$

3 a i x-intercepts $4, -2$,
  y-intercept -8
 ii axis of symm. $x = 1$
 iii vertex $(1, -9)$
 iv $y = x^2 - 2x - 8$

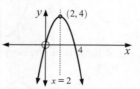

b i x-intercepts $0, -3$,
  y-intercept 0
 ii axis of symm. $x = -\frac{3}{2}$
 iii vertex $(-\frac{3}{2}, -\frac{9}{4})$
 iv $y = x^2 + 3x$

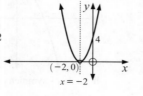

c i x-intercepts $0, 4$,
  y-intercept 0
 ii axis of symm. $x = 2$
 iii vertex $(2, 4)$
 iv $y = 4x - x^2$

d i x-intercept -2,
  y-intercept 4
 ii axis of symm. $x = -2$
 iii vertex $(-2, 0)$
 iv $y = x^2 + 4x + 4$

e i x-intercepts $-4, 1$,
  y-intercept -4
 ii axis of symm. $x = -\frac{3}{2}$
 iii vertex $(-\frac{3}{2}, -\frac{25}{4})$
 iv $y = x^2 + 3x - 4$

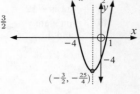

f i x-intercept 1,
  y-intercept -1
 ii axis of symm. $x = 1$
 iii vertex $(1, 0)$
 iv $y = -x^2 + 2x - 1$

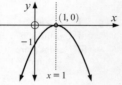

g i x-intercepts $-2, -4$,
  y-intercept -8
 ii axis of symmetry $x = -3$
 iii vertex $(-3, 1)$
 iv $y = -x^2 - 6x - 8$

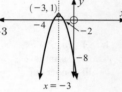

h i x-intercepts $1, 2$,
  y-intercept -2
 ii axis of symm. $x = \frac{3}{2}$
 iii vertex $(\frac{3}{2}, \frac{1}{4})$
 iv $y = -x^2 + 3x - 2$

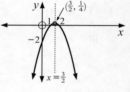

i i x-intercepts $\frac{1}{2}, -3$,
  y-intercept -3
 ii axis of symm. $x = -\frac{5}{4}$
 iii vertex $(-\frac{5}{4}, -\frac{49}{8})$
 iv $y = 2x^2 + 5x - 3$

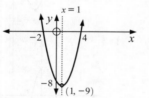

j i x-intercepts $-\frac{3}{2}, 4$,
  y-intercept -12
 ii axis of symm. $x = \frac{5}{4}$
 iii vertex $(\frac{5}{4}, -\frac{121}{8})$
 iv $y = 2x^2 - 5x - 12$

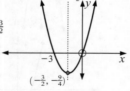

k i x-intercepts $\frac{2}{3}, -2$,
  y-intercept 4
 ii axis of symm. $x = -\frac{2}{3}$
 iii vertex $(-\frac{2}{3}, \frac{16}{3})$
 iv $y = -3x^2 - 4x + 4$

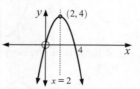

l i x-intercepts $0, 20$,
  y-intercept 0
 ii axis of symm. $x = 10$
 iii vertex $(10, 25)$
 iv $y = -\frac{1}{4}x^2 + 5x$

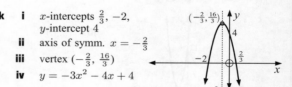
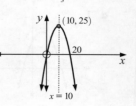
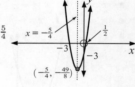

EXERCISE 9K

1 a $(1, 7)$ and $(2, 8)$ **b** $(4, 5)$ and $(-3, -9)$
c $(3, 0)$ (touching) **d** graphs do not meet
e graphs do not meet **f** $(4, 5)$ (touching)
2 a $(0.59, 5.59)$ and $(3.41, 8.41)$ **b** $(3, -4)$ touching
c graphs do not meet **d** $(-2.56, -18.81)$ and $(1.56, 1.81)$

EXERCISE 9L

1 a 9 seconds **b** 162 m **c** 18 seconds
2 a 12 **b** \$100 **c** \$244 **3 a** 21 **b** \$837 **c** \$45
4 500 m by 250 m **5 c** 100 m by 112.5 m
6 a $41\frac{2}{3}$ m by $41\frac{2}{3}$ m **b** 50 m by $31\frac{1}{4}$ m **7** 40 **8** 157

REVIEW SET 9A

1 a -12 **b** -17 **c** $x = 6$ or -3
2 a 0 **b** -15 **c** $-1\frac{1}{4}$
3 a i range is $\{y : y \geqslant -5\}$, domain is $\{x : x \in R\}$
ii x-intercepts are -1 and 5, y-intercept is $-\frac{25}{9}$
iii is a function
b i range is $\{y : y > 0 \text{ or } y \leqslant -1\}$,
domain is $\{x : x \in R, \ x \neq 0 \text{ and } x \neq 2\}$
ii no x-intercepts, no y-intercepts **iii** is a function
4 a domain is $\{x : x \geqslant -2\}$, range is $\{y : 1 \leqslant y < 3\}$
b domain is $\{x : x \in R\}$
range is $\{y : y = -1 \text{ or } y = 1 \text{ or } y = 2\}$
5 $a = -6$, $b = 13$ **6** $a = 1$, $b = -6$, $c = 5$
7 a i

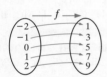

ii domain is $\{-2, -1, 0, 1, 2\}$
iii range is $\{1, 3, 5, 7, 9\}$
b i

ii domain is $\{-2, -1, 0, 1, 2\}$ **iii** range is $\{2, 4, 8\}$

REVIEW SET 9B

1 a -1 **b** -5 **c** $x = -2$ or $\frac{3}{2}$
2 a

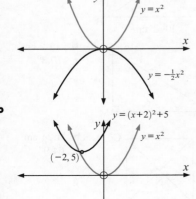

b

c

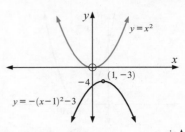

3 a i opens downwards **b**
ii 6
iii 1 and -3
iv $x = -1$

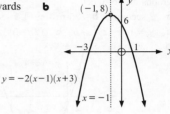

4 a i -15 **b**
ii 5 and -3
iii $x = 1$
iv $(1, -16)$

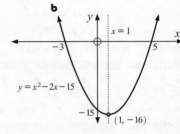

5 $(4, 6)$ touching **6 a** 2 seconds **b** 80 m **c** 6 seconds
7 a

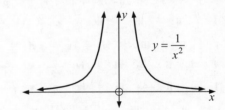

b domain is $\{x : x \in R, \ x \neq 0\}$, range is $\{y : y > 0\}$

REVIEW SET 9C

1 4 and 5
2 a

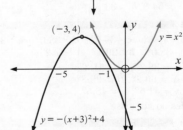

b

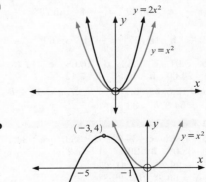

3 a i opens upwards **ii** 12 **iii** 2 (touching) **iv** $x = 2$

b

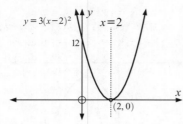

$y = 3(x-2)^2$ $x = 2$

$(2, 0)$

4 graphs do not meet

5 **a** 2 seconds **b** 25 m **c** 4.5 seconds

6 **a** $y = (x-2)^2 - 5$ **b** $(2, -5)$ **c** -1

    d

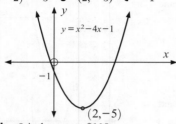

$y = x^2 - 4x - 1$

-1

$(2, -5)$

7 **a** $200 **b** 9 jackets **c** $119

EXERCISE 10A

1 **a** **i** JK **ii** JL **iii** KL **b** **i** RS **ii** TR **iii** ST
    c **i** AB **ii** BC **iii** AC

2 **a** b units **b** c units **c** c units **d** b units

EXERCISE 10B.1

1 **a** **i** $\frac{p}{r}$ **ii** $\frac{q}{r}$ **iii** $\frac{p}{q}$ **iv** $\frac{q}{r}$ **v** $\frac{p}{r}$ **vi** $\frac{q}{p}$

    b **i** $\frac{y}{x}$ **ii** $\frac{z}{x}$ **iii** $\frac{y}{z}$ **iv** $\frac{z}{x}$ **v** $\frac{y}{x}$ **vi** $\frac{z}{y}$

    c **i** $\frac{4}{5}$ **ii** $\frac{3}{5}$ **iii** $\frac{4}{3}$ **iv** $\frac{3}{5}$ **v** $\frac{4}{5}$ **vi** $\frac{3}{4}$

    d **i** $\frac{4}{\sqrt{41}}$ **ii** $\frac{5}{\sqrt{41}}$ **iii** $\frac{4}{5}$ **iv** $\frac{5}{\sqrt{41}}$ **v** $\frac{4}{\sqrt{41}}$ **vi** $\frac{5}{4}$

    e **i** $\frac{3}{\sqrt{13}}$ **ii** $\frac{2}{\sqrt{13}}$ **iii** $\frac{3}{2}$ **iv** $\frac{2}{\sqrt{13}}$ **v** $\frac{3}{\sqrt{13}}$ **vi** $\frac{2}{3}$

    f **i** $\frac{5}{6}$ **ii** $\frac{\sqrt{11}}{6}$ **iii** $\frac{5}{\sqrt{11}}$ **iv** $\frac{\sqrt{11}}{6}$ **v** $\frac{5}{6}$ **vi** $\frac{\sqrt{11}}{5}$

EXERCISE 10B.2

1 **a** $\sin 65^\circ = \frac{x}{a}$ **b** $\cos 32^\circ = \frac{x}{b}$ **c** $\tan 56^\circ = \frac{x}{c}$

    d $\cos 37^\circ = \frac{d}{x}$ **e** $\tan 49^\circ = \frac{e}{x}$ **f** $\tan 73^\circ = \frac{f}{x}$

    g $\sin 54^\circ = \frac{g}{x}$ **h** $\tan 27^\circ = \frac{x}{h}$ **i** $\cos 59^\circ = \frac{i}{x}$

2 **a** 17.62 **b** 8.91 **c** 6.63 **d** 16.68 **e** 14.43
    f 23.71 **g** 24.60 **h** 40.95 **i** 7.42

3 **a** $\theta = 62$, $a = 12.5$, $b = 6.7$ **b** $\theta = 27$, $a = 16.4$, $b = 7.4$
    c $\theta = 52$, $a = 54.8$, $b = 33.8$

EXERCISE 10B.3

1 **a** 51.3° **b** 41.8° **c** 39.6° **d** 38.0° **e** 34.1°
    f 48.6° **g** 38.0° **h** 29.1° **i** 37.2°

2 **a** $x = 6.2$, $\theta = 38.7^\circ$, $\phi = 51.3^\circ$
    b $x = 7.0$, $\alpha = 60.2^\circ$, $\beta = 29.8^\circ$
    c $x = 8.4$, $a = 41.5^\circ$, $b = 48.5^\circ$

4 The 3 triangles do not exist.

EXERCISE 10B.4

1 **a** $x = 4.125$ **b** $\alpha = 75.52$ **c** $\beta = 40.99$

2 **a** $\theta = 36.87$ **b** $r = 11.33$ **c** $\alpha = 61.93$

3 7.990 cm **4** 89.20° **5** 47.16° **6** 22.44°

7 11.83 cm **8** 119.5° **9** 36.49 cm

10 **a** $x = 3.441$ **b** $\alpha = 51.54$ **11** 129.5°

EXERCISE 10C

1 18.3 m **2** **a** 371 m **b** 1.62 km **3** 159 m **4** 1.58°

5 angle of elevation $= 26.4^\circ$, angle of depression $= 26.4^\circ$

6 418 m **7** 111 m **8** 72.0 m **9** 9.91 m

10 **a** 16.2 m/s **b** 11.5° **11** $\theta = 12.6$ **12** 9.56 m

13 2.80 km **14** 10.9 m **15** 786 m **16** 962 m

17 **a** 82.4 cm **b** 77.7 litres

EXERCISE 10D

1 **a** **i** $x = b\cos\alpha$ **b** **i** $x = b\tan\alpha$
        ii $\alpha = \cos^{-1}\left(\frac{x}{b}\right)$ **ii** $\alpha = \tan^{-1}\left(\frac{x}{b}\right)$

    c **i** $x = \frac{b}{\sin\alpha}$ **ii** $\alpha = \sin^{-1}\left(\frac{b}{x}\right)$

2 $\theta = \tan^{-1}\left(\frac{2}{x}\right) - \tan^{-1}\left(\frac{1}{x}\right)$

3 33.54 m from Q, angle of view is 7.65°

EXERCISE 10E.1

1 **a** 17.0 cm **b** 35.3°

2 **a** 10 cm **b** 31.0° **c** 12.6 cm **d** 25.4°

3 **a** 12.6 cm **b** 13.3° **4** 83.0 cm **5** 45°

EXERCISE 10E.2

1 **a** **i** FG **ii** GH **iii** FH **iv** HM
    b **i** CF **ii** DC **iii** DF **iv** CX **c** **i** MC **ii** MN

EXERCISE 10E.3

1 **a** **i** $\angle$AFE **ii** $\angle$AGE **iii** $\angle$BHF **iv** $\angle$BXF
    b **i** $\angle$QYR **ii** $\angle$QWR **iii** $\angle$PZS **iv** $\angle$PYS
    c **i** $\angle$ASX **ii** $\angle$AYX

2 **a**

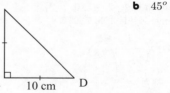

    b 45°

E, A, 10 cm, D

    c   **d** 35.3°

H, 10 cm, D, $10\sqrt{2}$ cm, B

3 **a**

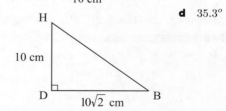

    b 11.2 cm
    c 70.3°

N, D, M

4 **a**

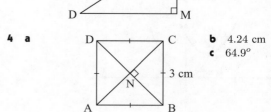

    b 4.24 cm
    c 64.9°

D, C, 3 cm, N, A, B

5 a 59.0° **b**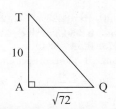
   c 49.7°

6 41.7°

EXERCISE 10F

1 a 28.9 cm^2 **b** 383.7 km^2 **c** 26.7 cm^2
2 $x = 19.0$ **3** 18.9 cm^2 **4 a** 71.616 m^2 **b** 8.43 m

EXERCISE 10G

1 a 28.8 cm **b** 3.38 km **c** 14.2 m
2 $\angle A = 52.0°$, $\angle B = 59.3°$, $\angle C = 68.7°$
3 112° **4 a** 40.3° **b** 107°

EXERCISE 10H.1

1 a $x = 28.4$ **b** $x = 13.4$ **c** $x = 3.79$
2 a $a = 21.3$ **b** $b = 76.9$ **c** $c = 5.09$

EXERCISE 10H.2

1 $\angle C = 62.1°$ or $\angle C = 117.9°$
   When $\angle C = 62.1°$, $a = 12.2$ cm, $\angle A = 77.9°$
   When $\angle C = 117.9°$, $a = 4.68$ cm, $\angle A = 22.1°$
2 a $\angle A = 49.5°$ **b** $\angle B = 72°$ or $108°$ **c** $\angle C = 44.3°$

EXERCISE 10I.1

1 17.7 m **2** 207 m **3** 23.9° **4** 77.5 m **5** 13.2°
6 a 38.0 m **b** 94.0 m

EXERCISE 10I.2

1 a 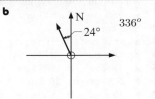 **b**

c

2 a 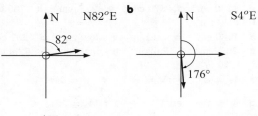 **b**

c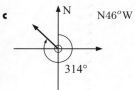

3 a 072° **b** 342° **c** 252° **d** 293°
4 a 000° **b** 060° **c** 120° **d** 180° **e** 240° **f** 300°
5 136 km **6** 12.6 km/h **7 a** 46.90 hectares **b** 2.635 km
8 a 8.76 km **b** 104° **9 a** 296° **b** 561 km
10 14.1 km on a bearing of 108.5° **11** 2 hours 17 minutes

REVIEW SET 10A

1 $\angle K = 66°$, KL = 6.91 cm, LM = 15.5 cm
2 a 36.7 cm **b** 60.6° **3** 22.3° **4** 25.2° **5** 9.78 cm
6 315 m
7 a 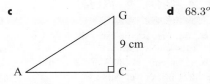 **b** 59.0°

c **d** 68.3°

8 21.1 km^2 **9** 36.8 cm^2 **10 a** $x = 41.5$ **b** $x = 15.4$
11 7.32 m **12 a** 10 600 m^2 **b** 1.06 hectares
13 a 203 km **b** bearing of 113°
14 179 km on a bearing 352°

REVIEW SET 10B

1 $\angle R = 52°$, PQ = 8.96 cm, PR = 11.4 cm
2 a 6.025 m **b** 6.13° **3** 6.20 cm **4** 51.3° **5** 369 km/h
6 a 22.6° **b** 50.2° **7** 25.4 km^2
8 a $x = 34.1$ **b** $x = 18.9$
9 $\angle R = 54.3°$, $\angle Q = 83.7°$, PR = 25.7 cm *or*
   $\angle R = 125.7°$, $\angle Q = 12.3°$, PR = 5.51 cm
10 113 cm^2 **11** 204 m **12** 530 m on a bearing of 077.2°
13 Yes, the ships are 69.2 km apart.

REVIEW SET 10C

1 15.8 km on a bearing of 147° **2** 7.28 m **3** 70.9 m
4 a 5 cm **b** 36.9° **5 a** 3.61 m **b** 33.7°
6 a **b** 11.2 cm
      c 57.9°

7 a 141 m **b** 45° **8** 44.9 cm^2
9 AC = 12.6 cm, $\angle A = 48.6°$, $\angle C = 57.8°$
10 17.7 km on a bearing of 239° **11** 15.6 cm or 25.5 cm
12 268° **13** 26.6 m^2

EXERCISE 11A.1

1 a 3.2 kg **b** 1870 kg **c** 0.047 835 kg **d** 4.653 g
   e 2 830 000 g **f** 63 200 g **g** 0.074 682 t
   h 1 700 000 000 mg **i** 91.275 kg
2 a 5.95 kg **b** 2000 pegs **c** 6400 bricks
3 1 500 000 sweets **4 a** 5.136 t **b** $1284

EXERCISE 11A.2

1 a 82.5 m **b** 29.5 cm **c** 6.25 km **d** 7380 cm
   e 0.2463 m **f** 0.009 761 km

2 a 4130 mm **b** 3 754 000 m **c** 482 900 cm
d 26 900 mm **e** 47 000 cm **f** 3 880 000 mm

3 11.475 km **4** 38 000

EXERCISE 11B.1

1 a 15 cm **b** 18 cm **c** 21 cm
2 a $3x$ units **b** $(5x + 5)$ units **c** $(3x + 3y + 3)$ units
3 a 12 m **b** 46 cm **c** 10.2 m
4 a

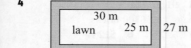

300 m, 220 m

b 1040 m

5 a 15.8 km **b** 13.5 km/h **6** 164 m
7 a 69.4 m **b** 146 m **8** $10.56
9 a 132 m **b** 660 m **c** 66 **d** $445.50
10 $939.70 **11 a** 20.2 m **b** $22.32

EXERCISE 11B.2

1 a 41.5 cm **b** 54.0 m **c** 361 m **d** 5.34 km
e 22.6 m **f** 1480 cm or 14.8 m
2 a 25.7 cm **b** 23.4 cm **c** 30.8 m **3** 240 m
4 a 48.6 m **b** 226 bricks **c** $384.20
5 a 11.9 m **b** 5.20 m **6** 31.8 m

EXERCISE 11C.1

1 a 0.23 cm^2 **b** 36 000 m^2 **c** 0.0726 m^2
d 7 600 000 mm^2 **e** 0.853 ha **f** 35 400 000 cm^2
g 1354 mm^2 **h** 4 320 000 cm^2 **i** 4820 mm^2
2 a 2100 chickens **b** 50 000 rectangles

EXERCISE 11C.2

1 a 32 cm^2 **b** $6a^2$ m^2 **c** 80 cm^2 **d** $9x^2$ m^2
2 a 3 cm^2 **b** 6 cm^2 **c** 640 m^2 **d** 12 cm^2
3 a a^2 cm^2 **b** $x(x + 1)$ m^2 **c** $6x$ m^2
4 a 24 cm^2 **b** 48 m^2 **c** 120 cm^2
5 a $x(x + 2)$ cm^2 **b** $2ab$ m^2 **c** $12x^2$ m^2
6 a 35 cm^2 **b** 30 cm^2 **c** 72 m^2
7 a $x(x + 2)$ cm^2 **b** $51x^2$ cm^2 **c** $\left(\frac{a+b}{2}\right) b$ m^2

EXERCISE 11C.3

1 a 78.5 cm^2 **b** 72.4 m^2 **c** $4\pi x^2$ m^2
2 a 63.6 cm^2 **b** $\frac{\pi x^2}{3}$ **c** 31.4 cm^2

EXERCISE 11C.4

1 a 83 cm^2 **b** 506 cm^2 **c** 143 cm^2
2 a $\frac{\pi r^2}{2}$ units^2 **b** $(\pi R^2 - \pi r^2)$ units^2
c $([x + 2a]^2 - \pi a^2)$ units^2 **d** $(2ab + \pi a^2)$ units^2
3 a 125 cm^2 **b** 76.0 cm^2 **c** 90 cm^2

EXERCISE 11D

1 a 280 km^2 **b** 910 m^2 **c** 66.7 m^2 **d** 1780 m^2 **e** 20.6 m^2
2 a 8 m **b** 1.8 cm **c** 2.4 m **3 a** 78 ha **b** $6162
4

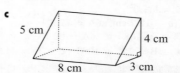

30 m, lawn, 25 m, 27 m, 32 m — area = 114 m^2

5 $28 800 **6** 12 flights

7 a 15 m^2 **b** 95.25 m^2 **c** 20 tiles **d** 1905 tiles **e** $4367
8 a 4 m **b** 50.3 m^2 **c** $160.85 **e** 150.80 m^2
f $678.60 **g** $840

9 medium, 1.358 cents per cm^2 **10** $239.10

11 a $BC = \dfrac{100}{x}$ **c** $x = 7.07$ **d**

7.07 m, A, B, 14.14 m, D, C

12 a $A = 3xy$
d $x = 11.55$ — 11.55 m
e

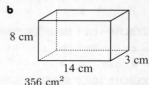

11.55 m, 17.32 m

EXERCISE 11E.1

1 a 106 cm^2 **b** 4600 cm^2 **c** $94x^2$ m^3 **2** 84 cm^2
3 a

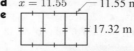

2.5 cm, cube
37.5 cm^2

b 8 cm, 14 cm, 3 cm
356 cm^2

c 5 cm, 4 cm, 8 cm, 3 cm
108 cm^2

4 a 44 m^2 **b** 538 m^2 **c** $2683

EXERCISE 11E.2

1 a 498 cm^2 **b** 267 cm^2 **c** 553 cm^2 **d** 396 cm^2
e 185 cm^2 **f** 845 cm^2
2 84.8 cm^2
3 a $8\pi x^2$ cm^2 **b** $16\pi x^2$ cm^2 **c** $12\pi x^2$ cm^2 **d** $24\pi x^2$ cm^2
4 a 603 cm^2 **b** 1070 cm^2 **c** 188 cm^2
5 a 19.4 m^2 **b** $883 **c** $21 201
6 a 5.39 m **b** 46.4 m^2 **c** $835
7 a 12.6 m^2 **b** $509 **8 a** $S_A = 4\pi r^2$ **b** 5.40 m
9 a $S_A = 3\pi r^2$ **b** 4.50 cm **c** 4.24 cm

EXERCISE 11F.1

1 a 8650 mm^3 **b** 86 cm^3 **c** 0.3 m^3 **d** 124 000 mm^3
e 0.3 cm^3 **f** 3 700 000 cm^3
2 a 7.5 m^3 **b** 47 400 sinkers

EXERCISE 11F.2

1 a 390 cm^2 **b** 5.54 m^3 **c** 381 cm^3 **d** 536 cm^3
e 1230 cm^3 **f** 373 cm^3 **g** 25.1 cm^3 **h** 765 cm^3
i 2940 cm^2
2 a 1180 cm^3 **b** 212 m^3 **c** 36.9 cm^3 **d** 232 cm^3
e 26.5 cm^3 **f** 2.62 m^3 **g** 4.60 cm^3 **h** 72.6 cm^3
i 1870 m^3
3 a $V = 6x^3$ **b** $V = \frac{1}{2}abc$ **c** $V = 4\pi r^3$
d $V = \pi x^3$ **e** $V = \dfrac{a^2 h}{3}$ **f** $V = 2\pi x^3$
4 a 1.36 cm **b** 6 cm
5 a

20 m, 1 m

b 65.97 m^2
c $\doteqdot$ 6.6 m^3

6 a 56.25 m^2 **b** 450 m^3 **7** 4 people

8 a

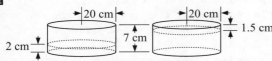

b 2513 cm^3 **c** 6912 cm^3 **d** 175.0%

9 a 0.5 m **b** 0.45 m **c** 0.373 m^3

10 a 7.18 m^3 **b** $\$972$ **11** 217 tonnes

12 a 2.67 cm **b** 3.24 cm **c** 4.46 cm **d** 2.60 m
e 1.74 cm **f** 5.60 m

EXERCISE 11G

1 a 3760 mL **b** 47.32 kL **c** 3500 L **d** 423 mL
e $54\,000 \text{ mL}$ **f** $0.058\,34 \text{ kL}$

2 a 1375 bottles **b** 9 tanks

3 a 83 m^3 **b** 3200 cm^3 **c** 2.3 L **d** $7\,154\,000 \text{ L}$
e 0.46 m^3 **f** $4\,600\,000 \text{ cm}^3$

4 a 25 mL **b** 32 m^3 **c** 7320 L

5 a 22.1 kL **b** 23.6 kL **c** 186 kL

6 a 24.4 mL **b** $\$80$ **c** $\$3.27$ per mL **7** 1.15 m

8 a 1.32 m^3 **b** 1.32 kL **c** 10.5 cm

9 a 39.3 mL **b** 2 cm **10** 78 mm **11** 35 truckloads

EXERCISE 11H

1 a 5 g/cm^3 **b** 1.33 g/cm^3 **2** 569 g **3** 312 cm^3

4 a 2.25 g/cm^3 **b** 20 g/cm^3 **5** 19.3 g/cm^3

6 4.90 cm **7** 417 600 beads **8** 4.24×10^6 tonnes

EXERCISE 11I

1 a 30 bricks, 8 layers

b

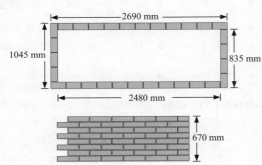

c 250, more than the 240 required to allow for breakages.

d 1.4 m^3 is required to fill the bed, since the soil will
compact when watered. **e** 78.1 litres

f Diameter (internal) of 2.24 m. It could hold 1.26 m^3 of
soil more than the rectangular bed. It could also hold
more plants because there is more space to plant them
(3.25 m^2 compared to 2.07 m^2).

g How many bricks are likely to be broken.

2 a 1.06 m **b** Cone: 2.66 m^2; Cylinder: 9.05 m^2;
Hemisphere: 4.02 m^2 **d** 3913 kg

3 a 16.8 m **b** 2.16 m^3 **c** 8.04 m^2

4 a 11.875 m^3 **b** 5.8 m **c** 1.357 m^3 more
d conical ends: 30.2 m^2, hemispherical ends: 32.8 m^2

e The hemispherical ends are better. As cost is proportional
to surface area, the higher the volume/surface area ratio
is, the better. This ratio is 0.4 for hemispheres and 0.396
for conical ends. Also, this design gives a shorter vehicle.

REVIEW SET 11A

1 a $15\,500 \text{ m}$ **b** $\$6510$ **2** 51.52 m

3 a 7.14 km^2 **b** 33.3 m^2

4 a 160 cm^3 **b** 137 m^3 **c** 1005 cm^3

5 a 120 cm^2 **b** 283 cm^2

6 a 7.24 km **b** 224 ha **c** 1.49 km **7** 11.8 cm

8 a 6.49 g/cm^3 **b** 812 g

9 a 5196 cans **b** 154 m^2 **10** 16 333 spikes

REVIEW SET 11B

1 a 24.2 m **b** 11 posts

2 a 17.3 m^2 **b** style 1: 900 cm^2, style 2: 800 cm^2
c 192 style 1 tiles, 216 style 2 tiles
d style 2 ($\$853.20$) is cheaper (style 1 $\$873.60$)

3 38.8 m **4 a** $A = bc + \left(\frac{a+b}{2}\right)h$ **b** $A = 4x^2 + \frac{\pi x^2}{2}$

5 a 67.45 m^2 **b** $\$222.75$

6 a 4.99 m^3 **b** 853 cm^3 **c** 0.452 m^3 **7** 5 tanks

8 68.4 mm **9 a** 5.31 g/cm^3 **b** 1520 g

10 a i 2.82 cm **ii** 94.0 cm^3 **b i** 1.86 m^3 **ii** 43.5 m^2

REVIEW SET 11C

1 a 565 m **b** $\$6927$

2 a $\frac{3}{8}$ **b** 52.3 cm **c** 170 cm^2

3 a 377 cm^2 **b** 340 cm^2 **c** 151 cm^2 **4** $82\,400 \text{ cm}^3$

5 3.22 m^3 **6** 1.03 m **7** 105 cm

8 a 62.8 cm^3 **b** 15 915 handles

9 a 10.1 m^3 **b** 24.5 m^3 **c** 34.6 m^3 **d** $\$4664$

10 98.8 tonnes

EXERCISE 12A

1 a 4, 13, 22, 31, ... **b** 45, 39, 33, 27, ...
c 2, 6, 18, 54, ... **d** 96, 48, 24, 12, ...

2 a Starts at 8 and each term is 8 more than the previous
term. Next two terms 40, 48.

b Starts at 2, each term is 3 more than the previous term;
14, 17.

c Starts at 36, each term is 5 less than the previous term;
16, 11.

d Starts at 96, each term is 7 less than the previous term;
68, 61.

e Starts at 1, each term is 4 times the previous term;
256, 1024.

f Starts at 2, each term is 3 times the previous term;
162, 486.

g Starts at 480, each term is half the previous term; 30, 15.

h Starts at 243, each term is $\frac{1}{3}$ of the previous term; 3, 1.

i Starts at 50 000, each term is $\frac{1}{5}$ of the previous term;
80, 16.

3 a 79, 75 **b** 1280, 5120 **c** 81, 90

4 a Each term is the square of the number of the term;
25, 36, 49.

b Each term is the cube of the number of the term;
125, 216, 343.

c Each term is $n \times (n+1)$ where n is the number of
the term; 30, 42, 56.

5 a 625, 1296 **b** 13, 21 **c** 9, 11 **d** 13, 17 (primes)
e 16, 22 **f** 16, 24

EXERCISE 12B

1 a 2, 4, 6, 8, 10 **b** 4, 6, 8, 10, 12 **c** 1, 3, 5, 7, 9
d $-1, 1, 3, 5, 7$ **e** 5, 7, 9, 11, 13 **f** 13, 15, 17, 19, 21

g 4, 7, 10, 13, 16   **h** 1, 5, 9, 13, 17
i 9, 14, 19, 24, 29

2 a 2, 4, 8, 16, 32   **b** 6, 12, 24, 48, 96
c 3, $1\frac{1}{2}$, $\frac{3}{4}$, $\frac{3}{8}$, $\frac{3}{16}$ **d** $-2, 4, -8, 16, -32$

3 17, 11, 23, -1, 47

EXERCISE 12C

1 a $u_1 = 6$, $d = 11$ **b** $u_n = 11n - 5$ **c** 545
d yes, u_{30} **e** no

2 a $u_1 = 87$, $d = -4$ **b** $u_n = 91 - 4n$ **c** -69 **d** no

3 b $u_1 = 1$, $d = 3$ **c** 169   **d** $u_{151} = 451$

4 b $u_1 = 32$, $d = -\frac{7}{2}$ **c** -227 **d** $n \geqslant 68$

5 a $k = 17\frac{1}{2}$ **b** $k = 4$ **c** $k = 4$ **d** $k = 0$
e $k = 3$, $k = -2$ **f** $k = 3$, $k = -1$

6 a $u_n = 6n - 1$ **b** $u_n = -\frac{3}{2}n + \frac{11}{2}$ **c** $u_n = -5n + 36$
d $u_n = -\frac{3}{2}n + \frac{1}{2}$

7 a 5, $6\frac{1}{4}$, $7\frac{1}{2}$, $8\frac{3}{4}$, 10
b -1, $3\frac{5}{7}$, $8\frac{3}{7}$, $13\frac{1}{7}$, $17\frac{6}{7}$, $22\frac{4}{7}$, $27\frac{2}{7}$, 32

8 a $u_1 = 36$, $d = -\frac{2}{3}$ **b** 100   **9** 100 006 (7692nd term)

EXERCISE 12D

1 a $b = 18$, $c = 54$ **b** $b = 2\frac{1}{2}$, $c = 1\frac{1}{4}$ **c** $b = 3$, $c = -1\frac{1}{2}$

2 a $u_1 = 5$, $r = 2$ **b** $u_n = 5 \times 2^{n-1}$, $u_{15} = 81920$

3 a $u_1 = 12$, $r = -\frac{1}{2}$ **b** $u_n = 12 \times (-\frac{1}{2})^{n-1}$, $u_{13} = \frac{3}{1024}$

4 a $u_1 = 8$, $r = -\frac{3}{4}$, $u_{10} = -0.60067749$

5 a $u_1 = 8$, $r = \frac{1}{\sqrt{2}}$, $u_n = 2^{\frac{7}{2} - \frac{n}{2}}$

6 a $k = \pm 14$ **b** $k = 2$ **c** $k = -2$ or 4

7 a $u_n = 3 \times 2^{n-1}$ **b** $u_n = 32 \times (-\frac{1}{2})^{n-1}$
c $u_n = 3 \times (\sqrt{2})^{n-1}$ **d** $u_n = 10 \times (\sqrt{2})^{1-n}$

8 a $u_9 = 13122$ **b** $u_{14} = 2916\sqrt{3} \doteqdot 5050.66$
c $u_{18} \doteqdot 0.00009155$

EXERCISE 12E

1 a $3993   **b** $993   **2** 11470.39 euro
3 a 43923 Yen   **b** 13923 Yen   **4** $23602.32
5 148024.43 Yen   **6** £51249.06   **7** $14976.01
8 £11477.02   **9** 19712.33 euro   **10** 19522.47 Yen

EXERCISE 12F

1 a i 1553 ants   **ii** 4823 ants   **b** 12.2 weeks
2 a 278 animals   **b** Year 2047
3 a i 73   **ii** 167   **b** 31 years   **4 a** 949364   **b** 18.1 months

EXERCISE 12G.1

1 a i $S_n = 3 + 11 + 19 + 27 + \dots + (8n - 5)$ **ii** 95
b i $S_n = 42 + 37 + 32 + \dots + (47 - 5n)$ **ii** 160
c i $S_n = 12 + 6 + 3 + 1\frac{1}{2} + \dots + 12(\frac{1}{2})^{n-1}$ **ii** $23\frac{1}{4}$
d i $S_n = 2 + 3 + 4\frac{1}{2} + 6\frac{3}{4} + \dots + 2(\frac{3}{2})^{n-1}$ **ii** $26\frac{3}{8}$
e i $S_n = 1 + \frac{1}{2} + \frac{1}{4} + \frac{1}{8} + \dots + \frac{1}{2^{n-1}}$ **ii** $1\frac{15}{16}$
f i $S_n = 1 + 8 + 27 + 64 + \dots + n^3$ **ii** 225

EXERCISE 12G.2

1 a 91   **b** 91   **c** 91
2 a 820   **b** 3087.5   **c** -1460 **d** -740
3 a 1749   **b** 2115   **c** $1410\frac{1}{2}$ **4** 203

5 -115.5 **6** 18   **7 a** 65   **b** 1914   **c** 47850
8 a 14025   **b** 71071   **c** 3266
10 a $u_n = 2n - 1$ **c** $S_1 = 1$, $S_2 = 4$, $S_3 = 9$, $S_4 = 16$
11 56, 49   **12** 10, 4, -2 or -2, 4, 10
13 2, 5, 8, 11, 14 or 14, 11, 8, 5, 2

EXERCISE 12G.3

1 a 23.9766   **b** $\doteqdot 189134$ **c** $\doteqdot 4.0001$ **d** $\doteqdot 0.5852$

2 a $S_n = \dfrac{3 + \sqrt{3}}{2}\left((\sqrt{3})^n - 1\right)$ **b** $S_n = 24(1 - (\frac{1}{2})^n)$
c $S_n = 1 - (0.1)^n$ **d** $S_n = \frac{40}{3}(1 - (-\frac{1}{2})^n)$

3 c $26361.59

4 a $\frac{1}{2}$, $\frac{3}{4}$, $\frac{7}{8}$, $\frac{15}{16}$, $\frac{31}{32}$ **b** $S_n = \dfrac{2^n - 1}{2^n}$
c $1 - (\frac{1}{2})^n = \dfrac{2^n - 1}{2^n}$ **d** as $n \to \infty$, $S_n \to 1$

5 b $S_n = 1 + 18(1 - (0.9)^{n-1})$ **c** 19 seconds

REVIEW SET 12A

1 a $\frac{1}{3}$, 1, 3, 9   **b** $\frac{5}{4}$, $\frac{8}{5}$, $\frac{11}{6}$, 2   **c** 5, -5, 35, -65
2 b $u_1 = 63$, $d = -5$ **c** -117 **d** $u_{54} = -202$
3 a $u_1 = 3$, $r = 4$ **b** $u_n = 3 \times 4^{n-1}$, $u_9 = 196608$
4 $k = -\frac{11}{2}$ **5** $u_n = 73 - 6n$, $u_{34} = -131$
6 b $u_1 = 6$, $r = \frac{1}{2}$ **c** 0.000183
7 $u_n = 33 - 5n$, $S_n = \frac{n}{2}(61 - 5n)$ **8** $k = \pm\frac{2\sqrt{3}}{3}$

REVIEW SET 12B

1 a 81   **b** $-1\frac{1}{2}$ **c** -486
2 23, 21, 19, 17, 15, 13, 11, 9
3 a $u_n = 89 - 3n$ **b** $u_n = \dfrac{2n + 1}{n + 3}$ **c** $u_n = 100(0.9)^{n-1}$
4 a 1587   **b** $47\frac{253}{256} \doteqdot 47.99$ **5** $t_{12} = 10240$
6 a 8415.31 euro   **b** 8488.67 euro   **c** 8505.75 euro
7 $u_n = \frac{1}{6} \times 2^{n-1}$
8 a 56275 euro   **b** It will reach 60000 euro after 7 completed years, i.e., in year 8.

REVIEW SET 12C

1 $u_n = (\frac{3}{4})2^{n-1}$ **a** 49152   **b** 24575.25   **2** 12
3 $u_{11} = \frac{8}{19683} \doteqdot 0.000406$ **4 a** 17   **b** $255\frac{511}{512} \doteqdot 256.0$
5 $k = -\frac{9}{5}$ or $k = 3$ **6 a** $18726.65   **b** $18885.74
7 $13972.28   **8 a** 3473 koalas   **b** Year 2009

EXERCISE 13A.1

1 a £303.35   **b** £530.87   **c** £1180.36
2 a 1349.7 kroner   **b** 3312.9 kroner   **c** 18773.1 kroner
3 a 412.32 krona   **b** 859 krona   **c** 2528.90 krona
4 a $1340.91 HK   **b** $7132.50 HK   **c** $48501 HK
5 a 2784.71 baht   **b** 32606.36 baht   **c** 136450.55 baht

EXERCISE 13A.2

1 $1293.25 NZ   **2** $1718.65 CAN
3 2943.83 Danish kroner   **4** 94.03 rupees
5 1693.99 ringgits   **6** 1080.69 euro
7 a i $312   **ii** $576   **b i** 167 pounds   **ii** 233 pounds

EXERCISE 13A.3

1 a i £7.50   **ii** $\doteqdot$ $923 US   **b i** £5.25   **ii** $\doteqdot$ 520 euro
c i £18   **ii** $\doteqdot$ $3360 NZ

2   **a** **i** $7.20 US   **ii** $\doteqdot$ £210   **b** **i** $12.60 US   **ii** $\doteqdot$ AUD $944
   c **i** $23.40 US   **ii** $\doteqdot$ 1027 euro

3   **a** $\doteqdot$ 1392 baht   **b** $\doteqdot$ 388 pesos   **c** $\doteqdot$ 12 pesos

4   **a** $\doteqdot$ 2837 rupees   **b** $\doteqdot$ 363 rand   **c** $\doteqdot$ 12 rand

EXERCISE 13A.4

1   **a** $2486.80 US   **b** 2813.20 Norwegian kroner
   c $18 759.95 NZ   **d** 22 994.60 Thai baht   **e** 110 830.85 Yen

EXERCISE 13B.1

1   **a** $480   **b** $1239.75   **c** $6933.33   **d** 402.07 Yen

2 $15 000 at 7% p.a. for 4 years is less by $221.25

3   **a** $2800   **b** £21 000   **4** $124 444.45

5   **a** $3\frac{1}{3}$% p.a.   **b** 8% p.a.   **6** $\doteqdot$ 7.09% p.a.

7   20% p.a.   **8**   **a** $2\frac{1}{2}$ years   **b** $6\frac{1}{2}$ years   **9**   6 years

EXERCISE 13B.2

1 $274.84   **2**   787.50 baht   **3** $1418.75   **4** $1660

5   Accept the loan from Lesley ($269.73 per month) compared to paying Rachel $338.22 per month.

EXERCISE 13C.1

1   **a** $5359.57   **b** $7293.04   **c** £9300.65

2   **a** 113.40 euro   **b** $1170.25   **c** $6663.24

3   **a** $17 496   **b** $2496

4   **a** 9663.60 kroner   **b** 1663.60 kroner

EXERCISE 13C.2

1 $9021.59   **2** $301.25

3   **a** $7650   **b** $8151.65   **c** $8243.80
   Clearly, compounding quarterly is $92.15 more than compounding half-yearly and $593.80 more than simple interest.

EXERCISE 13C.3

1   £13 373.53   **2** $546.01

3   **a** 4725 Yen   **b** 4940.81 Yen   **c** 4998.89 Yen
   d 5028.61 Yen
   e 5048.67 Yen - Clearly, interest compounded monthly gives the greatest return.

4   Bank A balance = $157 149.29, Bank B balance = $155 344.23
   Jai should invest with Bank A.

5   Simple interest earned = 8960 euro
   Interest compounded monthly = 9760.78 euro
   Interest compounded monthly is 800.78 euro more than simple interest.

EXERCISE 13C.4

1 $6631   **2** $4079.75   **3** $4159.10

4   **a** 308.70 rupees interest   **b** 636.36 rupees interest
   c 1351.20 rupees interest   Doubling the interest rate more than doubles the interest earned.

EXERCISE 13C.5

1   19 971.30 Yen   **2** $20 836.90   **3** $80 000

4   80 162.65 baht   **5** $1564.95

EXERCISE 13C.6

1 $\doteqdot$ 28 months (2 years, 4 months)   **2**   11 quarters (2.75 years)

3   159 months (13 years, 3 months)

4   2543 days (6 years, 353 days)

EXERCISE 13C.7

1   14.47% p.a.   **2**   6.0% p.a.   **3**   5.25% p.a.

4   Shares 13.05% p.a., house 9.48% p.a., i.e., the shares had the greater average annual percentage increase in value.

EXERCISE 13C.8

1   **a** $1676.03   **b** $1669.09 $\therefore$ **a** is more by $6.94

2   **a** Monthly: $863.16, half-yearly: $859.58
   b Monthly: $1148.08, half-yearly: $1143.33

3   **a** Derk should choose the 6.40% interest paid at maturity (i.e., annually).
   b $3200   **c** $1808

EXERCISE 13C.9

1   A is effectively 5.47% p.a., B is effectively 5.41% p.a., $\therefore$ A is the better rate.

2   P is effectively 7.87% p.a., Q is effectively 7.90% p.a., $\therefore$ Q is the better rate.

3   **a** 4.95% p.a.   **b** 5.01% p.a.
   c 4.9% p.a. compounded monthly

4   **a** 8.06% p.a.   **b** 8.11% p.a.
   c 7.95% p.a. compounded half-yearly

5   12 times (monthly)

6   **a** 6.379% p.a.   **b** 6.398% p.a.   **c** 6.399% p.a.   **d** 6.40% p.a.
   e Take interest at maturity to receive $2560 interest.

EXERCISE 13C.10

1   **a** 6.58% p.a.   **b** 5.76% p.a.   **c** 5.15% p.a.   **d** 4.69% p.a.

2   **a** 7.51% p.a.   **b** 4.30% p.a.   **c** 4.97% p.a.   **d** 3.62% p.a.

3   6.38% p.a.

EXERCISE 13D

1   **a**

Age (years)	Depreciation	Book value
0	$0	$15 000
1	$2250	$12 750
2	$1912.50	$10 837.50
3	$1625.63	$9211.87

   b **i** $2250   **ii** $1912.50   **iii** $1625.63

2   **a** $53 393.55   **b** $171 606.45

3   **a** $16 588.80   **b** $15 811.20   **4** $\doteqdot$ 24.79% p.a.

5   **a** $\doteqdot$ 18.37% p.a.   **6**   **a** $175 702.59   **b** $64 297.41

EXERCISE 13E

1   **a** $267   **b** $16 020   **c** $4020   **d** $11 704.53

2   **a** $308.80   **b** $11 116.80   **c** $1616.80   **d** $8817.04

3   **a** 2032.80 Yen   **b** 5250 Yen

4   **a** $7994   **b** $5110.40
   Becky should borrow from the Cash Credit Union as she pays $2883.60 less in interest (although her monthly repayments are more.)

EXERCISE 13F.1

1   **a** $66 354.49   **b** $56 020.65   **c** $10 333.84   **d** 3.44% p.a.

2   **a** $22 444.54   **b** $22 110.15   **c** $334.39   **d** 0.50% p.a.

3   Decreased by $20.49

4   **a** $1343.92   **b** $1806.11   **c** $2427.26

EXERCISE 13F.2

1   **a** $18 013.95   **b** $15 876.55

2   **a** $12 916.22   **b** **i** $10 090.13   **ii** $9156.55   **iii** $8317.11

3   **a** $7812.50   **b** **i** $21 049.33   **ii** $19 128.36

4   **a** $3060.45   **b** $2596.06
   c Jon needed to invest at greater than the inflation rate (i.e., 4.2% p.a.).

REVIEW SET 13A

1 $1199.90 US   **2**   248.60 euro   **3** $\doteqdot$ 5.43 years

4 $\div 5.77\%$ p.a.   **5** $395\,151.52$ **6** 6572.60

7 Bank $A = \$185\,250$,   Bank $B = \$231\,995.25$,
Bank $C = \$220\,787.17$.   Val should opt for Bank B.

8 a $21\,297.85$ **b** $26\,702.15$

9 a $29\,364.66$ **b** $26\,608.31$ **c** 2756.35 **d** 3.34%

REVIEW SET 13B

1 a 21.6 Swiss francs   **b** 802.14 euro   **2** $31\,772.57$

3 970.26 **4 a** 517.50 **b** $31\,050$ **c** 8050

5 a *Option 1* $26\,373.20$,   *Option 2* $26\,706$
   b *Option 1* 7873.20,   *Option 2* 8206
   Option 1 is cheaper by $332.80 and is over a shorter
   period, although a deposit of $1850 is required initially.

6 a £513   **b** £5780   **c** £23\,985.09

7 a $12\,116.18$ **b** $11\,055.79$

EXERCISE 14A

1 a 0.78   **b** 0.22   **2 a** 0.487   **b** 0.051   **c** 0.731

3 a 43 days   **b i** $\div 0.047$ **ii** $\div 0.186$ **iii** $\div 0.465$

4 a $\div 0.089$ **b** $\div 0.126$

EXERCISE 14C

1 a $\div 0.265$ **b** $\div 0.139$ **c** $\div 0.222$

2 a $\div 0.148$ **b** $\div 0.430$ **c** $\div 0.372$

3 a i $\div 0.189$ **ii** $\div 0.336$ **b** $\div 0.381$ **c** $\div 0.545$

EXERCISE 14D

1 a $\{A, B, C, D\}$ **b** $\{BB, BG, GB, GG\}$
   c $\{$ABCD, ABDC, ACBD, ACDB, ADBC, ADCB, BACD,
   BADC, BCAD, BCDA, BDAC, BDCA, CABD, CADB,
   CBAD, CBDA, CDAB, CDBA, DABC, DACB, DBAC,
   DBCA, DCAB, DCBA$\}$
   d $\{$GGG, GGB, GBG, BGG, GBB, BGB, BBG, BBB$\}$

2 a
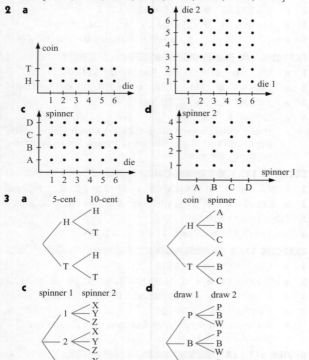

3 a

EXERCISE 14E

1 a $\frac{1}{5}$ **b** $\frac{1}{3}$ **c** $\frac{7}{15}$ **d** $\frac{4}{5}$ **e** $\frac{1}{5}$ **f** $\frac{8}{15}$

2 a 4   **b i** $\frac{2}{3}$ **ii** $\frac{1}{3}$ **3 a** 0   **b** $\frac{1}{4}$ **c** $\frac{19}{32}$ **d** $\frac{5}{32}$

4 a $\frac{1}{4}$ **b** $\frac{1}{9}$ **c** $\frac{4}{9}$ **d** $\frac{1}{36}$ **e** $\frac{1}{18}$ **f** $\frac{1}{6}$

5 a $\frac{1}{7}$ **b** $\frac{2}{7}$ **c** $\frac{124}{1461}$ **d** $\frac{237}{1461}$ {remember leap years}

6 {AKN, ANK, KAN, KNA, NAK, NKA}   **a** $\frac{1}{3}$ **b** $\frac{1}{3}$ **c** $\frac{1}{3}$ **d** $\frac{2}{3}$

7 a {GGG, GGB, GBG, BGG, GBB, BGB, BBG, BBB}
   b i $\frac{1}{8}$ **ii** $\frac{1}{8}$ **iii** $\frac{1}{8}$ **iv** $\frac{3}{8}$ **v** $\frac{1}{2}$ **vi** $\frac{7}{8}$

8 a {ABCD, ABDC, ACBD, ACDB, ADBC, ADCB,
   BACD, BADC, BCAD, BCDA, BDAC, BDCA,
   CABD, CADB, CBAD, CBDA, CDAB, CDBA,
   DABC, DACB, DBAC, DBCA, DCAB, DCBA}
   b i $\frac{1}{2}$ **ii** $\frac{1}{2}$ **iii** $\frac{1}{2}$ **iv** $\frac{1}{2}$

9 {HHHH, HHHT, HHTH, HTHH, THHH, HHTT, HTHT,
   THHT, THTH, HTTH, TTHH, TTTH, TTHT, THTT,
   HTTT, TTTT}
   a $\frac{1}{16}$ **b** $\frac{3}{8}$ **c** $\frac{5}{16}$ **d** $\frac{15}{16}$ **e** $\frac{1}{4}$

EXERCISE 14F

1
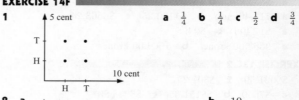
   a $\frac{1}{4}$ **b** $\frac{1}{4}$ **c** $\frac{1}{2}$ **d** $\frac{3}{4}$

2 a
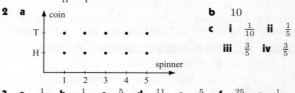
   b 10
   c i $\frac{1}{10}$ **ii** $\frac{1}{5}$
   iii $\frac{3}{5}$ **iv** $\frac{3}{5}$

3 a $\frac{1}{36}$ **b** $\frac{1}{18}$ **c** $\frac{5}{9}$ **d** $\frac{11}{36}$ **e** $\frac{5}{18}$ **f** $\frac{25}{36}$ **g** $\frac{1}{6}$
   h $\frac{5}{18}$ **i** $\frac{2}{9}$ **j** $\frac{13}{18}$

EXERCISE 14G.1

1 a $\frac{6}{7}$ **b** $\frac{36}{49}$ **c** $\frac{216}{343}$ **2 a** $\frac{1}{8}$ **b** $\frac{1}{8}$

3 a 0.0096   **b** 0.8096   **4 a** $\frac{1}{16}$ **b** $\frac{15}{16}$

5 a 0.56   **b** 0.06   **c** 0.14   **d** 0.24   **6 a** $\frac{8}{125}$ **b** $\frac{12}{125}$ **c** $\frac{27}{125}$

EXERCISE 14G.2

1 a $\frac{7}{15}$ **b** $\frac{7}{30}$ **c** $\frac{7}{15}$ **2 a** $\frac{14}{55}$ **b** $\frac{1}{55}$

3 a $\frac{3}{100}$ **b** $\frac{3}{100} \times \frac{2}{99} \div 0.0006$
   c $\frac{3}{100} \times \frac{2}{99} \times \frac{1}{98} \div 0.000\,006$ **d** $\frac{97}{100} \times \frac{96}{99} \times \frac{95}{98} \div 0.912$

4 a $\frac{4}{7}$ **b** $\frac{2}{7}$

EXERCISE 14H

1 a
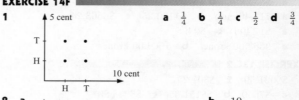

b i 0.42   **ii** 0.22

2 a
0.04 — F — 0.15 under 20 / 0.85 over 20
0.96 — M — 0.10 under 20 / 0.90 over 20

b i 0.096 **ii** 0.102 **c** 0.1305

3 a
0.6 — F — 0.2 violin / 0.8 not violin
0.4 — M — 0.3 violin / 0.7 not violin

b i 0.28 **ii** 0.24

4 a
0.45 — F — 0.5 had a takeaway meal / 0.5 did not
0.55 — M — 0.6 had a takeaway meal / 0.4 did not

b i 0.225 **ii** 0.445

5 a
0.22 — Navy — 0.19 officer / 0.81 other rank
0.47 — Army — 0.15 officer / 0.85 other rank
0.31 — Air Force — 0.21 officer / 0.79 other rank

b i 0.1774 **ii** 0.9582 **iii** 0.8644

6
$\frac{1}{5}$ — rain — $\frac{1}{2}$ win / $\frac{1}{2}$ lose
$\frac{4}{5}$ — no rain — $\frac{1}{20}$ win / $\frac{19}{20}$ lose

$Pr(win) = \frac{7}{50}$

7
0.4 — A — 0.05 spoiled / 0.95 not spoiled
0.6 — B — 0.02 spoiled / 0.98 not spoiled

0.032

8 $\frac{17}{40}$ **9 a** $\frac{11}{30}$ **b** $\frac{19}{30}$

EXERCISE 14I

1 a $\frac{20}{49}$ **b** $\frac{10}{21}$ **2 a** $\frac{3}{10}$ **b** $\frac{1}{10}$ **c** $\frac{3}{5}$

3 a $\frac{2}{9}$ **b** $\frac{5}{9}$ **4 a** $\frac{1}{3}$ **b** $\frac{2}{15}$ **c** $\frac{4}{15}$ **d** $\frac{4}{15}$

These cases cover all possibilities, so their probabilities must sum to 1.

5 a
egg 1 — egg 2
$\frac{6}{9}$ — S — $\frac{5}{8}$ S / $\frac{3}{8}$ D
$\frac{3}{9}$ — D — $\frac{6}{8}$ S / $\frac{2}{8}$ D

b $\frac{1}{12}$ **c** $\frac{5}{12}$

6 a chocolate 1 — chocolate 2
$\frac{10}{25}$ — H — $\frac{9}{24}$ H / $\frac{15}{24}$ S
$\frac{15}{25}$ — S — $\frac{10}{24}$ H / $\frac{14}{24}$ S

b $\frac{3}{20}$ **c** $\frac{7}{20}$

7 a $\frac{1}{5}$ **b** $\frac{3}{5}$ **c** $\frac{4}{5}$ **8** $\frac{19}{45}$ **9 a** $\frac{2}{100} \times \frac{1}{99} \doteq 0.0002$

b $\frac{98}{100} \times \frac{97}{99} \doteq 0.9602$ **c** $1 - \frac{98}{100} \times \frac{97}{99} \doteq 0.0398$

EXERCISE 14J

1 a **b**

c **d**

e **f**

2 a 29 **b** 17 **c** 26 **d** 5 **3 a** 65 **b** 9 **c** 4 **d** 52

4 a $\frac{19}{40}$ **b** $\frac{1}{2}$ **c** $\frac{4}{5}$ **d** $\frac{5}{8}$ **e** $\frac{13}{40}$ **f** $\frac{7}{20}$

5 a $\frac{19}{25}$ **b** $\frac{13}{25}$ **c** $\frac{6}{25}$ **d** $\frac{7}{19}$

6 a $\frac{7}{15}$ **b** $\frac{1}{15}$ **c** $\frac{2}{15}$ **d** $\frac{6}{7}$

7 a i $\dfrac{b+c}{a+b+c+d}$ **ii** $\dfrac{b}{a+b+c+d}$

iii $\dfrac{a+b+c}{a+b+c+d}$ **iv** $\dfrac{a+b+c}{a+b+c+d}$

b P(A or B) = P(A) + P(B) − P(A and B)

8 a $k = 5$ **b i** $\frac{7}{30}$ **ii** $\frac{11}{60}$ **iii** $\frac{7}{60}$ **iv** $\frac{53}{60}$ **v** $\frac{7}{60}$

vi $\frac{2}{15}$ **vii** $\frac{41}{60}$ **viii** $\frac{31}{60}$

EXERCISE 14K

1 a
22 study both

b i $\frac{9}{25}$ **ii** $\frac{11}{20}$

2 a $\frac{3}{8}$ **b** $\frac{7}{20}$ **c** $\frac{1}{5}$ **d** $\frac{15}{23}$

3 a $\frac{14}{25}$ **b** $\frac{4}{5}$ **c** $\frac{1}{5}$ **d** $\frac{5}{23}$ **e** $\frac{9}{14}$ **4 a** $\frac{22}{25}$ **b** $\frac{1}{2}$

5 $\frac{5}{6}$ **6 a** $\frac{13}{20}$ **b** $\frac{7}{20}$ **c** $\frac{11}{50}$ **d** $\frac{7}{25}$ **e** $\frac{4}{7}$ **f** $\frac{1}{4}$

7 a 0.93 **b** 0 **c** 0.18 **d** 0.02

8 a $\frac{3}{5}$ **b** $\frac{2}{3}$ **9 a** $\frac{23}{50}$ **b** $\frac{14}{23}$

10 a 95.68% **b** 0.95 **c** 0.99

d c, as people are "innocent until proven guilty"

e 0.7902 **f** 0.1978

11 a 0.97 **b** 0.95 **c** 0.962 **d** 0.605
12 a 0.015 **b** 0.019 **c** 0.316
13 a 0.023 69 **b i** 0.2875 **ii** 0.000 193 6

EXERCISE 14L

1 $P(A \cap B) = 0.2$ and $P(A) \times P(B) = 0.2$
$\therefore$ are independent events

2 a $\frac{7}{30}$ **b** $\frac{7}{12}$ **c** $\frac{7}{10}$ No, as $P(A \cap B) \neq P(A) \times P(B)$

3 a 0.35 **b** 0.85 **c** 0.15 **d** 0.15 **e** 0.5

4 $\frac{14}{15}$ **5 a** $\frac{91}{216}$ **b** 26

6 Hint: Show $P(A \cap B') = P(A) \, P(B')$
using a Venn diagram and $P(A \cap B)$

REVIEW SET 14A

1 a 39 days **b i** $\div 0.0256$ **ii** 0.282 **iii** 0.487
2 a $\div 0.402$ **b** $\div 0.571$ **c** $\div 0.830$
3 a $\frac{1}{9}$ **b** $\frac{4}{9}$ **c** $\frac{4}{9}$ **4 a** 0.60 **b** 0.40
5 a 0.21 **b** 0.15 **c** 32 logs **6** 0.657
7 a 0.63 **b** 0.03 **c** 0.97 **d** 0.07
8 a $\frac{19}{60}$ **b** $\frac{1}{20}$ **9 a** $a = 8$
b i $\frac{13}{30}$ **ii** $\frac{9}{20}$ **iii** $\frac{11}{60}$ **iv** $\frac{3}{5}$ **v** $\frac{7}{60}$ **vi** $\frac{1}{2}$ **vii** $\frac{1}{12}$

REVIEW SET 14B

1 ABCD, ABDC, ACBD, ACDB, ADBC, ADCB, BACD, BADC,
BCAD, BCDA, BDAC, BDCA, CABD, CADB, CBAD, CBDA,
CDAB, CDBA, DABC, DACB, DBAC, DBCA, DCAB, DCBA
a $\frac{1}{2}$ **b** $\frac{1}{3}$

2 a $\frac{3}{8}$ **b** $\frac{1}{8}$ **c** $\frac{5}{8}$ **3 a** $\frac{3}{25}$ **b** $\frac{24}{25}$ **c** $\frac{11}{12}$

4 P(N wins)
$= \frac{44}{125}$
$= 0.352$

5 a $\frac{2}{5}$ **b** $\frac{13}{15}$ **c** $\frac{4}{5}$

6 a $\frac{4}{500} \times \frac{3}{499} \times \frac{2}{498} \div 0.000\,000\,193$
b $1 - \frac{496}{500} \times \frac{495}{499} \times \frac{494}{498} \div 0.0239$

7
Day 1   Day 2
0.95 W
0.95 W
0.05 W'
0.9975
0.95 W
0.05 W'
0.05 W'

8 a $\frac{19}{30}$ **b** $\frac{6}{19}$

9 a

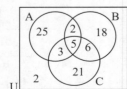

b 82 people
c i $\frac{25}{82}$ **ii** $\frac{5}{82}$
iii $\frac{55}{82}$ **iv** $\frac{1}{41}$
v $\frac{32}{41}$ **vi** $\frac{40}{41}$

REVIEW SET 14C

1 BBBB, BBBG, BBGB, BGBB, GBBB, BBGG, BGBG, BGGB,
GGBB, GBBG, GBGB, BGGG, GBGG, GGBG, GGGB, GGGG
$\frac{3}{8}$

2
die 1
a $\frac{2}{9}$
b $\frac{5}{12}$
die 2

3 a $\frac{1}{4}$ **b** $\frac{37}{40}$ **c** $\frac{10}{25} = \frac{2}{5}$ **4** $\frac{8}{33}$

5 a $\frac{5}{33}$ **b** $\frac{19}{66}$ **c** $\frac{5}{11}$ **d** $\frac{16}{33}$

6 a Two events are independent if the occurrence of one does
not affect the occurrence of the other.
For A and B independent, $P(A) \times P(B) = P(A \text{ and } B)$.
b Two events, A and B, are disjoint if they have no common
outcomes. $P(A \text{ or } B) = P(A) + P(B)$

7
0.36 W
0.25 R
0.64 W'
0.75 R'
0.36 W
0.64 W'
a 0.09
b 0.52

8 $1 - 0.9 \times 0.8 \times 0.7 = 0.496$ **9 a** $\frac{31}{70}$ **b** $\frac{21}{31}$

EXERCISE 15A.1

1 a proposition - False **b** proposition - False
c proposition - True **d** proposition - False
e proposition - True **f** proposition - True
g not a proposition **h** proposition - True
i not a proposition **j** proposition - False
k proposition - Indeterminate **l** not a proposition
m proposition - Indeterminate **n** proposition - True
o proposition - True **p** proposition - False

2 a All rectangles are not parallelograms - statement is true
b $\sqrt{5}$ is not an irrational number - statement is true
c 7 is not a rational number - statement true
d $23 - 14 \neq 12$ - negation true
e $52 \div 4 \neq 13$ - statement true
f The difference between any two odd numbers is not even
- statement true
g The product of consecutive integers is not always even
- statement true
h All obtuse angles are not equal - negation true
i All trapeziums are not parallelograms - negation true
j If a Δ has two equal angles it is not isosceles - statement
true

3 a $x \geqslant 5$ **b** $x < 3$ **c** $y \geqslant 8$ **d** $y > 10$

4 a not negation - Bic may have scored exactly 60%
b not negation - Jan could be elsewhere? **c** negation

5 Yes

6 a The football team lost or drew its game.
b She has two or more sisters.
c This suburb does not have a swimming pool.
d This garage is tidy.

7 a $x < 5$ for $x \in Z^+$ or $x \in \{1, 2, 3, 4\}$
b x is not a cow **c** $x < 0$ for $x \in Z$
d x is a female student **e** x is not a student.

EXERCISE 15A.2

1 a i $X = \{21, 24, 27\}$ **ii** $X = \{2, 4, 6, 8, 10\}$
iii $X = \{1, 2, 3, 6, 7, 14, 21, 42\}$

b i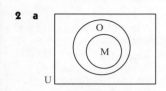

ii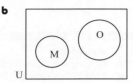

2 a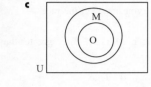

b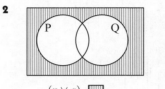

c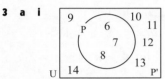

3 a i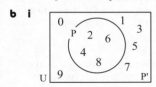

ii Let P be the truth set of p
$\neg p$......P$' = \{9, 10, 11, 12, 13, 14\}$

b i

ii Let P be the truth set of p
$\neg p$...... P$' = \{0, 1, 3, 5, 7, 9\}$

EXERCISE 15B

1 a The shirt is not new. **b** The shirt is not clean.
  c The shirt is new or clean. **d** The shirt is new and clean.
  e The shirt is new and not clean.
  f The shirt is not new and it is clean.
  g The shirt is new. **h** The shirt is clean.

2 a $\neg r$ **b** $\neg s$ **c** $r \wedge s$ **d** $\neg s \wedge \neg r$ **e** $s \vee r$

3 a $\neg x$ **b** $x \wedge y$ **c** $x \vee y$ **d** $\neg(x \wedge y)$ **e** $\neg x \wedge y$
  f $x \veebar y$

4 a p: Pam plays basketball **b** : Fisk enjoys drama
     q: Pam plays tennis   : Fisc enjoys comedy
  c : Seagulls are birds   **d** : $x > 2$
     : Seagulls are mammals (or not)   : $x \leqslant 10$
  e : $x \geqslant 2$
     : $x < 10$

5 Let P represent the truth set of p,
  Q represent the truth set of q.
  a P $= \{1, 3, 5, 7, 9\}$ Q $= \{2, 3, 5, 7, 11\}$
  truth set of $(p \wedge q) =$ P $\cap$ Q $= \{3, 5, 7\}$.
  b P $= \{3, 13\}$ Q $= \{1, 4, 9, 16, 25, 36, 49\}$
  $p \wedge q =$ P $\cap$ Q $= \{\ \}$ or $\varnothing$.

6 Let P represent truth set of p, Q represent truth set of q
  a P $= \{1, 2, 4, 5, 10, 20\}$ Q $= \{1, 3, 7, 21\}$

---

  truth set for $p \vee q$: $\{1, 2, 3, 4, 5, 7, 10, 20, 21\}$.
  b P $= \{7, 14, 21, 28, 35\}$ Q $= \{0, 10, 20, 30\}$
  truth set for $p \vee q$: $\{0, 7, 10, 14, 20, 21, 28, 30, 35\}$.

7 a True **b** False **c** False

8 a True **b** False **9 a** True **b** False

10 a i $x < 7$ **ii** $x < 3$, $x > 6$
  b i Either p, or q is a false statement (not both).
  ii Either p or q or both are false.

EXERCISE 15C

1

p	q	$p \vee q$	$\neg(p \vee q)$	$\neg p$	$\neg q$	$\neg p \wedge \neg q$
T	T	T	F	F	F	F
T	F	T	F	F	T	F
F	T	T	F	T	F	F
F	F	F	T	T	T	T

2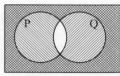

$\neg(p \vee q)$ ▤ $\neg p$ ▨ $\neg q$ ▧
$\neg p \wedge \neg q$ ▦

3 $\neg(p \vee q) = \neg p \wedge \neg q$; $\neg(p \wedge q) = \neg p \vee \neg q$
  a $x \neq -4$ and $x \neq 5$ **b** $x < 5$ or $x > 9$
  c $-1 \leqslant x \leqslant 7$
  d number does not end in 2 and does not end in 8
  e the enclosure is not an ellipse and not a circle
  f do not use the cosine rule and do not use the sine rule
  g lines PQ and RS and not parallel or not equal in length
  h $\triangle$KLM is not right angled or not an isosceles $\triangle$.

4 a

p	$\neg p$	$\neg(\neg p)$
T	F	T
T	F	T
F	T	F
F	T	F

b

p	$p \wedge p$
T	T
T	T
F	F
F	F

c

p	q	$p \vee q$	$\neg p \wedge q$	$p \vee (\neg p \wedge q)$
T	T	T	F	T
T	F	T	F	T
F	T	T	T	T
F	F	F	F	F

5 a

p	q	$\neg p$	$\neg q$	$(\neg p \wedge q)$	$(p \wedge \neg q)$	$(\neg p \wedge q) \vee (p \wedge \neg q)$
T	T	F	F	F	F	F
T	F	F	T	F	T	T
F	T	T	F	T	F	T
F	F	T	T	F	F	F

  b logically equivalent to the 'exclusive or'

6 a contradiction **b** neither **c** neither **d** tautology

7 a Tautologies have all true values $\therefore$ they are the same
    i.e., logically equivalent.
  b Logical contradictions have all false values
    $\therefore$ they are the same i.e., logically equivalent.

8 a a tautology **b** a logical contradiction
  c also a tautology

EXERCISE 15D

1 a

p	q	r	$\neg p$	$(q \wedge r)$	$\neg p \vee (q \wedge r)$
T	T	T	F	T	T
T	T	F	F	F	F
T	F	T	F	F	F
T	F	F	F	F	F
F	T	T	T	T	T
F	T	F	T	F	T
F	F	T	T	F	T
F	F	F	T	F	T

b

$(p \vee \neg q)$	$\wedge$	r
T	T	T
T	F	F
T	T	T
T	F	F
F	F	T
F	F	F
T	T	T
T	F	F

c

$(p \vee q)$	$\vee$	$(p \wedge \neg r)$
T	T	F
T	T	F
T	T	T
T	T	F
T	T	F
T	T	F
F	F	F
F	F	F

2 a

$(p \wedge q)$	$\wedge$	r	p	$\wedge$	$(q \wedge r)$
T	T	T	T	T	T
T	F	F	T	F	F
F	F	T	T	F	F
F	F	F	T	F	F
F	F	T	F	F	T
F	F	F	F	F	F
F	F	T	F	F	F
F	F	F	F	F	F

logical equivalent ∴ $(p \wedge q) \wedge r = p \wedge (q \wedge r)$

b

$(p \vee q)$	$\vee$	r	p	$\vee$	$(q \vee r)$
T	T	T	T	T	T
T	T	F	T	T	T
T	T	T	T	T	T
T	T	F	T	T	F
T	T	T	F	T	T
T	T	F	F	T	T
F	T	T	F	T	T
F	F	F	F	F	F

logical equivalent ∴ $(p \vee q) \vee r = p \vee (q \vee r)$

3 a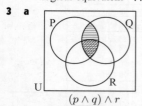
$(p \wedge q) \wedge r$

b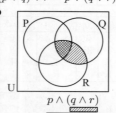
$p \wedge (q \wedge r)$

4 a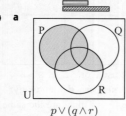
$p \vee (q \wedge r)$

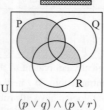
$(p \vee q) \wedge (p \vee r)$

b $p \vee (q \wedge r) = (p \vee q) \wedge (p \vee r)$
i.e., disjunction is distributive.

5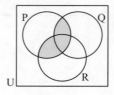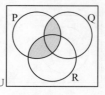
$p \wedge (q \vee r)$ $(p \wedge q) \vee (p \wedge r)$
∴ $p \wedge (q \vee r) = (p \wedge q) \vee (p \wedge r)$

EXERCISE 15E

1

	antecedent	conclusion
a	an integer ends in 0 or 5	an integer is divisible by 5
b	temp is low *enough*	the lake will freeze
c	$2x + 5 = 17$	$2x = 12$
d	you jump all 8 hurdles	you may win the race
e	$x^2 + 5 = 9$	$x = \pm 2$

2 a i If 17 is greater than 10 then 20, $(17 + 3)$, is greater than 13 $(10 + 3)$.   **ii** True

b i If $x > 4$ then $x^2 > 16$.   **ii** True

c i If $x^2 > 9$ then $x > 3$.   **ii** True

d i If $7 + 5 > 9$ then $\frac{7+5}{3} > 4$.   **ii** False

3 a $p \Rightarrow q$ **b** $q \Rightarrow p$ **c** $\neg q$ **d** $\neg p$ **e** $\neg p \Rightarrow \neg q$
f $p \Rightarrow \neg q$ **g** $\neg q \Rightarrow p$

4 a $p \Rightarrow \neg q$ **b** $\neg q \Rightarrow \neg p$ **c** $(p \wedge q) \Rightarrow p$

F	T	T
T	F	T
T	T	T
T	T	T

d $q \wedge (p \Rightarrow q)$ **e** $p \Leftrightarrow \neg q$ **f** $(p \Leftrightarrow q) \wedge \neg p$

T	F	T	F	F
F	T	F	F	F
T	T	F	F	T
F	F	T	T	T

g $p \Rightarrow (p \wedge \neg q)$ **h** $(p \Rightarrow q)$ $\Rightarrow$ $\neg p$

F	T	F	F
T	F	T	F
T	T	T	T
T	T	T	T

5 a $p \Leftrightarrow q$ **b** $(p \wedge q) \vee (\neg p \wedge \neg q)$

T	T
F	F
F	F
T	T

∴ $p \Leftrightarrow q = (p \wedge q) \vee (\neg p \wedge \neg q)$

6 None of the given set are logically equivalent to $\neg (q \Rightarrow p)$.

7 b and **c** are tautologies

EXERCISE 15F

1 *Converse*

a If $x = 3$ then $5x - 2 = 13$.
b If two triangles are equiangular then they are similar.
c If $x = \pm\sqrt{6}$ then $2x^2 = 12$.
d If the opposite sides are equal then the figure is a parallelogram.
e If the sides are equal in length then a triangle is equilateral.

Inverse

a If $5x - 2 \neq 13$ then $x \neq 3$.
b If two triangles are not similar they are not equiangular.

c If $2x^2 \neq 12$ then $x \neq \pm\sqrt{6}$.

d If the figure is not a parallelogram then its opposite sides are not equal.

e If a Δ is not equilateral then its sides are not equal in length.

2 a **i**, **iv** are true   **b** **iv**

3 a If x does not have thorns then x is not a rose bush.

b If x makes correct decisions all the time then x is not an umpire.

c If x has poor kicking skills then x is not a good soccer player.

d If x does not take the shape of the container then it is not liquid.

e If the person is not a doctor then the person is not fair nor clever.

4 a If x does not study Mathematics then x is not a high school student.

b **i** Kelly studies Maths.
   ii Tamra is not a high school student.
   iii Cannot say if Eli is a high school student.

5 a x^2 is not divisible by $\Rightarrow x$ is not divisible by 3.

b x is odd $\Rightarrow x$ is not a number ending in 2.

c PQ $\nparallel$ SR and PS $\nparallel$ QR $\Rightarrow$ PQRS is not a rectangle.

d $\angle$KML does not measure $60^o \Rightarrow$ KLM is not an equilateral Δ.

6 a **i** No weak student is in year 11
   ii No Year 11 student is weak.

b **i** If $x \in W$ then $x \notin E$.   **ii** If $x \in E$ then $x \notin W$.

c Statements are contrapositives.

EXERCISE 15G

1 a $(p \Rightarrow q) \wedge \neg q \Rightarrow \neg p$ **b** $(p \vee q) \wedge \neg p \Rightarrow q$

c $p \vee q \Rightarrow p$ **d** $(p \Rightarrow q) \wedge \neg p \Rightarrow \neg q$

e $(p \Rightarrow q) \wedge (q \Rightarrow p) \Rightarrow p$

2 a T   **b** T   **c** T   **d** T   **e** T
     T   T   T   T   T
     T   T   T   F   T
     T   T   T   T   F

a and **b** are valid arguments

3 a $q \wedge p$ **b** $\neg p$ **c** $q \wedge p \Rightarrow p$ **d** $(q \wedge p) \wedge \neg p \Rightarrow \neg q$

e tautology $\therefore$ valid

4 $\neg q \Rightarrow \neg p$ p q $\neg p$ $\neg q$ $\neg q \Rightarrow \neg p$
                                   T   T   F   F   T
                                   T   F   F   T   F *
                                   F   T   T   F   T
                                   F   F   T   T   T

Invalid when p is true, q is false.
Example: It is raining and I am not wearing my raincoat.

5 a valid **b** invalid **c** valid **d** invalid

6 a $(p \Rightarrow q) \wedge q \Rightarrow p$ **b** $(q \Rightarrow p) \wedge (p \Rightarrow q) \Rightarrow p$
          T   T
          T   T
          F   T
          T   F

EXERCISE 15H

1 a If it is sunny and warm then I feel happy.

b If it is sunny but not warm then I do not feel happy.

c If I am warm and happy then it is sunny.

2 a invalid **b** valid **c** invalid

3 Argument not valid when p, q and r are all true.   F
                                                                  T
                                                                  T
                                                                  T
                                                                  T
                                                                  T
                                                                  T
                                                                  T

4 a p : I do not like this subject
       q : I do not work hard
       r : I fail

b | $(p$ | $\Rightarrow$ | $q)$ | $\wedge$ | $(q$ | $\Rightarrow$ | $r)$ | $\wedge$ | $\neg r$ | $\Rightarrow$ | $\neg p$ |
|---|---|---|---|---|---|---|---|---|---|---|
| T | | T | T | T | | T | F | F | T | F |
| T | | F | F | F | | T | F | T | T | F |
| F | | F | T | F | | T | F | F | T | F |
| F | | F | T | F | | T | T | T | T | F |
| T | | T | T | T | | T | F | F | T | T |
| T | | F | F | F | | T | F | T | T | T |
| T | | T | T | T | | F | F | F | T | T |
| T | | T | T | T | | T | T | T | T | T |
| | | | | | | | * | | | * |

c Valid argument: conclusion is a result of correct reasoning.

REVIEW SET 15A

1 a proposition - True **b** not a proposition

c proposition - Indeterminate **d** not a proposition

e proposition - Indeterminate **f** proposition - True

g not a proposition **h** proposition - False

i proposition - Indeterminate **j** proposition - True

2 a x is not an even number.

b If x is an even number then x is divisible by 3.

c If x is an even number then x is not divisible by 3.

d x is either an even number or divisible by 3.

e x is not an even number and is divisible by 3.

f If x is not an even number then it is not divisible by 3.

g x is either an even number or is divisible by 3 but not both.

h x is either not an even number or is divisible by 3 but not both.

3 a $p \Rightarrow q$ **b** $\neg p$ **c** $q \wedge \neg p$ **d** $p \veebar q$ **e** $\neg (p \vee q)$

4 a **Implication:** If I love swimming then I live near the sea.
   Inverse: If I do not love swimming then I do not live near the sea.
   Converse: If I live near the sea then I love swimming.
   Contrapositive: If I do not live near the sea then I do not love swimming.

5 a   **b**

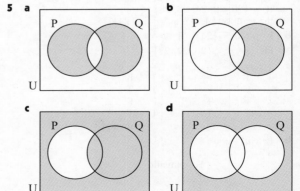

c   **d**

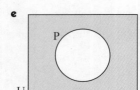

e, **f** (Venn diagrams)

6 Let P represent the truth set of p, Q the truth set of q
P = {1, 2, 3, 4, 6, 12}   Q = {1, 3, 5, 7, 9}
the truth set of $p \wedge q$ is {1, 3}
the truth set of $p \vee q$ is {1, 2, 3, 4, 5, 6, 7, 9, 12}

7 a True **b** False **c** True **d** True **e** True

REVIEW SET 15B

1 a I will pass. **b** The sky is not cloudy.
  c I will be late for school.
  d No single valid conclusion.

2 a Eddy is poor at football.
  b The maths class has 10 or less boys.
  c The writing is legible.
  d Ali owns either no car or an old car.

3 a If x is a bird then x has two legs.
  b If x is a snake then it is not a mammal.
  c If x is a rectangle then x does not have 5 sides.
  d If x is an equation then it has no real solutions.

4 a neither **b** $x \geqslant 0$, $x \in Q$

5 a $\neg(p \vee q)$ **b** $p \wedge \neg q$ **c** $\neg p \wedge q \wedge r$

6 a, **b** and **d** are logical equivalent

7 a
T
T
F
T
(Neither)

b
F
F
F
F
logical contradiction

c
F
T
T
F
(Neither)

d
T
F
T
F
(Neither)

e
T
F
T
T
T
F
T
F
tautology

f
T
T
T
T
tautology

REVIEW SET 15C

1 a Let P represent the truth set of p, Q the truth set of q, R the truth set of r. P = {20, 24, 28},
Q = {1, 2, 3, 4, 6, 8, 12, 24},
R = {20, 22, 24, 26, 28},
the truth set of $p \wedge q \wedge r$ is {24}.

  b i $p \wedge q$: {24} **ii** $q \wedge r$: {24}
  iii $p \wedge r$: {20, 24, 28}

2 a
$(p \Rightarrow q) \wedge \neg p \Rightarrow \neg q$
T F F T F
F F F T T
T T T F F
T T T T T
invalid

  b
$(p \vee q) \wedge \neg q \Rightarrow \neg p$
T
F invalid
T
T

c
$(p \Rightarrow q) \wedge (q \Rightarrow r) \Rightarrow r \vee p$
T
T
T
T
T invalid
T
T
F

3 a
$(p \Rightarrow q) \wedge p \Rightarrow q$ p: the sun is shining
T q: I wear my shorts
T
T valid
T

  b p: x is a teacher $(p \Rightarrow q) \wedge \neg p \Rightarrow \neg q$
  q: x works hard
T
T
F invalid
T

  c
$(p \Rightarrow q) \wedge (q \Rightarrow r) \wedge p \Rightarrow r$
T
T
T
T
T valid
T
T
T

4 **Implication:** The diagonals of a rhombus are equal.
Inverse: If x and y are not the diagonals of a rhombus then they are not equal.
Converse: If the diagonals are equal then they are the diagonals of a rhombus.
Contrapositive: If the diagonals are not equal then they are not the diagonals of a rhombus.

5 a i False **ii** True

(number line showing A at 5, B to the right)

  b i False **ii** True

(number line showing B from 6, A to 9)

6 a $\neg p \Rightarrow \neg q$ **b** $\neg p \Rightarrow q$ **c** $q \wedge \neg p$ **d** $\neg p \vee q$

7 a If the plane leaves from gate 5 then it does not leave this morning and does not leave from gate 2.
  b $r \Leftrightarrow (q \vee p)$

EXERCISE 16A

1 a 4 **b** 5 **c** 7 **d** $3\frac{1}{2}$ **e** $3\frac{1}{4}$

2 a 5 **b** 10 **c** 20 **d** $2\frac{1}{2}$ **e** $1\frac{1}{4}$

3 a 2 **b** 4 **c** 16 **d** 1 **e** $\frac{1}{4}$

4 a 1 **b** $\frac{1}{4}$ **c** $\frac{1}{16}$ **d** 4 **e** 16

5 a 6 **b** 2 **c** 18 **d** $\frac{2}{3}$ **e** 54

EXERCISE 16B

1 a 1.4 **b** 3.0 **c** 4.3

2 a i 1 **ii** 3 **iii** 1.7 **iv** 3.7

3 a i $x = 2$ **ii** $x = 1.6$ **iii** $x = 0$ **iv** $x = -0.3$

4 a i $x = 1$ **ii** $x = 1.3$ **iii** $x = 0$ **iv** $x = -0.8$

5 a

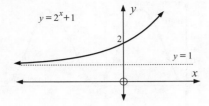

$y = 2^x + 1$
$y = 1$

b

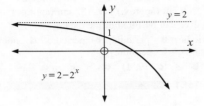

$y = 2$
$y = 2 - 2^x$

c

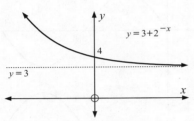

$y = 3 + 2^{-x}$
$y = 3$

d

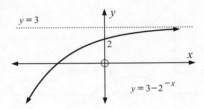

$y = 3$
$y = 3 - 2^{-x}$

EXERCISE 16C

1 a 500 ants **b i** 1553 ants **ii** 4823 ants

c

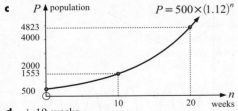

$P = 500 \times (1.12)^n$

d $\doteqdot$ 12 weeks

2 a 20 grams **b** $\doteqdot$ 23.64 g

c

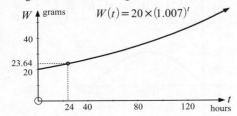

$W(t) = 20 \times (1.007)^t$

d $\doteqdot$ 231 hours

3 a 250 wasps **b i** 1073 **ii** 6532

c

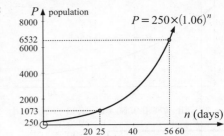

$P = 250 \times (1.06)^n$

d $\doteqdot$ 12 days

4 a

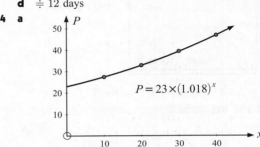

$P = 23 \times (1.018)^x$

b i $a = 23$ **ii** $b = 1.018$
c 1970 27 490; 1980 32 860; 1990 39 280
d i 56 120 **ii** 114 560
e

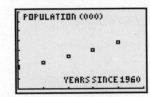

POPULATION (000)

YEARS SINCE 1960

f $P = 23.0 \times 1.018^x$
g 2010; 56 200 people,   2050; 114 600 people

EXERCISE 16D

1 a 100°C **b i** 50.0°C **ii** 25.0°C **iii** 12.5°C

c

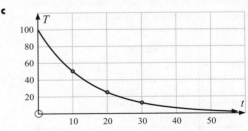

d i $\doteqdot$ 13 mins **ii** $\doteqdot$ 33 mins

2 a 250 g **b i** 112.2 g **ii** 50.4 g **iii** 22.6 g

c

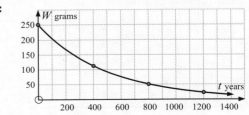

d $\doteqdot$ 346 years

3 a

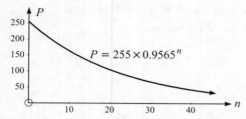

b i $a = 255$ **ii** $b = 0.9566$ **c** 1980 ÷ 204; **d** 2012
1985 ÷ 164;
1990 ÷ 131;
1995 ÷ 105

e

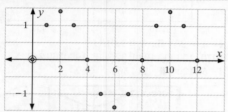

f $P = 254.8 \times 0.9565^n$ **g** ÷ July 2011

EXERCISE 16E

1 a

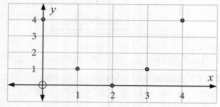

Data exhibits periodic behaviour.

b

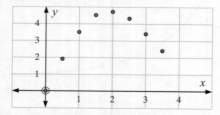

Not enough information to say data is periodic.
It may in fact be quadratic.

c

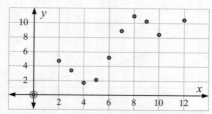

Not enough information to say data is periodic.
It may in fact be quadratic.

d

Not enough information to say data is periodic.

2 a

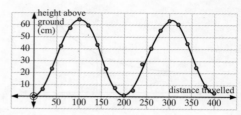

b The data is periodic.
i $y = 32$ (approx.) **ii** 64 cm (approx.)
iii 200 cm (approx.) **iv** 32 cm (approx.)
c A curve can be fitted to the data.

3 a periodic **b** periodic **c** periodic **d** not periodic
e periodic **f** periodic

4 a 2 **b** 8 **c** (2, 1) **d** 8 **e** $y = -1$

EXERCISE 16F.1

1 a

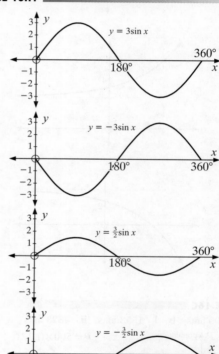

b

c

d

2 a

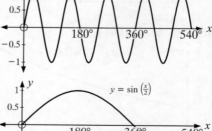

b

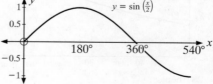

3 a 90° **b** 1080° **c** 600°

4 a $B = \frac{2}{5}$ **b** $B = 3$ **c** $B = \frac{1}{6}$

EXERCISE 16F.2

1 a $y = 3\sin x$ **b** $y = \sin x - 2$ **c** $y = -2\sin x - 1$

   d $y = \sin 2x$ **e** $y = -4\sin\left(\frac{x}{2}\right)$ **f** $y = \sin\left(\frac{x}{2}\right)$

   g $y = 2\sin(3x)$ **h** $y = 2\sin(2x) - 3$

2 a

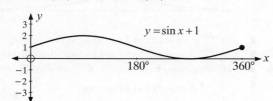

b

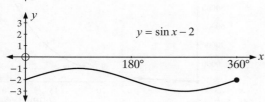

c

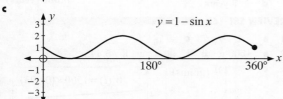

d

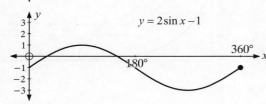

e

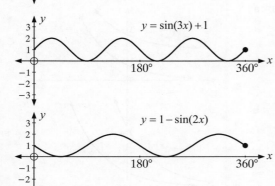

f

EXERCISE 16G

1 a $y = \cos x + 2$

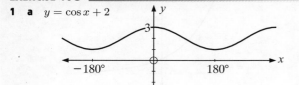

b $y = \cos x - 1$

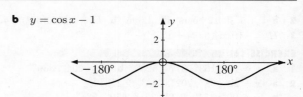

c $y = \frac{2}{3}\cos x$

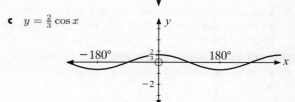

d $y = \frac{3}{2}\cos x$

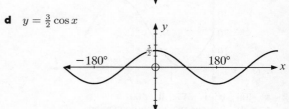

e $y = -\cos x$

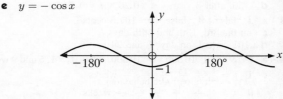

f $y = \cos(2x)$

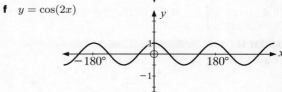

g $y = \cos\left(\frac{x}{2}\right)$

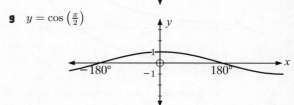

h $y = 3\cos(2x)$

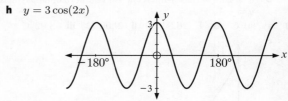

2 a 120° **b** 1080° **c** 720°

3 A: affects the amplitude, B: controls the period
   C: affects the vertical translation

4 a $y = 2\cos(2x)$ **b** $y = \cos\left(\frac{x}{2}\right) + 2$ **c** $y = 3\cos(4x)$

   d $y = -5\cos(2x)$

EXERCISE 16H

1 $T \doteqdot 4.5\cos(30t) + 11.5$

2 a i 1.5 **ii** 12 hours **iii** 4.5 m **b** $D \doteqdot 1.5\sin(30t) + 4.5$

3 $H = -10\cos(3.6t) + 12$

EXERCISE 16I

1 a $x \doteqdot 17°,\ 163°,\ 377°,\ 523°$ **b** $x \doteqdot 204°,\ 336°$

2 a $x \doteqdot 22°,\ 68°,\ 202°,\ 248°,\ 382°,\ 428°$

b $x \doteqdot 99°,\ 171°,\ 279°,\ 351°,\ 459°,\ 531°$

3 a $x \doteqdot 24.5°$ or $155.5°$ **b** $x \doteqdot 222.3°$ or $317.7°$

c no solutions exist **d** $x \doteqdot 4.7°,\ 85.3°,\ 184.7°,\ 265.3°$

e $x \doteqdot 74.6°,\ 285.4°$ **f** $x \doteqdot 104.3°,\ 165.7°$

EXERCISE 16J

1 a 20 m **b** at $t = \frac{3}{4}$ minute **c** 3 minutes

d

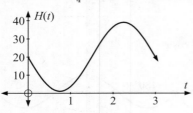

2 a 400 **b i** 577 **ii** 400

c 650   It is the maximum population.

d 150, after 3 years **e** $0.26 < t < 1.74$

3 a i false **ii** false **b** 101.5 cents/L

c on the 3rd, 13th and 19th days

d 101 cents/L on days 8 and 24

4 a i 7500 **ii** 10 500 **b** 10 500, when $t = 1$, 5 and 9 weeks

c i $t = \frac{1}{3},\ \frac{5}{3},\ \frac{13}{3},\ \frac{17}{3},\ \frac{25}{3},\ \frac{29}{3}$ weeks

ii $t = \frac{7}{3},\ \frac{11}{3},\ \frac{19}{3},\ \frac{23}{3},\ \frac{31}{3},\ \frac{35}{3}$ weeks

d $0.627 \leqslant t \leqslant 1.373$, $4.627 \leqslant t \leqslant 5.373$,
$8.627 \leqslant t \leqslant 9.373$

REVIEW SET 16A

1 a 3 **b** 24 **c** $\frac{3}{4}$

2

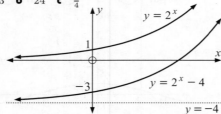

a $y = 2^x$ has y-intercept 1 and horizontal asymptote $y = 0$.

b $y = 2^x - 4$ has y-intercept -3 and horizontal asymptote
$y = -4$.

3 a 80°C **b i** 26.84°C **ii** 9.00°C **iii** 3.02°C

c

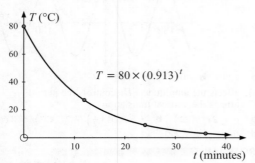

d $\doteqdot 12.8$ minutes

4 a $\frac{1}{2}$ **b** $\frac{1}{8}$ **c** 4

5

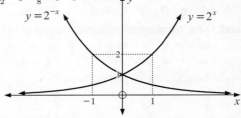

a $y = 2^x$ has y-intercept 1 and horizontal asymptote $y = 0$.

b $y = 2^{-x}$ has y-intercept 1 and horizontal asymptote $y = 0$.

6 a 1000 kangaroos **b i** $\doteqdot 4000$ **ii** $\doteqdot 16\,000$ **iii** $\doteqdot 64\,100$

c

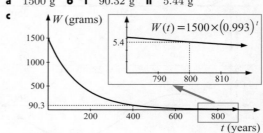

d $\doteqdot 7$ years

REVIEW SET 16B

1 a -1 **b** 7 **c** $-1\frac{8}{9}$

2 a 1500 g **b i** 90.32 g **ii** 5.44 g

c

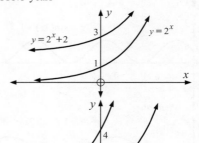

d $\doteqdot 385.5$ years

3 a

b

4 a

b

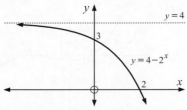

5 a 40 **b i** 113 **ii** 320 **iii** 2560

c

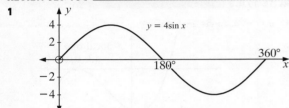

REVIEW SET 16C

1

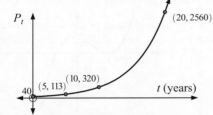

2

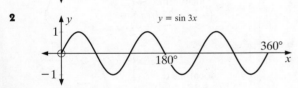

3 a $1080°$ **b** $90°$ **4 a** $x \doteq 24°, 156°$ **b** $x \doteq 192°, 348°$
5 $T \doteq 7.05 \sin(30t)° + 24.8$
6 a $x \doteq 22.5°, 157.5°, 382.5°$ **b** $x = 0°, 60°, 300°$ or $360°$
7 a 5000 **b** 3000, 7000
   c $0.5 \leqslant t \leqslant 2.5$ and $6.5 \leqslant t \leqslant 8$

REVIEW SET 16D

1

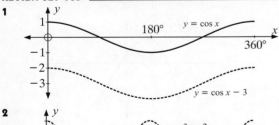

2

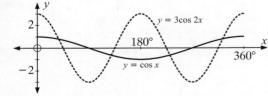

3 a $y = -4\cos(2x)°$ **b** $y = \cos\left(\frac{x}{2}\right)° + 2$
4 a $x \doteq 27°, 63°, 207°, 243°$ **b** $x \doteq 96°, 174°, 276°, 354°$
5 $x \doteq 64.0°, 296°, 424°$ **6 a** 28 units **b** 6 am Thursday
7 a $x \doteq 52.2°, 127.8°$ **b** $x = 30°, 150°, 270°$

EXERCISE 17A.1

1 a i $y = x^3 + 2$

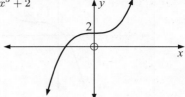

b i $y = -x^3 + 2$

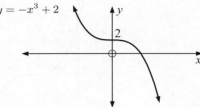

c $y = \frac{1}{2}x^3 + 3$

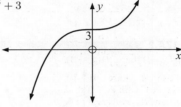

d $y = 2x^3 - 1$

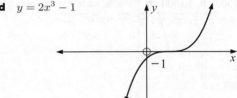

e $y = (x - 1)^3$

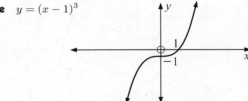

f $y = (x + 2)^3$

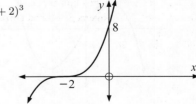

g $y = -(x + 1)^3$

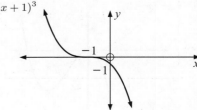

h $y = (x-1)^3 + 2$

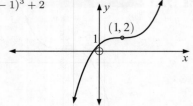

i $y = -(x-2)^3 - 5$

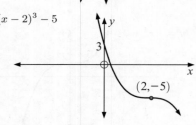

2 a $y = 3x^3 + 9x^2 + 9x + 1$
b $y = \frac{1}{2}x^3 + 3x^2 + 6x + 6$
c $y = -x^3 - 3x^2 - 3x + 4$

EXERCISE 17A.2

1 a $y = (x+1)(x-2)(x-3)$

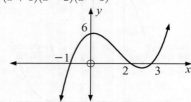

b $y = -2(x+1)(x-2)(x-\frac{1}{2})$

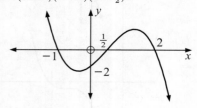

c $y = \frac{1}{2}x(x-4)(x+3)$

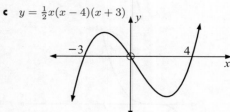

d $y = 2x^2(x-3)$

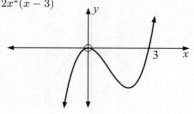

e $y = -\frac{1}{4}(x-2)^2(x+1)$

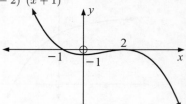

f $y = -3(x+1)^2(x-\frac{2}{3})$

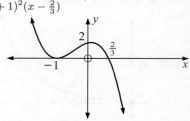

EXERCISE 17A.3

1 a $y = -2x^3 + 8x^2 + 6x - 36$

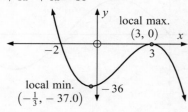

b $y = x^3 + 5x^2 + 2x - 8$

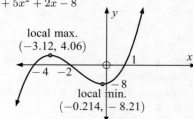

c $y = 3x^3 - 7x^2 - 28x + 10$

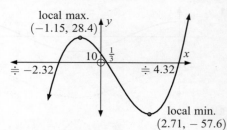

d $y = -3x^3 + 9x^2 + 27x + 15$

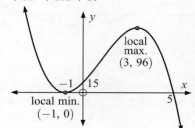

e $y = -3x^3 - 24x^2 - 48x$

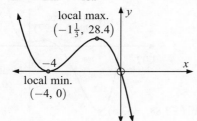

f $y = -x^3 + \frac{7}{2}x^2 + \frac{7}{2}x - 6$

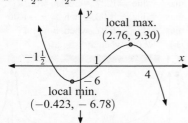

EXERCISE 17A.4

1 a $0 \leqslant x \leqslant 20$ **b** 13 **2** 92 items

3 a $a = 700$, the time at which the barrier has returned to its original position.
   b $k = \frac{85}{36\,000\,000}$, $f(1) = \frac{85}{36\,000\,000}t(t - 700)^2$
   c 120 mm   233.3 milliseconds

4 a 7.75 km^3 **b** 7.442 km^3 **c** 77.66% **d** 5418 km

5 February, March

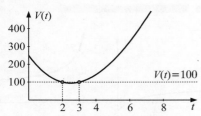

6 $V = x(32 - x)(40 - 2x)$,   4608 cm^3

EXERCISE 17B.1

1 a $y = (x + 4)(x + 1)(x - 2)(x - 3)$

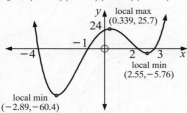

b $y = -x(x + 2)(x - 1)(x - 4)$

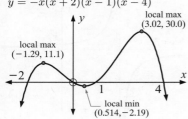

c $y = 2(x - 1)^2(x + 2)(2x - 5)$

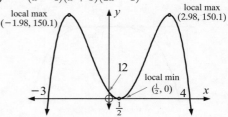

d $y = -(x - 4)(x + 3)(2x - 1)^2$

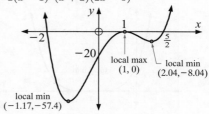

e $y = 3(x + 1)^2(x - 2)^2$

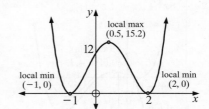

f $y = -2(x - 1)^2(x + 2)^2$

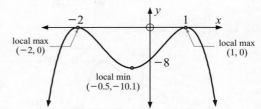

g $y = 4(x - 1)^3(2x + 3)$

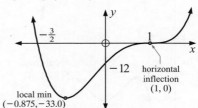

h $y = -x^3(2x - 3)$

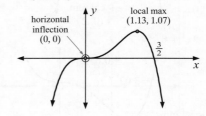

i $y = 3(x-1)^4$

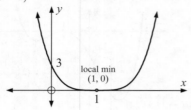

j $y = -2(x+2)^4$

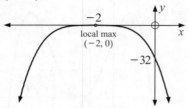

EXERCISE 17B.2

1 a $y = 2x^4 - 8x^3 + 7x + 2$

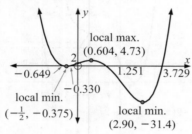

b $y = -2x^4 + 2x^3 + x^2 + x + 2$

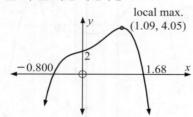

c $y = -\frac{1}{2}x^4 + 2x^2 + 2$

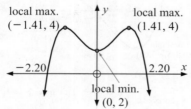

d $y = 2x^4 + 6x^3 + x + 2$

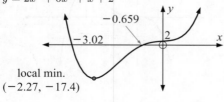

e $y = \frac{4}{5}x^4 + 5x^3 + 5x^2 + 2$

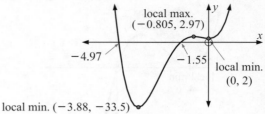

local min. $(-3.88, -33.5)$

2 a max. value $\doteq 27.95$, when $x \doteq 2.53$

b i max. value $\doteq 6.27$, when $x \doteq 1.17$

ii max. value $\doteq 4.03$, when $x \doteq -1.06$

iii max. value $\doteq 6.27$, when $x \doteq 1.17$

3 min. value $\doteq -2.03$, when $x \doteq -2.07$

EXERCISE 17C.1

1   **2**

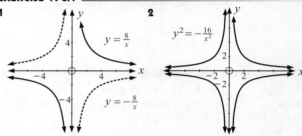

3 a $y = \dfrac{8}{x}$ **b** $y = \dfrac{3}{x}$ **c** $y = -\dfrac{12}{x}$

4 $(2, 2)$ and $(-2, -2)$

EXERCISE 17C.2

1 a $y = \dfrac{3}{x-1} + 2$

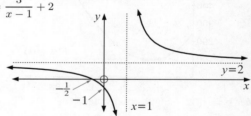

b $y = \dfrac{4}{x-3} + 1$

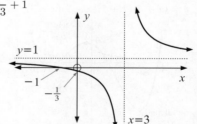

c $y = \dfrac{1}{x+2} + 3$

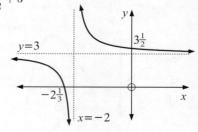

d $y = \dfrac{8}{x+2} + 1$

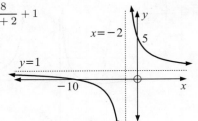

e $y = \dfrac{6}{x+2} - 4$

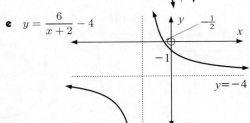

f $y = \dfrac{2}{x+4} - 1$

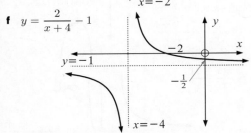

2 a i, ii and **iii** have no solution

b $\dfrac{ax+b}{cx+d} = \dfrac{a}{c}$ has no solution

c i $y = \dfrac{2x+1}{3x+4}$

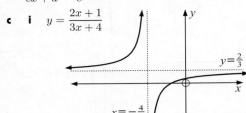

ii $y = \dfrac{5x-1}{2x+4}$

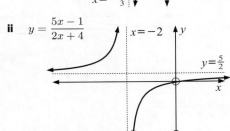

iii $y = \dfrac{3x-1}{-4x+3}$

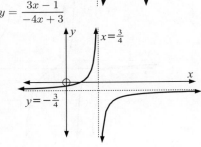

d HA is $y = \dfrac{a}{c}$, $c \neq 0$.

3 a $y = \dfrac{2x+3}{x-1}$

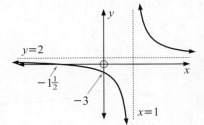

b $y = \dfrac{3x+1}{x+2}$

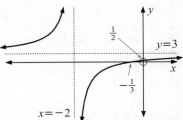

c $y = \dfrac{x-6}{x+3}$

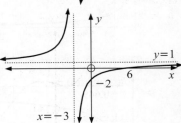

d $y = \dfrac{-2x+3}{x-2}$

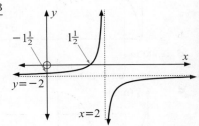

e $y = \dfrac{4x+2}{3x-4}$

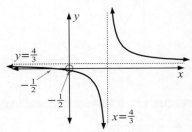

f $y = \dfrac{-5x+1}{2x-1}$

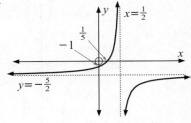

EXERCISE 17C.3

1 a 70 weeds   **b** 30 weeds   **c** 3 days

d $N = 20 + \dfrac{100}{t + 2}$

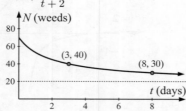

e no

2 a 250 amps
   b i 100 amps   **ii** 41.67 amps
   c 96 seconds

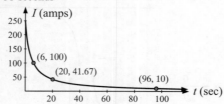

3 b i 2250 m   **ii** 2381 m
   c

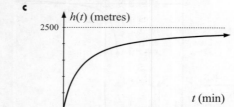

d 2500 m

4 a 16 m/s   **b** 0.429 secs   **c** 4 secs
   d

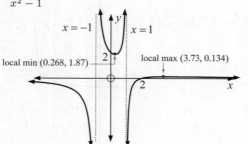

e $0 \leqslant t \leqslant 4$

EXERCISE 17D

1 a $y = \dfrac{x - 2}{x^2 - 1}$

local min (0.268, 1.87)   local max (3.73, 0.134)

b $y = \dfrac{x + 2}{x^2 - 2x}$

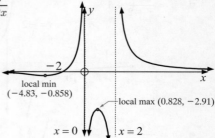

local min
$(-4.83, -0.858)$ local max $(0.828, -2.91)$

$x = 0$ $x = 2$

c $y = \dfrac{x + 3}{(x - 1)(x - 4)}$

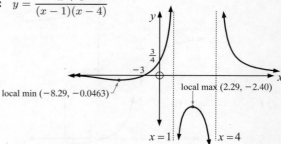

local min $(-8.29, -0.0463)$ local max $(2.29, -2.40)$

$x = 1$ $x = 4$

d $y = \dfrac{x(x - 2)}{x^2 - 1}$

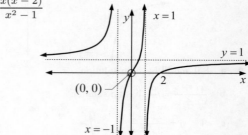

$x = 1$ $y = 1$

$(0, 0)$

$x = -1$

e $y = \dfrac{-x^2 + 2x}{x^2 - 3x + 2}$

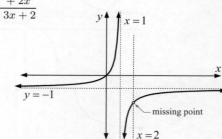

$x = 1$

$y = -1$

missing point

$x = 2$

f $y = \dfrac{(x - 1)(x + 4)}{(x + 1)(x - 4)}$

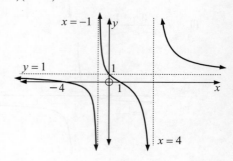

$x = -1$

$y = 1$

-4 1

$x = 4$

EXERCISE 17E

1

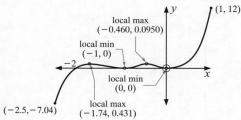

c Range is $\{y : -1 \leqslant y \leqslant 1\}$
d Is not periodic.

2

local max (0°, 2)   local max (360°, 2)

local min (180°, 0.5)

c Range is $\{y : 0.5 \leqslant y \leqslant 2\}$
d **i** The period is $360°$.
    ii The amplitude is meaningless here.

3 b $-2, -1$ and 0 **c** 0
c

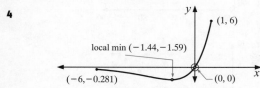

local max (−0.460, 0.0950)
local min (−1, 0)
local min (0, 0)
local max (−1.74, 0.431)
(−2.5, −7.04)
(1, 12)

4

local min (−1.44, −1.59)
(−6, −0.281)
(0, 0)
(1, 6)

c $y = 0$
d $f(-6) \doteq -0.281$, $f(1) = 6$
5 a local maxima at:
(97.3, 93.8), (451.2, 450.6), (810.6, 810.3)
b **i** There is no y-intercept. The graph seems to approach a value of 1.
    ii $f(0) = 0^0$ which is undefined.
c no
d

local max (810.6, 810.3)
local max (451.2, 450.6)
local max (97.3, 93.8)

EXERCISE 17F.1

1 a $(1, 7)$ and $(2, 8)$ **b** $(4, 5)$ and $(-3, -9)$

c $(3, 0)$ (touching) **d** graphs do not meet
2 a $(0.59, 5.59)$ and $(3.41, 8.41)$
b $(3, -4)$ touching **c** graphs do not meet
d $(-2.56, -18.81)$ and $(1.56, 1.81)$
3 a $(-1, 3)$ and $(-3, 1)$ **b** $(6, 1)$ and $(-1, -6)$
c $(2\frac{1}{2}, 4)$ and $(-2, -5)$ **d** do not meet
4 a $(-0.8508, -4.702)$ and $(2.351, 1.702)$ **b** do not meet
c $(-1.120, -5.359)$ and $(1.786, 3.359)$
d $(\frac{4}{3}, 9)$ and $(-3, -4)$ **e** $(-\frac{5}{3}, -3)$ and $(5, 1)$
f do not meet
5 b no
c The line cuts the curve at $(-2, -5)$ and $(-2\frac{1}{2}, -4)$.
6 a $(1, 4)$ and $(-\frac{3}{2}, \frac{1}{4})$ **b** $(\frac{3}{2}, 5)$ and $(-\frac{2}{3}, 5)$
c $(-\frac{1}{2}, -4)$ and $(\frac{8}{3}, 15)$

EXERCISE 17F.2

1 a $(-1, 5)$ **b** $(0, 1)$ and $(1, 3)$
c $(-1.69, 3.22)$ and $(3.03, 0.122)$
d $(-0.802, -0.414)$ and $(1.52, -5.29)$
e $(0.470, 2.13)$ **f** $(-0.838, 0.313)$ and $(0.632, 2.40)$
2 a $x \doteq \pm 1.19$ **b** Graphs meet at $(-1.19, -1.68)$ and $(1.19, 1.68)$. So, $x \doteq \pm 1.19$
c The graphing mode shows us how many solutions exist, but this takes a greater time to find.
3 a $(-1.81, 0.933)$ and $(1.42, 5.40)$
b $(0.293, 4.71)$ and $(1.51, 3.49)$
c $(-1.88, 0.272)$ and $(1.23, 2.34)$
d $(-2.86, -5.86)$ and $(2.45, -0.555)$
e $(30, 0.5)$, $(150, 0.5)$ and $(270, -1)$

REVIEW SET 17A

1 a $y = x^3 + 3$ **b** $y = -2x^2(x + 4)$

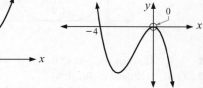

2 $y = 2x^3 - 6x^2 + 6x - 7$
3 a $y = (x - 2)^2(x + 1)(x - 3)$

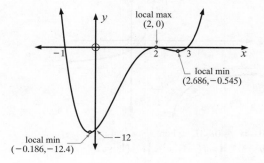

local max (2, 0)
local min (2.686, −0.545)
local min (−0.186, −12.4)

b $y = 3x^3 - x^2 + 2x - 1$

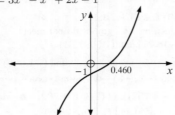

4 a $\doteqdot -1.62$ **b** 0 at $x = \pm 1$ **c** $\doteqdot -1.62$

5 $y = \dfrac{x-1}{(x+3)(x-2)}$

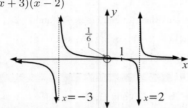

REVIEW SET 17B

1 a $y = -(x-1)(x+1)(x-3)^2$

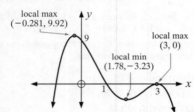

b $y = x^4 - 4x^3 + 6x^2 - 4x + 1$

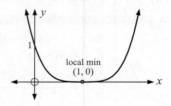

2 $\doteqdot -1.82$ when $x \doteqdot -0.474$

3 $(1.21, 1.42)$ and $(5.79, 10.6)$

4 a $x = -1$ **b** $y = -3$ **c** $-\frac{1}{3}$ **d** -1

$y = \dfrac{2}{x+1} - 3$

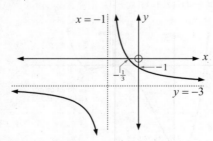

5 a 0 **b** -0.531, when $x \doteqdot -1.44$
   c $y = 0$ **d** $f(-4) = -\frac{1}{4}$, $f(1) = 2$

e $y = x \times 2^x$

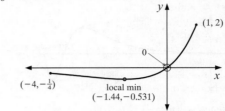

EXERCISE 18A

1 a i positive association **ii** roughly linear **iii** weak
   b i negative association **ii** linear **iii** strong
   c i no association **ii** non-linear **iii** zero
   d i negative association **ii** non-linear **iii** strong
   e i positive association **ii** linear **iii** moderate
   f i positive association **ii** non-linear **iii** moderate

2 a "If the variables x and y are positively associated then as x increases y increases."

   b "If there is negative correlation between the variables m and n then as m increases, n decreases."

   c "If there is no association between two variables then the points on the scatterplot appear to be randomly scattered."

3

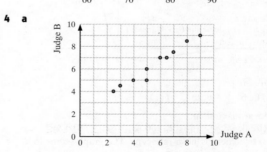

4 a

   b There appears to be strong, positive correlation between Judge A's scores and Judge B's scores. This means that as Judge A's scores increase, Judge B's scores increase.

5 a The independent variable is that which explains the relationship, while the dependent variable responds to values of the independent variable.

   b Explanatory and response, respectively. **c** Independent

6 a i no association **ii** zero **iii** non-linear
     iv no outliers

   b i positive association **ii** weak **iii** roughly linear
     iv no outliers

   c i negative association **ii** strong **iii** non-linear
     iv no outliers

d i positive association **ii** moderate **iii** linear
iv no outliers

e i negative association **ii** strong **iii** linear
iv no outliers

f i positive association **ii** moderate **iii** non-linear
iv no outliers

7 A causal relationship exists if the variables are related in such a way that if one is changed, the other changes.

EXERCISE 18B.1

1 a $r_A = 1$ $r_B = -1$ $r_C = 0$

b $r_A = 1$ confirms that there is a perfect linear positive correlation between the data points.

$r_B = -1$ confirms that there is a perfect linear negative correlation between the data points.

$r_C = 0$ confirms that there is no correlation between the points.

2 a $r \doteq 0.485$ **b** $r \doteq -0.0804$ **c** $r \doteq 0.985$

3 a

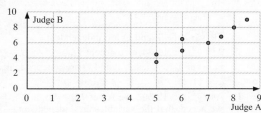

b There appears to be a strong, positive correlation between Judge A's scores and Judge B's scores. This means that as Judge A's scores increase, Judge B's scores increase.

c $r \doteq 0.943$, indicating a very strong linear relationship between Judge A's scores and Judge B's scores.

EXERCISE 18B.2

1 a $r^2 = 0.240$

b There is a very weak, positive association between the total number of crashes and the number of car crashes in which a casualty occurred.

2 a

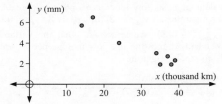

b $r \doteq -0.951$, $r^2 \doteq 0.904$

c The association between tread depth and number of km travelled is very strong, linear and negative.

3 a

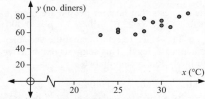

b $r = 0.743$, $r^2 = 0.552$

c There is a moderate, positive linear association between the number of diners and the noon temperature.

4 a

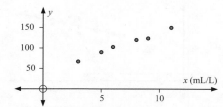

b $r = 0.994$ $r^2 = 0.989$

c There is a strong, positive linear association between yield and spray concentration.

5 a

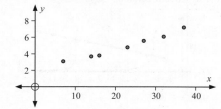

b $r = 0.989$ $r^2 = 0.978$

c There is a strong, positive linear association between cherry yield and the number of frosts.

6 a

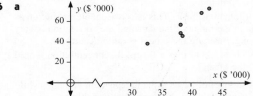

b $r = 0.921$, $r^2 = 0.848$

c There is a strong positive linear association between the starting salaries for Bachelor's degrees and the starting salaries for PhD's.

7 a

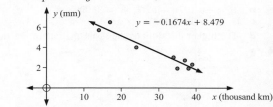

b $r = 0.840$ $r^2 = 0.706$
c moderate positive correlation

EXERCISE 18C

1 a tread depth **b** $y = -0.167x + 8.48$

c

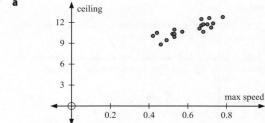

d 8.48 mm **e** 38 700 km

2 a temperature **b** Cov(X, Y) $\doteq 230$ (to 3 s.f.)
c $y = 2.24x + 6.23$

d

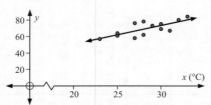

e Inaccurate, because high temperatures would eventually discourage people.

3 a Spray concentration is the independent variable, yield is the dependent variable.

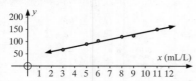

b $y = 9.93x + 39.5$

c The gradient indicates that an increase in concentration of 1 mL/L will increase the yield by roughly 9.9 tomatoes per bush. The intercept indicates that a bush can be expected to produce 39.5 tomatoes when there is no spraying.

d 109 tomatoes/bush. This is a reasonable prediction as it is in the range of the data and the correlation is very strong, meaning the model is quite accurate.

e No. It is likely that a 50 mL/L concentration would kill the bush.

4 a The number of frosts is the independent variable, while the cherry yield is the dependent variable.

b $y = 0.138x + 1.83$

c The gradient indicates that each extra frost increases the expected yield by 0.138 tonnes. The intercept indicates yields of 1.83 tonnes are expected when there are no frosts.

d 5.82 tonnes. This is reasonable as 29 frosts lies within the data range.

e 1.97 tonnes. This is unlikely to be accurate as the effect from the first few frosts may be negligible.

5 a

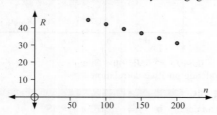

b $R = -0.107n + 52.7$, $r^2 = 0.999$

c i 43.1   **ii** 29.8

d The higher the altitude, the faster the rate of reaction.

6 a

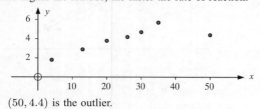

$(50, 4.4)$ is the outlier.

b i $\doteqdot 0.798$ **ii** $\doteqdot 0.993$

c i $y = 0.0672x + 2.22$ **ii** $y = 0.119x + 1.32$

d The one which excludes the outlier.

e Too much fertiliser often kills the plants (or makes them sick).

7 a $y = -0.795x + 10.8$ **b** $y = 0.160x + 42.5$

8 a i 10.2   **ii** 14.8   **b i** 4.80   **ii** 6.88

c Cov $(X, Y) \doteqdot 203.2$ **d** $y = 8.83x - 75.0$

EXERCISE 18D.1

1 a $\chi^2_{calc} \doteqdot 2.53$ **b** $\chi^2_{calc} \doteqdot 0.495$ **c** $\chi^2_{calc} \doteqdot 4.30$

d $\chi^2_{calc} \doteqdot 0.731$

3 a 1   **b** 1   **c** 2   **d** 3   **4** $df = 18$

EXERCISE 18D.2

1 $\chi^2_{calc} \doteqdot 10.5$, $df = 1$, $p = 0.00118$

As $\chi^2_{calc} > 3.84$, we reject H_0. So, the variables are not independent, i.e., they are related.

2 a $\chi^2_{calc} \doteqdot 5.20$, $df = 3$, $p = 0.158$

i As $\chi^2_{calc} < 7.81$, we accept H_0. So, at a 5% level, the variables P and Q are independent.

ii As $\chi^2_{calc} < 11.34$, we accept H_0, i.e., at a 10% level, P and Q are independent.

b $\chi^2_{calc} \doteqdot 11.25$, $df = 6$, $p \doteqdot 0.0809$

i As $\chi^2_{calc} < 12.59$, we reject H_0. So, at a 5% level, we accept that the variables are independent.

ii As $\chi^2_{calc} > 10.64$, we accept H_0, i.e., at a 10% level we accept that the variables are dependent.

3 $\chi^2_{calc} \doteqdot 8.58$, $df = 2$, $p \doteqdot 0.0137$

As $\chi^2_{calc} > 5.99$, we reject H_0 at a 5% level, that there is no association between age and the party to vote for.

4 Notice that the percentage of smokers is

$\frac{82}{294} \times 100\% = 27.9\%$ for low

$\frac{167}{835} \times 100\% = 20.0\%$ for average

$\frac{74}{502} \times 100\% = 14.7\%$ for high

$\frac{31}{199} \times 100\% = 15.6\%$ for very high

This data suggests low income people are more likely to smoke cigarettes. $\chi^2_{calc} \doteqdot 22.6$, $df = 3$, $p \doteqdot 0.00005$
As $\chi^2_{calc} > 6.25$ we reject H_0 that the variables are independent, suggesting that low income people are more likely to smoke cigarettes.

5 Percentage diabetics of

low $= \frac{11}{90} \times 100\% \doteqdot 12.2\%$

medium $= \frac{19}{87} \times 100\% \doteqdot 21.8\%$

heavy $= \frac{21}{95} \times 100\% \doteqdot 22.1\%$

obese $= \frac{38}{91} \times 100\% \doteqdot 41.8\%$

This data suggests that heavier people are more likely to get diabetes. $\chi^2_{calc} \doteqdot 22.6$, $df = 3$, $p \doteqdot 0.00005$
As $\chi^2_{calc} > 11.34$ we reject H_0 that the variables are independent at a 1% level. This suggests that heavier people are more likely to develop diabetes.

6 $\chi^2_{calc} = 12.6$, $df = 6$, $p \doteqdot 0.0499$

As $\chi^2_{calc} < 16.81$ (table) we accept H_0, that IQ and smoking are independent at a 1% level.

REVIEW SET 18A

1 a

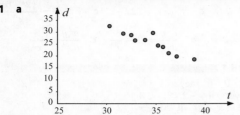

b $r = -0.9827$ Thus a strong negative linear relationship exists.

c $d = -1.64t + 82.3$, $50.0°C$

2 $\chi^2_{calc} \doteqdot 16.7$, $df = 2$, $p \doteqdot 0.0251$
As $\chi^2_{calc} > 9.21$ we reject H_0, that *wearing a seat belt and injury or death* are independent, at a 1% level.

3 a

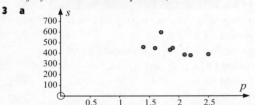

Yes, the point $(1.7, 597)$ is an outlier. It should not be deleted as there is no evidence that it is a mistake.

b $s = -116.5p + 665$
No, the prediction would not be accurate, as that much extrapolation is unreliable.

4 a number of waterings, n **b** $f = 34.0n + 19.3$

c

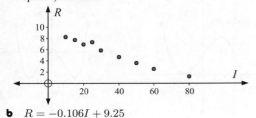

d i 104 (2 weeks), 359 (10 weeks)

ii Is unreliable as $n = 10$ is outside the poles.
Over watering could be a problem. $n = 2\frac{1}{2}$ is reliable.

5 a $\chi^2_{calc} \doteqdot 13.0$, $df = 6$, $p \doteqdot 0.00147$
As $\chi^2_{calc} > 12.59$, we reject H_0, i.e., we reject that P and Q are independent, at a 5% level.

b $\chi^2_{calc} \doteqdot 13.0$, $df = 6$, $p \doteqdot 0.00147$
As $\chi^2_{calc} > 16.81$, we accept H_0 that P and Q are independent, at a 1% level.

REVIEW SET 18B

1 a Income is the independent (explanatory) variable, while the peptic ulcer rate R is the dependent (response) variable.

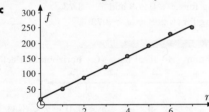

b $R = -0.106I + 9.25$

c The gradient of -0.1062 indicates that for each $1000 increase in income, the expected peptic ulcer rate decreases by 0.1062. The R-intercept of 9.2494 indicates that families of no income have an expected peptic ulcer rate of 9.2494.

d 4.5

e $87000. Predictions for incomes greater than this are negative, which is impossible (and hence inaccurate).

f i Point $(25, 7.3)$ as it lies well above the regression line.

ii $R = -0.104I + 9.08$ and $R(45) = 4.4$

2 $\chi^2_{calc} \doteqdot 42.1$, $df = 2$, $p \doteqdot 7.4 \times 10^{-10}$
$\chi^2_{calc} > 5.99$,   so we reject H_0 that *age of driver* and *increasing the speed limit* are independent, at a 5% level.

3 $\chi^2_{calc} \doteqdot 25.6$, $df = 9$, $p \doteqdot 0.00241$
$\chi^2_{calc} > 21.67$,   so we reject H_0 that *intelligence level* and *business success* are independent.

4 a

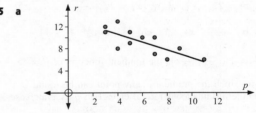

A linear model seems appropriate.

b $t = 0.0322v + 0.906$
$r^2 = 0.9997$, which indicates that this linear model is very close to the actual values.

c i 2.68 seconds   **ii** 4.45 seconds

d The driver's reaction time.

e Because the way in which the car brakes remains constant at all speeds.

5 a

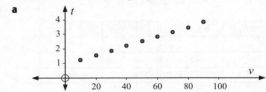

$r = -0.706p + 13.5$ dozen maidens

b Yes, there is a moderate-weak inverse linear relationship. This is indicated by $r = 0.763, r^2 = 0.582$.

c 9.2 dozen (111 maidens)

d This would predict that Silent Predator would abduct a negative number of maidens, which is unrealistic.

e r-int $= 13.5$, p-int $= 19.1$ **f** Silent Predator

EXERCISE 19A.1

1 a Every minute Karsten's heart beats 67 times.   **b** 4020

2 a 0.0015 errors/word   **b** 0.15 errors/100 words

3 Niko $12.35/h

4 a 1.77×10^{-4} mm/km   **b** 1.77 mm/10 000 km

5 a 76.1 km/h   **b** 21.1 m/s

EXERCISE 19A.2

1 a 96.2 km/h   **b** 65.5 km/h

2 a 800 m   **b** 125   **c** 125 m/min
d average walking speed   **e** 8 minutes
f 66.7 m/min   **g** 1.6 km

3 $\div 509000$ kL/day

4 a 1.19 kL/day **b** 1.03 kL/day **c** 1.05 kL/day

EXERCISE 19A.3

1 a 0.1 m/s **b** 0.9 m/s **c** 0.5 m/s

2 a i 3.2 beetles/gram **ii** 4.5 beetles/gram

b No effect 0 to 1 g, rapid decrease 1 to 8 grams,
rate of decrease decreases for 8 to 14 grams.

EXERCISE 19A.4

1 a 1 m/s **b** 3 km/h **c** $45/item sold **d** -5 bats/week

2 a 8200 L **b** 3000 L **c** 8200 L/h **d** 3000 L/h

EXERCISE 19D

1 a $f'(x) = 1$ **b** $f'(x) = 0$ **c** $f'(x) = 3x^2$ **d** $f'(x) = 4x^3$

2 a $f'(x) = 2$ **b** $f'(x) = 2x - 3$ **c** $f'(x) = 4x + 5$

3 a $f'(x) = -\dfrac{2}{x^3}$ **b** $f'(x) = -\dfrac{9}{x^4}$

4

Function	Derivative	Function	Derivative
x	1	x^{-1}	$-x^{-2}$
x^2	$2x^1$	x^{-2}	$-2x^{-3}$
x^3	$3x^2$	x^{-3}	$-3x^{-4}$
x^4	$4x^3$		

EXERCISE 19E

1 a $f'(x) = 3x^2$ **b** $f'(x) = 6x^2$ **c** $f'(x) = 14x$

d $f'(x) = 2x + 1$ **e** $f'(x) = -4x$ **f** $f'(x) = 2x + 3$

g $f'(x) = 3x^2 + 6x + 4$ **h** $f'(x) = 20x^3 - 12x$

i $f'(x) = 6x^{-2}$ **j** $f'(x) = -\dfrac{2}{x^2} + \dfrac{6}{x^3}$

k $f'(x) = 2x - \dfrac{5}{x^2}$ **l** $f'(x) = 2x + \dfrac{3}{x^2}$

2 a 4 **b** $-\dfrac{16}{729}$ **c** -7 **d** $\dfrac{13}{4}$ **e** $\dfrac{1}{8}$ **f** -11

3 a $\dfrac{dy}{dx} = 4 + \dfrac{3}{x^2}$, $\dfrac{dy}{dx}$ is the gradient function of $y = 4x - \dfrac{3}{x}$

from which the gradient at any point can be found.
It is the instantaneous rate of change in y as x increases.

b $\dfrac{dS}{dt} = 4t + 4$ metres per second, $\dfrac{dS}{dt}$ is the

instantaneous rate of change in position at the
time t, i.e., it is the velocity function.

c $\dfrac{dC}{dx} = 3 + 0.004x$ dollars per toaster

$\dfrac{dC}{dx}$ is the instantaneous rate of change in cost of
production as the number of toasters changes.

EXERCISE 19F

1 a $2x - y = 0$ **b** $3x + y = -5$

c $x - 2y = -4$ **d** $3x + y = -8$

2 a $(-\frac{3}{2}, \frac{11}{4})$ **b** $(-1, -1)$ and $(\frac{1}{3}, -\frac{23}{27})$

c $(2, 3)$ and $(-2, -1)$ **d** $(-\dfrac{b}{2a}, c - \dfrac{b^2}{4a})$

3 a $y = 21$ and $y = -6$ **b** $k = -5$

c $y = -3x + 1$ and $y = -3x + 5$

4 a $a = 2$ **b** $a = -12$ **5** $a = -4, b = 7$

EXERCISE 19G

1 a $f''(x) = 6$ **b** $f''(x) = 12x - 6$ **c** $f''(x) = \dfrac{12 - 6x}{x^4}$

2 a $\dfrac{d^2y}{dx^2} = -6x$ **b** $\dfrac{d^2y}{dx^2} = 2 - \dfrac{30}{x^4}$ **c** $\dfrac{d^2y}{dx^2} = \dfrac{8}{x^3}$

EXERCISE 19H.1

1 a i $x > 0$ **ii** never

b i never **ii** $-2 < x < 3$

c i $-2 < x < 0$ **ii** $0 < x < 2$

d i $x < 2$ **ii** $x > 2$

e i never **ii** all real x

f i all real x except $x = 0$ **ii** never

g i $1 < x < 5$ **ii** $x < 1, x > 5$

h i $2 < x < 4, x > 4$ **ii** $x < 0, 0 < x < 2$

i i $x < 0, 2 < x < 6$ **ii** $0 < x < 2, x > 6$

EXERCISE 19H.2

1 a increasing for $x > 0$, decreasing for $x < 0$

b decreasing for all $x \neq 0$

c increasing for $x > -\frac{3}{4}$, decreasing for $x < -\frac{3}{4}$

d increasing for $x < 0$ and $x > 4$,
decreasing for $0 < x < 4$

e increasing for $-0.816 < x < 0.816$,
decreasing for $x < -0.816$ and $x > 0.816$

f increasing for $-\frac{1}{2} < x < 3$,
decreasing for $x < -\frac{1}{2}$ and $x > 3$

g increasing for $x > -0.38$ and $x < -2.62$
decreasing for $-2.62 < x < -0.38$

h increasing for $x < 0.268$ and $x > 3.73$
decreasing for $0.268 < x < 3.73$

EXERCISE 19I

1 a A - local min B - local max C - horizontal inflection

b

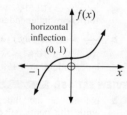

c i $x < -2, x > 3$ **ii** $-2 < x < 0, 0 < x < 3$

2 a $x = \frac{5}{4}$

b $f'(x) = 4x - 5, x = \frac{5}{4}$,
The gradient is 0 at the turning point (where $x = \frac{5}{4}$).

3 a

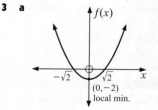

$(0, -2)$ local min.

b

horizontal inflection $(0, 1)$

c

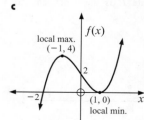

local max. $(-1, 4)$

$(1, 0)$ local min.

d

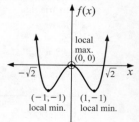

local max. $(0, 0)$

$(-1, -1)$ local min. $(1, -1)$ local min.

e

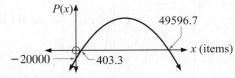

f

local max
(1.15, 3.08)

$f(x)$

local min
$(-1.15, -3.08)$

4 $x = -\dfrac{b}{2a}$, local min if $a > 0$, local max if $a < 0$

5 **a** $a = -12$, $b = -13$
  b local max is $(-2, 3)$, local min is $(2, -29)$

6 **a** greatest value $= 63$ (at $x = 5$)
   least value $= -18$ (at $x = 2$)
  b greatest value $= 4$ (at $x = 0$ and $x = 3$)
   least value $= -16$ (at $x = -2$)

7 Maximum hourly cost $= \$680.95$ when 150 hinges are made per hour. Minimum hourly cost $= \$529.80$ when 104 hinges are made per hour.

EXERCISE 19J

1 **a** $\dfrac{dM}{dt} = 3t^2 - 6t$ **b** $\dfrac{dR}{dt} = 8t + 4$

  c $\dfrac{dT}{dr} = 2r + \dfrac{100}{r^2}$ **d** $\dfrac{dA}{dh} = 2\pi + \tfrac{1}{2}h$

2 **a** cm^2/s **b** m^3/min **c** \$/items produced

3 **a** $B'(t) = 0.6t + 30$ is the rate at which the bacteria population is increasing.
  b $B'(3) = 31.8$ means that after 3 days, the bacteria population is increasing at 31.8 units per day
  c Since $B'(t)$ is positive.

4 **a** 2 m **b**

t (yr)	2	3	5	10	50
H (m)	11	14	16.4	18.2	19.64

  c $\dfrac{dH}{dt} = \dfrac{18}{t^2}$, m/yr **d** 18 m/yr, 2 m/yr, 0.18 m/yr

  e As $t^2 > 0$ for all $t \geqslant 1$, $\dfrac{dH}{dt} > 0$ for all $t \geqslant 0$.
   The height of the tree is increasing, i.e. it is growing.

5 **a** 0°C; 20, 20°C; 24, 40°C; 32
  b $\dfrac{dR}{dT} = \dfrac{1}{10} + \dfrac{T}{100}$
  c $\dfrac{dR}{dT} > 0$ (i.e., inc) for all $T > -10^\circ$C

6 **a** **i** \$4500 **ii** \$8250 **b** \$h/km
  c **i** increase of 100 \$h/km **ii** increase of 188.89 \$h/km

7 **a** The near part of the lake is 2 km from the sea, the furthest part is 3 km.

  c $x = \tfrac{1}{2}$; $\dfrac{dy}{dx} = 0.175$, height of hill is increasing as slope is positive
   $x = 1\tfrac{1}{2}$; $\dfrac{dy}{dx} = -0.225$, height of hill is decreasing as slope is negative
   $\therefore$ top of the hill is between $x = \tfrac{1}{2}$ and $x = 1\tfrac{1}{2}$.

8 **a** $C'(x) = 0.0009x^2 + 0.04x + 4$ dollars per pair

b $C'(220) = \$56.36$ per pair. This estimates the additional cost of making one more pair of jeans if 220 pairs are currently being made.

9 **a** $C'(x) = 0.000\,216x^2 - 0.001\,22x + 0.19$ dollars per item. $C'(x)$ is the rate at which the costs are increasing with respect to the number of items made.

  b $C'(300) = 19.26$ per item. This estimates the increase in cost to make 301 items, rather than 300.

  c \$19.32

10 **a**

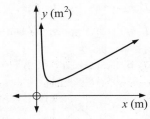

$P(x)$

49596.7

-20000 403.3 x (items)

   Profit for $404 < x < 49596$
  b $P'(x) = 5 - \frac{x}{5000}$. Profit increasing for $x < 25000$.

11 **a** $R(x) = 30x$ dollars
  b $P(x) = -0.002x^3 - 0.04x^2 + 20x - 3000$ dollars
  c $P'(x) = -0.006x^2 - 0.08x + 20$
  d $P'(50) = 1$ means that the profit from the 51st item is about \$1.

EXERCISE 19K

1 10 racquets **2** $61\tfrac{1}{4}$ m **3** 10 workers

4 **c** $C = \tfrac{x}{4}\left(108 - x^2\right)$ mL

  d $\dfrac{dC}{dx} = 27 - \dfrac{3x^2}{4}$, $\dfrac{dC}{dx} = 0$ when $x = \pm 6$.
  e 6 cm $\times$ 6 cm

5 5 cm $\times$ 5 cm squares **6** 50 fittings

7 **b** $L = 2x + \dfrac{100}{x}$

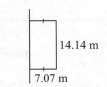

y (m^2)

x (m)

  c $L_{min} = 28.3$ m, **d**
   $x = 7.07$ m

14.14 m

7.07 m

8 **a** $2x$ cm **b** $V = 200 = 2x \times x \times h$
  c **Hint**: Show $h = \dfrac{100}{x^2}$ and substitute into the surface area equation.

  d $A = 4x^2 + \dfrac{600}{x}$

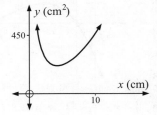

y (cm^2)

450

x (cm)

10

e $SA_{min} = 213.4$ cm^2,   **f**
   $x = 4.22$ cm

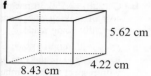

5.62 cm

8.43 cm   4.22 cm

9 a recall that $V_{cylinder} = \pi r^2 h$ and that
   1 L $= 1000$ cm^3
   b recall that $SA_{cylinder} = 2\pi r^2 + 2\pi rh$
   c $A = 2\pi r^2 + \dfrac{2000}{r}$

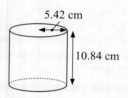

A (cm^2)

1500

x (cm)

15

d $A = 553$ cm^2,   **e**
   $r = 5.42$ cm

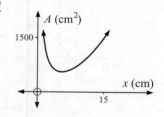

5.42 cm

10.84 cm

10 b 0.577 m wide, 0.816 m deep
11 Hint: Show that $C = 25x^2 + 200xy$ then use the result
   from **a**.   **c** 1.59 m $\times$ 1.59 m $\times$ 0.397 m

REVIEW SET 19A

1 -1
2 a $21x^2$ **b** $6x - 3x^2$ **c** $8x - 12$ **d** $7 + 4x$
3 $y = 4x + 2$
4 a $f''(x) = 6 - \dfrac{2}{x^3}$ **b** $f''(x) = 2$
5 a $a = -1$, $b = 2$
6 a 0 **b** local max at $(-1, 2)$
   local min at $(1, -2)$
   c

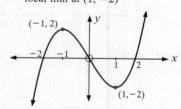

$(-1, 2)$

y

-2 -1

1 2

x

$(1, -2)$

7 a x intercepts are 0, 2 (touches) y intercept is 0.
   b local max at $(0.67, 1.19)$ local min at $(2, 0)$
   c

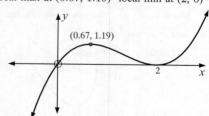

y

$(0.67, 1.19)$

2

x

8 6 cm $\times$ 6 cm
9 a $0 < x < \dfrac{200}{\pi}$ **c** $x \doteqdot 31.8$, $l = 100$

REVIEW SET 19B

1 $f'(x) = 3x^2 - 2$
2 a $\dfrac{dy}{dx} = 6x - 4x^3$ **b** $\dfrac{dy}{dx} = 1 + \dfrac{1}{x^2}$
3 $9x - y = 11$ **4** $(\frac{1}{\sqrt{2}}, \frac{4}{\sqrt{2}})$, $(-\frac{1}{\sqrt{2}}, -\frac{4}{\sqrt{2}})$
5 max is 21 when $x = 4$ min is 1 when $x = -1$ or 2.
6 a local max at $(-2, 51)$, local min at $(3, -74)$
   b increasing for $x < -2$ and $x > 3$
   decreasing for $-2 < x < 3$
   c

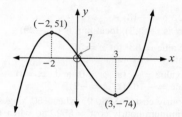

$(-2, 51)$

y

7

-2 3

x

$(3, -74)$

7 6 cm high, 12 cm wide
8 a $P = 2x + 2r + \pi r$ **b** $x = 100 - \dfrac{(2 + \pi)r}{2}$
   d $r = 28$ m, $x = 28$ m

REVIEW SET 19C

1 $f(x) = 2x + 2$
2 a $f'(x) = -6x^{-4} - 4x^{-5}$ **b** $f'(x) = -\dfrac{1}{x^2} + \dfrac{8}{x^3}$
   c $f'(x) = 2x - \dfrac{2}{x^3}$ **d** $f'(x) = -\dfrac{1}{4x^2}$
3 $y + 24x = 36$
4 a $A = -12$, $B = -13$
   b local max $(-2, 3)$, local min $(2, -29)$
5 a -4 **b** 1 is the only x-intercept.
   c there are none
   d $f(x) = x^3 + x^2 + 2x - 4$

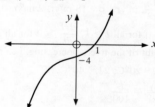

y

1

x

-4

6 a i \$312 **ii** \$1218.75
   b i 9.10 \$h/km **ii** 7.50 \$h/km
   c 3 km/h

7 a $y = \dfrac{1}{x^2}$ **c** 1.26 m $\times$ 1.26 m $\times$ 0.63 m

INDEX